Communication
Systems
Engineering

Communication Systems Engineering

John G. Proakis

Masoud Salehi

Northeastern University

 Prentice-Hall International, Inc.

© 1994 by Prentice-Hall, Inc.
A Paramount Communications Company
Englewood Cliffs, New Jersey 07632

Printed in the United States of America

10 9 8 7 6 5 4 3

ISBN 0-13-300625-5

Prentice-Hall International (UK) Limited, *London*
Prentice-Hall of Australia Pty. Limited, *Sydney*
Prentice-Hall Canada Inc., *Toronto*
Prentice-Hall Hispanoamericana, S.A., *Mexico*
Prentice-Hall of India Private Limited, *New Delhi*
Prentice-Hall of Japan, Inc., *Tokyo*
Simon & Schuster Asia Pte. Ltd., *Singapore*
Editora Prentice-Hall do Brasil, Ltda., *Rio de Janeiro*
Prentice-Hall, Inc., *Englewood Cliffs, New Jersey*

To Felia, George, and Elena (JGP)

To Fariba, Omid, and my parents (MS)

C

Contents

Preface   **xv**

1 Introduction   **1**

   1.1   Historical Review   2

   1.2   Elements of an Electrical Communication System   5

          1.2.1 Digital Communication System, 8
          1.2.2 Early Work in Digital Communications, 11

   1.3   Communication Channels and Their Characteristics   13

   1.4   Mathematical Models for Communication Channels   21

   1.5   Organization of the Book   23

   1.6   Further Reading   25

2 Signals and Linear Systems   26

2.1   Basic Concepts   27

    2.1.1   Classification of Signals, 27
    2.1.2   Some Important Signals and Their Properties, 36
    2.1.3   Classification of Systems, 42
    2.1.4   Analysis of LTI Systems in the Time Domain, 45

2.2   Fourier Series   47

    2.2.1   Signal Space Concepts, 48
    2.2.2   Orthogonal Expansion of Signals, 53
    2.2.3   Fourier Series and Its Properties, 55
    2.2.4   Response of LTI Systems to Periodic Signals, 65
    2.2.5   Parseval's Relation, 68

2.3   Fourier Transforms   70

    2.3.1   From Fourier Series to Fourier Transforms, 70
    2.3.2   Basic Properties of the Fourier Transform, 78
    2.3.3   Fourier Transform for Periodic Signals, 91
    2.3.4   Transmission over LTI Systems, 94

2.4   Power and Energy   97

    2.4.1   Energy-Type Signals, 98
    2.4.2   Power-Type Signals, 101

2.5   Sampling of Signals
    and Signal Reconstruction from Samples   104

    2.5.1   Ideal Sampling, 104
    2.5.2   Practical Sampling, 110

2.6   Bandpass Signals   112

    2.6.1   Properties of the Hilbert Transform, 122

2.7   Further Reading   125

    Problems   125

3 Random Processes   143

3.1   Probability and Random Variables   144

3.2   Random Processes: Basic Concepts   160

    3.2.1   Description of Random Processes, 162
    3.2.2   Statistical Averages, 164
    3.2.3   Stationary Processes, 167
    3.2.4   Random processes and linear systems, 175

3.3   Random Processes in the Frequency Domain   178

              3.3.1   Power Spectrum of Stochastic Processes, 179
              3.3.2   Transmission over LTI Systems, 185

   3.4   Gaussian and White Processes   188

              3.4.1   Gaussian Processes, 188
              3.4.2   White Processes, 190

   3.5   Bandlimited Processes and Sampling   193

   3.6   Bandpass Processes   196

   3.7   Further Reading   204

         Problems   204

4 Information Sources and Source Coding   221

   4.1   Modeling of Information Sources   222

              4.1.1   Measure of Information, 223
              4.1.2   Joint and Conditional Entropy, 225

   4.2   Source-Coding Theorem   228

   4.3   Source-Coding Algorithms   230

              4.3.1   The Huffman Source-Coding Algorithm, 230
              4.3.2   The Lempel-Ziv Source Coding Algorithm, 235

   4.4   Rate-Distortion Theory   237

              4.4.1   Mutual Information, 238
              4.4.2   Differential Entropy, 239
              4.4.3   Rate-Distortion Function, 240

   4.5   Quantization   246

              4.5.1   Scalar Quantization, 247
              4.5.2   Vector Quantization, 256

   4.6   Waveform Coding   259

              4.6.1   Pulse-Code Modulation, 259
              4.6.2   Differential Pulse-Code Modulation, 264
              4.6.3   Delta Modulation, 267

   4.7   Analysis-Synthesis Techniques   270

   4.8   Digital Audio Transmission and Digital Audio Recording   274

              4.8.1   Digital Audio in Telephone Transmission Systems, 274
              4.8.2   Digital Audio Recording, 276

   4.9   Further Reading   282

         Problems   282

5 Analog Signal Transmission and Reception   296

5.1   Introduction to Modulation   297

5.2   Amplitude Modulation   298

    5.2.1   Double-Sideband Suppressed Carrier AM, 298
    5.2.2   Conventional Amplitude Modulation, 306
    5.2.3   Single-Sideband AM, 310
    5.2.4   Vestigial-Sideband AM, 317
    5.2.5   Implementation of AM Modulators and Demodulators, 320
    5.2.6   Signal Multiplexing, 326

5.3   Angle Modulation   328

    5.3.1   Representation of FM and PM Signals, 329
    5.3.2   Spectral Characteristics of Angle Modulated Signals, 333
    5.3.3   Implementation of Angle Modulators and Demodulators, 343

5.4   Radio and Television Broadcasting   351

    5.4.1   AM Radio Broadcasting, 351
    5.4.2   FM Radio Broadcasting, 354
    5.4.3   Television Broadcasting, 356

5.5   Mobile Radio Systems   367

5.6   Further Reading   369

    Problems   370

6 Effect of Noise on Analog Communication Systems   385

6.1   Effect of Noise on Linear Modulation Systems   386

    6.1.1   Effect of Noise on a Baseband System, 386
    6.1.2   Effect of Noise on DSB-SC AM, 386
    6.1.3   Effect of Noise on SSB AM, 388
    6.1.4   Effect of Noise on Conventional AM, 389

6.2   Carrier Phase Estimation with a Phase-Locked Loop   394

    6.2.1   Effect of Additive Noise on Phase Estimation, 398

6.3   Effect of Noise on Angle Modulation   404

    6.3.1   Threshold Effect in Angle Modulation, 414
    6.3.2   Pre-emphasis and De-emphasis Filtering, 418

6.4   Comparison of Analog-Modulation Systems   421

6.5   Effects of Transmission Losses and Noise
in Analog Communication Systems   422

    6.5.1   Characterization of Thermal Noise Sources, 423

6.5.2 *Effective Noise Temperature and Noise Figure, 424*
6.5.3 *Transmission Losses, 427*
6.5.4 *Repeaters for Signal Transmission, 428*

6.6   Further Reading   431

       Problems   431

**7 Digital Transmission Through
   an Additive White Gaussian Noise Channel**   **437**

7.1   Pulse Modulation Signals
       and Their Geometric Representation   438

7.1.1 *Pulse Modulation Signals, 438*
7.1.2 *Geometric Representation of Signal Waveforms, 444*
7.1.3 *Geometric Representation of M-ary Pulse Modulation Signals, 448*
7.1.4 *Modulation Codes and Modulation Signals with Memory, 453*

7.2   Optimum Receiver for Pulse-Modulated Signals
       in Additive White Gaussian Noise   467

7.2.1 *Correlation-Type Demodulator, 469*
7.2.2 *Matched-Filter-Type Demodulator, 474*
7.2.3 *The Optimum Detector, 480*
7.2.4 *The Maximum-Likelihood Sequence Detector, 484*

7.3   Probability of Error for Signals
       in Additive White Gaussian Noise   488

7.3.1 *Probability of Error for Binary Modulation, 488*
7.3.2 *Probability of Error for M-ary Modulation, 492*
7.3.3 *Probability of Error for ML Sequence Detection, 502*
7.3.4 *Comparison of Modulation Methods, 504*

7.4   Regenerative Repeaters and Link Budget Analysis   506

7.4.1 *Regenerative Repeaters, 507*
7.4.2 *Link Budget Analysis for Radio Channels, 509*

7.5   Further Reading   512

       Problems   513

**8 Digital PAM Transmission
   Through Bandlimited AWGN Channels**   **530**

8.1   Digital Transmission
       Through Bandlimited Baseband Channels   531

8.1.1 *Digital PAM Transmission
       Through Bandlimited Baseband Channels, 535*

8.1.2   The Power Spectrum of a Digital PAM Signal, 537
8.1.3   The Power Spectrum of Digital Signals with Memory, 542

8.2   Signal Design for Bandlimited Channels   547

8.2.1   Design of Bandlimited Signals
        for Zero ISI—The Nyquist Criterion, 548
8.2.2   Design of Bandlimited Signals with Controlled ISI—
        Partial Response Signals, 555
8.2.3   Data Detection for Controlled ISI, 558

8.3   Probability of Error in Detection of Digital PAM   564

8.3.1   Probability of Error for Detection
        of Digital PAM with Zero ISI, 565
8.3.2   Probability of Error for Detection of Partial Response Signals, 566

8.4   System Design in the Presence of Channel Distortion   570

8.4.1   Design of Optimum Transmitting and Receiving Filters, 572
8.4.2   Channel Equalization, 577

8.5   Symbol Synchronization   595

8.5.1   Spectral-Line Methods, 597
8.5.2   Early–Late Gate Synchronizers, 600
8.5.3   Minimum Mean-Square-Error Method, 603
8.5.4   Maximum-Likelihood Methods, 603

8.6   Further Reading   606

      Problems   606

9   Digital Transmission via Carrier Modulation   617

9.1   Carrier-Amplitude Modulation   618

9.1.1   Amplitude Demodulation and Detection, 621
9.1.2   Probability of Error for PAM in an AWGN Channel, 624
9.1.3   Signal Demodulation in the Presence of Channel Distortion, 626

9.2   Carrier-Phase Modulation   628

9.2.1   Phase Demodulation and Detection, 633
9.2.2   Probability of Error for Phase Modulation in
        an AWGN Channel, 636
9.2.3   Carrier Phase Estimation, 640
9.2.4   Differential-Phase Modulation and Demodulation, 643
9.2.5   Probability of Error for DPSK in an AWGN Channel, 645

9.3   Quadrature Amplitude Modulation   646

9.3.1   Demodulation and Detection of QAM, 649
9.3.2   Probability of Error for QAM in AWGN Channel, 652

             9.3.3   Carrier-Phase Estimation, 658

9.4   Carrier-Frequency Modulation   659

             9.4.1   Frequency-Shift Keying, 659
             9.4.2   Demodulation and Detection of FSK Signals, 661
             9.4.3   Probability of Error for Noncoherent Detection of FSK, 667
             9.4.4   Continuous-Phase FSK, 671
             9.4.5   Spectral Characteristics of CPFSK Signals, 678

9.5   Continuous-Phase Carrier Modulation   681

             9.5.1   Demodulation and Detection of CPM Signals, 686
             9.5.2   Performance of CPM in an AWGN Channel, 691
             9.5.3   Spectral Characteristics of CPM Signals, 694

9.6   Digital Transmission on Fading Multipath Channels   695

             9.6.1   Channel Model for Time-Variant Multipath Channels, 697
             9.6.2   Signal Design for Fading Multipath Channels, 703
             9.6.3   Performance of Binary Modulation
                     in Rayleigh Fading Channels, 706

9.7   Further Reading   713

      Problems   714

10 Channel Capacity and Coding   **726**

    10.1   Modeling of Communication Channels   726

    10.2   Channel Capacity   729

             10.2.1   Gaussian Channel Capacity, 733

    10.3   Bounds on Communication   736

             10.3.1   Transmission of Analog Sources by PCM, 740

    10.4   Coding for Reliable Communication   742

             10.4.1   A Tight Bound on Error Probability of Orthogonal Signals, 743
             10.4.2   The Promise of Coding, 746

    10.5   Linear Block Codes   753

             10.5.1   Decoding and Performance of Linear Block Codes, 757
             10.5.2   Burst-Error-Correcting-Codes, 767

    10.6   Cyclic Codes   769

             10.6.1   The Structure of Cyclic Codes, 769

    10.7   Convolutional Codes   777

             10.7.1   Basic Properties of Convolutional Codes, 779

          10.7.2 Optimum Decoding
                    of Convolutional Codes—The Viterbi Algorithm, 784
          10.7.3 Other Decoding Algorithms for Convolutional Codes, 789
          10.7.4 Bounds on Error Probability of Convolutional Codes, 789

10.8   Coding for Bandwidth Constrained Channels   792

          10.8.1 Combined Coding and Modulation, 794
          10.8.2 Trellis Coded Modulation, 796

10.9   Practical Applications of Coding   803

          10.9.1 Coding for Deep-Space Communications, 803
          10.9.2 Coding for Telephone-Line Modems, 804
          10.9.3 Coding for Compact Disks, 805

10.10   Further Reading   809

        Problems   809

11 Spread-Spectrum Communication Systems   **822**

11.1   Model of
        a Spread-Spectrum Digital Communication Systems   823

11.2   Direct-Sequence Spread Spectrum Systems   825

          11.2.1 Probability of Error, 829
          11.2.2 Some Applications of DS Spread-Spectrum Signals, 837
          11.2.3 Generation of PN Sequences, 844

11.3   Frequency-Hopped Spread Spectrum   848

          11.3.1 Slow Frequency-Hopping Systems, 851
          11.3.2 Fast Frequency-Hopping Systems, 853
          11.3.3 Applications of FH Spread Spectrum, 855

11.4   Other Types of Spread-Spectrum Signals   856

11.5   Synchronization of Spread-Spectrum Systems   857

11.6   Further Reading   865

        Problems   865

**Appendix A: The Probability of Error
for Multichannel Reception of Binary Signals**   **869**

References   **873**

Index   **881**

P

Preface

The objective of this book is to provide an introduction to the basic principles in the analysis and design of communication systems. It is primarily intended for use as a text for a first course in communications, either at a senior level or at a first-year graduate level.

Broad Topical Coverage

Although we have placed a very strong emphasis on digital communications, we have provided abundant review of important mathematical foundational topics and a solid introduction to analog communications. The book is divided in three parts:

- A review of the fundamentals of system theory and probability (*Chapters 2–3*)

- An introduction to information sources and analog communications (*Chapters 4–6*)

- An introduction to digital communications (*Chapters 7–11*)

Emphasis on Digital Communications

Our motivation for emphasizing digital communications is due to the technological developments that have occurred during the past four decades. Today, digital communication systems are in common use and generally carry the bulk of our daily information transmission through a variety of communications media, such as wireline telephone channels, microwave radio, fiber-optic channels, and satellite channels. We are currently witnessing an explosive growth in the development of personal communication systems and ultra-high speed communication networks, which are based on digital transmission of the information, whether it is voice, still images, or video. We anticipate that, in the near future, we will witness a replacement of the current analog AM and FM radio and television broadcast by digital transmission systems.

The development of sophisticated, high-speed digital communication systems has been accelerated by concurrent developments in inexpensive high speed integrated circuits (IC) and programmable digital signal processing chips. The developments in microelecrtronic IC fabrication have made possible the implementation of high-speed, high precision A/D converters, of powerful error-correcting coders/decoders, and of complex digital modulation techniques. All of these technological developments point to a continuation in the trend toward increased use of digital communications as a means for transmitting information.

Overview of the Text

It is assumed that students using this book have a basic understanding of linear system theory, both continuous and discrete, including a working knowledge of Fourier series and Fourier transform techniques. Chapter 2 provides a review of basic material on signals and systems and establishes the necessary notation used in subsequent chapters. It is also assumed that students have had a first course in probability. Such courses are currently required in many undergraduate electrical engineering and computer engineering programs. Chapter 3 provides a review of probability and also covers the topic of random processes to the extent that is necessary for a first course in communications.

A logical beginning in the introduction of communication systems analysis and design is the characterization of information sources and source encoding. Chapter 4 is devoted to this topic. In this chapter we introduce the reader to the modeling of information sources, both discrete and continuous (analog) and the basic mathematical concepts of entropy and mutual information. Our discussion of source encoding for discrete sources includes the Huffman coding algorithm and the Lempel-Ziv algorithm. For the case of analog sources, we treat both scalar and vector quantization and describe the common waveform coding techniques, namely, PCM, DPCM and DM. We also describe the LPC-based source modeling method. As practical examples of the source coding methods described in this chapter we cite the digital speech transmission systems in the telephone plant and digital audio recording systems as embodied in the compact disc (CD) player.

Modulation and demodulation techniques are covered in Chapters 5 through 9. Chapter 5 is devoted to the treatment of analog modulation methods, specifically, AM, FM and PM, and their application to radio and television broadcasting.

Chapter 6 provides an analysis of the effects of noise in analog modulation systems. The phase-locked loops (PLL), which is used for estimating the phase of a sinusoidal carrier in both analog and digital communication systems is also described in Chapter 6. This chapter also includes a treatment of the effect of transmission losses and noise in analog modulation systems.

Digital modulation and demodulation techniques are described in Chapters 7, 8, and 9. We begin with the design of baseband pulse signals for the additive, white Gaussian noise (AWGN) channel in Chapter 7. Modulation codes are introduced and their applications to magnetic recording channels and CD players are illustrated. The demodulation and detection of baseband pulse modulation signals is described and the performance of the detectors for these signals is derived in terms of the probability of error. The chapter concludes with a discussion of regenerative repeaters and a description of a typical link budget analysis for radio channels.

Chapter 8 treats digital PAM transmission through bandlimited AWGN channels. In this chapter we derive the power spectral density of linearly modulated baseband signals and consider the problem of signal design for a bandlimited channel. We show that the effect of channel distortion is to introduce intersymbol interference (ISI), which can be eliminated or minimized by proper filter design at the transmitter and the receiver. The use of linear and nonlinear adaptive equalizers for reducing the effect of ISI is also described. The chapter concludes with a treatment of several methods for achieving symbol synchronization in synchronous digital communication systems.

Digital transmission via carrier modulation is the topic of Chapter 9. In this chapter we consider carrier amplitude modulation, carrier phase modulation, quadrature amplitude modulation, carrier frequency modulation, and continuous-phase carrier modulation. The optimum demodulation of these carrier modulation methods is described and their performance characteristics in transmission through an AWGN channel are evaluated. The last section of Chapter 9 treats the characterization of fading multipath channels and the performance of basic binary modulation techniques in a Rayleigh fading channel.

Chapter 10 treats the topic of channel coding and decoding. The capacity of a communication channel is first defined, and the capacity of the Gaussian channel is determined. Linear block codes and convolutional codes are introduced and appropriate decoding algorithm are described. The benefits of coding for bandwidth constrained channels are also described. The final section of this chapter presents three practical applications of coding.

The last chapter of this book treats spread spectrum modulation. The concept of spread spectrum signals was developed several decades ago to provide reliable military communication through radio channels that are generally vulnerable to jamming and other interference. Today, spread spectrum modulation is being considered for use in many wireless (radio) communication applications, including mobile cellular radio and for personal communications. This chapter emphasizes the two frequently used methods for generating spread spectrum signals, namely, direct-sequence and frequency-hopping spread spectrum.

Examples and Homework Problems

We have included a large number of carefully chosen examples and homework problems. The text contains over 180 worked-out examples and over 480 problems. Examples and problems range from simple exercises to more challenging and thought-provoking problems. A Solutions Manual is available from the publisher, which is provided in both typeset form and as a diskette formatted in LaTeX. This will enable instructors to print out solutions in any configuration easily.

Course Options

This book can serve as a text in either a one-semester or a two-semester course in communication system. An important consideration in the design of the course is whether or not the students have had a prior course in probability and random processes. Another important consideration is whether or not analog modulation and demodulation techniques are to be covered. Below, we outline three scenarios. Others are certainly possible.

  I. *A one-term course in analog and digital communication*
  Selected review sections in Chapter 2, all of Chapter 3, selected sections of Chapter 4, and all of Chapters 5, 6, 7, 8, 9, 10.

  II. *A one-term course in digital communication*
  Chapters 3 and 4 and Chapters 7 through 10

  III. *A two-term course sequence on analog and digital communications*
  1. Chapter 3 and Chapters 5, 6, 7 for the first course
  2. Chapters 8, 9, 4, 10, 11 for the second course.

Acknowledgments

We wish to express our appreciation to the undergraduate students at Northeastern University whose comments on several drafts of this book have helped to improve the presentation. We also wish to acknowledge the helpful comments of our faculty colleagues, Professors Brady, Kellner, and Raghavan, as well as the external reviewers, Professors Harry Leib of McGill University, James Cavers of Simon Fraser University, and Morton Kanefsky of the University of Pittsburgh, for their in-depth reviews and suggestions. Finally, we wish to thank Gloria Doukakis and Karen Blake for their assistance in the preparation of the manuscript.

John G. Proakis

Masoud Salehi

Communication Systems Engineering

1

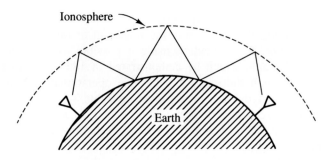

Ionosphere

Earth

Introduction

Every day, in our work and in our leisure time, we come in contact with and we use a variety of modern communication systems and communication media, the most common being the telephone, radio, and television. Through these media we are able to communicate (nearly) instantaneously with people on different continents, transact our daily business, and receive information about various developments and events of note that occur all around the world. Electronic mail and facsimile transmission have made it possible to rapidly communicate written messages across great distances.

Can you imagine a world without telephones, radio, and TV? Yet, when you think about it, most of these modern-day communication systems were invented and developed during the past century. Below, we present a brief historical review of major developments within the last two hundred years that have had a major role in the development of modern communication systems.

1.1 HISTORICAL REVIEW

Telegraphy and telephony. One of the earliest inventions of major significance to communications was the invention of the electric battery by Alessandro Volta in 1799. This invention made it possible for Samuel Morse to develop the electric telegraph, which he demonstrated in 1837. The first telegraph line linked Washington with Baltimore and became operational in May 1844. Morse devised the variable-length binary code given in Table 1.1, in which letters of the English alphabet are represented by a sequence of dots and dashes (code words). In this code, more frequently occurring letters are represented by short code words, while letters occurring less frequently are represented by longer code words.

The *Morse code* was the precursor to the variable-length source coding methods that are described in Chapter 4. It is remarkable that the earliest form of electrical communication that was developed by Morse, namely, *telegraphy,* was a binary digital communication system in which the letters of the English alpha-

TABLE 1-1   MORSE CODE

A	.—	N	—.		
B	—...	O	———		
C	—.—.	P	.——.		
D	—..	Q	——.—	1	.————
E	.	R	.—.	2	..———
F	..—.	S	...	3	...——
G	——.	T	—	4	....—
H	....	U	..—	5	.....
I	..	V	...—	6	—....
J	.———	W	.——	7	——...
K	—.—	X	—..—	8	———..
L	.—..	Y	—.——	9	————.
M	——	Z	——..	0	—————

(a) Letters   (b) Numbers

Period (.)	.—.—.—	Wait sign (AS)	.—...
Comma (,)	——..——	Double dash (break)	—...—
Interrogation (?)	..——..	Error sign	....
Quotation Mark (")	.—..—.	Fraction bar (/)	—..—.
Colon (:)	———...	End of message (AR)	.—.—.
Semicolon (;)	—.—.—.	End of transmission (SK)	...—.—
Parenthesis ( )	—.——.—		

(c) Punctuation and Special Characters

bet were efficiently encoded into corresponding variable-length code words having binary elements.

Nearly forty years later, in 1875, Emile Baudot developed a code for telegraphy in which each letter was encoded into fixed-length binary code words of length 5. In the *Baudot code,* the binary code elements were of equal length and designated as mark and space.

An important milestone in telegraphy was the installation of the first transatlantic cable in 1858 that linked the United States and Europe. This cable failed after about four weeks of operation. A second cable was laid a few years later and became operational in July 1866.

Telephony came into being with the invention of the telephone in the 1870's. Alexander Graham Bell patented his invention of the telephone in 1876 and in 1877 established the Bell Telephone Company. Early versions of telephone communication systems were relatively simple and provided service over several hundred miles. Significant advances in the quality and range of service during the first two decades of this century resulted from the invention of the carbon microphone and the induction coil.

The invention of the triode amplifier by Lee DeForest in 1906 made it possible to introduce signal amplification in telephone communication systems and thus to allow for telephone signal transmission over great distances. For example, transcontinental telephone transmission became operational in 1915.

Two world wars and the Great Depression during the 1930's must have been a deterrent to the establishment of transatlantic telephone service. It was not until 1953, when the first transatlantic cable was laid, that telephone service became available between the United States and Europe.

Automatic switching was another important advance in the development of telephony. The first automatic switch, developed by Strowger in 1897, was an electromechanical step-by-step switch. This type of switch was used for several decades. With the invention of the transistor, electronic (digital) switching became economically feasible. After several years of development at the Bell Telephone Laboratories, a digital switch was placed in service in Illinois in June 1960.

During the past thirty years, there have been numerous significant advances in telephone communications. Fiber optic cables are rapidly replacing copper wire in the telephone plant, and electronic switches have replaced the old electromechanical systems.

Wireless communications.   The development of wireless communications stems from the works of Oersted, Faraday, Gauss, Maxwell, and Hertz during the nineteenth century. In 1820, Oersted demonstrated that an electric current produces a magnetic field. On August 29, 1831, Michael Faraday showed that an induced current is produced by moving a magnet in the vicinity of a conductor. Thus, he demonstrated that a changing magnetic field produces an electric field. With this early work as background, James C. Maxwell in 1864 predicted the existence of electromagnetic radiation and formulated the basic theory that has been in use for over a century. Maxwell's theory was verified experimentally by Hertz in 1887.

In 1894, a sensitive device that could detect radio signals, called the *coherer*, was used by its inventor Oliver Lodge to demonstrate wireless communication over a distance of 150 yards at Oxford, England. Guglielmo Marconi is credited with the development of *wireless telegraphy*. Marconi demonstrated the transmission of radio signals at a distance of approximately 2 kilometers in 1895. Two years later, in 1897, he patented a radio telegraph system and established the Wireless Telegraph and Signal Company. On December 12, 1901, Marconi received a radio signal at Signal Hill in Newfoundland, which was transmitted from Cornwall, England—a distance of about 1700 miles.

The invention of the vacuum tube was especially instrumental in the development of radio communication systems. The vacuum diode was invented by John Fleming in 1904 and the vacuum triode amplifier was invented by Lee DeForest in 1906, as previously indicated. The invention of the triode made radio broadcast possible in the early part of the twentieth century. AM (amplitude modulation) broadcast was initiated in 1920 when radio station KDKA, Pittsburgh, went on the air. From that date, AM radio broadcasting grew very rapidly across the country and around the world. The *superheterodyne AM radio receiver*, as we know it today, was invented by Edwin Armstrong during World War I. Another significant development in radio communications was the invention of FM (frequency modulation), also by Armstrong. In 1933, Armstrong built and demonstrated the first FM communication system. However, the use of FM was slow to develop compared with AM broadcast. It was not until the end of World War II that FM broadcast gained in popularity and developed commercially.

The first television system was built in the United States by Vladimir Zworykin and demonstrated in 1929. Commercial television broadcasting began in London in 1936 by the British Broadcasting Corporation (BBC). Five years later, the Federal Communications Commission (FCC) authorized television broadcasting in the United States.

The past fifty years.   The growth in communications services over the past fifty years has been phenomenal. The invention of the transistor in 1947 by Walter Brattain, John Bardeen, and William Shockley; the integrated circuit in 1958 by Jack Kilby and Robert Noyce; and the laser by Townes and Schawlow in 1958, have made possible the development of small-size, low-power, low-weight, and high-speed electronic circuits which are used in the construction of satellite communication systems, wideband microwave radio systems, and lightwave communication systems using fiber optic cables. A satellite named Telstar I was launched in 1962 and used to relay TV signals between Europe and the United States. Commercial satellite communication services began in 1965 with the launching of the Early Bird satellite.

Currently, most of the wireline communication systems are being replaced by fiber optic cables which provide extremely high bandwidth and make possible the transmission of a wide variety of information sources, including voice, data, and video. Cellular radio has been developed to provide telephone service to people in automobiles. High-speed communication networks link computers and a variety of peripheral devices literally around the world.

Today we are witnessing a significant growth in the introduction and use of personal communications services, including voice, data, and video transmission. Satellite and fiber optic networks provide high-speed communication services around the world. Indeed, this is the dawn of the modern telecommunications era.

There are several historical treatments in the development of radio and telecommunications covering the past century. We cite books by McMahon, *The Making of a Profession—A Century of Electrical Engineering in America* (IEEE Press, 1984); Ryder and Fink, *Engineers and Electronics* (IEEE Press, 1984); and S. Millman, Ed., *A History of Engineering and Science in the Bell System—Communications Sciences* (1925-1980) (AT & T Bell Laboratories, 1984).

1.2 ELEMENTS OF AN ELECTRICAL COMMUNICATION SYSTEM

Electrical communication systems are designed to send messages or information from a source that generates the messages to one or more destinations. In general, a communication system can be represented by the functional block diagram shown in Figure 1.1.

The information generated by the source may be of the form of voice (speech source), a picture (image source), or plain text in some particular language, such as English, Japanese, German, French, etc. An essential feature of any source that generates information is that its output is described in probabilistic terms; that is, the output of a source is not deterministic. Otherwise, there would be no need to transmit the message.

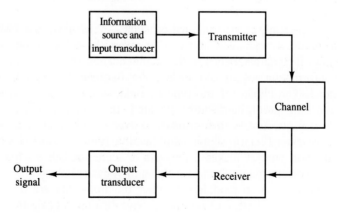

FIGURE 1.1.   Functional block diagram of a communication system.

A transducer is usually required to convert the output of a source into an electrical signal that is suitable for transmission. For example, a microphone serves as the transducer that converts an acoustic speech signal into an electrical signal, and a video camera converts an image into an electrical signal. At the destination, a similar transducer is required to convert the electrical signals that are received into a form that is suitable for the user; for example, acoustic signals, images, etc.

The heart of the communication system consists of three basic parts, namely, the transmitter, the channel, and the receiver. The functions performed by these three elements are described below.

The transmitter.   The transmitter converts the electrical signal into a form that is suitable for transmission through the physical channel or transmission medium. For example, in radio and TV broadcast, the Federal Communications Commission (FCC) specifies the frequency range for each transmitting station. Hence, the transmitter must translate the information signal to be transmitted into the appropriate frequency range that matches the frequency allocation assigned to the transmitter. Thus, signals transmitted by multiple radio stations do not interfere with one another. Similar functions are performed in telephone communication systems, where the electrical speech signals from many users are transmitted over the same wire.

In general, the transmitter performs the matching of the message signal to the channel by a process called *modulation*. Usually, modulation involves the use of the information signal to systematically vary the amplitude, frequency, or phase of a sinusoidal carrier. For example, in AM radio broadcast, the information signal that is transmitted is contained in the amplitude variations of the sinusoidal carrier, which is the center frequency in the frequency band allocated to the radio transmitting station. This is an example of *amplitude modulation*. In FM radio broadcast, the information signal that is transmitted is contained in the frequency variations of the sinusoidal carrier. This is an example of *frequency modulation. Phase modulation* (PM) is yet a third method for impressing the information signal on a sinusoidal carrier.

In general, carrier modulation such as AM, FM, and PM is performed at the transmitter, as indicated above, to convert the information signal to a form that matches the characteristics of the channel. Thus, through the process of modulation, the information signal is translated in frequency to match the allocation of the channel. The choice of the type of modulation is based on several factors, such as the amount of bandwidth allocated, the types of noise and interference that the signal encounters in transmission over the channel, and the electronic devices that are available for signal amplification prior to transmission. In any case, the modulation process makes it possible to accommodate the transmission of multiple messages from many users over the same physical channel.

In addition to modulation, other functions that are usually performed at the transmitter are filtering of the information-bearing signal, amplification of the modulated signal, and in the case of wireless transmission, radiation of the signal by means of a transmitting antenna.

The channel.   The communications channel is the physical medium that is used to send the signal from the transmitter to the receiver. In wireless transmission, the channel is usually the atmosphere (free space). On the other hand, telephone channels usually employ a variety of physical media, including wirelines, optical fiber cables, and wireless (microwave radio). Whatever the physical medium for signal transmission, the essential feature is that the transmitted signal is corrupted in a random manner by a variety of possible mechanisms. The most common form of signal degradation comes in the form of additive noise, which is generated at the front end of the receiver, where signal amplification is performed. This noise is often called *thermal noise*. In wireless transmission, additional additive disturbances are man-made noise and atmospheric noise picked up by a receiving antenna. Automobile ignition noise is an example of man-made noise, and electrical lightning discharges from thunderstorms is an example of atmospheric noise. Interference from other users of the channel is another form of additive noise that often arises in both wireless and wireline communication systems.

In some radio communication channels, such as the ionospheric channel that is used for long-range, short-wave radio transmission, another form of signal degradation is multipath propagation. Such signal distortion is characterized as a nonadditive signal disturbance which manifests itself as time variations in the signal amplitude, usually called *fading*. This phenomenon is described in more detail in Section 1.3.

Both additive and nonadditive signal distortions are usually characterized as random phenomena and described in statistical terms. The effect of these signal distortions must be taken into account in the design of the communication system.

In the design of a communication system, the system designer works with mathematical models that statistically characterize the signal distortion encountered on physical channels. Often, the statistical description that is used in a mathematical model is a result of actual empirical measurements obtained from experiments involving signal transmission over such channels. In such case, there is a physical justification for the mathematical model used in the design of communication systems. On the other hand, in some communication system designs, the statistical characteristics of the channel may vary significantly with time. In such cases, the system designer may design a communication system that is robust to the variety of signal distortions. This can be accomplished by having the system adapt some of its parameters to the channel distortion encountered.

The receiver.   The function of the receiver is to recover the message signal contained in the received signal. If the message signal is transmitted by carrier modulation, the receiver performs *carrier demodulation* to extract the message from the sinusoidal carrier. Since the signal demodulation is performed in the presence of additive noise and possibly other signal distortion, the demodulated message signal is generally degraded to some extent by the presence of these distortions in the received signal. As we shall see, the fidelity of the received message signal is a function of the type of modulation, the strength of the additive noise, the type and strength of any other additive interference, and the type of any nonadditive interference.

Besides performing the primary function of signal demodulation, the receiver also performs a number of peripheral functions, including signal filtering and noise suppression.

1.2.1 Digital Communication System

Up to this point, we have described an electrical communication system in rather broad terms based on the implicit assumption that the message signal is a continuous time-varying waveform. We refer to such continuous-time signal waveforms as *analog signals* and to the corresponding information sources that produce such signals as *analog sources.* Analog signals can be transmitted directly via carrier modulation over the communication channel and demodulated accordingly at the receiver. We call such a communication system an *analog communication system.*

Alternatively, an analog source output may be converted into a digital form and the message can be transmitted via digital modulation and demodulated as a digital signal at the receiver. There are some potential advantages to transmitting an analog signal by means of digital modulation. The most important reason is that signal fidelity is better controlled through digital transmission than analog transmission. In particular, digital transmission allows us to regenerate the digital signal in long-distance transmission, thus eliminating effects of noise at each regeneration point. In contrast, the noise added in analog transmission is amplified along with the signal when amplifiers are used periodically to boost the signal level in long-distance transmission. Another reason for choosing digital transmission over analog is that the analog message signal may be highly redundant. With digital processing, redundancy may be removed prior to modulation, thus conserving channel bandwidth. Yet a third reason may be that digital communication systems are often cheaper to implement.

In some applications, the information to be transmitted is inherently digital, e.g., in the form of English text, computer data, etc. In such cases, the information source that generates the data is called a *discrete (digital) source.*

In a digital communication system, the functional operations performed at the transmitter and receiver must be expanded to include message signal discretization at the transmitter and message signal synthesis or interpolation at the receiver. Additional functions include redundancy removal, and channel coding and decoding.

Figure 1.2 illustrates the functional diagram and the basic elements of a digital communication system. The source output may be either an analog signal, such as audio or video signal, or a digital signal, such as the output of a teletype machine which is discrete in time and has a finite number of output characters. In

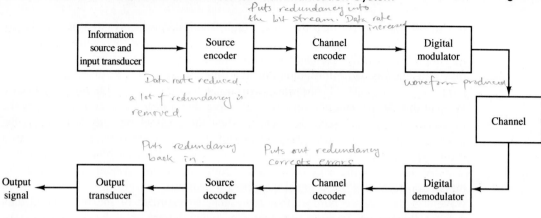

FIGURE 1.2.   Basic elementes of a digital communication system.

a digital communication system, the messages produced by the source are usually converted into a sequence of binary digits. Ideally, we would like to represent the source output (message) by as few binary digits as possible. In other words, we seek an efficient representation of the source output that results in little or no redundancy. The process of efficiently converting the output of either an analog or a digital source into a sequence of binary digits is called *source encoding* or *data compression*. We shall describe source encoding methods in Chapter 4.

The sequence of binary digits from the source encoder, which we call the *information sequence,* is passed to the *channel encoder*. The purpose of the channel encoder is to introduce in a controlled manner some redundancy in the binary information sequence which can be used at the receiver to overcome the effects of noise and interference encountered in the transmission of the signal through the channel. Thus, the added redundancy serves to increase the reliability of the received data and improves the fidelity of the received signal. In effect, redundancy in the information sequence aids the receiver in decoding the desired information sequence. For example, a (trivial) form of encoding of the binary information sequence is simply to repeat each binary digit m times, where m is some positive integer. More sophisticated (nontrivial) encoding involves taking k information bits at a time and mapping each k-bit sequence into a unique n-bit sequence, called a *code word*. The amount of redundancy introduced by encoding the data in this manner is measured by the ratio n/k. The reciprocal of this ratio, namely, k/n, is called the rate of the code or, simply, the *code rate*.

The binary sequence at the output of the channel encoder is passed to the digital modulator, which serves as the interface to the communications channel. Since nearly all of the communication channels encountered in practice are capable of transmitting electrical signals (waveforms), the primary purpose of the digital modulator is to map the binary information sequence into signal waveforms. To elaborate on this point, let us suppose that the coded information sequence is to be transmitted one bit at a time at some uniform rate R bits/s. The digital modulator may simply map the binary digit 0 into a waveform $s_0(t)$ and the binary digit 1 into a waveform $s_1(t)$. In this manner, each bit from the channel encoder is transmitted separately. We call this *binary modulation*. Alternatively, the modulator may transmit b coded information bits at a time by using $M = 2^b$ distinct waveforms $s_i(t)$, $i = 0, 1, \ldots, M-1$, one waveform for each of the 2^b possible b-bit sequences. We call this *M-ary modulation* $(M > 2)$. Note that a new b-bit sequence enters the modulator every b/R seconds. Hence, when the channel bit rate R is fixed, the amount of time available to transmit one of the M waveforms corresponding to a b-bit sequence is b times the time period in a system that uses binary modulation.

At the receiving end of a digital communications system, the digital demodulator processes the channel-corrupted transmitted waveform and reduces each waveform to a single number that represents an estimate of the transmitted data symbol (binary or M-ary). For example, when binary modulation is used, the demodulator may process the received waveform and decide on whether the transmitted bit is a 0 or 1. In such a case, we say the *demodulator has made a binary decision*. As one alternative, the demodulator may make a ternary decision; that is, it decides that the transmitted bit is either a 0 or 1 or it makes no decision at all, depending on the apparent quality of the received signal. When no decision is made on a particular bit, we say that *the demodulator has inserted an erasure in the demodulated data*. Using the redundancy in the transmitted data, the decoder attempts to fill in the positions where erasures occurred. Viewing the decision process performed by the demodulator as a form of quantization, we observe that binary and ternary decisions are special cases of a demodulator that quantizes to Q levels, where $Q \geq 2$. In general, if the digital communications system employs M-ary modulation, where $m = 0, 1, \ldots, M-1$ represent the M possible transmitted symbols, each corresponding to $k = \log_2 M$ bits, the demodulator may make a Q-ary decision, where $Q \geq M$. In the extreme case where no quantization is performed, $Q = \infty$.

When there is no redundancy in the transmitted information, the demodulator must decide which of the M waveforms was transmitted in any given time interval. Consequently, $Q = M$, and since there is no redundancy in the transmitted information, no discrete channel decoder is used following the demodulator. On the other hand, when there is redundancy introduced by a discrete channel encoder at the transmitter, the Q-ary output from the demodulator occurring every k/R seconds is fed to the decoder, which attempts to reconstruct the original information sequence from knowledge of the code used by the channel encoder and the redundancy contained in the received data.

A measure of how well the demodulator and encoder perform is the frequency with which errors occur in the decoded sequence. More precisely, the average probability of a bit-error at the output of the decoder is a measure of the performance of the demodulator-decoder combination. In general, the probability of error is a function of the code characteristics, the types of waveforms used to transmit the information over the channel, the transmitter power, the characteristics of the channel (i.e., the amount of noise), the nature of the interference, etc., and the method of demodulation and decoding. These items and their effect on performance will be discussed in detail in subsequent chapters.

As a final step, when an analog output is desired, the source decoder accepts the output sequence from the channel decoder, and from knowledge of the source encoding method used, attempts to reconstruct the original signal from the source. Due to channel decoding errors and possible distortion introduced by the source encoder and, perhaps, the source decoder, the signal at the output of the source decoder is an approximation to the original source output. The difference or some function of the difference between the original signal and the reconstructed signal is a measure of the distortion introduced by the digital communications system.

1.2.2 Early Work in Digital Communications

Although Morse is responsible for the development of the first electrical digital communication system (telegraphy), the beginnings of what we now regard as modern digital communications stem from the work of Nyquist (1924), who investigated the problem of determining the maximum signaling rate that can be used over a telegraph channel of a given bandwidth without intersymbol interference. He formulated a model of a telegraph system in which a transmitted signal has the general form

$$s(t) = \sum_n a_n g(t - nT)$$

where $g(t)$ represents a basic pulse shape and $\{a_n\}$ is the binary data sequence of $\{\pm 1\}$ transmitted at a rate of $1/T$ bits per second. Nyquist set out to determine the optimum pulse shape that was bandlimited to W Hz and maximized the bit rate $1/T$ under the constraint that the pulse caused no intersymbol interference at the sampling times k/T, $k = 0$, ± 1, ± 2, ... His studies led him to conclude that the maximum pulse rate $1/T$ is $2W$ pulses per second. This rate is now called the *Nyquist rate*. Moreover, this pulse rate can be achieved by using the pulses $g(t) = (\sin 2\pi Wt)/2\pi Wt$. This pulse shape allows the recovery of the data without intersymbol interference at the sampling instants. Nyquist's result is equivalent to a version of the sampling theorem for bandlimited signals, which was later stated precisely by Shannon (1948). The sampling theorem states that a signal of bandwidth W can be reconstructed from samples taken at the Nyquist rate of $2W$ samples per second using the interpolation formula

$$s(t) = \sum_n s\left(\frac{n}{2W}\right) \frac{\sin 2\pi W(t - n/2W)}{2\pi W(t - n/2W)}$$

In light of Nyquist's work, Hartley (1928) considered the issue of the amount of data that can be transmitted reliably over a bandlimited channel when multiple amplitude levels are used. Due to the presence of noise and other interference, Hartley postulated that the receiver can reliably estimate the received signal amplitude to some accuracy, say A_δ. This investigation led Hartley to conclude that there is a maximum data rate that can be communicated reliably over a bandlimited channel, when the maximum signal amplitude is limited to $A_{\max}$ (fixed power constraint), and the amplitude resolution is A_δ.

Another significant advance in the development of communications was the work of Wiener (1942) who considered the problem of estimating a desired signal waveform $s(t)$ in the presence of additive noise $n(t)$, based on observation of the received signal $r(t) = s(t) + n(t)$. This problem arises in signal demodulation. Wiener determined the linear filter whose output is the best mean-square approximation to the desired signal $s(t)$. The resulting filter is called the *optimum linear (Wiener) filter*.

Hartley's and Nyquist's results on the maximum transmission rate of digital information were precursors to the work of Shannon (1948 a,b) who established the mathematical foundations for information theory and derived the fundamental limits for digital communication systems. In his pioneering work, Shannon formulated the basic problem of reliable transmission of information in statistical terms, using probabilistic models for information sources and communication channels. Based on such a statistical formulation, he adopted a logarithmic measure for the information content of a source. He also demonstrated that the effect of a transmitter power constraint, a bandwidth constraint, and additive noise can be associated with the channel and incorporated into a single parameter, called the *channel capacity*. For example, in the case of an additive white (spectrally flat) Gaussian noise interference, an ideal bandlimited channel of bandwidth W has a capacity C given by

$$C = W \log_2 \left(1 + \frac{P}{W N_0} \right) \text{ bits/s}$$

where P is the average transmitted power and $N_0/2$ is the power spectral density of the additive noise. The significance of the channel capacity is as follows: If the information rate R from the source is less than C $(R < C)$, then it is theoretically possible to achieve reliable (error-free) transmission through the channel by appropriate coding. On the other hand, if $R > C$, reliable transmission is not possible regardless of the amount of signal processing performed at the transmitter and receiver. Thus, Shannon established basic limits on communication of information and gave birth to a new field that is now called *information theory*.

Initially the fundamental work of Shannon had a relatively small impact on the design and development of new digital communications systems. In part, this was due to the small demand for digital information transmission during the 1950's. Another reason was the relatively large complexity and, hence, the high cost of digital hardware required to achieve the high efficiency and high reliability predicted by Shannon's theory.

Another important contribution to the field of digital communications is the work of Kotelnikov (1947) which provided a coherent analysis of the various digital communication systems based on a geometrical approach. Kotelnikov's approach was later expanded by Wozencraft and Jacobs (1965).

The increase in the demand for data transmission during the last three decades, coupled with the development of more sophisticated integrated circuits, has led to the development of very efficient and more reliable digital communications systems. In the course of these developments, Shannon's original results and the generalization of his results on maximum transmission limits over a channel and on bounds on the performance achieved have served as benchmarks for any given communications system design. The theoretical limits derived by Shannon and other researchers that contributed to the development of information theory serve as an ultimate goal in the continuing efforts to design and develop more efficient digital communications systems.

Following Shannon's publications came the classic work of Hamming (1950) on error detecting and error-correcting codes to combat the detrimental effects of channel noise. Hamming's work stimulated many researchers in the years that followed, and a variety of new and powerful codes were discovered, many of which are used today in the implementation of modern communication systems.

1.3 COMMUNICATION CHANNELS AND THEIR CHARACTERISTICS

As indicated in the preceding discussion, the communication channel provides the connection between the transmitter and the receiver. The physical channel may be a pair of wires that carry the electrical signal, or an optical fiber that carries the information on a modulated light beam, or an underwater ocean channel in which the information is transmitted acoustically, or free space over which the information-bearing signal is radiated by use of an antenna. Other media that can be characterized as communication channels are data storage media, such as magnetic tape, magnetic disks, and optical disks.

One common problem in signal transmission through any channel is additive noise. In general, additive noise is generated internally by components such as resistors and solid-state devices used to implement the communication system. This is sometimes called *thermal noise*. Other sources of noise and interference may arise externally to the system, such as interference from other users of the channel. When such noise and interference occupy the same frequency band as the desired signal, its effect can be minimized by proper design of the transmitted signal and its demodulator at the receiver. Other types of signal degradations that may be encountered in transmission over the channel are signal attenuation, amplitude and phase distortion, and multipath distortion.

The effects of noise may be minimized by increasing the power in the transmitted signal. However, equipment and other practical constraints limit the power level in the transmitted signal. Another basic limitation is the available channel bandwidth. A bandwidth constraint is usually due to the physical limitations of the medium and the electronic components used to implement the transmitter and the receiver. These two limitations result in constraining the amount of data that can be transmitted reliably over any communications channel. Shannon's basic results relate the channel capacity to the available transmitted power and channel bandwidth.

Below, we describe some of the important characteristics of several communication channels.

Wireline channels. The telephone network makes extensive use of wire lines for voice signal transmission, as well as data and video transmission. Twisted-pair wire lines and coaxial cable are basically guided electromagnetic channels which provide relatively modest bandwidths. Telephone wire generally used to connect a customer to a central office has a bandwidth of several hundred kilohertz (kHz). On the other hand, coaxial cable has a usable bandwidth of several megahertz (MHz). Figure 1.3 illustrates the frequency range of guided electromagnetic channels, which includes waveguides and optical fibers.

Signals transmitted through such channels are distorted in both amplitude and phase and further corrupted by additive noise. Twisted-pair wireline channels are also prone to crosstalk interference from physically adjacent channels. Because wireline channels carry a large percentage of our daily communications around the country and the world, much research has been performed on the characterization of their transmission properties and on methods for mitigating the amplitude and phase distortion encountered in signal transmission. In Chapter 8, we describe methods for designing optimum transmitted signals and their demodulation, including the design of channel equalizers that compensate for amplitude and phase distortion.

Fiber optic channels. Optical fibers offer the communications system designer a channel bandwidth that is several orders of magnitude larger than coaxial cable channels. During the past decade, optical fiber cables have been developed that have a relatively low signal attenuation, and highly reliable photonic devices have been developed for signal generation and signal detection. These technological advances have resulted in a rapid deployment of optical fiber channels, both in domestic telecommunication systems as well as for trans-Atlantic and trans-Pacific communications. With the large bandwidth available on fiber optic channels it is possible for telephone companies to offer subscribers a wide array of telecommunication services, including voice, data, facsimile, and video.

The transmitter or modulator in a fiber optic communication system is a light source, either a light-emitting diode (LED) or a laser. Information is transmitted by varying (modulating) the intensity of the light source with the message signal. The light propagates through the fiber as a light wave and is amplified periodically (in the case of digital transmission, it is detected and regenerated by repeaters) along the transmission path to compensate for signal attenuation. At the receiver, the light intensity is detected by a photodiode, whose output is an electrical signal that varies in direct proportion to the power of the light impinging on the photodiode.

It is envisioned that optical fiber channels will replace nearly all wireline channels in the telephone network by the turn of the century.

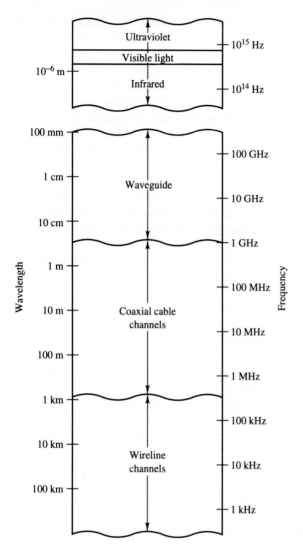

FIGURE 1.3.   Frequency range for guided wire channel.

Wireless electromagnetic channels.   In wireless communication systems, electromagnetic energy is coupled to the propagation medium by an antenna which serves as the radiator. The physical size and the configuration of the antenna depend primarily on the frequency of operation. To obtain efficient radiation of electromagnetic energy, the antenna must be longer than 1/10 of the wavelength. Consequently, a radio station transmitting in the AM frequency band, say at 1 MHz (corresponding to a wavelength of $\lambda = c/f_c = 300$ m), requires an antenna of at least 30 meters. Other important characteristics and attributes of antennas for wireless transmission are described in Chapter 7.

Figure 1.4 illustrates the various frequency bands of the electromagnetic spectrum. The mode of propagation of electromagnetic waves in the atmosphere and in free space may be subdivided into three categories, namely, ground-wave propagation, sky-wave propagation, and line-of-sight (LOS) propagation. In the VLF and ELF frequency bands, where the wavelengths exceed 10 km, the earth and the ionosphere act as a waveguide for electromagnetic wave propagation. In these frequency ranges, communication signals practically propagate around the globe. For this reason, these frequency bands are primarily used to provide navigational aids from shore to ships around the world. The channel bandwidths available in these frequency bands are relatively small (usually from 1% to 10% of the center frequency), and hence, the information that is transmitted through these channels is relatively slow speed and, generally, confined to digital transmission. A dominant type of noise at these frequencies is generated from thunderstorm activity around the globe, especially in tropical regions. Interference results from the many users of these frequency bands.

Ground-wave propagation, as illustrated in Figure 1.5, is the dominant mode of propagation for frequencies in the MF band (0.3 to 3 MHz). This is the frequency band used for AM broadcasting and maritime radio broadcasting. In AM broadcasting, the range with ground-wave propagation of even the more powerful radio stations is limited to about 100 miles. Atmospheric noise, man-made noise, and thermal noise from electronic components at the receiver are dominant disturbances for signal transmission of MF.

Sky-wave propagation, as illustrated in Figure 1.6, results from transmitted signals being reflected (bent or refracted) from the ionosphere, which consists of several layers of charged particles ranging in altitude from 30 to 250 miles above the surface of the earth. During the daytime hours, the heating of the lower atmosphere by the sun causes the formation of the lower layers at altitudes below 75 miles. These lower layers, especially the D-layer, serve to absorb frequencies below 2 MHz, thus severely limiting sky-wave propagation of AM radio broadcast. However, during the night-time hours, the electron density in the lower layers of the ionosphere drops sharply and the frequency absorption that occurs during the daytime is significantly reduced. As a consequence, powerful AM radio broadcast stations can propagate over large distances via sky wave over the F-layer of the ionosphere, which ranges from 90 miles to 250 miles above the surface of the earth.

A frequently occurring problem with electromagnetic wave propagation via sky wave in the HF frequency range is *signal multipath*. Signal multipath occurs when the transmitted signal arrives at the receiver via multiple propagation paths at different delays. Signal multipath generally results in intersymbol interference in a digital communication system. Moreover, the signal components arriving via different propagation paths may add destructively, resulting in a phenomenon called *signal fading,* which most people have experienced when listening to a distant radio station at night when sky wave is the dominant propagation

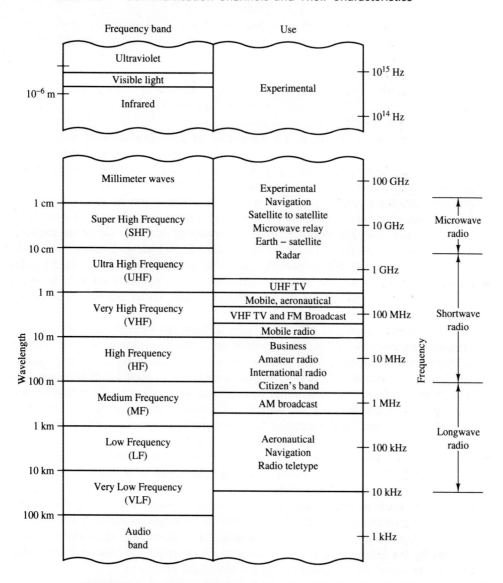

FIGURE 1.4. Frequency range for wireless electromagnetic channels. (*Adapted from Carlson (1975), Sec. Ed., © McGraw-Hill Book Compay Co. Reprinted with permission of the publisher.*)

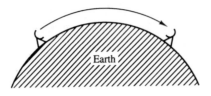

FIGURE 1.5. Illustration of ground-wave propagation.

mode. Additive noise at HF is a combination of atmospheric noise and thermal voice.

Sky-wave ionospheric propagation ceases to exist at frequencies above approximately 30 MHz, which is the end of the HF band. However, it is possible to have ionospheric scatter propagation at frequencies in the range of 30 MHz to 60MHz, resulting from signal scattering from the lower ionosphere. It is also possible to communicate over distances of several hundred miles by use of tropospheric scattering at frequencies in the range of 40 MHz to 300 MHz. Troposcatter results from signal scattering due to particles in the atmosphere at altitudes of 10 miles or less. Generally, ionospheric scatter and tropospheric scatter involve large signal propagation losses and require a large amount of transmitter power and relatively large antennas.

Frequencies above 30 MHz propagate through the ionosphere with relatively little loss and make satellite and extraterrestrial communications possible. Hence, at frequencies in the VHF band and higher, the dominant mode of electromagnetic propagation is line-of-sight (LOS) propagation. For terrestrial communication systems, this means that the transmitter and receiver antennas must be in direct LOS with relatively little or no obstruction. For this reason, television stations transmitting in the VHF and UHF frequency bands mount their antennas on high towers to achieve a broad coverage area.

In general, the coverage area for LOS propagation is limited by the curvature of the earth. If the transmitting antenna is mounted at a height h feet above the surface of the earth, the distance to the radio horizon, assuming no physical obstructions such as mountains, is approximately $d = \sqrt{2h}$ miles. For example, a TV antenna mounted on a tower of 1000 ft. in height provides a coverage of approximately 50 miles. As another example, microwave radio relay systems used extensively for telephone and video transmission at frequencies above 1 GHz have antennas mounted on tall towers or on the top of tall buildings.

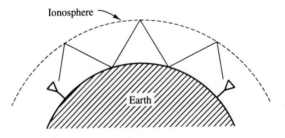

FIGURE 1.6. Illustration of sky-wave propagation.

The dominant noise limiting the performance of communication systems in the VHF and UHF frequency ranges is thermal noise generated in the receiver front end and cosmic noise picked up by the antenna. At frequencies in the SHF band above 10 GHz, atmospheric conditions play a major role in signal propagation. Figure 1.7 illustrates the signal attenuation in dB/mile due to precipitation for frequencies in the range of 10 to 100 GHz. We observe that heavy rain introduces extremely high propagation losses that can result in service outages (total breakdown in the communication system).

At frequencies above the EHF band, we have the infrared and visible light regions of the electromagnetic spectrum which can be used to provide LOS optical

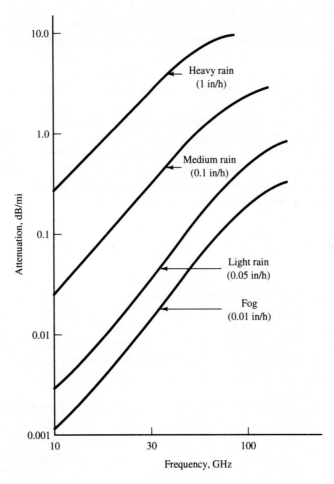

FIGURE 1.7. Signal attenuation due to precipitation (*From Ziemer and Tranter (1990), © Houghton Mifflin Co. Reprinted with permission of the publisher.*

communication in free space. To date, these frequency bands have been used in experimental communication systems, such as satellite-to-satellite links.

Underwater acoustic channels. Over the past few decades, ocean exploration activity has been steadily increasing. Coupled with this increase is the need to transmit data collected by sensors placed under water to the surface of the ocean. From there it is possible to relay the data via a satellite to a data collection center.

Electromagnetic waves do not propagate over long distances under water except at extremely low frequencies. However, the transmission of signals at such low frequencies is prohibitively expensive because of the large and powerful transmitters required. The attenuation of electromagnetic waves in water can be expressed in terms of the *skin depth,* which is the distance a signal is attenuated by $1/e$. For sea water, the skin depth $\delta = 250/\sqrt{f}$, where f is expressed in Hz and δ is in meters. For example, at 10 kHz, the skin depth is 2.5 meters. In contrast, acoustic signals propagate over distances of tens and even hundreds of kilometers.

An underwater acoustic channel is characterized as a multipath channel due to signal reflections from the surface and the bottom of the sea. Because of wave motion, the signal multipath components undergo time-varying propagation delays which result in signal fading. In addition, there is frequency-dependent attenuation, which is approximately proportional to the square of the signal frequency.

Ambient ocean acoustic noise is caused by shrimp, fish, and various mammals. Near harbors, there is also man-made acoustic noise in addition to the ambient noise. In spite of this hostile environment, it is possible to design and implement efficient and highly reliable underwater acoustic communication systems for transmitting digital signals over large distances.

Storage channels. Information storage and retrieval systems constitute a very significant part of data-handling activities on a daily basis. Magnetic tape, including digital audio tape and video tape, magnetic disks used for storing large amounts of computer data, optical disks used for computer data storage, and compact disks are examples of data storage systems that can be characterized as communication channels. The process of storing data on a magnetic tape or a magnetic or optical disk is equivalent to transmitting a signal over a telephone or a radio channel. The readback process and the signal processing involved in storage systems to recover the stored information is equivalent to the functions performed by a receiver in a telephone or radio communication system to recover the transmitted information.

Additive noise generated by the electronic components and interference from adjacent tracks is generally present in the readback signal of a storage system, just as is the case in a telephone or a radio communication system.

The amount of data that can be stored is generally limited by the size of the disk or tape and the density (number of bits stored per square inch) that can be achieved by the write/read electronic systems and heads. For example, a packing density of 10^9 bits per square inch has been recently demonstrated in an experimental magnetic disk storage system. (Current commercial magnetic storage products achieve a much lower density.) The speed at which data can be written on a disk or tape and the speed at which it can be read back is also limited by the associated mechanical and electrical subsystems that constitute an information storage system.

Channel coding and modulation are essential components of a well-designed digital magnetic or optical storage system. In the readback process, the signal is demodulated and the added redundancy introduced by the channel encoder is used to correct errors in the readback signal.

1.4 MATHEMATICAL MODELS FOR COMMUNICATION CHANNELS

In the design of communication systems for transmitting information through physical channels, we find it convenient to construct mathematical models that reflect the most important characteristics of the transmission medium. Then, the mathematical model for the channel is used in the design of the channel encoder and modulator at the transmitter and the demodulator and channel decoder at the receiver. Below, we provide a brief description of the channel models that are frequently used to characterize many of the physical channels that we encounter in practice.

The additive noise channel.   The simplest mathematical model for a communication channel is the additive noise channel, illustrated in Figure 1.8. In this model, the transmitted signal $s(t)$ is corrupted by an additive random noise process $n(t)$. Physically, the additive noise process may arise from electronic components and amplifiers at the receiver of the communication system, or from interference encountered in transmission as in the case of radio signal transmission.

If the noise is introduced primarily by electronic components and amplifiers at the receiver, it may be characterized as thermal noise. This type of noise is characterized statistically as a *Gaussian noise process*. Hence, the resulting mathematical model for the channel is usually called the *additive Gaussian noise channel*. Because this channel model applies to a broad class of physical communication channels and because of its mathematical tractability, this is the predominant channel model used in our communication system analysis and design. Channel attenuation is easily incorporated into the model. When the signal undergoes

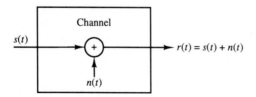

FIGURE 1.8.   The additive noise channel.

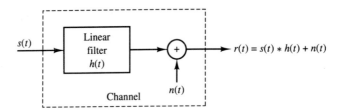

FIGURE 1.9.   The linear filter channel with additive noise.

attenuation in transmission through the channel, the received signal is

$$r(t) = \alpha s(t) + n(t) \tag{1.4.1}$$

where α represents the attenuation factor.

The linear filter channel.   In some physical channels such as wireline telephone channels, filters are used to ensure that the transmitted signals do not exceed specified bandwidth limitations and thus do not interfere with one another. Such channels are generally characterized mathematically as linear filter Channels with additive noise, as illustrated in Figure 1.9. Hence, if the channel input is the signal $s(t)$, the channel output is the signal

$$r(t) = s(t) \star h(t) + n(t)$$

$$= \int_{-\infty}^{+\infty} h(\tau)s(t - \tau)\,d\tau + n(t) \tag{1.4.2}$$

where $h(t)$ is the impulse response of the linear filter and $\star$ denotes convolution.

The linear time-variant filter channel.   Physical channels such as underwater acoustic channels and ionospheric radio channels which result in time-variant multipath propagation of the transmitted signal may be characterized mathematically as time-variant linear filters. Such linear filters are characterized by a time-variant channel impulse response $h(\tau; t)$, where $h(\tau; t)$ is the response of the channel at time t due to an impulse applied at time $t - \tau$. Thus, τ represents the "age" (elapsed-time) variable. The linear time-variant filter channel with additive noise is illustrated Figure 1.10. For an input signal $s(t)$, the channel output signal is

$$r(t) = s(t) \star h(\tau; t) + n(t)$$

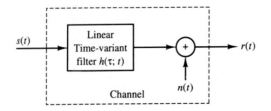

FIGURE 1.10.   Linear time-variant filter channel with additive noise.

$$= \int_{-\infty}^{+\infty} h(\tau;t)s(t-\tau)\,d\tau + n(t) \qquad (1.4.3)$$

A good model for multipath signal propagation through physical channels, such as the ionosphere (at frequencies below 30 MHz) and mobile cellular radio channels, is a special case of (1.4.3) in which the time-variant impulse response has the form

$$h(\tau;t) = \sum_{k=1}^{L} a_k(t)\delta(\tau - \tau_k) \qquad (1.4.4)$$

where the $\{a_k(t)\}$ represents the possibly time-variant attenuation factors for the L multipath propagation paths. If (1.4.4) is substituted into (1.4.3), the received signal has the form

$$r(t) = \sum_{k=1}^{L} a_k(t)s(t-\tau_k) + n(t) \qquad (1.4.5)$$

Hence, the received signal consists of L multipath components, where each component is attenuated by $\{a_{k(t)}\}$ and delayed by $\{\tau_k\}$.

The three mathematical models described above adequately characterize a large majority of physical channels encountered in practice. These three channel models are used in this text for the analysis and design of communication systems.

1.5 ORGANIZATION OF THE BOOK

Before we embark on the analysis and design of communication systems, it is important to develop the mathematical tools that are used throughout the book. We begin with a review of basic concepts in signal theory and linear systems in Chapter 2. The methods of signal representation in both time and frequency are presented. Emphasis is placed on the Fourier series and the Fourier transform representation of signals and the use of transforms in linear systems analysis. The process of sampling a bandlimited analog signal is also considered. Finally, we discuss the representation of bandpass signals and systems, which are widely used in the analysis and design of communication systems.

In Chapter 3 we present a review of the basic definitions and concepts in probability and random processes. These topics are particularly important in our study of electrical communications, because information sources produce random signals at their output and communication channels generally corrupt the transmitted signals in a random manner through the addition of noise and other channel distortion. Special emphasis is placed on Gaussian random processes, which provide mathematically tractable models for additive noise disturbances. Both time domain and frequency domain representations of random signals are presented.

Our study of communication system analysis and design begins in Chapter 4, which is devoted to information sources and source coding. In this chapter, we introduce a measure for the information content of a discrete source and describe two algorithms for efficient encoding of the source output. The major part of the chapter is devoted to the problem of encoding the outputs of analog sources. Several waveform encoding methods are described, including pulse code modulation (PCM), differential PCM, and delta modulation (DM). We also describe a model-based approach to analog source encoding, based on linear prediction.

In Chapters 5 through 9 we treat the subjects of modulation and demodulation. We begin with modulation and demodulation of analog information-bearing signals. Chapter 5 is devoted entirely to this topic, with detailed descriptions of amplitude modulation (AM), frequency modulation (FM), and phase modulation (PM). In Chapter 6, we evaluate the effects of additive noise in the demodulation of analog signals.

Chapter 7 is concerned with baseband digital signal transmission and reception. In baseband transmission, the frequency of the transmitted signal is concentrated in a band of frequencies around $f = 0$ (dc). Various types of binary and nonbinary (M-ary) signal types are considered for signal transmission. The optimum form of the signal demodulator is derived when the channel additive noise is a Gaussian random process. Finally, the performance of the communication system, as measured by the probability of error, is obtained for the various types of signals used to transmit the information.

In Chapter 8, we continue our treatment of baseband transmission, and consider the design of the transmitter and the receiver when the channel and, hence, the transmitted signal is constrained in bandwidth. The effect of channel distortion, in the form of amplitude and phase distortion, is also considered in the context of optimizing the transmitted signal and its demodulation. In this context, we describe the design of an adaptive equalizer for mitigating the effects of amplitude and phase distortion that is encountered in signal transmission.

Chapter 9 deals with carrier modulation methods for transmitting digital information over a communication channel. These include amplitude-shift keying (ASK); phase-shift keying (PSK); frequency-shift keying (FSK); combined ASK-PSK, including quadrature amplitude modulation (QAM); and continuous-phase modulation (CPM). The error probability performance achieved with these modulation methods is also derived. The topic of signal fading is also considered in this chapter.

Channel coding and decoding is the topic of Chapter 10. In this chapter, we describe the concept of channel capacity and present the capacity of an additive white Gaussian noise channel. Linear block codes and convolutional codes are considered for enhancing the performance of a digital communication system in the presence of additive noise. Decoding algorithms for both block codes and convolutional codes are also described. The final topic of this chapter provides a treatment of trellis-coded modulation, which is widely used in the implementation of high speed modems.

The last chapter of the book, Chapter 11, is devoted to an introduction of spread-spectrum signals for digital communications. Spread-spectrum signals have found widespread use in military communication systems over the past four decades.

They are particularly effective in providing resistance to jamming (intentional interference). They have also been used to provide covert communications by means of signal transmission at very low power levels. Their jam-resistant property makes spread-spectrum signals very attractive for commercial use, especially for mobile (cellular) radio communications and low-power radio communications within large buildings.

In an introductory book of this level, we have not attempted to provide a large number of references to the technical literature. However, we have included in each chapter several supplementary references for further reading, including textbooks and basic or tutorial treatments of important topics. References are cited by giving the author's name with the year of publication in parentheses, for example, Nyquist (1924).

1.6   FURTHER READING

We have already cited several historical treatments of radio and telecommunications during the past century. These include the book by McMahon (1984), Ryder and Fink (1984) and Millman (1984). The classical works of Nyquist (1924), Hartley (1928), Kotelnikov (1947), Shannon (1948), and Hamming (1950) are particularly important because they lay the groundwork of modern communication systems engineering.

2

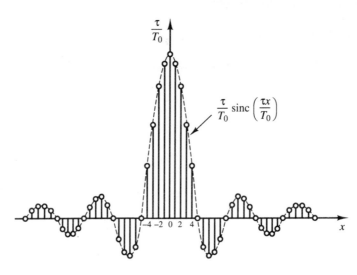

$$\frac{\tau}{T_0} \operatorname{sinc}\left(\frac{\tau x}{T_0}\right)$$

Signals and Linear Systems

In this chapter, we review the basics of signals and linear systems. The motivation for studying these fundamental concepts stems from the basic role they play in modeling various types of communication systems. In particular, signals are used to transmit information over a communication channel. Such signals are usually called *information-bearing signals.*

In the transmission of an information-bearing signal over a communication channel, the shape of the signal is changed, or *distorted,* by the channel. In other words, the output of the communication channel, which is called the *received signal,* is not an exact replica of the channel input due to the channel distortion. The communication channel is an example of a system, that is, an entity that produces an output signal when excited by an input signal. A large number of communication channels can be modeled closely by a subclass of systems called *linear systems.* Linear systems are a large subclass of systems which arise naturally in many practical applications and are rather easy to deal with. We have devoted this entire chapter to the study of the basics of signals and linear systems.

2.1 BASIC CONCEPTS

This section includes the basic definitions of signals and linear systems.

2.1.1 Classification of Signals

A signal is any function that carries some information. We will only consider
signals whose independent variable is time. The symbol used for a signal is the
symbol used for mathematical functions, e.g., $x(t)$ or $f(t)$. An example of a signal
is a speech signal. A *sample waveform* (the term usually used to denote a realization
of a signal) of speech is shown in Figure 2.1.

   Continuous-Time and Discrete-Time Signals.   Based on the range of the
independent variable, signals can be divided into two classes, *discrete-time* signals
and *continuous-time* signals.

   Definition 2.1.1.   A continuous-time signal is a signal $x(t)$ for which the
independent variable t takes real numbers. A discrete-time signal, denoted by $x[n]$,
is a signal for which the independent variable n takes its values in the set of
integers.

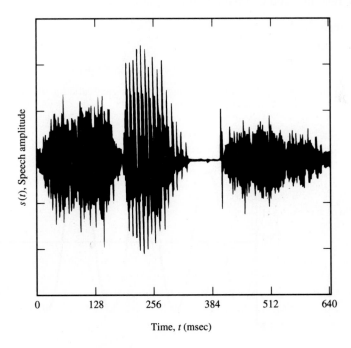

FIGURE 2.1.   A sample speech waveform.

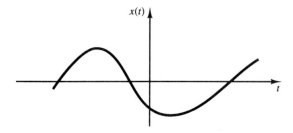

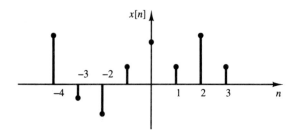

FIGURE 2.2. Examples of discrete-time and continuous-time signals.

By sampling a continuous-time signal $x(t)$ at time instants separated by T_0 we can define the discrete-time signal $x[n] = x(nT_0)$. Figure 2.2 shows examples of discrete-time and continuous-time signals.

Example 2.1.1

Let

$$x(t) = A\cos(2\pi f_0 t + \theta)$$

This is an example of a continuous-time signal called a *sinusoidal* signal. A sketch of this signal is given in Figure 2.3.

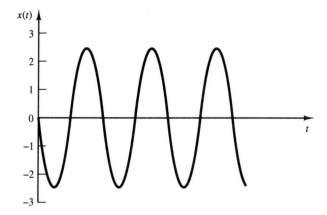

FIGURE 2.3. Sinusoidal signal.

Example 2.1.2

Let

$$x[n] = A\cos(2\pi f_0 n + \theta)$$

where $n \in \mathbb{Z}$ ($\mathbb{Z}$ is the set of integers). A sketch of this discrete-time signal is given in Figure 2.4.

Classifying signals into two major classes of continuous- and discrete-time signals is one of the ways the signals can be classified. There are other types of classification.

Real and Complex Signals.   Signals are functions, and functions at a given value of their independent variables are just numbers, which can be either real or complex.

Definition 2.1.2.   A *real signal* takes its values in the set of real numbers, i.e., $x(t) \in \mathbb{R}$. A *complex signal* takes its values in the set of complex numbers, i.e., $x(t) \in \mathbb{C}$.

Complex signals are usually used in communications to model signals that convey amplitude and phase information. Like complex numbers, a complex signal can be represented by two real signals. These two real signals can be either the real and imaginary parts or the absolute value (or *modulus* or *magnitude*) and phase. A graph of a complex signal can be given in either of these representations. However, the magnitude and phase graphs are commonly used.

Example 2.1.3

The signal $x(t)$ defined by

$$x(t) = Ae^{j(2\pi f_0 t + \theta)}, \quad A > 0$$

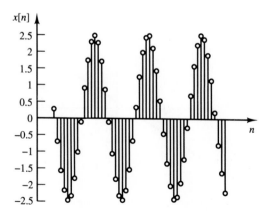

FIGURE 2.4.   Discrete sinusoidal signal.

is a complex signal. Its real part is

$$x_r(t) = A\cos(2\pi f_0 t + \theta)$$

and its imaginary part

$$x_i(t) = A\sin(2\pi f_0 t + \theta)$$

We could equivalently describe this signal in terms of its modulus and phase. The absolute value of $x(t)$ is

$$|x(t)| = \sqrt{x_r^2(t) + x_i^2(t)} = A$$

and its phase is

$$\angle x(t) = 2\pi f_0 t + \theta$$

Graphs of these functions are given in Figure 2.5.

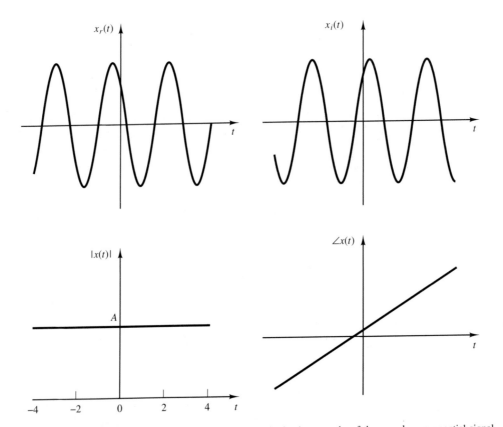

FIGURE 2.5. Real-imaginary and magnitude-phase graphs of the complex exponential signal.

The real and complex components, and the modulus and phase of any complex signal are related by the following relations

$$x_r(t) = |x(t)| \cos\left(\angle x(t)\right) \tag{2.1.1}$$

$$x_i(t) = |x(t)| \sin\left(\angle x(t)\right) \tag{2.1.2}$$

$$|x(t)| = \sqrt{x_r^2(t) + x_i^2(t)} \tag{2.1.3}$$

$$\angle x(t) = \begin{cases} \arctan \dfrac{x_i(t)}{x_r(t)}, & x_r(t) > 0 \\[2mm] \arctan \dfrac{x_i(t)}{x_r(t)} + \pi, & x_r(t) < 0,\; x_i(t) > 0 \\[2mm] \arctan \dfrac{x_i(t)}{x_r(t)} - \pi, & x_r(t) < 0,\; x_i(t) < 0 \end{cases} \tag{2.1.4}$$

Deterministic and Random Signals.

Definition 2.1.3.   A deterministic signal is a signal for which at any time instant t the value of $x(t)$ is given as a real or complex number. A random or *stochastic* signal is a signal for which at any time instant t the value of $x(t)$ is a random variable, i.e., it is defined by a probability density function.

All examples given so far have been deterministic signals. Random signals are discussed in Chapter 3.

Periodic and Nonperiodic Signals.   A periodic signal has the property that it repeats itself in time, and hence, it is sufficient to specify the signal in the basic interval called the *period*.

Definition 2.1.4.   A periodic signal is a signal $x(t)$ that satisfies the property

$$x(t + kT_0) = x(t)$$

for all t, all integers k, and some positive real number T_0 called the *period* of the signal. For discrete-time periodic signals we have

$$x[n + kN_0] = x[n]$$

for all integers n, all integers k, and a positive integer N_0 called the *period*. A signal that does not satisfy the conditions of periodicity is called *nonperiodic*.

Example 2.1.4

$$x(t) = A\cos(2\pi f_0 t + \theta)$$

and

$$x(t) = Ae^{j(2\pi f_0 t + \theta)}$$

are examples of real and complex periodic signals. The signal $u_{-1}(t)$ defined by

$$u_{-1}(t) = \begin{cases} 1 & t > 0 \\ \dfrac{1}{2} & t = 0 \\ 0 & t < 0 \end{cases} \qquad (2.1.5)$$

and illustrated in Figure 2.6 is an example of a nonperiodic signal. This signal is known as the *unit step* signal.

Example 2.1.5

The discrete-time sinusoidal signal shown in Figure 2.4 is not periodic for all values of f_0. The condition for it to be periodic is that

$$2\pi f_0(n + kN_0) + \theta = 2\pi f_0 n + \theta + 2m\pi \qquad (2.1.6)$$

for all integers n and k, some positive integer N_0, and some integer m. From this we conclude that

$$2\pi f_0 k N_0 = 2\pi m$$

or

$$f_0 = \frac{m}{kN_0}$$

i.e., the discrete sinusoidal signal is periodic *only for rational values of f_0.*

Causal and Noncausal Signals.   The concept of causality is an important concept in classifying *systems*. This concept has a close relation to realizability of a system. We will discuss this issue later in our treatment of various types of systems. Here we define the concept of causal signals, which is closely related to the concept of causal systems.

Definition 2.1.5.   A signal $x(t)$ is called *causal* if for all $t < 0$, we have $x(t) = 0$; otherwise, the signal is *noncausal*. Equivalently, a discrete-time signal is a causal signal if it is identically equal to zero for $n < 0$.

Example 2.1.6

The signal $x(t)$ defined by

$$x(t) = \begin{cases} A\cos(2\pi f_0 t + \theta) & \text{for } t \geq 0 \\ 0 & \text{otherwise} \end{cases}$$

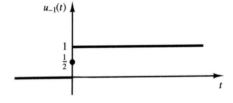

FIGURE 2.6.   The unit step signal.

is a causal signal. Its graph is shown in Figure 2.7.

Similarly, we can define *anti-causal* signals as signals whose time inverse is causal. Therefore, an anti-causal signal is identically equal to zero for $t > 0$.

Even and Odd Signals.   Evenness and oddness are expressions of various types of symmetry present in signals. A signal $x(t)$ is even if it has mirror symmetry with respect to the vertical axis. A signal is odd if it is symmetric with respect to the origin.

Definition 2.1.6.   The signal $x(t)$ is even if and only if for all t we have

$$x(-t) = x(t)$$

and is odd if and only if for all t

$$x(-t) = -x(t)$$

Figure 2.8 shows graphs of even and odd signals. In general, any signal $x(t)$ can be written as the sum of its even and odd parts.

$$x(t) = x_e(t) + x_o(t) \tag{2.1.7}$$

$$x_e(t) = \frac{x(t) + x(-t)}{2} \tag{2.1.8}$$

$$x_o(t) = \frac{x(t) - x(-t)}{2} \tag{2.1.9}$$

Example 2.1.7

The sinusoidal signal $x(t) = A\cos(2\pi f_0 t + \theta)$ is, in general, neither even nor odd. However, the special cases $\theta = 0$ and $\theta = \pm\frac{\pi}{2}$ correspond to even and odd signals, respectively. In general,

$$x(t) = \frac{A}{2}\cos(\theta)\cos(2\pi f_0 t) - \frac{A}{2}\sin(\theta)\sin(2\pi f_0 t)$$

Since $\cos(2\pi f_0 t)$ is even and $\sin(2\pi f_0 t)$ is odd, we conclude that

$$x_e(t) = \frac{A}{2}\cos(\theta)\cos(2\pi f_0 t)$$

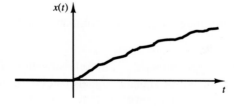

FIGURE 2.7.   An example of a causal signal.

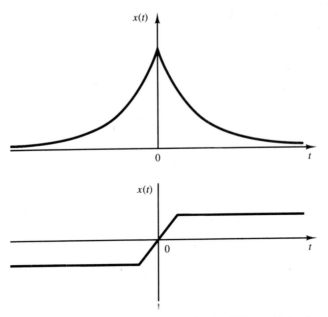

FIGURE 2.8. Examples of (a) an even signal and (b) an odd signal.

and

$$x_o(t) = -\frac{A}{2}\sin(\theta)\sin(2\pi f_0 t)$$

Example 2.1.8

From Figure 2.5 it is seen that for $\theta = 0$, $x(t) = Ae^{j2\pi f_0 t}$, the real part and the magnitude are even and the imaginary part and the phase are odd.

Hermitian Symmetry for Complex Signals. For complex signals, another form of symmetry, called Hermitian symmetry, is defined as follows:

Definition 2.1.7. A complex signal $x(t)$ is called Hermitian if its real part is even and its imaginary part is odd.

It can be easily shown that the magnitude of a Hermitian signal is even and the phase is odd.

Energy-Type and Power-Type Signals. This classification deals with the energy content and the power content of signals. Signals can be classified based on the finiteness of their energy content and power content.

Definition 2.1.8.   For any signal $x(t)$, the *energy content* of the signal, or simply the energy of the signal, E_x is defined by

$$E_x = \int_{-\infty}^{+\infty} |x(t)|^2 dt = \lim_{T \to \infty} \int_{-T/2}^{T/2} |x(t)|^2 dt \tag{2.1.10}$$

The *power content,* or power, of a signal is denoted by P_x and defined by

$$P_x = \lim_{T \to \infty} \frac{1}{T} \int_{-T/2}^{T/2} |x(t)|^2 dt \tag{2.1.11}$$

For real signals, $|x(t)|^2$ is replaced by $x^2(t)$.

Definition 2.1.9.   A signal $x(t)$ is an *energy-type* signal if and only if E_x is well defined and finite. A signal is a *power-type* signal if and only if P_x is well defined and

$$0 < P_x < \infty$$

Example 2.1.9

The energy content of $A \cos(2\pi f_0 t + \theta)$ is

$$E_x = \lim_{T \to \infty} \int_{-T/2}^{T/2} A^2 \cos^2(2\pi f_0 t + \theta) \, dt = \infty$$

Therefore, this signal is not an energy-type signal. However, the power of this signal is

$$
\begin{aligned}
P_x &= \lim_{T \to \infty} \frac{1}{T} \int_{-T/2}^{T/2} A^2 \cos^2(2\pi f_0 t + \theta) \, dt \\
&= \lim_{T \to \infty} \frac{1}{T} \int_{-T/2}^{T/2} \frac{A^2}{2} \left[1 + \cos(4\pi f_0 t + 2\theta) \right] dt \\
&= \lim_{T \to \infty} \left[\frac{A^2 T}{2T} + \left[\frac{A^2}{8\pi f_0 T} \sin(4\pi f_0 t + 2\theta) \right]_{-T/2}^{T/2} \right] \\
&= \frac{A^2}{2} < \infty
\end{aligned}
\tag{2.1.12}
$$

Hence, $x(t)$ is a power-type signal, and its power is $\frac{A^2}{2}$.

Example 2.1.10

For any periodic signal with period T_0, the energy is

$$E_x = \lim_{T \to \infty} \int_{-T/2}^{T/2} |x(t)|^2 dt$$

$$= \lim_{n\to\infty} \int_{-\frac{nT_0}{2}}^{+\frac{nT_0}{2}} |x(t)|^2 dt$$

$$= \lim_{n\to\infty} n \int_{-\frac{T_0}{2}}^{+\frac{T_0}{2}} |x(t)|^2 dt$$

$$= \infty \tag{2.1.13}$$

Therefore, periodic signals are not typically energy-type. The power content of any periodic signal is

$$P_x = \lim_{T\to\infty} \frac{1}{T} \int_{-T/2}^{T/2} |x(t)|^2 dt$$

$$= \lim_{n\to\infty} \frac{1}{nT_0} \int_{-\frac{nT_0}{2}}^{+\frac{nT_0}{2}} |x(t)|^2 dt$$

$$= \lim_{n\to\infty} \frac{n}{nT_0} \int_{-\frac{T_0}{2}}^{+\frac{T_0}{2}} |x(t)|^2 dt$$

$$= \frac{1}{T_0} \int_{-\frac{T_0}{2}}^{+\frac{T_0}{2}} |x(t)|^2 dt \tag{2.1.14}$$

This means that the power content of a periodic signal is equal to the average power in one period.

2.1.2 Some Important Signals and Their Properties

In our study of communication systems, certain signals appear frequently. Here we briefly introduce these signals and give some of their properties.

The Sinusoidal Signal. The sinusoidal signal is defined, in general, by

$$x(t) = A \cos(2\pi f_0 t + \theta)$$

The parameters A, f_0, and θ are called amplitude, frequency, and phase of the signal. A sinusoidal signal is periodic with period $T_0 = 1/f_0$. For a graph of this signal, see Figure 2.3.

The Complex Exponential Signal. The complex exponential signal is defined by $x(t) = Ae^{j(2\pi f_0 t + \theta)}$. Here again, A, f_0, and θ are called amplitude, frequency, and phase of the signal. This signal is shown in Figure 2.5.

The Unit Step Signal. The unit step signal defined in Section 2.1.1 is another frequently encountered signal. The unit step multiplied by any signal produces a "causal version" of the signal. The unit step signal is shown in Figure 2.6

The Rectangular Pulse.   This signal is denoted by $\Pi(t)$ and defined as

$$\Pi(t) = \begin{cases} 1 & -\frac{1}{2} < t < \frac{1}{2} \\ \dfrac{1}{2} & t = \pm\frac{1}{2} \\ 0 & \text{otherwise} \end{cases} \tag{2.1.15}$$

The graph of the rectangular pulse is shown in Figure 2.9. It is obvious that

$$\Pi(t) = u_{-1}\left(t + \frac{1}{2}\right) - u_{-1}\left(t - \frac{1}{2}\right)$$

The Triangular Signal.   This signal is denoted by $\Lambda(t)$ and defined as

$$\Lambda(t) = \begin{cases} t+1 & -1 \leq t \leq 0 \\ -t+1 & 0 \leq t \leq 1 \\ 0 & \text{otherwise} \end{cases} \tag{2.1.16}$$

(see Figure 2.10). It is not difficult to verify that[†]

$$\Lambda(t) = \Pi(t) \star \Pi(t) \tag{2.1.17}$$

The Sinc Signal.   The sinc signal is defined as

$$\text{sinc}(t) = \begin{cases} \sin(\pi t)/(\pi t) & t \neq 0 \\ 1 & t = 0 \end{cases} \tag{2.1.18}$$

The waveform corresponding to this signal is shown in Figure 2.11. From this figure it is seen that the sinc signal achieves its maximum at $t = 0$, and this maximum

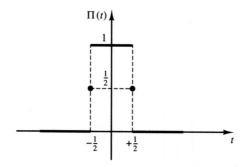

FIGURE 2.9.   The rectangular pulse.

[†] $x(t) \star y(t)$ denotes the convolution of two signals, which is defined by

$$x(t) \star y(t) = \int_{-\infty}^{+\infty} x(\tau) y(t-\tau)\, d\tau = \int_{-\infty}^{+\infty} x(t-\tau) y(\tau)\, d\tau$$

For more details see Section 2.1.4.

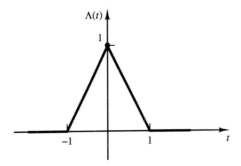

FIGURE 2.10.   The triangular pulse.

value equals 1. The zeros of the sinc signal are at $t = \pm1, \pm2, \pm3, \ldots$. Also note that local extrema of the sinc signal are *not* at $t = \pm\frac{3}{2}, \pm\frac{5}{2}, \ldots$.

The Sign or the Signum Signal.   The sign or the signum signal denotes the sign of the independent variable t and is defined by

$$\text{sgn}(t) = \begin{cases} 1 & t > 0 \\ -1 & t < 0 \\ 0 & t = 0 \end{cases} \tag{2.1.19}$$

This signal is shown in Figure 2.12. The signum signal can be expressed as the limit of the signal $x_n(t)$ defined by

$$x_n(t) = \begin{cases} e^{-\frac{t}{n}} & t > 0 \\ -e^{\frac{t}{n}} & t < 0 \\ 0 & t = 0 \end{cases} \tag{2.1.20}$$

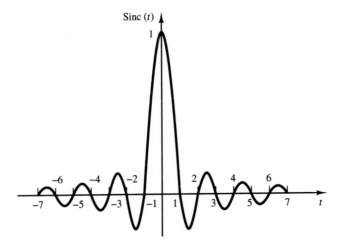

FIGURE 2.11.   The sinc signal.

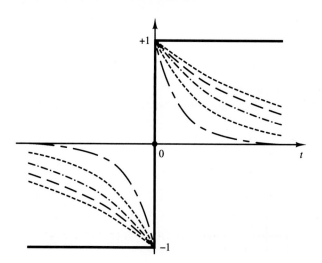

FIGURE 2.12.   The signum signal as the limit of $x_n(t)$.

when $n \to \infty$. We will later use this definition of the signum signal to find its Fourier transform. This limiting behavior is also shown in Figure 2.12.

The Impulse or Delta Signal.   The impulse or delta signal is a mathematical model for representing physical phenomena that take place in a very small time duration, so small that it is beyond the resolution of the measuring instruments involved and, for all practical purposes, their duration can be assumed to equal zero. Examples of such phenomena are a hammer blow, a very narrow voltage or current pulse, etc. In the precise mathematical sense, the impulse signal, denoted by $\delta(t)$, is not a function (or signal), it is a *distribution* or a *generalized function.* A distribution is defined in terms of its effect on another function (usually called "test function") under the integral sign. The impulse distribution (or signal) can be defined by its effect on the "test function" $\phi(t)$, which is assumed to be continuous at the origin, by the following relation:

$$\int_{-\infty}^{+\infty} \phi(t)\delta(t)\,dt = \phi(0) \qquad (2.1.21)$$

This property is called the *sifting* property of the impulse signal. In other words, the effect of the impulse signal on the "test function" $\phi(t)$ under the integral sign is to extract or *sift* its value at the origin. As it is seen, $\delta(t)$ is defined in terms of its action on $\phi(t)$ and not in terms of its value for different values of t.

Sometimes it is helpful to visualize $\delta(t)$ as the limit of certain known signals. The most commonly used forms are

$$\delta(t) = \lim_{\epsilon \downarrow 0} \frac{1}{\epsilon} \Pi\left(\frac{t}{\epsilon}\right) \qquad (2.1.22)$$

and

$$\delta(t) = \lim_{\epsilon \downarrow 0} \frac{1}{\epsilon} \operatorname{sinc}\left(\frac{t}{\epsilon}\right) \tag{2.1.23}$$

See Figure 2.13 for graphs of these signals.

The following properties follow from the definition of the impulse signal:

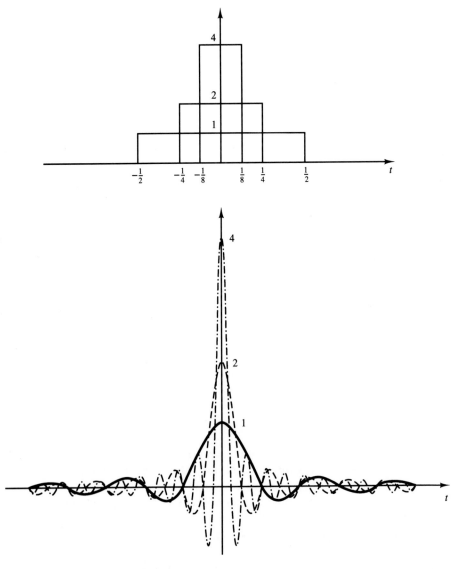

FIGURE 2.13. The impulse signal as a limit.

1. $\delta(t) = 0$ for all $t \neq 0$ and $\delta(0) = \infty$.

2. For any $\phi(t)$ continuous at t_0, we have

$$\int_a^b \phi(t)\delta(t - t_0)\, dt = \begin{cases} \phi(t_0) & a < t_0 < b \\ \dfrac{1}{2}\phi(t_0) & t_0 = a \text{ or } t_0 = b \\ 0 & \text{otherwise} \end{cases} \qquad (2.1.24)$$

3. For any $\phi(t)$ continuous at the origin,

$$\int_a^b \phi(t + t_0)\delta(t)\, dt = \begin{cases} \phi(t_0) & a < 0 < b \\ \dfrac{1}{2}\phi(t_0) & a = 0 \text{ or } b = 0 \\ 0 & \text{otherwise} \end{cases} \qquad (2.1.25)$$

4. For all $a \neq 0$,

$$\delta(at) = \frac{1}{|a|}\delta(t) \qquad (2.1.26)$$

5. $\delta(t)$ is the unitary element of the convolution operation, i.e., the result of convolution of any signal with the impulse signal is the original signal

$$x(t) \star \delta(t) = x(t) \qquad (2.1.27)$$

and

$$x(t) \star \delta(t - t_0) = x(t - t_0) \qquad (2.1.28)$$

6. The unit step signal is the integral of the impulse signal, and the impulse signal is the (*generalized* or *in distribution sense*) derivative of the unit step signal, i.e.,

$$u_{-1}(t) = \int_{-\infty}^t \delta(\tau)\, d\tau \qquad (2.1.29)$$

and

$$\delta(t) = \frac{d}{dt} u_{-1}(t) \qquad (2.1.30)$$

7. Similar to the definition of $\delta(t)$, we can define $\delta'(t)$, $\delta''(t)$, $\ldots$, $\delta^{(n)}(t)$, the *generalized derivatives* of $\delta(t)$ by the following equation:

$$\int_{-\infty}^{+\infty} \delta^{(n)}(t)\phi(t)\, dt = (-1)^n \left. \frac{d^n}{dt^n}\phi(t) \right|_{t=0} \qquad (2.1.31)$$

We can generalize this result to

$$\int_{-\infty}^{+\infty} \delta^{(n)}(t - t_0)\phi(t)\, dt = (-1)^n \left. \frac{d^n}{dt^n}\phi(t) \right|_{t=t_0} \qquad (2.1.32)$$

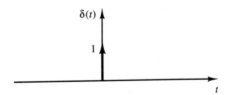

FIGURE 2.14. The impulse signal.

8. For even values of n, $\delta^{(n)}(t)$ is even, and for odd values of n, it is odd.

A schematic representation of the impulse signal is given in Figure 2.14

2.1.3 Classification of Systems

A system is an interconnection of various elements or devices that, from a certain viewpoint, behave as a whole. A system from a communication point of view is an entity that is excited by an *input signal,* and as a result of this excitation, produces an *output signal.* From a communication engineer's viewpoint, a system is a law that assigns output signals to various input signals. An electric circuit with some voltage source as the input and some current in a certain branch is an example of a system. The important point in the definition of a system is that *its output must be uniquely defined for any legitimate input.* This definition can be written mathematically as

$$y(t) = T\big[x(t)\big] \tag{2.1.33}$$

where $x(t)$ is the input, $y(t)$ is the output, and T is the operation performed by the system. Figure 2.15 shows a pictorial representation of a system.

Example 2.1.11

The input-output relationship $y(t) = ax(t) + bx^2(t)$ defines a system. For any input $x(t)$, the output $y(t)$ is uniquely determined.

A system is defined by (1) the operation that describes the system and (2) the set of legitimate input signals. We will denote the operation that describes the system by the operator T and the space of legitimate inputs to the system by X.

Example 2.1.12

The system described by the input-output relationship

$$y(t) = T\big[x(t)\big] = \frac{d}{dt}x(t) \tag{2.1.34}$$

for which X is the space of all differentiable signals, describes a system. This system is referred to as the *differentiator.*

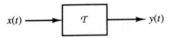

FIGURE 2.15. System, input and output.

The space X is usually defined by the system operation and therefore is not usually given explicitly.

Systems, like signals, can be classified according to their properties. Based on this point of view, various classifications are possible. Here we briefly introduce some of the fundamental system classifications.

Discrete-Time and Continuous-Time Systems.   Systems are defined in terms of their operation on input signals to produce the corresponding output signal. Systems can accept either discrete-time or continuous-time signals as their inputs and outputs. This is the basis of system classification into discrete-time and continuous-time systems.

Definition 2.1.10.   A *discrete-time system* accepts discrete-time signals as the input and produces discrete-time signals at the output. In a *continuous-time system,* both input and output signals are continuous-time signals.

Example 2.1.13

The systems described so far have all been continuous-time systems. An example of a discrete-time system is the system described by

$$y[n] = x[n] - x[n - 1] \qquad (2.1.35)$$

This system is called a *discrete-time differentiator.*

Linear and Nonlinear Systems.   Linear systems are systems for which the *superposition* property is satisfied; that is, the response of the system to a linear combination of the inputs is the linear combination of the responses to the corresponding inputs.

Definition 2.1.11.   A system $\mathcal{T}$ is *linear* if and only if for any two legitimate input signals, $x_1(t)$ and $x_2(t)$, and any two scalars, α and β, the linear combination $\alpha x_1(t) + \beta x_2(t)$ is also a legitimate input (in other words, the input space is a linear space), and

$$\mathcal{T}\big[\alpha x_1(t) + \beta x_2(t)\big] = \alpha \mathcal{T}\big[x_1(t)\big] + \beta \mathcal{T}\big[x_2(t)\big] \qquad (2.1.36)$$

A system that does not satisfy this relation is called *nonlinear.*

Linearity is a very important property. In a linear system, we can decompose the input into a linear combination of some fundamental signals whose outputs can be derived easily, then find the linear combination of the corresponding outputs.

Example 2.1.14

The differentiator described earlier is an example of a linear system because if $x_1(t)$ and $x_2(t)$ are differentiable, so is $\alpha x_1(t) + \beta x_2(t)$ for any choice of α and β, and

$$\frac{d}{dt}\big[\alpha x_1(t) + \beta x_2(t)\big] = \alpha x_1'(t) + \beta x_2'(t)$$

FIGURE 2.16. The input–output relation for the delay system.

The system described by

$$y(t) = ax^2(t)$$

is nonlinear because its response to $2x(t)$ is

$$\mathcal{T}\left[2x(t)\right] = 4x^2(t) \neq 2x^2(t) = 2\mathcal{T}\left[x(t)\right]$$

Example 2.1.15

A delay system is defined by $y(t) = x(t - \Delta)$, that is, the output is a delayed version of the input (see Figure 2.16). If $x(t) = \alpha x_1(t) + \beta x_2(t)$, then the response of the system is obviously $\alpha x_1(t - \Delta) + \beta x_2(t - \Delta)$. Therefore, the system is linear.

We denote the operation of linear systems by $\mathcal{L}$ rather than $\mathcal{T}$.

Time-Invariant and Time-Varying Systems. A system is called time-invariant if its input-output relationship does not change with time. This means that a delayed version of an input results in a delayed version of the output.

Definition 2.1.12. A system is *time-invariant* if and only if for all $x(t)$ and all values of t_0, its response to $x(t - t_0)$ is $y(t - t_0)$, where $y(t)$ is the response of the system to $x(t)$ (see Figure 2.17).

Example 2.1.16

The differentiator is a time-invariant system, since

$$\frac{d}{dt}x(t - t_0) = x'(t)\big|_{t=t-t_0}$$

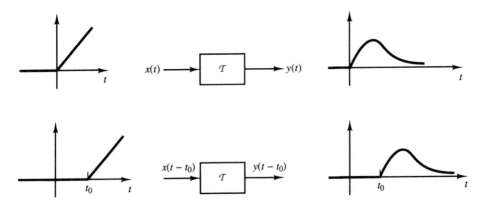

FIGURE 2.17. A time-invariant system.

Example 2.1.17

The *modulator*, defined by $y(t) = x(t)\cos 2\pi f_0 t$, is an example of a time-varying system. If $y(t) = x(t)\cos(2\pi f_0 t)$ is the response to $x(t)$, the response to $x(t - t_0)$ is

$$x(t - t_0)\cos(2\pi f_0 t)$$

which is not equal to $y(t - t_0)$.

The class of *linear time-invariant systems* (LTI for short) is particularly important. The response of these systems to inputs can be derived simply by finding the convolution of the input and the impulse response of the system. We will discuss this property in more detail in Section 2.1.4.

Causal and Noncausal Systems.   Causality deals with physical realizability of systems. Since no physical system can predict what values its input signal will assume in the future, it is sound to assume that in a physically realizable system the output at any time depends only on the values of the input signal up to that time and does not depend on the future values of the input.

Definition 2.1.13.   A system is *causal* if its output at any time-instant t_0 depends on the input at times prior to t_0, i.e.,

$$y(t) = \mathcal{T}\big[x(t) : t \le t_0\big]$$

A necessary and sufficient condition for an LTI system to be causal is that its impulse response $h(t)$ (which will be defined in next section) be a causal signal, i.e., for $t < 0$ we must have $h(t) = 0$. For noncausal systems, the value of the output at t_0 also depends on the values of the input at times after t_0. Noncausal systems are encountered in situations where signals are not processed in real time.

Example 2.1.18

A differentiator is an example of a causal system. A modulator is also a causal but time-varying system. A delay system defined in Example 2.1.15 is causal for $\Delta \ge 0$ and noncausal for $\Delta < 0$.

2.1.4 Analysis of LTI Systems in the Time Domain

The class of LTI systems play an important role both in communication and system theory. For this class of systems, the input–output relation is particularly simple and can be expressed in terms of the convolution integral. To develop this relationship, we first introduce the concept of the *impulse response* of a system.

Definition 2.1.14.   The impulse response $h(t)$ of a system is the response of the system to a unit impulse input $\delta(t)$,

$$h(t) = \mathcal{T}\big[\delta(t)\big]$$

The response of the system to a unit impulse applied at time τ, i.e., $\delta(t - \tau)$, is denoted by $h(t, \tau)$. For time-invariant systems obviously $h(t, \tau) = h(t - \tau)$.†

The convolution integral. Here we will derive the output $y(t)$ of an LTI system to any input signal $x(t)$. We will show that $y(t)$ can be expressed in terms of the input $x(t)$ and the impulse response $h(t)$ of the system.

We have seen in Section 2.1.2 that for any signal $x(t)$ we have

$$x(t) = x(t) \star \delta(t) = \int_{-\infty}^{+\infty} x(\tau)\delta(t - \tau)\,d\tau \qquad (2.1.37)$$

Now, if we denote the response of the LTI system to the input $x(t)$ by $y(t)$, we can write

$$y(t) = L\big[x(t)\big]$$

$$= L\left[\int_{-\infty}^{+\infty} x(\tau)\delta(t - \tau)\,d\tau\right]$$

$$\overset{a}{=} \int_{-\infty}^{+\infty} x(\tau)L\big[\delta(t - \tau)\big]\,d\tau$$

$$\overset{b}{=} \int_{-\infty}^{+\infty} x(\tau)h(t - \tau)\,d\tau$$

$$= x(t) \star h(t) \qquad (2.1.38)$$

where (a) follows from linearity of the system and (b) follows from time-invariance. This shows that the response to $x(t)$ is the convolution of $x(t)$ and the impulse response $h(t)$. Therefore, for the class of LTI systems, the impulse response completely characterizes the system. This means that the impulse response is all the information we need to describe the system behavior.

Example 2.1.19

The system described by

$$y(t) = \int_{-\infty}^{t} x(\tau)\,d\tau \qquad (2.1.39)$$

is called an *integrator*. Since integration is linear, this system is linear. Also, the response to $x(t - t_0)$ is

$$y_1(t) = \int_{-\infty}^{t} x(\tau - t_0)\,d\tau$$

---

†Note that here we are sloppy in notation, in the sense that h denotes two different functions, $h(t, \tau)$, which is a function of two variables, and $h(t)$, which is a function of one variable.

$$= \int_{-\infty}^{t-t_0} x(u)\,du$$

$$= y(t - t_0) \tag{2.1.40}$$

where we have used the change of variables $u = \tau - t_0$. It is seen that the system is LTI. The impulse response is obtained by applying an impulse at the input; i.e.,

$$h(t) = \int_{-\infty}^{t} \delta(\tau)\,d\tau = u_{-1}(t)$$

Example 2.1.20

Let a linear time-invariant system have impulse response $h(t)$. Assume this system has a complex exponential signal as input, $x(t) = Ae^{j(2\pi f_0 t + \theta)}$. The response to this input can be obtained by

$$y(t) = \int_{-\infty}^{+\infty} h(\tau) A e^{j\left(2\pi f_0 (t-\tau) + \theta\right)} d\tau$$

$$= A e^{j\theta} e^{j2\pi f_0 t} \int_{-\infty}^{+\infty} h(\tau) e^{-j2\pi f_0 \tau} d\tau$$

$$= A |H(f_0)| e^{j\left(2\pi f_0 t + \theta + \angle H(f_0)\right)} \tag{2.1.41}$$

where

$$H(f_0) = |H(f_0)| e^{j\angle H(f_0)} = \int_{-\infty}^{+\infty} h(\tau) e^{-j2\pi f_0 \tau} d\tau \tag{2.1.42}$$

This shows that the response of an LTI system to the complex exponential with frequency f_0 is a complex exponential with the same frequency. The amplitude of the response can be obtained by multiplying the amplitude of the input by $|H(f_0)|$, and its phase is obtained by adding $\angle H(f_0)$ to the input phase. $H(f_0)$ is a function of the impulse response and the input frequency. The family of complex exponentials are called *eigenfunctions* of the class of linear time-invariant systems. The eigenfunctions of a system are the set of inputs for which the output is a (possibly complex) scaling of the input. Because of this important property, finding the response of the systems to their eigenfunctions is particularly simple, and therefore, it is desirable to find ways to express arbitrary signals in terms of complex exponentials. We will later explore ways to do this.

2.2 FOURIER SERIES

A large number of building blocks in a communication system can be modeled by linear time-invariant systems. LTI systems provide good and accurate models for a large class of communication channels. Some basic components of transmitters and receivers, such as filters, amplifiers, and equalizers, are LTI systems.

Our main objective in this and subsequent sections is to develop methods and tools necessary to analyze linear time-invariant systems. By analyzing a system, we mean finding the output corresponding to a given input and, at the same time, having insight into the behavior of the system. We have already seen that the input and output of an LTI system are related by the convolution integral, given by

$$y(t) = \int_{-\infty}^{+\infty} h(\tau)x(t-\tau)\,d\tau = \int_{-\infty}^{+\infty} h(t-\tau)x(\tau)\,d\tau \qquad (2.2.1)$$

where $h(t)$ denotes the impulse response of the system. The convolution integral provides the basic tool for analyzing LTI systems. However, there are major drawbacks in direct application of the convolution integral. First, finding the response of an LTI system using the convolution integral, although straightforward, is not always an easy task. Second, even when the convolution integral can be performed with reasonable effort, it does not provide good insight into what the system really does.

In this and the subsequent section, we will develop another approach to analyzing LTI systems. The basic idea is *to expand the input as a linear combination of some basic signals whose output can be obtained easily, and then employing the linearity properties of the system to obtain the corresponding output.* This approach is much easier than direct computation of the convolution integral and at the same time provides better insight into the behavior of LTI systems. This method is based on the close similarity between the expansion of signals in terms of a basic signal set and the expansion of vectors in Euclidean space in terms of unit vectors.

2.2.1 Signal Space Concepts

In many respects, signals behave like vectors. Signals can be added and subtracted like vectors, and the result will be a new signal. Signals can be multiplied by scalars to yield new signals, and linear combinations of signals are new signals. Like vectors, signals can be represented by some fundamental signals that form a basis for the "signal space." Like vectors, we can define inner product and norm for signals, and like vector spaces, we can express a linear transformation on signals by their effect on the basis of the signal space. We start our discussion with some review of the basic properties of vectors and vector spaces.

Vectors in space. Vectors in three-dimensional space are represented in terms of their projections on the unit vectors, or "basis" of the space. A vector $\mathbf{a}$ can be expressed as $\mathbf{a} = a_1\mathbf{e}_1 + a_2\mathbf{e}_2 + a_3\mathbf{e}_3$, where $\mathbf{e}_1$, $\mathbf{e}_2$, and $\mathbf{e}_3$ are the unit vectors in the three basic directions of the space. The vector $\mathbf{a}$ can be equally represented by the triplet (a_1, a_2, a_3). In an n-dimensional space with basis $(\mathbf{e}_1, \mathbf{e}_2, \ldots, \mathbf{e}_n)$, the representation of any vector $\mathbf{a} = \sum_{i=1}^{n} a_i\mathbf{e}_i$ is similarly $(a_1, a_2, \ldots, a_n)$. The dimension of the space, n, is the number of unit vectors necessary and sufficient for representation of any vector in the space. Vectors $\mathbf{a}_1, \mathbf{a}_1, \ldots, \mathbf{a}_n$ are independent if none of them can be represented as a linear combination of the others.

The *inner product* of two vectors $\mathbf{a} = \sum_{i=1}^{n} a_i \mathbf{e}_i$ and $\mathbf{b} = \sum_{i=1}^{n} b_i \mathbf{e}_i$ is defined by

$$\mathbf{a} \cdot \mathbf{b} = \sum_{i=1}^{n} a_i b_i \tag{2.2.2}$$

Two vectors are orthogonal if $\mathbf{a} \cdot \mathbf{b} = 0$. The *norm* of a vector $\mathbf{a}$, denoted by $\|\mathbf{a}\|$, is a nonnegative number defined by

$$\|\mathbf{a}\| = (\mathbf{a} \cdot \mathbf{a})^{1/2} = \sqrt{\sum_{i=1}^{n} a_i^2} \tag{2.2.3}$$

A set of vectors are called orthonormal if they are mutually orthogonal and all have unity norm. So far we have assumed that our basis is an orthonormal basis. For an orthonormal basis, $a_i = \mathbf{a} \cdot \mathbf{e}_i$, and therefore

$$\mathbf{a} = \sum_{i=1}^{n} (\mathbf{a} \cdot \mathbf{e}_i) \mathbf{e}_i \tag{2.2.4}$$

The *triangle inequality* for any vectors $\mathbf{a}$ and $\mathbf{b}$ is

$$\|\mathbf{a} + \mathbf{b}\| \leq \|\mathbf{a}\| + \|\mathbf{b}\|, \tag{2.2.5}$$

with equality if and only if $\mathbf{a}$ and $\mathbf{b}$ are in the same direction, i.e.,

$$\mathbf{a} = k\mathbf{b} \tag{2.2.6}$$

for some positive real scalar k. From the triangle inequality we can derive the Cauchy–Schwartz inequality, given as

$$|\mathbf{a} \cdot \mathbf{b}| \leq \|\mathbf{a}\| \cdot \|\mathbf{b}\| \tag{2.2.7}$$

with equality if $\mathbf{a} = k\mathbf{b}$. For two vectors $\mathbf{a}$ and $\mathbf{b}$ we have

$$\|\mathbf{a} + \mathbf{b}\|^2 = (\mathbf{a} + \mathbf{b})^2 = \|\mathbf{a}\|^2 + \|\mathbf{b}\|^2 + 2\mathbf{a} \cdot \mathbf{b} \tag{2.2.8}$$

If $\mathbf{a}$ and $\mathbf{b}$ are orthogonal, then $\mathbf{a} \cdot \mathbf{b} = 0$ and

$$\|\mathbf{a} + \mathbf{b}\|^2 = (\mathbf{a} + \mathbf{b})^2 = \|\mathbf{a}\|^2 + \|\mathbf{b}\|^2 \tag{2.2.9}$$

This is the *Pythagorean relation*.

A linear transformation in an n-dimensional space is a transformation L that satisfies the superposition equality

$$L[\alpha \mathbf{a} + \beta \mathbf{b}] = \alpha L[\mathbf{a}] + \beta L[\mathbf{b}] \tag{2.2.10}$$

for any vectors $\mathbf{a}$ and $\mathbf{b}$ and any scalars α and β. Since any vector $\mathbf{a}$ can be represented as $\sum_{i=1}^{n} a_i \mathbf{e}_i$, we have

$$L[\mathbf{a}] = \sum_{i=1}^{n} a_i L[\mathbf{e}_i] \tag{2.2.11}$$

This means that *it is enough to know the effect of the linear transformation on the basis of the space.* From this information we can find its effect on any arbitrary vector in the space.

Eigenvectors of a linear transformation L are vectors **a** which do not change direction when transformed by L, i.e., the result of the transformation is a scaling of the original vector (compare with the definition of the eigenfunctions of a system given in Example 2.1.20). Therefore, for vector **a** to be an eigenvector of the linear transformation L, we must have

$$L[\mathbf{a}] = \lambda \mathbf{a} \qquad (2.2.12)$$

If this relation holds, **a** is an eigenvector of L and λ is called the *eigenvalue* corresponding to the eigenvector **a**. Also note the similarity between the concepts of linear transformations and linear systems.

Gram–Schmidt orthogonalization process.   Given n linearly independent vectors $\mathbf{a}_1, \mathbf{a}_2, \ldots, \mathbf{a}_n$, we want to find an orthonormal basis for representation of these vectors. The process by which we can find a set of orthonormal basis is known as the Gram–Schmidt orthogonalization process.

We start with one of the vectors, say $\mathbf{a}_1$, and scale it such that its norm becomes unity. This can be done by dividing $\mathbf{a}_1$ by its norm. Hence, we define vectors $\mathbf{v}_1$ and $\mathbf{i}_1$ as

$$\mathbf{v}_1 = \mathbf{a}_1$$
$$\mathbf{i}_1 = \frac{\mathbf{v}_1}{\|\mathbf{v}_1\|} \qquad (2.2.13)$$

To find $\mathbf{i}_2$, we first subtract from $\mathbf{a}_2$ its projection on $\mathbf{i}_1$ to find $\mathbf{v}_2 = \mathbf{a}_2 - (\mathbf{a}_2 \cdot \mathbf{i}_1)\mathbf{i}_1$. Then we normalize $\mathbf{v}_2$ to obtain $\mathbf{i}_2$.

$$\mathbf{v}_2 = \mathbf{a}_2 - (\mathbf{a}_2 \cdot \mathbf{i}_1)\mathbf{i}_1$$
$$\mathbf{i}_2 = \frac{\mathbf{v}_2}{\|\mathbf{v}_2\|} \qquad (2.2.14)$$

Since $\mathbf{v}_2$ has no projection on $\mathbf{i}_1$, both $\mathbf{v}_2$ and $\mathbf{i}_2$ are orthogonal to $\mathbf{i}_1$. To find $\mathbf{i}_3$, we have to first subtract from $\mathbf{a}_3$ its projection on $\mathbf{i}_1$ and $\mathbf{i}_2$, and then normalize the resulting vector. Thus,.

$$\mathbf{v}_3 = \mathbf{a}_3 - (\mathbf{a}_3 \cdot \mathbf{i}_1)\mathbf{i}_1 - (\mathbf{a}_3 \cdot \mathbf{i}_2)\mathbf{i}_2$$
$$\mathbf{i}_3 = \frac{\mathbf{v}_3}{\|\mathbf{v}_3\|} \qquad (2.2.15)$$

Again $\mathbf{v}_3$, and therefore $\mathbf{i}_3$, have no projection on $\mathbf{i}_1$ and $\mathbf{i}_2$ and therefore are orthogonal to both $\mathbf{i}_1$ and $\mathbf{i}_2$. In general, for all $1 \leq m \leq n$, we can determine $\mathbf{v}_m$

and $\mathbf{i}_m$ by

$$\mathbf{v}_m = \mathbf{a}_m - \sum_{j=1}^{m-1} (\mathbf{a}_m \cdot \mathbf{i}_j)\mathbf{i}_j$$

$$\mathbf{i}_m = \frac{\mathbf{v}_m}{\|\mathbf{v}_m\|} \qquad\qquad (2.2.16)$$

It is easy to verify that $\mathbf{i}_1, \mathbf{i}_2, \ldots, \mathbf{i}_n$, obtained by this process constitute an orthonormal basis. From the process just described it is obvious that depending on which vector we start with, we obtain a different orthonormal basis. Therefore, the Gram–Schmidt process does not provide a unique orthonormal basis. Figure 2.18 shows the Gram–Schmidt orthogonalization process in three-dimensional space.

Example 2.2.1

Let $\mathbf{a}_1 = 2\mathbf{e}_1 + \mathbf{e}_2$, $\mathbf{a}_2 = \mathbf{e}_1 + \mathbf{e}_2 + \mathbf{e}_3$, and $\mathbf{a}_3 = \mathbf{e}_2 + \mathbf{e}_3$ be vectors in three-dimensional space. Using the Gram–Schmidt process, find an orthonormal basis for the space spanned by these vectors.

Solution   Using the Gram–Schmidt orthogonalization process, we have

$$\mathbf{v}_1 = 2\mathbf{e}_1 + \mathbf{e}_2$$
$$\|\mathbf{v}_1\| = \sqrt{1+4} = \sqrt{5}$$

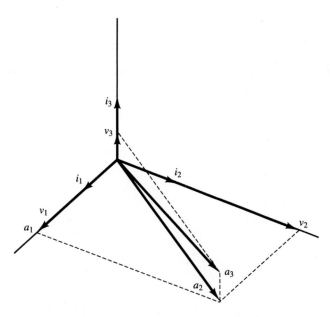

FIGURE 2.18.   Gram–Schmidt orthogonalization process

$$\mathbf{i}_1 = \frac{\mathbf{v}_1}{\|\mathbf{v}_1\|} = \frac{1}{\sqrt{5}}(2\mathbf{e}_1 + \mathbf{e}_2)$$

To find $\mathbf{i}_2$, we have

$$\mathbf{a}_2 \cdot \mathbf{i}_1 = (\mathbf{e}_1 + \mathbf{e}_2 + \mathbf{e}_3) \cdot (\frac{1}{\sqrt{5}}(2\mathbf{e}_1 + \mathbf{e}_2)) = \frac{3}{\sqrt{5}}$$

$$\mathbf{v}_2 = \mathbf{a}_2 - (\mathbf{a}_2 \cdot \mathbf{i}_1)\mathbf{i}_1$$

$$= (\mathbf{e}_1 + \mathbf{e}_2 + \mathbf{e}_3) - \frac{3}{\sqrt{5}}(\frac{2}{\sqrt{5}}\mathbf{e}_1 + \frac{1}{\sqrt{5}}\mathbf{e}_2)$$

$$= -\frac{1}{5}\mathbf{e}_1 + \frac{2}{5}\mathbf{e}_2 + \mathbf{e}_3$$

Now we can proceed to determine $\mathbf{i}_2$ as follows:

$$\mathbf{i}_2 = \frac{\mathbf{v}_2}{\|\mathbf{v}_2\|}$$

$$\|\mathbf{v}_2\| = \sqrt{\frac{1}{25} + \frac{4}{25} + 1} = \frac{\sqrt{30}}{5}$$

$$\mathbf{i}_2 = \frac{5}{\sqrt{30}}(-\frac{1}{5}\mathbf{e}_1 + \frac{2}{5}\mathbf{e}_2 + \mathbf{e}_3)$$

$$= -\frac{1}{\sqrt{30}}\mathbf{e}_1 + \frac{2}{\sqrt{30}}\mathbf{e}_2 + \frac{5}{\sqrt{30}}\mathbf{e}_3$$

To find $\mathbf{i}_3$, we start with

$$\mathbf{a}_3 \cdot \mathbf{i}_1 = (\mathbf{e}_2 + \mathbf{e}_3) \cdot (\frac{1}{\sqrt{5}}(2\mathbf{e}_1 + \mathbf{e}_2)) = \frac{1}{\sqrt{5}}$$

$$\mathbf{a}_3 \cdot \mathbf{i}_2 = (\mathbf{e}_2 + \mathbf{e}_3) \cdot (-\frac{1}{\sqrt{30}}\mathbf{e}_1 + \frac{2}{\sqrt{30}}\mathbf{e}_2 + \frac{5}{\sqrt{30}}\mathbf{e}_3)$$

$$= \frac{1}{\sqrt{30}}(2 + 5) = \frac{7}{\sqrt{30}}$$

From this we have

$$\mathbf{v}_3 = \mathbf{a}_3 - (\mathbf{a}_3 \cdot \mathbf{i}_1)\mathbf{i}_1 - (\mathbf{a}_3 \cdot \mathbf{i}_2)\mathbf{i}_2$$

$$= (\mathbf{e}_2 + \mathbf{e}_3) - \frac{1}{\sqrt{5}}(\frac{1}{\sqrt{5}}(2\mathbf{e}_1 + \mathbf{e}_2))$$

$$- \frac{7}{\sqrt{30}}(-\frac{1}{\sqrt{30}}\mathbf{e}_1 + \frac{2}{\sqrt{30}}\mathbf{e}_2 + \frac{5}{\sqrt{30}}\mathbf{e}_3)$$

$$= -\frac{1}{6}\mathbf{e}_1 + \frac{1}{3}\mathbf{e}_2 - \frac{1}{6}\mathbf{e}_3$$

Then, the norm of $\mathbf{v}_3$ is

$$\|\mathbf{v}_3\| = \sqrt{\frac{1}{36} + \frac{1}{9} + \frac{1}{36}} = \frac{1}{\sqrt{6}}$$

and, therefore,

$$\mathbf{i}_3 = \frac{\mathbf{v}_3}{\|\mathbf{v}_3\|} = -\frac{1}{\sqrt{6}}\mathbf{e}_1 + \frac{2}{\sqrt{6}}\mathbf{e}_2 - \frac{1}{\sqrt{6}}\mathbf{e}_3$$

It is easy to verify that $\mathbf{i}_1, \mathbf{i}_2, \mathbf{i}_3$ obtained above constitute an orthonormal basis.

2.2.2 Orthogonal Expansion of Signals

We have already seen the similarities that exist between the space of signals and the space of vectors. The fundamental similarity comes from the fact that any linear combination of signals is itself a signal, as is the case with vectors. This, in mathematical terms, means that the space of signals is a *vector space,* or a *linear space.*

If we consider the signals defined on the interval $[a, b]$, the inner product of the two signals $x(t)$ and $y(t)$, denoted by $\langle x(t), y(t) \rangle$, is defined by

$$\langle x(t), y(t) \rangle = \int_a^b x(t)y^*(t)\, dt \tag{2.2.17}$$

This definition has the basic properties of the inner product of vectors. Note that $\langle x(t), y(t) \rangle$ is, in general, a complex number and we have

$$\langle x(t), y(t) \rangle = \langle y(t), x(t) \rangle^* \tag{2.2.18}$$

The norm of a signal can be defined by the square root of the inner product of the signal with itself

$$\|x(t)\| = \langle x(t), x(t) \rangle^{1/2} = \left(\int_a^b |x(t)|^2 dt \right)^{1/2} \tag{2.2.19}$$

Obviously $\|x(t)\|$ is always a nonnegative real number.

A sequence of signals $\{\psi_n(t)\}$ is called orthonormal if and only if

$$\langle \psi_n(t), \psi_m(t) \rangle = \delta_{mn} = \begin{cases} 1 & m = n \\ 0 & m \neq n \end{cases} \tag{2.2.20}$$

where δ_{mn} denotes the Kronecker's delta. If it happens that all the elements of a signal space can be expanded in terms of the orthonormal basis $\psi_n(t)$, then this basis is called a *complete basis* for the space. To find the coefficients of expansion of $x(t)$ in terms of $\{\psi_n(t)\}$, we have to find the inner products of $x(t)$ and $\{\psi_n(t)\}$, i.e., $x(t)$ can be expanded as

$$x(t) = \sum_n x_n \psi_n(t) = \sum_n \langle x(t), \psi_n(t) \rangle \psi_n(t) \tag{2.2.21}$$

This means that the coefficients of the expansion x_n can be determined from

$$x_n = \langle x(t), \psi_n(t) \rangle \tag{2.2.22}$$

The equivalent of the Pythagorean theorem for signals is

$$\|x(t)\|^2 = \int_a^b |x(t)|^2 dt = \sum_n |x_n|^2 = \sum_n |\langle x(t), \psi_n(t)\rangle|^2 \qquad (2.2.23)$$

and the equivalent of the Cauchy–Schwartz inequality becomes

$$\left| \int_a^b x(t) y^*(t)\, dt \right| \leq \left[\int_a^b |x(t)|^2 dt \right]^{1/2} \cdot \left[\int_a^b |y(t)|^2 dt \right]^{1/2} \qquad (2.2.24)$$

with equality if and only if $y(t) = k x(t)$ for some complex k. For a proof and some variations of the Cauchy–Schwartz inequality see Problem 2.32.

Example 2.2.2

We can apply the Gram–Schmidt orthogonalization process to find an orthonormal basis for a space of signals. Let us assume we want to find an orthonormal basis for representation of polynomials in the $[-1, +1]$ interval. To do this, we start with the set $f_n(t) = t^n$, $n = 0, 1, 2, \ldots$, and apply the Gram–Schmidt process.

$$\|f_0(t)\| = \left[\int_{-1}^{+1} 1\, dt \right]^{1/2} = \sqrt{2}$$

$$\psi_0(t) = \frac{f_0(t)}{\|f_0(t)\|} = \frac{1}{\sqrt{2}} \qquad -1 \leq t \leq 1$$

$$\langle f_1(t), \psi_0(t)\rangle = \int_{-1}^{+1} \frac{t}{\sqrt{2}}\, dt = 0$$

$$v_1(t) = f_1(t) - \langle f_1(t), \psi_0(t)\rangle \psi_0(t) = t, \quad -1 \leq t \leq 1$$

$$\|v_1(t)\| = \left[\int_{-1}^{+1} t^2 dt \right]^{1/2} = \sqrt{2/3}$$

$$\psi_1(t) = \frac{v_1(t)}{\|v_1(t)\|} = \sqrt{3/2}\, t \qquad -1 \leq t \leq 1$$

For $n = 2, 3, \ldots$ we can find $\psi_n(t)$ similarly.

There are numerous choices for the set of orthonormal signals $\{\psi_n(t)\}$. Assuming that a signal $x(t)$ is expanded in terms of the orthonormal basis $\{\psi_n(t)\}$, let us find the response of an LTI system to this signal. We have

$$x(t) = \sum_n x_n \psi_n(t)$$

$$y(t) = L[x(t)] = L\left[\sum_n x_n \psi_n(t) \right] = \sum_n x_n L[\psi_n(t)] \qquad (2.2.25)$$

If the set $\{\psi_n(t)\}$ is appropriately chosen, then finding the response of the LTI system to $\psi_n(t)$ is easy, and consequently, finding the response to any $x(t)$ will be a simple task. Therefore, finding a complete orthonormal basis, for which the response of linear systems is straightforward, is essential. We have already seen in Example 2.1.20 that complex exponentials constitute such a set of signals. Therefore, it is desirable to find ways of expanding arbitrary signals in terms of complex exponentials. The Fourier series and the Fourier transform provide the means of such an expansion.

2.2.3 Fourier Series and Its Properties

The set of complex exponentials are the eigenfunctions of LTI systems. The response of an LTI system to a complex exponential is a complex exponential with the same frequency and with a change in amplitude and phase. It was shown in Example 2.1.20 that the change in phase and amplitude are functions of the frequency of the complex exponential and the impulse response of the LTI system. Now the natural question is: Which signals can be expanded in terms of complex exponentials? The following theorem, known as the Fourier series theorem, gives a set of sufficient conditions for expansion of periodic signals in terms of complex exponentials. The expansion of nonperiodic signals will be discussed later.

Theorem 2.2.1 [Fourier Series].   Let the signal $x(t)$ be a periodic signal with period T_0. If the following conditions (known as the Dirichlet conditions) are satisfied,

1. $x(t)$ is absolutely integrable over its period, i.e.,

$$\int_0^{T_0} |x(t)|dt < \infty,$$

2. the number of maxima and minima of $x(t)$ in each period is finite,
3. the number of discontinuities of $x(t)$ in each period is finite,

then $x(t)$ can be expanded in terms of the complex exponential signals $\{e^{j2\pi \frac{n}{T_0} t}\}_{n=-\infty}^{+\infty}$ as

$$x_{\pm}(t) = \sum_{n=-\infty}^{+\infty} x_n e^{j2\pi \frac{n}{T_0} t} \qquad (2.2.26)$$

where

$$x_n = \frac{1}{T_0} \int_{\alpha}^{\alpha+T_0} x(t) e^{-j2\pi \frac{n}{T_0} t} dt \qquad (2.2.27)$$

for some arbitrary α and

$$x_\pm(t) = \begin{cases} x(t) & \text{if } x(t) \text{ is continuous at } t \\ \dfrac{x(t^+) + x(t^-)}{2} & \text{if } x(t) \text{ is discontinuous at } t \end{cases} \qquad (2.2.28)$$

Some observations concerning this theorem are in order.

- The coefficients x_n are called the Fourier series coefficients of the signal $x(t)$. These are, in general, complex numbers.
- The parameter α in the limits of the integral is arbitrary. It can be chosen to simplify computation of the integral. Usually $\alpha = 0$ or $\alpha = -T_0/2$ are good choices.
- For all practical purposes, $x_\pm(t)$ equals $x(t)$. From now on, we will use $x(t)$ instead of $x_\pm(t)$ but will keep in mind that the Fourier series expansion at points of discontinuity of $x(t)$ gives the midpoint between right and left limits of the signal.
- The Dirichlet conditions are only sufficient conditions for the existence of the Fourier series expansion. For some signals that do not satisfy these conditions we can still find the Fourier series expansion.
- The quantity $f_0 = \frac{1}{T_0}$ is called the *fundamental frequency* of the signal $x(t)$. We observe that the frequencies of the complex exponential signals are multiples of this fundamental frequency.
- Conceptually this is a very important result. It states that the periodic signal $x(t)$ can be described by the period T_0 (or fundamental frequency f_0) and the sequence of complex numbers $\{x_n\}$. In other words, to describe $x(t)$, it is sufficient to specify a *countable* set of (in general, complex) numbers. This is a considerable reduction in complexity for describing $x(t)$, since to define $x(t)$ for all t, we have to specify its values on an *uncountable* set of points.
- The Fourier series expansion can be expressed in terms of the angular frequency $\omega_0 = 2\pi f_0$ by

$$x_n = \frac{\omega_0}{2\pi} \int_{\alpha}^{\alpha + \frac{2\pi}{\omega_0}} x(t) e^{-jn\omega_0 t} \, dt \qquad (2.2.29)$$

and

$$x(t) = \sum_{n=-\infty}^{+\infty} x_n e^{jn\omega_0 t} \qquad (2.2.30)$$

- In general, $x_n = |x_n| e^{j\angle x_n}$. $|x_n|$ gives the magnitude of the nth harmonic and $\angle x_n$ gives its phase. Figure 2.19 shows a graph of the magnitude and phase of various harmonics in $x(t)$. This type of graph is called the *discrete spectrum* of the periodic signal $x(t)$.

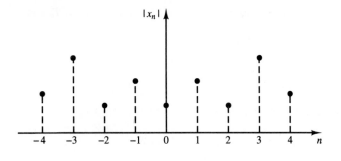

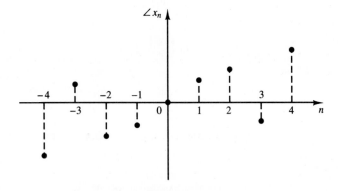

FIGURE 2.19. The discrete spectrum of $x(t)$.

Example 2.2.3

Let $x(t)$ denote the periodic signal depicted in Figure 2.20 and described analytically by

$$x(t) = \sum_{n=-\infty}^{+\infty} \Pi\left(\frac{t - nT_0}{\tau}\right) \qquad (2.2.31)$$

Determine the Fourier series expansion for this signal.

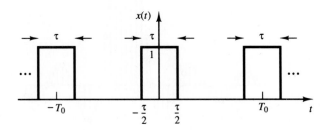

FIGURE 2.20. Periodic signal $x(t)$.

Solution   We first observe that the period of the signal is T_0 and

$$
\begin{aligned}
x_n &= \frac{1}{T_0} \int_{-\frac{T_0}{2}}^{+\frac{T_0}{2}} x(t) e^{-jn\frac{2\pi t}{T_0}}\, dt \\[2mm]
&= \frac{1}{T_0} \int_{-\frac{\tau}{2}}^{+\frac{\tau}{2}} e^{-jn\frac{2\pi t}{T_0}}\, dt \\[2mm]
&= \frac{1}{T_0} \frac{T_0}{-jn2\pi} \left[e^{-jn\frac{\pi\tau}{T_0}} - e^{+jn\frac{\pi\tau}{T_0}} \right] \\[2mm]
&= \frac{1}{\pi n} \sin\left(\frac{n\pi\tau}{T_0} \right) \\[2mm]
&= \frac{\tau}{T_0} \operatorname{sinc}\left(\frac{n\tau}{T_0} \right)
\end{aligned}
\tag{2.2.32}
$$

Therefore,

$$
x(t) = \sum_{n=-\infty}^{+\infty} \frac{\tau}{T_0} \operatorname{sinc}\left(\frac{n\tau}{T_0} \right) e^{jn\frac{2\pi t}{T_0}}
\tag{2.2.33}
$$

A graph of the Fourier series coefficients is shown in Figure 2.21

Example 2.2.4

Determine the Fourier series expansion for the signal $x(t)$ shown in Figure 2.22 and described by

$$
x(t) = \sum_{n=-\infty}^{+\infty} (-1)^n \Pi(t - n)
\tag{2.2.34}
$$

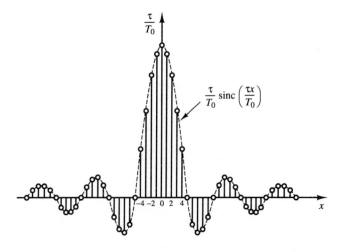

FIGURE 2.21.   The discrete spectrum of the rectangular pulse train.

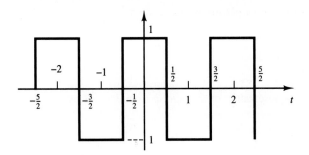

FIGURE 2.22. Signal $x(t)$.

Solution   Here $T_0 = 2$ and it is convenient to choose $\alpha = -\frac{1}{2}$. Then,

$$x_n = \frac{1}{2} \int_{-\frac{1}{2}}^{\frac{3}{2}} x(t) e^{-jn\pi t} dt$$

$$= \frac{1}{2} \int_{-\frac{1}{2}}^{\frac{1}{2}} e^{-jn\pi t} dt - \frac{1}{2} \int_{\frac{1}{2}}^{\frac{3}{2}} e^{-jn\pi t} dt$$

$$= -\frac{1}{j2\pi n} \left[e^{-j\frac{n\pi}{2}} - e^{j\frac{n\pi}{2}} \right] - \frac{1}{-j2\pi n} \left[e^{-jn\frac{3\pi}{2}} - e^{-jn\frac{\pi}{2}} \right]$$

$$= \frac{1}{n\pi} \sin\left(\frac{n\pi}{2}\right) - \frac{1}{n\pi} e^{-jn\pi} \sin\left(\frac{n\pi}{2}\right)$$

$$= \frac{1}{n\pi} (1 - \cos(n\pi)) \sin\left(\frac{n\pi}{2}\right)$$

$$= \begin{cases} \dfrac{2}{n\pi} & n = 4k + 1 \\[2mm] -\dfrac{2}{n\pi} & n = 4k + 3 \\[2mm] 0 & n \text{ even} \end{cases} \qquad (2.2.35)$$

From these values of x_n we have the following Fourier series expansion for $x(t)$:

$$x(t) = \frac{2}{\pi} (e^{j\pi t} + e^{-j\pi t}) - \frac{2}{3\pi} (e^{j3\pi t} + e^{-j3\pi t}) + \frac{2}{5\pi} (e^{j5\pi t} + e^{-j5\pi t}) - \cdots$$

$$= \frac{4}{\pi} \cos(\pi t) - \frac{4}{3\pi} \cos(3\pi t) + \frac{4}{5\pi} \cos(5\pi t) - \cdots$$

$$= \frac{4}{\pi} \sum_{k=0}^{\infty} \frac{(-1)^k}{2k + 1} \cos(2k + 1)\pi t \qquad (2.2.36)$$

Some nice and useful mathematical series can be obtained from Fourier series expansions. For instance, from the preceding example we can substitute $t = 0$ in both sides of the above relationship to obtain

$$1 - \frac{1}{3} + \frac{1}{5} - \frac{1}{7} + \frac{1}{9} - \cdots = \frac{\pi}{4} \tag{2.2.37}$$

Example 2.2.5

What is the Fourier series representation of an *impulse train* denoted by

$$x(t) = \sum_{n=-\infty}^{+\infty} \delta(t - nT_0) \tag{2.2.38}$$

and shown in Figure 2.23.

Solution   We have

$$x_n = \frac{1}{T_0} \int_{-\frac{T_0}{2}}^{+\frac{T_0}{2}} x(t) e^{-j2\pi \frac{n}{T_0} t} dt$$

$$= \frac{1}{T_0} \int_{-\frac{T_0}{2}}^{+\frac{T_0}{2}} \delta(t) e^{-j2\pi \frac{n}{T_0} t} dt$$

$$= \frac{1}{T_0} \tag{2.2.39}$$

With these coefficients we have the following expansion:

$$\sum_{n=-\infty}^{+\infty} \delta(t - nT_0) = \frac{1}{T_0} \sum_{n=-\infty}^{+\infty} e^{j2\pi \frac{n}{T_0} t} \tag{2.2.40}$$

This is a very useful relation and we will employ it frequently.

Fourier series for real signals.   If the signal $x(t)$ is a real signal satisfying the conditions of the Fourier series theorem, then there exist alternative ways to expand the signal. For real $x(t)$ we have

$$x_{-n} = \frac{1}{T_0} \int_{\alpha}^{\alpha+T_0} x(t) e^{j2\pi \frac{n}{T_0} t} dt$$

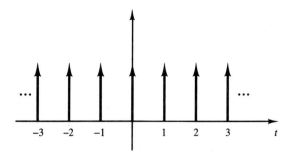

FIGURE 2.23.   An impulse train.

$$= \left[\frac{1}{T_0} \int_{\alpha}^{\alpha+T_0} x(t) e^{-j2\pi \frac{n}{T_0} t} dt \right]^*$$

$$= x_n^* \tag{2.2.41}$$

This means that for real $x(t)$ the positive and negative coefficients are conjugates. Hence, $|x_n|$ has even symmetry and $\angle x_n$ has odd symmetry with respect to the $n = 0$ axis. An example of the discrete spectrum for a real signal is shown in Figure 2.24. From $x_{-n} = x_n^*$ it follows that if we denote $x_n = \frac{a_n - jb_n}{2}$, then $x_{-n} = \frac{a_n + jb_n}{2}$ and, therefore, for $n \geq 1$,

$$x_n e^{j2\pi \frac{n}{T_0} t} + x_{-n} e^{-j2\pi \frac{n}{T_0} t} = \frac{a_n - jb_n}{2} e^{j2\pi \frac{n}{T_0} t} + \frac{a_n + jb_n}{2} e^{-j2\pi \frac{n}{T_0} t}$$

$$= a_n \cos\left(2\pi \frac{n}{T_0} t\right) + b_n \sin\left(2\pi \frac{n}{T_0} t\right)$$

Since x_0 is real and given as $x_0 = \frac{a_0}{2}$, we conclude that

$$x(t) = \frac{a_0}{2} + \sum_{n=1}^{\infty} \left[a_n \cos\left(2\pi \frac{n}{T_0} t\right) + b_n \sin(2\pi \frac{n}{T_0} t) \right] \tag{2.2.42}$$

This relation, which only holds for a real periodic signal $x(t)$, is called the *trigonometric Fourier series expansion*. To obtain a_n and b_n, we have

$$x_n = \frac{a_n - jb_n}{2} = \frac{1}{T_0} \int_{\alpha}^{\alpha+T_0} x(t) e^{-j2\pi \frac{n}{T_0} t} dt,$$

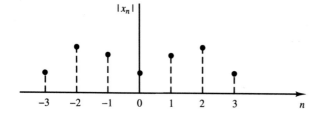

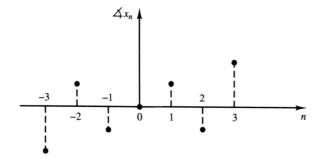

FIGURE 2.24. Discrete spectrum for a real-valued signal.

and therefore,

$$\frac{a_n - jb_n}{2} = \frac{1}{T_0} \int_\alpha^{\alpha+T_0} x(t) \cos\left(2\pi \frac{n}{T_0} t\right) dt - \frac{j}{T_0} \int_\alpha^{\alpha+T_0} x(t) \sin\left(2\pi \frac{n}{T_0} t\right) dt$$

From this we obtain

$$a_n = \frac{2}{T_0} \int_\alpha^{\alpha+T_0} x(t) \cos\left(2\pi \frac{n}{T_0} t\right) dt \tag{2.2.43}$$

$$b_n = \frac{2}{T_0} \int_\alpha^{\alpha+T_0} x(t) \sin\left(2\pi \frac{n}{T_0} t\right) dt \tag{2.2.44}$$

There exists still a third way to represent the Fourier series expansion of a real signal. Noting that

$$x_n e^{j2\pi \frac{n}{T_0} t} + x_{-n} e^{-j2\pi \frac{n}{T_0} t} = 2|x_n| \cos\left(2\pi \frac{n}{T_0} t + \angle x_n\right) \tag{2.2.45}$$

we have

$$x(t) = x_0 + 2\sum_{n=1}^{\infty} |x_n| \cos\left(2\pi \frac{n}{T_0} t + \angle x_n\right) \tag{2.2.46}$$

In summary, for a real periodic signal $x(t)$, we have three alternative ways to represent the Fourier series expansion

$$x(t) = \sum_{n=-\infty}^{+\infty} x_n e^{j2\pi \frac{n}{T_0} t} \tag{2.2.47}$$

$$= \frac{a_0}{2} + \sum_{n=1}^{\infty} \left[a_n \cos\left(2\pi \frac{n}{T_0} t\right) + b_n \sin\left(2\pi \frac{n}{T_0} t\right)\right] \tag{2.2.48}$$

$$= x_0 + 2\sum_{n=1}^{\infty} |x_n| \cos\left(2\pi \frac{n}{T_0} t + \angle x_n\right) \tag{2.2.49}$$

where the corresponding coefficients are obtained from

$$x_n = \frac{1}{T_0} \int_\alpha^{\alpha+T_0} x(t) e^{-j2\pi \frac{n}{T_0} t} dt = \frac{a_n}{2} - j\frac{b_n}{2} \tag{2.2.50}$$

$$a_n = \frac{2}{T_0} \int_\alpha^{\alpha+T_0} x(t) \cos\left(2\pi \frac{n}{T_0} t\right) dt \tag{2.2.51}$$

$$b_n = \frac{2}{T_0} \int_\alpha^{\alpha+T_0} x(t) \sin\left(2\pi \frac{n}{T_0} t\right) dt \tag{2.2.52}$$

$$|x_n| = \frac{1}{2} \sqrt{a_n^2 + b_n^2} \tag{2.2.53}$$

$$\angle x_n = -\arctan\left(\frac{b_n}{a_n}\right) \tag{2.2.54}$$

Fourier series expansion for even and odd signals.   If in addition to being real, a signal is either even or odd, then the Fourier series expansion can be further simplified. For even $x(t)$ we have

$$b_n = \frac{2}{T_0}\int_{-\frac{T_0}{2}}^{\frac{T_0}{2}} x(t)\sin\left(2\pi\frac{n}{T_0}t\right)dt = 0 \tag{2.2.55}$$

Since $x(t)\sin(2\pi\frac{n}{T_0}t)$ is the product of an even and an odd signal, it will be odd and its integral will be zero. Therefore, for even signals the Fourier series expansion has only cosine terms, i.e., we have

$$x(t) = \frac{a_0}{2} + \sum_{n=1}^{\infty} a_n\cos\left(2\pi\frac{n}{T_0}t\right) \tag{2.2.56}$$

Equivalently, since $x_n = \frac{a_n - jb_n}{2}$, we conclude that for an even signal all x_n's are real (or all phases are either 0 or π depending on the sign of x_n).

For odd signals we can conclude in a similar way that all a_n's vanish and therefore the Fourier series expansion only contains the sine terms, or equivalently, all x_n's are imaginary. In this case, we have

$$x(t) = \sum_{n=1}^{\infty} b_n\sin\left(2\pi\frac{n}{T_0}t\right) \tag{2.2.57}$$

Odd harmonic and even harmonic signals.   A signal $x(t)$ is called an odd harmonic signal if in its Fourier series expansion only the odd-numbered harmonics are present. An even harmonic signal is one for which only the even-numbered harmonics are present. For an odd harmonic signal we have

$$x(t) = \sum_{n=-\infty}^{+\infty} x_{2n+1}e^{j2\pi\frac{2n+1}{T_0}t} \tag{2.2.58}$$

Therefore,

$$x\left(t + \frac{T_0}{2}\right) = \sum_{n=-\infty}^{+\infty} x_{2n+1}e^{j2\pi\frac{2n+1}{T_0}(t+\frac{T_0}{2})}$$

$$= \sum_{n=-\infty}^{+\infty} x_{2n+1}e^{j2\pi\frac{2n+1}{T_0}t}\cdot e^{j\pi(2n+1)}$$

$$= -\sum_{n=-\infty}^{+\infty} x_{2n+1}e^{j2\pi\frac{2n+1}{T_0}t}$$

$$= -x(t) \tag{2.2.59}$$

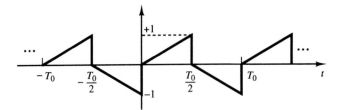

FIGURE 2.25. An example of an odd harmonic signal.

In other words, in odd harmonic signals the relation $x\left(t + \frac{T_0}{2}\right) = -x(t)$ always holds. Figure 2.25 shows an example of an odd harmonic signal. It can be easily shown that if the relation $x\left(t + \frac{T_0}{2}\right) = -x(t)$ holds for a signal, then the coefficients of all the even harmonics in the Fourier series expansion of the signal vanish, and therefore the signal will be an odd harmonic signal. This condition is both a necessary and a sufficient condition.

Note that the definition of an odd harmonic signal does not depend on the choice of the time origin. Therefore, any shift of an odd harmonic signal results in an odd harmonic signal. This is in contrast to the definition of even and odd signals, where the choice of the time origin is critical. This is also the reason why the concept of evenness and oddness is related to the concept of *phase* in the Fourier series expansion coefficients, while the concepts of even harmonic and odd harmonic are related to the concept of *frequency*.

For even harmonic signals we can similarly show that $x\left(t + \frac{T_0}{2}\right) = x(t)$, i.e., the period of an even harmonic signal is $\frac{T_0}{2}$ and not T_0.

Example 2.2.6

Assuming $T_0 = 2$, determine the Fourier series expansion of the signal shown in Figure 2.25.

Solution   We have

$$x_n = \frac{1}{T_0} \int_{-\frac{T_0}{2}}^{\frac{T_0}{2}} x(t) e^{-j2\pi \frac{n}{T_0} t}\, dt$$

$$= \frac{1}{2} \int_{-1}^{1} x(t) e^{-j\pi n t}\, dt$$

$$= \frac{1}{2}\left[\int_{-1}^{0} (-t - 1) e^{-j\pi n t}\, dt + \int_{0}^{+1} t e^{-j\pi n t}\, dt \right]$$

Now using $\int t e^{\alpha t}\, dt = \frac{1}{\alpha} t e^{\alpha t} - \frac{1}{\alpha^2} e^{\alpha t}$, we have

$$\int t e^{-j\pi n t}\, dt = \frac{j}{\pi n} t e^{-j\pi n t} + \frac{1}{\pi^2 n^2} e^{-j\pi n t}$$

and, therefore,

$$x_n = -\frac{j}{2\pi n} t e^{-j\pi nt} \Big]_{-1}^{0} - \frac{1}{2\pi^2 n^2} e^{-j\pi nt} \Big]_{-1}^{0}$$

$$- \frac{j}{2\pi n} e^{-j\pi nt} \Big]_{-1}^{1} + \frac{j}{2\pi n} t e^{-j\pi nt} \Big]_{0}^{1} + \frac{1}{2\pi^2 n^2} e^{-j\pi nt} \Big]_{0}^{1}$$

$$= (\cos(\pi n) - 1)\left(\frac{1}{\pi^2 n^2} + \frac{j}{2\pi n}\right)$$

$$= \begin{cases} -\dfrac{2}{\pi^2 n^2} - \dfrac{j}{\pi n} & n \text{ odd} \\ 0 & n \text{ even} \end{cases}$$

Noting that $x_n = \frac{a_n - jb_n}{2}$, we have

$$a_n = \begin{cases} -\dfrac{4}{\pi^2 n^2} & n \text{ odd} \\ 0 & n \text{ even} \end{cases}$$

and

$$b_n = \begin{cases} \dfrac{2}{\pi n} & n \text{ odd} \\ 0 & n \text{ even} \end{cases}$$

The Fourier series expansion is

$$x(t) = \sum_{n=0}^{\infty} \left[\frac{2}{\pi(2n+1)} \sin(2n+1)\pi t - \frac{4}{\pi^2(2n+1)^2} \cos(2n+1)\pi t\right]$$

Now if we substitute $t = 1$ in both sides and note that this is a discontinuity point of $x(t)$ and $x_{\pm}(1) = \frac{x(1^+) + x(1^-)}{2} = \frac{1}{2}$, we obtain

$$\frac{1}{2} = -\frac{4}{\pi^2} \sum_{n=0}^{\infty} \frac{1}{(2n+1)^2} \cos(2n+1)\pi = \frac{4}{\pi^2} \sum_{n=0}^{\infty} \frac{1}{(2n+1)^2}$$

or

$$\frac{\pi^2}{8} = 1 + \frac{1}{9} + \frac{1}{25} + \frac{1}{49} + \cdots + \frac{1}{(2n+1)^2} + \cdots \tag{2.2.60}$$

2.2.4 Response of LTI Systems to Periodic Signals

As we have already observed, the response of an LTI system to a complex exponential is a complex exponential with the same frequency and a change in amplitude and phase. In particular, if $h(t)$ is the impulse response of the system, then from Example 2.1.20 we know that the response to the exponential $e^{j2\pi f_0 t}$ is $H(f_0)e^{j2\pi f_0 t}$ where

$$H(f) = \int_{-\infty}^{+\infty} h(t)e^{-j2\pi ft}\, dt$$

Now let us assume that $x(t)$, the input to the LTI system, is periodic with period T_0 having a Fourier series representation

$$x(t) = \sum_{n=-\infty}^{+\infty} x_n e^{j2\pi \frac{n}{T_0} t}$$

Then we have

$$y(t) = L[x(t)]$$

$$= L\left[\sum_{-\infty}^{+\infty} x_n e^{j2\pi \frac{n}{T_0} t} \right]$$

$$= \sum_{-\infty}^{+\infty} x_n L\left[e^{j2\pi \frac{n}{T_0} t} \right]$$

$$= \sum_{-\infty}^{+\infty} x_n H\left(\frac{n}{T_0} \right) e^{j2\pi \frac{n}{T_0} t} \tag{2.2.61}$$

where $H(n/T_0) = H(f)|_{f=n/T_0}$. From this relation we can conclude that:

- If the input to an LTI system is periodic with period T_0, then the output is also periodic (what is the period of the output?). The output has a Fourier series expansion given by

$$y(t) = \sum_{n=-\infty}^{+\infty} y_n e^{j2\pi \frac{n}{T_0} t}$$

  where $y_n = x_n H\left(\frac{n}{T_0} \right)$. This is equivalent to $|y_n| = |x_n| \cdot \left| H\left(\frac{n}{T_0} \right) \right|$ and $\angle y_n = \angle x_n + \angle H\left(\frac{n}{T_0} \right)$.

- Only the frequency components that are present at the input can be present at the output. This means that *an LTI system cannot introduce new frequency components in the output other than those already present at the input*. In other words, all systems capable of introducing new frequency components are either nonlinear and/or time varying.

- The amount of change in amplitude, $\left| H\left(\frac{n}{T_0} \right) \right|$, and phase, $\angle H\left(\frac{n}{T_0} \right)$, are functions of n, the harmonic order, and $h(t)$, the impulse response of the system. The function

$$H(f) = \int_{-\infty}^{+\infty} h(t) e^{-j2\pi ft} dt \tag{2.2.62}$$

is called the *frequency response* or *frequency characteristic* of the LTI system. In general, $H(f)$ is a complex function that can be described by its magnitude

$|H(f)|$ and phase $\angle H(f)$. Having $H(f)$, or equivalently, $h(t)$, is all the information needed to find the output of an LTI system to a given periodic input.

Example 2.2.7

Let $x(t)$ denote a signal as shown in Figure 2.22 but with a period equal to $T_0 = 10^{-5}$ seconds. This signal is passed through a filter with a frequency response depicted in Figure 2.26. Determine the output of the filter.

Solution We first start with the Fourier series expansion of the input. This can be easily obtained as

$$x(t) = \frac{4}{\pi} \sum_{n=0}^{\infty} \frac{(-1)^n}{2n+1} \cos\left(2\pi(2n+1)10^5 t\right)$$

$$= \frac{2}{\pi} e^{j2\pi 10^5 t} + \frac{2}{\pi} e^{-j2\pi 10^5 t}$$

$$- \frac{2}{3\pi} e^{j6\pi 10^5 t} - \frac{2}{3\pi} e^{-j6\pi 10^5 t}$$

$$+ \frac{2}{5\pi} e^{j10\pi 10^5 t} + \frac{2}{5\pi} e^{-j10\pi 10^5 t} - \cdots \qquad (2.2.63)$$

To find the output corresponding to each frequency, we must multiply the coefficient of each frequency component by $H(f)$ corresponding to that frequency. Therefore, we have

$$H(10^5) = 1e^{j\frac{\pi}{2}} = e^{j\frac{\pi}{2}}$$

$$H(-10^5) = 1e^{-j\frac{\pi}{2}} = e^{-j\frac{\pi}{2}}$$

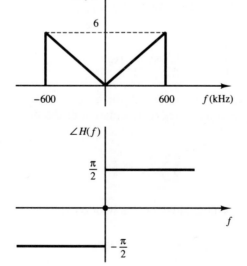

FIGURE 2.26. Frequency response of the filter.

$$H(3 \times 10^5) = 3e^{j\frac{\pi}{2}}$$
$$H(-3 \times 10^5) = 3e^{-j\frac{\pi}{2}}$$
$$H(5 \times 10^5) = 5e^{j\frac{\pi}{2}}$$
$$H(-5 \times 10^5) = 5e^{-j\frac{\pi}{2}} \tag{2.2.64}$$

For higher frequencies, $H(f) = 0$. Therefore, we have

$$
y(t) = \frac{2}{\pi} e^{(j2\pi 10^5 t + \frac{\pi}{2})} + \frac{2}{\pi} e^{(-j2\pi 10^5 t - \frac{\pi}{2})}
$$
$$
- \frac{2}{\pi} e^{(j6\pi 10^5 t + \frac{\pi}{2})} - \frac{2}{\pi} e^{(-j6\pi 10^5 t - \frac{\pi}{2})}
$$
$$
+ \frac{2}{\pi} e^{(j10\pi 10^5 t + \frac{\pi}{2})} + \frac{2}{\pi} e^{(-j10\pi 10^5 t - \frac{\pi}{2})}
$$

or equivalently,

$$
y(t) = -\frac{4}{\pi} \sin(2\pi 10^5 t) + \frac{4}{\pi} \sin(6\pi 10^5 t) - \frac{4}{\pi} \sin(10\pi 10^5 t) \tag{2.2.65}
$$

2.2.5 Parseval's Relation

Parseval's relation is a special form of the Pythagorean relation for the Fourier series expansion of periodic signals. Basically, Parseval's relation says that the power content of a periodic signal is the sum of the power contents of its component in the Fourier series representation of that signal. This relation is a consequence of the orthogonality of the basis that is used for the Fourier series expansion, that is, the exponential signals. Recall that the Pythagorean relation says that if the signal $x(t)$ can be expanded in terms of the orthonormal basis $\{\psi_n(t)\}$ as $x(t) = \sum_n x_n \psi_n(t)$, then

$$
\int_a^b |x(t)|^2 dt = \sum_n |x_n|^2
$$

where the interval $[a, b]$ is the interval on which the basis $\{\psi_n(t)\}$ is orthonormal.

Let us assume that the Fourier series representation of the periodic signal $x(t)$ is given by

$$
x(t) = \sum_{n=-\infty}^{+\infty} x_n e^{j2\pi \frac{n}{T_0} t}
$$

Then, the complex conjugate of both sides of this relation is

$$
x^*(t) = \sum_{n=-\infty}^{+\infty} x_n^* e^{-j2\pi \frac{n}{T_0} t}
$$

By multiplying the two equations, we obtain

$$|x(t)|^2 = \sum_{n=-\infty}^{+\infty} \sum_{m=-\infty}^{+\infty} x_n x_m^* e^{j2\pi \frac{n-m}{T_0} t}$$

Next we integrate both sides over one period and note

$$\int_\alpha^{\alpha+T_0} e^{j2\pi \frac{n-m}{T_0} t} dt = T_0 \delta_{m,n} = \begin{cases} T_0 & n=m \\ 0 & n \neq m \end{cases}$$

Thus, we obtain

$$\int_\alpha^{\alpha+T_0} |x(t)|^2 dt = \sum_{n=-\infty}^{+\infty} \sum_{m=-\infty}^{+\infty} x_n x_m^* T_0 \delta_{m,n}$$

$$= T_0 \sum_{n=-\infty}^{+\infty} |x_n|^2 \qquad (2.2.66)$$

or finally,

$$\frac{1}{T_0} \int_\alpha^{\alpha+T_0} |x(t)|^2 dt = \sum_{n=-\infty}^{+\infty} |x_n|^2 \qquad (2.2.67)$$

This is the formal statement of Parseval's relation. Note that the left-hand side of this relation is P_x, the power content of the signal $x(t)$, and $|x_n|^2$ is the power content of $x_n e^{j2\pi \frac{n}{T_0}}$, the nth harmonic. Therefore, Parseval's relation says that the power content of the periodic signal is the sum of the power contents of its harmonics.

If we substitute $x_n = \frac{a_n - jb_n}{2}$ in Parseval's relation, we obtain

$$P_x = \frac{1}{T_0} \int_\alpha^{\alpha+T_0} |x(t)|^2 dt = \frac{a_0^2}{4} + \frac{1}{2} \sum_{n=1}^{\infty} (a_n^2 + b_n^2) \qquad (2.2.68)$$

Noting that the power content of $a_n \cos\left(2\pi \frac{n}{T_0}\right)$ and $b_n \sin\left(2\pi \frac{n}{T_0}\right)$ are $\frac{a_n^2}{2}$ and $\frac{b_n^2}{2}$, respectively, we see that here again the power content of $x(t)$ is the sum of the power contents of its harmonics.

Example 2.2.8

Determine the power contents of the input and output signals in Example 2.2.7.

Solution We have

$$P_x = \frac{1}{2} \int_{-\frac{1}{2}}^{\frac{3}{2}} |x(t)|^2 dt = \frac{1}{2}\left[\int_{-\frac{1}{2}}^{\frac{1}{2}} 1 dt + \int_{\frac{1}{2}}^{\frac{3}{2}} 1 dt \right] = 1$$

We could employ Parseval's relation to obtain the same result. To do this, we have

$$P_x = \sum_{n=1}^{\infty} \frac{a_n^2}{2} = \frac{8}{\pi^2}\left(1 + \frac{1}{9} + \frac{1}{25} + \frac{1}{49} + \cdots\right)$$

Using relation (2.2.60), we obtain $P_x = 1$. To find the output power, we have

$$P_y = \frac{1}{2} \int_\alpha^{\alpha+2} |y(t)|^2 dt = \sum_{n=1}^\infty \frac{b_n^2}{2} = \frac{1}{2} \left[\frac{16}{\pi^2} + \frac{16}{\pi^2} + \frac{16}{\pi^2} \right] = \frac{24}{\pi^2}$$

2.3 FOURIER TRANSFORMS

2.3.1 From Fourier Series to Fourier Transforms

The Fourier series is a means for expanding a periodic signal in terms of complex exponentials. This expansion considerably decreases the complexity of the description of the signal, and simultaneously, the expansion in terms of complex exponentials is particularly useful when analyzing LTI systems. This is because the complex exponentials are the eigenfunctions of LTI systems. Equivalent to any periodic signal is the sequence $\{x_n\}$ of the Fourier series expansion coefficients which, together with the fundamental frequency f_0, completely describe the signal.

In this section, we will extend the idea of the Fourier series representation of periodic signals to the case of nonperiodic signals. We will show that expansion of a nonperiodic signal in terms of complex exponentials is still possible. However, the resulting spectrum is not discrete anymore. In other words, the spectrum of nonperiodic signals covers a continuous range of frequencies.

Let the signal $x(t)$ be a nonperiodic signal and let us assume that we are looking for an expansion of this signal which is valid in the interval $-\frac{T_0}{2} < t \le \frac{T_0}{2}$. The value of T_0 is quite arbitrary, and therefore this assumption does not cause any loss of generality in our treatment. To find the appropriate expansion, we introduce the truncated signal

$$x_{T_0}(t) = \begin{cases} x(t) & -\frac{T_0}{2} < t \le \frac{T_0}{2} \\ 0 & \text{otherwise} \end{cases}$$

i.e., a signal that coincides with the original $x(t)$ in the desired interval and vanishes outside the desired interval. Now we define the *periodic* signal

$$x_{T_0}^p(t) = \sum_{n=-\infty}^{+\infty} x_{T_0}(t - nT_0) \tag{2.3.1}$$

This signal is also equal to $x(t)$ in the desired interval. Figure 2.27 shows a sketch of $x(t)$, $x_{T_0}(t)$ and $x_{T_0}^p(t)$. It is obvious that if we find an expansion for $x_{T_0}^p(t)$, this will serve as an expansion for $x(t)$ in the desired interval, because in the desired interval both signals are identical. Since $x_{T_0}^p(t)$ is periodic with period T_0, we can find the Fourier series representation of it as

$$x_n^p = \frac{1}{T_0} \int_{-\frac{T_0}{2}}^{\frac{T_0}{2}} x_{T_0}^p(\tau) e^{-j2\pi \frac{n}{T_0} \tau} d\tau$$

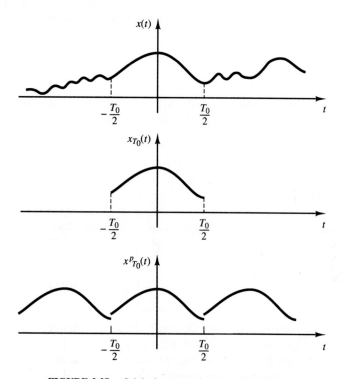

FIGURE 2.27.   Original, truncated, and replicated signals.

$$= \frac{1}{T_0} \int_{-\frac{T_0}{2}}^{\frac{T_0}{2}} x(\tau) e^{-j2\pi \frac{n}{T_0} \tau} d\tau$$

$$x(t) = x_{T_0}^P(t) = \sum_{n=-\infty}^{+\infty} x_n^P e^{j2\pi \frac{n}{T_0} t} \qquad -\frac{T_0}{2} < t \le \frac{T_0}{2}$$

Therefore, for all $-\frac{T_0}{2} < t \le \frac{T_0}{2}$ we can write

$$x(t) = \sum_{n=-\infty}^{+\infty} \left(\frac{1}{T_0} \int_{-\frac{T_0}{2}}^{\frac{T_0}{2}} x(\tau) e^{-j2\pi \frac{n}{T_0} \tau} d\tau \right) e^{j2\pi \frac{n}{T_0} t} \qquad (2.3.2)$$

Now if we let $T_0 \to \infty$, we obtain an expansion that is valid for all values of t. To do this, we introduce the new variables

$$f = \frac{n}{T_0} = nf_0$$

$$\Delta f = (n+1)f_0 - nf_0 = f_0 = \frac{1}{T_0}$$

Then, upon substitution into (2.3.2), we obtain

$$x(t) = \sum_{n=-\infty}^{+\infty} \left(\Delta f \int_{-\frac{T_0}{2}}^{\frac{T_0}{2}} x(\tau) e^{-j2\pi f \tau} \, d\tau \right) e^{j2\pi f t} \tag{2.3.3}$$

Letting $T_0 \to \infty$ (and $f_0 \to 0$), $\Delta f \to 0$, the above sum becomes an integral and the limits of the inside integration tend to infinity. Therefore, we have

$$x(t) = \int_{-\infty}^{+\infty} \left(\int_{-\infty}^{+\infty} x(\tau) e^{-j2\pi f \tau} d\tau \right) e^{j2\pi f t} df \tag{2.3.4}$$

Now, defining

$$X(f) = \int_{-\infty}^{+\infty} x(t) e^{-j2\pi f t} dt \tag{2.3.5}$$

we obtain

$$x(t) = \int_{-\infty}^{+\infty} X(f) e^{j2\pi f t} df \tag{2.3.6}$$

This is the equivalent of the Fourier series expansion for nonperiodic signals. The function $X(f)$ plays the role of $\{x_n\}$ in the Fourier series expansion and gives the magnitude and phase of various frequency components present in $x(t)$. Also note that unlike the Fourier series expansion for periodic signals, in general, the spectrum of $x(t)$ is not discrete. Rather, it covers a continuous range of frequencies.

With this (admittedly handwaving) argument, we can formally state the Fourier transform theorem as follows.

Theorem 2.3.1 [Fourier Transform]. If the signal $x(t)$ satisfies certain conditions known as the Dirichlet conditions, namely,

1. $x(t)$ is absolutely integrable on the real line, that is,

$$\int_{-\infty}^{+\infty} |x(t)| dt < \infty$$

2. The number of maxima and minima of $x(t)$ in any finite interval on the real line is finite,

3. The number of discontinuities of $x(t)$ in any finite interval on the real line is finite,

then, the Fourier transform (or Fourier integral) of $x(t)$, defined by

$$X(f) = \int_{-\infty}^{+\infty} x(t) e^{-j2\pi f t} dt \tag{2.3.7}$$

exists and the original signal can be obtained from its Fourier transform by

$$x_{\pm}(t) = \int_{-\infty}^{+\infty} X(f)e^{j2\pi ft}df \tag{2.3.8}$$

where $x_{\pm}(t)$ is defined as in Theorem 2.2.1.

We make the following observations concerning the Fourier transform:

- $X(f)$ is, in general, a complex function. Its magnitude $|X(f)|$ and phase $\angle X(f)$ represent the amplitude and phase of various frequency components in $x(t)$. The function $X(f)$ is sometimes referred to as the *spectrum*[†] of the signal $x(t)$.
- To denote that $X(f)$ is the Fourier transform of $x(t)$, the following notation is frequently employed

$$X(f) = \mathcal{F}[x(t)]$$

To denote that $x(t)$ is the *inverse Fourier transform* of $X(f)$, the following notation is used

$$x(t) = \mathcal{F}^{-1}[X(f)]$$

Sometimes the following notation is used as a shorthand for both relations

$$x(t) \Leftrightarrow X(f)$$

- If the variable in the Fourier transform is chosen to be ω rather than f, then we have

$$X(\omega) = \int_{-\infty}^{+\infty} x(t)e^{-j\omega t}\,dt$$

and

$$x(t) = \frac{1}{2\pi} \int_{-\infty}^{+\infty} X(\omega)e^{j\omega t}\,d\omega$$

- The Fourier transform and the inverse Fourier transform relations can be written as

$$x(t) = \int_{-\infty}^{+\infty} \left[\int_{-\infty}^{+\infty} x(\tau)e^{-j2\pi f\tau}\,d\tau \right] e^{j2\pi ft}df$$

$$= \int_{-\infty}^{+\infty} \left[\int_{-\infty}^{+\infty} e^{j2\pi f(t-\tau)}\,df \right] x(\tau)\,d\tau \tag{2.3.9}$$

[†]Sometimes $X(f)$ is referred to as *voltage spectrum*, as opposed to *power spectrum* to de defined later.

On the other hand,

$$x(t) = \int_{-\infty}^{+\infty} \delta(t - \tau)x(\tau)\,d\tau \tag{2.3.10}$$

Comparing (2.3.9) with (2.3.10), we obtain

$$\delta(t - \tau) = \int_{-\infty}^{+\infty} e^{j2\pi f(t-\tau)}\,df \tag{2.3.11}$$

or, in general,

$$\delta(t) = \int_{-\infty}^{+\infty} e^{j2\pi ft}\,df \tag{2.3.12}$$

Example 2.3.1

Determine the Fourier transform of the signal $\Pi(t)$.

Solution We have

$$\begin{aligned}
\mathcal{F}[\Pi(t)] &= \int_{-\infty}^{+\infty} \Pi(t)e^{-j2\pi ft}\,dt \\
&= \int_{-\frac{1}{2}}^{+\frac{1}{2}} e^{-j2\pi ft}\,dt \\
&= \frac{1}{-j2\pi f}\left[e^{-j\pi f} - e^{j\pi f}\right] \\
&= \frac{\sin \pi f}{\pi f} \\
&= \operatorname{sinc}(f) \tag{2.3.13}
\end{aligned}$$

Therefore,

$$\mathcal{F}[\Pi(t)] = \operatorname{sinc}(f)$$

Figure 2.28 illustrates the Fourier transform relationship for this signal.

Example 2.3.2

Find the Fourier transform of an impulse signal $x(t) = \delta(t)$.

Solution The Fourier transform can be obtained by

$$\begin{aligned}
\mathcal{F}[\delta(t)] &= \int_{-\infty}^{+\infty} \delta(t)e^{-j2\pi ft}\,dt \\
&= 1 \tag{2.3.14}
\end{aligned}$$

where we have used the sifting property of $\delta(t)$. This shows that all frequencies are present in the spectrum of $\delta(t)$ with unity magnitude and zero phase. The graphs of $\delta(t)$ and its Fourier transform are given in Figure 2.29. Similarly, from the relation

$$\int_{-\infty}^{+\infty} \delta(f)e^{j2\pi ft}\,dt = 1$$

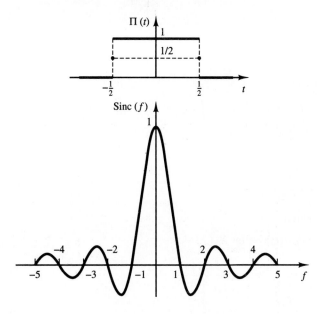

FIGURE 2.28. $\Pi(t)$ and its Fourier transform.

we conclude that

$$\mathcal{F}[1] = \delta(f)$$

Example 2.3.3

Determine the Fourier transform of signal $\text{sgn}(t)$.

Solution   We begin with the definition of $\text{sgn}(t)$ as a limit of an exponential, as shown in Figure 2.12, and given by

$$x_n(t) = \begin{cases} e^{-\frac{t}{n}} & t > 0 \\ -e^{\frac{t}{n}} & t < 0 \\ 0 & t = 0 \end{cases} \qquad (2.3.15)$$

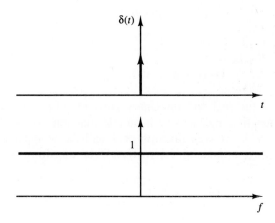

FIGURE 2.29. Impulse signal and its spectrum.

For this signal, the Fourier transform is

$$X_n(f) = \mathcal{F}[x_n(t)]$$

$$= \int_{-\infty}^{0} (-e^{\frac{t}{n}})e^{-j2\pi ft}dt + \int_{0}^{+\infty} e^{-\frac{t}{n}} e^{-j2\pi ft}dt$$

$$= -\int_{-\infty}^{0} e^{t(\frac{1}{n} - j2\pi f)}dt + \int_{0}^{+\infty} e^{-t(\frac{1}{n} + j2\pi f)}dt$$

$$= -\frac{1}{\frac{1}{n} - j2\pi f} + \frac{1}{\frac{1}{n} + j2\pi f}$$

$$= \frac{-j4\pi f}{\frac{1}{n^2} + 4\pi^2 f^2} \tag{2.3.16}$$

Now letting $n \to \infty$, we obtain

$$\mathcal{F}[\text{sgn}(t)] = \lim_{n\to\infty} X_n(f)$$

$$= \lim_{n\to\infty} \frac{-j4\pi f}{\frac{1}{n^2} + 4\pi^2 f^2}$$

$$= \frac{1}{j\pi f} \tag{2.3.17}$$

The graphs of sgn(t) and its spectrum are shown in Figure 2.30.

Fourier transform of real, even and odd signals.   The Fourier transform relation can be written in general as

$$\mathcal{F}[x(t)] = \int_{-\infty}^{+\infty} x(t)e^{-j2\pi ft}dt$$

$$= \int_{-\infty}^{+\infty} x(t)\cos(2\pi ft)\,dt - j\int_{-\infty}^{+\infty} x(t)\sin(2\pi ft)\,dt$$

For real $x(t)$, both integrals

$$\int_{-\infty}^{+\infty} x(t)\cos(2\pi ft)\,dt$$

and

$$\int_{-\infty}^{+\infty} x(t)\sin(2\pi ft)\,dt$$

are real and therefore denote the real and imaginary parts of $X(f)$, respectively. Since the cosise is an even function and the sine is an odd function, we see that for real $x(t)$ the real part of $X(f)$ is an even function of f and the imaginary part is an odd function of f. Therefore, in general, for real $x(t)$, the transform $X(f)$ is a Hermitian function

$$X(-f) = X^*(f)$$

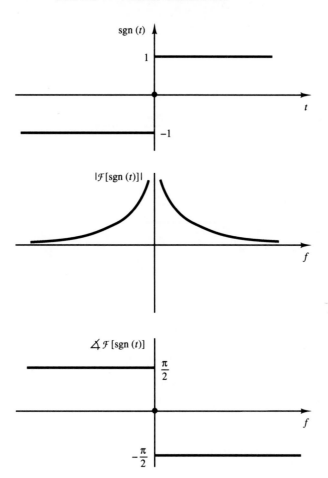

FIGURE 2.30. The signum signal and its spectrum.

This is equivalent to the following relations:

$$\text{Re}[X(-f)] = \text{Re}[X(f)]$$

$$\text{Im}[X(-f)] = -\text{Im}[X(f)]$$

$$|X(-f)| = |X(f)|$$

$$\angle X(-f) = -\angle X(f)$$

Typical plots of $|X(f)|$ and $\angle X(f)$ for a real $x(t)$ are given in Figure 2.31. If in addition to being real, $x(t)$ is an even signal, then the integral

$$\int_{-\infty}^{+\infty} x(t)\sin(2\pi ft)\,dt$$

vanishes because the integrand is the product of an even and odd signal and therefore odd. Hence, the Fourier transform $X(f)$ will be real and even. Similarly, if

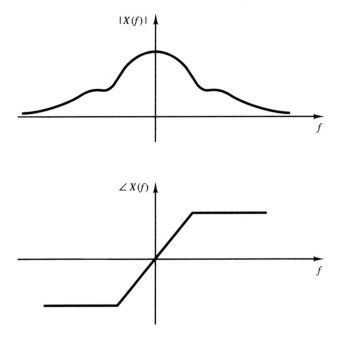

FIGURE 2.31. Magnitude and phase of the spectrum of a real signal.

$x(t)$ is real and odd, the real part of its Fourier transform vanishes and $X(f)$ will be imaginary and odd.

2.3.2 Basic Properties of the Fourier Transform

In this section, we will develop the basic properties of the Fourier transform. With each theorem, we present examples of applications of these theorems.

     Theorem 2.3.2 [Linearity]. The Fourier transform operation is linear. That is, if $x_1(t)$ and $x_2(t)$ are signals possessing Fourier transforms $X_1(f)$ and $X_2(f)$, respectively, the Fourier transform of $\alpha x_1(t) + \beta x_2(t)$ is $\alpha X_1(f) + \beta X_2(f)$, where α and β are two arbitrary (real or complex) scalars.

    Proof. This is a direct result of the linearity of integration.

Example 2.3.4

    Determine the Fourier transform of $u_{-1}(t)$, the unit step signal.

Solution It can be easily seen that

$$u_{-1}(t) = \frac{1}{2} + \frac{1}{2}\,\text{sgn}(t) = \frac{1}{2} \times 1 + \frac{1}{2}\,\text{sgn}(t)$$

By applying the linearity theorem, we obtain

$$\mathcal{F}[u_{-1}(t)] = \mathcal{F}\left[\frac{1}{2} \times 1 + \frac{1}{2} \operatorname{sgn}(t)\right]$$

$$= \frac{1}{j2\pi f} + \frac{1}{2}\delta(f) \tag{2.3.18}$$

Theorem 2.3.3 [Duality].   If

$$X(f) = \mathcal{F}[x(t)]$$

then

$$x(f) = \mathcal{F}[X(-t)]$$

and

$$x(-f) = \mathcal{F}[X(t)]$$

Proof.   Beginning with the inverse Fourier transform relation

$$x(t) = \int_{-\infty}^{+\infty} X(f)e^{j2\pi ft}\,dt$$

we let $u = -f$, then we obtain

$$x(t) = \int_{-\infty}^{+\infty} X(-u)e^{-j2\pi ut}\,du$$

Letting $t = f$, we have

$$x(f) = \int_{-\infty}^{+\infty} X(-u)e^{-j2\pi uf}\,du$$

and finally, substituting t for u, we obtain

$$x(f) = \int_{-\infty}^{+\infty} X(-t)e^{-j2\pi tf}\,dt$$

or

$$x(f) = \mathcal{F}[X(-t)] \tag{2.3.19}$$

Using the same technique once more, we obtain

$$x(-f) = \mathcal{F}[X(t)] \tag{2.3.20}$$

Example 2.3.5

Determine the Fourier transform of $\operatorname{sinc}(t)$.

Solution   Noting that $\Pi(t)$ is an even signal, and therefore, $\Pi(-f) = \Pi(f)$, and using the duality theorem, we obtain

$$\mathcal{F}[\operatorname{sinc}(t)] = \Pi(-f) = \Pi(f) \tag{2.3.21}$$

Example 2.3.6

Determine the Fourier transform of $\frac{1}{t}$.

Solution   Here again, we apply the duality theorem to the transform pair

$$\mathcal{F}[\text{sgn}(t)] = \frac{1}{j\pi f}$$

to obtain

$$\mathcal{F}[\frac{1}{j\pi t}] = \text{sgn}(-f) = -\text{sgn}(f)$$

Using the linearity theorem, we have

$$\mathcal{F}\left[\frac{1}{t}\right] = -j\pi \, \text{sgn}(f) \tag{2.3.22}$$

Theorem 2.3.4 [Shift in Time Domain].   A shift of t_0 in the time origin causes a phase shift of $-2\pi f t_0$ in the frequency domain. In other words,

$$\mathcal{F}[x(t - t_0)] = e^{-j2\pi f t_0} \mathcal{F}[x(t)]$$

Proof.   We start with the Fourier transform of $x(t - t_0)$,

$$\mathcal{F}[x(t - t_0)] = \int_{-\infty}^{+\infty} x(t - t_0)e^{-j2\pi f t}dt$$

With a change of variable of $u = t - t_0$, we obtain

$$\mathcal{F}[x(t - t_0)] = \int_{-\infty}^{+\infty} x(u)e^{-j2\pi f t_0}e^{-j2\pi f u}du$$

$$= e^{-j2\pi f t_0} \int_{-\infty}^{+\infty} x(u)e^{-j2\pi f u}du$$

$$= e^{-j2\pi f t_0} \mathcal{F}[x(t)] \tag{2.3.23}$$

Note that a change in the time origin does not change the magnitude of the transform. It only introduces a phase shift linearly proportional to the time shift (or delay).

Example 2.3.7

Determine the Fourier transform of the signal shown in Figure 2.32.

Solution   We have

$$x(t) = \Pi(t - \frac{3}{2})$$

By applying the shift theorem, we obtain

$$\mathcal{F}[x(t)] = e^{-j2\pi f \times \frac{3}{2}} = e^{-j3\pi f} \text{sinc}(f) \tag{2.3.24}$$

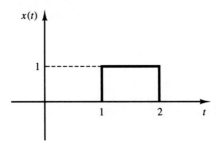

FIGURE 2.32. Signal $x(t)$.

Example 2.3.8

Determine the Fourier transform of the impulse train

$$\sum_{n=-\infty}^{+\infty} \delta(t - nT_0)$$

Solution Applying the shift theorem we have

$$\mathcal{F}\left[\delta(t - nT_0)\right] = e^{-j2\pi fnT_0} \mathcal{F}[\delta(t)] = e^{-j2\pi fnT_0}$$

Therefore,

$$\mathcal{F}\left[\sum_{n=-\infty}^{+\infty} \delta(t - nT_0)\right] = \sum_{n=-\infty}^{+\infty} e^{-j2\pi fnT_0}$$

Now using (2.2.40),

$$\sum_{n=-\infty}^{+\infty} \delta(t - nT_0) = \frac{1}{T_0} \sum_{n=-\infty}^{+\infty} e^{j2\pi \frac{n}{T_0} t}$$

and substituting f for t and $\frac{1}{T_0}$ for T_0, we obtain

$$\sum_{n=-\infty}^{+\infty} \delta\left(f - \frac{n}{T_0}\right) = T_0 \sum_{n=-\infty}^{+\infty} e^{j2\pi nfT_0} = T_0 \sum_{n=-\infty}^{+\infty} e^{-j2\pi nfT_0} \qquad (2.3.25)$$

or

$$\sum_{n=-\infty}^{+\infty} e^{-j2\pi nfT_0} = \frac{1}{T_0} \sum_{n=-\infty}^{+\infty} \delta\left(f - \frac{n}{T_0}\right) \qquad (2.3.26)$$

Using this relation, we obtain

$$\mathcal{F}\left[\sum_{n=-\infty}^{+\infty} \delta(t - nT_0)\right] = \frac{1}{T_0} \sum_{n=-\infty}^{+\infty} \delta\left(f - \frac{n}{T_0}\right) \qquad (2.3.27)$$

The case of $T_0 = 1$ is particularly interesting. For this case we have

$$\mathcal{F}\left[\sum_{n=-\infty}^{+\infty} \delta(t - n)\right] = \sum_{n=-\infty}^{+\infty} \delta(f - n) \qquad (2.3.28)$$

That is, the Fourier transform of $\sum_{n=-\infty}^{+\infty} \delta(t - n)$ is itself after substituting f for t.

Theorem 2.3.5 [Scaling].   For any real $a \neq 0$, we have

$$\mathcal{F}[x(at)] = \frac{1}{|a|} X\left(\frac{f}{a}\right) \tag{2.3.29}$$

Proof.   We start with

$$\mathcal{F}[x(at)] = \int_{-\infty}^{+\infty} x(at)e^{-j2\pi ft} dt \tag{2.3.30}$$

and make the change in variable $u = at$, then,

$$\mathcal{F}[x(at)] = \frac{1}{|a|} \int_{-\infty}^{+\infty} x(u)e^{-j2\pi fu/a} du$$

$$= \frac{1}{|a|} X\left(\frac{f}{a}\right) \tag{2.3.31}$$

where we have treated the cases $a > 0$ and $a < 0$ separately.

Note that in the above expression if $a > 1$, then $x(at)$ is a contracted form of $x(t)$, whereas if $a < 1$, $x(at)$ is an expanded version of $x(t)$. This means that if we expand a signal in the time domain, its frequency domain representation (Fourier transform) contracts, and if we contract a signal in the time domain, its frequency domain representation expands. This is exactly what one expects since contracting a signal in the time domain makes the changes in the signal more abrupt, thus increasing its frequency content.

Example 2.3.9

Determine the Fourier transform of the signal $\cos(2\pi f_0 t)$.

Solution   From Problem 2.43 we know that $\mathcal{F}[\cos \pi t] = \frac{1}{2}\delta(f + \frac{1}{2}) + \frac{1}{2}\delta(f - \frac{1}{2})$. Letting $a = 2f_0$ in the scaling theorem and applying it to this result, we obtain

$$\mathcal{F}[\cos 2\pi f_0 t] = \frac{1}{2f_0}\left[\frac{1}{2}\delta\left(\frac{f}{2f_0} - \frac{1}{2}\right) + \frac{1}{2}\delta\left(\frac{f}{2f_0} + \frac{1}{2}\right)\right]$$

$$= \frac{1}{2}\delta(f - f_0) + \frac{1}{2}\delta(f + f_0) \tag{2.3.32}$$

where we have used the relation $\delta(at) = \frac{1}{|a|}\delta(t)$ in the last step.

Theorem 2.3.6 [Convolution].   If the signals $x(t)$ and $y(t)$ both possess Fourier transforms, then

$$\mathcal{F}[x(t) \star y(t)] = \mathcal{F}[x(t)] \cdot \mathcal{F}[y(t)] = X(f) \cdot Y(f) \tag{2.3.33}$$

Proof.   We have

$$\mathcal{F}[x(t) \star y(t)] = \int_{-\infty}^{+\infty}\left[\int_{-\infty}^{+\infty} x(\tau)y(t - \tau)\, d\tau\right] e^{-j2\pi ft} dt$$

$$= \int_{-\infty}^{+\infty} x(\tau) \left[\int_{-\infty}^{+\infty} y(t - \tau) e^{-j2\pi f(t-\tau)} \, dt \right] e^{-j2\pi f\tau} d\tau$$

Now with the change of variable $u = t - \tau$, we have

$$\int_{-\infty}^{+\infty} y(t - \tau) e^{-j2\pi f(t-\tau)} dt = \int_{-\infty}^{+\infty} y(u) e^{-j2\pi fu} du$$

$$= \mathcal{F}[y(t)]$$

$$= Y(f)$$

and therefore,

$$\mathcal{F}[x(t) \star y(t)] = \int_{-\infty}^{+\infty} x(\tau) Y(f) e^{-j2\pi f\tau} d\tau$$

$$= X(f) \cdot Y(f) \tag{2.3.34}$$

This theorem is very important and is a direct result of the fact that the complex exponentials are eigenfunctions of LTI systems (or equivalently, eigenfunctions of the convolution operation). We see that finding the response of an LTI system to a given input is much simpler in the frequency domain compared to the time domain. This theorem is the basis of the frequency domain analysis of LTI systems.

Example 2.3.10

Determine the Fourier transform of the signal $\Lambda(t)$, shown in Figure 2.10.

Solution It is enough to note that $\Lambda(t) = \Pi(t) \star \Pi(t)$ and use the convolution theorem. Thus, we obtain

$$\mathcal{F}[\Lambda(t)] = \mathcal{F}[\Pi(t)] \cdot \mathcal{F}[\Pi(t)] = \text{sinc}^2(f) \tag{2.3.35}$$

Theorem 2.3.7 [Modulation]. The Fourier transform of $x(t)e^{j2\pi f_0 t}$ is $X(f - f_0)$.

Proof. We have

$$\mathcal{F}[x(t)e^{j2\pi f_0 t}] = \int_{-\infty}^{+\infty} x(t) e^{j2\pi f_0 t} e^{-j2\pi ft} \, dt$$

$$= \int_{-\infty}^{+\infty} x(t) e^{-j2\pi t(f-f_0)} dt$$

$$= X(f - f_0) \tag{2.3.36}$$

This theorem is the dual of the time-shift theorem. The time-shift theorem says that a shift in the time domain results in a multiplication by a complex exponential in the frequency domain. The modulation theorem states that a multiplication in the time domain by a complex exponential results in a shift in the frequency domain. A shift in the frequency domain is usually called modulation.

Example 2.3.11

Determine the Fourier transform of the signal

$$x(t)\cos(2\pi f_0 t)$$

Solution   We have

$$\mathcal{F}[x(t)\cos(2\pi f_0 t)] = \mathcal{F}\left[\frac{1}{2}x(t)e^{j2\pi f_0 t} + \frac{1}{2}x(t)e^{-j2\pi f_0 t}\right]$$

$$= \frac{1}{2}X(f - f_0) + \frac{1}{2}X(f + f_0) \qquad (2.3.37)$$

Figure 2.33 shows a pictorial representation of this relation.

Example 2.3.12

Determine the Fourier transform of the signal

$$x(t) = \begin{cases} \cos(\pi t) & |t| \leq \frac{1}{2} \\ 0 & \text{otherwise} \end{cases} \qquad (2.3.38)$$

shown in Figure 2.34.

Solution $x(t)$ can be expressed as

$$x(t) = \Pi(t)\cos(\pi t) \qquad (2.3.39)$$

Therefore,

$$\mathcal{F}[\Pi(t)\cos(\pi t)] = \frac{1}{2}\,\text{sinc}\left(f - \frac{1}{2}\right) + \frac{1}{2}\,\text{sinc}\left(f + \frac{1}{2}\right) \qquad (2.3.40)$$

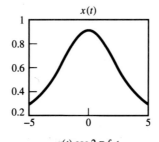

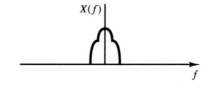

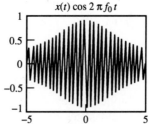

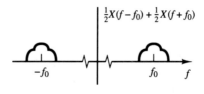

FIGURE 2.33.   Effect of modulation in time and frequency domain.

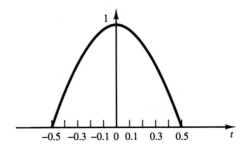

FIGURE 2.34. Signal $x(t)$.

where we have used the result of Example 2.3.11, with $f_0 = \frac{1}{2}$.

Theorem 2.3.8 [Parseval's relation].   If the Fourier transforms of the signals $x(t)$ and $y(t)$ are denoted by $X(f)$ and $Y(f)$, respectively, then

$$\int_{-\infty}^{+\infty} x(t)y^*(t)\,dt = \int_{-\infty}^{+\infty} X(f)Y^*(f)\,df \tag{2.3.41}$$

Proof

$$\int_{-\infty}^{+\infty} x(t)y^*(t)\,dt = \int_{-\infty}^{+\infty} \left[\int_{-\infty}^{+\infty} X(f)e^{j2\pi ft}df\right]\left[\int_{-\infty}^{+\infty} Y(f')e^{j2\pi f't}df'\right]^* dt$$

$$= \int_{-\infty}^{+\infty} \left[\int_{-\infty}^{+\infty} X(f)e^{j2\pi ft}df\right]\left[\int_{-\infty}^{+\infty} Y^*(f')e^{-j2\pi f't}df'\right] dt$$

$$= \int_{-\infty}^{+\infty} X(f)\left[\int_{-\infty}^{+\infty} Y^*(f')\left[\int_{-\infty}^{+\infty} e^{j2\pi t(f-f')}\,dt\right]df'\right]df$$

Now using (2.3.11), we have

$$\int_{-\infty}^{+\infty} e^{j2\pi t(f-f')}dt = \delta(f - f')$$

and therefore,

$$\int_{-\infty}^{+\infty} x(t)y^*(t)\,dt = \int_{-\infty}^{+\infty} X(f)\left[\int_{-\infty}^{+\infty} Y^*(f')\delta(f - f')\,df'\right]df$$

$$= \int_{-\infty}^{+\infty} X(f)Y^*(f)\,df \tag{2.3.42}$$

where we have employed the sifting property of the impulse signal in the last step.

Note that if we let $y(t) = x(t)$, we obtain

$$\int_{-\infty}^{+\infty} |x(t)|^2 dt = \int_{-\infty}^{+\infty} |X(f)|^2 df \tag{2.3.43}$$

This is known as *Rayleigh's theorem* and is similar to Parseval's relation for periodic signals.

Example 2.3.13

Using Parseval's theorem, determine the values of the integrals

$$\int_{-\infty}^{+\infty} \operatorname{sinc}^4(t)\, dt$$

and

$$\int_{-\infty}^{+\infty} \operatorname{sinc}^3(t)\, dt$$

Solution   We have $\mathcal{F}[\operatorname{sinc}^2(t)] = \Lambda(t)$. Therefore, using Rayleigh's theorem, we have

$$\int_{-\infty}^{+\infty} \operatorname{sinc}^4(t)\, dt = \int_{-\infty}^{+\infty} \Lambda^2(f)\, df$$

$$= \int_{-1}^{0} (f+1)^2 df + \int_{0}^{1} (-f+1)^2 df$$

$$= \frac{2}{3}$$

For the second integral we note that $\mathcal{F}[\operatorname{sinc}(t)] = \Pi(f)$, and therefore, by Parseval's theorem we have

$$\int_{-\infty}^{+\infty} \operatorname{sinc}^2(t)\operatorname{sinc}(t)\, dt = \int_{-\infty}^{+\infty} \Pi(f)\Lambda(f)\, df$$

Figure 2.35 shows the product of $\Pi(f)$ and $\Lambda(f)$. From this figure we see that

$$\int_{-\infty}^{+\infty} \Pi(f)\Lambda(f)\, df = 1 \times \frac{1}{2} + \frac{1}{2} \times \frac{1}{2}$$

and therefore,

$$\int_{-\infty}^{+\infty} \operatorname{sinc}^3(t)\, dt = \frac{3}{4}$$

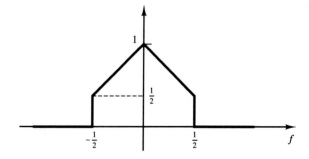

FIGURE 2.35. Product of $\Pi(f)$ and $\Lambda(f)$.

Theorem 2.3.9 [Autocorrelation].   The (*time*) *autocorrelation function* of the signal $x(t)$ is denoted by $R_x(\tau)$ and is defined by

$$R_x(\tau) = \int_{-\infty}^{+\infty} x(t)x^*(t - \tau)\, dt \qquad (2.3.44)$$

The autocorrelation theorem states that

$$\mathcal{F}[R_x(\tau)] = |X(f)|^2 \qquad (2.3.45)$$

Proof.   Note that $R_x(\tau) = x(\tau) \star x^*(-\tau)$. By using the convolution theorem, the autocorrelation theorem follows easily.

From this theorem we conclude that the Fourier transform of the autocorrelation of a signal is always a real-valued positive function.

Theorem 2.3.10 [Differentiation].   The Fourier transform of the derivative of a signal can be obtained from the relation

$$\mathcal{F}\left[\frac{d}{dt}x(t)\right] = j2\pi f X(f) \qquad (2.3.46)$$

Proof.   We have

$$\frac{d}{dt}x(t) = \frac{d}{dt} \int_{-\infty}^{+\infty} X(f)e^{j2\pi ft}\, df$$

$$= \int_{-\infty}^{+\infty} j2\pi f X(f)e^{j2\pi ft}\, df \qquad (2.3.47)$$

From this we conclude that

$$\mathcal{F}^{-1}[j2\pi f X(f)] = \frac{d}{dt}x(t)$$

or

$$\mathcal{F}\left[\frac{d}{dt}x(t)\right] = j2\pi f X(f)$$

Corollary.   With repeated application of the differentiation theorem, we obtain the relation

$$\mathcal{F}\left[\frac{d^n}{dt^n}x(t)\right] = (j2\pi f)^n X(f) \qquad (2.3.48)$$

Example 2.3.14

Determine the Fourier transform of the signal shown in Figure 2.36.

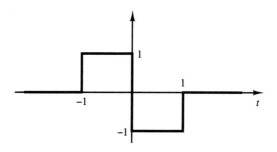

FIGURE 2.36. Signal $x(t)$.

Solution It is easily seen that $x(t) = \frac{d}{dt} \Lambda(t)$. Therefore, applying the differentiation theorem, we have

$$\mathcal{F}[x(t)] = \mathcal{F}\left[\frac{d}{dt}\Lambda(t)\right]$$

$$= j2\pi f\, \mathcal{F}[\Lambda(t)]$$

$$= j2\pi f\, \text{sinc}^2(f) \tag{2.3.49}$$

Theorem 2.3.11 [Differentiation in Frequency Domain]

$$\mathcal{F}[tx(t)] = \frac{j}{2\pi}\frac{d}{df}X(f) \tag{2.3.50}$$

Proof.   The proof is basically the same as the differentiation theorem in the time domain and is left as an exercise.

Corollary.   Repeated use of this theorem yields

$$\mathcal{F}[t^n x(t)] = (\frac{j}{2\pi})^n \frac{d^n}{df^n}X(f) \tag{2.3.51}$$

Example 2.3.15

Determine the Fourier transform of $x(t) = t$.

Solution We have

$$\mathcal{F}[t] = \mathcal{F}[t \times 1]$$

$$= \frac{j}{2\pi}\frac{d}{df}\mathcal{F}[1]$$

$$= \frac{j}{2\pi}\frac{d}{df}\delta(f)$$

$$= \frac{j}{2\pi}\delta'(f) \tag{2.3.52}$$

Theorem 2.3.12 [Integration].   The Fourier transform of the integral of a signal can be determined from the relation

$$\mathcal{F}\left[\int_{-\infty}^{t} x(\tau)\,d\tau\right] = \frac{X(f)}{j2\pi f} + \frac{1}{2}X(0)\delta(f) \qquad (2.3.53)$$

Proof.   Using the result of Problem 2.11, we have

$$\int_{-\infty}^{t} x(\tau)\,d\tau = x(t) \star u_{-1}(t)$$

Now using the convolution theorem and the Fourier transform of $u_{-1}(t)$, we have

$$\mathcal{F}\left[\int_{-\infty}^{t} x(\tau)\,d\tau\right] = X(f)\left[\frac{1}{j2\pi f} + \frac{1}{2}\delta(f)\right]$$

$$= \frac{X(f)}{j2\pi f} + \frac{1}{2}X(0)\delta(f) \qquad (2.3.54)$$

Example 2.3.16

Determine the Fourier transform of the signal shown in Figure 2.37.

Solution   It is seen that

$$x(t) = \int_{-\infty}^{t} \Pi(\tau)\,d\tau$$

Therefore, by using the integration theorem, we obtain

$$\mathcal{F}[x(t)] = \frac{\operatorname{sinc}(f)}{j2\pi f} + \frac{1}{2}\operatorname{sinc}(0)\delta(f)$$

$$= \frac{\operatorname{sinc}(f)}{j2\pi f} + \frac{1}{2}\delta(f) \qquad (2.3.55)$$

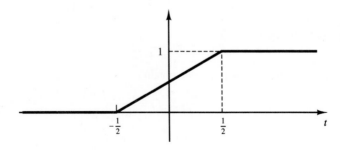

FIGURE 2.37. Signal $x(t)$.

Theorem 2.3.13 [Moments]. If $\mathcal{F}[x(t)] = X(f)$, then $\int_{-\infty}^{+\infty} t^n x(t)\,dt$, and the nth moment of $x(t)$ can be obtained from the relation

$$\int_{-\infty}^{+\infty} t^n x(t)\,dt = \left(\frac{j}{2\pi}\right)^n \frac{d^n}{df^n} X(f)\Big|_{f=0} \tag{2.3.56}$$

Proof. Using the result of the corollary to the differentiation in the frequency domain theorem, we have

$$\mathcal{F}[t^n x(t)] = \left(\frac{j}{2\pi}\right)^n \frac{d^n}{df^n} X(f) \tag{2.3.57}$$

This means that

$$\int_{-\infty}^{+\infty} t^n x(t) e^{-j2\pi ft}\,dt = \left(\frac{j}{2\pi}\right)^n \frac{d^n}{df^n} X(f)$$

Now letting $f = 0$ on both sides, we obtain the desired result.

Corollary. For the special case of $n = 0$, we obtain this simple relation for finding the area under a signal, i.e.,

$$\int_{-\infty}^{+\infty} x(t)\,dt = X(0) \tag{2.3.58}$$

This theorem and corollary provide a simple method for expressing the moments of a signal in terms of its Fourier transform.

Example 2.3.17

Let $\alpha > 0$ and $x(t) = e^{-\alpha t} u_{-1}(t)$. Find the nth moment of $x(t)$.

Solution   First we find $X(f)$. We have

$$X(f) = \int_{-\infty}^{+\infty} e^{-\alpha t} u_{-1}(t) e^{-j2\pi ft}\,dt$$

$$= \int_{0}^{\infty} e^{-t(\alpha + j2\pi ft)}\,dt$$

$$= -\frac{1}{\alpha + j2\pi f}(0 - 1)$$

$$= \frac{1}{\alpha + j2\pi f} \tag{2.3.59}$$

By differentiating n times, we obtain

$$\frac{d^n}{df^n} X(f) = \frac{(-j2\pi)^n n!}{(\alpha + j2\pi f)^{n+1}} \tag{2.3.60}$$

and hence,

$$\int_{-\infty}^{+\infty} t^n e^{-\alpha t} u_{-1}(t)\,dt = \left(\frac{j}{2\pi}\right)^n (-j2\pi)^n \cdot \frac{n!}{\alpha^{n+1}}$$

$$= \frac{n!}{\alpha^{n+1}} \qquad (2.3.61)$$

Example 2.3.18

Determine the Fourier transform of $x(t) = e^{-\alpha|t|}$, where $\alpha > 0$ (see Figure 2.38).

Solution   We can write

$$x(t) = e^{-\alpha t}u_{-1}(t) + e^{\alpha t}u_{-1}(-t) = x_1(t) + x_1(-t) \qquad (2.3.62)$$

We have already seen that

$$\mathcal{F}[x_1(t)] = \mathcal{F}\left[e^{-\alpha t}u_{-1}(t)\right] = \frac{1}{\alpha + j2\pi f}$$

Then, by using the scaling theorem with $a = -1$, we obtain

$$\mathcal{F}[x_1(-t)] = \mathcal{F}\left[e^{\alpha t}u_{-1}(-t)\right] = \frac{1}{\alpha - j2\pi f}$$

Hence, by the linearity property we have

$$\mathcal{F}\left[e^{-\alpha|t|}\right] = \frac{1}{\alpha + j2\pi f} + \frac{1}{\alpha - j2\pi f}$$

$$= \frac{2\alpha}{\alpha^2 + 4\pi^2 f^2} \qquad (2.3.63)$$

Table 2.1 gives a collection of frequently used Fourier transform pairs.

2.3.3 Fourier Transform for Periodic Signals

In this section, we extend the results already obtained to develop methods for finding the Fourier transform of periodic signals. We have already obtained the

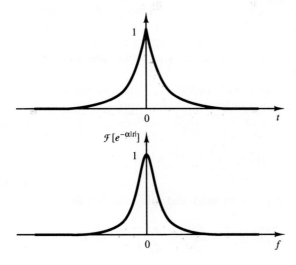

FIGURE 2.38.   Signal $e^{-\alpha|t|}$ and its Fourier transform.

TABLE 2-1   TABLE OF FOURIER TRANSFORMS

Time Domain	Frequency Domain
$\delta(t)$	1
1	$\delta(f)$
$\delta(t - t_0)$	$e^{-j2\pi f t_0}$
$e^{j2\pi f_0 t}$	$\delta(f - f_0)$
$\cos(2\pi f_0 t)$	$\frac{1}{2}\delta(f - f_0) + \frac{1}{2}\delta(f + f_0)$
$\sin(2\pi f_0 t)$	$-\frac{1}{2j}\delta(f + f_0) + \frac{1}{2j}\delta(f - f_0)$
$\Pi(t)$	$\text{sinc}(f)$
$\text{sinc}(t)$	$\Pi(f)$
$\Lambda(t)$	$\text{sinc}^2(f)$
$\text{sinc}^2(t)$	$\Lambda(f)$
$e^{-\alpha t}u_{-1}(t), \alpha > 0$	$\frac{1}{\alpha + j2\pi f}$
$te^{-\alpha t}u_{-1}(t), \alpha > 0$	$\frac{1}{(\alpha + j2\pi f)^2}$
$e^{-\alpha|t|}$	$\frac{2\alpha}{\alpha^2 + (2\pi f)^2}$
$e^{-\pi t^2}$	$e^{-\pi f^2}$
$\text{sgn}(t)$	$1/(j\pi f)$
$u_{-1}(t)$	$\frac{1}{2}\delta(f) + \frac{1}{j2\pi f}$
$\delta'(t)$	$j2\pi f$
$\delta^{(n)}(t)$	$(j2\pi f)^n$
$\frac{1}{t}$	$-j\pi\,\text{sgn}(f)$
$\sum_{n=-\infty}^{n=+\infty}\delta(t - nT_0)$	$\frac{1}{T_0}\sum_{n=-\infty}^{n=+\infty}\delta(f - \frac{n}{T_0})$

Fourier transform for some periodic signals (see Table 2.1). These include $e^{j2\pi f_0 t}$, $\cos(2\pi f_0 t)$, $\sin(2\pi f_0 t)$, and $\sum_{n=-\infty}^{n=+\infty}\delta(t - nT_0)$. The common property of the Fourier transform of all these periodic signals is that the Fourier transform consists of impulse functions in the frequency domain. We will show in this section that this property holds for all periodic signals and in fact there exists a close relationship between the Fourier transform of a periodic signal and the Fourier series representation of the signal.

Let $x(t)$ be a periodic signal with period T_0, satisfying the Dirichlet conditions. Let $\{x_n\}$ denote the Fourier series coefficients corresponding to this signal. Then

$$x(t) = \sum_{n=-\infty}^{+\infty} x_n e^{j2\pi \frac{n}{T_0} t}$$

By taking the Fourier transform of both sides and using the fact that

$$\mathcal{F}\left[e^{j2\pi \frac{n}{T_0} t}\right] = \delta\left(f - \frac{n}{T_0}\right)$$

we obtain

$$X(f) = \sum_{n=-\infty}^{+\infty} x_n \delta\left(f - \frac{n}{T_0}\right) \tag{2.3.64}$$

From this relation we observe that the Fourier transform of a periodic signal $x(t)$ consists of a sequence of impulses in frequency at multiples of the fundamental frequency of the periodic signal. The weights of the impulses are just the Fourier series coefficients of the periodic signal. This relation gives a shortcut for computing Fourier series coefficients of a signal using the properties of the Fourier transform. If we define the truncated signal $x_{T_0}(t)$ as

$$x_{T_0}(t) = \begin{cases} x(t) & -\frac{T_0}{2} < t \leq \frac{T_0}{2} \\ 0 & \text{otherwise} \end{cases} \tag{2.3.65}$$

we see that

$$x(t) = \sum_{n=-\infty}^{+\infty} x_{T_0}(t - nT_0) \tag{2.3.66}$$

Noting that $x_{T_0}(t - nT_0) = x_{T_0}(t) \star \delta(t - nT_0)$, we have

$$x(t) = x_{T_0}(t) \star \sum_{n=-\infty}^{+\infty} \delta(t - nT_0) \tag{2.3.67}$$

Therefore, using the convolution theorem and Table 2.1, we obtain

$$X(f) = X_{T_0}(f)\left[\frac{1}{T_0} \sum_{n=-\infty}^{+\infty} \delta\left(f - \frac{n}{T_0}\right)\right] \tag{2.3.68}$$

which simplifies to

$$X(f) = \frac{1}{T_0} \sum_{n=-\infty}^{+\infty} X_{T_0}\left(\frac{n}{T_0}\right) \delta\left(f - \frac{n}{T_0}\right) \tag{2.3.69}$$

Comparing this result with

$$X(f) = \sum_{n=-\infty}^{+\infty} x_n \delta\left(f - \frac{n}{T_0}\right) \tag{2.3.70}$$

we conclude that

$$x_n = \frac{1}{T_0} X_{T_0}\left(\frac{n}{T_0}\right) \tag{2.3.71}$$

This equation gives an alternative way to find the Fourier series coefficients. Given the periodic signal $x(t)$, we carry out the following steps to find x_n:

1. First, we find the truncated signal $x_{T_0}(t)$.

2. Then, we determine the Fourier transform of the truncated signal using Table 2.1 and the Fourier transform theorems and properties.
3. Finally, we evaluate the Fourier transform of the truncated signal at $f = \frac{n}{T_0}$ to obtain the nth harmonic and multiply by $\frac{1}{T_0}$.

Example 2.3.19

Determine the Fourier series coefficients of the signal $x(t)$ shown in Figure 2.20.

Solution   We follow the preceding steps. The truncated signal is

$$x_{T_0}(t) = \Pi(\frac{t}{\tau})$$

and its Fourier transform is

$$X_{T_0}(f) = \tau \operatorname{sinc}(\tau f)$$

Therefore,

$$x_n = \frac{\tau}{T_0} \operatorname{sinc}\left(\frac{n\tau}{T_0}\right)$$

2.3.4 Passage through LTI Systems

The convolution theorem is the basis for the analysis of LTI systems in the frequency domain. We have seen that the output of an LTI system is equal to the convolution of the input and the impulse response. If we translate this relationship in the frequency domain using the convolution theorem, we observe that if $X(f)$, $H(f)$, and $Y(f)$ are the Fourier transforms of the input, system impulse response, and the output, respectively, then

$$Y(f) = X(f)H(f)$$

Hence, the input-output relation for an LTI system in the frequency domain is much simpler than the corresponding relation in the time domain. In the time domain we have the convolution integral, whereas in the frequency domain we have a simple multiplication. This simplicity is a direct consequence of the fact that the Fourier transform of a signal is a representation of it in terms of the eigenfunctions of LTI systems. Therefore, to find the output of an LTI system to a given input, it suffices to find the Fourier transform of the input and the Fourier transform of the system impulse response and multiply them to obtain the Fourier transform of the output. To obtain the time-domain representation of the output, we can find the inverse Fourier transform of the result. In most cases, finding the inverse Fourier transform is not necessary, and the frequency domain representation of the output gives enough information about the output.

Example 2.3.20

Let the input to an LTI system be the signal

$$x(t) = \operatorname{sinc}(W_1 t)$$

and the impulse response of the system be

$$h(t) = \mathrm{sinc}(W_2 t)$$

Determine the output signal.

Solution   First we transform the signals to the frequency domain. Thus, we obtain

$$X(f) = \frac{1}{W_1} \Pi\left(\frac{f}{W_1}\right)$$

and

$$H(f) = \frac{1}{W_2} \Pi\left(\frac{f}{W_2}\right)$$

Figure 2.39 shows $X(f)$ and $H(f)$. To obtain the output in the frequency domain, we have

$$Y(f) = X(f)H(f)$$

$$= \frac{1}{W_1 W_2} \Pi\left(\frac{f}{W_1}\right) \Pi\left(\frac{f}{W_2}\right)$$

$$= \begin{cases} \dfrac{1}{W_1 W_2} \Pi\left(\dfrac{f}{W_1}\right) & W_1 \le W_2 \\[2mm] \dfrac{1}{W_1 W_2} \Pi\left(\dfrac{f}{W_2}\right) & W_1 > W_2 \end{cases} \qquad (2.3.72)$$

From this result we obtain

$$y(t) = \begin{cases} \dfrac{1}{W_2} \mathrm{sinc}(W_1 t) & W_1 \le W_2 \\[2mm] \dfrac{1}{W_1} \mathrm{sinc}(W_2 t) & W_1 > W_2 \end{cases}$$

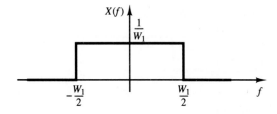

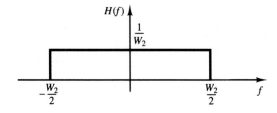

FIGURE 2.39. Lowpass signal and lowpass filter.

Signals such as $x(t)$ in the above example, whose frequency domain representation contains frequencies around the zero frequency and does not contain frequencies beyond some W_1, are called *lowpass signals.* Similarly, an LTI system that can pass all frequencies less than some W_1 and rejects all frequencies beyond W_1 is called an *ideal lowpass filter.* The frequency response of an ideal lowpass filter is 1 for all frequencies $-W \leq f \leq W$ and is 0 outside this interval. W is called the *bandwidth* of the filter. Similarly, we can have *ideal highpass filters,* for which $H(f)$ is unity outside the interval $-W \leq f \leq W$ and zero inside, and *ideal bandpass filters,* whose frequency response is unity in some interval $W_1 \leq |f| \leq W_2$ and 0 otherwise. In this case, the bandwidth of the filter is $W_2 - W_1$. Figure 2.40 shows frequency response functions of various filter types. For nonideal lowpass or bandpass filters, the bandwidth is usually defined as the band of frequencies at which the power transfer ratio of the filter is at least half of the maximum power transfer ratio. This bandwidth is usually called the 3-dB bandwidth of the filter, because reducing the power by a factor of 2 is equivalent to decreasing it by 3-dB on the logarithmic scale. Figure 2.41 shows the 3-dB bandwidth of filters.

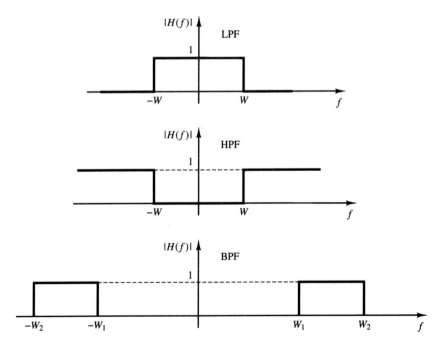

FIGURE 2.40. Various filter types.

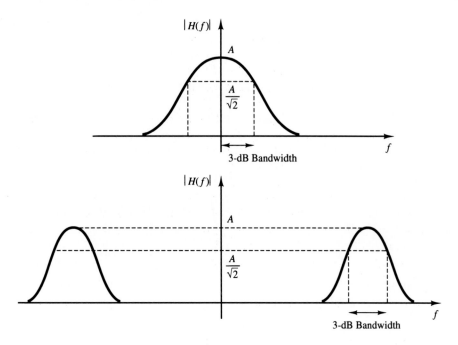

FIGURE 2.41.   3-dB bandwidth of filters.

2.4 POWER AND ENERGY

The concepts of power and energy and power-type and energy-type signals were defined in Section 2.1.1. In this section, we expand these concepts both in the time and frequency domains.

The energy and power of a signal are representatives of the energy or power delivered by the signal when the signal is interpreted as a voltage or current source feeding a 1 ohm resistor. The energy content of a signal $x(t)$, denoted by E_x, is defined as

$$E_x = \int_{-\infty}^{+\infty} |x(t)|^2 dt$$

and the power content of a signal is

$$P_x = \lim_{T \to \infty} \frac{1}{T} \int_{-\frac{T}{2}}^{+\frac{T}{2}} |x(t)|^2 dt$$

A signal is energy-type if $E_x < \infty$ and is power-type if $0 < P_x < \infty$. A signal cannot be both power- and energy-type, because for energy-type signals $P_x = 0$ and for power-type signals $E_x = \infty$. A signal can be neither energy-type nor power-type. An example of such a signal is given in Problem 2.5. However, most of the signals of interest are either energy-type or power-type. Almost all periodic signals are power-type and have power

$$P_x = \frac{1}{T_0} \int_{\alpha}^{\alpha+T_0} |x(t)|^2 dt$$

where T_0 is the period and α is any arbitrary real number.

2.4.1 Energy-Type Signals

For an energy-type signal $x(t)$ we define the autocorrelation function $R_x(\tau)$ as

$$R_x(\tau) = x(\tau) \star x^*(-\tau)$$

$$= \int_{-\infty}^{+\infty} x(t)x^*(t-\tau)\, dt$$

$$= \int_{-\infty}^{+\infty} x(t+\tau)x^*(t)\, dt \tag{2.4.1}$$

By setting $\tau = 0$ in the definition of the autocorrelation function of $x(t)$, we obtain the energy content of it; i.e.,

$$E_x = \int_{-\infty}^{+\infty} |x(t)|^2 dt$$

$$= R_x(0) \tag{2.4.2}$$

Now using the autocorrelation theorem of the Fourier transform (see Section 2.3.2), we find the Fourier transform of $R_x(\tau)$ to be $|X(f)|^2$. Using this result, or equivalently, by employing Rayleigh's theorem, we have

$$E_x = \int_{-\infty}^{+\infty} |x(t)|^2 dt$$

$$= \int_{-\infty}^{+\infty} |X(f)|^2 df \tag{2.4.3}$$

If we pass the signal $x(t)$ through a filter with (in general, complex) impulse response $h(t)$ and frequency response $H(f)$, the output will be $y(t) = x(t) \star h(t)$ and in the frequency domain $Y(f) = X(f)H(f)$. To find the energy content of the output signal $y(t)$, we have

$$E_y = \int_{-\infty}^{+\infty} |y(t)|^2 dt$$

$$= \int_{-\infty}^{+\infty} |Y(f)|^2 df$$

$$= \int_{-\infty}^{+\infty} |X(f)|^2 |H(f)|^2 df$$

$$= R_y(0) \tag{2.4.4}$$

where $R_y(\tau) = y(\tau) \star y^*(-\tau)$ is the autocorrelation function of the output. The inverse Fourier transform of $|Y(f)|^2$ is

$$R_y(\tau) = \mathcal{F}^{-1}\left[|Y(f)|^2\right]$$

$$= \mathcal{F}^{-1}\left[|X(f)|^2 |H(f)|^2\right]$$

$$\stackrel{a}{=} \mathcal{F}^{-1}\left[|X(f)|^2\right] \star \mathcal{F}^{-1}\left[|H(f)|^2\right]$$

$$\stackrel{b}{=} R_x(\tau) \star R_h(\tau) \tag{2.4.5}$$

where (a) follows from the convolution theorem and (b) follows from the autocorrelation theorem. Now let us assume that

$$H(f) = \begin{cases} 1 & W < f < W + \Delta W \\ 0 & \text{otherwise} \end{cases}$$

Then

$$|Y(f)|^2 = \begin{cases} |X(f)|^2 & W < f < W + \Delta W \\ 0 & \text{otherwise} \end{cases}$$

and

$$E_y = \int_{-\infty}^{+\infty} |Y(f)|^2 df$$

$$\approx |X(W)|^2 \Delta W \tag{2.4.6}$$

This filter passes the frequency components in a small interval around $f = W$ and rejects all the other components. Therefore, the output energy is representative of how much energy is located in the vicinity of $f = W$ in the input signal. This means that $|X(W)|^2 \Delta W$ is the amount of energy in $x(t)$ located in the bandwidth $[W, W + \Delta W]$. From this we see that

$$|X(W)|^2 = \frac{\text{Energy in } [W, W + \Delta W] \text{ bandwidth}}{\Delta W}$$

This is why $|X(f)|^2$ is called the *energy spectral density* of a signal $x(t)$, and represents the amount of energy per hertz of bandwidth present in the signal at various frequencies. From this we define the energy spectral density (or energy spectrum) of the signal $x(t)$ as

$$\mathcal{G}_x(f) = |X(f)|^2$$

$$= \mathcal{F}\left[R_x(\tau)\right] \tag{2.4.7}$$

To summarize:

1. For any energy-type signal $x(t)$ we define the autocorrelation function $R_x(\tau) = x(\tau) \star x^*(-\tau)$.

2. The energy spectral density of $x(t)$ is denoted by $G_x(f)$ and is the Fourier transform of $R_x(\tau)$ which is equal to $|X(f)|^2$.

3. E_x, the energy content of $x(t)$, is the value of the autocorrelation function evaluated at $\tau = 0$, or equivalently, the integral of the energy spectral density over all frequencies; i.e.,

$$E_x = R_x(0)$$

$$= \int_{-\infty}^{+\infty} G_x(f)\, df \tag{2.4.8}$$

4. If $x(t)$ is passed through a filter with impulse response $h(t)$ and the output is denoted by $y(t)$, we have

$$y(t) = x(t) \star h(t)$$

$$R_y(t) = R_x(t) \star R_h(t)$$

$$G_y(f) = G_x(f) G_h(f) = |X(f)|^2 |H(f)|^2$$

Example 2.4.1

Determine the autocorrelation function, energy spectral density, and energy content of the signal $x(t) = e^{-\alpha t} u_{-1}(t)$, $\alpha > 0$.

Solution   First we find the Fourier transform of $x(t)$. From Table 2.1 we have

$$X(f) = \frac{1}{\alpha + j2\pi f}$$

Hence,

$$G_x(f) = |X(f)|^2 = \frac{1}{\alpha^2 + (2\pi f)^2}$$

and

$$R_x(\tau) = \mathcal{F}^{-1}\left[|X(f)|^2\right] = \frac{1}{2\alpha} e^{-\alpha|\tau|}$$

To find the energy content, it is enough to find the value of the autocorrelation function at zero

$$E_x = R_x(0) = \frac{1}{2\alpha}$$

Example 2.4.2

If the signal in the above example is passed through a filter with impulse response $h(t) = e^{-\beta t} u_{-1}(t)$, $\beta > 0$, $\beta \neq \alpha$, determine the autocorrelation function, the power spectral density, and the energy content of the signal at the output.

Solution The frequency response of the filter is

$$H(f) = \frac{1}{\beta + j2\pi f}$$

Therefore,

$$|Y(f)|^2 = |X(f)|^2|H(f)|^2$$

$$= \frac{1}{(\alpha^2 + 4\pi^2 f^2)(\beta^2 + 4\pi^2 f^2)}$$

$$= \frac{1}{\beta^2 - \alpha^2}\left[\frac{1}{\alpha^2 + 4\pi^2 f^2} - \frac{1}{\beta^2 + 4\pi^2 f^2}\right]$$

From this result we obtain

$$R_y(\tau) = \frac{1}{\beta^2 - \alpha^2}\left[\frac{1}{2\alpha}e^{-\alpha|\tau|} - \frac{1}{2\beta}e^{-\beta|\tau|}\right]$$

and

$$E_y = R_y(0)$$

$$= \frac{1}{2\alpha\beta(\alpha + \beta)}$$

2.4.2 Power-Type Signals

For the class of power-type signals a similar development is possible. In this case, we define the *time-average autocorrelation function* of the power-type signal $x(t)$ as

$$R_x(\tau) = \lim_{T\to\infty}\frac{1}{T}\int_{-\frac{T}{2}}^{+\frac{T}{2}} x(t)x^*(t - \tau)\,dt \qquad (2.4.9)$$

Now, obviously, the power content of the signal can be obtained from

$$P_x = \lim_{T\to\infty}\frac{1}{T}\int_{-\frac{T}{2}}^{+\frac{T}{2}} |x(t)|^2 dt$$

$$= R_x(0) \qquad (2.4.10)$$

We define $S_x(f)$, the *power spectral density* or the *power spectrum* of the signal $x(t)$ to be the Fourier transform of the time-average autocorrelation function.

$$S_x(f) = \mathcal{F}[R_x(\tau)] \qquad (2.4.11)$$

Subsequently, we will justify this definition. Now we can express the power content of the signal $x(t)$ in terms of $S_x(f)$ by noting that $R_x(0) = \int_{-\infty}^{+\infty} S_x(f)\,df$, i.e.,

$$P_x = R_x(0)$$

$$= \int_{-\infty}^{+\infty} S_x(f)\,df \qquad (2.4.12)$$

If a power-type signal $x(t)$ is passed through a filter with impulse response $h(t)$, the output is

$$y(t) = \int_{-\infty}^{+\infty} x(\tau)h(t - \tau) \, d\tau$$

and the time-average autocorrelation function for the output signal is

$$R_y(\tau) = \lim_{T \to \infty} \frac{1}{T} \int_{-\frac{T}{2}}^{+\frac{T}{2}} y(t)y^*(t - \tau) \, dt$$

Substituting for $y(t)$, we obtain

$$R_y(\tau) = \lim_{T \to \infty} \frac{1}{T} \int_{-\frac{T}{2}}^{+\frac{T}{2}} \left[\int_{-\infty}^{+\infty} h(u)x(t - u) \, du \right] \left[\int_{-\infty}^{+\infty} h^*(v)x^*(t - \tau - v) \, dv \right] dt$$

By making a change of variables $w = t - u$ and, changing the order of integration, we obtain

$$R_y(\tau) = \lim_{T \to \infty} \frac{1}{T} \int_{-\frac{T}{2}}^{+\frac{T}{2}} \int_{-\infty}^{+\infty} \int_{-\infty}^{+\infty} h(u)h^*(v) \times$$

$$\times \left[x(w)x^*(u + w - \tau - v) \, dw \right] du \, dv \, dt$$

$$\stackrel{a}{=} \int_{-\infty}^{+\infty} \int_{-\infty}^{+\infty} R_x(\tau + v - u)h(u)h^*(v) \, du \, dv$$

$$\stackrel{b}{=} \int_{-\infty}^{+\infty} \left[R_x(\tau + v) \star h(\tau + v) \right] h^*(v) \, dv$$

$$\stackrel{c}{=} R_x(\tau) \star h(\tau) \star h^*(-\tau) \tag{2.4.13}$$

where in (a) we have used the definition of R_x and in (b) and (c) we have used the definition of the convolution integral. Taking the Fourier transform of both sides of this equation, we obtain

$$S_y(f) = S_x(f)H(f)H^*(f)$$

$$= S_x(f)|H(f)|^2 \tag{2.4.14}$$

This relation between the input-output power spectral densities is the same as the relation between the energy spectral densities at the input and the output of a filter. Now using the same arguments used for the case of energy spectral density, we conclude that $S_x(f)$, as defined above, represents the amount of power at various frequencies. This justifies the definition of the power spectral density as the Fourier transform of the time-average autocorrelation function.

We have already seen that periodic signals are power-type signals. For periodic signals, the time-average autocorrelation function and the power spectral density simplify considerably. Let us assume that the signal $x(t)$ is a periodic signal with period T_0 having the Fourier series coefficients $\{x_n\}$. To find the time-average autocorrelation function we have

$$R_x(\tau) = \lim_{T \to \infty} \frac{1}{T} \int_{-\frac{T}{2}}^{+\frac{T}{2}} x(t)x^*(t - \tau)\, dt$$

$$= \lim_{k \to \infty} \frac{1}{kT_0} \int_{-\frac{kT_0}{2}}^{+\frac{kT_0}{2}} x(t)x^*(t - \tau)\, dt$$

$$= \lim_{k \to \infty} \frac{k}{kT_0} \int_{-\frac{T_0}{2}}^{+\frac{T_0}{2}} x(t)x^*(t - \tau)\, dt$$

$$= \frac{1}{T_0} \int_{-\frac{T_0}{2}}^{+\frac{T_0}{2}} x(t)x^*(t - \tau)\, dt \qquad (2.4.15)$$

This relation gives the time-average autocorrelation function for a periodic signal. If we substitute the Fourier series expansion of the periodic signal in this relation, we obtain

$$R_x(\tau) = \frac{1}{T_0} \int_{-\frac{T_0}{2}}^{+\frac{T_0}{2}} \sum_{n=-\infty}^{+\infty} \sum_{m=-\infty}^{+\infty} x_n x_m^* e^{j2\pi \frac{m}{T_0} \tau} e^{j2\pi \frac{n-m}{T_0} t}\, dt \qquad (2.4.16)$$

Now using the fact that

$$\frac{1}{T_0} \int_{-\frac{T_0}{2}}^{+\frac{T_0}{2}} e^{j2\pi \frac{n-m}{T_0} t}\, dt = \delta_{m,n}$$

we obtain

$$R_x(\tau) = \sum_{n=-\infty}^{+\infty} |x_n|^2 e^{j2\pi \frac{n}{T_0} \tau} \qquad (2.4.17)$$

From this relation we see that the time-average autocorrelation function of a periodic signal is itself periodic with the same period as the original signal, and its Fourier series coefficients are magnitude squares of the Fourier series coefficients of the original signal.

To determine the power spectral density of a periodic signal, it is enough to find the Fourier transform of $R_x(\tau)$. Since we are dealing with a periodic function, the Fourier transform consists of impulses in the frequency domain. This is what we expect because a periodic signal consists of a sum of sinusoidal (or exponential) signals, and therefore the power is concentrated at discrete frequencies (the harmonics). Therefore, the power spectral density of a periodic signal is given by

$$S_x(f) = \sum_{n=-\infty}^{+\infty} |x_n|^2 \delta\left(f - \frac{n}{T_0}\right) \qquad (2.4.18)$$

To find the power content of a periodic signal, we have to integrate this relation over the whole frequency spectrum. When we do this, we obtain

$$P_x = \sum_{n=-\infty}^{+\infty} |x_n|^2$$

This is the same relation we obtained in Section 2.2.5. If this periodic signal passes through an LTI system with frequency response $H(f)$, the output will be periodic and the power spectral density of the output can be obtained by employing the relation between the power spectral densities of the input and the output of a filter. Thus,

$$S_y(f) = |H(f)|^2 \sum_{n=-\infty}^{+\infty} |x_n|^2 \delta\left(f - \frac{n}{T_0}\right)$$

$$= \sum_{n=-\infty}^{+\infty} |x_n|^2 \left|H\left(\frac{n}{T_0}\right)\right|^2 \delta\left(f - \frac{n}{T_0}\right) \tag{2.4.19}$$

and the power content of the output signal is

$$P_y = \sum_{n=-\infty}^{+\infty} |x_n|^2 \left|H\left(\frac{n}{T_0}\right)\right|^2$$

Table 2.2 shows a summary of our results concerning energy and power of signals.

2.5 SAMPLING OF SIGNALS AND SIGNAL RECONSTRUCTION FROM SAMPLES

The sampling theorem is one of the most important results in the analysis of signals and has widespread applications in communications and signal processing. This theorem and the numerous applications that it provides clearly show how much can be gained by employing the frequency domain methods and the insight provided by frequency domain signal analysis. Many modern signal processing techniques and the whole family of digital communication methods are based on the validity of this theorem and the insight provided by it. In fact, this theorem, together with results from signal quantization techniques, provide a bridge that connects the analog world to digital techniques.

2.5.1 Ideal Sampling

The idea leading to the sampling theorem is very simple and quite intuitive. Let us assume that we have two signals $x_1(t)$ and $x_2(t)$ as shown in Figure 2.42. $x_1(t)$ is a smooth signal, it varies very slowly, and its main frequency content is at low frequencies. In contrast, $x_2(t)$ is a signal with rapid changes due to the presence of

TABLE 2-2 SUMMARY OF ENERGY AND POWER RELATIONS

	Autocorrelation function	Energy/power-spectral density	Input-output relation
Energy-type signals	$R_x(\tau) = \int_{-\infty}^{+\infty} x(t)x^*(t-\tau)\,dt$	$G_x(f) = \mathcal{F}[R_x(\tau)] = |X(f)|^2$	$G_y = G_x(f)|H(f)|^2$ $R_y(\tau) = R_x(\tau) \star h(\tau) \star h^*(-\tau)$
Power-type signals	$R_x(\tau) = \lim_{T\to\infty} \frac{1}{T} \int_{-T/2}^{+T/2} x(t)x^*(t-\tau)\,dt$	$S_x(f) = \mathcal{F}[R_x(\tau)] = \lim_{T\to\infty} \frac{|X_T(f)|^2}{T}$	$S_y(f) = S_x(f)|H(f)|^2$ $R_y(\tau) = R_x(\tau) \star h(\tau) \star h^*(-\tau)$
Periodic signals	$R_x(\tau) = \frac{1}{T_0} \int_{a}^{a+T_0} x(t)x^*(t-\tau)\,dt$ $R_x(\tau) = \sum_{n=-\infty}^{+\infty} |x_n|^2 e^{j2\pi(n/T_0)\tau}$	$S_x(f) = \mathcal{F}[R_x(\tau)]$ $S_x(f) = \sum_{n=-\infty}^{+\infty} |x_n|^2 \delta(f - \frac{n}{T_0})$	$S_y(f) = S_x(f)|H(f)|^2$ $S_y(f) = \sum_{n=-\infty}^{+\infty} |x_n|^2 |H(\frac{n}{T_0})|^2 \delta(f - \frac{n}{T_0})$

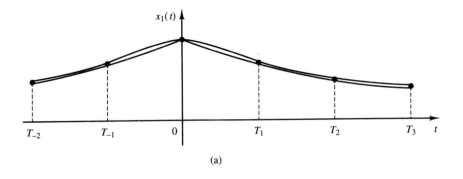

(a)

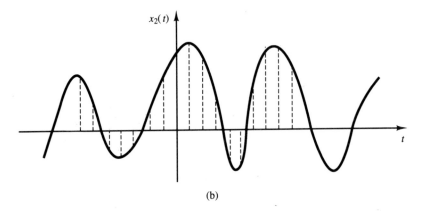

(b)

FIGURE 2.42.　Sampling of a (a) smooth and (b) rapidly varying signal.

the high frequency components. We are to approximate these signals with samples taken at regular intervals T_1 and T_2, respectively. To obtain an approximation of the original signal, we can use, for example, linear interpolation of the sampled values. It is obvious that the sampling interval for the signal $x_1(t)$ can be much larger than the sampling interval necessary to reconstruct signal $x_2(t)$ with comparable distortion. This is a direct consequence of the smoothness of the signal $x_1(t)$ compared to $x_2(t)$. Therefore, the sampling interval for the signals of smaller bandwidth can be made larger, or the sampling frequency can be made smaller. The sampling theorem is, in fact, a precise statement of this intuitive reasoning. It basically states that

1. If the signal $x(t)$ is bandlimited to W; i.e., if $X(f) \equiv 0$ for $|f| \geq W$, then it is sufficient to sample at intervals $T_s = \frac{1}{2W}$.

2. If we are allowed to employ more sophisticated interpolating signals compared to linear interpolation, we are able to obtain the exact original signal back from the samples as long as the previous condition is satisfied.

As seen here, the importance of the sampling theorem lies in the fact that it provides both a method of reconstruction of the original signal from the sampled values and also gives a precise upper bound on the sampling interval (or equivalently, a lower bound on the sampling frequency) required for distortionless reconstruction.

Theorem 2.5.1 [Sampling].   Let the signal $x(t)$ be bandlimited with bandwidth W, i.e., let $X(f) \equiv 0$ for $|f| \geq W$. Let $x(t)$ be sampled at multiples of some basic sampling interval T_s, where $T_s \leq \frac{1}{2W}$ to yield the sequence $\{x(nT_s)\}_{n=-\infty}^{+\infty}$. Then it is possible to reconstruct the original signal $x(t)$ from the sampled values by the reconstruction formula

$$x(t) = \sum_{n=-\infty}^{+\infty} 2W'T_s x(nT_s)\, \text{sinc}[2W'(t - nT_s)] \tag{2.5.1}$$

where W' is any arbitrary number that satisfies

$$W \leq W' \leq \frac{1}{T_s} - W$$

In the special case where $T_s = \frac{1}{2W}$, the reconstruction relation simplifies to

$$x(t) = \sum_{n=-\infty}^{+\infty} x(nT_s)\, \text{sinc}\left(\frac{t}{T_s} - n\right)$$

Proof.   Let $x_\delta(t)$ denote the result of sampling the original signal by impulses at nT_s time instants. Then

$$x_\delta(t) = \sum_{n=-\infty}^{+\infty} x(nT_s)\delta(t - nT_s) \tag{2.5.2}$$

We can write $x_\delta(t)$ as

$$x_\delta(t) = x(t) \sum_{n=-\infty}^{+\infty} \delta(t - nT_s) \tag{2.5.3}$$

Now if we find the Fourier transform of both sides of the above relation and apply the dual of the convolution theorem to the right-hand side, we obtain

$$X_\delta(f) = X(f) \star \mathcal{F}\left[\sum_{n=-\infty}^{+\infty} \delta(t - nT_s)\right] \tag{2.5.4}$$

Now using Table 2.1 to find the Fourier transform of $\sum_{n=-\infty}^{+\infty} \delta(t - nT_s)$, we obtain

$$\mathcal{F}\left[\sum_{n=-\infty}^{+\infty} \delta(t - nT_s)\right] = \frac{1}{T_s} \sum_{n=-\infty}^{+\infty} \delta\left(f - \frac{n}{T_s}\right) \tag{2.5.5}$$

By substituting (2.5.5) into (2.5.4) we obtain

$$X_\delta(f) = X(f) \star \frac{1}{T_s} \sum_{n=-\infty}^{+\infty} \delta\left(f - \frac{n}{T_s}\right)$$

$$= \frac{1}{T_s} \sum_{n=-\infty}^{+\infty} X\left(f - \frac{n}{T_s}\right) \tag{2.5.6}$$

where in the last step we have employed the convolution property of the impulse signal. This relation shows that $X_\delta(f)$, the Fourier transform of the impulse-sampled signal, is a replication of the Fourier transform of the original signal at a $\frac{1}{T_s}$ rate. Figure 2.43 shows this situation. Now if $T_s > \frac{1}{2W}$, then the replicated spectrum of $x(t)$ overlaps, and reconstruction of the original signal is not possible. The type of distortion that results from undersampling is known as *aliasing error* or *aliasing distortion*. However, if $T_s \leq \frac{1}{2W}$, no overlap occurs, and by employing an appropriate filter, we can reconstruct the original signal back. To obtain the original signal back, it is sufficient to filter the sampled signal by a lowpass filter with frequency response characteristic

1. $H(f) = T_s$ for $|f| < W$.
2. $H(f) = 0$ for $|f| \geq \frac{1}{T_s} - W$

For $W \leq |f| < \frac{1}{T_s} - W$, the filter can have any characteristics that make its implementation easy. Of course, one obvious (though not practical) choice is an ideal

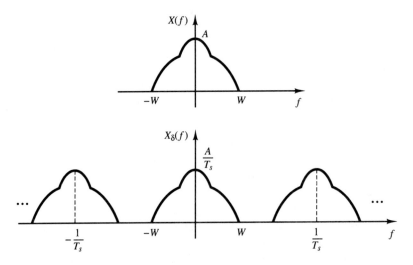

FIGURE 2.43. Frequency domain representation of the sampled signal.

lowpass filter with bandwidth W' where W' satisfies $W \le W' < \frac{1}{T_s} - W$; i.e.,

$$H(f) = T_s \Pi \left(\frac{f}{2W'} \right) \tag{2.5.7}$$

With this choice we have

$$X(f) = X_\delta(f) T_s \Pi \left(\frac{f}{2W'} \right) \tag{2.5.8}$$

Taking the inverse Fourier transform of both sides, we obtain

$$x(t) = x_\delta(t) \star 2W' T_s \operatorname{sinc}(2W't)$$

$$= \left(\sum_{n=-\infty}^{+\infty} x(nT_s)\delta(t - nT_s) \right) \star 2W' T_s \operatorname{sinc}(2W't)$$

$$= \sum_{n=-\infty}^{+\infty} 2W' T_s x(nT_s) \operatorname{sinc}\left(2W'(t - nT_s) \right) \tag{2.5.9}$$

This relation shows that if we use sinc functions for interpolation of the sampled values, we can reconstruct the original signal perfectly. The sampling rate $f_s = \frac{1}{2W}$ is the minimum sampling rate at which no aliasing occurs. This sampling rate is known as the *Nyquist sampling rate*. If sampling is done at the Nyquist rate, then the only choice for the reconstruction filter is an ideal lowpass filter and $W' = W = \frac{1}{2T_s}$. Then,

$$x(t) = \sum_{n=-\infty}^{+\infty} x \left(\frac{n}{2W} \right) \operatorname{sinc}(2Wt - n)$$

$$= \sum_{n=-\infty}^{+\infty} x(nT_s) \operatorname{sinc}\left(\frac{t}{T_s} - n \right) \tag{2.5.10}$$

In practical systems, sampling is done at a rate higher than the Nyquist rate. This allows for the reconstruction filter to be realizable and easier to build. In such cases, the distance between two adjacent replicated spectra in the frequency domain, i.e., $(\frac{1}{T_s} - W) - W = f_s - 2W$ is known as the *guard band*.

   Note that there exists a strong similarity between our development of the sampling theorem and our previous development of the Fourier transform for periodic signals (or Fourier series). In the Fourier transform for periodic signals, we started with a (time) periodic signal and showed that its Fourier transform consists of a sequence of impulses. Therefore, to define the signal, it was enough to give the weights of these impulses (Fourier series coefficients). In the sampling theorem, we started with an impulse-sampled signal, or a sequence of impulses in the time domain, and showed that the Fourier transform is a periodic function in the frequency domain. Here again, the values of the samples were enough to define the signal completely. This similarity is a consequence of the duality between the time and frequency domains and the fact that both the Fourier series expansion and reconstruction from samples are orthogonal expansions, one in terms of the exponential signals and the other in terms of the sinc functions. This fact will be further explored in the problems.

Example 2.5.1

In the above development, we assumed that samples are taken at multiples of T_s. What happens if we sample regularly with T_s as the sampling interval but the first sample is taken at some $0 < t_0 < T_s$?

Solution   We define a new signal $y(t) = x(t + t_0)$. Then $y(t)$ is bandlimited with $Y(f) = e^{j2\pi f t_0} X(f)$ and the samples of $y(t)$ at $\{kT_s\}_{k=-\infty}^{\infty}$ are equal to the samples of $x(t)$ at $\{t_0 + kT_s\}_{k=-\infty}^{\infty}$. Applying the sampling theorem to the reconstruction of $y(t)$, we have

$$y(t) = \sum_{k=-\infty}^{\infty} y(kT_s) \operatorname{sinc}\left(2W(t - kT_s)\right)$$

$$= \sum_{k=-\infty}^{\infty} x(t_0 + kT_s) \operatorname{sinc}\left(2W(t - kT_s)\right)$$

and, hence,

$$x(t + t_0) = \sum_{k=-\infty}^{\infty} x(t_0 + kT_s) \operatorname{sinc}\left(2W(t - kT_s)\right)$$

Substituting $t = -t_0$, we obtain the following important interpolation relation.

$$x(0) = \sum_{k=-\infty}^{\infty} x(t_0 + kT_s) \operatorname{sinc}\left(2W(t_0 + kT_s)\right) \qquad (2.5.11)$$

even function

Example 2.5.2

A bandlimited signal has a bandwidth equal to 3400 Hz. What sampling rate should be used to guarantee a guard band of 1200 Hz?

Solution   We have

$$\text{guard band} = f_s - 2W = 1200$$

Therefore

$$f_s = 1200 + 2 \times 3400 = 8000$$

2.5.2 Practical Sampling

Although the preceding sampling theorem establishes the basic limit on the sampling frequency to make distortionless reconstruction possible, sampling by impulses is hardly practical. In practice, two other types of sampling are frequently used. These two schemes are *switched sampling* and *zero-order-hold sampling*.

Switched sampling.   In switched sampling, the bandlimited signal is sampled by closing a switch at regular intervals and keeping it closed for some fixed

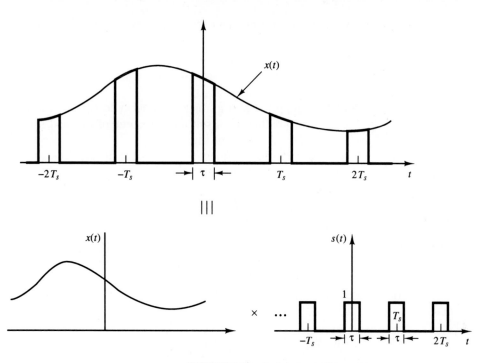

FIGURE 2.44.   Switched sampling system.

duration. Figure 2.44 shows an example of a switched sampling system. The switching action can be described by a switching signal

$$s(t) = \sum_{n=-\infty}^{+\infty} \Pi\left(\frac{t - nT_s}{\tau}\right) \quad\text{rectangular function.}$$ (2.5.12)

and the sampled signal $x_s(t)$ can be written as

$$x_s(t) = x(t)s(t)$$ (2.5.13)

In the frequency domain, we have

$$X_s(f) = X(f) \star S(f)$$ (2.5.14)

The Fourier transform of the signal $s(t)$ may be obtained from the Fourier series coefficients given in Section 2.3.3, and is expressed as

$$S(f) = \frac{\tau}{T_s} \sum_{n=-\infty}^{+\infty} \text{sinc}\left(\frac{n\tau}{T_s}\right) \delta\left(f - \frac{n}{T_s}\right)$$ (2.5.15)

Substituting (2.5.15) into (2.5.14), we obtain

$$X_s(f) = \sum_{n=-\infty}^{+\infty} \frac{\tau}{T_s} \text{sinc}\left(\frac{n\tau}{T_s}\right) X\left(f - \frac{n}{T_s}\right)$$ (2.5.16)

As seen here, again we have a replication in the frequency domain, but this time with different coefficients. For $n = 0$ the corresponding part is $\frac{\tau}{T_s} X(f)$. To reconstruct the signal, we have to employ, for example, an ideal filter with gain $\frac{T_s}{\tau}$. Figure 2.45 on page 113 illustrates the reconstruction process.

Zero-Order-Hold Sampling. Flat-top, or zero-order-hold (ZOH), sampling is done by sampling at nT_s and holding the result until the time of the next sample, i.e., $(n + 1)T_s$. Figure 2.46 shows this type of sampling.

Zero-order-hold sampling can be considered as a cascade of impulse-sampling with a zero-order-hold filter; i.e., a filter with impulse response

$$h(t) = \Pi\left(\frac{t - \frac{T_s}{2}}{T_s} \right)$$

and frequency response

$$H(f) = T_s e^{-j\pi T_s f} \operatorname{sinc}(T_s f)$$

We have already seen that the output of the impulse sampler in the frequency domain is

$$X_\delta(f) = \frac{1}{T_s} \sum_{n=-\infty}^{+\infty} X\left(f - \frac{n}{T_s} \right)$$

By passing this signal through a zero-order-hold filter, we obtain

$$X_{\text{ZOH}}(f) = \sum_{n=-\infty}^{+\infty} e^{-j\pi f T_s} \operatorname{sinc}(T_s f) X\left(f - \frac{n}{T_s} \right) \qquad (2.5.17)$$

To obtain the original signal back, a filter with frequency response

$$H(f) = \frac{e^{j\pi T_s f}}{\operatorname{sinc}(T_s f)} \Pi\left(\frac{f}{2W} \right) \qquad (2.5.18)$$

can be employed. If the sampling frequency is quite high, that is, if $f_s \gg 2W$, then for all $|f| < W$, $T_s f \ll 1$ and $\operatorname{sinc}(T_s f) \approx 1$. In this case an ideal filter can reconstruct the signal with good accuracy.

2.6 BANDPASS SIGNALS

In this section, we examine time domain and frequency domain characteristics of a class of signals frequently encountered in communication system analysis. This class of signals is the class of *bandpass* or *narrowband* signals. The concept of bandpass signals is a generalization of the concept of monochromatic signals, and our analysis of the properties of these signals follows that used in analyzing monochromatic signals.

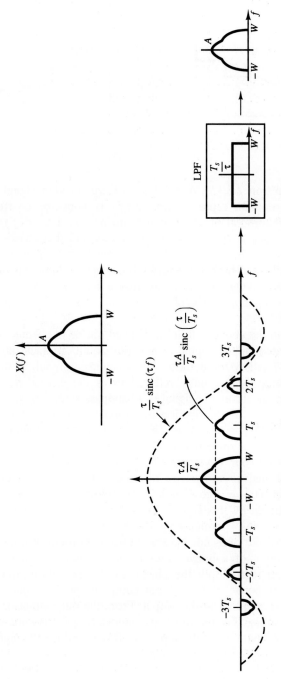

FIGURE 2.45. Reconstruction of the original signal in switched sampling.

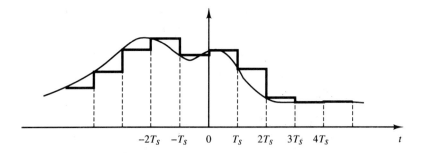

FIGURE 2.46.   Zero-order-hold sampling.

Definition 2.6.1.   A *bandpass* or *narrowband* signal is a signal $x(t)$ whose frequency domain representation $X(f)$ is nonzero for frequencies in a usually small neighborhood of some high frequency f_0, i.e., $X(f) \equiv 0$ for $|f - f_0| \geq W$, where $W < f_0$. A *bandpass system* is a system that passes signals with frequency components in the neighborhood of some high frequency f_0, i.e., $H(f) = 1$ for $|f - f_0| \leq W$ and highly attenuates frequency components outside of this frequency band. Alternatively, we may say that a bandpass system is one whose impulse response is a bandpass signal.

Note that in the above definition f_0 need not be the center of the signal bandwidth, or be located in the signal bandwidth at all. In fact, all the spectra shown in Figure 2.47 satisfy the definition of a bandpass signal. With the above precautions, the frequency f_0 is usually referred to as the *central frequency* of the bandpass signal. A monochromatic signal is a bandpass signal for which $W = 0$. A large class of signals used for information transmission, the *modulated signals,* are examples of bandpass signals or at least closely represented by bandpass signals. Throughout this section, we assume that the bandpass signal $x(t)$ is real-valued.

To begin our development of bandpass signals, let us start with the tools used in the analysis of systems or circuits driven by monochromatic (or sinusoidal) signals. Let $x(t) = A\cos(2\pi f_0 t + \theta)$ denote a monochromatic signal. To analyze a circuit driven by this signal, we first introduce the *phasor* corresponding to this signal as $X = Ae^{j\theta}$, which contains the information about the amplitude and phase of the signal but does not have any information concerning the frequency of it. To find the output of a linear time-invariant circuit driven by this sinusoidal signal, it is enough to multiply the phasor of the excitation signal by the value of the frequency response of the system computed at the input frequency to obtain the phasor corresponding to the output. From the output phasor we can find the output signal by noting that the input and output frequencies are the same. To obtain the phasor corresponding to the input, we first introduce the signal $z(t)$ as

$$z(t) = Ae^{j(2\pi f_0 t + \theta)}$$

$$= A\cos(2\pi f_0 t + \theta) + jA\sin(2\pi f_0 t + \theta)$$

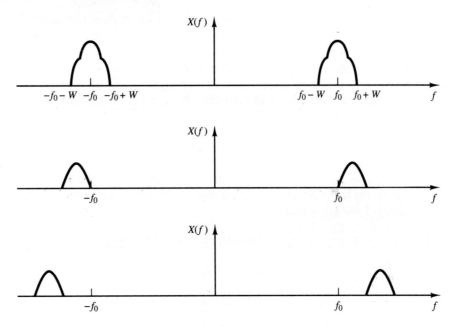

FIGURE 2.47.   Examples of narrowband signals.

$$= x(t) + jx_q(t)$$

where $x_q(t) = A \sin(2\pi f_0 t + \theta)$ is a 90° phase shift version of the original signal and the subscript stands for *quadrature*. Note that $z(t)$ represents a vector rotating at an angular frequency equal to $2\pi f_0$ as shown in Figure 2.48, and X, the phasor, is obtained from $z(t)$ by deleting the rotation at the angular frequency of $2\pi f_0$, or equivalently by rotating the vector corresponding to $z(t)$ at an angular frequency equal to $2\pi f_0$ in the opposite direction, which is equivalent to multiplying by $e^{-j2\pi f_0 t}$, or

$$X = z(t)e^{-j2\pi f_0 t}$$

In the frequency domain, this is equivalent to shifting $Z(f)$ to the left by f_0. Also

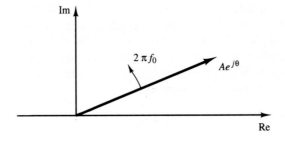

FIGURE 2.48.   Phasor of a monochromatic signal.

note that the frequency domain representation of $Z(f)$ is obtained by deleting the negative frequencies from $X(f)$ and multiplying the positive frequencies by 2.

To obtain a parallel development for narrowband signals, we start with a signal corresponding to $z(t)$. We define $z(t)$ for narrowband signals in the same way that $z(t)$ was defined for monochromatic signals, i.e., by multiplying the positive frequencies in $X(f)$ by 2 and deleting the negative frequencies. By doing this, we have

$$Z(f) = 2u_{-1}(f)X(f) \qquad (2.6.1)$$

The signal $z(t)$ defined by the above relation is called the *analytic signal* corresponding to $x(t)$, or *pre-envelope* of $x(t)$. To obtain the time domain representation of $z(t)$, we first start with finding a signal whose Fourier transform is $u_{-1}(f)$. From Table 2.1 we know that

$$\mathcal{F}[u_{-1}(t)] = \frac{1}{2}\delta(f) + \frac{1}{j2\pi f}$$

Applying the duality theorem, we obtain

$$\mathcal{F}\left[\frac{1}{2}\delta(t) + \frac{j}{2\pi t}\right] = u_{-1}(f) \qquad (2.6.2)$$

Now, using the convolution theorem, we have

$$z(t) = \left(\delta(t) + \frac{j}{\pi t}\right) \star x(t)$$

$$= x(t) + j\frac{1}{\pi t} \star x(t)$$

$$= x(t) + j\hat{x}(t) \qquad (2.6.3)$$

where

$$\hat{x}(t) = \frac{1}{\pi t} \star x(t) \qquad (2.6.4)$$

Comparing this result with the corresponding monochromatic result

$$z(t) = \underbrace{A\cos(2\pi f_0 t + \theta)}_{x(t)} + j\underbrace{A\sin(2\pi f_0 t + \theta)}_{\hat{x}(t)} \qquad (2.6.5)$$

we see that $\hat{x}(t)$ plays the same role as $A\sin(2\pi f_0 t + \theta)$. $\hat{x}(t)$ is called the *Hilbert transform* of $x(t)$. The name "transform" is somewhat misleading because there is no change of domain involved (as is the case for Fourier, Laplace, and Z transforms, for example). In fact, the Hilbert transform is a simple filter (seen from the fact that it can be expressed in terms of a convolution integral). To see what the Hilbert transform does in the frequency domain, we note that

$$\mathcal{F}\left[\frac{1}{\pi t}\right] = -j\,\text{sgn}(f)$$

$$= \begin{cases} -j & f > 0 \\ 0 & f = 0 \\ +j & f < 0 \end{cases}$$

$$= \begin{cases} e^{-j\frac{\pi}{2}} & f > 0 \\ 0 & f = 0 \\ e^{j\frac{\pi}{2}} & f < 0 \end{cases}$$

$$= e^{-j\frac{\pi}{2}\,\text{sgn}(f)}, \quad f \neq 0 \tag{2.6.6}$$

This means that the Hilbert transform is equivalent to a $-\frac{\pi}{2}$ phase shift for positive frequencies and $+\frac{\pi}{2}$ phase shift for negative frequencies and can be represented by a filter with transfer function $H(f) = -j\,\text{sgn}(f)$. This filter is called a *quadrature filter*, emphasizing its role in providing a 90° phase shift. Later in this section, we will investigate some of the most important properties of the Hilbert transform.

To obtain the equivalent of a "phasor" for the bandpass signal, we have to shift the spectrum of $z(t)$, i.e., $Z(f)$, to the left by f_0 to obtain a signal denoted by $x_l(t)$ and called the *lowpass representation of the bandpass signal* $x(t)$.

$$X_l(f) = Z(f + f_0) = 2u_{-1}(f + f_0)X(f + f_0) \tag{2.6.7}$$

and

$$x_l(t) = z(t)e^{-j2\pi f_0 t} \tag{2.6.8}$$

Figure 2.49 shows $Z(f)$ and $X_l(f)$ corresponding to a bandpass signal $x(t)$. As seen, $x_l(t)$ is a lowpass signal, meaning that its frequency components are located around the zero frequency, or $X_l(f) \equiv 0$ for $|f| \geq W$ where $W < f_0$. $x_l(t)$ plays the role of the phasor for bandpass signals. In general, $x_l(t)$ is a complex signal having $x_c(t)$ and $x_s(t)$ as its real and imaginary parts, respectively, i.e.,

$$x_l(t) = x_c(t) + jx_s(t) \tag{2.6.9}$$

$x_c(t)$ and $x_s(t)$ are lowpass signals, called *in-phase* and *quadrature* components of the bandpass signal $x(t)$. Substituting for $x_l(t)$ and rewriting $z(t)$, we obtain

$$\begin{aligned} z(t) &= x(t) + j\hat{x}(t) \\ &= x_l(t)e^{j2\pi f_0 t} \\ &= (x_c(t) + jx_s(t))e^{j2\pi f_0 t} \\ &= (x_c(t)\cos(2\pi f_0 t) - x_s(t)\sin(2\pi f_0 t)) + \\ &\quad + j(x_c(t)\sin(2\pi f_0 t) + x_s(t)\cos(2\pi f_0 t)) \end{aligned} \tag{2.6.10}$$

Equating the real and imaginary parts, we have

$$x(t) = x_c(t)\cos(2\pi f_0 t) - x_s(t)\sin(2\pi f_0 t) \tag{2.6.11}$$

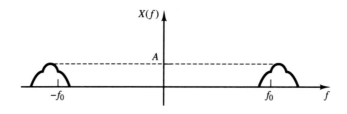

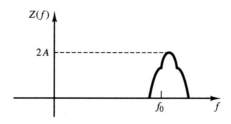

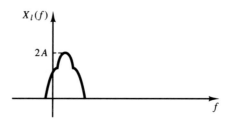

FIGURE 2.49. $Z(f)$ and $X_l(f)$ corresponding to $x(t)$.

and

$$\hat{x}(t) = x_c(t) \sin(2\pi f_0 t) + x_s(t) \cos(2\pi f_0 t) \qquad (2.6.12)$$

These relations give $x(t)$ and $\hat{x}(t)$ in terms of two lowpass quadrature component signals $x_c(t)$ and $x_s(t)$ and are known as *bandpass to lowpass transformation
relations.*

If we define $V(t)$, the envelope of $x(t)$, as

$$V(t) = \sqrt{x_c^2(t) + x_s^2(t)} \qquad (2.6.13)$$

and $\Theta(t)$, the phase of $x(t)$, as

$$\Theta(t) = \arctan \frac{x_s(t)}{x_c(t)} \qquad (2.6.14)$$

we can write

$$x_l(t) = V(t)e^{j\Theta(t)} \qquad (2.6.15)$$

which looks more like the familiar phasor relation $X = Ae^{j\theta}$. The only difference is that in this case the envelope ($V(t)$) and phase ($\Theta(t)$) are both (slowly) time-varying functions. Therefore, in contrast to the monochromatic phasor which has constant amplitude and phase, the envelope and phase of a bandpass signal vary slowly with time, and therefore the vector representation of it moves on a curve in the complex plane (see Figure 2.50). Substituting $x_l(t) = V(t)e^{j\Theta(t)}$ in $z(t)$ in (2.6.10), we obtain

$$
\begin{aligned}
z(t) &= x(t) + j\hat{x}(t) \\
&= x_l(t)e^{j2\pi f_0 t} \\
&= V(t)e^{j\Theta(t)}e^{j2\pi f_0 t} \\
&= V(t)\cos(2\pi f_0 t + \Theta(t)) + jV(t)\sin(2\pi f_0 t + \Theta(t)) \quad (2.6.16)
\end{aligned}
$$

from which we have

$$
x(t) = V(t)\cos(2\pi f_0 t + \Theta(t)) \quad (2.6.17)
$$

and

$$
\hat{x}(t) = V(t)\sin(2\pi f_0 t + \Theta(t)) \quad (2.6.18)
$$

These relations show why $V(t)$ and $\Theta(t)$ are called envelope and phase of the signal $x(t)$. Figure 2.51 shows the relation between $x(t)$, $V(t)$, and $\Theta(t)$.

Example 2.6.1

Show that $X(f)$ can be written in terms of $X_l(f)$ as

$$
X(f) = \frac{1}{2}[X_l(f - f_0) + X_l^*(-f - f_0)] \quad (2.6.19)
$$

Solution   To obtain $X(f)$ from $X_l(f)$, we do exactly the inverse of what we did to get $X_l(f)$ from $X(f)$. First, we shift $X_l(f)$ to the right by f_0 to get $Z(f)$. Then we have

$$
Z(f) = X_l(f - f_0)
$$

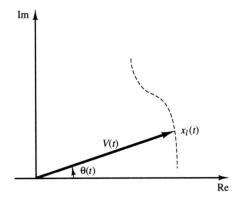

FIGURE 2.50.   The phasor representation of a bandpass signal.

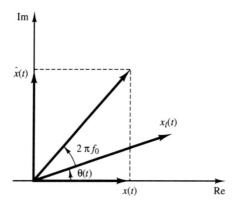

FIGURE 2.51. The envelope and phase of a bandpass signal.

To get $X(f)$ from $Z(f)$, we have to multiply the positive frequencies by a factor of $\frac{1}{2}$ and reconstruct the negative frequencies. Since $x(t)$ is assumed to be real-valued, its Fourier transform has Hermitian symmetry. Therefore, if we write

$$X(f) = X_+(f) + X_-(f)$$

where $X_+(f)$ denotes the positive frequency part of $X(f)$ and $X_-(f)$ denotes the negative frequency part, we have

$$X_-(f) = X_+^*(-f)$$

Since

$$X_+(f) = \frac{1}{2} X_l(f - f_0)$$

we obtain

$$X(f) = \frac{1}{2} [X_l(f - f_0) + X_l^*(-f - f_0)]$$

The relation between the various signals discussed in this section are summarized in Table 2.3.

Transmission of bandpass signals through bandpass systems.   In the same way that phasors make analysis of systems driven by monochromatic signals easier, lowpass equivalents of bandpass signals can be employed to find the outputs of bandpass systems driven by bandpass signals. Let $x(t)$ be a bandpass signal with center frequency f_0, and let $h(t)$ be the impulse response of an LTI system. Let us assume that $h(t)$ is narrowband with the same center frequency as $x(t)$. To find $y(t)$, the output of the system when driven by $x(t)$, we use frequency domain analysis. In the frequency domain we have $Y(f) = X(f)H(f)$. The signal $y(t)$ is obviously a bandpass signal, and therefore it has a lowpass equivalent $y_l(t)$. To obtain $Y_l(f)$, we have

$$Y_l(f) = 2u_{-1}(f + f_0)Y(f + f_0)$$

$$= 2u_{-1}(f + f_0)X(f + f_0)H(f + f_0) \qquad (2.6.20)$$

TABLE 2-3   SUMMARY OF BANDPASS TO
LOWPASS TRANSLATION RELATIONS

$$\begin{cases} x(t) = x_c(t)\cos(2\pi f_0 t) - x_s(t)\sin(2\pi f_0 t) \\ \hat{x}(t) = x_c(t)\sin(2\pi f_0 t) + x_s(t)\cos(2\pi f_0 t) \end{cases}$$

$$\begin{cases} x(t) = \text{Re}[x_l(t)e^{j2\pi f_0 t}] \\ \hat{x}(t) = \text{Im}[x_l(t)e^{j2\pi f_0 t}] \end{cases}$$

$$\begin{cases} x(t) = \text{Re}[z(t)] \\ \hat{x}(t) = \text{Im}[z(t)] \end{cases}$$

$$\begin{cases} x(t) = V(t)\cos(2\pi f_0 t + \Theta(t)) \\ \hat{x}(t) = V(t)\sin(2\pi f_0 t + \Theta(t)) \end{cases}$$

$$\begin{cases} x_c(t) = x(t)\cos(2\pi f_0 t) + \hat{x}(t)\sin(2\pi f_0 t) \\ x_s(t) = \hat{x}(t)\cos(2\pi f_0 t) - x(t)\sin(2\pi f_0 t) \end{cases}$$

$$\begin{cases} x_c(t) = \text{Re}[x_l(t)] \\ x_s(t) = \text{Im}[x_l(t)] \end{cases}$$

$$\begin{cases} x_c(t) = \text{Re}[z(t)e^{-j2\pi f_0 t}] \\ x_s(t) = \text{Im}[z(t)e^{-j2\pi f_0 t}] \end{cases}$$

$$\begin{cases} x_c(t) = V(t)\cos\Theta(t) \\ x_s(t) = V(t)\sin\Theta(t) \end{cases}$$

$$\begin{cases} x_l(t) = (x(t) + j\hat{x}(t))e^{-j2\pi f_0 t} \\ x_l(t) = x_c(t) + jx_s(t) \\ x_l(t) = z(t)e^{-j2\pi f_0 t} \\ x_l(t) = V(t)e^{j\Theta(t)} \end{cases}$$

$$\begin{cases} z(t) = x(t) + j\hat{x}(t) \\ z(t) = (x_c(t) + jx_s(t))e^{j2\pi f_0 t} \\ z(t) = x_l(t)e^{j2\pi f_0 t} \\ z(t) = V(t)e^{j(2\pi f_0 t + \Theta(t))} \end{cases}$$

$$\begin{cases} V(t) = \sqrt{x^2(t) + \hat{x}^2(t)} \\ \Theta(t) = \arctan\frac{\hat{x}(t)}{x(t)} - 2\pi f_0 t \end{cases}$$

$$\begin{cases} V(t) = \sqrt{x_c^2(t) + x_s^2(t)} \\ \Theta(t) = \arctan\frac{x_s(t)}{x_c(t)} \end{cases}$$

$$\begin{cases} V(t) = |x_l(t)| \\ \Theta(t) = \angle x_l(t) \end{cases}$$

$$\begin{cases} V(t) = |z(t)| \\ \Theta(t) = \angle z(t) - 2\pi f_0 t \end{cases}$$

By writing $H(f)$ and $X(f)$ in terms of their lowpass equivalents, we obtain

$$X_l(f) = 2u_{-1}(f + f_0)X(f + f_0)$$

and

$$H_l(f) = 2u_{-1}(f + f_0)H(f + f_0)$$

By multiplying these two relations and noting that $u_{-1}^2(f) = u_{-1}(f)$, we have

$$X_l(f)H_l(f) = 4u_{-1}(f + f_0)X(f + f_0)H(f + f_0)$$

Finally, by substituting in the relation for $Y_l(f)$, we obtain

$$Y_l(f) = \frac{1}{2}X_l(f)H_l(f) \tag{2.6.21}$$

or, in time domain, we have

$$y_l(t) = \frac{1}{2}x_l(t) \star h_l(t) \tag{2.6.22}$$

This relation shows that to obtain the output $y(t)$ we can carry out the convolution at low frequencies to obtain $y_l(t)$ and then transform to higher frequencies using the relation

$$y(t) = \text{Re}[y_l(t)e^{j2\pi f_0 t}] \tag{2.6.23}$$

2.6.1 Properties of the Hilbert Transform

In this section, we will explore some further properties of the Hilbert transform. We have already seen that performing the Hilbert transform on a signal is equivalent to a 90° phase shift in all frequency components of the signal. Therefore, it is obvious that the only change that the Hilbert transform performs on a signal is changing its phase. In particular, no change is done to the amplitude of the frequency components, and therefore, the energy and power of the signal do not change by performing the Hilbert transform operation. On the other hand, since performing the Hilbert transform changes cosines into sines, it is not surprising that the Hilbert transform of a signal is orthogonal to it. Also, since the Hilbert transform introduces a 90° phase shift, carrying it out twice causes a total 180° phase shift or a sign reversal of the original signal. In all the following theorems, we assume that $X(f)$ does not have any impulses at zero frequency.

Theorem 2.6.1. The Hilbert transform of an even signal is odd, and the Hilbert transform of an odd signal is even.

Proof. If $x(t)$ is even, then $X(f)$ is a real and even function and therefore $-j \, \text{sgn}(f)X(f)$ is an imaginary and odd function. Hence, its inverse Fourier transform $\hat{x}(t)$ will be odd. If $x(t)$ is odd, then $X(f)$ is imaginary and odd and $-j \, \text{sgn}(f)X(f)$ is real and even, and therefore, $\hat{x}(t)$ is even.

Theorem 2.6.2. Applying the Hilbert transform operation to a signal twice causes a change of sign of the signal, i.e.,

$$\hat{\hat{x}}(t) = -x(t) \tag{2.6.24}$$

Proof. It is enough to note that

$$\mathcal{F}[\hat{\hat{x}}(t)] = (-j \, \text{sgn}(f))^2 X(f) \tag{2.6.25}$$

and hence,

$$\mathcal{F}[\hat{\hat{x}}(t)] = -X(f) \tag{2.6.26}$$

where we have used the fact that $X(f)$ does not contain any impulses at the origin.

Theorem 2.6.3. The energy content of a signal is equal to the energy content of its Hilbert transform.

Proof. Using Rayleigh's theorem of the Fourier transform, we have

$$E_x = \int_{-\infty}^{+\infty} |x(t)|^2 dt = \int_{-\infty}^{+\infty} |X(f)|^2 df \tag{2.6.27}$$

and

$$E_{\hat{x}} = \int_{-\infty}^{+\infty} |\hat{x}(t)|^2 dt = \int_{-\infty}^{+\infty} |-j \, \text{sgn}(f) X(f)|^2 df \tag{2.6.28}$$

Noting that $|-j \, \text{sgn}(f)|^2 = 1$ except for $f = 0$ and that $X(f)$ does not contain any impulses at the origin completes the proof.

Theorem 2.6.4. The signal $x(t)$ and its Hilbert transform are orthogonal.

Proof. Here we use Parseval's theorem of the Fourier transform to obtain

$$\int_{-\infty}^{+\infty} x(t)\hat{x}(t) \, dt = \int_{-\infty}^{+\infty} X(f)[-j \, \text{sgn}(f) X(f)]^* df$$

$$= -j \int_{-\infty}^{0} |X(f)|^2 df + j \int_{0}^{+\infty} |X(f)|^2 df$$

$$= 0 \tag{2.6.29}$$

where in the last step we have used the fact that $X(f)$ is Hermitian and therefore $|X(f)|^2$ is even.

Theorem 2.6.5. Let $x(t)$ represent a bandpass signal and $m(t)$ denote a lowpass signal with nonoverlapping spectra. Then the Hilbert transform of $c(t) = m(t)x(t)$ is equal to $m(t)\hat{x}(t)$.

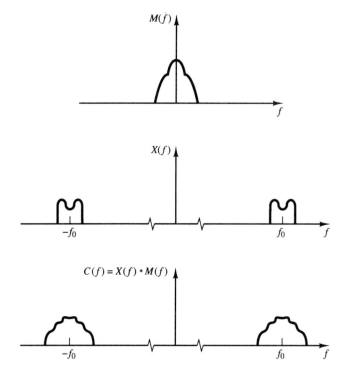

FIGURE 2.52. Frequency domain representation of $M(f)$, $X(f)$, and $C(f)$.

Proof.   Figure 2.52 shows the representation of $M(f)$, $X(f)$, and $C(f)$. Note that $C(f) = M(f) \star X(f)$. From the assumption on the bandwidth of $m(t)$ and $x(t)$, we see that $C(f)$ consists of two separate positive frequency and negative frequency regions that do not overlap. Let us denote these regions by $C_+(f)$ and $C_-(f)$, respectively. It can be seen that

$$C_+(f) = M(f) \star X_+(f)$$

and

$$C_-(f) = M(f) \star X_-(f)$$

To find the Hilbert transform of $c(t)$, we note that

$$\mathcal{F}[\hat{c}(t)] = -j\,\text{sgn}(f)C(f)$$
$$= -jC_+(f) + jC_-(f)$$
$$= -jM(f) \star X_+(f) + jM(f) \star X_-(f)$$
$$= M(f) \star [-jX_+(f) + jX_-(f)]$$
$$= M(f) \star [-j\,\text{sgn}(f)X(f)]$$

$$= M(f) \star \mathcal{F}[\hat{x}(t)] \tag{2.6.30}$$

Returning to the time domain, we obtain

$$\hat{c}(t) = m(t)\hat{x}(t) \tag{2.6.31}$$

Example 2.6.2

Determine the Hilbert transform of $x(t) = A\cos(2\pi f_0 t + \theta)$.

Solution   The Fourier transform of $x(t)$ is

$$X(f) = \frac{1}{2}Ae^{j\theta}\delta(f - f_0) + \frac{1}{2}Ae^{-j\theta}\delta(f + f_0)$$

Therefore,

$$-j\,\text{sgn}(f)X(f) = -\frac{j}{2}Ae^{j\theta}\delta(f - f_0) + \frac{j}{2}Ae^{-j\theta}\delta(f + f_0)$$

Taking the inverse transform, we obtain

$$\hat{x}(t) = A\sin(2\pi f_0 t + \theta) \tag{2.6.32}$$

Example 2.6.3

Let $x(t)$ denote a lowpass signal. Determine the Hilbert transform of $x(t)\cos(2\pi f_0 t)$ where f_0 is much larger than the bandwidth of $x(t)$.

Solution   Here the conditions of the last theorem apply. Therefore,

$$\widehat{x(t)\cos(2\pi f_0 t)} = x(t)\sin(2\pi f_0 t) \tag{2.6.33}$$

2.7 FURTHER READING

There exist numerous references covering the analysis of LTI systems in the time and frequency domains. Oppenheim, Willsky, and Young (1983) contains a wide coverage of time and frequency domain analysis of both discrete-time and continuous-time systems. Papoulis (1962) and Bracewell (1965) provide an in-depth analysis of the Fourier series and transform techniques. A more advanced treatment of linear systems based on linear operator theory can be found in Franks (1969).

PROBLEMS

2.1 Two signals $x_1(t) = 1$ and $x_2(t) = \cos 2\pi t$ are sampled at $t = 0, \pm 1, \pm 2, \ldots$. The resulting discrete-time signals are denoted by $x_1[n]$ and $x_2[n]$. Verify that $x_1[n] = x_2[n]$. What can you conclude from this observation?

2.2 Show that the sum of two discrete periodic signals is periodic, whereas the sum of two continuous periodic signals is not necessarily periodic. Under what condition is the sum of two continuous periodic signals periodic?

2.3 Classify the following signals into even, odd, and neither even nor odd signals. In the last case, find the even and odd parts of the signals.

1. $x_1(t) = \begin{cases} e^{-t} & t > 0 \\ -e^t & t < 0 \\ 0 & t = 0 \end{cases}$

2. $x_2(t) = e^{-|t|}$

3. $x_3(t) = \begin{cases} \dfrac{t}{|t|} & t \neq 0 \\ 0 & t = 0 \end{cases}$

4. $x_4(t) = \begin{cases} t & t \geq 0 \\ 0 & t < 0 \end{cases}$

5. $x_5(t) = \sin t + \cos t$

6. $x_6(t) = x_1(t) - x_2(t)$, where $x_1(t)$ is even and $x_2(t)$ is odd

2.4 Classify the following signals into energy-type, power-type, and neither energy-type nor power-type signals. For energy-type and power-type signals, find the energy or the power content of the signal.

1. $x_1(t) = e^{-t} \cos t\, u_{-1}(t)$
2. $x_2(t) = e^{-t} \cos t$
3. $x_3(t) = \text{sgn}(t)$
4. $x_4(t) = A \cos 2\pi f_1 t + B \cos 2\pi f_2 t$

2.5 Using the definitions of power-type and energy-type signals,

1. Show that $x(t) = A e^{j(2\pi f_0 t + \theta)}$ is a power-type signal and its power content is A^2.
2. Show that the unit step signal $u_{-1}(t)$ is a power-type signal and find its power content.
3. Show that the signal

$$x(t) = \begin{cases} Kt^{-\frac{1}{4}} & t > 0 \\ 0 & t \leq 0 \end{cases}$$

   is neither an energy- nor a power-type signal.

2.6 Find the even and odd parts of the signal $x(t) = \Lambda(t) u_{-1}(t)$.

2.7 Based on the definition of even and odd signals,

1. Show that the decomposition of a signal into even and odd parts is unique.
2. Show that the product of two even or two odd signals is even, whereas the product of an even and an odd signal is odd.

2.8 Give a plot of the following signals.

1. $x_1(t) = \Pi(t) + \Pi(-t)$
2. $x_2(t) = \Pi(t) - \Pi(t-1)$
3. $x_3(t) = \Lambda(t)\Pi(t)$
4. $x_4(t) = \sum_{n=-\infty}^{+\infty} \Lambda(t-2n)$
5. $x_5(t) = \sum_{n=-\infty}^{+\infty}(-1)^n\Lambda(t-n)$
6. $x_6(t) = \mathrm{sgn}(t) + \mathrm{sgn}(1-t)$
7. $x_7(t) = 1 + \mathrm{sgn}(t)$
8. $x_8(t) = \mathrm{sgn}^2(t)$
9. $x_9(t) = \mathrm{sinc}(t)\,\mathrm{sgn}(t)$
10. $x_{10}(t) = \sum_{n=-\infty}^{+\infty}(-1)^n n\delta(t-n)$
11. $x_{11}(t) = \sum_{n=1}^{\infty} \frac{1}{2^n}\Pi(\frac{t}{n})$

2.9 Using the properties of the impulse function, find the values of these expressions.

1. $x_1(t) = \mathrm{sinc}(t)\,\delta(t)$
2. $x_2(t) = \Lambda(t) \star \sum_{n=-\infty}^{+\infty}\delta(t-2n)$
3. $x_3(t) = \Lambda(t) \star \delta'(t)$
4. $x_4(t) = \cos t\,\delta(3t)$
5. $x_5(t) = \delta(5t) \star \delta(4t)$
6. $x_6(t) = \delta(5t) \star \delta'(3t)$
7. $x_7(t) = \cos t\,\delta'(t)$
8. $\int_{-\infty}^{+\infty} \Pi(t)\,\delta(2t-1)\,dt$
9. $\int_{-\infty}^{+\infty} \mathrm{sinc}(t)\,\delta(t)\,dt$
10. $\int_{-\infty}^{+\infty} \mathrm{sinc}(t+1)\,\delta(t)\,dt$
11. $\int_{-\infty}^{+\infty} \left[\sum_{n=1}^{\infty} \frac{1}{2^n}\Pi(\frac{t}{n})\right]\delta(t)\,dt$
12. $\int_{-\infty}^{+\infty} \cos t\left[\sum_{n=1}^{\infty}\delta(2^n t)\right]\,dt$

2.10 Show that the impulse signal is even. What can you say about the evenness or oddness of its nth derivative?

2.11 We have seen that $x(t) \star \delta(t) = x(t)$. Show that

$$x(t) \star \delta^{(n)}(t) = \frac{d^n}{dt^n}x(t)$$

and

$$x(t) \star u_{-1}(t) = \int_{-\infty}^{t} x(\tau)\,d\tau$$

2.12 Classify these systems into linear and nonlinear.

1. $y(t) = \begin{cases} 1 & x(t) > 0 \\ 0 & x(t) \le 0 \end{cases}$

2. $y(t) = \begin{cases} \dfrac{x(t)}{|x(t)|} & x(t) \ne 0 \\ 0 & x(t) = 0 \end{cases}$

3. $y(t) = \begin{cases} \dfrac{x^2(t)}{|x(t)|} & x(t) \ne 0 \\ 0 & \text{otherwise} \end{cases}$

4. $y(t) = e^{-t} x(t)$
5. $y(t) = x(t) u_{-1}(t)$
6. $y(t) = x(t) \delta(t)$
7. $y(t) = x(t) \sum_{n=-\infty}^{+\infty} \delta(t - nT)$

8. $y(t) = \begin{cases} \dfrac{d}{dt} x(t) & t > 0 \\ x(t) & t < 0 \\ 0 & t = 0 \end{cases}$

9. $y(t) = x(t) + y(t - 1)$
10. $y(t) =$ Algebraic sum of jumps in $x(t)$ in the interval $(-\infty, t]$.

2.13 Prove that a system is linear if and only if

1. It is *homogeneous;* i.e., for all input signals $x(t)$ and all real numbers α, we have $T[\alpha x(t)] = \alpha T[x(t)]$.
2. It is *additive;* i.e., for all input signals $x_1(t)$ and $x_2(t)$, we have

$$T[x_1(t) + x_2(t)] = T[x_1(t)] + T[x_2(t)]$$

2.14 Prove that if a system satisfies the additivity property described in Problem 2.13, then it is homogeneous for all *rational* α.

2.15 Show that the system described by

$$y(t) = \begin{cases} \dfrac{x^2(t)}{x'(t)} & x'(t) \ne 0 \\ 0 & x'(t) = 0 \end{cases}$$

is homogeneous but nonlinear. Can you give another example of such a system?

2.16 Show that the response of a linear system to identically zero input is an output that is identically zero.

2.17 The system defined by the input-output relation

$$y(t) = x(t)\cos(2\pi f_0 t)$$

where f_0 is a constant, is called a *modulator.* Is this system linear? Is it time-invariant?

2.18 Comment on the truth or falseness of these statements.

1. A system whose components are nonlinear is necessarily nonlinear.
2. A system whose components are time-varying is necessarily a time-varying system.
3. The response of a causal system to a causal signal is itself causal.

2.19 Classify the following systems into time-varying and time-invariant.

1. $y(t) = x(-t)$
2. $y(t) = x(t)u_{-1}(t)$
3. $y(t) = x(t)\delta(t)$
4. $y(t) = x(t)\sum_{n=-\infty}^{+\infty}\delta(t - nT)$
5. $y(t) = \int_{-\infty}^{t} x(\tau)\,d\tau$
6. $y(t) = x(t) + y(t - 1)$
7. $y(t) = \begin{cases} \dfrac{x(t)}{|x(t)|} & x(t) \neq 0 \\ 0 & x(t) = 0 \end{cases}$

2.20 The response of a linear time-invariant system to the input $x(t) = e^{-\alpha t}u_{-1}(t)$ is $\delta(t)$. Using time-domain analysis, find the impulse response of this system. What is the response of the system to a general input $x(t)$?

2.21 Let a system be defined by

$$y(t) = \frac{1}{2T}\int_{t-T}^{t+T} x(\tau)\,d\tau$$

Is this system causal?

2.22 For an LTI system to be causal, it is required that $h(t)$ be zero for $t < 0$. Give an example of a nonlinear system that is causal but whose impulse response is nonzero for $t < 0$.

2.23 Using the convolution integral, show that the response of an LTI system to $u_{-1}(t)$ is given by $\int_{-\infty}^{t} h(\tau)\,d\tau$.

2.24 What is the impulse response of a differentiator? Find the output of this system to an arbitrary input $x(t)$ by finding the convolution of the input and the impulse response. Repeat for the delay system.

2.25 Show that in an LTI system if the response to $x(t)$ is $y(t)$, then the response to $x'(t)$ is $y'(t)$.

2.26 The system defined by

$$y(t) = \int_{t-T}^{t} x(\tau)\, d\tau$$

(T is a constant) is a finite-time integrator. Is this system LTI? If yes, find the impulse response.

2.27 Compute the following convolution integrals.

1. $e^{-t}u_{-1}(t) \star e^{-t}u_{-1}(t)$
2. $e^{-t}u_{-1}(t) \star u_{-1}(t)$
3. $\Pi(t) \star \Lambda(t)$
4. $(\Lambda(t)\,\mathrm{sgn}(t)) \star u_{-1}(t)$
5. $\Lambda(t) \star \mathrm{sgn}(t)$
6. $(\Lambda(t)u_{-1}(t)) \star \Pi(t)$

2.28 Show that in a causal LTI system the convolution integral reduces to

$$y(t) = \int_{0}^{+\infty} x(t-\tau)h(\tau)\, d\tau = \int_{-\infty}^{t} x(\tau)h(t-\tau)\, d\tau$$

2.29 Using the Gram–Schmidt orthogonalization process, determine an orthonormal basis for the space spanned by vectors $\mathbf{e}_1$, $\mathbf{e}_1 + \mathbf{e}_2$, and $\mathbf{e}_1 + \mathbf{e}_2 + \mathbf{e}_3$.

2.30 Show that the set of signals $\psi_n(t) = \sqrt{\frac{1}{T_0}}\, e^{j2\pi \frac{n}{T_0} t}$ constitute an orthonormal set of signals on the interval $[\alpha, \alpha + T_0]$, where α is arbitrary.

2.31 Let the inner product of two functions $x(t)$ and $y(t)$ be defined by $\int_{0}^{1} x(t)y(t)\, dt$. Using the Gram–Schmidt orthogonalization procedure, find an orthonormal basis for the space of polynomial functions $x_1(t) = 1$, $x_2(t) = t$, $x_3(t) = t^2$, $x_4(t) = t^3$, ...

2.32 In this problem, we present the proof of the Cauchy–Schwartz inequality.

1. Show that for real $\{\alpha_i\}_{i=1}^{n}$ and $\{\beta_i\}_{i=1}^{n}$,

$$\sum_{i=1}^{n} \alpha_i \beta_i \leq \left[\sum_{i=1}^{n} \alpha_i^2 \right]^{\frac{1}{2}} \left[\sum_{i=1}^{n} \beta_i^2 \right]^{\frac{1}{2}}$$

What are the conditions for equality?

2. Let $\{x_i\}_{i=1}^n$ and $\{y_i\}_{i=1}^n$ be complex numbers. Show that

$$\left|\sum_{i=1}^n x_i y_i^*\right| \leq \sum_{i=1}^n |x_i y_i^*| = \sum_{i=1}^n |x_i||y_i^*|$$

What are the conditions for equality?

3. From (1) and (2) conclude that

$$\left|\sum_{i=1}^n x_i y_i^*\right| \leq \left[\sum_{i=1}^n |x_i|^2\right]^{\frac{1}{2}} \left[\sum_{i=1}^n |y_i|^2\right]^{\frac{1}{2}}$$

What are the conditions for equality?

4. Generalize the above results to integrals and prove the Cauchy–Schwartz inequality

$$\left|\int_{-\infty}^{+\infty} x(t)y^*(t)\, dt\right| \leq \left[\int_{-\infty}^{+\infty} |x(t)|^2\, dt\right]^{\frac{1}{2}} \left[\int_{-\infty}^{+\infty} |y(t)|^2\, dt\right]^{\frac{1}{2}}$$

What are the conditions for equality?

2.33 Let $\{\phi_i(t)\}_{i=1}^N$ be an orthonormal set on N signals; i.e.,

$$\int_{-\infty}^{+\infty} \phi_i(t)\phi_j^*(t)\, dt = \begin{cases} 1 & i \neq j \\ 0 & i = j \end{cases} \quad 1 \leq i, j \leq N$$

and let $x(t)$ be an arbitrary signal. Let $\hat{x}(t) = \sum_{i=1}^N \alpha_i \phi_i(t)$ be a linear approximation of $x(t)$ in terms of $\{\phi_i(t)\}_{i=1}^N$. We are interested in finding α_i's such that

$$\epsilon^2 = \int_{-\infty}^{+\infty} |x(t) - \hat{x}(t)|^2\, dt$$

is minimized.

1. Show that the minimizing α_i's satisfy

$$\alpha_i = \int_{-\infty}^{+\infty} x(t)\phi_i^*(t)\, dt$$

2. Show that with the above choice of α_i's we have

$$\epsilon_{\min}^2 = \int_{-\infty}^{+\infty} |x(t)|^2\, dt - \sum_{i=1}^N |\alpha_i|^2$$

2.34 Determine the Fourier series expansion of the following signals.

1. $x_1(t) = \sum_{n=-\infty}^{+\infty} \Lambda(t - 2n)$

2. $x_2(t) = \sum_{n=-\infty}^{+\infty} \Lambda(t-n)$

3. $x_3(t) = e^{t-n}$ for $n \le t < n+1$

4. $x_4(t) = \cos t + \cos 2.5t$

5. $x_5(t) = \sum_{n=-\infty}^{+\infty} \Lambda(t-n)u_{-1}(t-n)$

6. $x_6(t) = \sum_{n=-\infty}^{+\infty} (-1)^n \delta(t-nT)$

7. $x_7(t) = \sum_{n=-\infty}^{+\infty} \delta'(t-nT)$

8. $x_8(t) = |\cos 2\pi f_0 t|$ (full-wave rectifier output)

9. $x_9(t) = \cos 2\pi f_0 t + |\cos 2\pi f_0 t|$ (half-wave rectifier output)

2.35 Show that for real $x(t)$ we have

$$x_e(t) = \frac{a_0}{2} + \sum_{n=1}^{\infty} a_n \cos\left(2\pi \frac{n}{T_0} t\right)$$

$$x_o(t) = \sum_{n=1}^{\infty} a_n \sin\left(2\pi \frac{n}{T_0} t\right)$$

where $x_e(t)$ and $x_o(t)$ denote the even and odd parts of $x(t)$.

2.36 Determine the Fourier series expansion of each of the periodic signals shown in Figure P-2.36. For each signal, also determine the trigonometric Fourier series.

2.37 In the previous problem, determine the output of each of the following filters when $T = 1$ and each of the signals in the previous problem is the input

1. $H(f) = 10\Pi(\frac{f}{4})$.

2. $H(f) = \begin{cases} -j & 0 < f \le 4 \\ j & -4 \le f < 0 \\ 0 & \text{otherwise} \end{cases}$

2.38 Let x_n and y_n represent the Fourier series coefficients of $x(t)$ and $y(t)$, respectively. Assuming the period of $x(t)$ is T_0, express y_n in terms of x_n in each of the following cases.

1. $y(t) = x(t-t_0)$

2. $y(t) = x(t)e^{j2\pi f_0 t}$

3. $y(t) = x(at),\ a \ne 0$

4. $y(t) = \frac{d}{dt} x(t)$

2.39 Let $x(t)$ and $y(t)$ be two periodic signals with period T_0, and let x_n and y_n denote the Fourier series coefficients of these two signals. Show that

$$\frac{1}{T_0} \int_{\alpha}^{\alpha+T_0} x(t)y^*(t)\, dt = \sum_{n=-\infty}^{+\infty} x_n y_n^*$$

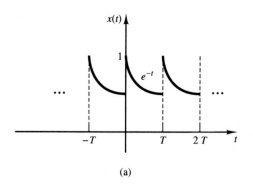

(a)

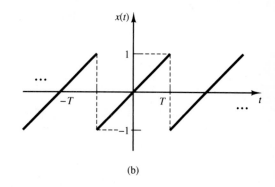

(b)

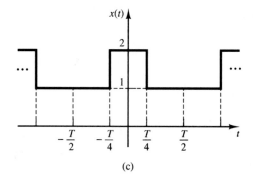

(c)

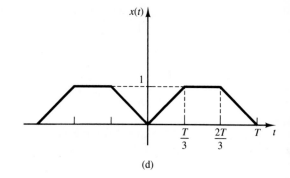

(d)

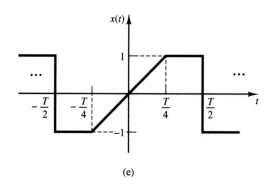

(e)

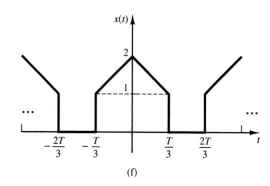

(f)

FIGURE P-2.36

2.40 Show that for all periodic physical signals that have finite power, the coefficients of the Fourier series expansion x_n tend to zero as $n \to \infty$.

2.41 Determine the power content of the signal shown in Figure 2.25. Use Parseval's relation to prove the following identity.

$$1 + \frac{1}{3^4} + \frac{1}{5^4} + \cdots + \frac{1}{(2n+1)^4} + \cdots = \frac{\pi^4}{96}$$

2.42 Determine the Fourier transform of each of the following signals (α is positive).

1. $x(t) = \frac{1}{1+t^2}$
2. $\Pi(t-3) + \Pi(t+3)$
3. $\Lambda(2t+3) + \Lambda(3t-2)$
4. $\text{sinc}^3 t$
5. $t \, \text{sinc} \, t$
6. $t \cos 2\pi f_0 t$
7. $e^{-\alpha|t|} \cos(\beta t)$
8. $te^{-\alpha|t|} \cos(\beta t)$

2.43 Show that the Fourier transform of $\frac{1}{2}\delta\left(t + \frac{1}{2}\right) + \frac{1}{2}\delta\left(t - \frac{1}{2}\right)$ is $\cos(\pi f)$. Prove the following transform pairs.

$$\mathcal{F}[\cos(\pi t)] = \frac{1}{2}\delta\left(f + \frac{1}{2}\right) + \frac{1}{2}\delta\left(f - \frac{1}{2}\right)$$

and

$$\mathcal{F}[\sin(\pi t)] = \frac{j}{2}\delta\left(f + \frac{1}{2}\right) - \frac{j}{2}\delta\left(f - \frac{1}{2}\right)$$

2.44 Determine the Fourier transform of the signals shown in Figure P-2.44:

2.45 Use the convolution theorem to show that

$$\text{sinc}(t) \star \text{sinc}(t) = \text{sinc}(t)$$

2.46 Prove that convolution in the frequency domain is equivalent to multiplication in the time domain, that is,

$$\mathcal{F}[x(t)y(t)] = X(f) \star Y(f)$$

2.47 Let $x(t)$ be an arbitrary signal and define $x_1(t) = \sum_{n=-\infty}^{+\infty} x(t - nT_0)$.

1. Show that $x_1(t)$ is a periodic signal.
2. How can you write $x_1(t)$ in terms of $x(t)$ and $\sum_{n=-\infty}^{+\infty} \delta(t - nT_0)$?
3. Find the Fourier transform of $x_1(t)$ in terms of the Fourier transform of $x(t)$.

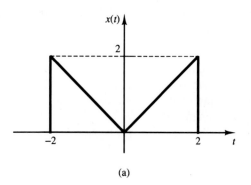

(a)

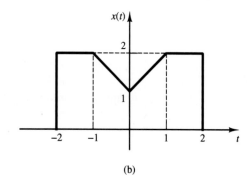

(b)

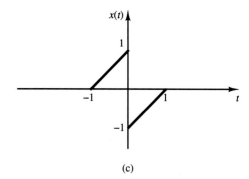

(c)

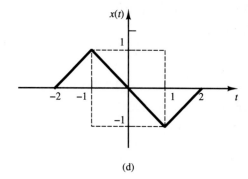

(d)

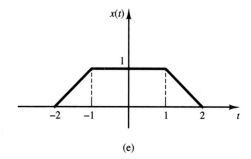

(e)

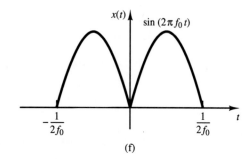

(f)

FIGURE P-2.44

2.48 Using the properties of the Fourier transform, evaluate the following integrals (α is positive)

1. $\int_{-\infty}^{+\infty} \text{sinc}^5(t)\, dt$

2. $\int_{0}^{\infty} e^{-\alpha t} \text{sinc}(t)\, dt$

3. $\int_0^\infty e^{-\alpha t} \operatorname{sinc}^2(t)\, dt$

4. $\int_0^\infty e^{-\alpha t} \cos(\beta t)\, dt$

2.49 A linear time-invariant system with impulse response $h(t) = e^{-\alpha t} u_{-1}(t)$ is driven by the input $x(t) = e^{-\beta t} u_{-1}(t)$. It is assumed that $\alpha, \beta > 0$. Using frequency domain analysis, find the output of the system. Is the output power-type or energy-type? Find its power or energy.

2.50 Let $x(t)$ be periodic with period T_0, and let $0 \le \alpha < T_0$. Define

$$x_\alpha(t) = \begin{cases} x(t) & \alpha \le t < \alpha + T_0 \\ 0 & \text{otherwise} \end{cases}$$

and let $X_\alpha(f)$ denote the Fourier transform of $x_\alpha(t)$. Prove that $X_\alpha(\frac{n}{T_0})$ is independent of the choice of α.

2.51 Using the identity

$$\sum_{n=-\infty}^{+\infty} \delta(t - nT_s) = \frac{1}{T_s} \sum_{n=-\infty}^{+\infty} e^{jn \frac{2\pi t}{T_s}}$$

show that for any signal $x(t)$ and any T_s, the following identity holds

$$\sum_{n=-\infty}^{+\infty} x(t - nT_s) = \frac{1}{T_s} \sum_{n=-\infty}^{+\infty} X\left(\frac{n}{T_s}\right) e^{jn \frac{2\pi t}{T_s}}$$

From this conclude the following relation known as *Poisson's sum formula.*

$$\sum_{n=-\infty}^{+\infty} x(nT_s) = \frac{1}{T_s} \sum_{n=-\infty}^{+\infty} X\left(\frac{n}{T_s}\right)$$

2.52 Using Poisson's sum formula, show that

1. $\sum_{n=-\infty}^{+\infty} \frac{2\alpha}{\alpha^2 + 4\pi^2 n^2} = \sum_{n=-\infty}^{+\infty} e^{-\alpha|n|}$

2. $\sum_{n=-\infty}^{+\infty} \operatorname{sinc}(\frac{n}{K}) = K$ for all $K \in \{1, 2, \ldots\}$

3. $\sum_{n=-\infty}^{+\infty} \operatorname{sinc}^2(\frac{n}{K}) = K$ for all $K \in \{1, 2, \ldots\}$

2.53 The response of an LTI system to $e^{-\alpha t} u_{-1}(t)$, $(\alpha > 0)$, is $\delta(t)$. Using frequency domain analysis techniques, find the response of the system to $x(t) = e^{-\alpha t} \cos(\beta t) u_{-1}(t)$.

2.54 Find the output of an LTI system with impulse response $h(t)$ when driven by the input $x(t)$ in each of the following cases.

1. $h(t) = \delta(t) + \delta'(t)$ $x(t) = e^{-\alpha|t|}$, $(\alpha > 0)$

2. $h(t) = e^{-\alpha t} u_{-1}(t)$ $x(t) = e^{-\beta t} u_{-1}(t)$, $(\alpha, \beta > 0)$

3. $h(t) = e^{-\alpha t} \cos(\beta t) u_{-1}(t)$ $x(t) = e^{-\gamma t} u_{-1}(t)$, $(\alpha, \beta, \gamma > 0)$

4. $h(t) = e^{-\alpha|t|}$ $x(t) = e^{-\beta t}u_{-1}(t)$, $(\alpha, \beta > 0)$ (Treat the special case $\alpha = \beta$ separately.)

5. $h(t) = \text{sinc}(t)$ $x(t) = \text{sinc}^2(t)$

2.55 Can the response of an LTI system to $x(t) = \text{sinc}(t)$ be $y(t) = \text{sinc}^2(t)$? Justify your answer.

2.56 Let the response of an LTI system to $\Pi(t)$ be $\Lambda(t)$.

1. Can you find the response of this system to $x(t) = \cos 2\pi t$ from the information provided above?

2. Show that $h_1(t) = \Pi(t)$ and $h_2(t) = \Pi(t) + \cos 2\pi t$ can both be impulse responses of this system, and therefore having the response of a system to $\Pi(t)$ does not uniquely determine the system.

3. Does the response of an LTI system to $u_{-1}(t)$ uniquely determine the system? How about the response to $e^{-\alpha t}u_{-1}(t)$ for some $\alpha > 0$? In general, what conditions must the input $x(t)$ satisfy so that the system can be uniquely determined by knowing its corresponding output?

2.57 Determine whether these signals are energy-type or power-type. In each case, find the energy or power spectral density and the energy or power content of the signal.

1. $x(t) = e^{-\alpha t}u_{-1}(t)$ $\alpha > 0$
2. $x(t) = \text{sinc}(t)$
3. $x(t) = \sum_{n=-\infty}^{+\infty} \Lambda(t - 2n)$
4. $x(t) = u_{-1}(t)$.
5. $x(t) = \frac{1}{t}$

2.58 Find the energy spectral density and the energy content, or power spectral density and the power content, of the output of the following LTI systems when driven by the signals of the previous problem.

1. $h(t) = e^{-\gamma t}u_{-1}(t)$, $\gamma > 0$
2. $h(t) = \text{sinc}(6t)$
3. $h(t) = \frac{1}{\pi t}$

2.59 Show that if $x_T(t)$ denotes the truncated signal corresponding to the power-type signal $x(t)$, that is,

$$x_T(t) = \begin{cases} x(t) & -\frac{T}{2} < t \leq \frac{T}{2} \\ 0 & \text{otherwise} \end{cases}$$

and if $G_{x_T}(f)$ denotes the energy spectral density of $x_T(t)$, then $S_x(f)$, the power spectral density of $x(t)$, can be expressed as

$$S_x(f) = \lim_{T \to \infty} \frac{G_{x_T}(f)}{T}$$

2.60 Show that if the input to an LTI system satisfies $\int_{-\infty}^{+\infty} \int_{-\infty}^{+\infty} |x(t - \tau)|^2 \, dt \, d\tau < \infty$ and the impulse response of the system is energy-type, then the output is also energy-type. Can you give an example of an LTI system with an energy-type signal as its input such that the corresponding output signal is not energy-type?

2.61 For a lowpass signal with a bandwidth of 6000 Hz, what is the minimum sampling frequency for perfect reconstruction of the signal? What is the minimum required sampling frequency if a guard band of 2000 Hz is required? What is the minimum required sampling frequency and the value of K for perfect reconstruction if the reconstruction filter has the following frequency response.

$$H(f) = \begin{cases} K & |f| < 7000 \\ K - K \dfrac{|f| - 7000}{3000} & 7000 < |f| < 10000 \\ 0 & \text{otherwise} \end{cases}$$

2.62 Let the signal $x(t) = A \operatorname{sinc}(1000t)$ be sampled with a sampling frequency of 2000 samples per second. Determine the most general class of reconstruction filters for perfect reconstruction of this signal.

2.63 The lowpass signal $x(t)$ with a bandwidth of W is sampled with a sampling interval of T_s and the signal

$$x_p(t) = \sum_{n=-\infty}^{+\infty} x(nT_s) p(t - nT_s)$$

is generated, where $p(t)$ is an arbitrary shaped pulse (not necessarily time-limited to the interval $[0, T_s]$).

1. Find the Fourier transform of $x_p(t)$.
2. Find the conditions for perfect reconstruction of $x(t)$ from $x_p(t)$.
3. Determine the required reconstruction filter.

2.64 The lowpass signal $x(t)$ with a bandwidth of W is sampled at the Nyquist rate and the signal

$$x_1(t) = \sum_{n=-\infty}^{+\infty} (-1)^n x(nT_s) \delta(t - nT_s)$$

is generated.

1. Find the Fourier transform of $x_1(t)$.
2. Can $x(t)$ be reconstructed from $x_1(t)$ by using an LTI system? Why?
3. Can $x(t)$ be reconstructed from $x_1(t)$ by using a linear time-varying system? How?

2.65 A lowpass signal $x(t)$ with bandwidth W is sampled with a sampling interval T_s and the sampled values are denoted by $x(nT_s)$. A new signal $x_1(t)$ is generated by linear interpolation of the sampled values; i.e.,

$$x_1(t) = x(nT_s) + \frac{t - nT_s}{T_s} (x((n+1)T_s) - x(nT_s)) \qquad nT_s \le t \le (n+1)T_s$$

1. Find the power spectrum of $x_1(t)$.
2. Under what conditions can the original signal be reconstructed from the sampled signal, and what is the required reconstruction filter?

2.66 A lowpass signal $x(t)$ with bandwidth of 50 Hz is sampled at the Nyquist rate and the resulting sampled values are

$$x(nT_s) = \begin{cases} -1 & -4 \le n < 0 \\ 1 & 0 < n \le 4 \\ 0 & \text{otherwise} \end{cases}$$

1. Find $x(.005)$.
2. Is this signal power-type or energy-type? Find its power or energy content.

2.67 Let W be arbitrary and $x(t)$ be a lowpass signal with bandwidth W.

1. Show that the set of signals $\{\phi_n(t)\}_{n=-\infty}^{\infty}$ where $\phi_n = \text{sinc}(2Wt - n)$ represent an orthogonal signal set. How should these signals be weighted to generate an orthonormal set?
2. Conclude that the reconstruction from the samples relation

$$x(t) = \sum_{n=-\infty}^{+\infty} x(nT_s) \, \text{sinc}(2Wt - n)$$

is an orthogonal expansion relation.
3. From the above, show that for all n,

$$\int_{-\infty}^{+\infty} x(t) \, \text{sinc}(2Wt - n) \, dt = K x(nT_s)$$

and find K.

2.68 Prove that

$$x_c(t) = x(t)\cos(2\pi f_0 t) + \hat{x}(t)\sin(2\pi f_0 t)$$
$$x_s(t) = \hat{x}(t)\cos(2\pi f_0 t) - x(t)\sin(2\pi f_0 t)$$

2.69 Show that

$$X_c(f) = \frac{1}{2}[X_l(f) + X_l^*(-f)]$$

2.70 A lowpass signal $x(t)$ has a Fourier transform shown in Figure P-2.70(a). This signal is applied to the system shown in Figure P-2.70(b). The blocks marked by $\mathcal{H}$ represent Hilbert transform blocks, and it is assumed that $W \ll f_0$. Determine the signals $x_i(t)$ for $1 \le i \le 7$ and plot $X_i(f)$ for $1 \le i \le 7$.

2.71 Show that the Hilbert transform of $A\sin(2\pi f_0 t + \theta)$ is $-A\cos(2\pi f_0 t + \theta)$.

2.72 Show that the Hilbert transform of the signal $e^{j2\pi f_0 t}$ is equal to $-j\,\text{sgn}(f_0)e^{j2\pi f_0 t}$.

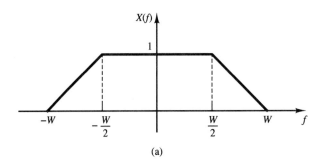

(a)

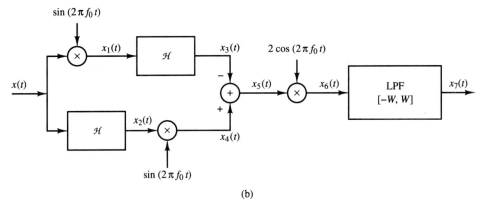

(b)

FIGURE P-2.70

2.73 Show that

$$\mathcal{F}[\widehat{\frac{d}{dt}x(t)}] = 2\pi |f| \mathcal{F}[x(t)]$$

2.74 Show that the Hilbert transform of the derivative of a signal is equal to the derivative of its Hilbert transform.

2.75 The bandpass signal $x(t) = \operatorname{sinc} t \cos(2\pi f_0 t)$ is passed through a bandpass filter with impulse response $h(t) = \operatorname{sinc}^2(t) \sin 2\pi f_0 t$. Using the lowpass equivalents of both the input and the impulse response, find the lowpass equivalent of the output, and from it find the output $y(t)$.

2.76 The real narrowband signal $x(t)$ whose frequency components are in the neighborhood of some f_0 (and $-f_0$) is passed through a filter with transfer function $H(f)$, and the output is denoted by $y(t)$. The magnitude of the transfer function is denoted by $A(f)$, and its phase is denoted by $\theta(f)$. It is assumed that the transfer function of the filter is smooth enough such that in the bandwidth of the input signal the magnitude of the transfer function is essentially constant, and its phase can be approximated by its first-order Taylor series expansion, i.e.,

$$A(f) \approx A(f_0)$$

$$\theta(f) \approx \theta(f_0) + (f - f_0)\theta'(f)|_{f=f_0}$$

1. Show that $Y_l(f)$, the Fourier transform of the lowpass equivalent of the output, can be written as

$$Y_l(f) \approx X_l(f)A(f_0)e^{j\left(\theta(f_0) + f\theta'(f)|_{f=f_0}\right)}$$

2. Conclude that

$$y(t) \approx A(f_0)V_x(t - t_g)\cos(2\pi f_0(t - t_p) + \theta_x(t - t_g))$$

where $V_x(t)$ is the envelope of the input $x(t)$ and

$$t_g = -\frac{1}{2\pi}\frac{d\theta(f)}{df}\bigg|_{f=f_0}$$

$$t_p = -\frac{1}{2\pi}\frac{\theta(f)}{f}\bigg|_{f=f_0}$$

3. The quantities t_g and t_p are called *envelope delay* (or *group delay*) and *phase delay*, respectively. Can you interpret their role and justify this nomenclature?

2.77 We have seen that the Hilbert transform introduces a 90° phase shift in the components of a signal, and the transfer function of a quadrature filter can be written as

$$H(f) = \begin{cases} e^{-j\frac{\pi}{2}} & f > 0 \\ 0 & f = 0 \\ e^{j\frac{\pi}{2}} & f < 0 \end{cases}$$

We can generalize this concept to a new transform that introduces a phase shift of θ in the frequency components of a signal by introducing

$$H_\theta(f) = \begin{cases} e^{-j\theta} & f > 0 \\ 0 & f = 0 \\ e^{j\theta} & f < 0 \end{cases}$$

and denote the result of this transform by $x_\theta(t)$, i.e., $X_\theta(f) = X(f)H_\theta(f)$, where $X_\theta(f)$ denotes the Fourier transform of $x_\theta(t)$. Throughout this problem, assume that the signal $x(t)$ does not contain any DC components.

1. Find $h_\theta(t)$, the impulse response of the filter representing the transform described above.
2. Show that $x_\theta(t)$ is a linear combination of $x(t)$ and its Hilbert transform.
3. Show that if $x(t)$ is an energy-type signal, $x_\theta(t)$ will also be an energy-type signal and its energy content will be equal to the energy content of $x(t)$.

2.78 Let $m(t) = \mathrm{sinc}^2(t)$ and let $x(t) = m(t)\cos 2\pi f_0 t - \hat{m}(t)\sin 2\pi f_0 t$ represent a bandpass signal.

1. Find the pre-envelope, $z(t)$, and the lowpass equivalent signal to $x(t)$.
2. Determine and plot the Fourier transform of the signal $x(t)$. What is the bandwidth of $x(t)$?
3. Repeat step 2 for $x(t) = m(t)\cos 2\pi f_0 t + \hat{m}(t)\sin 2\pi f_0 t$

3

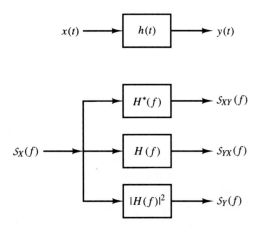

Random Processes

This chapter is devoted to the study of random processes and their properties. Random processes provide good models for both information sources and noise. When a signal is transmitted through a communication channel, there are two types of imperfections that cause the received signal to be different from the transmitted signal. One class of imperfections are deterministic in nature, such as linear and nonlinear distortion, intersymbol interference, etc. The second class is nondeterministic (such as addition of noise, multipath fading, etc.). For a quantitative study of these phenomena, we model them as random processes.

The information that is to be transmitted is also, by its nature, best modeled as a random process. This is due to the fact that any signal that conveys information must have some uncertainty in it, otherwise its transmission is of no interest. We will explore this aspect later in greater detail in Chapter 4. In this chapter, after a brief review of the basics of probability theory and random variables, we introduce the concept of a random process and the basic tools used in the mathematical analysis of random processes.

3.1 PROBABILITY AND RANDOM VARIABLES

In this section we give a brief review of some basics of probability theory that are needed for our treatment of random processes. It is assumed that the reader already has been exposed to probability theory elsewhere, and therefore our treatment in this section will be brief.

Probability space.   The fundamental concept in any probabilistic model is the concept of a *random experiment,* which is any experiment whose outcome, for some reason, cannot be predicted with certainty. Flipping a coin, throwing a die, and drawing a card from a deck are examples of random experiments. What is common in all these cases is that the result (or outcome) of the experiment is uncertain. A random experiment has certain *outcomes* that are the elementary results of the experiment. In the flipping of a coin, "head" and "tail" are the possible outcomes. In throwing a die, 1, 2, 3, 4, 5, 6 are the possible outcomes. The set of all possible outcomes is called the *sample space* and is denoted by Ω. Outcomes are denoted by ω's, and certainly each ω lies in Ω, i.e., $\omega \in \Omega$.

A sample space is *discrete* if the number of its elements are finite or countably infinite, otherwise it is a *nondiscrete* sample space. All the random experiments given above have discrete sample spaces. If one chooses randomly a number between 0 and 1, then the sample space corresponding to this random experiment is nondiscrete.

Events are subsets of the sample space for which a probability can be defined. For discrete sample spaces, any subset of the sample space is an event, that is, a probability can be defined for it. For instance, in throwing a die various events such as "the outcome is even," "the outcome is greater than 3," and "the outcome divides 3" can be considered. For a nondiscrete sample space, not every subset of Ω can be assigned a probability without sacrificing basic intuitive properties of probability. To overcome this difficulty, we define a σ-field $\mathcal{B}$ on the sample space Ω as a collection of subsets of Ω such that the following conditions are satisfied:

1. $\Omega \in \mathcal{B}$.
2. If the subset (event) $E \in \mathcal{B}$ then $E^c \in \mathcal{B}$ where E^c denotes the complement of E.
3. If $E_i \in \mathcal{B}$ for all i, then $\cup_{i=1}^{\infty} E_i \in \mathcal{B}$.

We define a *probability measure* P on $\mathcal{B}$ as a set function assigning nonnegative values to all events E in $\mathcal{B}$ such that the following conditions are satisfied:

1. $0 \le P(E) \le 1$ for all $E \in \mathcal{B}$.
2. $P(\Omega) = 1$.

3. For disjoint events E_1, E_2, E_3, $\cdots$ (i.e., events for which $E_i \cap E_j = \emptyset$ for all $i \neq j$, where $\emptyset$ is the null set), we have $P(\cup_{i=1}^{\infty} E_i) = \sum_{i=1}^{\infty} P(E_i)$.

The triple $(\Omega, \mathcal{B}, P)$ is called a *probability space.*

Some basic properties of the probability measure follow easily from the set theoretical properties of events together with the basic properties of probability measure. We list some of the most important properties here.

1. $P(E^c) = 1 - P(E)$.
2. $P(\emptyset) = 0$.
3. $P(E_1 \cup E_2) = P(E_1) + P(E_2) - P(E_1 \cap E_2)$.
4. If $E_1 \subset E_2$ then $P(E_1) \leq P(E_2)$.

Conditional probability.   Let us assume that the two events E_1 and E_2 are defined on the same probability space with corresponding probabilities $P(E_1)$ and $P(E_2)$. Then, if an observer receives the information that the event E_2 has in fact occurred, his probability about event E_1 will not be $P(E_1)$ anymore. In fact, the information that he received changes the probabilities of various events, and new probabilities, called *conditional probabilities,* are defined. The conditional probability of the event E_1 given the event E_2 is defined by

$$
P(E_1|E_2) = \begin{cases} \frac{P(E_1 \cap E_2)}{P(E_2)}, & P(E_2) \neq 0 \\ 0, & \text{otherwise} \end{cases}
$$

If it happens that $P(E_1|E_2) = P(E_1)$, then knowledge of E_2 does not change the probability of the occurrence of E_1. In this case, the events E_1 and E_2 are said to be *statistically independent.* For statistically independent events, $P(E_1 \cap E_2) = P(E_1)P(E_2)$.

Example 3.1.1

In throwing a fair die, the probability of

$$A = \{\text{the outcome is greater than 3}\}$$

is

$$P(A) = P(4) + P(5) + P(6) = \frac{1}{2} \tag{3.1.1}$$

The probability of

$$B = \{\text{the outcome is even}\} \tag{3.1.2}$$

is

$$P(B) = P(2) + P(4) + P(6) = \frac{1}{2} \tag{3.1.3}$$

In this case

$$P(A|B) = \frac{P(A \cap B)}{P(B)} = \frac{P(4) + P(6)}{\frac{1}{2}} = \frac{2}{3}$$

If the events $\{E_i\}_{i=1}^n$ make a partition of the sample space Ω, if the following two conditions are satisfied

$$\cup_{i=1}^n E_i = \Omega$$

$$E_i \cap E_j = \emptyset \quad \text{for all } 1 \le i, j \le n \text{ and } i \ne j$$

Then, if for an event A we have the conditional probabilities $\{P(A|E_i)\}_{i=1}^n$, $P(A)$ can be obtained by applying the *total probability theorem* stated as

$$P(A) = \sum_{i=1}^n P(E_i)P(A|E_i)$$

Bayes rule gives the conditional probabilities $P(E_i|A)$ by the following relation

$$P(E_i|A) = \frac{P(E_i)P(A|E_i)}{\sum_{j=1}^n P(E_j)P(A|E_j)}$$

Random variables. A (real) *random variable* is a mapping from the sample space Ω to the set of real numbers. A schematic diagram representing a random variable is given in Figure 3.1. Random variables are denoted by capital letters X, Y, etc., and individual values of the random variable X are $X(\omega)$. A random variable is discrete if the range of its values is either finite or countably infinite. This range is usually denoted by $\{x_i\}$.

The *cumulative distribution function* (c.d.f.) of a random variable X is defined as

$$F_X(x) = P(\omega \in \Omega : X(\omega) \le x)$$

which can be simply written as

$$F_X(x) = P(X \le x)$$

and has the following properties:

1. $0 \le F_X(x) \le 1$.
2. $F_X(x)$ is nondecreasing.

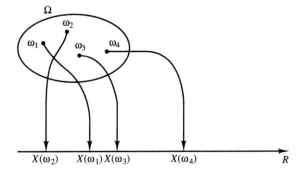

$X(\omega_2)$ $X(\omega_1) X(\omega_3)$ $X(\omega_4)$ R

FIGURE 3.1. Random variable as a mapping from Ω to $\mathbb{R}$.

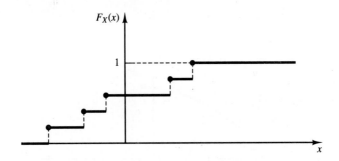

FIGURE 3.2.   The c.d.f. for a discrete random variable.

3. $\lim_{x \to -\infty} F_X(x) = 0$ and $\lim_{x \to +\infty} F_X(x) = 1$.

4. $F_X(x)$ is continuous from the right, i.e., $\lim_{\epsilon \downarrow 0} F(x + \epsilon) = F(x)$.

5. $P(a < X \le b) = F_X(b) - F_X(a)$.

6. $P(X = a) = F_X(a) - F_X(a^-)$.

For discrete random variables, $F_X(x)$ is a stair-case function. A random variable is called *continuous* if $F_X(x)$ is a continuous function. A random variable is called *mixed* if it is neither discrete nor continuous. Examples of c.d.f.'s for discrete, continuous, and mixed random variables are shown in Figures 3.2, 3.3, and 3.4, respectively.

The *probability density function,* or p.d.f., of a random variable X is defined as the derivative of $F_X(x)$, i.e.,

$$f_X(x) = \frac{d}{dx} F_X(x)$$

In case of discrete or mixed random variables, the p.d.f. involves impulses. The basic properties of p.d.f. are listed below:

1. $f_X(x) \ge 0$.

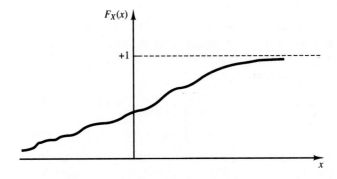

FIGURE 3.3.   The c.d.f. for a continuous random variable.

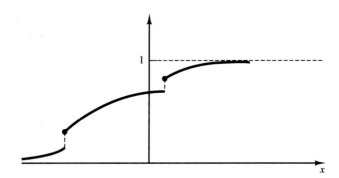

FIGURE 3.4.   The c.d.f. for a mixed random variable.

2. $\int_{-\infty}^{+\infty} f_X(x)\,dx = 1$.

3. $\int_{a^+}^{b^+} f_X(x)\,dx = P(a < X \le b)$.

4. In general, $P(X \in A) = \int_A f_X(x)\,dx$.

5. $F_X(x) = \int_{-\infty}^{x^+} f_X(u)\,du$.

For discrete random variables, it is more common to define the *probability mass function,* or p.m.f., which is defined as $\{p_i\}$ where $p_i = P(X = x_i)$. Obviously for all i we have $p_i \ge 0$ and $\sum_i p_i = 1$.

Important random variables.   The most commonly used random variables in communications are:

Bernoulli Random Variable.   This is a discrete random variable that takes two values 1 and 0 with probabilities p and $1 - p$. A Bernoulli random variable is a good model for a binary data generator. Also, when binary data are transmitted over a communication channel, some bits are received in error. We can model an error by modulo-2 addition of a 1 to the input bit, thus changing a 0 into a 1 and a 1 into a 0. Therefore, a Bernoulli random variable can be employed to model the channel errors.

Binomial Random Variable.   This is a discrete random variable that gives the number of 1's in a sequence of n independent Bernoulli trials. The p.m.f. is given by

$$P(X = k) = \begin{cases} \dbinom{n}{k} p^k (1-p)^{n-k}, & 0 \le k \le n \\[2ex] 0, & \text{otherwise} \end{cases}$$

This random variable models, for example, the total number of bits received in error when a sequence of n bits is transmitted over a channel with bit-error probability of p.

Uniform Random Variable.   This is a continuous random variable that takes values between a and b with equal probabilities over intervals of equal length. The density function is given by

$$f_X(x) = \begin{cases} \frac{1}{b-a}, & a < x < b \\ 0, & \text{otherwise} \end{cases}$$

This is a model for continuous random variables whose range is known, but nothing else is known about the likelihood of various values that the random variable can assume. For example, when the phase of a sinusoid is random, it is usually modeled as a uniform random variable between 0 and 2π.

Gaussian or Normal Random Variable.   This is a continuous random variable that is described by the density function

$$f_X(x) = \frac{1}{\sqrt{2\pi}\,\sigma}\, e^{-\frac{(x-m)^2}{2\sigma^2}}$$

The Gaussian random variable is the most important and frequently encountered random variable in communications. The reason is that thermal noise, which is the major source of noise in communication systems, has a Gaussian distribution. The properties of Gaussian noise will be investigated in more detail later in this chapter. Graphs of the p.d.f.'s and p.m.f.'s of the above random variables are given in Figures 3.5–3.8. The c.d.f. for the Gaussian random variable with $m = 0$ and $\sigma = 1$ is denoted by $\Phi(x)$ and given by

$$\Phi(x) = P(X \le x) = \int_{-\infty}^{x} \frac{1}{\sqrt{2\pi}}\, e^{-\frac{t^2}{2}}\, dt$$

A closely related function is $Q(x) = 1 - \Phi(x)$ giving $P(X > x)$. This function is well tabulated and frequently used in communications. It is easy to see that $Q(x)$ satisfies the following relations:

$$Q(-x) = 1 - Q(x) \tag{3.1.4}$$

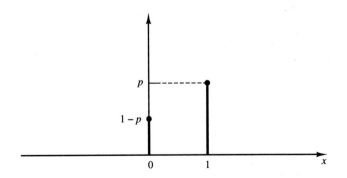

FIGURE 3.5.   The p.m.f. for the Bernoulli random variable.

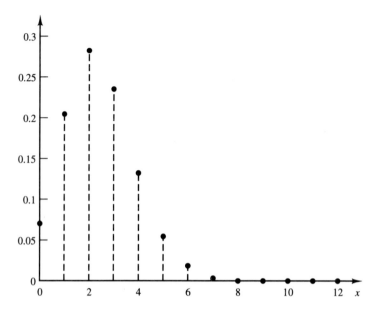

FIGURE 3.6. The p.m.f for the binomial random variable.

$$Q(0) = \frac{1}{2} \tag{3.1.5}$$

$$Q(\infty) = 0. \tag{3.1.6}$$

Table 3.1 gives the values of this function for various values of x.

There exist certain bounds on the Q function that are used widely to find bounds on error probability of various communication systems. These bounds are investigated in the problems at the end of this chapter. The two mostly used upper bounds are

$$Q(x) \le \frac{1}{2} e^{-\frac{x^2}{2}} \quad \text{for all } x \ge 0 \tag{3.1.7}$$

and

$$Q(x) < \frac{1}{\sqrt{2\pi}x} e^{-\frac{x^2}{2}} \quad \text{for all } x \ge 0 \tag{3.1.8}$$

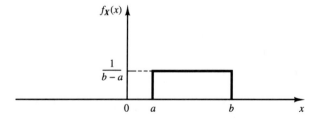

FIGURE 3.7. The p.d.f. for the uniform random variable.

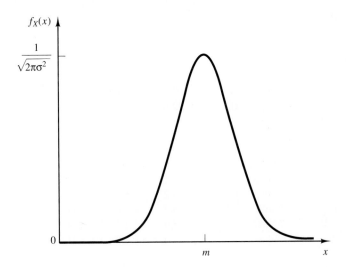

FIGURE 3.8.   The p.d.f. for the Gaussian random variable.

The frequently used lower bound is

$$Q(x) > \frac{1}{\sqrt{2\pi}x}\left(1 - \frac{1}{x^2}\right)e^{-\frac{x^2}{2}} \quad \text{for all } x \geq 0 \tag{3.1.9}$$

A plot of $Q(x)$ and these bounds is given in Figure 3.9.

A Gaussian variable can be described in terms of its two parameters m and σ by $\mathcal{N}(m, \sigma^2)$. For this random variable, a simple change of variable in the integral that computes $P(X > x)$ results in $P(X > x) = Q\left(\frac{x-m}{\sigma}\right)$. This gives the so-called *tail* probability in a Gaussian random variable.

Functions of a random variable.   A function of a random variable $Y = g(X)$ is itself a random variable. To find its c.d.f. we can use the definition of the c.d.f. to obtain

$$F_Y(y) = P(\omega \in \Omega : g(X(\omega)) \leq y)$$

In the special case that, for all y, the equation $g(x) = y$ has a countable number of solutions $\{x_i\}$, and for all these solutions $g'(x_i)$ exists and is nonzero, we can use the following relation to find the p.d.f of the random variable $Y = g(X)$.

$$f_Y(y) = \sum_i \frac{f_X(x_i)}{|g'(x_i)|} \tag{3.1.10}$$

Example 3.1.2

Assuming X is a Gaussian random variable with $m = 0$ and $\sigma = 1$, find the probability density function of the random variable Y given by $Y = aX + b$.

TABLE 3-1   TABLE OF $Q(x)$ VALUES

x	$Q(x)$	x	$Q(x)$	x	$Q(x)$
0	5.000000e−01	2.4	8.197534e−03	4.8	7.933274e−07
0.1	4.601722e−01	2.5	6.209665e−03	4.9	4.791830e−07
0.2	4.207403e−01	2.6	4.661189e−03	5.0	2.866516e−07
0.3	3.820886e−01	2.7	3.466973e−03	5.1	1.698268e−07
0.4	3.445783e−01	2.8	2.555131e−03	5.2	9.964437e−06
0.5	3.085375e−01	2.9	1.865812e−03	5.3	5.790128e−08
0.6	2.742531e−01	3.0	1.349898e−03	5.4	3.332043e−08
0.7	2.419637e−01	3.1	9.676035e−04	5.5	1.898956e−08
0.8	2.118554e−01	3.2	6.871378e−04	5.6	1.071760e−08
0.9	1.840601e−01	3.3	4.834242e−04	5.7	5.990378e−09
1.0	1.586553e−01	3.4	3.369291e−04	5.8	3.315742e−09
1.1	1.356661e−01	3.5	2.326291e−04	5.9	1.817507e−09
1.2	1.150697e−01	3.6	1.591086e−04	6.0	9.865876e−10
1.3	9.680049e−02	3.7	1.077997e−04	6.1	5.303426e−10
1.4	8.075666e−02	3.8	7.234806e−05	6.2	2.823161e−10
1.5	6.680720e−02	3.9	4.809633e−05	6.3	1.488226e−10
1.6	5.479929e−02	4.0	3.167124e−05	6.4	7.768843e−11
1.7	4.456546e−02	4.1	2.065752e−05	6.5	4.016001e−11
1.8	3.593032e−02	4.2	1.334576e−05	6.6	2.055790e−11
1.9	2.871656e−02	4.3	8.539898e−06	6.7	1.042099e−11
2.0	2.275013e−02	4.4	5.412542e−06	6.8	5.230951e−12
2.1	1.786442e−02	4.5	3.397673e−06	6.9	2.600125e−12
2.2	1.390345e−02	4.6	2.112456e−06	7.0	1.279813e−12
2.3	1.072411e−02	4.7	1.300809e−06		

Solution   In this case $g(x) = ax + b$, therefore $g'(x) = a$. The equation $ax + b = y$ has only one solution given by $x_1 = \frac{y-b}{a}$. Using these results we obtain

$$f_Y(y) = \frac{f_X(\frac{y-b}{a})}{|a|}$$

$$= \frac{1}{\sqrt{2\pi a^2}} e^{-\frac{(y-b)^2}{2a^2}} \qquad (3.1.11)$$

It is observed that Y is a Gaussian random variable $\mathcal{N}(b, a^2)$.

From the above example, we arrive at the important conclusion that *a linear function of a Gaussian random variable is itself a Gaussian random variable.*

Statistical averages.   The *expected value* of the random variable X is defined as

$$E(X) = \int_{-\infty}^{+\infty} x f_X(x)\, dx$$

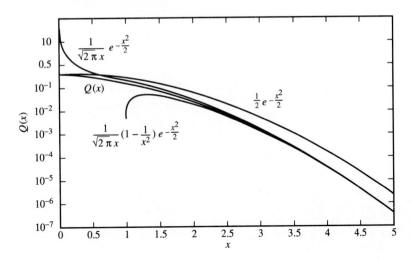

FIGURE 3.9. Bounds on Q-function.

and is usually denoted by m_X. The expected value of a random variable is a measure of the average of the value that the random variable takes in a large number of experiments. In general, the nth moment of a random variable X is defined as

$$m_X^{(n)} = \int_{-\infty}^{+\infty} x^n f_X(x)\, dx \qquad (3.1.12)$$

The expected value of $Y = g(X)$ is

$$E(g(X)) = \int_{-\infty}^{+\infty} g(x) f_X(x)\, dx$$

For discrete random variables, these equations become

$$E(X) = \sum_i x_i P(X = x_i)$$

and

$$E(g(X)) = \sum_i g(x_i) P(X = x_i)$$

In the special case where $g(X) = (X - E(X))^2$, $E(Y)$ is called the *variance* of X, which is a measure of spread of the density function of X. The variance is denoted by σ_X^2 and its square root, σ_X, is called the *standard deviation*. The relation for the variance can be written as $\sigma_X^2 = E(X^2) - (E(X))^2$. For any constant c, the following hold:

1. $E(cX) = cE(X)$.
2. $E(c) = c$.

3. $E(X + c) = E(X) + c$.

It is also easy to verify the following properties of the variance

1. $\text{VAR}(cX) = c^2 \text{VAR}(X)$.
2. $\text{VAR}(c) = 0$.
3. $\text{VAR}(X + c) = \text{VAR}(X)$.

Characteristic functions.   The characteristic function of a random variable X is denoted by $\Psi_X(v)$ and defined as

$$\Psi_X(v) = \int_{-\infty}^{+\infty} f_X(x) e^{jvx} \, dx \tag{3.1.13}$$

As it is observed from the above definition, the characteristic function of a random variable is closely related to the Fourier transform of its density function. The characteristic function of a random variable provides a simple way to find its various moments. This can be easily shown by using the moments theorem of the Fourier transform. To obtain the nth moment of the random variable X, we can use the relation

$$m_X^{(n)} = \frac{1}{j^n} \frac{d^n}{dv^n} \Psi_X(v) \Big|_{v=0} \tag{3.1.14}$$

The characteristic function of a Gaussian random variable with mean m and variance σ^2 is given by

$$\Psi_X(v) = \int_{-\infty}^{+\infty} \left[\frac{1}{\sqrt{2\pi\sigma^2}} \exp\left[-\frac{(x-m)^2}{2\sigma^2} \right] \right] e^{jvx} \, dx$$

$$= e^{jvm - \frac{v^2\sigma^2}{2}} \tag{3.1.15}$$

Multiple random variables.   Let X and Y be two random variables defined on the same sample space Ω. For these two random variables, we can define the *joint c.d.f.* as

$$F_{X,Y}(x, y) = P(\omega \in \Omega : X(\omega) \le x, Y(\omega) \le y)$$

or simply as

$$F_{X,Y}(x, y) = P(X \le x, Y \le y)$$

and the *joint p.d.f.* as

$$f_{X,Y}(x, y) = \frac{\partial^2}{\partial x \partial y} F_{X,Y}(x, y)$$

The following relations summarize the basic properties of the joint and marginal c.d.f.'s and p.d.f.'s.

1. $F_X(x) = F_{X,Y}(x, \infty)$.
2. $F_Y(y) = F_{X,Y}(\infty, y)$.
3. $f_X(x) = \int_{-\infty}^{+\infty} f_{X,Y}(x, y)\, dy$.
4. $f_Y(y) = \int_{-\infty}^{+\infty} f_{X,Y}(x, y)\, dx$.
5. $\int_{-\infty}^{+\infty} \int_{-\infty}^{+\infty} f_{X,Y}(x, y)\, dx\, dy = 1$.
6. $P((X, Y) \in A) = \iint_{(x,y) \in A} f_{X,Y}(x, y)\, dx\, dy$.
7. $F_{X,Y}(x, y) = \int_{-\infty}^{x} \int_{-\infty}^{y} f_{X,Y}(u, v)\, du\, dv$.

The *conditional p.d.f.* of the random variable Y, given that the value of the random variable X is equal to x, is denoted by $f_{Y|X}(y|x)$ and defined as

$$f_{Y|X}(y|x) = \begin{cases} \frac{f_{X,Y}(x,y)}{f_X(x)}, & f_X(x) \neq 0 \\ 0, & \text{otherwise} \end{cases}$$

If it happens that the density function after knowledge of X is the same as the density function before the knowledge of X, then the random variables are said to be statistically independent. For statistically independent random variables,

$$f_{X,Y}(x, y) = f_X(x) f_Y(y)$$

The expected value of $g(X, Y)$, where $g(X, Y)$ is any function of X and Y, is obtained from

$$E(g(X, Y)) = \int_{-\infty}^{+\infty} \int_{-\infty}^{+\infty} g(x, y) f_{X,Y}(x, y)\, dx\, dy$$

In the special case where $g(X, Y) = XY$, we obtain $E(XY)$, which is called the *correlation* of X and Y. In the case where $g(X, Y) = (X - m_X)(Y - m_Y)$, we obtain $E(XY) - m_X m_Y$, which is called the *covariance* of X and Y. Note that if $X = Y$, then $\text{COV}(X, Y) = \text{VAR}(X)$. The normalized version of the covariance, called the *correlation coefficient*, is denoted by $\rho_{X,Y}$ and is defined by

$$\rho_{X,Y} = \frac{\text{COV}(X, Y)}{\sigma_X \sigma_Y}$$

Using the Cauchy-Schwartz inequality (see 2.2.24), it can be shown that $|\rho_{X,Y}| \leq 1$ and $\rho = \pm 1$ indicates a first-order relationship between X and Y, i.e., a relation of the form $Y = aX + b$. The case $\rho = 1$ corresponds to a positive a, and $\rho = -1$ corresponds to a negative a. Also, it is easy to verify that if X and Y are independent, then $\text{COV}(X, Y) = \rho_{X,Y} = 0$. That is, independence implies lack of correlation ($\rho = 0$). It should be noted that lack of correlation does not in general imply independence. That is, ρ might be zero and the random variables may still be statistically dependent.

Some properties of the expected value and variance applied to multiple random variables are listed below:

1. $E(\sum_i c_i X_i) = \sum_i c_i E(X_i)$.
2. $\text{VAR}(\sum_i c_i X_i) = \sum_i c_i^2 \text{VAR}(X_i) + \sum_i \sum_{j \neq i} c_i c_j COV(X_i, X_j)$.

where c_i's are constants.

Multiple functions of multiple random variables. If we define two functions of the random variables X and Y by

$$Z = g(X, Y)$$
$$W = h(X, Y)$$

then the joint c.d.f. and p.d.f. of Z and W can be obtained directly by applying the definition of the c.d.f. However, if it happens that for all z and w the set of equations

$$g(x, y) = z$$
$$h(x, y) = w$$

has a countable number of solutions $\{x_i, y_i\}$, and at these points the determinant of the Jacobian matrix

$$\mathbf{J}(x, y) = \begin{bmatrix} \frac{\partial z}{\partial x} & \frac{\partial z}{\partial y} \\ \frac{\partial w}{\partial x} & \frac{\partial w}{\partial y} \end{bmatrix}$$

is nonzero, then we have

$$f_{Z,W}(z, w) = \sum_i \frac{f(x_i, y_i)}{|\det \mathbf{J}(x_i, y_i)|} \qquad (3.1.16)$$

where $\det \mathbf{J}$ denotes the determinant of the matrix $\mathbf{J}$.

Example 3.1.3

The two random variables X and Y are independent and identically distributed, each with a Gaussian density function with mean equal to zero and variance equal to σ^2. If these two random variables denote the coordinates of a point in the plane, find the p.d.f. of the magnitude and the phase of that point in polar coordinates.

Solution   First we have to find the joint p.d.f. of X and Y. Since X and Y are independent, their joint p.d.f. is the product of their marginal p.d.f.'s, that is,

$$f_{X,Y}(x, y) = f_X(x) f_Y(y)$$

$$= \frac{1}{2\pi\sigma^2} \exp\left(-\frac{x^2 + y^2}{2\sigma^2}\right) \qquad (3.1.17)$$

The magnitude of the point with coordinates (X, Y) in the polar plane is given by $V = \sqrt{X^2 + Y^2}$ and its phase is given by $\Theta = \arctan\frac{Y}{X}$. We begin by deriving the joint p.d.f. of V and Θ. In this case, $g(X, Y) = \sqrt{X^2 + Y^2}$ and $h(X, Y) = \arctan\frac{Y}{X}$.

The Jacobian matrix is given by

$$\mathbf{J}(x, y) = \begin{bmatrix} \dfrac{x}{\sqrt{x^2 + y^2}} & \dfrac{y}{\sqrt{x^2 + y^2}} \\ -\dfrac{y}{x^2 + y^2} & \dfrac{x}{x^2 + y^2} \end{bmatrix} \tag{3.1.18}$$

The determinant of the Jacobian matrix can be easily determined to be equal to

$$|\det \mathbf{J}(x, y)| = \frac{1}{\sqrt{x^2 + y^2}}$$

$$= \frac{1}{v} \tag{3.1.19}$$

The set of equations

$$\begin{cases} \sqrt{x^2 + y^2} = v \\ \arctan \dfrac{y}{x} = \theta \end{cases} \tag{3.1.20}$$

has only one solution given by

$$\begin{cases} x = v \cos \theta \\ y = v \sin \theta \end{cases} \tag{3.1.21}$$

Substituting these results into (3.1.16), we obtain the joint p.d.f. of the magnitude and the phase as

$$f_{V,\Theta}(v, \theta) = v f_{X,Y}(v \cos \theta, v \sin \theta)$$

$$= \frac{v}{2\pi \sigma^2} \exp \left(-\frac{v^2}{2\sigma^2} \right) \tag{3.1.22}$$

To derive the marginal p.d.f. for the magnitude and the phase, we have to integrate the joint p.d.f. To obtain the p.d.f. of the phase, we have

$$f_\Theta(\theta) = \int_0^\infty f_{V,\Theta}(v, \theta) \, dv$$

$$= \frac{1}{2\pi} \int_0^\infty \frac{v}{\sigma^2} \exp \left(-\frac{v^2}{2\sigma^2} \right) dv$$

$$= \frac{1}{2\pi} \left[-\exp \left(-\frac{v^2}{2\sigma^2} \right) \right]_0^\infty$$

$$= \frac{1}{2\pi} \tag{3.1.23}$$

Hence, the phase is uniformly distributed on $[0, 2\pi)$. To obtain the marginal p.d.f. for the magnitude we have

$$f_V(v) = \int_0^{2\pi} f_{V,\Theta}(v, \theta) \, d\theta$$

$$= \frac{v}{\sigma^2} e^{-\frac{v^2}{2\sigma^2}} \tag{3.1.24}$$

The above relation holds only for positive v. For negative v, of course, $f_V(v) = 0$. Therefore,

$$f_V(v) = \begin{cases} \dfrac{v}{\sigma^2} e^{-\frac{v^2}{2\sigma^2}}, & v \geq 0 \\ 0, & v < 0 \end{cases} \tag{3.1.25}$$

This p.d.f. is known as the *Rayleigh probability density function* and has widespread applications in the study of the fading communication channels. It is also interesting to note that in the above example

$$f_{V,\Theta}(v, \theta) = f_V(v) f_\Theta(\theta) \tag{3.1.26}$$

and, therefore, the magnitude and the phase are independent random variables.

Jointly Gaussian random variables. Jointly Gaussian or binormal random variables X and Y are defined by the joint density function

$$f_{X,Y}(x, y) = \frac{1}{2\pi\sigma_1\sigma_2\sqrt{(1 - \rho^2)}} \exp\left\{ -\frac{1}{2(1 - \rho^2)} \right.$$

$$\times \left[\frac{(x - m_1)^2}{\sigma_1^2} + \frac{(y - m_2)^2}{\sigma_2^2} - \frac{2\rho(x - m_1)(y - m_2)}{\sigma_1\sigma_2} \right] \right\}$$

When the two random variables X and Y are distributed according to a binormal distribution, not only are X and Y themselves normal random variables, but the conditional densities $f(x|y)$ and $f(y|x)$ are also Gaussian. This property is the main difference between jointly Gaussian random variables and two random variables each having a Gaussian distribution. Also, it is straightforward to show that if X and Y are jointly Gaussian, X is Gaussian with mean m_1 and variance σ_1^2, Y is Gaussian with mean m_2 and variance σ_2^2, and the correlation coefficient between X and Y is ρ.

The definition of two jointly Gaussian random variables can be extended to n random variables $(X_1, X_2, X_3, \cdots, X_n)$. If we define the random vector $\mathbf{X} = (X_1, X_2, \cdots, X_n)$, the vector of the means $\mathbf{m} = (m_1, m_2, \cdots, m_n)$, and the $n \times n$ covariance matrix $\mathbf{C}$ such that $C_{i,j} = \text{COV}(X_i, X_j)$, then the random variables $\{X_i\}$ are jointly Gaussian if

$$f(x_1, x_2, \cdots, x_n) = \frac{1}{\sqrt{(2\pi)^n \det(\mathbf{C})}} \exp\{-\frac{1}{2}(\mathbf{x} - \mathbf{m})\mathbf{C}^{-1}(\mathbf{x} - \mathbf{m})^t\}$$

Here are the main properties of jointly Gaussian random variables:

1. If n random variables are jointly Gaussian, any subset of them is also distributed according to a jointly Gaussian distribution of the appropriate size. In particular all individual random variables are Gaussian.

2. Jointly Gaussian random variables are completely characterized by their mean vector **m** and covariance matrix **C**. These so-called *second-order* properties completely describe the random variables.

3. Any subset of the jointly Gaussian random variables conditioned on any other subset of the original random variables makes a jointly Gaussian distribution of the appropriate size.

4. Any set of linear combinations of $(X_1, X_2, \cdots, X_n)$ are themselves jointly Gaussian. In particular, any linear combination of X_i's is a Gaussian random variable.

5. Two uncorrelated jointly Gaussian random variables are independent. Therefore, *for jointly Gaussian random variables, independence and uncorrelatedness are equivalent.* As we have seen before, this is not true in general for non-Gaussian random variables.

Sums of random variables.   If we have a sequence of random variables $(X_1, X_2, \cdots, X_n)$ with basically the same properties, then the behavior of their average $Y = \frac{1}{n} \sum_{i=1}^{n} X_i$ is expected to be "less random" than each X_i. The law of large numbers and the central limit theorem are precise statements of this intuitive fact.

The *weak law of large numbers* (WLLN) states that if the sequence of random variables $X_1, X_2, \cdots, X_n$ are uncorrelated with the same mean m_X and variance $\sigma_X^2 < \infty$, then for any $\epsilon > 0$, $\lim_{n \to \infty} P(|Y - m_X| > \epsilon) = 0$, where $Y = \frac{1}{n} \sum_{i=1}^{n} X_i$. This means that the average converges (in probability) to the expected value.

The *central limit theorem* not only states the convergence of the average to the mean but also gives some insight into the distribution of the average. This theorem states that if $(X_1, X_2, \cdots, X_n)$ are independent with means $(m_1, m_2, \cdots, m_n)$ and variances $(\sigma_1^2, \sigma_2^2, \cdots, \sigma_n^2)$, then the c.d.f. of the random variable $\frac{1}{\sqrt{n}} \sum_{i=1}^{n} \frac{X_i - m_i}{\sigma_i}$ converges to the c.d.f. of a Gaussian random variable with mean 0 and variance 1. In the special case that the X_i's are i.i.d. (independent and identically distributed), this theorem says that the c.d.f. of $Y = \frac{1}{n} \sum_{i=1}^{n} X_i$ converges to the c.d.f. of a $\mathcal{N}\left(m, \frac{\sigma^2}{n}\right)$. Note that, although from the central limit theorem we can conclude that the average converges to the expected value, we cannot say that the law of large numbers follows from the central limit theorem. This is because the requirements for the central limit theorem to hold are much stronger. For the central limit theorem to hold, we need the random variables to be independent, whereas the law of large numbers holds under the less restrictive condition of uncorrelatedness of the random variables.

This concludes our brief review of the basics of the probability theory. References at the end of this chapter provide sources for further study.

3.2 RANDOM PROCESSES: BASIC CONCEPTS

A random process is a natural extension of the concept of random variable when dealing with signals. In analyzing communication systems, we are basically dealing with time-varying signals. Chapter 2 was devoted entirely to analysis of linear systems when driven by time-varying signals. In our development so far, we have assumed that all the signals are deterministic. In many situations the deterministic assumption on time-varying signals is not a valid assumption, and it is more appropriate to model signals as random rather than as deterministic functions. One such example is the case of thermal noise in electronic circuits. This type of noise is due to the random movement of electrons as a result of thermal agitation and, therefore, the resulting current and voltage can only be described statistically. Another example is the reflection of radio waves from different layers of the ionosphere that makes long-range broadcasting of short-wave radio possible. Due to randomness of this reflection, the received signal can again be modeled as a random signal. These two examples show that random signals are suitable for describing certain phenomena in signal transmission.

Another situation where modeling by random processes proves useful is in the characterization of information sources. An information source, such as a speech source, generates time-varying signals whose contents are not known in advance; otherwise, there would be no need to transmit them. Therefore, random processes provide a natural way to model information sources as well.

A *random process,* a *stochastic process,* or a *random signal* can be viewed in two different, although closely related, ways. One way is to view a random process as a collection of time functions, or signals, corresponding to various outcomes of a random experiment. From this viewpoint, corresponding to each outcome ω_i in a probability space $(\Omega, \mathcal{B}, P)$, there exists a signal $x(t; \omega_i)$. This description is very similar to the description of random variables in which a real number is assigned to each outcome ω_i. Figure 3.10 depicts this characterization of random processes. Thus, for each ω_i there exists a deterministic time function $x(t; \omega_i)$, which is called a *sample function,* or a *realization* of the random process. At each time instant t_0, and for each $\omega_i \in \Omega$, we have the number $x(t_0; \omega_i)$. For the different outcomes ω_i's at a fixed time t_0, the numbers $x(t_0; \omega_i)$ constitute a random variable denoted by $X(t_0)$. After all, a random variable is nothing but an assignment of real numbers to outcomes of a random experiment. This is a very important observation and a bridge that connects the concept of a random process to the more familiar concept of a random variable. In other words, *at any time instant the value of a random process is a random variable.*

Alternatively, we may view the random signal at $t_1, t_2, \ldots$, or in general, all $t \in \mathbb{R}$ as a collection of random variables $\{X(t_1), X(t_2), \ldots\}$, i.e., $\{X(t), t \in \mathbb{R}\}$. From this viewpoint, a random process is represented as a collection of random variables indexed by some index set (e.g., $\mathbb{R}$ in the latter case). If the index set is the set of real numbers, the random process is called a *continuous-time random process;* and if it is the set of all integers, then the random process is a *discrete-*

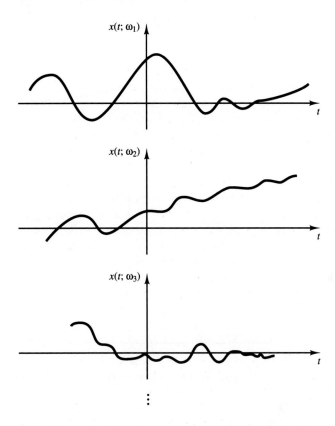

FIGURE 3.10.   Sample functions of a random process.

time random process. A discrete-time random process is nothing but a sequence of random variables $\{X_i\}_{i=-\infty}^{+\infty}$. This second view of random processes, although less intuitive, is more appropriate for precise mathematical development of the theory of random processes.

Example 3.2.1

Let $(\Omega, \mathcal{B}, P)$ denote the probability space corresponding to the random experiment of throwing a die. Obviously, in this case $\Omega = \{1, 2, 3, 4, 5, 6\}$. For all ω_i, let $x(t; \omega_i) = \omega_i e^{-t} u_{-1}(t)$ denote a random process. Then $X(1)$ is a random variable taking values $e^{-1}, 2e^{-1}, \ldots, 6e^{-1}$ each with probability $\frac{1}{6}$. Sample functions of this random process are shown in Figure 3.11. For this example, the first viewpoint of random processes is the preferred view.

Example 3.2.2

Let ω_i denote the outcome of a random experiment consisting of independent drawings from a Gaussian random variable distributed according to $\mathcal{N}(0, 1)$. Let the discrete

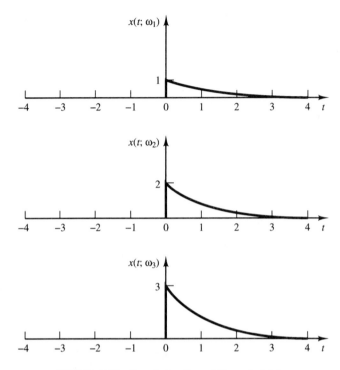

FIGURE 3.11.   Sample functions of Example 3.2.1.

time random process $\{X_n\}_{n=0}^{\infty}$ be defined by: $X_0 = 0$ and $X_n = X_{n-1} + \omega_n$ for all $n \geq 1$. It follows from the basic properties of the Gaussian random variables that for all $i \geq 1$ and $j \geq 1$, $i < j$, $\{X_n\}_i^j$ is a $j - i + 1$ dimensional Gaussian vector. For this example the second view, i.e., interpreting the random process as a collection of random variables, is more appropriate.

3.2.1 Description of Random Processes

Based on the adopted viewpoint, there are two types of descriptions possible for random processes. If the random process is viewed as a collection of signals, the *analytic description* may be appropriate. In this description, analytic expressions are given for each sample in terms of one or more random variables, i.e., the random process is given as $X(t) = f(t; \theta)$ where $\theta = (\theta_1, \theta_2, \ldots, \theta_n)$ is in general a random vector with a given joint probability density function. This is a very informative description of a random process because it completely describes the analytic form of various realizations of the process. For real-life processes, it is hardly possible to give such a complete description. If an analytic description is not possible, a *statistical description* may be appropriate. Such a description is based on the second viewpoint of random processes—as a collection of random variables indexed by some index set.

Definition 3.2.1. A *complete statistical description* of a random process $X(t)$ is known if for any integer n and any choice of $(t_1, t_2, \ldots, t_n) \in \mathbb{R}^n$, the joint p.d.f. of $(X(t_1), X(t_2), \ldots, X(t_n))$ is given.

If the complete statistical description of the process is given, for any n the joint density function of $(X(t_1), X(t_2), \ldots, X(t_n))$ is given by $f_{X(t_1), X(t_2), \ldots, X(t_n)}(x_1, x_2, \ldots, x_n)$.

Definition 3.2.2. A process $X(t)$ is described by its Mth *order statistics* if for all $n \leq M$ and all $(t_1, t_2, \ldots, t_n) \in \mathbb{R}^n$, the joint p.d.f. of $(X(t_1), X(t_2), \ldots, X(t_n))$ is given.

A very important special case, in the study of communication systems, is the case of $M = 2$, in which second-order statistics are known. This simply means that, at each time instant t, we have the density function of $X(t)$, and for all choices of (t_1, t_2), the joint density function of $(X(t_1), X(t_2))$ is given.

Example 3.2.3

A random process is defined by $X(t) = A \cos(2\pi f_0 t + \Theta)$ where Θ is a random variable uniformly distributed on $[0, 2\pi)$. In this case, we have an analytic description of the random process. Note that by having the analytic description, we can find the complete statistical description. Figure 3.12 shows some samples of this process.

Example 3.2.4

The process $X(t)$ is defined by $X(t) = X$, where X is a random variable uniformly distributed on $[-1, 1]$. In this case again, an analytic description of the random process is given. For this random process, each sample is a constant signal. Samples of this process are shown in Figure 3.13.

Example 3.2.5

The process $X(t), t > 0$, is defined by the property that for any n and any $(t_1, t_2, \ldots, t_n) \in \mathbb{R}^{+n}$, the joint density function of $\{X(t_i)\}_{i=1}^n$ is a jointly Gaussian vector with mean 0 and covariance matrix described by

$$C_{i,j} = \text{COV}(X(t_i), X(t_j)) = \sigma^2 \min(t_i, t_j)$$

This is a complete statistical description of the random process $X(t)$. A sample of this process is shown in Figure 3.14.

Note that in the last example, although a complete statistical description of the process is given, little insight can be obtained in the shape of each realization of the process.

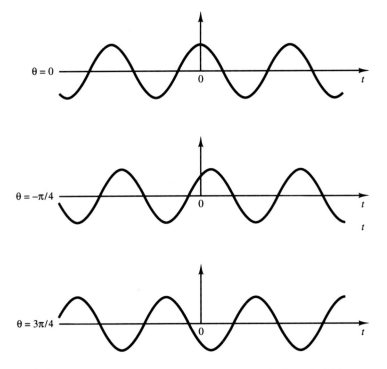

FIGURE 3.12.   Samples of the random process given in Example 3.2.3.

3.2.2 Statistical Averages

The fact that a random process defines a random variable at any given time and a random vector at any given set of times enables us to define various statistical averages for the process via statistical averages of the corresponding random variables.

Definition 3.2.3.   The *mean*, or *expectation* of the random process $X(t)$ is a deterministic function of time $m_X(t)$ that at each time instant t_0 equals the mean of the random variable $X(t_0)$. That is, $m_X(t) = E[X(t)]$ for all t.

Since at any t_0, the random variable $X(t_0)$ is well defined with a p.d.f. $f_{X(t_0)}(x)$, we have

$$E[X(t_0)] = m_X(t_0) = \int_{-\infty}^{+\infty} x f_{X(t_0)}(x)dx$$

Figure 3.15 gives a pictorial description of this definition.

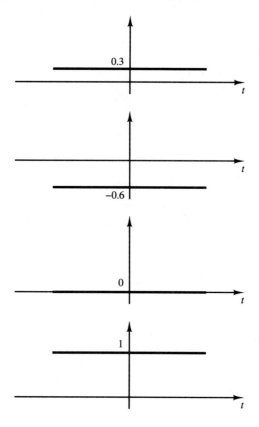

FIGURE 3.13. Samples of the random process given in Example 3.2.4.

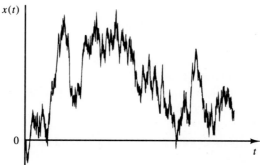

FIGURE 3.14. Samples of the process given in Example 3.2.5.

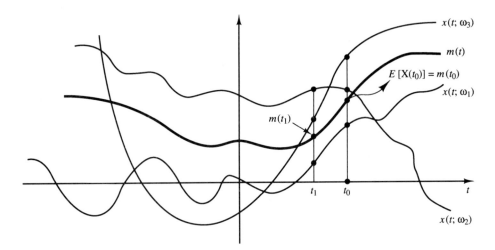

FIGURE 3.15. The mean of a random process.

Example 3.2.6

The mean of the random process in Example 3.2.3 is obtained by noting that

$$f_\Theta(\theta) = \begin{cases} \frac{1}{2\pi} & 0 \le \theta < 2\pi \\ 0, & \text{otherwise} \end{cases}$$

Hence

$$E[X(t)] = \int_0^{2\pi} A\cos(2\pi f_0 t + \theta)\, \frac{1}{2\pi}\, d\theta = 0$$

We observe that in this case $m_X(t)$ is independent of t.

Another statistical average that plays a very important role in our study of random processes is the *autocorrelation function*. The autocorrelation function is especially important because it completely describes the power-spectral density and the power content of a large class of random processes.

Definition 3.2.4. The *autocorrelation function* of the random process $X(t)$, denoted as $R_{XX}(t_1, t_2)$, is defined by $R_{XX}(t_1, t_2) = E[X(t_1)X(t_2)]$.

From the above definition it is clear that $R_{XX}(t_1, t_2)$ is a deterministic function of two variables, t_1 and t_2, given by

$$R_{XX}(t_1, t_2) = \int_{-\infty}^{+\infty} \int_{-\infty}^{+\infty} x_1 x_2 f_{X(t_1), X(t_2)}(x_1, x_2)\, dx_1\, dx_2$$

The autocorrelation function is usually denoted by $R_X(t_1, t_2)$ for short.

Example 3.2.7

The autocorrelation function of the random process in Example 3.2.4 is

$$R_X(t_1, t_2) = E[A\cos(2\pi f_0 t_1 + \Theta) A\cos(2\pi f_0 t_2 + \Theta)]$$

$$= A^2 E\left[\frac{1}{2}\cos 2\pi f_0(t_1 - t_2) + \frac{1}{2}\cos(2\pi f_0(t_1 + t_2) + 2\Theta)\right]$$

$$= \frac{A^2}{2}\cos 2\pi f_0(t_1 - t_2)$$

Note that

$$E[\cos(2\pi f_0(t_1 + t_2) + 2\Theta)] = \int_0^{2\pi} \cos[2\pi f_0(t_1 + t_2) + 2\theta] \frac{1}{2\pi} d\theta = 0$$

Example 3.2.8

For the random process given in Example 3.2.4, we have

$$R_X(t_1, t_2) = E(X^2) = \int_{-1}^{+1} \frac{x^2}{2} dx = \frac{1}{3}$$

So far, we have given two methods, analytical and statistical, to describe random processes. A statistical description can be a complete description or an Mth-order description. Sometimes, even having a second-order statistical description of a random variable is not practical. In these cases, it may be possible to find various statistical averages by averaging over different realizations; in particular, it may be possible to find $m_X(t)$ and $R_X(t_1, t_2)$. Although this information is much less than a complete statistical description of the random process, it may be adequate. As we will see later, the mean and autocorrelation function provide a complete statistical description for the important class of Gaussian random processes.

3.2.3 Stationary Processes

In a complete statistical description of a random process, for any n and any $(t_1, t_2, \ldots, t_n)$, the joint p.d.f. $f_{X(t_1), X(t_2), \ldots, X(t_n)}(x_1, x_2, \ldots, x_n)$ is given. This joint p.d.f. in general depends on the choice of the time origin. In a very important class of random processes, the joint density function is independent of the choice of the time origin. These processes whose statistical properties are time independent are called *stationary processes*. There are different notions of stationarity. A *strictly stationary process* is a process in which for all n and all $(t_1, t_2, \ldots, t_n)$, $f_{X(t_1), X(t_2), \ldots, X(t_n)}(x_1, x_2, \ldots, x_n)$ depends only on the relative position of $t_1, t_2, \ldots, t_n$ and not on their values directly. In other words, a shift in the time origin does not change the statistical properties of the process. A formal definition of a strictly stationary process is the following:

Definition 3.2.5. A *strictly stationary process* is a process in which for all n, all $(t_1, t_2, \ldots, t_n)$, and all Δ

$$f_{X(t_1), X(t_2), \ldots, X(t_n)}(x_1, x_2, \ldots, x_n) = f_{X(t_1+\Delta), X(t_2+\Delta), \ldots, X(t_n+\Delta)}(x_1, x_2, \ldots, x_n)$$

A process is called M^{th}-order stationary if the above condition holds for all $n \leq M$.

It is obvious that for both strictly stationary processes and M^{th}-order stationary processes the density function of $X(t)$ is time independent.

Example 3.2.9

Let $X(t)$ denote a process such that for all n and all $(t_1, t_2, \ldots, t_n)$, the vector $(X(t_1), X(t_2), \ldots, X(t_n))$ is an n-dimensional Gaussian random variable with mean zero and covariance matrix $\mathbf{I}_n$ (n-dimensional identity matrix). Then clearly $X(t)$ is a strictly stationary process.

Strict stationarity is a very strong condition that only a few physical processes may satisfy. A less restrictive definition of stationarity is often more useful.

Definition 3.2.6. A process $X(t)$ is *wide-sense stationary* (WSS) if the following conditions are satisfied:

1. $m_X(t) = E[X(t)]$ is independent of t.
2. $R_X(t_1, t_2)$ depends only on the time difference $\tau = t_1 - t_2$ and not on t_1 and t_2 individually.

Hereafter the term *stationary* with no adjectives means WSS, and for these processes the mean and autocorrelation are denoted by m_X and $R_{XX}(\tau)$ or simply $R_X(\tau)$.

Example 3.2.10

For the random process in Exercise 3.2.3, we have already seen that $m_X = 0$ and $R_X(t_1, t_2) = \frac{A^2}{2} \cos 2\pi f_0(t_1 - t_2)$. Therefore, the process is WSS.

A class of processes closely related to stationary processes is the class of *cyclostationary processes*. In cyclostationary processes, the statistical properties are not time independent, but periodic with time. We will give the formal definition of the cyclostationary processes here, but their importance in communication system analysis will become more apparent in Chapter 7.

Definition 3.2.7 A random process $X(t)$ with mean $m_X(t)$ and autocorrelation function $R_X(t + \tau, t)$ is called *cyclostationary* if both the mean and the autocorrelation are periodic in t with some period T_0, i.e., if

$$m_X(t + kT_0) = m_X(t) \tag{3.2.1}$$

and

$$R_X(t + \tau + kT_0, t + kT_0) = R_X(t + \tau, t) \tag{3.2.2}$$

for all t, τ, and k.

Example 3.2.11

Let $Y(t) = X(t)\cos(2\pi f_0 t)$ where $X(t)$ is a stationary random process with mean m and autocorrelation $R_X(\tau)$. Then

$$m_Y(t) = E\left[X(t)\cos(2\pi f_0 t)\right] = m_X \cos(2\pi f_0 t) \tag{3.2.3}$$

and

$$R_Y(t + \tau, t) = E\left[X(t + \tau)\cos(2\pi f_0(t + \tau))X(t)\cos(2\pi f_0 t)\right]$$

$$= R_X(\tau)\left[\frac{1}{2}\cos(2\pi f_0 \tau) + \frac{1}{2}\cos(4\pi f_0 t + 2\pi f_0 \tau)\right] \tag{3.2.4}$$

It is seen that both $m_Y(t)$ and $R_Y(t + \tau, t)$ are periodic with period $T_0 = \frac{1}{f_0}$. Therefore, the process is cyclostationary.

The following theorem gives the basic properties of the autocorrelation function of a stationary process.

Theorem 3.2.1. $R_X(\tau)$, the autocorrelation function of the stationary process $X(t)$, has the following properties:

1. $R_X(\tau)$ is an even function, i.e., $R_X(-\tau) = R_X(\tau)$.
2. The maximum absolute value of $R_X(\tau)$ is achieved at $\tau = 0$, i.e., $|R_X(\tau)| \leq R_X(0)$.
3. If for some T_0 we have $R_X(T_0) = R_X(0)$, then for all integers k, $R_X(kT_0) = R_X(0)$.

Proof

1. By definition

$$R_X(\tau) = E[X(t)X(t - \tau)] = E[X(t - \tau)X(t)] = R_X(-\tau)$$

This proves the first part of the theorem.
2. By noting that

$$E[X(t) \pm X(t - \tau)]^2 \geq 0$$

and expanding the result, we have

$$E[X^2(t)] + E[X^2(t - \tau)] \pm 2E[X(t)X(t - \tau)] \geq 0$$

Hence,

$$\pm R_X(\tau) \leq R_X(0)$$

which means that $|R_X(\tau)| \leq R_X(0)$.

3. The proof is by mathematical induction. Assume that $R_X(iT_0) = R_X(0)$, for $i = 1, 2, \ldots, k - 1$. Expanding

$$E[X(t) - X(t - kT_0)]^2$$
$$= E[(X(t) - X(t - (k-1)T_0)) + (X(t - (k-1)T_0) - X(t - kT_0))]^2$$

we obtain

$$2R_X(0) - 2R_X(kT_0)$$
$$= E[X(t) - X(t - (k-1)T_0)]^2$$
$$+ E[X(t - (k-1)T_0) - X(t - kT_0)]^2$$
$$+ 2E[(X(t) - X(t - (k-1)T_0))(X(t - (k-1)T_0) - X(t - kT_0))]$$

Now using the relation $R_X(iT_0) = R_X(0)$ for $i \leq k - 1$,

$$E[X(t) - X(t - (k-1)T_0)]^2 = 2R_X((k-1)T_0) - 2R_X(0) = 0$$

and

$$E[X(t - (k-1)T_0) - X(t - kT_0)]^2 = 2R_X(T_0) - 2R_X(0) = 0$$

By applying the Cauchy-Schwartz inequality, we obtain

$$|E[(X(t) - X(t - (k-1)T_0))(X(t - (k-1)T_0) - X(t - kT_0))]|$$
$$\leq \sqrt{E[X(t) - X(t - (k-1)T_0)]^2 E[X(t - (k-1)T_0) - X(t - kT_0)]^2}$$
$$= \sqrt{4(R_X((k-1)T_0) - R_X(0))(R_X(T_0) - R_X(0))} = 0$$

from which we conclude $R_X(kT_0) = R_X(0)$ for all integers k.

Ergodic processes.   For a strictly stationary process $X(t)$ and for any function $g(x)$, we can define two types of averages:

1. By looking at a given time t_0 and different realizations of the process, we have a random variable $X(t_0)$ with density function $f_{X(t_0)}(x)$, which is independent of t_0 since the process is strictly stationary. For this random variable, we can find the statistical average (or *ensemble average*) of any function $g(X)$ as

$$E[g(X(t_0))] = \int_{-\infty}^{+\infty} g(x) f_{X(t_0)}(x)\, dx$$

This value is of course independent of t_0.

2. By looking at an individual realization, we have a deterministic function of time $x(t; \omega_i)$. Based on this function, we can find the time average for a function $g(x)$, defined as

$$\langle g(x) \rangle_i = \lim_{T \to \infty} \frac{1}{T} \int_{-T/2}^{+T/2} g(x(t; \omega_i))\, dt$$

$\langle g(x) \rangle_i$ is, of course, a real number independent of t, but, in general, it is dependent on the particular realization chosen (ω_i or i). Therefore, for each ω_i, we have a corresponding real number $\langle g(x) \rangle_i$. Hence, $\langle g(x) \rangle_i$ is the value assumed by a random variable. We denote this random variable by $\langle g(X) \rangle$.

If it happens that for all functions $g(x)$, $\langle g(x) \rangle_i$ is independent of i and equals $E[g(X(t_0))]$, then the process is called *ergodic*.[†]

Definition 3.2.8. A stationary process $X(t)$ is also *ergodic* if for all functions $g(x)$ and all $\omega_i \in \Omega$

$$\lim_{T \to \infty} \frac{1}{T} \int_{-T/2}^{+T/2} g(x(t; \omega_i)) \, dt = E[g(X(t))]$$

In other words, if all time averages are equal to the corresponding statistical averages, then the stationary process is ergodic.

A natural consequence of ergodicity is that in measuring various statistical averages (mean and autocorrelation, for example), it is sufficient to look at one realization of the process and find the corresponding time average, rather than considering a large number of realizations and averaging over them.

Example 3.2.12

For the process given in Example 3.2.3, for any value of $0 \le \theta < 2\pi$ (i.e. for any realization of the process), we have

$$\langle g(X(t; \theta)) \rangle = \lim_{T \to \infty} \frac{1}{T} \int_{-T/2}^{+T/2} g(A \cos(2\pi f_0 t + \theta)) \, dt$$

$$= \lim_{N \to \infty} \frac{1}{2NT_0} \int_{-NT_0}^{NT_0} g(A \cos(2\pi f_0 t + \theta)) \, dt$$

$$= \frac{1}{T_0} \int_0^{T_0} g(A \cos(2\pi f_0 t + \theta)) \, dt$$

$$\overset{2\pi f_0 t + \theta = u}{=} \frac{1}{T_0} \int_\theta^{2\pi + \theta} g(A \cos u) \frac{du}{2\pi f_0}$$

$$= \frac{1}{2\pi} \int_0^{2\pi} g(A \cos u) \, du$$

[†]Our notion of ergodicity and the definition that follows are not precise. A precise definition of ergodicity requires knowledge of measure theory and is not treated here. In fact, our definition of ergodicity is one of the properties satisfied by stationarity and ergodic processes and sometimes is referred to as the ergodic property. For most engineering applications, however, this notion is adequate. For a precise definition of ergodicity and some related misconceptions, the reader is referred to Gray and Davisson (1986).

where $T_0 = \frac{1}{f_0}$. On the other hand,

$$E[X(t)] = \int_0^{2\pi} g(A\cos(2\pi f_0 t + \theta)) \frac{1}{2\pi} d\theta$$

$$\overset{2\pi f_0 t + \theta = u}{=} \frac{1}{2\pi} \int_{2\pi f_0 t}^{2\pi f_0 t + 2\pi} g(A\cos(u)) du$$

$$= \frac{1}{2\pi} \int_0^{2\pi} g(A\cos u) du$$

and, therefore, the process is ergodic.

Example 3.2.13

In the process given in Example 3.2.4, each sample has a different constant value, and, therefore, the time average for each ω_i depends on i. This means that the process is not ergodic.

Example 3.2.14

We have already seen that a sequence of random variables defines a discrete-time random process. Let us assume that the sequence $\{X_i\}_{i=-\infty}^{\infty}$ is a sequence of i.i.d. random variables. Then for any realization, by the law of large numbers, we have $\lim_{N \to \infty} \frac{1}{2N} \sum_{i=-N}^{N} X_i = E[X]$. This means that i.i.d. processes are ergodic.

Ergodicity is a very strong property. Unfortunately, there exists no simple test for ergodicity. For the important class of Gaussian processes, however, there exists a simple test that we will discuss later.

Power and energy.   In our discussion of power and energy for deterministic signals, we defined two types of signals: energy-type and power-type.

We can extend these notions to random processes. Let $X(t)$ be a random process with sample functions $x(t, \omega_i)$. The energy and power of each sample function then are defined as

$$E_i = \int_{-\infty}^{+\infty} x^2(t, \omega_i) dt$$

and

$$P_i = \lim_{T \to \infty} \frac{1}{T} \int_{-T/2}^{T/2} x^2(t, \omega_i) dt$$

This shows that for each $\omega_i \in \Omega$, we have real numbers E_i and P_i, denoting energy and power, respectively, and, therefore, it is clear that both energy and power are random variables, which we denote[†] by $\mathcal{E}_X$ and $\mathcal{P}_X$. It makes sense to define the

---

[†]Note that here we are using script letters to denote random variables.

expected values of $\mathcal{P}_X$ and $\mathcal{E}_X$ as a measure of the power and energy content of the process. These quantities are defined as follows:

Definition 3.2.9.   The power content P_X and the energy content E_X of the random process $X(t)$ are defined as

$$P_X = E[\mathcal{P}_X]$$

and

$$E_X = E[\mathcal{E}_X]$$

where

$$\mathcal{E}_X = \int_{-\infty}^{+\infty} X^2(t)\,dt$$

and

$$\mathcal{P}_X = \lim_{T \to \infty} \frac{1}{T} \int_{-T/2}^{T/2} X^2(t)\,dt$$

From this definition,

$$E_X = E\left[\int_{-\infty}^{+\infty} X^2(t)\,dt \right]$$

$$= \int_{-\infty}^{+\infty} E[X^2(t)]\,dt$$

$$= \int_{-\infty}^{+\infty} R_X(t,t)\,dt \qquad (3.2.5)$$

and

$$P_X = E\left[\lim_{T \to \infty} \frac{1}{T} \int_{-T/2}^{T/2} X^2(t)\,dt \right]$$

$$= \lim_{T \to \infty} \frac{1}{T} \int_{-T/2}^{T/2} E[X^2(t)]\,dt$$

$$= \lim_{T \to \infty} \frac{1}{T} \int_{-T/2}^{T/2} R_X(t,t)\,dt \qquad (3.2.6)$$

If the process is stationary, then $R_X(t,t) = R_X(0)$ is independent of t and we have

$$P_X = R_X(0)$$

and

$$E_X = \int_{-\infty}^{+\infty} R_X(0)\,dt$$

It is seen that for stationary processes, if $E_X < \infty$, i.e., if the process is "energy-type," then we have $R_X(0) = E[X^2(t)] = 0$. This means that for all t, $X(t)$ is zero with probability one. This shows that for the case of stationary processes, only power-type processes are of theoretical and practical interest.

If, in addition to being stationary, the process is also ergodic, then $\mathcal{P}_X$ is not random anymore, and for *each sample function of the process* we have

$$P_i = P_X = R_X(0)$$

Example 3.2.15

For the process given in Example 3.2.3, which is both stationary and ergodic, we have

$$P_X = R_X(0) = \left. \frac{A^2}{2} \cos(2\pi f_0 \tau) \right|_{\tau=0} = \frac{A^2}{2}$$

This is, in fact, the power content of each sample function in this process since each realization is a sinusoidal waveform.

Example 3.2.16

For the process in Example 3.2.4, which is stationary but not ergodic, we have $P_X = R_X(0) = \frac{1}{3}$. In this case, for each sample, which is a constant waveform x where $-1 \leq x \leq 1$, the power is x^2 and the power content of the process is simply $E[X^2]$.

Example 3.2.17

For the process of Example 3.2.5, which is neither stationary nor ergodic, we have

$$P_X = \lim_{T \to \infty} \frac{1}{T} \int_0^T R_X(t, t) \, dt$$

But

$$R_X(t, t) = \sigma^2 \min(t, t) = \sigma^2 t \quad \text{for all } t > 0.$$

Hence,

$$P_X = \lim_{T \to \infty} \frac{1}{T} \int_0^T \sigma^2 t \, dt = \infty$$

Multiple random processes. Multiple random processes arise naturally when dealing with statistical properties of two or more random processes defined on the same probability space. For example, take the case where we are dealing with a random process $X(t)$ and we pass it through a linear time-invariant system. For each sample function input $x(t; \omega_i)$, we have a sample function output defined by $y(t; \omega_i) = x(t; \omega_i) \star h(t)$, where $h(t)$ denotes the impulse response of the system. It is seen that for each $\omega_i \in \Omega$, we have two signals $x(t; \omega_i)$ and $y(t; \omega_i)$. Therefore, we are dealing with two random processes $X(t)$ and $Y(t)$ defined on the same probability space. Although we can define many random processes on the same probability space, we hardly need anything beyond the case of two random processes in communication systems and, therefore, we will only treat this case. When dealing with two random processes, a natural question is the dependence between the random processes under consideration. To this end, we define independence of two random processes.

Definition 3.2.10.   Two random processes $X(t)$ and $Y(t)$ are *independent* if for all t_1, t_2, the random variables $X(t_1)$ and $Y(t_2)$ are independent. Similarly, $X(t)$ and $Y(t)$ are *uncorrelated* if $X(t_1)$ and $Y(t_2)$ are uncorrelated for all t_1, t_2.

From the properties of random variables we know that independence of random processes results in their being uncorrelated, whereas, in general, uncorrelatedness does not imply independence, except for the important class of Gaussian processes for which the two properties are equivalent. We define the correlation function for the case of two random processes as follows:

Definition 3.2.11.   The *crosscorrelation* between two random processes $X(t)$ and $Y(t)$ is defined as

$$R_{XY}(t_1, t_2) = E[X(t_1)Y(t_2)]$$

It is seen from the above definition that, in general,

$$R_{XY}(t_1, t_2) = R_{YX}(t_2, t_1) \tag{3.2.7}$$

The concept of stationarity can also be generalized to joint stationarity for the case of two random processes as follows.

Definition 3.2.12.   Two random processes $X(t)$ and $Y(t)$ are *jointly wide-sense stationary*, or simply *jointly stationary*, if both $X(t)$ and $Y(t)$ are individually stationary and the cross correlation $R_{XY}(t_1, t_2)$ depends only on $\tau = t_1 - t_2$.

Example 3.2.18

Assuming that the two random processes $X(t)$ and $Y(t)$ are jointly stationary, determine the autocorrelation of the process $Z(t) = X(t) + Y(t)$.

Solution   By definition

$$\begin{aligned} R_Z(t+\tau, t) &= E[Z(t+\tau)Z(t)] \\ &= E[(X(t+\tau) + Y(t+\tau))(X(t) + Y(t))] \\ &= R_X(\tau) + R_Y(\tau) + R_{XY}(\tau) + R_{XY}(-\tau) \end{aligned}$$

3.2.4 Random processes and linear systems

In the section on multiple random processes, we saw that when a random process passes through a linear time-invariant system the output is also a random process defined on the original probability space. In this section, we will study the properties of the output process based on the knowledge of the input process. We are assuming that a stationary process $X(t)$ is the input to a linear time-invariant system with impulse response $h(t)$, and the output process is denoted by $Y(t)$ as shown in Figure 3.16. The following questions are of interest: Under what conditions will the output process be stationary? Under what conditions will the input and output processes be jointly stationary? How can we obtain the mean and autocorrelation of the output process and the crosscorrelation between the input and output processes? The following theorem answers these questions.

Theorem 3.2.2. If a stationary process $X(t)$ with mean m_X and auto-correlation function $R_X(\tau)$ is passed through a linear time-invariant system with impulse response $h(t)$, the input and output processes $X(t)$ and $Y(t)$ will be jointly stationary with

$$m_Y = m_X \int_{-\infty}^{+\infty} h(t)\,dt \tag{3.2.8}$$

$$R_{XY}(\tau) = R_X(\tau) \star h(-\tau) \tag{3.2.9}$$

$$R_Y(\tau) = R_X(\tau) \star h(\tau) \star h(-\tau) \tag{3.2.10}$$

Proof. By noting that $Y(t) = \int_{-\infty}^{+\infty} X(\tau)h(t-\tau)\,d\tau$, we see that

$$m_Y(t) = E\left[\int_{-\infty}^{+\infty} X(\tau)h(t-\tau)\,d\tau\right]$$

$$= \int_{-\infty}^{+\infty} E[X(\tau)]h(t-\tau)\,d\tau$$

$$= \int_{-\infty}^{+\infty} m_X h(t-\tau)\,d\tau$$

$$\overset{u=t-\tau}{=} m_X \int_{-\infty}^{+\infty} h(u)\,du$$

This proves that m_Y is independent of t and concludes the first part of the theorem. For the second part, we have

$$R_{XY}(t_1, t_2) = E[X(t_1)Y(t_2)]$$

$$= E\left[X(t_1)\int_{-\infty}^{+\infty} X(s)h(t_2-s)\,ds\right]$$

$$= \int_{-\infty}^{+\infty} E[X(t_1)X(s)]h(t_2-s)\,ds$$

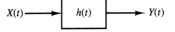

FIGURE 3.16. A random process passing through a linear time-invariant system.

$$= \int_{-\infty}^{+\infty} R_X(t_1 - s)h(t_2 - s)\, ds$$

$$\overset{u=s-t_2}{=} \int_{-\infty}^{+\infty} R_X(t_1 - t_2 - u)h(-u)\, du$$

$$= \int_{-\infty}^{+\infty} R_X(\tau - u)h(-u)\, du$$

$$= R_X(\tau) \star h(-\tau)$$

The last relation shows that $R_{XY}(t_1, t_2)$ depends only on τ. Therefore, if we prove the third part of the theorem, we also have proved that the input and the output are jointly stationary.

To prove the third part (and hence the second part) of the theorem, we use the result we obtained above. We observe that

$$R_Y(t_1, t_2) = E[Y(t_1)Y(t_2)]$$

$$= E\left[\left(\int_{-\infty}^{+\infty} X(s)h(t_1 - s)\, ds \right) Y(t_2) \right]$$

$$= \int_{-\infty}^{+\infty} R_{XY}(s - t_2)h(t_1 - s)\, ds$$

$$\overset{u=s-t_2}{=} \int_{-\infty}^{+\infty} R_{XY}(u)h(t_1 - t_2 - u)\, du$$

$$= R_{XY}(\tau) \star h(\tau)$$

$$= R_X(\tau) \star h(-\tau) \star h(\tau)$$

where in the last step we have used the result of the preceding step. This shows that R_Y, the autocorrelation function of the output depends only on τ and, hence, the output process is stationary. Therefore, the input and output processes are jointly stationary.

Example 3.2.19

Assume a stationary process passes through a differentiator. What are the mean and autocorrelation function of the output and what is the crosscorrelation between the input and output?

Solution   In a differentiator $h(t) = \delta'(t)$. Since $\delta'(t)$ is odd, it follows that

$$m_Y = m_X \int_{-\infty}^{+\infty} \delta'(t)\, dt = 0$$

and

$$R_{XY} = R_X(\tau) \star \delta'(-\tau) = -R_X(\tau) \star \delta'(\tau) = -\frac{d}{d\tau} R_X(\tau)$$

and

$$R_Y(\tau) = -\frac{d}{d\tau} R_X(\tau) \star \delta'(\tau) = -\frac{d^2}{d\tau^2} R_X(\tau)$$

Example 3.2.20

Repeat the previous example for the case where the LTI system is a quadrature filter defined by $h(t) = \frac{1}{\pi t}$ and, therefore, $H(f) = -j \operatorname{sgn}(f)$. The output of the filter is the Hilbert transform of the input in this case (see Section 2.6).

Solution   We have

$$m_Y = m_X \int_{-\infty}^{+\infty} \frac{1}{\pi t} \, dt = 0$$

because $\frac{1}{\pi t}$ is an odd function. Also, we have

$$R_{XY}(\tau) = R_X(\tau) \star \frac{1}{-\pi \tau} = -\hat{R}_X(\tau)$$

and

$$R_Y(\tau) = R_{XY}(\tau) \star \frac{1}{\pi \tau} = -\hat{\hat{R}}_X(\tau) = R_X(\tau)$$

where we have used the fact that $\hat{\hat{x}}(t) = -x(t)$ and assumed that the $R_X(\tau)$ has no dc component (see Section 2.6.1).

3.3   RANDOM PROCESSES IN THE FREQUENCY DOMAIN

In the last part of the previous section, we dealt with the passage of random processes through linear systems. We saw that if a stationary process passes through a LTI system, the input and output will be jointly stationary. We also found a relation between the input and output autocorrelation functions and the crosscorrelation function between the input and the output. In Chapter 2, we have seen that using frequency domain techniques greatly simplifies the input–output relation of linear systems. A natural question now is: How can frequency-domain analysis techniques be applied to the case of LTI systems with random inputs? Our main objective in this section is to develop the necessary techniques that can be used in the frequency-domain analysis of LTI systems when driven by random inputs.

A first attempt would be to define Fourier transforms for each sample function of the random process $x(t; \omega_i)$, and, thus, to define a new process with variable f on the original probability space. The problem with this approach is that there is no guarantee that all the sample functions of the random process possess Fourier transforms. In fact, for many processes we cannot define a Fourier transform for individual sample functions.

Another approach would be to look at the input–output power spectra. This approach is inspired by the convolution integral relation that exists between input and output autocorrelation functions when a system is driven by a random input signal and also by the close relationship between the power spectrum of deterministic signals and their time-average autocorrelation functions as seen in Section 2.4. But prior to a formal definition, we have to see what we mean by the power spectral density of a stochastic process.

3.3.1 Power Spectrum of Stochastic Processes

The power spectrum of a stochastic process is a natural extension of the definition of the power spectrum for deterministic signals when the statistical nature of the process is also taken into account.

Let $X(t)$ denote a random process and let $x(t; \omega_i)$ denote a sample function of this process. To define the power-spectral density for this sample function, we truncate it by defining

$$x_T(t; \omega_i) = x_{T_i}(t) = \begin{cases} x(t; \omega_i), & |t| < T/2 \\ 0, & \text{otherwise} \end{cases}$$

By truncating the signal, we make sure that the result is an energy-type signal and, therefore, possesses a Fourier transform, which we denote by $X_{T_i}(f)$. From the definition of energy-spectral density for energy-type signals, we know that the energy-spectral density for this signal is simply $|X_{T_i}(f)|^2$. Having the energy-spectral density, we can define the power-spectral density as the average-energy-spectral density per unit of time, i.e., $|X_{T_i}(f)|^2/T$. Now, by letting T become arbitrarily large, we define the power-spectral density for the sample function being considered, and we can denote it by $S_{x_i}(f)$.

It is obvious that, in general, various sample functions give rise to various $S_{x_i}(f)$'s, i.e., for each f we have a random variable denoting the amount of power at that frequency in each sample function. It makes sense to define the power spectrum as the ensemble average of these values, i.e.,

$$S_X(f) \overset{\text{def}}{=} E\left[\lim_{T \to \infty} \frac{|X_T(f)|^2}{T} \right] = \lim_{T \to \infty} \frac{E[|X_T(f)|^2]}{T} \tag{3.3.1}$$

The above definition of the power-spectral density is a straightforward generalization of the corresponding relation for the case of deterministic signals. Although this definition is quite intuitive, using it to find the power-spectral density is not always easy.

Example 3.3.1

Find the power-spectral density for the process defined in Example 3.2.4.

Solution Let X denote a random variable uniformly distributed on $[-1, 1]$. Then the truncated random signal is simply

$$X_T(t) = X \Pi \left(\frac{t}{T} \right)$$

Hence,

$$X_T(f) = XT\,\text{sinc}(Tf)$$

and

$$S_X(f) = \lim_{T\to\infty} E(X^2)T\,\text{sinc}^2(Tf)$$

Noting that $E(X^2) = 1/3$, we have to find $\lim_{T\to\infty} T\,\text{sinc}^2(Tf)$. But $T\,\text{sinc}^2(Tf)$ is the Fourier transform of $\Lambda\left(\frac{t}{T}\right)$, and as T goes to infinity, this function goes to 1. Therefore,

$$\lim_{T\to\infty} T\,\text{sinc}^2(Tf) = \delta(f)$$

and $S_X(f) = \frac{1}{3}\delta(f)$.

There exists a very important theorem known as the *Wiener-Khinchin theorem* that gives the power-spectral density of a random process in terms of its autocorrelation function.

Theorem 3.3.1 [Wiener-Khinchin]   If for all finite τ and any interval $\mathcal{A}$ of length $|\tau|$, the autocorrelation function of the random process $X(t)$ satisfies the condition

$$\left|\int_{\mathcal{A}} R_X(t+\tau, t)\,dt\right| < \infty \tag{3.3.2}$$

then the power-spectral density of $X(t)$ is the Fourier transform of $\langle R_X(t+\tau, t)\rangle$ where

$$\langle R_X(t+\tau, t)\rangle = \lim_{T\to\infty} \frac{1}{T}\int_{-T/2}^{T/2} R_X(t+\tau, t)\,dt \tag{3.3.3}$$

Proof.   We begin with the definition of the power-spectral density of a random process

$$S_X(f) = \lim_{T\to\infty} \frac{1}{T} E|X_T(f)|^2$$

where

$$X_T(f) = \int_{-T/2}^{T/2} X(t)e^{-j2\pi ft}\,dt$$

Substituting for $X_T(f)$, we have

$$S_X(f) = \lim_{T\to\infty} \frac{1}{T} E\left[\int_{-T/2}^{T/2} X(s)e^{-j2\pi fs}\,ds \int_{-T/2}^{T/2} X(t)e^{+j2\pi ft}\,dt\right]$$

$$= \lim_{T\to\infty} \frac{1}{T}\int_{-T/2}^{T/2}\int_{-T/2}^{T/2} R_X(s, t)e^{-j2\pi f(s-t)}\,dt\,ds$$

Now we proceed by finding the inverse Fourier transform of the above and showing that it equals $\langle R_X(t+\tau, t)\rangle$. We start with

$$\mathcal{F}^{-1}[S_X(f)] = \lim_{T \to \infty} \frac{1}{T} \int_{-\infty}^{+\infty} e^{+j2\pi ft} \int_{-T/2}^{T/2} \int_{-T/2}^{T/2} R_X(s, t) e^{-j2\pi f(s-t)} \, dt \, ds \, df$$

$$= \lim_{T \to \infty} \frac{1}{T} \int_{-T/2}^{T/2} \int_{-T/2}^{T/2} R_X(s, t) \, dt \, ds \int_{-\infty}^{+\infty} e^{j2\pi f[\tau - (s-t)]} \, df$$

Using the fact that $\mathcal{F}^{-1}[1] = \delta(t)$, we have

$$\int_{-\infty}^{+\infty} e^{j2\pi f[\tau - (s-t)]} \, df = \delta(\tau - s + t)$$

Substituting this result in the inverse Fourier transform, we obtain

$$\mathcal{F}^{-1}[S_X(f)] = \lim_{T \to \infty} \frac{1}{T} \int_{-T/2}^{T/2} dt \int_{-T/2}^{T/2} R_X(s, t) \delta(\tau - s + t) \, ds$$

But,

$$\int_{-T/2}^{T/2} R_X(s, t) \delta(\tau - s + t) \, ds = \begin{cases} R_X(t+\tau, t), & -T/2 < t+\tau < T/2 \\ 0, & \text{otherwise} \end{cases}$$

Hence,

$$\mathcal{F}^{-1}[S_X(f)] = \lim_{T \to \infty} \frac{1}{T} \int_{-T/2}^{T/2} \begin{cases} R_X(t+\tau, t), & -T/2 < t+\tau < T/2 \\ 0, & \text{otherwise} \end{cases} \, dt$$

This can be written as

$$\mathcal{F}^{-1}[S_X(f)] = \begin{cases} \lim_{T \to \infty} \frac{1}{T} \int_{-T/2}^{T/2-\tau} R_X(t+\tau, t) \, dt, & \tau > 0 \\ \lim_{T \to \infty} \frac{1}{T} \int_{-T/2-\tau}^{T/2} R_X(t+\tau, t) \, dt, & \tau < 0 \end{cases}$$

or

$$\mathcal{F}^{-1}[S_X(f)] = \begin{cases} \lim_{T \to \infty} \frac{1}{T} \left[\int_{-T/2}^{T/2} R_X(t+\tau, t) \, dt - \int_{T/2-\tau}^{T/2} R_X(t+\tau, t) \, dt \right], \\ \hspace{10cm} \tau > 0 \\ \lim_{T \to \infty} \frac{1}{T} \left[\int_{-T/2}^{T/2} R_X(t+\tau, t) \, dt - \int_{-T/2}^{-T/2-\tau} R_X(t+\tau, t) \, dt \right], \\ \hspace{10cm} \tau < 0 \end{cases}$$

Since $\left| \int_{\mathcal{A}} R_X(t+\tau, t) \, dt \right| < \infty$ for all $\mathcal{A}$ of length τ, as $T \to \infty$ the second terms in the above brackets make no contribution. Therefore, by taking the Fourier transform of both sides we have

$$S_X(f) = \mathcal{F} \left[\lim_{T \to \infty} \frac{1}{T} \int_{-T/2}^{T/2} R_X(t+\tau, t) \, dt \right] \tag{3.3.4}$$

This concludes the proof of the Wiener-Khinchin theorem.

The following corollaries are immediate results of the Wiener-Khinchin theorem.

Corollary. If $X(t)$ is a stationary process in which $\tau R_X(\tau)$ remains finite for all finite τ, then

$$S_X(f) = \mathcal{F}[R_X(\tau)] \tag{3.3.5}$$

Proof. In this case, the condition of the preceding theorem is satisfied and

$$\langle R_X(t+\tau, t)\rangle = \lim_{T\to\infty} \frac{1}{T}\int_{-T/2}^{T/2} R_X(\tau)\,dt = R_X(\tau)$$

Applying the Wiener-Khinchin theorem gives the desired result.

Corollary In a cyclostationary process, if

$$\left|\int_0^{T_0} R_X(t+\tau, t)\,dt\right| < \infty$$

then the condition of the Wiener-Khinchin theorem is satisfied (why?) and the power-spectral density is obtained from by

$$S_X(f) = \mathcal{F}[\bar{R}_X(\tau)] \tag{3.3.6}$$

where

$$\bar{R}_X(\tau) = \frac{1}{T_0}\int_{-T_0/2}^{T_0/2} R_X(t+\tau, t)\,dt \tag{3.3.7}$$

and T_0 is the period of the cyclostationary process.

Proof. The proof is immediate by noting that for a cyclostationary process

$$\lim_{T\to\infty} \frac{1}{T}\int_{-T/2}^{T/2} R_X(t+\tau, t)\,dt = \frac{1}{T_0}\int_{-T_0/2}^{T_0/2} R_X(t+\tau, t)\,dt$$

Example 3.3.2

It was shown in Example 3.2.11, that if $X(t)$ is stationary, then $Y(t) = X(t)\cos(2\pi f_0 t)$ is a cyclostationary process with

$$R_Y(t+\tau, t) = R_X(\tau)\left[\frac{1}{2}\cos(2\pi f_0 \tau) + \frac{1}{2}\cos(4\pi f_0 t + 2\pi f_0 \tau)\right] \tag{3.3.8}$$

Using the above corollary, we have

$$\bar{R}_Y(\tau) = \frac{1}{T_0}\int_{-T_0/2}^{T_0/2} R_Y(t+\tau, t)\,dt$$

$$= \frac{R_X(\tau)}{2}\cos(2\pi f_0 \tau) \tag{3.3.9}$$

Therefore,

$$S_Y(f) = \mathcal{F}\left[\tilde{R}_Y(\tau)\right]$$

$$= \frac{1}{4}\left[S_X(f + f_0) + S_X(f - f_0)\right] \tag{3.3.10}$$

Based on the Wiener-Khinchin theorem, the following observations are in order:

- To find the total power in the process it is enough to integrate the power-spectral density

$$P_X = \int_{-\infty}^{+\infty} S_X(f)\,df$$

$$= \left. \langle R_X(t + \tau, t) \rangle \right|_{\tau=0}$$

$$= \left. \lim_{T \to \infty} \frac{1}{T} \int_{-T/2}^{T/2} E[X(t + \tau)X(t)]\,dt \right|_{\tau=o}$$

$$= E\left[\lim_{T \to \infty} \frac{1}{T} \int_{-T/2}^{T/2} X^2(t)\,dt \right]$$

which is exactly the power content of a random process as defined before.

- For stationary and ergodic processes, each sample function $x(t; \omega_i)$ is a deterministic function for which we can define the power-spectral density as given in Section 2.4. To this end, we define the power-spectral density as the Fourier transform of the (time-average) autocorrelation function of $x(t; \omega_i)$ defined by

$$R_{x(t;\omega_i)}(\tau) = \lim_{T \to \infty} \frac{1}{T} \int_{-T/2}^{T/2} x(t; \omega_i)x(t - \tau; \omega_i)\,dt$$

and

$$S_{x(t;\omega_i)} = \mathcal{F}\left[R_{x(t;\omega_i)}(\tau)\right]$$

Now, since the process is ergodic, the time average in the equation defining $R_{x(t;\omega_i)}(\tau)$ is equal to the corresponding statistical average for all samples $x(t; \omega_i)$, namely,

$$\lim_{T \to \infty} \frac{1}{T} \int_{-T/2}^{T/2} x(t; \omega_i)x(t - \tau; \omega_i)\,dt = E[X(t)X(t - \tau)] = R_X(\tau)$$

and, therefore,

$$S_X(f) = \mathcal{F}[R_X(\tau)] = S_{x(t;\omega_i)}$$

This shows that *for stationary and ergodic processes the power-spectral density of each sample function is equal to the power-spectral density of the process.* In other words, each sample function is a good representative of the spectral characteristics of the entire process. Obviously, the power content of the process is also equal to the power content of each sample function.

- The power-spectral density was originally defined as

$$\lim_{T \to \infty} \frac{E[|X_T(f)|^2]}{T}$$

From this definition, it is obvious that the power-spectral density is always a real, nonnegative, and even function of f. The fact that the power-spectral density is real and even is a direct consequence of the autocorrelation function being real and even, but nonnegativeness of the power-spectral density is not a direct consequence of the properties of the autocorrelation function that we have studied so far.[†] Therefore, we can add this property to the already mentioned properties of the autocorrelation function of a stationary process as given in Section 3.2.3.

Example 3.3.3

For the stationary and ergodic random process in Example 3.2.3, we had

$$R_X(\tau) = \frac{A^2}{2} \cos(2\pi f_0 \tau)$$

Hence,

$$S_X(f) = \frac{A^2}{4} [\delta(f - f_0) + \delta(f + f_0)]$$

The power spectral density is shown in Figure 3.17. All the power content of the process is located at f_0 and $-f_0$, and this is expected because the sample functions of this process are sinusoidals with their power at those frequencies.

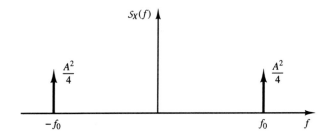

FIGURE 3.17. Power-spectral density of the random process of Example 3.2.3.

[†]The fact that $S_x(f)$ is nonnegative is a consequence of the fact that $R_X(\tau)$ is *positive semidefinite* in the sense that for any signal $g(t)$, we have $\int_{-\infty}^{+\infty} \int_{-\infty}^{+\infty} g(t) R_X(t - s) g(s) \, dt \, ds \geq 0$.

Example 3.3.4

The process of Example 3.2.4 is stationary but not ergodic. In this case,

$$R_X(\tau) = E[X(t+\tau)X(t)] = E[X^2] = \frac{1}{3}$$

Hence,

$$S_X(f) = \mathcal{F}[\frac{1}{3}] = \frac{1}{3}\delta(f)$$

as was shown before without using the Wiener-Khinchin theorem. It is obvious that, in this case, for each realization of the process, we have a different power spectrum.

Example 3.3.5

The process given in Example 3.2.11 is cyclostationary with

$$R_X(t+\tau, t) = \frac{1}{6}[\cos(2\pi f_0(2t+\tau)) + \cos(2\pi f_0\tau)]$$

from which we obtain

$$\bar{R}_X(\tau) = \frac{1}{T_0}\int_{-T_0/2}^{T_0/2} R_X(t+\tau, t)\, dt = \frac{1}{6}\cos(2\pi f_0\tau)$$

Therefore,

$$S_X(f) = \frac{1}{12}[\delta(f-f_0) + \delta(f+f_0)]$$

3.3.2 Passage through LTI Systems

We have already seen that, when a stationary random process with mean m_x and autocorrelation function $R_X(\tau)$ passes through a linear-time invariant system with impulse response $h(t)$, the output process will be also stationary with mean

$$m_Y = m_X \int_{-\infty}^{+\infty} h(t)\, dt$$

and autocorrelation

$$R_Y(\tau) = R_X(\tau) \star h(\tau) \star h(-\tau)$$

We have also seen that $X(t)$ and $Y(t)$ will be jointly stationary with crosscorrelation function

$$R_{XY}(\tau) = R_X(\tau) \star h(-\tau)$$

Translation of these relations into the frequency domain is straightforward. By noting that $\mathcal{F}[h(-\tau)] = H^*(f)$ and $\int_{-\infty}^{+\infty} h(t)\, dt = H(0)$, we have

$$m_Y = m_X H(0) \tag{3.3.11}$$

$$S_Y(f) = S_X(f)|H(f)|^2 \tag{3.3.12}$$

The first equation above (3.3.11) says that, since the mean of a random process is basically its dc value, the mean value of the response of the system only depends on the value of $H(f)$ at $f = 0$ (dc response). The second equation (3.3.12) says that, when dealing with the power spectrum, the phase of $H(f)$ is irrelevant; only magnitude of $H(f)$ affects the output power spectrum. This is also intuitive because power depends on the amplitude, and not the phase, of the signal.

We can also define a frequency domain relation for the autocorrelation function. Let us define the *cross-spectral density* $S_{XY}(f)$ as

$$S_{XY}(f) = \mathcal{F}[R_{XY}(\tau)] \tag{3.3.13}$$

Then

$$S_{XY}(f) = S_X(f) H^*(f) \tag{3.3.14}$$

and since $R_{YX}(\tau) = R_{XY}(-\tau)$, we have

$$S_{YX}(f) = S^*_{XY}(f) = S_X(f) H(f) \tag{3.3.15}$$

Note that, although $S_X(f)$ and $S_Y(f)$ are real nonnegative functions, $S_{XY}(f)$ and $S_{YX}(f)$ can, in general, be complex functions. Figure 3.18 shows how the above quantities are related.

Example 3.3.6

If the process in Example 3.2.3 passes through a differentiator, we have $H(f) = j2\pi f$ and, therefore,

$$S_Y(f) = 4\pi^2 f^2 \left[\frac{A^2}{4} (\delta(f - f_0) + \delta(f + f_0)) \right] = A^2 \pi^2 f_0^2 [\delta(f - f_0) + \delta(f + f_0)]$$

and

$$S_{XY}(f) = (-j2\pi f)S_X(f) = \frac{jA^2\pi f_0}{2} \left[\delta(f + f_0) - \delta(f - f_0) \right]$$

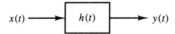

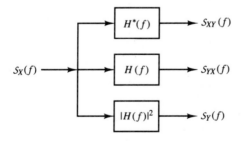

FIGURE 3.18. Input–output relations for power-spectral density and cross-spectral density.

Example 3.3.7

Passing the process in Example 3.2.4 through a differentiator results in

$$S_Y(f) = 4\pi^2 f^2 (\frac{1}{3}\delta(f)) = 0$$

$$S_{XY}(f) = (-j2\pi f)(\frac{1}{3}\delta(f)) = 0$$

These results are intuitive because the sample functions of this process are constant and differentiating them results in zero output.

Power-spectral density of a sum process.   In practice, we often encounter the sum of two random processes. For example, in the case of communication over a channel with additive noise, the noise process is added to the signal process. Below, we determine the power-spectral density for the sum of two jointly stationary processes.

Let us assume that $Z(t) = X(t) + Y(t)$ where $X(t)$ and $Y(t)$ are jointly stationary random processes. We already know that $Z(t)$ is a stationary process with

$$R_Z(\tau) = R_X(\tau) + R_Y(\tau) + R_{XY}(\tau) + R_{YX}(\tau) \qquad (3.3.16)$$

Taking the Fourier transform of both sides of this equation and using the result of Problem 3.50, we obtain

$$S_Z(f) = S_X(f) + S_Y(f) + S_{XY}(f) + \underbrace{S_{YX}(f)}_{S_{XY}^*(f)}$$

$$= S_X(f) + S_Y(f) + 2\,\mathrm{Re}\,[S_{XY}(f)] \qquad (3.3.17)$$

The above relation shows that the power-spectral density of the sum process is the sum of the power spectra of the individual processes plus a third term that depends on the correlation between the two processes.

If the two processes $X(t)$ and $Y(t)$ are uncorrelated, then

$$R_{XY}(\tau) = m_X m_Y$$

Now, if at least one of the processes is zero mean, we will have $R_{XY}(\tau) = 0$ and

$$S_Z(f) = S_X(f) + S_Y(f) \qquad (3.3.18)$$

Example 3.3.8

Let $X(t)$ represent the process in Example 3.2.3 and let $Z(t) = X(t) + \frac{d}{dt} X(t)$. Then

$$S_{XY}(f) = \frac{j A^2 \pi f_0}{2} \left[\delta(f + f_0) - \delta(f - f_0)\right]$$

and, therefore,

$$\mathrm{Re}[S_{XY}(f)] = 0$$

Hence,

$$S_Z(f) = S_X(f) + S_Y(f) = A^2(\frac{1}{4} + \pi^2 f_0^2)[\delta(f - f_0) + \delta(f + f_0)]$$

3.4 GAUSSIAN AND WHITE PROCESSES

Gaussian processes play an important role in communication systems. The fundamental reason for their importance is that thermal noise in electronic devices, which is produced by random movement of electrons due to thermal agitation, can be closely modeled by a Gaussian process. The reason for the Gaussian behavior of thermal noise is that the current introduced by movement of electrons in an electric circuit can be regarded as the sum of small currents of a very large number of sources, namely individual electrons. It can be assumed that at least a majority of these sources behave independently and, therefore, the total current is the sum of a large number of i.i.d. random variables. Now by applying the central limit theorem, this total current has a Gaussian distribution.

Apart from thermal noise, Gaussian processes provide rather good models for some information sources as well. Some interesting properties of the Gaussian processes, which will be discussed in this section, make these processes mathematically tractable and easy to deal with.

3.4.1 Gaussian Processes

We start our discussion with a formal definition of Gaussian processes.

Definition 3.4.1.   A random process $X(t)$ is a *Gaussian process* if for all n and all $(t_1, t_2, \ldots, t_n)$, the random variables $\{X(t_i)\}_{i=1}^n$ have a jointly Gaussian density function.

From the above definition it is seen that, in particular, at any time instant t_0, the random variable $X(t_0)$ is Gaussian, and at any two points t_1, t_2, random variables $(X(t_1), X(t_2))$ are distributed according to a two-dimensional Gaussian random variable. Moreover, since a complete statistical description of $\{X(t_i)\}_{i=1}^n$ depends only on **m** and **C**, the mean and autocorrelation matrices, we have the following theorem.

Theorem 3.4.1.   For Gaussian processes, knowledge of the mean and autocorrelation, i.e., $m_X(t)$ and $R_X(t_1, t_2)$ gives a complete statistical description of the process.

The following theorem is of fundamental importance in dealing with Gaussian processes.

Theorem 3.4.2.   If the Gaussian process $X(t)$ is passed through an LTI system, then the output process $Y(t)$ also will be a Gaussian process.

Proof.   To prove that $Y(t)$ is Gaussian, we have to prove that for all n and all $\{t_i\}_{i=1}^n$, the vector $(Y(t_1), Y(t_2), \ldots, Y(t_n))$ is a Gaussian vector. In general, we have

$$Y(t_i) = \int_{-\infty}^{+\infty} X(\tau)h(t_i - \tau)\, d\tau = \lim_{N \to \infty} \lim_{\Delta \to 0} \sum_{j=-N}^{j=N} X(j\Delta)h(t_i - j\Delta)\Delta$$

Hence,

$$\begin{cases} Y(t_1) & = & \lim_{N\to\infty} \lim_{\Delta\to 0} \sum_{j=-N}^{j=N} X(j\Delta)h(t_1 - j\Delta)\Delta \\ Y(t_2) & = & \lim_{N\to\infty} \lim_{\Delta\to 0} \sum_{j=-N}^{j=N} X(j\Delta)h(t_2 - j\Delta)\Delta \\ & \vdots & \\ Y(t_n) & = & \lim_{N\to\infty} \lim_{\Delta\to 0} \sum_{j=-N}^{j=N} X(j\Delta)h(t_n - j\Delta)\Delta \end{cases}$$

Now, since $\{X(j\Delta)\}_{j=-N}^N$ is a Gaussian vector and random variables $(Y(t_1), Y(t_2), \ldots, Y(t_n))$ are linear combinations of random variables $\{X(j\Delta)\}_{j=-N}^N$, we conclude that they are also jointly Gaussian.

This theorem is a very important result and demonstrates one of the nice properties of Gaussian processes that makes them attractive. For a non-Gaussian process, knowledge of the statistical properties of the input process does not easily lead to the statistical properties of the output process. For Gaussian processes, we know that the output process of an LTI system will also be Gaussian. Hence, a complete statistical description of the output process requires only knowledge of the mean and autocorrelation functions of it. Therefore, it only remains to find the mean and the autocorrelation function of the output process and, as we have already seen in Section 3.2.4, this is an easy task.

Note that the above results hold for all Gaussian processes regardless of stationarity. Since a complete statistical description of Gaussian processes depends only on $m_X(t)$ and $R_X(t_1, t_2)$, we have also the following theorem.

Theorem 3.4.3.   For Gaussian processes, WSS and strict stationarity are equivalent.

We also state the following theorem without proof. This theorem gives sufficient conditions for the ergodicity of zero-mean stationary Gaussian processes. For a proof, see Wong and Hajek (1985).

Theorem 3.4.4.   A sufficient condition for the ergodicity of the stationary zero-mean Gaussian process $X(t)$ is that

$$\int_{-\infty}^{+\infty} |R_X(\tau)|\, d\tau < \infty$$

Parallel to the definition of jointly Gaussian random variables, we can define *jointly Gaussian random processes.*

Definition 3.4.2. The random processes $X(t)$ and $Y(t)$ are *jointly Gaussian* if for all n, m and all $(t_1, t_2, \ldots, t_n)$ and $(\tau_1, \tau_2, \ldots, \tau_m)$, the random vector $(X(t_1), X(t_2), \ldots, X(t_n), Y(\tau_1)Y(\tau_2), \ldots, Y(\tau_m))$ is distributed according to an $n+m$ dimensional jointly Gaussian distribution.

We have also the following important theorem:

Theorem 3.4.5. For jointly Gaussian processes, uncorrelatedness and independence are equivalent.

Proof. This is also a straightforward consequence of the basic properties of Gaussian random variables as outlined in our discussion of jointly Gaussian random variables.

3.4.2 White Processes

The term *white process* is used to denote the processes in which all frequency components appear with equal power, i.e., the power-spectral density is constant for all frequencies. This parallels the notion of "white light" in which all colors exist.

Definition 3.4.3. A process $X(t)$ is called a *white process* if it has a flat spectral density, i.e., if $S_X(f)$ is a constant for all f.

The importance of white processes in practice stems from the fact that thermal noise can be closely modeled as a white process over a wide range of frequencies. Also a wide range of processes used to describe a variety of information sources can be modeled as the output of LTI systems driven by a white process. Figure 3.19 shows the power spectral density of a white process.

If we find the power content of a white process using $S_X(f) = C$, a constant, we will have

$$P_X = \int_{-\infty}^{+\infty} S_X(f)\,df = \int_{-\infty}^{+\infty} C\,df = \infty$$

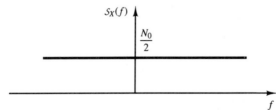

FIGURE 3.19. Power spectral density of a white process.

Obviously, no real physical process can have infinite power, and, therefore, a white process is not a meaningful physical process. However, quantum mechanical analysis of the thermal noise shows that it has a power-spectral density given by

$$S_n(f) = \frac{\hbar f}{2(e^{\frac{\hbar f}{kT}} - 1)} \qquad (3.4.1)$$

in which $\hbar$ denotes *Planck's constant* (equal to 6.6×10^{-34} Joules $\times$ sec) and k is *Boltzmann's constant* (equal to 1.38×10^{-23} Joules/Kelvin). T denotes the temperature in degrees Kelvin. This power spectrum is shown in Figure 3.20.

The above spectrum achieves its maximum at $f = 0$ and the value of this maximum is $\frac{kT}{2}$. The spectrum goes to zero as f goes to infinity, but the rate of convergence to zero is very slow. For instance, at room temperature ($T = 300°$K) $S_n(f)$ drops to 90% of its maximum at about $f \approx 2 \times 10^{12}$Hz, which is beyond the frequencies employed in conventional communication systems. From this we conclude that thermal noise, though not precisely white, for all practical purposes can be modeled as a white process with the power spectrum equaling $\frac{kT}{2}$. The value kT is usually denoted by N_0 and, therefore, the power-spectral density of thermal noise is usually given as $S_n(f) = \frac{N_0}{2}$, and sometimes referred to as the *two-sided power-spectral density,* emphasizing that this spectrum extends to both positive and negative frequencies. We will avoid this terminology throughout and simply use *power spectrum* or *power-spectral density.*

Looking at the autocorrelation function for a white process, we see that

$$R_n(\tau) = \mathcal{F}^{-1}\left[\frac{N_0}{2}\right] = \frac{N_0}{2}\delta(\tau)$$

This shows that for all $\tau \neq 0$, we have $R_X(\tau) = 0$, i.e., if we sample a white process at two points t_1 and t_2 ($t_1 \neq t_2$), the resulting random variables will be uncorrelated. If, in addition to being white, the random process is also Gaussian, the sampled random variables will also be independent.

In short, the thermal noise that we will use in subsequent chapters is assumed to be a stationary, ergodic, zero mean, white Gaussian process whose power spectrum is $N_0/2 = kT/2$.

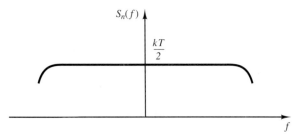

FIGURE 3.20.   Power spectrum of thermal noise.

Noise-equivalent bandwidth.   When white Gaussian noise passes through a filter, the output process, although still Gaussian, will not be white anymore. The filter characteristics determine the spectral properties of the output process, and we have

$$S_Y(f) = S_X(f)|H(f)|^2 = \frac{N_0}{2}|H(f)|^2$$

Now, if we want to find the power content of the output process, we have to integrate $S_Y(f)$. Thus,

$$P_Y = \int_{-\infty}^{+\infty} S_Y(f)\,df = \frac{N_0}{2}\int_{-\infty}^{+\infty}|H(f)|^2\,df$$

Therefore, to determine the output power, we have to evaluate the integral $\int_{-\infty}^{+\infty}|H(f)|^2\,df$. To do this calculation, we define B_{neq}, the *noise-equivalent bandwidth* of a filter with frequency response $H(f)$, as

$$B_{neq} = \frac{\int_{-\infty}^{+\infty}|H(f)|^2\,df}{2\,H_{max}^2} \tag{3.4.2}$$

where H_{max} denotes the maximum of $|H(f)|$ in the passband of the filter. Figure 3.21 shows H_{max} and B_{neq} for a lowpass filter.

Using the above definition, we have

$$P_Y = \frac{N_0}{2}\int_{-\infty}^{+\infty}|H(f)|^2\,df$$

$$= \frac{N_0}{2}\times 2B_{neq}\,H_{max}^2$$

$$= N_0\,B_{neq}\,H_{max}^2 \tag{3.4.3}$$

Therefore, by having B_{neq}, finding the output noise power becomes a simple task. The noise-equivalent bandwidth of filters and amplifiers is usually provided by the manufacturer.

Example 3.4.1

Find the noise-equivalent bandwidth of a lowpass RC filter.

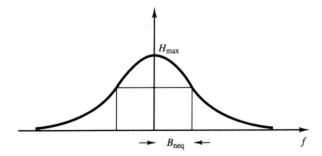

FIGURE 3.21. Noise-equivalent bandwidth of a lowpass filter.

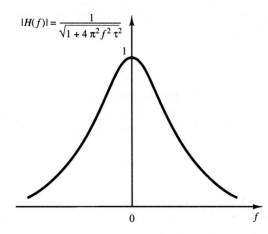

FIGURE 3.22. Frequency response of a lowpass RC filter.

Solution $H(f)$ for this filter is

$$H(f) = \frac{1}{1 + j2\pi f RC}$$

and is shown in Figure 3.22. Defining $\tau = RC$, we have

$$|H(f)| = \frac{1}{\sqrt{1 + 4\pi^2 f^2 \tau^2}}$$

and therefore $H_{\max} = 1$. We also have

$$\int_{-\infty}^{+\infty} |H(f)|^2 \, df = 2 \int_0^{\infty} \frac{1}{1 + 4\pi^2 f^2 \tau^2} \, df$$

$$\overset{u = 2\pi f \tau}{=} 2 \int_0^{\infty} \frac{1}{1 + u^2} \times \frac{du}{2\pi \tau}$$

$$= \frac{1}{\pi \tau} \times \frac{\pi}{2}$$

$$= \frac{1}{2\tau}$$

Hence,

$$B_{neq} = \frac{\frac{1}{2\tau}}{2 \times 1} = \frac{1}{4RC}$$

3.5 BANDLIMITED PROCESSES AND SAMPLING

A bandlimited process is a random process whose power-spectral density occupies a finite bandwidth. In other words, a bandlimited process is a process with the property that for all $|f| \geq W$, we have $S_X(f) \equiv 0$ where W is the bandwidth of the process.

Almost all of the processes encountered in nature are bandlimited because there is a limit on the bandwidth of all physical systems. Bandlimited processes also arise when passing random processes through bandlimited linear systems. The output, usually called a *filtered process,* is a bandlimited process. The power-spectral density of a typical bandlimited process is shown in Figure 3.23.

We have already seen in Section 2.5 that for bandlimited signals there exists the powerful sampling theorem that states that the signal can be perfectly reconstructed from its sampled values as long as the sampling rate is more than twice the highest frequency component in the signal, i.e., the bandlimited signal can be expressed in terms of its samples taken at regular intervals T_s, where $T_s \leq \frac{1}{2W}$, by the relation

$$x(t) = \sum_{k=-\infty}^{\infty} 2WT_s\, x(kT_s)\, \text{sinc}\left(2W(t - kT_s)\right)$$

In the special case where $T_s = \frac{1}{2W}$, the above relation simplifies to

$$x(t) = \sum_{k=-\infty}^{\infty} x(\frac{k}{2W})\, \text{sinc}(2W(t - \frac{k}{2W}))$$

One wonders if such a relation exists for bandlimited random processes. In other words, is it possible to express a bandlimited process in terms of its sampled values? The following theorem, which is the sampling theorem for the bandlimited random processes, shows that this is in fact true.

Theorem 3.5.1   Let $X(t)$ be a stationary bandlimited process, i.e., $S_X(f) \equiv 0$ for $|f| \geq W$. Then the following relation holds.

$$E\left|X(t) - \sum_{k=-\infty}^{\infty} X(kT_s)\, \text{sinc}(2W(t - kT_s))\right|^2 = 0 \qquad (3.5.1)$$

where $T_s = \frac{1}{2W}$ denotes the sampling interval.

Proof.   Let us start by expanding the above relation. The left-hand side becomes

$$E\left|X(t) - \sum_{k=-\infty}^{\infty} X(kT_s)\, \text{sinc}(2W(t - kT_s))\right|^2$$

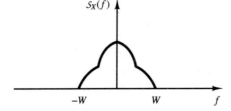

FIGURE 3.23.   Power spectrum of a bandlimited process.

$$= R_X(0) - 2 \sum_{k=-\infty}^{\infty} R_X(t - kT_s) \operatorname{sinc}(2W(t - kT_s))$$

$$+ \sum_{k=-\infty}^{\infty} \sum_{l=-\infty}^{\infty} R_X((k - l)T_s) \operatorname{sinc}(2W(t - kT_s)) \operatorname{sinc}(2W(t - lT_s))$$

Introducing the change of variable $m = l - k$ in the last line of the above relation, we have

$$\sum_{k=-\infty}^{\infty} \sum_{l=-\infty}^{\infty} R_X((k - l)T_s) \operatorname{sinc}(2W(t - kT_s)) \operatorname{sinc}(2W(t - lT_s))$$

$$= \sum_{k=-\infty}^{\infty} \sum_{m=-\infty}^{\infty} R_X(-mT_s) \operatorname{sinc}(2W(t - kT_s)) \operatorname{sinc}(2W(t - kT_s - mT_s))$$

$$= \sum_{k=-\infty}^{\infty} \operatorname{sinc}(2W(t - kT_s)) \sum_{m=-\infty}^{\infty} R_X(mT_s) \operatorname{sinc}(2W(t - kT_s - mT_s))$$

where we have used the fact that $R_X(-mT_s) = R_X(mT_s)$.

The fact that the process is bandlimited means that the Fourier transform of $R_X(\tau)$ is bandlimited to W and, therefore, for $R_X(\tau)$ we have the expansion

$$R_X(t) = \sum_{k=-\infty}^{\infty} R_X(kT_s) \operatorname{sinc}[2W(t - kT_s)]$$

from which we obtain

$$\sum_{m=-\infty}^{\infty} R_X(mT_s) \operatorname{sinc}[2W(t - kT_s - mT_s)] = R_X(t - kT_s)$$

Therefore,

$$E \left| X(t) - \sum_{k=-\infty}^{\infty} X(kT_s) \operatorname{sinc}[2W(t - kT_s)] \right|^2$$

$$= R_X(0) - 2 \sum_{k=-\infty}^{\infty} R_X(t - kT_s) \operatorname{sinc}[2W(t - kT_s)] +$$

$$+ \sum_{k=-\infty}^{\infty} R_X(t - kT_s) \operatorname{sinc}[2W(t - kT_s)] =$$

$$= R_X(0) - \sum_{k=-\infty}^{\infty} R_X(t - kT_s) \operatorname{sinc}[2W(t - kT_s)]$$

Now we can apply the result of Example 2.5.1 to $R_X(\tau)$ to obtain

$$R_X(0) = \sum_{k=-\infty}^{\infty} R_X(t - kT_s) \operatorname{sinc}[2W(t - kT_s)]$$

Substituting this result in the expression for $E[X(t)]$, we obtain

$$E \left| X(t) - \sum_{k=-\infty}^{\infty} X(kT_s) \operatorname{sinc}[2W(t - kT_s)] \right|^2 = 0$$

This concludes the proof of the theorem.

This result is parallel to the sampling theorem for deterministic signals developed in Section 2.5. Note that, due to the random nature of the entities involved in this case, the equality of $X(t)$ and $\sum_{k=-\infty}^{\infty} X(kT_s) \operatorname{sinc}[2W(t - kT_s)]$ is not pointwise, and we can only say that

$$E \left| X(t) - \sum_{k=-\infty}^{\infty} X(kT_s) \operatorname{sinc}[2W(t - kT_s)] \right|^2 = 0$$

This is usually called *equality in quadratic mean* or *equality in the mean-squared sense* and denoted by

$$X(t) \stackrel{\text{q.m.}}{=} \sum_{k=-\infty}^{\infty} X(kT_s) \operatorname{sinc}[2W(t - kT_s)] \qquad (3.5.2)$$

Now that we have seen that a bandlimited process can be recovered from its sampled values taken at $\frac{1}{2W}$ intervals, an interesting question is whether or not these samples are uncorrelated. It can be shown that a necessary and sufficient condition for the uncorrelatedness of these samples is that the power spectrum of the process be flat over the passband, i.e.,

$$S_X(f) = \begin{cases} K & |f| < W \\ 0, & \text{otherwise} \end{cases}$$

Unless this condition is satisfied, the samples of the process will be correlated, and this correlation can be exploited to make their transmission easier. This fact will be further explored in Chapter 4.

3.6 BANDPASS PROCESSES

Bandpass random processes are the equivalents of bandpass deterministic signals. A bandpass process has all its power located in the neighborhood of some central frequency f_0.

Definition 3.6.1. $X(t)$ is a *bandpass* or *narrowband* process if $S_X(f) \equiv 0$ for $|f - f_0| \geq W$ where $W < f_0$.

Although the term *narrowband process* is usually used for the case where $W \ll f_0$, we will use both terms *bandpass process* and *narrowband process* interchangeably. Figure 3.24 shows the power spectrum of a bandpass process. Note that, as in Section 2.6, there is no need for f_0 to be the mid-band frequency, or, in general, there is even no need for f_0 to be in the frequency band of the process.

Bandpass processes are suitable for modeling modulated signals. A random process is usually modulated on a carrier for transmission over a communication channel and the resulting process is a bandpass process. The noise that passes through a bandpass filter at the front end of a receiver is also a bandpass process.

As we have seen in Section 2.6, there are certain ways to express bandpass signals in terms of equivalent lowpass signals. The purpose of this section is to generalize those results to the case of random processes. As we will see, many of the results of Section 2.6 can be generalized in a straightforward manner to bandpass processes.

Let $X(t)$ be a bandpass process as defined above. Then $R_X(\tau)$ is a deterministic bandpass signal whose Fourier transform $S_X(f)$ is nonzero in the neighborhood of f_0. If $X(t)$ is passed through a quadrature filter with impulse response $\frac{1}{\pi t}$ and transfer function $H(f) = -j \, \text{sgn}(f)$, the output process is the Hilbert transform of the input process, and according to Example 3.2.20, we have

$$R_{X\hat{X}}(\tau) = -\hat{R}_X(\tau) \qquad\qquad (3.6.1)$$

$$R_{\hat{X}}(\tau) = R_X(\tau) \qquad\qquad (3.6.2)$$

Now, parallel to the deterministic case, let us define two new processes $X_c(t)$ and $X_s(t)$ by

$$X_c(t) = X(t)\cos(2\pi f_0 t) + \hat{X}(t)\sin(2\pi f_0 t) \qquad\qquad (3.6.3)$$

$$X_s(t) = \hat{X}(t)\cos(2\pi f_0 t) - X(t)\sin(2\pi f_0 t) \qquad\qquad (3.6.4)$$

As in the deterministic case, $X_c(t)$ and $X_s(t)$ are called *in-phase* and *quadrature* components of the process $X(t)$. Throughout the rest of this chapter, we will assume that the process $X(t)$ is stationary with zero mean and, based on this assumption, will explore the properties of its in-phase and quadrature components.

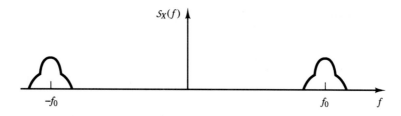

FIGURE 3.24.   Power spectrum of a bandpass process.

Theorem 3.6.1. If $X(t)$ is a zero-mean stationary bandpass process, the processes $X_c(t)$ and $X_s(t)$ will be a zero-mean jointly stationary process.

Proof. The fact that both $X_c(t)$ and $X_s(t)$ are zero mean is trivial. It remains to be proven that $R_{X_c}(t+\tau, t)$, $R_{X_s}(t+\tau, t)$, and $R_{X_c X_s}(t+\tau, t)$ are functions of τ only. We have

$$R_{X_c}(t+\tau, t)$$

$$= E[X_c(t+\tau)X_c(t)]$$

$$= E\left[\left(X(t+\tau)\cos(2\pi f_0(t+\tau)) + \hat{X}(t+\tau)\sin(2\pi f_0(t+\tau))\right)\right.$$

$$\left. \times \left(X(t)\cos(2\pi f_0 t) + \hat{X}(t)\sin(2\pi f_0 t)\right)\right]$$

$$= R_X(\tau)\cos(2\pi f_0 t)\cos(2\pi f_0(t+\tau)) - \hat{R}_X(\tau)\sin(2\pi f_0 t)\cos(2\pi f_0(t+\tau))$$

$$\quad + \hat{R}_X(\tau)\cos(2\pi f_0 t)\sin(2\pi f_0(t+\tau)) + R_X(\tau)\sin(2\pi f_0 t)\sin(2\pi f_0(t+\tau))$$

$$= R_X(\tau)\cos(2\pi f_0\tau) + \hat{R}_X(\tau)\sin(2\pi f_0\tau)$$

where we have used the fact that $\hat{R}_X(\tau)$ is an odd function, because it is the Hilbert transform of an even function (see Theorem 2.6.1).

It can be proved similarly that

$$R_{X_s}(\tau) = R_X(\tau)\cos(2\pi f_0\tau) + \hat{R}_X(\tau)\sin(2\pi f_0\tau)$$

and

$$R_{X_c X_s}(\tau) = R_X(\tau)\sin(2\pi f_0\tau) - \hat{R}_X(\tau)\cos(2\pi f_0\tau)$$

In conclusion, we have the following relations to which we will refer frequently

$$\begin{cases} R_{X_c}(\tau) & = & R_{X_s}(\tau) = R_X(\tau)\cos(2\pi f_0\tau) + \hat{R}_X(\tau)\sin(2\pi f_0\tau) \\ R_{X_c X_s}(\tau) & = & R_X(\tau)\sin(2\pi f_0\tau) - \hat{R}_X(\tau)\cos(2\pi f_0\tau) \end{cases} \tag{3.6.5}$$

This concludes the proof of the theorem.

Theorem 3.6.2. $X_c(t)$ and $X_s(t)$ are lowpass processes, i.e., their power spectra vanishes for $|f| \geq W$.

Proof. Note that $R_{X_c}(\tau)$ and $R_{X_s}(\tau)$ are deterministic signals. Comparing

$$R_{X_c}(\tau) = R_{X_s}(\tau) = R_X(\tau)\cos(2\pi f_0\tau) + \hat{R}_X(\tau)\sin(2\pi f_0\tau)$$

with the corresponding relations for deterministic signals as developed in Section 2.6, we observe that the relations describing both $R_{X_c}(\tau)$ and $R_{X_s}(\tau)$ are exactly the same relations that describes $x_c(t)$, the in-phase component of a deterministic bandpass signal $x(t)$. This means that $R_{X_c}(\tau)$ and $R_{X_s}(\tau)$ are both lowpass signals.

This in turn means that the Fourier transform of these signals, i.e., $S_{X_c}(f)$ and $S_{X_s}(f)$ are bandlimited to $|f| < W$, and therefore the theorem is proved.

From the above theorems it is seen that the processes $X_c(t)$ and $X_s(t)$ play the same role as the in-phase and quadrature components for deterministic signals, and therefore they are called the in-phase and quadrature processes. Furthermore, we have seen that these processes are jointly stationary with the *same power-spectral density*. Parallel to the deterministic case, we can also define the envelope and phase processes by

$$V(t) = \sqrt{X_c^2(t) + X_s^2(t)} \tag{3.6.6}$$

$$\Theta(t) = \arctan \frac{X_s(t)}{X_c(t)} \tag{3.6.7}$$

To find the power spectrum of the in-phase and quadrature components, we have

$$S_{X_c}(f) = S_{X_s}(f) =$$

$$= \mathcal{F}[R_X(\tau)\cos(2\pi f_0\tau) + \hat{R}_X(\tau)\sin(2\pi f_0\tau)]$$

$$= \frac{S_X(f - f_0)}{2} + \frac{S_X(f + f_0)}{2}$$

$$+ [-j\,\mathrm{sgn}(f)\,S_X(f)] \star [\frac{\delta(f - f_0) - \delta(f + f_0)}{2j}]$$

$$= \frac{S_X(f - f_0)}{2} + \frac{S_X(f + f_0)}{2}$$

$$+ \frac{1}{2}\,\mathrm{sgn}(f + f_0)S_X(f + f_0) - \frac{1}{2}\,\mathrm{sgn}(f - f_0)S_X(f - f_0)$$

$$= \frac{S_X(f - f_0)}{2}[1 - \mathrm{sgn}(f - f_0)] + \frac{S_X(f + f_0)}{2}[1 + \mathrm{sgn}(f + f_0)]$$

$$= \begin{cases} S_X(f - f_0) & f < -f_0 \\ S_X(f - f_0) + \frac{1}{2}S_X(f + f_0) & f = -f_0 \\ S_X(f - f_0) + S_X(f + f_0) & |f| < f_0 \\ S_X(f + f_0) + \frac{1}{2}S_X(f - f_0) & f = f_0 \\ S_X(f + f_0) & f > f_0 \end{cases}$$

From the fact that the original process $X(t)$ is bandpass, it follows that

$$S_{X_c}(f) = S_{X_s}(f) = \begin{cases} S_X(f - f_0) + S_X(f + f_0) & |f| < f_0 \\ 0, & \text{otherwise} \end{cases} \tag{3.6.8}$$

This simply means that in order to obtain the power spectrum of both $X_c(t)$ and $X_s(t)$, it is enough to shift the positive-frequency part of $S_X(f)$ to the left by f_0 and to shift the negative-frequency part of it to the right by f_0 and add the result. Figure 3.25 shows the relation between $S_X(f)$, $S_{X_c}(f)$, and $S_{X_s}(f)$.

To obtain the cross spectral density, we have

$$S_{X_c X_s}(f) = \mathcal{F}[R_{X_c X_s}(\tau)]$$

$$= S_X(f) \star \frac{\delta(f - f_0) - \delta(f + f_0)}{2j}$$

$$-[-j\,\mathrm{sgn}(f)S_X(f)] \star \frac{\delta(f - f_0) + \delta(f + f_0)}{2}$$

$$= \frac{j}{2}[S_X(f + f_0) - S_X(f - f_0)]$$

$$+ \frac{j}{2}[\mathrm{sgn}(f + f_0)S_X(f + f_0) + \mathrm{sgn}(f - f_0)S_X(f - f_0)]$$

$$= \frac{j}{2}S_X(f + f_0)[1 + \mathrm{sgn}(f + f_0)] - \frac{j}{2}S_X(f - f_0)[1 - \mathrm{sgn}(f - f_0)]$$

$$= \begin{cases} -jS_X(f - f_0) & f < -f_0 \\ \frac{j}{2}S_X(f + f_0) - jS_X(f - f_0) & f = -f_0 \\ j[S_X(f + f_0) - S_X(f - f_0)] & |f| < f_0 \\ -\frac{j}{2}S_X(f - f_0) + jS_X(f + f_0) & f = f_0 \\ jS_X(f + f_0) & f > f_0 \end{cases}$$

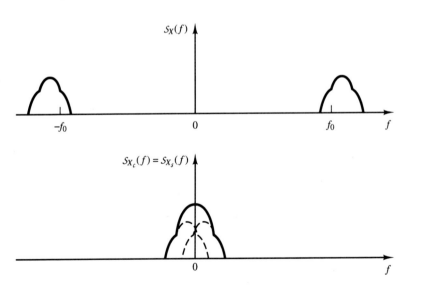

FIGURE 3.25. Power spectrum of in-phase and quadrature components.

Again, since the process is bandpass, the above relation simplifies to

$$S_{X_c X_s}(f) = \begin{cases} j[S_X(f + f_0) - S_X(f - f_0)] & |f| < f_0 \\ 0, & \text{otherwise} \end{cases} \qquad (3.6.9)$$

Figure 3.26 shows the relation between $S_{X_c X_s}(f)$ and $S_X(f)$.

Since the crosscorrelation

$$R_{X_c X_s}(\tau) = R_X(\tau) \sin(2\pi f_0 \tau) - \hat{R}_X(\tau) \cos(2\pi f_0 \tau)$$

is an odd function, it follows that $R_{X_c X_s}(0) = 0$. This means that at any time instant t_0, the random variables $X_c(t_0)$ and $X_s(t_0)$ are uncorrelated. Of course, from this we cannot conclude that the random processes $X_c(t)$ and $X_s(t)$ are uncorrelated because, in general there is no guarantee that for arbitrary t_1 and t_2 the random variables $X(t_1)$ and $X(t_2)$ are uncorrelated. However, if the symmetry condition

$$S_X(f + f_0) = S_X(f - f_0) \qquad (3.6.10)$$

is satisfied for all $|f| < f_0$, then $S_{X_c X_s}(f) \equiv 0$ and therefore $R_{X_c X_s}(\tau) \equiv 0$ for all τ. This means that $X_c(t_1)$ and $X_s(t_2)$ are uncorrelated for arbitrary t_1 and t_2, which in turn means that the random processes $X_c(t)$ and $X_s(t)$ are uncorrelated. If $S_X(f)$ happens to be symmetric around f_0, then the above condition is satisfied, and the processes $X_c(t)$ and $X_s(t)$ will be uncorrelated.

If the zero-mean, stationary, and bandpass process $X(t)$ is Gaussian as well, then $X_c(t)$ and $X_s(t)$ will be jointly Gaussian (why?). In this case, uncorrelated-ness is equivalent to independence, and, therefore, under the above conditions the processes $X_c(t)$ and $X_s(t)$ will be independent.

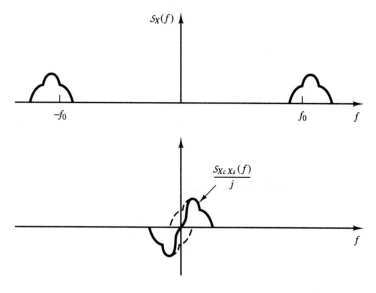

FIGURE 3.26. Cross-spectral density of in-phase and quadrature components.

Example 3.6.1

The white Gaussian noise $N(t)$ with power spectrum $\frac{N_0}{2}$ passes through an ideal bandpass filter with frequency response

$$H(f) = \begin{cases} 1 & |f - f_c| < W \\ 0, & \text{otherwise} \end{cases}$$

where $W \ll f_c$. The output process is denoted by $X(t)$. Find the power spectrum and the cross-spectral density of the in-phase and quadrature components in the following two cases:

1. f_0 is chosen to be equal to f_c.
2. f_0 is chosen to be equal to $f_c - W$.

Solution   The process $X(t)$ is obviously a bandpass process whose power-spectral density is given by

$$S_X(f) = \begin{cases} \frac{N_0}{2}, & |f - f_c| < W \\ 0, & \text{otherwise} \end{cases}$$

If we choose f_c as the central frequency, then to obtain power spectra of $X_c(t)$ and $X_s(t)$ we have to shift the positive-frequency part of $S_X(f)$ to the left by $f_0 = f_c$ and the negative-frequency part of it to the right by the same amount and add the results. If we do this, the resulting spectrum will be

$$S_{X_c}(f) = S_{X_s}(f) = \begin{cases} N_0 & |f| < W \\ 0, & \text{otherwise} \end{cases}$$

Because we are choosing f_0 to be the axis of symmetry of the spectrum of $X(t)$, the process $X_c(t)$ and $X_s(t)$ will be independent with this choice and $S_{X_c X_s}(f) = 0$. Figure 3.27 shows the power spectra in this case.

   If we choose $f_0 = f_c - W$ as the central frequency,[†] the results will be quite different. In this case, shifting the positive and negative components of $X(t)$ by f_0 to the left and right and adding them results in

$$S_{X_c}(f) = S_{X_s}(f) = \begin{cases} \frac{N_0}{2} & |f| < 2W \\ 0, & \text{otherwise} \end{cases}$$

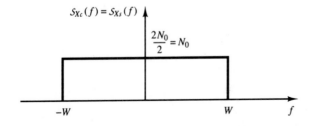

$S_{X_c}(f) = S_{X_s}(f)$

$\frac{2N_0}{2} = N_0$

$-W$ W f

FIGURE 3.27.   In-phase and quadrature power spectra in first part of Example 3.6.1.

[†]*Central frequency* does not, of course, mean the center of the bandwidth. It means the frequency with respect to which we are expanding the process.

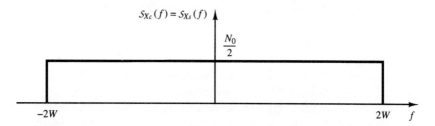

FIGURE 3.28. In-phase and quadrature components for the second part of Example 3.6.1.

and for the cross-spectral density, we will have

$$S_{X_c X_s}(f) = \begin{cases} -j\,\frac{N_0}{2} & -2W < f < 0 \\ j\,\frac{N_0}{2} & 0 < f < 2W \\ 0, & \text{otherwise} \end{cases}$$

Of course, in this case the processes $X_c(t)$ and $X_s(t)$ will not be independent. See Figure 3.28 for this case.

So far we have defined $X_c(t)$ and $X_s(t)$ in terms of $X(t)$ and $\hat{X}(t)$ by

$$\begin{cases} X_c(t) &= X(t)\cos(2\pi f_0 t) + \hat{X}(t)\sin(2\pi f_0 t) \\ X_s(t) &= \hat{X}(t)\cos(2\pi f_0 t) - X(t)\sin(2\pi f_0 t) \end{cases}$$

and have shown that $X_c(t)$ and $X_s(t)$ are jointly stationary lowpass processes. Obtaining $X(t)$ and $\hat{X}(t)$ in terms of the in-phase and quadrature components is straightforward, and we can summarize the result as

$$\begin{cases} X(t) &= X_c(t)\cos(2\pi f_0 t) - X_s(t)\sin(2\pi f_0 t) \\ \hat{X}(t) &= X_c(t)\sin(2\pi f_0 t) + X_s(t)\cos(2\pi f_0 t) \end{cases} \tag{3.6.11}$$

These relations are frequently used in subsequent chapters to express bandpass processes in terms of the corresponding lowpass processes.

The power contents of the in-phase and quadrature components of $X(t)$ can be obtained easily. From

$$R_{X_c}(\tau) = R_{X_s}(\tau) = R_X(\tau)\cos(2\pi f_0 \tau) + \hat{R}_X(\tau)\sin(2\pi f_0 \tau)$$

we have

$$P_{X_c} = P_{X_s} = R_{X_c}(\tau)|_{\tau=0} = R_X(0) = P_X \tag{3.6.12}$$

In other words, the power content of the in-phase and quadrature components is equal to the power content of the original bandpass process.

3.7 FURTHER READING

Elementary treatments of probability and random processes for engineers can be found in books by Leon-Garcia (1989) and Helstrom (1991). More extensive treatments can be found in books by Davenport and Root (1987) and Papoulis (1991). Gray and Davisson (1991) is particularly interesting because it covers random processes for electrical engineers without compromising mathematical vigor. Advanced treatment of the material in this chapter, based on measure theory concepts, can be found in the book by Wong and Hajek (1985).

PROBLEMS

3.1 A random experiment consists of drawing a ball from an urn that contains four red balls numbered 1, 2, 3, 4 and three black balls numbered 1, 2, 3. Precisely state what outcomes are contained in the following events.

    1. $E_1 =$ The number on the ball is even.

    2. $E_2 =$ The color of the ball is red and its number is greater than 1.

    3. $E_3 =$ The number on the ball is less than 3.

    4. $E_4 = E_1 \cup E_3$.

    5. $E_5 = E_1 \cup (E_2 \cap E_3)$.

3.2 If all balls in the preceding problem are equally likely to be drawn, find the probabilities of E_i, $1 \le i \le 5$.

3.3 In a certain city three car brands A, B, and C have 20%, 30%, and 50% of the market share, respectively. The probability that a car needs major repair during its first year of purchase for the three brands are 5%, 10%, and 15%, respectively.

    1. What is the probability that a car in this city needs major repair during its first year of purchase?

    2. If a car in this city needs major repair during its first year of purchase, what is the probability that it is made by manufacturer A?

3.4 Under what conditions can two disjoint events A and B be independent?

3.5 An information source produces 0 and 1 with probabilities 0.3 and 0.7, respectively. The output of the source is transmitted via a channel that has a probability of error (turning a 1 into a 0 or a 0 into a 1) equal to 0.2.

    1. What is the probability that at the output a 1 is observed?

    2. What is the probability that a 1 was the output of the source if at the output of the channel a 1 is observed?

3.6 A coin is flipped three times and the random variable X denotes the total number of heads that show up. The probability of a head in one flip of this coin is denoted by p.

    1. What values can the random variable X take?
    2. What is the p.m.f. of the random variable X?
    3. Derive and plot the c.d.f. of X.
    4. What is the probability that X exceeds 1?

3.7 Coin A has a probability of head equal to 0.25 and probability of tail equal to 0.75, and coin B is a fair coin. Each coin is flipped four times. Let the random variable X denote the number of heads resulting from coin A and Y denote the resulting number of heads from coin B.

    1. What is the probability that $X = Y = 2$?
    2. What is the probability that $X = Y$?
    3. What is the probability that $X > Y$?
    4. What is the probability that $X + Y \leq 5$?

3.8 A random variable X is defined by the c.d.f.

$$F_X(x) = \begin{cases} 0 & x < 0 \\ \dfrac{1}{2}x & 0 \leq x < 1 \\ K & x \geq 1 \end{cases}$$

    1. Find the value of K.
    2. Is this random variable discrete, continuous, or mixed?
    3. What is the probability that $\frac{1}{2} < X \leq 1$?
    4. What is the probability that $\frac{1}{2} < X < 1$?
    5. What is the probability that X exceeds 2?

3.9 Random variable X is distributed according to $f_X(x) = \Lambda(x)$.

    1. Find the c.d.f. of X.
    2. Determine $P(X > \frac{1}{2})$.
    3. Determine $P(X > 0 | X < \frac{1}{2})$.
    4. What is $f_X(x | X > \frac{1}{2})$?
    5. What is $E[X | X > \frac{1}{2}]$?

3.10 The noise voltage in an electric circuit can be modeled as a Gaussian random variable with mean equal to zero and variance equal to 10^{-8}.

1. What is the probability that the value of the noise exceeds 10^{-4}? What is the probability that it exceeds 4×10^{-4}? What is the probability that the noise value is between -2×10^{-4} and 10^{-4}?

2. Given that the value of the noise is positive, what is the probability that it exceeds 10^{-4}?

3. This noise passes through a half-wave rectifier with characteristics

$$
g(x) = \begin{cases} x & x > 0 \\ 0 & x \le 0 \end{cases}
$$

   Find the p.d.f. of the rectified noise by first finding its c.d.f. Why can we not use the general expression in (3.1.10) here?

4. Find the expected value of the rectified noise in the previous part.

5. Now assume that the noise passes through a full-wave rectifier defined by $g(x) = |x|$. Find the density function of the rectified noise in this case. What is the expected value of the output noise in this case?

3.11 X is a $\mathcal{N}(0, \sigma^2)$ random variable. This random variable is passed through a system whose input–output relation is given by $y = g(x)$. Find the p.d.f. or the p.m.f. of the output random variable Y in each of the following cases.

1. Square-law device, $g(x) = ax^2$.
2. Limiter,

$$
g(x) = \begin{cases} -b & x \le -b \\ b & x \ge b \\ x & |x| < b \end{cases}
$$

3. Hard limiter,

$$
g(x) = \begin{cases} a & x > 0 \\ 0 & x = 0 \\ b & x < 0 \end{cases}
$$

4. Quantizer, $g(x) = x_n$ for $a_n \le x < a_{n+1}$, $1 \le n \le N$, where x_n lies in the interval $[a_n, a_{n+1}]$ and the sequence $\{a_1, a_2, \ldots, a_{N+1}\}$ satisfies the conditions $a_1 = -\infty$, $a_{N+1} = \infty$ and for $i > j$ we have $a_i > a_j$.

3.12 The random variable Φ is uniformly distributed on the interval $[-\frac{\pi}{2}, \frac{\pi}{2}]$. Find the p.d.f. of $X = \tan \Phi$. Find the mean and the variance of X.

3.13 Let Y be a positive valued random variable, i.e., $f_Y(y) = 0$ for $y < 0$.

1. Let α be any positive constant; show that $P(Y > \alpha) \leq \frac{E[Y]}{\alpha}$ (Markov inequality).

2. Let X be any random variable with variance σ^2 and define $Y = (X - E[X])^2$ and $\alpha = \epsilon^2$ for some ϵ; obviously, the conditions of the problem are satisfied for Y and α as chosen here. Derive the Chebychev inequality

$$P\left(|X - E[X]| > \epsilon\right) \leq \frac{\sigma^2}{\epsilon^2}$$

3.14 Show that for a binomial random variable the mean is given by np and the variance is given by $np(1 - p)$.

3.15 Show that for a Poisson random variable defined by the p.m.f. $P(X = k) = \frac{\lambda^k}{k!} e^{-k}$ $k = 0, 1, 2, \ldots$, the characteristic function is given by $\Psi_X(v) = e^{\lambda(e^{jv} - 1)}$. Use this result to show that $E(X) = \lambda$ and $\text{VAR}(X) = \lambda$.

3.16 Let X denote a Gaussian random variable with mean equal to zero and variance equal to σ^2. Show that

$$E[X^n] = \begin{cases} 0 & n = 2k + 1 \\ 1 \times 3 \times 5 \times \cdots \times (n-1)\sigma^n & n = 2k \end{cases}$$

(Hint: Differentiate the identity $\int_{-\infty}^{+\infty} e^{-\frac{x^2}{2\sigma^2}} \, dx = \sqrt{2\pi\sigma^2}$, k times.)

3.17 Two random variables X and Y are distributed according to

$$f_{X,Y}(x, y) = \begin{cases} K(x + y) & 0 \leq x, y \leq 1 \\ 0, & \text{otherwise} \end{cases}$$

1. Find K.
2. What is the probability that $X + Y > 1$?
3. Find $P(X > Y)$.
4. What is $P(X > Y | X + 2Y > 1)$?
5. Find $P(X = Y)$.
6. What is $P(X > 0.5 | X = Y)$?
7. Find $f_X(x)$ and $f_Y(y)$.
8. Find $f_X(x | X + 2Y > 1)$ and $E[X | X + 2Y > 1]$.

3.18 Let $X_1, X_2, \ldots, X_n$ denote independent and identically distributed random variables, each with p.d.f. $f_X(x)$,

1. If $Y = \min\{X_1, X_2, \ldots, X_n\}$, find the p.d.f. of Y.
2. If $Z = \max\{X_1, X_2, \ldots, X_n\}$, find the p.d.f. of Z.

3.19 Show that for a Rayleigh density function

$$f_X(x) = \begin{cases} \dfrac{x}{\sigma^2} e^{-\frac{x^2}{2\sigma^2}} & x > 0 \\ 0, & \text{otherwise} \end{cases}$$

we have $E[X] = \sigma\sqrt{\frac{\pi}{2}}$ and $\text{VAR}(X) = \left(2 - \frac{\pi}{2}\right)\sigma^2$.

3.20 Let X and Y be independent random variables with

$$f_X(x) = \begin{cases} \alpha e^{-\alpha x} & x > 0 \\ 0, & \text{otherwise} \end{cases}$$

and

$$f_Y(y) = \begin{cases} \beta e^{-\beta x} & x > 0 \\ 0, & \text{otherwise} \end{cases}$$

where α and β are assumed to be positive constants. Find the p.d.f. of $X + Y$ and treat the special case $\alpha = \beta$ separately.

3.21 Two random variables X and Y are distributed according to

$$f_{X,Y}(x, y) = \begin{cases} K e^{-x-y} & x \geq y \geq 0 \\ 0, & \text{otherwise} \end{cases}$$

1. Find the value of the constant K.
2. Find the marginal density functions of X and Y.
3. Are X and Y independent?
4. Find $f_{X|Y}(x|y)$.
5. Find $E[X|Y = y]$.
6. Find $\text{COV}(X, Y)$ and $\rho_{X,Y}$.

3.22 Let Θ be uniformly distributed on $[0, \pi)$, and let random variables X and Y be defined by $X = \cos\Theta$ and $Y = \sin\Theta$. Show that X and Y are uncorrelated but not independent.

3.23 Let X and Y be two independent Gaussian random variables, each with mean zero and variance 1. Define the two events, $E_1(r) = \{X > r \text{ and } Y > r\}$ and $E_2(r) = \{\sqrt{X^2 + Y^2} > \sqrt{2}r\}$, where r is any nonnegative constant.

1. Show that $E_1(r) \subseteq E_2(r)$ and therefore $P\left(E_1(r)\right) \leq P\left(E_2(r)\right)$.
2. Show that $P\left(E_1(r)\right) = Q^2(r)$.

3. Use rectangular to polar transformation relations to find $P\left(E_2(r)\right)$ and conclude with the bound

$$Q(r) \le \frac{1}{2}e^{-\frac{r^2}{2}}$$

on the Q function.

3.24 It can be shown that the Q function can be well approximated by

$$Q(x) \approx \frac{e^{-\frac{x^2}{2}}}{\sqrt{2\pi}}\left(b_1 t + b_2 t^2 + b_3 t^3 + b_4 t^4 + b_5 t^5\right)$$

where $t = \frac{1}{1+px}$ and

$$p = 0.2316419$$
$$b_1 = 0.31981530$$
$$b_2 = -0.356563782$$
$$b_3 = 1.781477937$$
$$b_4 = -1.821255978$$
$$b_5 = 1.330274429$$

Using this relation, write a computer program to compute the Q function at any given value of its argument. Compute $Q(x)$ for $x = 1, 1.5, 2, 2.5, 3, 3.5, 4, 4.5, 5$ and compare the results with those obtained from the table of the Q function.

3.25 Let the random vector $\mathbf{X} = (X_1, X_2, \ldots, X_n)$ be jointly Gaussian distributed with mean $\mathbf{m}$ and covariance matrix $\mathbf{C}$. Define a new random vector $\mathbf{Y} = \mathbf{A}\mathbf{X}^t + \mathbf{b}$, where $\mathbf{Y}$ is an n-dimensional random vector and $\mathbf{A}$ and $\mathbf{b}$ are constant matrices. Using the fact that linear functions of jointly Gaussian random variables are themselves jointly Gaussian, find the mean and covariance matrices of $\mathbf{Y}$.

3.26 Let X and Y be independent Gaussian random variables, each distributed according to $\mathcal{N}(0, \sigma^2)$.

1. Find the joint density function of the random variables $Z = X + Y$ and $W = 2X - Y$. What is the correlation coefficient between these two random variables?

2. Find the p.d.f. of the random variable $R = \frac{X}{Y}$. What is the mean and the variance of R?

3.27 Random variables X and Y are jointly Gaussian with

$$\mathbf{m} = \begin{bmatrix} 1 & 2 \end{bmatrix}$$

$$C = \begin{bmatrix} 4 & -4 \\ -4 & 9 \end{bmatrix}$$

1. Find the correlation coefficient between X and Y.
2. If $Z = 2X + Y$ and $W = X - 2Y$, find $\text{COV}(X, Y)$.
3. Find the p.d.f. of Z.

3.28 Let X and Y be two jointly Gaussian random variables with means m_X and m_Y, variances σ_X^2 and σ_Y^2 and correlation coefficient $\rho_{X,Y}$. Show that $f_{X|Y}(x|y)$ is a Gaussian distribution with mean $m_X + \rho \frac{\sigma_X}{\sigma_Y}(y - m_Y)$ and variance $\sigma_X^2(1 - \rho_{X,Y}^2)$. What happens if $\rho = 0$? What if $\rho = \pm 1$?

3.29 X and Y are zero-mean jointly Gaussian random variables, each with variance σ^2. The correlation coefficient between Y and Y is denoted by ρ. Random variables Z and W are defined by

$$\begin{cases} Z = X \cos\theta + Y \sin\theta \\ W = -X \sin\theta + Y \cos\theta \end{cases}$$

where θ is a constant angle.

1. Show that Z and W are jointly Gaussian random variables and determine their joint p.d.f.
2. For what values of θ are the random variables Z and W independent?

3.30 Two random variables X and Y are distributed according to

$$f_{X,Y}(x, y) = \begin{cases} \dfrac{K}{\pi} e^{-\frac{x^2+y^2}{2}} & \text{if } xy \geq 0 \\ 0 & \text{if } xy < 0 \end{cases}$$

1. Find K.
2. Show that X and Y are each Gaussian random variables.
3. Show that X and Y are not jointly Gaussian.
4. Are X and Y independent?
5. Are X and Y uncorrelated?
6. Find $f_{X|Y}(x|y)$. Is this a Gaussian distribution?

3.31 Let X and Y be two independent Gaussian random variables with common variance σ^2. The mean of X is m and Y is a zero-mean random variable. We define the random variable V as $V = \sqrt{X^2 + Y^2}$. Show that

$$f_V(v) = \begin{cases} \dfrac{v}{\sigma^2} I_0\left(\dfrac{mv}{\sigma^2}\right) e^{-\frac{v^2+m^2}{2\sigma^2}} & v > 0 \\ 0 & v \le 0 \end{cases}$$

where

$$I_0(x) = \frac{1}{2\pi} \int_0^{2\pi} e^{x\cos u}\, du = \frac{1}{2\pi} \int_{-\pi}^{\pi} e^{x\cos u}\, du$$

is called the *modified Bessel function of the first kind and zero order*. The distribution of V is known as the *Rician distribution*. Show that, in the special case of $m = 0$, the Rician distribution simplifies to the Rayleigh distribution.

3.32 A coin, whose probability of a head is 0.25, is flipped 2,000 times.

1. Using the law of large numbers, find a lower bound to the probability that the total number of heads lies between 480 and 520.
2. Using the central limit theorem, find the probability that the total number of heads lies between 480 and 520.

3.33 Find the covariance matrix of the random vector X in Example 3.2.2.

3.34 Find $m_X(t)$ for the random process $X(t)$ given in Example 3.2.4. Is it independent of t?

3.35 Let the random process $X(t)$ be defined by $X(t) = A + Bt$ where A and B are independent random variables each uniformly distributed on $[-1, 1]$. Find $m_X(t)$ and $R_X(t_1, t_2)$.

3.36 What is the autocorrelation function of the random process given in Example 3.2.5?

3.37 Show that the process given in Example 3.2.4 is a stationary process.

3.38 Is the process given in Example 3.2.5 wide-sense stationary?

3.39 Show that any M^{th}-order stationary process, $M \ge 2$, is WSS.

3.40 Which one of the following functions can be the autocorrelation function of a random process and why?

1. $f(\tau) = \sin(2\pi f_0 \tau)$
2. $f(\tau) = \tau^2$
3. $f(\tau) = \begin{cases} 1 - |\tau| & |\tau| \le 1 \\ 1 + |\tau| & |\tau| > 1 \end{cases}$
4. $f(\tau)$ as shown in Figure P-3.40.

3.41 Is the process given in Example 3.2.5 an ergodic process?

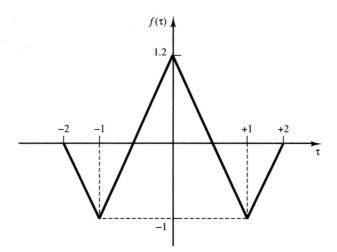

FIGURE P-3.40

3.42 Is the process of Example 3.2.1 power-type or energy-type? Is this process stationary?

3.43 A random process $Z(t)$ takes values 0,1. A transition from 0 to 1 or from 1 to 0 occurs randomly, and the probability of having n transitions in a time interval of duration τ, $(\tau > 0)$, is given by

$$p_N(n) = \frac{1}{1+\alpha\tau}\left(\frac{\alpha\tau}{1+\alpha\tau}\right)^n, \qquad n = 0, 1, 2, \ldots$$

where $\alpha > 0$ is a constant. We further assume that at $t = 0$, $Z(0)$ is equally likely to be 0 or 1.

   1. Find $m_Z(t)$.
   2. Find $R_Z(t+\tau, t)$. Is $Z(t)$ stationary? Is it cyclostationary?
   3. Determine the power-spectral density of $Z(t)$.

3.44 The random process $X(t)$ is defined by

$$X(t) = X\cos 2\pi f_0 t + Y \sin 2\pi f_0 t$$

where X and Y are two zero-mean independent Gaussian random variables each with variance σ^2.

   1. Find $m_X(t)$.
   2. Find $R_X(t+\tau, t)$. Is $X(t)$ stationary? Is it cyclostationary?
   3. Find the power-spectral density of $X(t)$.
   4. Answer the above questions for the case where $\sigma_X^2 \neq \sigma_Y^2$.

3.45 Let $\{A_k\}_{k=-\infty}^{\infty}$ be a sequence of random variables with $E[A_k] = m$ and $E[A_k A_j] = R_A(k - j)$. We further assume that $R_A(k - j) = R_A(j - k)$. Let $p(t)$ be any deterministic signal whose Fourier transform is $P(f)$, and define the random process

$$X(t) = \sum_{k=-\infty}^{+\infty} A_k p(t - kT)$$

where T is a constant.

1. Find $m_X(t)$.

2. Find $R_X(t + \tau, t)$.

3. Show that this process is cyclostationary with period T.

4. Show that

$$\bar{R}_X(\tau) = \frac{1}{T} \int_0^T R_X(t + \tau, t) \, dt = \frac{1}{T} \sum_{n=-\infty}^{+\infty} R_A(n) R_p(\tau - nT)$$

where $R_p(\tau) = p(\tau) \star p(-\tau)$ is the (deterministic) autocorrelation function of $p(t)$.

5. Show that the power-spectral density of $X(t)$ is given by

$$S_X(f) = \frac{|P(f)|^2}{T} \left[R_A(0) + 2 \sum_{k=1}^{\infty} R_A(k) \cos 2\pi k f T \right]$$

3.46 Using the result of Problem 3.45 find the power-spectral density of the random process

$$X(t) = \sum_{n=-\infty}^{+\infty} A_n p(t - nT)$$

in the following cases

1. A_n's are independent random variables each taking values ± 1 with equal probability and

$$p(t) = \begin{cases} 1 & 0 \leq t \leq T \\ 0, & \text{otherwise} \end{cases}$$

2. A_n's take values 0 and 1 with equal probability, all the other conditions as in part 1.

3. Solve parts 1 and 2 for the case where

$$p(t) = \begin{cases} 1 & 0 \leq t \leq 3T \\ 0, & \text{otherwise} \end{cases}$$

3.47 Let A_n's denote a sequence of independent binary valued random variables, each taking values ± 1 with equal probability. Random variables B_n are defined according to $B_n = A_n + A_{n-1}$ and the random process $X(t)$ is defined as $X(t) = \sum_{n=-\infty}^{+\infty} B_n p(t - nT)$.

   1. Using the results of Problem 3.45, determine the power-spectral density of $X(t)$.
   2. Assuming that

$$
p(t) = \begin{cases} 1 & 0 \leq t \leq T \\ 0, & \text{otherwise} \end{cases}
$$

   plot a sample function of $X(t)$. Find the power-spectral density of $X(t)$ and plot it.
   3. Let $B_n = A_n + \alpha A_{n-1}$ and find the power-spectral density of $X(t)$.

3.48 Let $X(t)$ be a cyclostationary process with period T. From Corollary 3.3.1 we have seen that, to find the power spectral density of $X(t)$, we first determine $\bar{R}(\tau) = \frac{1}{T} \int_0^T R_X(t+\tau, t)\, dt$ and then find $\mathcal{F}[\bar{R}(\tau)]$. We now obtain this result by another approach.

   1. Let Θ be a random variable, independent of $X(t)$ and uniformly distributed on $[0, T)$. Show that $Y(t) = X(t + \Theta)$ is stationary, and its autocorrelation function is given as

$$
R_Y(\tau) = \frac{1}{T} \int_0^T R_X(t + \tau, t)\, dt
$$

   2. Show that $Y(t)$ and $X(t)$ have equal power-spectral densities.
   3. Conclude that

$$
S_X(f) = \mathcal{F}\left[\frac{1}{T} \int_0^T R_X(t + \tau, t)\, dt \right]
$$

3.49 The RMS bandwidth of a process is defined as

$$
W_{\text{RMS}} = \frac{\int_{-\infty}^{+\infty} f^2 S_X(f)\, df}{\int_{-\infty}^{+\infty} S_X(f)\, df}
$$

Show that for a stationary process we have

$$
W_{\text{RMS}} = -\frac{1}{4\pi^2 R_X(0)} \frac{d^2}{d\tau^2} R_X(\tau)\Big|_{\tau=0}
$$

3.50 Show that for jointly stationary processes $X(t)$ and $Y(t)$, we have $R_{XY}(\tau) = R_{YX}(-\tau)$. From this, conclude that $S_{XY}(f) = S_{YX}^*(f)$.

3.51 A zero-mean white Gaussian noise with power-spectral density of $\frac{N_0}{2}$ passes through an ideal lowpass filter with bandwidth B.

   1. Find the autocorrelation of the output process $Y(t)$.
   2. Assuming $\tau = \frac{1}{2B}$, find the joint probability-density function of the random variables $Y(t)$ and $Y(t + \tau)$. Are these random variables independent?

3.52 Find the output autocorrelation function for a delay line with delay Δ when the input is a stationary process with autocorrelation $R_X(\tau)$. Interpret the result.

3.53 We have proved that when the input to a LTI system is stationary, the output is also stationary. Is the converse of this theorem also true? That is, if we know that the output process is stationary, can we conclude that the input process is necessarily stationary?

3.54 It was shown in this chapter that, if a stationary random process $X(t)$ with autocorrelation function $R_X(\tau)$ is applied to a LTI system with impulse response $h(t)$, the output $Y(t)$ is also stationary with autocorrelation function $R_Y(\tau) = R_X(\tau) \star h(\tau) \star h(-\tau)$. In this problem we show that a similar relation holds for cyclostationary processes.

   1. Let $X(t)$ be a cyclostationary process applied to a LTI system with impulse response $h(t)$. Show that the output process is also cyclostationary.
   2. Show that

$$\bar{R}_Y(t) = \bar{R}_X(t) \star h(\tau) \star h(-\tau)$$

   3. Conclude that the relation

$$S_Y(f) = S_X(f)|H(f)|^2$$

   is true for both stationary and cyclostationary processes.

3.55 Generalize the result of Example 3.3.8 to show that, if $X(t)$ is stationary,

   1. $X(t)$ and $\frac{d}{dt}X(t)$ are uncorrelated processes.
   2. The power spectrum of $Z(t) = X(t) + \frac{d}{dt}X(t)$ is the sum of the power spectra of $X(t)$ and $\frac{d}{dt}X(t)$.

Express the power spectrum of the sum in terms of the power spectrum of $X(t)$.

3.56 $X(t)$ is a stationary process with power-spectral density $S_X(f)$. This process passes through the system shown in Figure P-3.56.

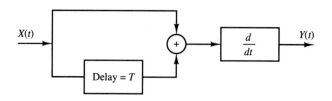

FIGURE P-3.56

1. Is $Y(t)$ stationary? Why?
2. What is the power-spectral density of $Y(t)$?
3. What frequency components can *not* be present in the output process and why?

3.57 The stationary random process $X(t)$ has a power-spectral density denoted by $S_X(f)$.

1. What is the power-spectral density of $Y(t) = X(t) - X(t - T)$?
2. What is the power-spectral density of $Z(t) = X'(t) - X(t)$?
3. What is the power-spectral density of $W(t) = Y(t) + Z(t)$?

3.58 Show that for two jointly stationary processes $X(t)$ and $Y(t)$, we have

$$|R_{XY}(\tau)| \le R_X^{\frac{1}{2}}(0) R_Y^{\frac{1}{2}}(0) \le \frac{1}{2}[R_X(0) + R_Y(0)]$$

3.59 The stationary process $X(t)$ with autocorrelation function $R_X(t) = e^{-\alpha|t|}$ is passed through a LTI system and the output process is denoted by $Y(t)$. Find the output autocorrelation function and the crosscorrelation function between the input and the output processes in each of the following cases.

1. A delay system with delay Δ.
2. A system with $h(t) = \frac{1}{t}$.
3. A system described by the differential equation

$$\frac{d}{dt} Y(t) + Y(t) = \frac{d}{dt} X(t) - X(t)$$

4. A finite-time averager defined by the input–output relation

$$y(t) = \frac{1}{2T} \int_{t-T}^{t+T} x(\tau) \, d\tau$$

where T is a constant.

3.60 Give an example of two processes $X(t)$ and $Y(t)$ for which $R_{XY}(t+\tau, t)$ is a function of τ but $X(t)$ and $Y(t)$ are not stationary.

3.61 For each of the following processes, find the power-spectral density.

1. $X(t) = A\cos(2\pi f_0 t + \Theta)$, where A is a constant and Θ is a random variable uniformly distributed on $[0, \frac{\pi}{4}]$.
2. $X(t) = X + Y$, where X and Y are independent, X is uniform on $[-1, 1]$, and Y is uniform on $[0, 1]$.

3.62 $X(t)$ is a stationary random process with autocorrelation function $R_X(\tau) = e^{-\alpha|\tau|}$, $\alpha > 0$. This process is applied to a LTI system with $h(t) = e^{-\beta t}u(t)$, where $\beta > 0$. Find the power-spectral density of the output process $Y(t)$. Treat the cases $\alpha \neq \beta$ and $\alpha = \beta$ separately.

3.63 Let $Y(t) = X(t) + N(t)$, where $X(t)$ and $N(t)$ are signal and noise processes. It is assumed that $X(t)$ and $N(t)$ are jointly stationary with autocorrelation functions $R_X(\tau)$ and $R_N(\tau)$ and crosscorrelation function $R_{XN}(\tau)$. It is desired to separate the signal from the noise by passing $Y(t)$ through a LTI system with impulse response $h(t)$ and transfer function $H(f)$. The output process is denoted by $\hat{X}(t)$, which is desired to be as close to $X(t)$ as possible.

1. Find the crosscorrelation between $\hat{X}(t)$ and $X(t)$ in terms of $h(\tau)$, $R_X(\tau)$, $R_N(\tau)$, and $R_{XN}(\tau)$.
2. Show that the LTI system that minimizes $E[X(t) - \hat{X}(t)]^2$ has a transfer function

$$H(f) = \frac{S_X(f) + S_{XN}(f)}{S_X(f) + S_N(f) + 2\,\mathrm{Re}[S_{XN}(f)]}$$

3. Now assume that $X(t)$ and $N(t)$ are independent and $N(t)$ is a zero-mean white Gaussian process with power-spectral density $\frac{N_0}{2}$. Find the optimal $H(f)$ under these conditions. What is the corresponding value of $E[X(t) - \hat{X}(t)]^2$ in this case?
4. In the special case of $S_N(f) = 1$, $S_X(f) = \frac{1}{1+f^2}$, and $S_{XN}(f) = 0$, find the optimal $H(f)$.

3.64 In this problem, we examine the estimation of a random process from the observation of another random process. Let $X(t)$ and $Z(t)$ be two jointly stationary random processes. We are interested in designing a LTI system with impulse response $h(t)$ such that when $Z(t)$ is passed through it the output process $\hat{X}(t)$ is as close to $X(t)$ as possible. In other words, we are interested in the best *linear* estimate of $X(t)$ based on the observation of $Y(t)$ that minimizes $E[X(t) - \hat{X}(t)]^2$.

1. Let us assume we have two LTI systems with impulse responses $h(t)$ and $g(t)$. $Z(t)$ is applied to both systems and the outputs are denoted by $\hat{X}(t)$ and $\tilde{X}(t)$, respectively. The first filter is designed such that its output satisfies the condition

$$E\left[(X(t) - \hat{X}(t))Z(t - \tau)\right] = 0$$

for all values of τ and t where the second filter does not satisfy this property. Show that

$$E[X(t) - \tilde{X}(t)]^2 \geq E[X(t) - \hat{X}(t)]^2$$

i.e., the necessary and sufficient condition for an optimal filter is that its output satisfy the *orthogonality condition* as given by

$$E\left[(X(t) - \hat{X}(t))Z(t - \tau)\right] = 0$$

which simply means that the estimation error $\epsilon(t) = X(t) - \hat{X}(t)$ must be orthogonal to the observable process $Z(t)$ at all times.

2. Show that the optimal $h(t)$ must satisfy

$$R_{XZ}(\tau) = R_Z(\tau) \star h(\tau)$$

3. Show that the optimal filter satisfies

$$H(f) = \frac{S_{XZ}(f)}{S_Z(f)}$$

4. Derive an expression for $E[\epsilon^2(t)]$ when the optimal filter is employed.

3.65 The random process $X(t)$ is defined by

$$X(t) = \sum_{n=-\infty}^{+\infty} A_n \operatorname{sinc} 2W(t - nT)$$

where A_n's are independent random variables with mean zero and common variance σ^2.

1. Use the result of Problem 3.45 to obtain the power-spectral density of $X(t)$.

2. In the special case of $T = \frac{1}{2W}$ what is the power content of $X(t)$?

3. Now let $X_1(t)$ be a zero-mean stationary process with power-spectral density $S_{X_1}(f) = \frac{N_0}{2} \Pi\left(\frac{f}{2W}\right)$, and let $A_n = X_1(nT)$ where $T = \frac{1}{2W}$. Determine the power-spectral density of $X(t)$ and its power content. What is the relation between $X_1(t)$ and $X(t)$?

3.66 What is the noise-equivalent bandwidth of an ideal bandpass filter with bandwidth W?

3.67 Work out Example 3.6.1 with $f_0 = f_c - \frac{W}{2}$.

3.68 Verify (3.6.12) for both parts of Example 3.6.1.

3.69 A zero-mean white Gaussian noise, $n_w(t)$ with power-spectral density $\frac{N_0}{2}$, is passed through an ideal filter whose passband is 3–11 kHz. The output process is denoted by $n(t)$.

1. If $f_0 = 7$ kHz, find $S_{n_c}(f)$, $S_{n_s}(f)$, and $R_{n_c n_s}(\tau)$ where $n_c(t)$ and $n_s(t)$ are the in-phase and quadrature components of $n(t)$.

2. Repeat part 1 with $f_0 = 6$ kHz.

3.70 Let $p(t)$ be a bandpass signal with in-phase and quadrature components $p_c(t)$ and $p_s(t)$ and let $X(t) = \sum_{n=-\infty}^{+\infty} A_n p(t - nT)$ where A_n's are independent random variables. Express $X_c(t)$ and $X_s(t)$ in terms of $p_c(t)$ and $p_s(t)$.

3.71 Let $X(t)$ be a bandpass process and let $V(t)$ denote its envelope. Show that for all choices of the center frequency f_0, $V(t)$ remains unchanged.

3.72 Let $n_w(t)$ be a zero-mean white Gaussian noise with power-spectral density $\frac{N_0}{2}$, and let this noise be passed through an ideal bandpass filter with bandwidth $2W$ centered at frequency f_c. Denote the output process by $n(t)$.

1. Assuming $f_0 = f_c$, find the power content of the in-phase and quadrature components of $n(t)$.

2. Find the density function of $V(t)$, the envelope of $n(t)$.

3. Now assume $X(t) = A \cos 2\pi f_0 t + n(t)$ where A is a constant. What is the density function of the envelope of $X(t)$?

3.73 A noise process has a power-spectral density given by

$$S_n(f) = \begin{cases} 10^{-8}\left(1 - \dfrac{|f|}{10^8}\right) & |f| < 10^8 \\ 0 & |f| > 10^8 \end{cases}$$

This noise is passed through an ideal bandpass filter with a bandwidth of 2 MHz centered at 50 MHz.

1. Find the power content of the output process.

2. Write the output process in terms of the in-phase and quadrature components and find the power in each component. Assume $f_0 = 50$ MHz.

3. Find the power-spectral density of the in-phase and quadrature components.

4. Now assume that the filter is not an ideal filter and is described by

$$|H(f)|^2 = \begin{cases} |f| - 49 \times 10^6 & 49 \text{ MHz} < |f| < 51 \text{ MHz} \\ 0, & \text{otherwise} \end{cases}$$

Repeat parts 1, 2, and 3 with this assumption.

4

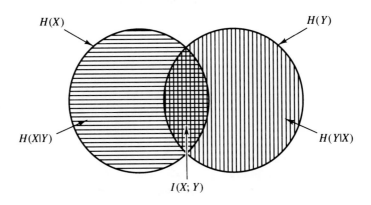

$H(X)$ $H(Y)$

$H(X|Y)$ $H(Y|X)$

$I(X; Y)$

Information Sources and Source Coding

Communication systems are designed to transmit information. In any communication system there exists an information source that produces the information, and the purpose of the communication system is to transmit the output of the source to its destination. In radio broadcasting, for instance, the information source is either a speech source or a music source. In TV broadcasting, the information source is a video source whose output is a moving image. In FAX transmission, the information source produces a still image. In communication between computers, either binary data or ASCII characters are transmitted, and, therefore, the source can be modeled as a binary or ASCII source. In the storage of binary data on a computer disk, the source is again a binary source.

Everyone has an intuitive notion of the meaning of information. However, performance analysis of communication systems can hardly be conceived without a quantitative measure of information and mathematical modeling of information sources. Hartley, Nyquist, and Shannon were the pioneers in defining quantitative measures for information. In this chapter, we investigate mathematical modeling of information sources and provide a measure of information. Then we will see how the output of an information source can be made more compact and, therefore, easier to transmit or store.

4.1 MODELING OF INFORMATION SOURCES

The intuitive and common notion of information refers to any new knowledge about something. One can obtain information via hearing, seeing, or other means of perception. The information source, therefore, produces outputs that are of interest to the receiver of information, who does not know these outputs in advance. The role of the communication-system designer is to make sure that this information is transmitted to the receiver correctly. Because the output of the information source is a time-varying unpredictable function (if predictable, there is no need to transmit it), it can be modeled as a random process. We have seen already in Chapter 3 that in communication channels the existence of noise causes stochastic dependence between the input and output of the channel. Therefore, the communication-system designer designs a system that transmits the output of a random process (information source) to a destination via a random medium (channel) and ensures low distortion.

Information sources can be modeled by random processes, and the properties of the random process depend on the nature of the information source. For example, when modeling speech signals, the resulting random process has all its power in a frequency band of approximately 300–3400 Hz. Therefore, the power-spectral density of the speech signal also occupies this band of frequencies. A typical power-spectral density for the speech signal is shown in Figure 4.1.

Video signals are rastered from a still or moving image, and, therefore, the bandwidth depends on the required resolution. For TV transmission, depending on the system employed (NTSC, PAL, or SECAM), this band is typically between 0–4.5 MHz and 0–6.5 MHz. For telemetry data the bandwidth, of course, depends on the rate of change of data.

What is common in all these processes is that they are bandlimited processes and, therefore, can be sampled at the Nyquist rate or faster and reconstructed from the sampled values. Therefore, it makes sense to confine ourselves to discrete-time random processes in this chapter because all information sources of interest can be modeled by such a process. The mathematical model for an information source is shown in Figure 4.2. Here the source is modeled by a discrete-time random process $\{X_i\}_{i=-\infty}^{\infty}$. The alphabet over which the random variables X_i are defined can be either discrete (in transmission of binary data, for instance) or continuous (e.g., sampled speech). The statistical properties of the discrete-time random process depend on the nature of the information source.

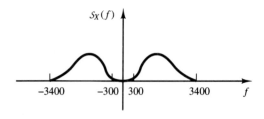

FIGURE 4.1. Typical power spectrum of speech signal.

In this chapter we will study only rather simple models for information sources. Study of more complicated models is mathematically demanding and beyond the scope of our treatment. However, even simple models enable us to define precisely a measure of information and bounds on compression of information.

The simplest model for the information source that we study is the *discrete memoryless source* (d.m.s.). A d.m.s. is a discrete-time discrete-amplitude random process in which all X_i's are generated independently and with the same distribution. Therefore, a d.m.s. generates a sequence of independent and identically distributed (i.i.d.) random variables taking values in a discrete set.

Let set $\mathcal{A} = \{a_1, a_2, \ldots, a_N\}$ denote the set in which the random variable X takes its values, and let the probability mass function for the discrete random variable X be denoted by $p_i = p(X = a_i)$ for all $i = 1, 2, \ldots, N$. A full description of the d.m.s. is given by the set $\mathcal{A}$, called the alphabet, and the probabilities $\{p_i\}_{i=1}^{N}$.

Example 4.1.1

An information source is described by the alphabet $\mathcal{A} = \{0, 1\}$ and $p(X_i = 1) = 1 - p(X_i = 0) = p$. This is an example of a d.m.s. In the special case where $p = .5$, the source is called a *binary symmetric source,* or b.s.s.

4.1.1 Measure of Information

To give a quantitative measure of information, we will start with the basic model of an information source and try to define the information content of the source to satisfy certain intuitive properties. Let us assume that the source that we are considering is a discrete source. Let the outputs of this source be revealed to an interested party. Let a_1 be the most-likely and a_N be the least-likely output. For example, one could imagine the source to represent both the weather condition and air pollution in a certain city (in the northern hemisphere) during July. In this case, $\mathcal{A}$ represents various combinations of various weather conditions and pollution such as hot and polluted, hot and lightly polluted, cold and highly polluted, cold and mildly polluted, very cold and lightly polluted, etc. The question is: which output conveys more information, a_1 or a_N (the most-probable or the least-probable one)? Intuitively, revealing a_N (or equivalently, very cold and lightly polluted in the previous example) reveals the most information. From this it follows that a rational measure of information for an output of an information source should be a decreasing function of the probability of that output. A second intuitive property of a measure of information is that a small change in the probability of a certain output should not change the information delivered by that output by a large amount. In other words, the information measure should be a decreasing and continuous function of the probability of the source output.

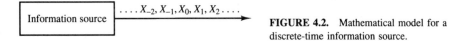

FIGURE 4.2.   Mathematical model for a discrete-time information source.

Now let us assume that the information about output a_j can be broken into two independent parts, say a_{j1} and a_{j2}, i.e., $X_j = (X^{(j1)}, X^{(j2)})$, $a_j = \{a^{j1}, a^{j2}\}$ and $p(X = a_j) = p(X^{(j1)} = a^{j1})p(X^{(j2)} = a^{j2})$. This can happen, for example, if we assume that the temperature and pollution were almost independent, and, therefore, each source output can be broken into two independent components. Because the components are independent, revealing information about one component (temperature) does not provide any information about the other component (pollution) and, therefore intuitively, the amount of information provided by revealing a_j is the sum of the informations obtained by revealing a^{j1} and a^{j2}. From the above discussion, we can conclude that the amount of information revealed about an output a_j with probability p_j must satisfy the following four conditions:

1. The information content of output a_j depends only on the probability of a_j and not on the value of a_j. We denote this function by $I(p_j)$ and call it *self-information.*

2. Self-information is a continuous function of p_j, i.e., $I(\cdot)$ is a continuous function.

3. Self-information is a decreasing function of its argument.

4. If $p_j = p^{(j1)}p^{(j2)}$, then $I(p_j) = I(p^{(j1)}) + I(p^{(j2)})$.

It can be proved that the only function that satisfies all the above properties is the logarithmic function, i.e., $I(x) = -\log(x)$. The base of logarithm is not important and defines the unit by which the information is measured. If the base is 2, the information is expressed in *bits,* and if the natural logarithm is employed, the unit is *nats.*

Now that the information revealed about each source output a_i is defined as the self information of that output, given by $-\log(p_i)$, we can define the information content of the source as the weighted average of self-information of all source outputs. This is justified by the fact that various source outputs appear with their corresponding probabilities. Therefore, the information revealed by an unidentified source output is the weighted average of the self-information of the various source outputs, i.e., $\sum_{i=1}^{N} p_i I(p_i) = \sum_{i=1}^{N} -p_i \log p_i$. The information content of the information source is known as the *entropy* of the source and is denoted by $H(X)$.

Definition 4.1.1. The entropy of a discrete random variable X is a function of its p.m.f. and is defined by

$$H(X) = -\sum_{i=1}^{N} p_i \log p_i = \sum_{i=1}^{N} p_i \log\left(\frac{1}{p_i}\right) \tag{4.1.1}$$

Note that there exists a slight abuse of notation here. One would expect $H(X)$ to denote a function of the random variable X and hence be a random variable itself. However, $H(X)$ is a function of the p.m.f. of the random variable X and is therefore a number.

Example 4.1.2

In the d.m.s. with probabilities p and $1 - p$, respectively, we have

$$H(X) = -p \log p - (1 - p) \log(1 - p) \tag{4.1.2}$$

This function, denoted by $H_b(p)$, is known as the *binary entropy function,* and a plot of it is given in Figure 4.3.

Example 4.1.3

A source with bandwidth 4000 Hz is sampled at the Nyquist rate. Assuming that the resulting sequence can be approximately modeled by a d.m.s. with alphabet $\mathcal{A} = \{-2, -1, 0, 1, 2\}$ and with corresponding probabilities $\{\frac{1}{2}, \frac{1}{4}, \frac{1}{8}, \frac{1}{16}, \frac{1}{16}\}$, determine the rate of the source in bits/sec.

Solution   We have

$$H(X) = \frac{1}{2} \log 2 + \frac{1}{4} \log 4 + \frac{1}{8} \log 8 + 2 \times \frac{1}{16} \log 16 = \frac{15}{8} \text{ bits/sample}$$

and because we have 8000 samples/sec, the source produces information at a rate of 15,000 bits/sec.

4.1.2 Joint and Conditional Entropy

When dealing with two or more random variables, exactly in the same way that joint and conditional probabilities are introduced, one can introduce joint and conditional

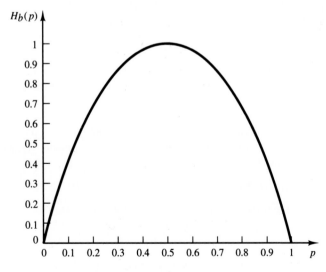

FIGURE 4.3.   The binary entropy function.

entropies. These concepts are especially important when dealing with sources with memory.

Definition 4.1.2. The *joint entropy* of two discrete random variables (X, Y) is defined by

$$H(X, Y) = -\sum_{x,y} p(x, y) \log p(x, y) \qquad (4.1.3)$$

For the case of n random variables $\mathbf{X} = (X_1, X_2, \ldots, X_n)$, we have

$$H(\mathbf{X}) = -\sum_{x_1, x_2, \ldots, x_n} p(x_1, x_2, \ldots, x_n) \log p(x_1, x_2, \ldots, x_n) \qquad (4.1.4)$$

As seen, the joint entropy is simply the entropy of a vector-valued random variable.

The conditional entropy of the random variable X, given the random variable Y, can be defined by noting that if $Y = y$, then the p.m.f. of the random variable X will be $p(x|y)$, and the corresponding entropy is $H(X|Y = y) = -\sum_x p(x|y) \log p(x|y)$, which is intuitively the amount of uncertainty in X when one knows $Y = y$. The weighted average of the above quantities over all y is the uncertainty in X when Y is known. This quantity is known as the conditional entropy and defined as follows:

Definition 4.1.3. The *conditional entropy* of the random variable X given the random variable Y is defined by

$$H(X|Y) = -\sum_{x,y} p(x, y) \log p(x|y) \qquad (4.1.5)$$

In general we have

$$H(X_n|X_1, \ldots, X_{n-1}) = -\sum_{x_1, \ldots, x_n} p(x_1, \ldots, x_n) \log p(x_n|x_1, \ldots, x_{n-1}) \qquad (4.1.6)$$

Example 4.1.4

Using chain rule for p.m.f.'s, $p(x, y) = p(y)p(x|y)$ show that $H(X, Y) = H(Y) + H(X|Y)$. Generalize this result to the case of n random variables to show the following chain rule for entropies

$$H(\mathbf{X}) = H(X_1) + H(X_2|X_1) + \cdots + H(X_n|X_1, X_2, \ldots, X_{n-1}) \qquad (4.1.7)$$

Solution From the definition of the joint entropy of two random variables, we have

$$H(X, Y) = -\sum_{x,y} p(x, y) \log p(x, y)$$

$$= -\sum_{x,y} p(x, y) \log[p(y)p(x|y)]$$

$$= -\sum_{x,y} p(x, y) \log p(y) - \sum_{x,y} p(x, y) \log p(x|y)$$

$$= -\sum_{y} p(y) \log p(y) - \sum_{x,y} p(x, y) \log p(x|y)$$

$$= H(Y) + H(X|Y) \tag{4.1.8}$$

where in the last step we have used

$$\sum_{x} p(x, y) = p(y) \tag{4.1.9}$$

This relation says that the information content of the pair (X, Y) is equal to the information content of Y plus the information content of X after Y is known. Equivalently, it says that the same information is transferred by either revealing the pair (X, Y), or by first revealing Y and then revealing the remaining information in X. The proof for general n is similar and is left as an exercise. In the case where the random variables $(X_1, X_2, \ldots, X_n)$ are independent, this relation reduces to

$$H(\mathbf{X}) = \sum_{i=1}^{n} H(X_i) \tag{4.1.10}$$

If the random variable X_n denotes the output of a discrete (not necessarily memoryless) source at time n, then $H(X_2|X_1)$ denotes the fresh information provided by source output X_2 to someone who already knows source output X_1. In the same way, $H(X_n|X_1, X_2, \ldots, X_{n-1})$ denotes the fresh information in X_n for an observer who has observed the sequence $(X_1, X_2, \ldots, X_{n-1})$. The limit of the above conditional entropy as n tends to infinity is known as the *entropy rate* of the random process.

Definition 4.1.4.   The *entropy rate* of a stationary discrete-time random process is defined by

$$H = \lim_{n \to \infty} H(X_n|X_1, X_2, \ldots, X_{n-1})$$

Stationarity ensures the existence of the limit, and it can be proven that an alternative definition of the entropy rate for sources with memory is given by

$$H = \lim_{n \to \infty} \frac{1}{n} H(X_1, X_2, \ldots, X_n).$$

Entropy rate plays the role of entropy for sources with memory. It is basically a measure of the uncertainty per output symbol of the source.

4.2 SOURCE-CODING THEOREM

The source-coding theorem is one of the three fundamental theorems of information theory introduced by Shannon (1948a, 1948b). The source-coding theorem establishes a fundamental limit on the rate at which the output of an information source can be compressed without causing a large error probability. We have seen already that the entropy of an information source is a measure of the uncertainty or, equivalently, the information content of the source. Therefore, it is natural that in the statement of the source coding theorem the entropy of the source plays a major role.

The entropy of an information source has a very intuitive meaning. Let us assume that we are observing outputs of length n of a d.m.s. where n is very large. Then according to the law of large numbers (see Chapter 3), in this sequence, with high probability (that goes to 1 as $n \to \infty$) letter a_1 is repeated approximately np_1 times, letter a_2 is repeated approximately np_2 times,..., and letter a_N is repeated approximately np_N times. This means that for n large enough, with probability approaching 1, every sequence from the source has the same composition and therefore the same probability. To put it in another way, asymptotically "almost everything is almost equally probable."[†]. The sequences $\mathbf{x}$ that have the above structure are called *typical sequences*. The probability of a typical sequence is given by

$$p(\mathbf{X} = \mathbf{x}) \approx \Pi_{i=1}^{N} p_i^{np_i}$$
$$= \Pi_{i=1}^{N} 2^{np_i \log p_i}$$
$$= 2^{n \sum_{i=1}^{N} p_i \log p_i}$$
$$= 2^{-nH(X)}$$

(In this chapter all the logarithms are in base 2, and all entropies are in bits unless otherwise specified). This means that for large n almost all the output sequences of length n of the source are equally probable with probability $\approx 2^{-nH(X)}$. These are called typical sequences. On the other hand, the probability of the set of nontypical sequences is negligible.

Because the probability of the typical sequences is almost 1 and each typical sequence has a probability of almost $2^{-nH(X)}$, the total number of typical sequences is almost $2^{nH(X)}$. Therefore, although a source of alphabet size N can produce N^n sequences of length n, the effective number of outputs is $2^{nH(X)}$. By "effective number of outputs" we mean that almost nothing is lost by neglecting the other outputs, and the probability of having lost anything goes to zero as n goes to infinity. Figure 4.4 gives a schematic diagram of the property mentioned above. This is a very important result, which tells us that for all practical purposes, it is enough to consider the set of typical sequences rather than the set of all possible

---

[†]Borrowed from Cover and Thomas (1991).

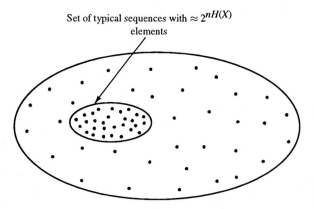

Set of typical sequences with $\approx 2^{nH(X)}$ elements

FIGURE 4.4.   The set of typical and nontypical sequences.

outputs of the source. The error introduced in ignoring nontypical sequences can be made smaller than any given $\epsilon > 0$ by choosing n to be large enough. This is the essence of *data compression,* the practice of representing the output of the source with a smaller number of sequences compared to the number of the outputs that the source really produces.

From the above result and since ϵ is an arbitrary positive number, it is seen that we can only represent the typical source outputs without introducing considerable error. Because the total number of typical sequences is roughly $2^{nH(X)}$, we need $nH(X)$ bits to represent them. However, these bits are used to represent source outputs of length n. Therefore, on the average, any source output requires $H(X)$ bits for an essentially error-free representation. This, once again, justifies the notion of entropy as the amount of information per source output.

So far we have assumed that the source is discrete and memoryless and can be represented by an i.i.d. random variable. Such a source can be compressed only if its p.m.f. is not uniform. For X uniformly distributed, we have $H(X) = \log N$, and therefore $2^{nH(X)} = 2^{n \log N} = N^n$. This means that the effective number of source outputs of length n is equal to the total number of source outputs, and no compression is possible.

We have not considered the case where the source has memory. For sources with memory, the outputs of the source are not independent, and previous outputs therefore reveal some information about the future ones. This means that the rate at which fresh information is produced decreases as more and more source outputs are revealed. A classic example of such a case is the English alphabet, which shows a lot of dependency between letters and words (a "q" is almost always followed by a "u," a single letter between two spaces is either "I" or "a," etc.). The entropy per letter for a large text of English is roughly the limit of $H(X_n|X_1, X_2, \ldots, X_{n-1})$ as n becomes large (the entropy rate defined in Section 4.1.2). In general stationary sources, the entropy rate has the same significance as the entropy for the case of memoryless sources and defines the number of effective source outputs for n large enough, i.e., 2^{nH} where H is the entropy rate.

Studies with statistical models of the English language show that the entropy rate converges rather quickly, and for $n = 10$ we are very close to the limit. These studies show that for $n = 1$, i.e., a memoryless source model, we have $H(X) = 4.03$ bits/letter. As the memory increases, the size of the space over which conditional probabilities are computed increases rapidly, and it is not easy to find the conditional probabilities required to compute the entropy rate. Some methods for estimating these conditional probabilities have been proposed in the literature, and, based on these methods, the entropy of English is estimated to be around 1.3 bits/letter. It is worth mentioning here that in these studies only the 26 letters of the English alphabet and the space mark (a total of 27) have been considered.

So far we have given an informal description of the source-coding theorem and justified it. A formal statement of the theorem, without proof, is given below. The interested reader is referred to the references at the end of this chapter for a proof.

Theorem 4.2.1 [Source-Coding Theorem]. A source with entropy rate H can be encoded with arbitrarily small error probability at any rate R (bits/source output) as long as $R > H$. Conversely, if $R < H$, the error probability will be bounded away from zero, independent of the complexity of the encoder and the decoder employed.

This theorem, first proved by Shannon (1948a), only gives necessary and sufficient conditions for the existence of source codes. It does not provide any algorithm for the design of codes that achieve the performance predicted by this theorem. In the next section, we present two algorithms for compression of information sources. One is due to Huffman (1952), and the second is due to Lempel and Ziv (1978).

4.3   SOURCE-CODING ALGORITHMS

In the preceding section, we observed that H, the entropy of a source, gives a sharp bound on the rate at which a source can be compressed for reliable reconstruction. This means that at rates above entropy it is possible to design a code with an error probability as small as desired, whereas at rates below entropy such a code does not exist. This important result, however, does not provide specific algorithms to design codes approaching this bound. In this section, we will introduce two algorithms to design codes that perform very close to the entropy bound. These coding methods are the Huffman source-coding algorithm and the Lempel-Ziv source-coding algorithm.

4.3.1 The Huffman Source-Coding Algorithm

In Huffman coding, fixed-length blocks of the source output are mapped to variable-length binary blocks. This is called *fixed- to variable-length* coding. The idea is to map the more-frequently occurring fixed-length sequences to shorter binary sequences and the less-frequently occurring ones to longer binary sequences. In variable-length coding, synchronization is a problem. This means that there should be one and only one way to break the binary-received sequence into code words. The following example clarifies this point.

Example 4.3.1

Let us assume that the possible outputs of an information source are $\{a_1, a_2, a_3, a_4, a_5\}$, and consider the following three codes for this source.

		Codewords			
Letter	Probability	Code 1	Code 2	Code 3	Code 4
a_1	$p_1 = \frac{1}{2}$	1	1	0	00
a_2	$p_2 = \frac{1}{4}$	01	10	10	01
a_3	$p_3 = \frac{1}{8}$	001	100	110	10
a_4	$p_4 = \frac{1}{16}$	0001	1000	1110	11
a_5	$p_5 = \frac{1}{16}$	00001	10000	1111	110

In the first code each codeword ends with a 1. Therefore, as soon as the decoder observes a 1, it knows that the codeword has ended and a new codeword will start. This means that the code is a *self-synchronizing code*. In the second code each codeword starts with a 1. Therefore, upon observing a 1, the decoder knows that a new codeword has started and, hence, the previous bit was the last bit of the previous codeword. This code is again self-synchronizing but not as desirable as the first code because with this code we have to wait to receive the first bit of the next codeword to recognize that a new codeword has started, whereas in code 1 we recognize the last bit without having to receive the first bit of the next codeword. Both codes 1 and 2 therefore are *uniquely decodable*. However, only code 1 is *instantaneous*. Codes 1 and 3 have the nice property that no codeword is the prefix of another codeword; it is said that they satisfy the *prefix condition*. It can be proved that a necessary and sufficient condition for a code to be uniquely decodable and instantaneous is that it satisfy the prefix condition. This means that both codes 1 and 3 are uniquely decodable and instantaneous. However, code 3 has the advantage of having a smaller average codeword length. In fact, for code 1 the average codeword length is

$$E[L] = 1 \times \frac{1}{2} + 2 \times \frac{1}{4} + 3 \times \frac{1}{8} + 4 \times \frac{1}{16} + 5 \times \frac{1}{16} = \frac{31}{16}$$

and for code 3

$$E[L] = 1 \times \frac{1}{2} + 2 \times \frac{1}{4} + 3 \times \frac{1}{8} + 4 \times \frac{1}{16} + 4 \times \frac{1}{16} = \frac{30}{16}$$

Code 4 has a major disadvantage because it is not uniquely decodable. For example, the sequence 110110 can be decoded in two ways, as a_5a_5 or as $a_4a_2a_3$. Codes that are not uniquely decodable are not desirable and should be avoided in practice. From the discussion above it is seen that the most desirable of the above four codes is code 3, which is uniquely decodable, is instantaneous, and has the least average codeword length. This code is an example of a *Huffman code,* to be discussed shortly.

As already mentioned, the idea in Huffman coding is to choose codeword lengths such that more-probable sequences have shorter codewords. If we can map each source output of probability p_i to a codeword of length approximately $\log \frac{1}{p_i}$ and at the same time ensure unique decodability, we can achieve an average codeword length of approximately $H(X)$. Huffman codes are uniquely decodable instantaneous codes with minimum average codeword length. In this sense they are optimal. By optimal we mean that, among all codes that satisfy the prefix condition (and therefore are uniquely decodable and instantaneous), Huffman codes have the minimum-average codeword length. We present below the algorithm for the design of the Huffman code. From the algorithm, it is obvious that the resulting code satisfies the prefix condition. The proof of the optimality is omitted, and the interested reader is referred to the references at the end of this chapter.

Huffman encoding algorithm.

1. Sort source outputs in decreasing order of their probabilities.
2. Merge the two least-probable outputs into a single output whose probability is the sum of the corresponding probabilities.
3. If the number of remaining outputs is 2, then go to the next step; otherwise go to step 1.
4. Arbitrarily assign 0 and 1 as codewords for the two remaining outputs.
5. If an output is the result of the merger of two outputs in a preceding step, append the current codeword with a 0 and a 1 to obtain the codeword for the preceding outputs and repeat step 5. If no output is preceded by another output in a preceding step, then stop.

Figure 4.5 shows a flow chart of this algorithm.

Example 4.3.2

Design a Huffman code for the source given in the preceding example.

Solution　The tree diagram summarizes the design steps for code construction and the resulting codewords.

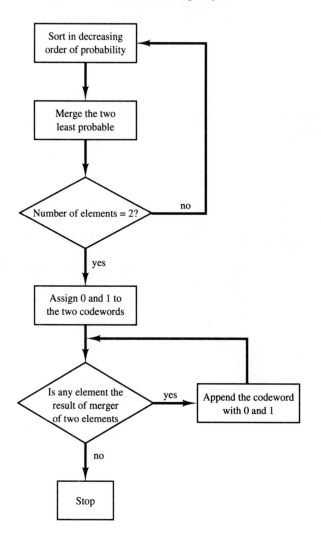

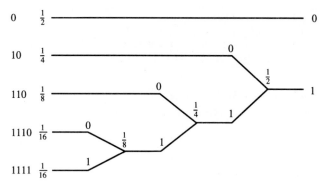

FIGURE 4.5. Huffman coding algorithm.

It can be shown that the average length of a Huffman code, defined by

$$\overline{R} = \sum_{x \in X} p(x)l(x) \qquad (4.3.1)$$

where $l(x)$ is the length of the codeword corresponding to the source output x, satisfies the following inequality

$$H(X) \leq \overline{R} < H(X) + 1 \qquad (4.3.2)$$

Now, if instead of single-source letters, the Huffman code was designed for sequences of source letters of length n (n^{th} extension of the source), we would have

$$H(\mathbf{X}^n) \leq \overline{R_n} < H(\mathbf{X}^n) + 1$$

where $\overline{R}_n$ denotes the average codeword length for the extended-source sequence and, therefore, $\overline{R} = \frac{1}{n} \overline{R}_n$. In case the source we are dealing with is memoryless, we also have $H(\mathbf{X}^n) = nH(X)$. Substituting these in the above equation and dividing by n, we have

$$H(X) \leq \overline{R} < H(X) + \frac{1}{n} \qquad (4.3.3)$$

Therefore, for n large enough $\overline{R}$ can be made as close to $H(X)$ as desired. It is also obvious that for discrete sources with memory, $\overline{R}$ approaches the entropy rate of the source.

Example 4.3.3

A d.m.s. with equiprobable outputs and alphabet $\mathcal{A} = \{a_1, a_2, a_3\}$ has the Huffman code designed below.

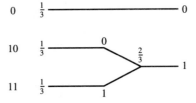

The entropy of the source is $H(X) = \log 3 = 1.585$ and $\overline{R} = \frac{5}{3} = 1.667$. If we use sequences of two letters, we will have the source

$$\mathcal{A}^2 = \{(a_1, a_1), (a_1, a_2), (a_1, a_3), \dots, (a_3, a_2), (a_3, a_3)\}$$

with the probability vector $\mathbf{p}^{(2)} = \{\frac{1}{9}, \frac{1}{9}, \frac{1}{9}, \frac{1}{9}, \frac{1}{9}, \frac{1}{9}, \frac{1}{9}, \frac{1}{9}, \frac{1}{9}\}$. A Huffman code for this source is designed in the following tree diagram.

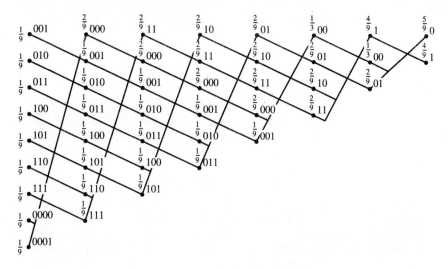

Here the average codeword length is $\overline{R}_2 = 3.222$ bits/pair of source outputs or 1.611 bits/each source output. Comparing this with the previous example, we see that the average length in this case is closer to the entropy of the source.

In our treatment, only binary Huffman codes were treated. The algorithm for design of binary Huffman codes can be easily generalized to an algorithm for the design of general M-ary Huffman codes. We will examine the general algorithm in a problem at the end of this chapter.

4.3.2 The Lempel-Ziv Source Coding Algorithm

We have seen already that Huffman codes are optimal in the sense that for a given source they provide a prefix code with minimum-average block length. Nevertheless, there exist major problems implementing Huffman codes. One problem is that Huffman codes depend strongly on the source probabilities (statistics). The source statistics have to be known in advance to design a Huffman code. If we can observe only the source outputs, then we have to do the coding in two passes. In the first pass, we estimate the statistics of the source (which, in the case of sources with memory and in cases where we want to apply Huffman coding to the extension of the source, becomes quite time consuming) and in the second pass coding is done. The other problem with Huffman codes is that if the code is designed for source blocks of length 1, it only employs variations in the frequency of the source outputs and not the source memory. If one wants to employ the source memory as well, one has to design the code for blocks of length 2 or more and this exponentially increases the complexity of the algorithm. For instance, encoding of ASCII characters with a block length of 1 requires a tree with 256 terminal nodes, but if a block length of 2 is desired the size of the tree and therefore the complexity of coding becomes much higher. In certain applications such as storage in magnetic or optical media, where the high-transfer rates are desirable, the complexity and speed of Huffman coding becomes a bottleneck.

The Lempel-Ziv algorithm belongs to the class of *universal source-coding algorithms,* i.e., algorithms that are independent of the source statistics. This algorithm is a *variable- to fixed-length* coding scheme. This means that any sequence of source outputs is uniquely parsed into phrases of varying length, and these phrases are encoded using codewords of equal length. Parsing is done by identifying phrases of the smallest length that have not appeared so far. To this end, the parser observes the source output, and as long as the new source output sequence after the last phrase coincides with one of the existing phrases, no new phrase is introduced and another letter from the source is considered. As soon as the new output sequence is different from the previous phrases, it is recognized as a new phrase and encoded. The encoding scheme is simple. The new phrase is the concatenation of a previous phrase and a new source output. To encode it, *the binary expansion of the lexicographic ordering of the previous phrase and the new bit are concatenated.* For example, let us assume that we want to parse and encode the following sequence:

$$0100001100001010000010100000110000010100001001001$$

Parsing the sequence by the rules explained before results in the following phrases:

$$0, 1, 00, 001, 10, 000, 101, 0000, 01, 010, 00001, 100, 0001, 0100, 0010, 01001$$

It is seen that all the phrases are different and each phrase is a previous phrase concatenated with a new source output. The number of phrases is 16. This means that for each phrase we need 4 bits, plus an extra bit to represent the new source output. The above sequence is encoded by

$$0000\ 0, 0000\ 1, 0001\ 0, 0011\ 1, 0010\ 0, 0011\ 0, 0101\ 1, 0110\ 0,$$
$$0001\ 1, 1001\ 0, 1000\ 1, 0101\ 0, 0110\ 1, 1010\ 0, 0100\ 0, 1110\ 1$$

Table 4.3.2 summarizes this procedure.

This representation can hardly be called a data-compression scheme because a sequence of length 49 has been mapped into a sequence of length 80, but as the length of the original sequence is increased the compression role of this algorithm becomes more apparent. It can be proved that for a stationary and ergodic source, as the length of the sequence increases, the number of bits in the compressed sequence approaches $nH(X)$, where $H(X)$ is the entropy rate of the source. The decompression of the encoded sequence is straightforward and can be done very easily.

One problem with the Lempel-Ziv algorithm is how the number of phrases should be chosen. Here we have chosen 16 phrases and therefore 4 bits to represent each phrase. In general, any fixed number of phrases will sooner or later become too small and cause an overflow. For example, if we were to continue coding the above source for additional input letters, we could not add the new phrases to our dictionary because we have assigned 4 bits for representation of the elements of the

TABLE 4-1   SUMMARY OF LEMPEL-ZIV
EXAMPLE

Dictionary Location		Dictionary Contents	Codeword	
1	0001	0	0000	0
2	0010	1	0000	1
3	0011	00	0001	0
4	0100	001	0011	1
5	0101	10	0010	0
6	0110	000	0011	0
7	0111	101	0101	1
8	1000	0000	0110	0
9	1001	01	0001	1
10	1010	010	1001	0
11	1011	00001	1000	1
12	1100	100	0101	0
13	1101	0001	0110	1
14	1110	0010	1010	0
15	1111	0010	0100	0
16			1110	1

dictionary and we have already 16 phrases in the dictionary. To solve this problem, the encoder and decoder must purge from their dictionaries those elements that are no longer useful and substitute new elements for them. The purging method should, of course, be a method on which the encoder and the decoder have agreed.

The Lempel-Ziv algorithm is used widely in practice to compress computer files. The "compress" and "uncompress" utilities under the UNIX© operating system and numerous algorithms under the MS-DOS© operating system (ZIP, ZOO, LZH, ARJ, etc.) are implementations of various versions of this algorithm.

4.4 RATE-DISTORTION THEORY

In Section 4.2, we saw that it is possible to encode the output of a memoryless information source at a rate arbitrarily close to its entropy and still recover the output reliably. Reliable reconstruction of the source means that the error probability can be made to approach zero as the block length approaches infinity. In many cases, however, transmitting at rates close to entropy is not possible. For example, if one has limited storage space and the entropy of the information source is greater than the storage capacity, error-free retrieval of the source output from the stored data is impossible. In cases such as this, some *lossy* compression technique has to be employed and some *distortion* will be introduced.

An example of such a case is the encoding of the output of an analog source. The samples of an analog source are real numbers, and to represent a real number an infinite number of bits is required. Therefore, in a practical digital system there exists no way to transmit or store analog data without any loss in precision. To encode analog samples, the samples must be quantized, and the quantized values are then encoded using one of the source-encoding algorithms.

In this section we will study the case of data compression subject to a fidelity criterion. We first introduce some quantities that are needed for quantitative analysis of this case and then define the rate-distortion function.

4.4.1 Mutual Information

For discrete random variables $H(X|Y)$ denotes the entropy (or uncertainty) of the random variable X after random variable Y is known. Therefore, if the starting entropy of the random variable X is $H(X)$, then $H(X) - H(X|Y)$ denotes the amount of uncertainty of X that has been removed by revealing random variable Y. In other words, $H(X) - H(X|Y)$ is the amount of information provided by the random variable Y about random variable X. This quantity plays an important role in both source and channel coding and is called the *mutual information* between two random variables.

Definition 4.4.1. The *mutual information* between two discrete random variables X and Y is denoted by $I(X; Y)$ and defined by

$$I(X; Y) = H(X) - H(X|Y). \tag{4.4.1}$$

Example 4.4.1

Let X and Y be binary random variables with $p(X = 0, Y = 0) = \frac{1}{3}$, $p(X = 1, Y = 0) = \frac{1}{3}$, and $p(X = 0, Y = 1) = \frac{1}{3}$. Find $I(X; Y)$ in this case.

Solution From above, $p(X = 0) = p(Y = 0) = \frac{2}{3}$, and, therefore, $H(X) = H(Y) = H_b(\frac{2}{3}) = 0.919$. On the other hand, the (X, Y) pair is a random vector uniformly distributed on three values $(0, 0)$, $(1, 0)$, and $(0, 1)$. Therefore, $H(X, Y) = \log 3 = 1.585$. From this, we have $H(X|Y) = H(X, Y) - H(Y) = 1.585 - 0.919 = 0.666$, and $I(X; Y) = H(X) - H(X|Y) = 0.919 - 0.666 = 0.253$

Mutual information has certain properties that are explored in problems and summarized here.

1. $I(X; Y) \geq 0$ with equality if and only if X and Y are independent.
2. $I(X; Y) \leq \min(H(X), H(Y))$.
3. $I(X; Y) = \sum_{x, y} p(x, y) \log \frac{p(x, y)}{p(x)p(y)}$.
4. $I(X; Y) = H(X) + H(Y) - H(X, Y)$.
5. $I(X; Y|Z)$ is the conditional mutual information and defined by $I(X; Y|Z) = H(X|Z) - H(X|Y, Z)$.
6. $I(X; Y|Z) = \sum_z p(z) I(X; Y|Z = z)$.
7. $I(XY; Z) = I(X; Z) + I(Y; Z|X)$. This is the chain rule for mutual information.

8. In general, $I(X_1, \ldots, X_n; Y) = I(X_1; Y) + I(X_2; Y|X_1) + \ldots + I(X_n; Y|X_1, \ldots, X_{n-1})$.

Figure 4.6 represents the relation among entropy, conditional entropy and mutual information quantities.

4.4.2 Differential Entropy

So far we have defined entropy and mutual information for discrete sources. If we are dealing with a discrete-time continuous-alphabet source whose outputs are real numbers, nothing exists that has the intuitive meaning of entropy. In the continuous case, another quantity that resembles entropy, called *differential entropy,* is defined. However, it does not have the intuitive meaning of entropy. In fact, to reconstruct the output of a continuous source reliably, an infinite number of bits per source output are required because any output of the source is a real number and the binary expansion of a real number has infinitely many bits.

Definition 4.4.2.   The *differential entropy* of a continuous random variable X with p.d.f. $f_X(x)$ is denoted by $h(X)$ and defined by

$$h(X) = -\int_{-\infty}^{+\infty} f_X(x) \log f_X(x) dx \qquad (4.4.2)$$

where $0 \log 0 = 0$.

Example 4.4.2

Determine the differential entropy of a random variable X uniformly distributed on $[0, a]$.

Solution   Using the definition of differential entropy

$$h(X) = -\int_0^a \frac{1}{a} \log \frac{1}{a} dx = \log a$$

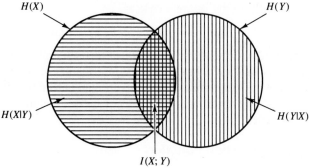

FIGURE 4.6.   Entropy, conditional entropy, and mutual information.

It is seen that for $a < 1$, we have $h(X) < 0$, which is in contrast to the nonnegativity of the discrete entropy. Also, for $a = 1$, $h(X) = 0$ without X being deterministic. This is again in contrast to the properties of discrete entropy.

Example 4.4.3

Determine the differential entropy of a Gaussian random variable with mean zero and variance σ^2.

Solution   The p.d.f. is $f(x) = \frac{1}{\sqrt{2\pi\sigma^2}} e^{-\frac{x^2}{2\sigma^2}}$. Therefore, using natural logarithms, we find the differential entropy in nats

$$h(X) = -\int_{-\infty}^{+\infty} \ln\left(\frac{1}{\sqrt{2\pi\sigma^2}}\right) f(x)dx - \int_{-\infty}^{+\infty} \ln\left(e^{-\frac{x^2}{2\sigma^2}}\right) f(x)dx$$

$$= \ln\left(\sqrt{2\pi\sigma^2}\right) + \frac{\sigma^2}{2\sigma^2} =$$

$$= \frac{1}{2} \ln\left(2\pi e\sigma^2\right) \quad \text{nats} \tag{4.4.3}$$

where we have used $\int_{-\infty}^{+\infty} f(x)dx = 1$ and $\int_{-\infty}^{+\infty} x^2 f(x)dx = \sigma^2$. Changing the logarithm base to 2, we have

$$h(X) = \frac{1}{2} \log(2\pi e\sigma^2) \quad \text{bits} \tag{4.4.4}$$

Extensions of the definition of differential entropy to joint random variables and conditional differential entropy are straightforward. For two random variables, we have

$$h(X, Y) = -\int_{-\infty}^{+\infty} \int_{-\infty}^{+\infty} f(x, y) \log f(x, y)dx\, dy \tag{4.4.5}$$

and

$$h(X|Y) = h(X, Y) - h(Y) \tag{4.4.6}$$

The mutual information between two continuous random variables X and Y is defined similarly to the discrete case as

$$I(X; Y) = h(Y) - h(Y|X) = h(X) - h(X|Y) \tag{4.4.7}$$

Although differential entropy does not have the intuitive interpretation of discrete entropy, it can be shown that the mutual information of continuous random variables has basically the same interpretation as the mutual information of discrete random variables, i.e., the information provided by one random variable about the other random variable.

4.4.3 Rate-Distortion Function

Returning to our original problem of representing a continuous source with finite number of bits/symbol, the question is: if the source output is to be compressed and represented by a certain number of bits/symbol, how close can the compressed version and the original version be? This question is not applicable only to continuous sources. A similar question can be asked for discrete sources. Let there be a discrete source with entropy $H(X)$. This means that we can transmit the output of this source using $H(X)$ bits/source output and with arbitrarily small-error probability. Now let us assume that using $H(X)$ bits/source symbol is too high and that we cannot tolerate it (e.g., in magnetic disk storage, the disk space required becomes huge). Now that the number of available bits/source output is less than $H(X)$, error-free restoration of the source is not possible and some errors will be inevitable. The question is; At a given number of bits/source symbol, what is the minimum achievable error rate and, of course, how can this minimum be achieved? The question can also be asked in the opposite direction, i.e., what is the minimum number of bits/source output required to reproduce the source at a certain level of distortion?

To formally present the above discussion, we have to define the notion of distortion. Distortion in reproduction of a source is a measure of fidelity or closeness of the reproduction to the original source output. In a high-fidelity reproduction, the reproduced signal is very close to the original signal and distortion is low, whereas in a low-fidelity reproduction there exists more "distance" between the original signal and the reproduction and, therefore, a higher distortion. A *distortion measure* is a measure of how far apart the signal and its reproduction are. One could think of many distortion measures between a signal $x(t)$ and its reproduction $\hat{x}(t)$, for instance, $\max_t |x(t) - \hat{x}(t)|$, $\lim_{T \to \infty} \frac{1}{T} \int_{-T/2}^{T/2} |x(t) - \hat{x}(t)| dt$, and $\lim_{T \to \infty} \frac{1}{T} \int_{-T/2}^{T/2} (x(t) - \hat{x}(t))^2 dt$ are three distortion measures.

A good distortion measure must satisfy two properties. First, it has to be a good approximation to the perception process, and second, it has to be simple enough to be mathematically tractable. The first property simply says that, for example, if we are dealing with speech transmission and in perception of speech the phase of the waveform is not a very crucial factor, the distortion measure should not be heavily dependent on exact phase reproduction. On the other hand, if we are dealing with image perception, phase plays an important role and this must therefore be reflected in our choice of the distortion measure. Finding a distortion measure that meets both of these requirements is usually not an easy task.

In general, a distortion measure is a distance between x and its reproduction $\hat{x}$, denoted by $d(x, \hat{x})$. In the discrete case a commonly used distortion measure is the *Hamming distortion,* defined by

$$d_H(x, \hat{x}) = \begin{cases} 1 & x \neq \hat{x} \\ 0, & \text{otherwise} \end{cases}$$

In the continuous case the *squared-error distortion,* defined by

$$d(x, \hat{x}) = (x - \hat{x})^2$$

is frequently used. It is also assumed that we are dealing with a *per letter distortion measure,* meaning that the distortion between sequences is the average of the distortion between their components, i.e.,

$$d(\mathbf{x}^n, \hat{\mathbf{x}}^n) = \frac{1}{n} \sum_{i=1}^{n} d(x_i, \hat{x}_i)$$

This assumption simply means that the position of the "error" in reproduction is not important and the distortion is not context dependent.

Now, since the source output is a random process, $d(\mathbf{X}^n, \hat{\mathbf{X}}^n)$ is a random variable. We define the distortion for the source as the expected value of this random variable,

$$D = E[d(\mathbf{X}^n, \hat{\mathbf{X}}^n)] = \frac{1}{n} \sum_{i=1}^{n} E[d(X_i, \hat{X}_i)] = E[d(X, \hat{X})]$$

where in the last step we have used the stationarity assumption on the source (independence of the distributions from the index i).

Example 4.4.4

Show that with Hamming distortion, D represents the error probability.

Solution

$$D = E[d_H(X, \hat{X})] = 1 \times p[X \neq \hat{X}] + 0 \times p[X = \hat{X}]$$
$$= p[X \neq \hat{X}] = p[\text{error}]$$

With all these definitions our original question can be restated as follows: given a memoryless information source with alphabet X and probability distribution $p(x)$, a reproduction alphabet $\hat{X}$, and a distortion measure $d(x, \hat{x})$ defined for all $x \in X$ and $\hat{x} \in \hat{X}$, what is R, the minimum number of bits/source output required to guarantee that the average distortion between the source-output sequence and the corresponding reproduction-output sequence does not exceed some given D? It is obvious that R is a decreasing function of D, i.e., if we need high-fidelity reproduction (low D), we require a high R. The relation between R and D is expressed via the *rate-distortion function*. The following theorem gives the general form of the rate-distortion function. For a proof, the reader is referred to the references at the end of this chapter.

Theorem 4.4.1 [Rate-Distortion]. The minimum number of bits/source output required to reproduce a memoryless source with distortion less than or equal to D is called the *rate-distortion function*, denoted by $R(D)$ and given by

$$R(D) = \min_{p(\hat{x}|x):Ed(X,\hat{X})\leq D} I(X; \hat{X}) \qquad (4.4.8)$$

Figure 4.7 is a schematic representation of this theorem. The space of source outputs of length n, i.e., X^n is divided into 2^{nR} regions. If the output of the source

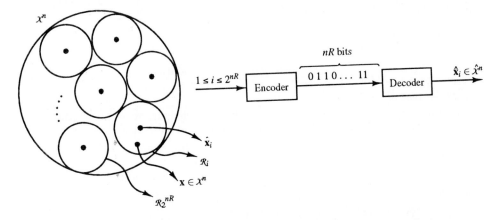

FIGURE 4.7.   Schematic representation of the rate-distortion theorem.

$\mathbf{x}^n$ falls in region i, the binary representation of i is transmitted to the decoder. Since $1 \leq i \leq 2^{nR}$, the binary representation is of length nR, and coding is done therefore at a rate of R bits/source output. The decoder, after receiving the binary representation of i, generates a predetermined sequence $\hat{\mathbf{x}}^n$ such that its average distance (distortion) from the $\mathbf{x}$ sequences in region i is minimum. This is the best representation of the region i sequences. Note that for large values of R, we have a large number of regions, and, therefore, our representation is very precise (fine quantization), whereas for small R, the number of regions is small and the distortion is large (coarse quantization). There exist two extreme cases. The first happens when there is only one region ($R = 0$). In this case, the representation point is in some sense (which will be clarified later) the centroid of the whole input space. The second extreme case is when each region consists of a single source output. In this case, R is taking its maximum value, and the distortion is zero.[†] From the rate-distortion theorem, it is seen that if the rate-distortion function is given for a source and a distortion measure, we know the minimum number of bits/source symbol required to reconstruct the source with any given distortion measure. Given any rate, we can determine the minimum achievable distortion if a code of that rate is used. It should be emphasized that, as was the case with the source-coding theorem, the results indicated by the rate-distortion function are "fundamental limits" in the sense that they can only be achieved asymptotically and with increasing complexity of the encoding–decoding schemes.

[†]The distortion is zero when the source alphabet and the representation alphabet are the same. In general, the minimum distortion is given by $D_{\min} = E[\min_{\hat{x}} d(X, \hat{x})] = \sum_x p(x) \min_{\hat{x}} d(x, \hat{x})$.

244 Information Sources and Source Coding Chap. 4

Example 4.4.5

For a binary memoryless source with $p(X_i = 1) = 1 - p(X_i = 0) = p$, and with Hamming distortion, it can be shown that the rate-distortion function is given by

$$R(D) = \begin{cases} H_b(p) - H_b(D) & 0 \leq D \leq \min\{p, 1-p\} \\ 0 & \text{otherwise} \end{cases} \qquad (4.4.9)$$

1. Assuming $p = 0.5$, how many bits/source output are required to transmit this information source with a probability of error at most equal to 0.25?
2. With $p = 0.5$ and a channel that can transmit 0.75 bits/each source output, what is the minimum achievable error probability?

Solution

1. Recall that for the Hamming distortion, the error probability and the average distortion coincide (see Example 4.4.4). Therefore, $P_e = D = 0.25$ and since $p = 0.5$, we are dealing with the case where $0 \leq D \leq \min\{p, 1-p\}$. This means that $R(0.25) = H_b(0.5) - H_b(0.25)$, which results in $R \approx 0.189$.
2. For $R = 0.75$, we must solve the equation $H_b(p) - H_b(D) = 0.75$, where $H_b(p) = H_b(0.5) = 1$ and, therefore, $H_b(D) = 0.25$, which gives $P_e = D = 0.042$.

A plot of the rate-distortion function for a binary source and with a Hamming distortion measure is given in Figure 4.8. Note that for zero distortion (zero-error

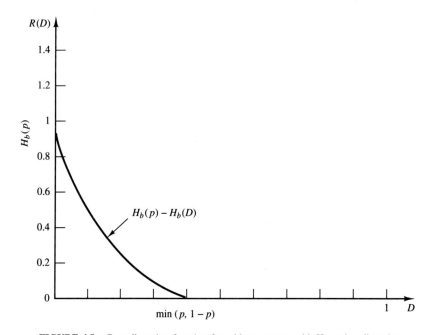

FIGURE 4.8. Rate-distortion function for a binary source with Hamming distortion.

probability), we have $R(D) = H_b(p)$, which is in agreement with the source-coding theorem. Assuming $p < 0.5$, for $D = p$ we have $R(D) = 0$, i.e., we can reproduce the source at a distortion of p with no transmission at all, by setting the reproduction vector to be the all-zero vector. This means that $D = p_e = P[X \neq \hat{X}] = P[X \neq 0] = P[X = 1] = p$.

It can also be shown that for a zero-mean Gaussian source with variance σ^2 and with the squared-error distortion measure, the rate-distortion function is given by

$$R(D) = \begin{cases} \frac{1}{2} \log \frac{\sigma^2}{D} & 0 \leq D \leq \sigma^2 \\ 0, & \text{otherwise} \end{cases} \qquad (4.4.10)$$

A plot of this rate-distortion function is given in Figure 4.9. An interesting question now is: By what factor is the distortion decreased if the rate of the coding of the source is increased by 1 bit? The answer, of course, depends on the rate-distortion function of the source (which in turn depends on the source statistics and the distortion measure). Let us consider the case of a zero-mean Gaussian d.m.s. with squared-error distortion. Since for $0 \leq D \leq \sigma^2$, $R(D) = \frac{1}{2} \log \frac{\sigma^2}{D}$, we can express the *distortion-rate function* as $D(R) = \sigma^2 2^{-2R}$. Obviously, increasing R by 1 will

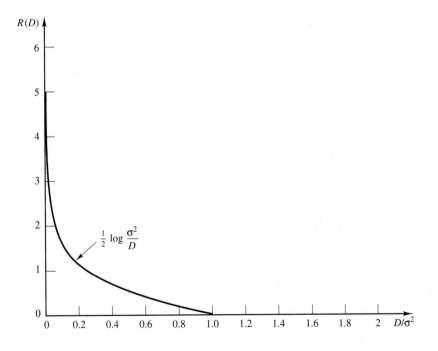

FIGURE 4.9. Rate-distortion function for a Gaussian source with squared-error distortion.

decrease D by a factor of 4, or equivalently by 6 dB. This means that every 1 bit of transmission/source output reduces the distortion by 6 dB.

Example 4.4.6

In the representation of a zero-mean unit-variance Gaussian source, what is the minimum attainable distortion if 8 bits/source output are employed? By what factor is the distortion decreased if we use 16 bits/source output?

Solution   Using the relation $D(R) = \sigma^2 2^{-2R}$ with $R = 8$ and $\sigma = 1$, we have $D = \frac{1}{2^{16}} \approx$ 1.52×10^{-5}. If instead of 8, 16 bits are employed, the distortion is reduced by 48 dB, or a factor of 4^8.

4.5 QUANTIZATION

In the previous section, we saw that when dealing with analog sources precise description of the source requires an infinite number of bits/source output, which is not an achievable goal. Therefore, in transmission of analog sources some distortion is always present, and the goal is to minimize this distortion. We also introduced the rate-distortion function, which gives a fundamental limit on the trade off between the code rate and the distortion. In this section, we will investigate practical schemes to represent the output of an analog source at low rates and, at the same time, without introducing excessive distortion. As we have already seen, the fundamental limit promised by the rate-distortion function can be approached only asymptotically, that is, by using very complex encoders and decoders. The encoder observes source outputs of length n, $\mathbf{x} \in X^n$ and maps them into representation sequences of length n, $\hat{\mathbf{x}}^n \in \hat{X}^n$. The number of the latter sequences is 2^{nR}, and, therefore, R bits/source output are required for their transmission. The larger the value of n, the closer to the rate-distortion limit the system operates. This means that an effective quantization scheme should work on blocks of source outputs rather than each output separately. Quantizers that operate on blocks of source output are called *vector quantizers* as opposed to *scalar quantizers*, which quantize each output separately. In this section, we will study both scalar and vector quantizers.

Aside from classifying quantizers as scalar and vector, one can classify quantizers (or, in general, source coders) on the basis of their general method for compressing data, as either *waveform coders* or *analysis-synthesis coders*. In waveform coding for data compression, the output of the source, which is a waveform, is compressed using one of several compression schemes. In this approach the mechanism by which the waveform is generated is not important, and the only important factors are the characteristics of the source output as a waveform, i.e., its bandwidth, power-spectral density, statistical properties, etc. Because, the mechanism by which the waveform is generated is not important, the results are very robust and can be applied to all sources regardless of their nature. In analysis-synthesis coders, the waveform is not directly compressed and transmitted. Instead, a model for production of the waveform is adopted, and the main parameters of that model are compressed and transmitted. For example, in speech coding the mechanism by which speech is produced can be modeled as a time-varying filter exited by either white noise or a sequence of impulses. In the analysis-synthesis approach to speech coding, parameters of the time-varying filter and its inputs are quantized (by scalar or vector quantization) and transmitted. At the receiving end, a filter that simulates the behavior of the vocal tract is generated and then exited by the appropriate input and, thus, a close replica of the waveform is generated. This approach is certainly a model-based approach and does not have the generality of the waveform-coding approach. On the positive side, model-based quantization schemes achieve better compression ratios compared to waveform coders.

4.5.1 Scalar Quantization

In scalar quantization each single-source output is quantized into a number of levels and these levels then are encoded into a binary sequence. In general, each source output is a real number, but transmission of real numbers requires an infinite number of bits. Therefore, it is required to map the set of real numbers into a finite set and at the same time minimize the distortion introduced. In scalar quantization the set of real numbers $\mathcal{R}$ is partitioned into N disjoint subsets denoted by $\mathcal{R}_k$, $1 \le k \le N$. Corresponding to each subset $\mathcal{R}_k$, a representation point $\hat{x}_k$, which usually belongs to $\mathcal{R}_k$, is chosen. If the source output at time i, x_i, belongs to $\mathcal{R}_k$, then it is represented by $\hat{x}_k$, which is the *quantized version* of x. $\hat{x}_k$ is then represented by a binary sequence and transmitted. Because there are N possibilities for the quantized levels, $\log N$ bits are enough to encode these levels into binary sequences (N is generally chosen to be a power of 2). Therefore, the number of bits required to transmit each source output is $R = \log N$ bits. The price that we have paid for a decrease in rate from infinity to $\log N$ is, of course, the introduction of distortion.

Figure 4.10 shows an example of an 8-level quantization scheme. In this scheme the 8 regions are defined as $\mathcal{R}_1 = (-\infty, a_1]$, $\mathcal{R}_2 = (a_1, a_2]$, ..., $\mathcal{R}_8 = (a_7, +\infty)$. The representation point (or quantized value) in each region is denoted by $\hat{x}_i$ and shown in the figure. The quantization function Q is defined by

$$Q(x) = \hat{x}_i \qquad \text{for all } x \in \mathcal{R}_i \tag{4.5.1}$$

This function is also shown in figure 4.10. As seen, the quantization function is a nonlinear function that is noninvertible. This is because all points in $\mathcal{R}_i$ are mapped into a single point $\hat{x}_i$. Because the quantization function is noninvertible, some information is lost in the process of quantization and this lost information is not recoverable.

If we are using the squared error distortion measure, then

$$d(x, \hat{x}) = (x - Q(x))^2 = \tilde{x}^2$$

where $\tilde{x} = x - \hat{x} = x - Q(x)$. Since X is a random variable, so are $\hat{X}$ and $\tilde{X}$; therefore

$$D = E[d(X, \hat{X})] = E[X - Q(X)]^2$$

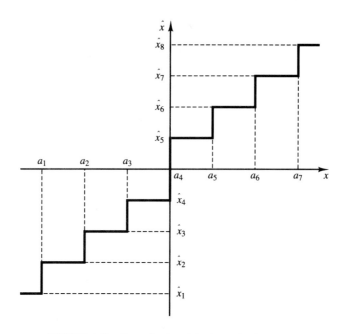

FIGURE 4.10.   Example of an 8-level quantization scheme.

Example 4.5.1

The source $X(t)$ is a stationary Gaussian source with mean zero and power-spectral density

$$S_x(f) = \begin{cases} 2 & |f| < 100 \text{ Hz} \\ 0 & \text{otherwise} \end{cases}$$

The source is sampled at Nyquist rate, and each sample is quantized using the 8-level quantizer shown in Figure 4.10 with $a_1 = -60, a_2 = -40, a_3 = -20, a_4 = 0, a_5 = 20, a_6 = 40, a_7 = 60$, and $\hat{x}_1 = -70, \hat{x}_2 = -50, \hat{x}_3 = -30, \hat{x}_4 = -10, \hat{x}_5 = 10, \hat{x}_6 = 30, \hat{x}_7 = 50, \hat{x}_8 = 70$. What is the resulting distortion and rate?

Solution   The sampling frequency is $f_s = 200$ Hz. Each sample is a zero-mean Gaussian random variable with variance

$$\sigma^2 = E(X_i^2) = R_X(\tau)|_{\tau=0} = \int_{-\infty}^{+\infty} S_X(f)df = \int_{-100}^{100} 2df = 400$$

Because each sample is quantized into 8 levels, then $\log 8 = 3$ bits are required per sample and, therefore, the required rate is

$$R = 3f_s = 600 \text{ bits/sec.}$$

To find the distortion we have to evaluate $E(X - \hat{X})^2$ for each sample. We will prove in a problem at the end of this chapter (see Problem 4.45) that, as long as the process is stationary, this is equal to $P_{\tilde{X}(t)}$, the power content of the process $\tilde{X}(t)$, or equivalently, the time-average distortion introduced by quantization. But

$$D = E[X - \hat{X}]^2 = \int_{-\infty}^{+\infty} (x - Q(x))^2 f_X(x)dx$$

where $f_X(x)$ denotes the p.d.f. of the random variable X. From here, we have

$$D = \sum_{i=1}^{8} \int_{\mathcal{R}_i} (x - Q(x))^2 f_X(x)dx$$

or, equivalently,

$$D = \int_{-\infty}^{a_1} (x - \hat{x}_1)^2 f_X(x)dx + \sum_{i=2}^{7} \int_{a_{i-1}}^{a_i} (x - \hat{x}_i)^2 f_X(x)dx$$

$$+ \int_{a_7}^{\infty} (x - \hat{x}_8)^2 f_X(x)\,dx \qquad (4.5.2)$$

where $f_X(x)$ is $\frac{1}{\sqrt{2\pi 400}} \exp(-x^2/800)$. Substituting $\{a_i\}_{i=1}^{7}$ and $\{\hat{x}_i\}_{i=1}^{8}$ in the above integral and evaluating the result with the Q-function table, we obtain $D \approx 33.38$. Note that if we were to use 0 bits/source output, then the best strategy would be to set the reconstructed signal equal to zero. In this case we would have a distortion of $D = E(X - 0)^2 = \sigma^2 = 400$. This quantization scheme and the transmission of 3 bits/source output has enabled us to reduce the distortion to 33.38. It is also interesting to compare the above result with the result predicted by the rate-distortion bound. Substituting $R = 3$ and $\sigma = 20$, in

$$R = \frac{1}{2} \log \frac{\sigma^2}{D}$$

we obtain $D = 6.25$. Obviously, the simple quantization scheme shown in Figure 4.10 is far from optimal. The reason for this poor performance is threefold. First and foremost, the rate-distortion bound is an asymptotic bound and holds for optimal mapping of *blocks* of source outputs when the length of the block tends to infinity, whereas, in this example, we have employed a scalar quantizer operating on single-source outputs. The second reason is that, even as a scalar quantizer with 8 levels (3 bits/source output), no attempt has been made to design an optimal quantizer by choosing the $\{a_i\}$ and $\hat{x}_i$'s appropriately. The third reason is that after quantization the 8 outputs $\{\hat{x}\}_{i=1}^{8}$ are not equiprobable and can be further compressed. In this example

$$p(\hat{x}_i) = \int_{a_{i-1}}^{a_i} \frac{1}{\sqrt{2\pi 400}} \exp(-x^2/800)dx \quad \text{for } 2 \le i \le 7$$

and

$$p(\hat{x}_1) = p(\hat{x}_8) = \int_{a_7}^{\infty} \frac{1}{\sqrt{2\pi 400}} \exp(-x^2/800)dx$$

which results in $p(\hat{x}_1) = p(\hat{x}_8) = .0014$, $p(\hat{x}_2) = p(\hat{x}_7) = 0.0214$, $p(\hat{x}_3) = p(\hat{x}_6) = 0.1359$, $p(\hat{x}_4) = p(\hat{x}_5) = 0.3414$. Using the source-coding theorem, we see that the output of the quantizer can be compressed further to $H(\hat{X}) = 2.105$ bits/source output. This means that it makes more sense to compare 33.38 with the value of the rate-distortion function at $R = 2.105$, which results in $D = 21.61$ as opposed to 6.25.

In the above example, we have chosen $E[X - Q(X)]^2$, which is called *mean-squared distortion* or *quantization noise,* as the measure of performance. A more meaningful measure of performance is a normalized version of the quantization noise, normalized with respect to the power of the original signal.

Definition 4.5.1.   If the random variable X is quantized to $Q(X)$, the *signal-to-quantization-noise ratio* (SQNR) is defined by

$$\text{SQNR} = \frac{E[X^2]}{E[X - Q(X)]^2} \tag{4.5.3}$$

When dealing with signals, the quantization-noise power is

$$P_{\tilde{X}} = \lim_{T \to \infty} \frac{1}{T} \int_{-\frac{T}{2}}^{\frac{T}{2}} E[X(t) - Q(X(t))]^2 dt \tag{4.5.4}$$

and the signal power is

$$P_X = \lim_{T \to \infty} \frac{1}{T} \int_{-\frac{T}{2}}^{\frac{T}{2}} E[X^2(t)] dt \tag{4.5.5}$$

Hence, the SQNR is

$$\text{SQNR} = \frac{P_X}{P_{\tilde{X}}} \tag{4.5.6}$$

It can be shown (see Problem 4.45 at the end of this chapter) that if $X(t)$ is stationary, then the above relation simplifies to Equation 4.5.3 where X is the random variable representing $X(t)$ at any point.

Uniform quantization.   Uniform quantizers are the simplest examples of scalar quantizers. In a uniform quantizer the entire real line is partitioned into N regions. All regions except $\mathcal{R}_1$ and $\mathcal{R}_N$ are of equal length, which is denoted by Δ. This means that for all $1 \le i \le N - 1$, we have $a_{i+1} - a_i = \Delta$. Figure 4.10 is an example of an 8-level uniform quantizer. In a uniform quantizer the distortion is given by

$$D = \int_{-\infty}^{a_1} (x - \hat{x}_1)^2 f_X(x) dx + \sum_{i=1}^{N-2} \int_{a_1+(i-1)\Delta}^{a_1+i\Delta} (x - \hat{x}_{i+1})^2 f_X(x) dx$$

$$+ \int_{a_1+(N-2)\Delta}^{\infty} (x - \hat{x}_N)^2 f_X(x) dx \tag{4.5.7}$$

It is seen from above that D is a function of $N + 2$ design parameters, namely a_1, Δ, and $\{\hat{x}_i\}_{i=1}^N$. To design the optimal uniform quantizer, one has to differentiate D with respect to the above variables and find the values that minimize D.

Further assumptions simplify the above relation to some extent. If we assume that $f_X(x)$ is an even function of x (symmetric density function), then the best quantizer will also have symmetry properties. This means that for even N, we will have $a_i = -a_{N-i} = -(\frac{N}{2} - i)\Delta$ for all $1 \le i \le \frac{N}{2}$ (which means $a_{\frac{N}{2}} = 0$), and $\hat{x}_i = -\hat{x}_{N+1-i}$ for $1 \le i \le \frac{N}{2}$. In this case we have

$$D = 2 \int_{-\infty}^{(-\frac{N}{2}+1)\Delta} (x - \hat{x}_1)^2 f_X(x)\,dx$$

$$+ 2 \sum_{i=1}^{\frac{N}{2}-1} \int_{(-\frac{N}{2}+i)\Delta}^{(-\frac{N}{2}+i+1)\Delta} (x - \hat{x}_{i+1})^2 f_X(x)\,dx \qquad (4.5.8)$$

When N is odd, we have a situation such as the one shown in Figure 4.11. In this case $a_i = -a_{N-i} = (-\frac{N}{2} + i)\Delta$ for $1 \le i \le \frac{N-1}{2}$ and $\hat{x}_i = -\hat{x}_{N+1-i}$ for $1 \le i \le \frac{N+1}{2}$, which means $\hat{x}_{\frac{N+1}{2}} = 0$. The distortion is given by

$$D = 2 \int_{-\infty}^{(-\frac{N}{2}+1)\Delta} (x - \hat{x}_1)^2 f_X(x)\,dx$$

$$+ 2 \sum_{i=1}^{\frac{N-3}{2}} \int_{(-\frac{N}{2}+i)\Delta}^{(-\frac{N}{2}+i+1)\Delta} (x - \hat{x}_{i+1})^2 f_X(x)\,dx$$

$$+ \int_{-\frac{\Delta}{2}}^{\frac{\Delta}{2}} x^2 f_X(x)\,dx \qquad (4.5.9)$$

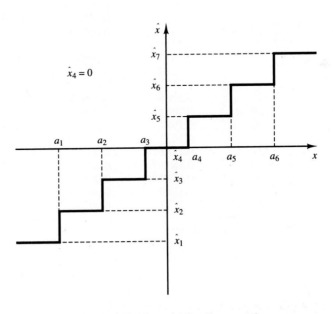

FIGURE 4.11.   Seven-level uniform quantizer.

TABLE 4-2   OPTIMAL UNIFORM QUANTIZER FOR A
GAUSSIAN SOURCE

No. output levels N	Output-level spacing Δ	Mean-squared error D	Informational entropy $H(\hat{x})$
1	—	1.000	0.0
2	1.596	0.3634	1.000
3	1.224	0.1902	1.536
4	0.9957	0.1188	1.904
5	0.8430	0.08218	2.183
6	0.7334	0.06065	2.409
7	0.6508	0.04686	2.598
8	0.5860	0.03744	2.761
9	0.5338	0.03069	2.904
10	0.4908	0.02568	3.032
11	0.4546	0.02185	3.148
12	0.4238	0.01885	3.253
13	0.3972	0.01645	3.350
14	0.3739	0.01450	3.440
15	0.3534	0.01289	3.524
16	0.3352	0.01154	3.602
17	0.3189	0.01040	3.676
18	0.3042	0.009430	3.746
19	0.2909	0.008594	3.811
20	0.2788	0.007869	3.874
21	0.2678	0.007235	3.933
22	0.2576	0.006678	3.990
23	0.2482	0.006185	4.045
24	0.2396	0.005747	4.097
25	0.2315	0.005355	4.146
26	0.2240	0.005004	4.194
27	0.2171	0.004687	4.241
28	0.2105	0.004401	4.285
29	0.2044	0.004141	4.328
30	0.1987	0.003905	4.370
31	0.1932	0.003688	4.410
32	0.1881	0.003490	4.449
33	0.1833	0.003308	4.487
34	0.1787	0.003141	4.524
35	0.1744	0.002986	4.560
36	0.1703	0.002843	4.594

From Max (1960) © IEEE

Minimization of distortion in these cases, although much simpler compared to the general case, is still a tedious task and is done mainly by numerical techniques. Table 4.2 gives the optimal quantization level spacing for a zero-mean, unit variance Gaussian random variable when the $\hat{x}$'s are chosen to be the mid-points of the quantization regions.

Nonuniform quantization.   If we relax the condition that the quantization regions (except for the first and the last one) be of equal length, we are then minimizing the distortion with less constraints and the resulting nonuniform quantizer will perform better compared to a uniform quantizer with the same number of levels. Let us assume that we are interested in designing the optimal quantizer with N levels of quantization with no other constraint on the regions. The distortion will be given by

$$D = \int_{-\infty}^{a_1} (x - \hat{x}_1)^2 f_X(x)dx + \sum_{i=1}^{N-2} \int_{a_i}^{a_{i+1}} (x - \hat{x}_{i+1})^2 f_X(x)dx$$

$$+ \int_{a_{N-1}}^{\infty} (x - \hat{x}_N)^2 f_X(x)dx \qquad (4.5.10)$$

There exist a total of $2N - 1$ variables in the above expression and the minimization of D is to be done with respect to these variables. Differentiating with respect to a_i yields

$$\frac{\partial}{\partial a_i} D = f_X(a_i)[(a_i - \hat{x}_i)^2 - (a_i - \hat{x}_{i+1})^2] = 0 \qquad (4.5.11)$$

which results in

$$a_i = \frac{1}{2}(\hat{x}_i + \hat{x}_{i+1}) \qquad (4.5.12)$$

This result simply means that, in an optimal quantizer, *the boundaries of the quantization regions are the midpoints of the quantized values.* Because quantization is done on a minimum distance basis, each x value is quantized to the nearest $\{\hat{x}_i\}_{i=1}^{N}$.

To determine the quantized values $\hat{x}_i$, we differentiate D with respect to $\hat{x}_i$ and define $a_0 = -\infty$ and $a_N = +\infty$. Thus, we obtain

$$\frac{\partial}{\partial \hat{x}_i} D = \int_{a_{i-1}}^{a_i} 2(x - \hat{x}_i) f_X(x)dx = 0 \qquad (4.5.13)$$

which results in

$$\hat{x}_i = \frac{\int_{a_{i-1}}^{a_i} x f_X(x)dx}{\int_{a_{i-1}}^{a_i} f_X(x)dx}$$

$$= \frac{\int_{a_{i-1}}^{a_i} x f_X(x)dx}{p(a_{i-1} < X \le a_i)}$$

$$= \int_{a_{i-1}}^{a_i} x \, \frac{f_X(x)}{p(a_{i-1} < X \le a_i)} \, dx$$

$$= \int_{-\infty}^{+\infty} x f_X(x|a_{i-1} < X \le a_i) dx$$

$$= E[X|a_{i-1} < X \le a_i] \qquad (4.5.14)$$

where we have used the definition of the conditional density function

$$f_X(x|a_{i-1} < X \le a_i) = \begin{cases} \frac{f_X(x)}{p(a_{i-1} < X \le a_i)} & a_{i-1} < x \le a_{i-1} \\ 0, & \text{otherwise} \end{cases}$$

Equation (4.5.14) shows that in an optimal quantizer *the quantized value (or representation point) for a region should be chosen to be the centroid (conditional expected value) of that region.* Equations (4.5.12) and (4.5.14) give the necessary conditions for a scalar quantizer to be optimal and are known as the Lloyd-Max conditions. The criteria for optimal quantization can then be summarized as

1. The boundaries of the quantization regions are the midpoints of the corresponding quantized values (nearest-neighbor rule).
2. The quantized values are the centroids of the quantization regions.

Although these rules are very simple, they do not result in analytical solutions to the optimal quantizer design. The usual method of designing the optimal quantizer is to start with a set of quantization regions and then find the quantized values using the second criterion. Then we design new quantization regions for the new quantized values and alternate between the two steps until the distortion does not change much from one step to the next. Based on this method one can design the optimal quantizer for various source statistics. Table 4.3 shows the optimal nonuniform quantizers for various values of N for a zero-mean unit variance Gaussian source.

Example 4.5.2

How would the results of Example 4.5.1 change if instead of the uniform quantizer shown in Figure 4.10, we used an optimal nonuniform quantizer with the same number of levels?

Solution We can find the quantization regions and the quantized values from Table 4.3 with $N = 8$. It should be noted that this table is designed for a unit variance Gaussian source, and to obtain the values for the source under study, all the values read from the table should be multiplied by the σ of the source, in this case 20. This gives us the values $a_1 = -a_7 = -34.96, a_2 = -a_6 = -21, a_3 = -a_5 = -10.012, a_4 = 0$ and $\hat{x}_1 = -\hat{x}_8 = -43.04, \hat{x}_2 = -\hat{x}_7 = -26.88, \hat{x}_3 = -\hat{x}_6 = -15.12, \hat{x}_4 = -\hat{x}_5 = -4.902$. Based on these values, the distortion is evaluated to be $D = 13.816$. The SQNR is

$$\text{SQNR} = \frac{400}{13.816} = 28.95 \sim 14.62 \text{ dB}$$

TABLE 4-3   OPTIMAL NONUNIFORM QUANTIZER FOR A GAUSSIAN SOURCE

N	$\pm a_i$	$\pm \hat{x}_i$	D	$H(\hat{X})$
1	—	0	1	0
2	0	0.7980	0.3634	1
3	0.6120	0, 1.224	0.1902	1.536
4	0, 0.9816	0.4528, 1.510	0.1175	1.911
5	0.3823, 1.244	0, 0.7646, 1.724	0.07994	2.203
6	0, 0.6589, 1.447	0.3177, 1.000, 1.894	0.05798	2.443
7	0.2803, 0.8744, 1.611	0, 0.5606, 1.188, 2.033	0.04400	2.647
8	0, 0.5006, 1.050, 1.748	0.2451, 0.7560, 1.344, 2.152	0.03454	2.825
9	0.2218, 0.6812, 1.198, 1.866	0, 0.4436, 0.9188, 1.476, 2.255	0.02785	2.983
10	0, 0.4047, 0.8339, 1.325, 1.968	0.1996, 0.6099, 1.058, 1.591, 2.345	0.02293	3.125
11	0.1837, 0.5599, 0.9656, 1.436, 2.059	0, 0.3675, 0.7524, 1.179, 1.693, 2.426	0.01922	3.253
12	0, 0.3401, 0.6943, 1.081, 1.534, 2.141	0.1684, 0.5119, 0.8768, 1.286, 1.783, 2.499	0.01634	3.372
13	0.1569, 0.4760, 0.8126, 1.184, 1.623, 2.215	0, 0.3138, 0.6383, 0.9870, 1.381, 1.865, 2.565	0.01406	3.481
14	0, 0.2935, 0.5959, 0.9181, 1.277, 1.703, 2.282	0.1457, 0.4413, 0.7505, 1.086, 1.468, 1.939, 2.625	0.01223	3.582
15	0.1369, 0.4143, 0.7030, 1.013, 1.361, 1.776, 2.344	0, 0.2739, 0.5548, 0.8512, 1.175, 1.546, 2.007, 2.681	0.01073	3.677
16	0, 0.2582, 0.5224, 0.7996, 1.099, 1.437, 1.844, 2.401	0.1284, 0.3881, 0.6568, 0.9424, 1.256, 1.618, 2.069, 2.733	0.009497	3.765
17	0.1215, 0.3670, 0.6201, 0.8875, 1.178, 1.508, 1.906, 2.454	0, 0.2430, 0.4909, 0.7493, 1.026, 1.331, 1.685, 2.127, 2.781	0.008463	3.849
18	0, 0.2306, 0.4653, 0.7091, 0.9680, 1.251, 1.573, 1.964, 2.504	0.1148, 0.3464, 0.5843, 0.8339, 1.102, 1.400, 1.746, 2.181, 2.826	0.007589	3.928
19	0.1092, 0.3294, 0.5551, 0.7908, 1.042, 1.318, 1.634, 2.018, 2.55	0, 0.2184, 0.4404, 0.6698, 0.9117, 1.173, 1.464, 1.803, 2.232, 2.869	0.006844	4.002
20	0, 0.2083, 0.4197, 0.6375, 0.8661, 1.111, 1.381, 1.690, 2.068, 2.594	0.1038, 0.3128, 0.5265, 0.7486, 0.9837, 1.239, 1.524, 1.857, 2.279, 2.908	0.006203	4.074
21	0.09918, 0.2989, 0.5027, 0.7137, 0.9361, 1.175, 1.440, 1.743, 2.116, 2.635	0, 0.1984, 0.3994, 0.6059, 0.8215, 1.051, 1.300, 1.579, 1.908, 2.324, 2.946	0.005648	4.141
22	0, 0.1900, 0.3822, 0.5794, 0.7844, 1.001, 1.235, 1.495, 1.793, 2.160, 2.674	0.09469, 0.2852, 0.4793, 0.6795, 0.8893, 1.113, 1.357, 1.632, 1.955, 2.366, 2.982	0.005165	4.206
23	0.09085, 0.2736, 0.4594, 0.6507, 0.8504, 1.062, 1.291, 1.546, 1.841, 2.203, 2.711	0, 0.1817, 0.3654, 0.5534, 0.7481, 0.9527, 1.172, 1.411, 1.682, 2.000, 2.406, 3.016	0.004741	4.268
24	0, 0.1746, 0.3510, 0.5312, 0.7173, 0.9122, 1.119, 1.344, 1.595, 1.885, 2.243, 2.746	0.08708, 0.2621, 0.4399, 0.6224, 0.8122, 1.012, 1.227, 1.462, 1.728, 2.042, 2.444, 3.048	0.004367	4.327

TABLE 4-3   CONTINUED

N	$\pm a_i$	$\pm \hat{x}_i$	D	$H(\hat{X})$
25	0.08381, 0.2522, 0.4231, 0.5982, 0.7797, 0.9702, 1.173, 1.394, 1.641, 1.927, 2.281, 2.779	0, 0.1676, 0.3368, 0.5093, 0.6870, 0.8723, 1.068, 1.279, 1.510, 1.772, 2.083, 2.480, 3.079	0.004036	4.384
26	0, 0.1616, 0.3245, 0.4905, 0.6610, 0.8383, 1.025, 1.224, 1.442, 1.685, 1.968, 2.318, 2.811	0.08060, 0.2425, 0.4066, 0.5743, 0.7477, 0.9289, 1.121, 1.328, 1.556, 1.814, 2.121, 2.514, 3.109	0.003741	4.439
27	0.07779, 0.2340, 0.3921, 0.5587, 0.7202, 0.8936, 1.077, 1.273, 1.487, 1.727, 2.006, 2.352, 2.842	0, 0.1556, 0.3124, 0.4719, 0.6354, 0.8049, 0.9824, 1.171, 1.374, 1.599, 1.854, 2.158, 2.547, 3.137	0.003477	4.491
28	0, 0.1503, 0.3018, 0.04556, 0.6132, 0.7760, 0.9460, 1.126, 1.319, 1.529, 1.766, 2.042, 2.385, 2.871	0.07502, 0.2256, 0.3780, 0.5333, 0.6930, 0.8589, 1.033, 1.118, 1.419, 1.640, 1.892, 2.193, 2.578, 3.164	0.003240	4.542

From Max (1960) © IEEE

If one computes the probability of each quantized value using the relation

$$p(\hat{x}_i) = \int_{a_{i-1}}^{a_i} f_X(x)\,dx$$

and finds the entropy of the random variable $\hat{X}$, one would obtain $H(\hat{X}) = 2.825$ bits/source output. The distortion obtained from the rate-distortion function with $R = 2.825$ and with $\sigma = 20$ is $D = 7.966$. The difference between 13.816 and 7.966 is purely the difference between a quantization scheme based on single outputs (scalar quantization) and blocks of source outputs (vector quantization).

4.5.2 Vector Quantization

In scalar quantization each output of the discrete-time source (which is usually the result of a sampling of a continuous-time source) is quantized separately and then encoded. For example, if we are using a 4-level scalar quantizer and encoding each level into 2 bits, we are using 2 bits/each source output. This quantization scheme is shown in Figure 4.12.

Now if we consider two samples of the source at each time, and interpret these two samples as a point in the plane, the scalar quantizer partitions the entire plane into 16 quantization regions as shown in Figure 4.13. It is seen that the regions in the two-dimensional space are all of rectangular shape. If we allow 16 regions of any shape in the two-dimensional space, we are capable of obtaining better results. This means that we are quantizing 2 source outputs at a time using 16 regions, which is equivalent to 4 bits/2 source outputs or 2 bits/each source output. Therefore, the number of bits/source output for quantizing two samples at a time is equal to the number of bits/source output obtained in the scalar case. Because we

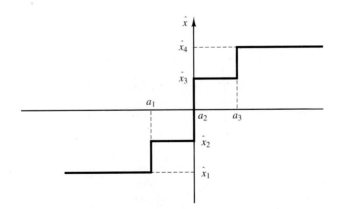

FIGURE 4.12.   Four-level scalar quantizer.

are relaxing the requirement of having rectangular regions, the performance will be improved. Now, if we take three samples at a time and quantize the entire three-dimensional space into 64 regions, we will have even less distortion with the same number of bits/source output. The idea of vector quantization is to take blocks of source outputs of length n and design the quantizer in the n-dimensional Euclidean space rather than doing the quantization based on single samples in one-dimensional space.

Let us assume that the quantization regions in the n-dimensional space are denoted by $\mathcal{R}_i$, $1 \leq i \leq K$. These K regions partition the n-dimensional space. Each block of source output of length n is denoted by $\mathbf{x} \in \mathbb{R}^n$ and if $\mathbf{x} \in \mathcal{R}_i$, it is quantized to $Q(\mathbf{x}) = \hat{\mathbf{x}}_i$. Figure 4.14 shows this quantization scheme for $n = 2$ and $K = 37$. Because there are a total of K quantized values, $\log K$ bits are enough to represent these values. This means that we require $\log K$ bits/n source outputs or, the rate of the source code is

$$R = \frac{\log K}{n} \text{ bits/source output} \qquad (4.5.15)$$

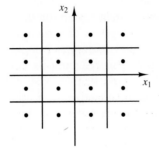

FIGURE 4.13.   Scalar 4-level quantization applied to two samples.

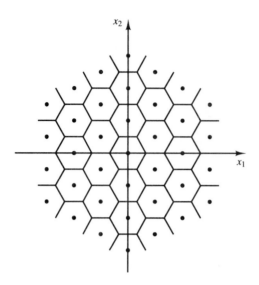

FIGURE 4.14.   Vector quantization in two dimensions.

The optimal vector quantizer of dimension n and number of levels K is the one that chooses the region $\mathcal{R}_i$'s and the quantized values $\hat{\mathbf{x}}_i$'s such that the resulting distortion is minimized. Applying the same procedure that we used for the case of scalar quantization, we obtain the following criteria for an optimal vector quantizer design:

1. Region $\mathcal{R}_i$ is the set of all points in the n-dimensional space that are closer to $\hat{\mathbf{x}}_i$ than any other $\hat{\mathbf{x}}_j$, for all $j \neq i$.

$$\mathcal{R}_i = \{\mathbf{x} \in \mathcal{R}^n : \|\mathbf{x} - \hat{\mathbf{x}}_i\| < \|\mathbf{x} - \hat{\mathbf{x}}_j\|, \; \forall j \neq i\}$$

2. $\hat{\mathbf{x}}_i$ is the centroid of the region $\mathcal{R}_i$.

$$\hat{\mathbf{x}}_i = \frac{1}{p(\mathbf{X} \in \mathcal{R}_i)} \int \int \ldots \int_{\mathcal{R}_i} \mathbf{x} f_{\mathbf{X}}(\mathbf{x}) d\mathbf{x}$$

A practical approach to designing optimal vector quantizers is based on the same approach employed in designing optimal scalar quantizers. Starting from a given set of quantization regions, we derive the optimal quantized vectors for these regions using the second criterion, then repartition the space using the first criterion, and going back and forth until changes in distortion are negligible.

Vector quantization has found widespread applications in speech and image coding, and numerous algorithms for reducing its computational complexity have been proposed. It can be also proved that for stationary and ergodic sources, the performance of the optimal vector quantizer approaches the optimal performance given by the rate-distortion function as n becomes large.

4.6 WAVEFORM CODING

Waveform-coding schemes are designed to reproduce the waveform output of the source at the destination with as small a distortion as possible. In these techniques no attention is paid to the mechanism that produces the waveform, and all attempts are directed at reproduction of the source output at the destination with high fidelity. Because the structure of the source plays no role in the design of waveform coders and only properties of the waveform affect the design, waveform coders are robust and can be used with a variety of sources as long as the waveforms produced by the sources have certain similarities. In this section we study some basic waveform-coding methods that are widely applied to a variety of sources.

4.6.1 Pulse-Code Modulation

Pulse-code modulation (PCM) is the simplest and oldest waveform-coding scheme. A pulse-code modulator consists of three basic sections: a sampler, a quantizer, and an encoder. A functional block diagram of a PCM system is shown in Figure 4.15.

The waveform entering the sampler is a bandlimited waveform with bandwidth W. Usually there exists a filter with bandwidth W prior to the sampler to prevent any components beyond W from entering the sampler. This filter is called the *presampling filter*. The sampling is done at a rate higher than the Nyquist rate to allow for some guard-band. The sampled values then enter a scalar quantizer. The quantizer is either a uniform quantizer, which results in a uniform PCM system, or a nonuniform quantizer. The choice of the quantizer is based on the characteristics of the source output. The output of the quantizer is then encoded into a binary sequence of length ν where $N = 2^{\nu}$ is the number of quantization levels.

Uniform PCM.   In uniform PCM applications, it is assumed that the range of the input samples is $[-x_{\max}, +x_{\max}]$ and the number of quantization levels N is a power of 2, $N = 2^{\nu}$. From this, the length of each quantization region is given by

$$\Delta = \frac{2x_{\max}}{N} = \frac{x_{\max}}{2^{\nu-1}} \tag{4.6.1}$$

The quantized values in uniform PCM are chosen to be the midpoints of the quantization regions, and therefore the error $\tilde{x} = x - Q(x)$ is a random variable taking values in the interval $(-\frac{\Delta}{2}, +\frac{\Delta}{2}]$. In ordinary PCM applications, the number of levels (N) is usually high, and the range of variations of the input signal (amplitude variations $x_{\max}$) is small. This means that the length of each quantization region (Δ) is small and, under these assumptions, in each quantization region the error

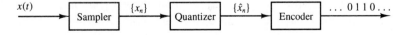

FIGURE 4.15.   Block diagram of a PCM system.

$\check{X} = X - Q(X)$ can be well approximated by a uniformly distributed random variable on $(-\frac{\Delta}{2}, +\frac{\Delta}{2}]$. The distortion introduced by quantization (quantization noise) is therefore

$$E[\tilde{X}^2] = \int_{-\frac{\Delta}{2}}^{+\frac{\Delta}{2}} \frac{1}{\Delta} \tilde{x}^2 d\tilde{x} = \frac{\Delta^2}{12} = \frac{x_{max}^2}{3N^2} = \frac{x_{max}^2}{3 \times 4^\nu} \qquad (4.6.2)$$

where ν is the number of bits per source sample. The SQNR then becomes

$$SQNR = \frac{\overline{X^2}}{\overline{\tilde{X}^2}} = \frac{3 \times N^2 \overline{X^2}}{x_{max}^2} = \frac{3 \times 4^\nu \overline{X^2}}{x_{max}^2} \qquad (4.6.3)$$

If we denote the normalized X by $\check{X}$, that is, $\check{X} = \frac{X}{x_{max}}$, then

$$SQNR = 3 \times N^2 \overline{\check{X}^2} = 3 \times 4^\nu \overline{\check{X}^2} \qquad (4.6.4)$$

Note that by definition $|\check{X}| \le 1$ and therefore $\overline{\check{X}^2} \le 1$. This means that $3N^2 = 3 \times 4^\nu$ is an upperbound to the SQNR in uniform PCM. This also means that the SQNR in uniform PCM deteriorates as the dynamic range of the source increases because an increase in the dynamic range of the source results in a decrease in $\overline{\check{X}^2}$. In a problem at the end of this chapter (Problem 4.57), we will see that this sensitivity to the source dynamic range can be improved by employing nonuniform PCM.

Expressing the SQNR in dB produces

$$SQNR\Big|_{dB} \approx P_{\check{X}}\Big|_{dB} + 6\nu + 4.8 \qquad (4.6.5)$$

It is seen that each extra bit (increase in ν by one) increases the SQNR by 6 dB. This increase is comparable to that of an optimal system, as was shown in Section 4.4.

Example 4.6.1

What is the resulting SQNR for a signal uniformly distributed on $[-1, 1]$ when uniform PCM with 256 levels is employed.

Solution   Because $x_{max} = 1$, then $\check{X} = X$ and $P_{\check{X}} = \int_{-1}^{1} \frac{1}{2} x^2 dx = \frac{1}{3}$. Therefore, using $\nu = \log 256 = 8$, we have

$$SQNR = 3 \times 4^\nu \overline{\check{X}^2} = 4^\nu = 65536 \sim 48.16 \text{ dB}$$

The issue of bandwidth requirements of pulse-transmission systems, of which PCM is an example, is dealt with in detail in Chapter 8. Here we briefly discuss some results concerning the bandwidth requirements of a PCM system. If a signal has a bandwidth of W, then the minimum number of samples for perfect reconstruction of the signal is given by the sampling theorem and is equal to $2W$ samples/sec. If some guard-band is required, then the number of samples/sec is f_s, which is more than $2W$. For each sample ν bits are used, therefore a total of νf_s bits/sec are required for transmission of the PCM signal. In the case of sampling at the

Nyquist rate, this is equal to $2\nu W$ bits/sec. The minimum bandwidth requirement for transmission of R bits/sec (or, more precisely, R pulses per second) is $\frac{R}{2}$ (see Chapter 8).[†] Therefore the minimum bandwidth requirement of a PCM system is

$$\mathrm{BW} = \frac{\nu f_s}{2} \tag{4.6.6}$$

which, in the case of sampling at the Nyquist rate, gives the absolute minimum bandwidth requirement as

$$\mathrm{BW} = \nu W \tag{4.6.7}$$

This means that a PCM system expands the bandwidth of the original signal by a factor of at least ν.

Nonuniform PCM.   As long as the statistics of the input signal are close to the uniform distribution, uniform PCM works fine. However, in coding of certain signals such as speech, the input distribution is far from being uniformly distributed. For a speech waveform, in particular, there exists a higher probability for smaller amplitudes and lower probability for larger amplitudes. Therefore, it makes sense to design a quantizer with more quantization regions at lower amplitudes and less quantization regions at larger amplitudes. The resulting quantizer will be a nonuniform quantizer having quantization regions of various sizes.

The usual method for performing nonuniform quantization[†] is to first pass the samples through a nonlinear element that compresses the large amplitudes (reduces dynamic range of the signal) and then perform a uniform quantization on the output. At the receiving end, the inverse (expansion) of this nonlinear operation is applied to obtain the sampled value. This technique is called *companding* (*comp*ressing-exp*anding*). A block diagram of this system is shown in Figure 4.16.

There are two types of companders that are widely used for speech coding. The μ-law compander used in the United States and Canada employs the logarithmic function at the transmitting side, with $|x| \leq 1$,

$$g(x) = \frac{\log(1 + \mu|x|)}{\log(1 + \mu)} \, \mathrm{sgn}(x) \tag{4.6.8}$$

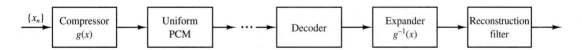

FIGURE 4.16.   Block diagram of a nonuniform PCM system.

[†] A more practical bandwidth requirement is $\frac{R}{\alpha}$, where $1 < \alpha < 2$.
[†] Sometimes the term *nonlinear quantization* is used, which is misleading because all quantization schemes, uniform or nonuniform, are nonlinear.

The parameter μ controls the amount of compression and expansion. The standard PCM system in the United States and Canada employs a compressor with $\mu = 255$, followed by a uniform quantizer with 128 levels (7 bits/sample). Use of a compander in this system improves the performance of the system by about 24 dB. A plot of the μ-law compander characteristics is shown in Figure 4.17.

The second widely used logarithmic compressor is the A-law compander. The characteristic of this compander is shown in Figure 4.18 and is given by

$$g(x) = \begin{cases} \dfrac{1 + \ln A|x|}{1 + \ln A} \, \mathrm{sgn}(x), & \frac{1}{A} \le |x| \le 1 \\[3mm] \dfrac{A|x|}{1 + \ln A} \, \mathrm{sgn}(x), & 0 \le |x| \le \frac{1}{A} \end{cases} \tag{4.6.9}$$

where A is chosen to be 87.56.

Optimal compander design.   In Section 4.5.1 we studied the criteria for optimal quantizer design. We approach the problem of optimum compander design in the same manner. This approach gives us an approximation to the characteristics of the optimal compressor that, when followed by a uniform quantizer, gives close

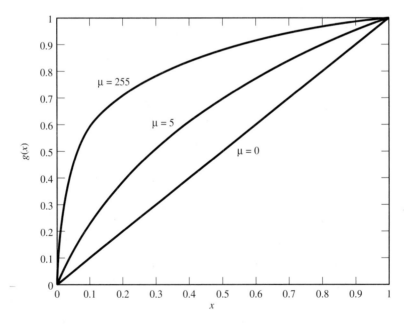

FIGURE 4.17. μ-law compander characteristics.

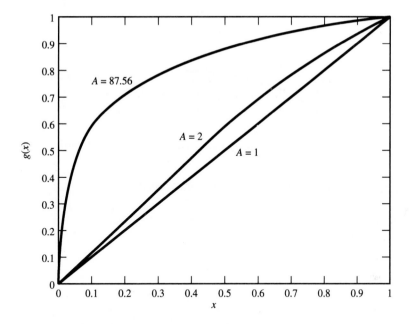

FIGURE 4.18.   *A*-law compander characteristics.

to optimal performance. By defining $a_0 = -x_{\max}$ and $a_N = +x_{\max}$, we have

$$D = \sum_{i=1}^{N} \int_{a_{i-1}}^{a_i} (x - \hat{x}_i)^2 f_X(x)\,dx \qquad (4.6.10)$$

If the number of quantization regions is large and the density function is smooth enough, one can assume a uniform distribution in each region and, therefore, $\hat{x}_i = \frac{a_{i-1}+a_i}{2}$. After substituting for $\hat{x}_i$ in the above relation, we have

$$D \approx \sum_{i=1}^{N} f_X(a_{i-1}) \frac{\Delta_i^3}{12} \qquad (4.6.11)$$

where $\Delta_i = a_i - a_{i-1}$. Noting that the input to the compressor is a nonuniformly quantized sample with quantization regions of size Δ_i and the output is a uniformly quantized sample with quantization regions of equal size Δ, from Figure 4.19 we see that

$$g'(a_{i-1}) \approx \frac{\Delta}{\Delta_i} \qquad (4.6.12)$$

We are assuming that the function $g(x)$ maps the interval $[-x_{\max}, x_{\max}]$ into

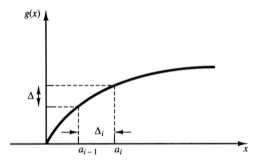

FIGURE 4.19. Compander input–output relation.

$[-y_{max}, y_{max}]$, and therefore the boundary conditions are

$$\begin{cases} g(-x_{max}) = -y_{max} \\ g(x_{max}) = y_{max} \end{cases} \tag{4.6.13}$$

Substituting from (4.6.12) into (4.6.11), we have

$$D \approx \frac{1}{12} \sum_{i=1}^{N} f_X(a_{i-1})[\frac{\Delta}{g'(a_{i-1})}]^2 \Delta_i \tag{4.6.14}$$

Using the relation $\Delta = \frac{y_{max}}{2^{\nu-1}}$, and assuming that N is very large, one obtains

$$D \approx \frac{y_{max}^2}{3 \times 4^{\nu}} \int_{-\infty}^{+\infty} \frac{f_X(x)}{[g'(x)]^2} \, dx \tag{4.6.15}$$

where $g(x)$ denotes the characteristics of the compressor. The above relation gives the distortion when a compressor with characteristics $g(x)$ is employed. One can minimize D to obtain the optimal compressor. The resulting optimal compressor has the form

$$g(x) = y_{max} \left[\frac{2 \int_{-\infty}^{x} [f_X(\eta)]^{\frac{1}{3}} \, d\eta}{\int_{-\infty}^{+\infty} [f_X(\eta)]^{\frac{1}{3}} \, d\eta} - 1 \right] \tag{4.6.16}$$

and the resulting distortion will be

$$D = \frac{1}{12 \times 4^{\nu}} \left[\int_{-\infty}^{+\infty} [f_X(\eta)]^{\frac{1}{3}} \, d\eta \right]^3 \tag{4.6.17}$$

4.6.2 Differential Pulse-Code Modulation

In a PCM system, after sampling the information signal, each sample is quantized independently using a scalar quantizer. This means that previous sample values have no effect on the quantization of the new samples. However, as was mentioned in Section 3.5, when a bandlimited random process is sampled at the Nyquist rate or faster, the sampled values are usually correlated random variables. The exception is when the spectrum of the process is flat within its bandwidth. This means that the previous samples give some information about the next sample, and this information can be employed to improve the performance of the PCM system. For instance, if the previous sample values were small, with high probability the next sample value will be small as well and, hence, it is not necessary to quantize a wide range of values to achieve a good performance.

In the simplest form of differential pulse-code modulation (DPCM), the difference between two adjacent samples is quantized. Because two adjacent samples are highly correlated, their difference has small variations and, therefore, to achieve a certain level of performance fewer bits are required to quantize it. This means that DPCM can achieve performance levels compared to PCM at lower bit rates.

Figure 4.20 shows a block diagram of this simple DPCM scheme. As seen in the figure, the input to the quantizer is not simply $X_n - X_{n-1}$ but rather $X_n - \hat{Y}'_{n-1}$. We will see that $\hat{Y}'_{n-1}$ is closely related to X_{n-1}, and the above choice has the advantage that accumulation of quantization noise is prevented. The input to the quantizer, Y_n, is quantized by a scalar quantizer (uniform or nonuniform) to produce $\hat{Y}_n$. Using the relations

$$Y_n = X_n - \hat{Y}'_{n-1} \tag{4.6.18}$$

and

$$\hat{Y}'_n = \hat{Y}_n + \hat{Y}'_{n-1} \tag{4.6.19}$$

we obtain the quantization error between the input and the output of the quantizer as

$$\hat{Y}_n - Y_n = \hat{Y}_n - (X_n - \hat{Y}'_{n-1})$$

$$= \hat{Y}_n - X_n + \hat{Y}'_{n-1}$$

$$= \hat{Y}'_n - X_n \tag{4.6.20}$$

At the receiving end, we have

$$\hat{X}_n = \hat{Y}_n + \hat{X}_{n-1} \tag{4.6.21}$$

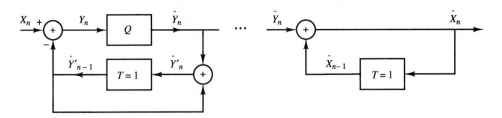

FIGURE 4.20. A simple DPCM scheme.

By comparing (4.6.19) and (4.6.21), it is seen that $\hat{Y}'_n$ and $\hat{X}_n$ satisfy the same difference equation with the same excitation function ($\hat{Y}_n$). Therefore, if the initial conditions of $\hat{Y}'_n$ and $\hat{X}_n$ are chosen to be the same, they will be equal. For instance, if we let $\hat{Y}'_{-1} = \hat{X}_{-1} = 0$, then for all n we will have $\hat{Y}'_n = \hat{X}_n$. Substituting this in (4.6.20), one obtains

$$\hat{Y}_n - Y_n = \hat{X}_n - X_n \tag{4.6.22}$$

This shows that the quantization error between X_n and its reproduction $\hat{X}_n$ is the same as the quantization error between the input and the output of the quantizer. However, the range of variations of Y_n is usually much smaller compared to that of X_n and, therefore, Y_n can be quantized with fewer bits.

In a more complex version of DPCM, instead of using only the previous sample, the last p samples are used to predict the value of the next sample. Then the difference between the sample X_n and its predicted value is quantized. Usually a linear predictor of the form $\sum_{i=1}^{p} a_i X_{n-i}$ is employed, and the coefficients of the predictor a_i are chosen to minimize the mean-squared error between the sample X_n and its predicted value

$$D = E[X_n - \sum_{i=1}^{p} a_i X_{n-i}]^2 \tag{4.6.23}$$

Expanding and assuming the process X_n to be stationary, one obtains

$$D = R_X(0) - 2 \sum_{i=1}^{p} a_i R_X(i) + \sum_{i=1}^{p} \sum_{j=1}^{p} a_i a_j R_X(i-j) \tag{4.6.24}$$

To minimize D, we differentiate with respect to the a_i's and find the roots. After differentiating we have

$$\sum_{i=1}^{p} a_i R_X(i-j) = R_X(j) \qquad 1 \le j \le p \tag{4.6.25}$$

where R_X denotes the autocorrelation function of the process X_n. Solving the above set of equations (usually referred to as *Yule-Walker equations*), one can find the optimal set of predictor coefficients $\{a_i\}_{i=1}^{p}$.

Figure 4.21 shows the block diagram of a general DPCM system. This block diagram is quite similar to the block diagram shown in Figure 4.20. The only difference is that the delay $T = 1$ has been substituted with the prediction filter $\sum_{i=1}^{p} a_i X_{n-i}$. Exactly the same analysis shows that

$$\hat{Y}_n - Y_n = \hat{X}_n - X_n \tag{4.6.26}$$

Because we are using a p-step predictor, we are using more information in predicting X_n and, therefore, the range of variations of Y_n will be less. This in turn means that even lower bit rates are possible here. Differential PCM systems find wide applications in speech and image compression.

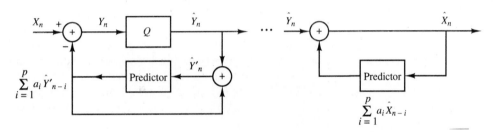

FIGURE 4.21.   A general DPCM scheme.

4.6.3 Delta Modulation

Delta modulation (ΔM) is a simplified version of the simple DPCM scheme shown in Figure 4.20. In ΔM the quantizer is a 1-bit (2-level) quantizer with magnitudes $\pm\Delta$. A block diagram of a ΔM system is shown in Figure 4.22. The same analysis that was applied to the simple DPCM system is valid here.

Because in ΔM only 1 bit/sample is employed, the quantization noise will be high unless the dynamic range of Y_n is very low. This in turn means that X_n and X_{n-1} must have a very high correlation coefficient. To have high correlation between X_n and X_{n-1}, one has to sample at rates much higher than the Nyquist rate. Therefore, in ΔM the sampling rate is usually much higher than the Nyquist rate but, because the number of bits/sample is only 1, the total number of bits/sec required to transmit a waveform is lower than that of a PCM system.

A major advantage of ΔM is the very simple structure of the system. At the receiving end we have the following relation for the reconstruction of $\hat{X}_n$

$$\hat{X}_n - \hat{X}_{n-1} = \hat{Y}_n \tag{4.6.27}$$

By solving this equation for $\hat{X}_n$ and assuming zero initial conditions, one obtains

$$\hat{X}_n = \sum_{i=0}^{n} \hat{Y}_i \tag{4.6.28}$$

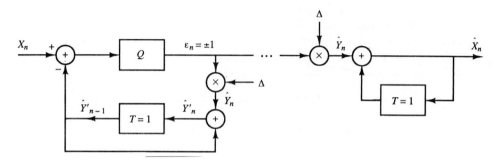

FIGURE 4.22.   Delta modulation.

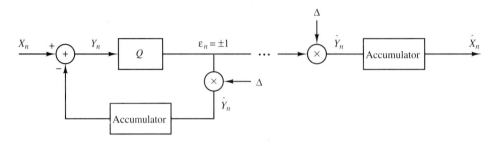

FIGURE 4.23. Delta modulation with integrators.

This means that to obtain $\hat{X}_n$, one only has to accumulate the values of $\hat{Y}_n$. If the sampled values are represented by impulses, the accumulator will be a simple integrator. This simplifies the block diagram of a ΔM system to Figure 4.23

The step-size Δ is a very important parameter in designing a ΔM system. Large values of Δ cause the modulator to follow rapid changes in the input signal but at the same time cause excessive quantization noise when input changes slowly. This case is shown in Figure 4.24. As seen in this figure, for large Δ, when the input varies slowly a large quantization noise occurs, which is known as *granular noise*. The case of a too small Δ is shown in Figure 4.25. In this case we have problems with rapid changes in the input. When the input changes rapidly (high-input slope), it takes a rather long time for the output to follow the input and an excessive quantization noise is caused in this period. This type of distortion, which is caused by high slope of the input waveform, is called *slope-overload distortion.*

Adaptive delta modulation.   We have seen that too large a step size causes granular noise and too small a step size results in sample-overload distortion. This means that a good choice for Δ is a "medium" value, but in some cases the performance of the best medium value (i.e., the one minimizing the mean-squared distortion) is not satisfactory. An approach that works well in these cases is to change the step size according to changes in the input. If the input tends to change rapidly, the step size is chosen to be large such that the output can follow the input quickly and no sample-overload distortion results. When the input is more or less

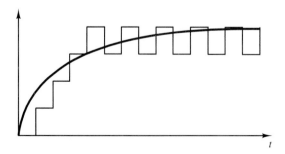

FIGURE 4.24. Large Δ and granular noise.

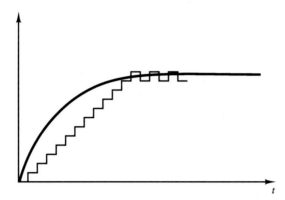

FIGURE 4.25.   Small Δ and
slope-overload distortion.

flat (slowly varying), the step size is changed to a small value to prevent granular
noise. Such changes in the step size are shown in Figure 4.26.

   To adaptively change the step size one has to design a mechanism for recog-
nizing large and small input slopes. If the slope of the input is small, the output
of the quantizer $\hat{Y}$ alternates between Δ and $-\Delta$, as shown in Figure 4.26. This is
the case where granular noise is the main source of noise and one has to decrease
the step size. However, in the case of slope overload, the output cannot follow the
input rapidly and the output of the quantizer will be a succession of $+\Delta$'s or $-\Delta$'s.
From the above it is seen that the sign of two successive $\hat{Y}_n$'s is a good criterion for
changing the step size. If the two successive outputs have the same sign, the step
size should be increased, and if they are of opposite signs, it should be decreased.

   A particularly simple rule to change the step size is given by

$$\Delta_n = \Delta_{n-1} K^{\epsilon_n \times \epsilon_{n-1}} \tag{4.6.29}$$

where ϵ_n is the output of the quantizer before being scaled by the step size and K
is some constant larger than one. It has been verified that in the 20–60 kbits/sec
range, with a choice of $K = 1.5$, the performance of adaptive ΔM systems is 5–10
dB better than the performance of ΔM when applied to speech sources.

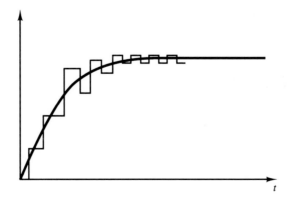

FIGURE 4.26.   Performance of adaptive
ΔM.

4.7 ANALYSIS-SYNTHESIS TECHNIQUES

In contrast to waveform coding, analysis-synthesis techniques are methods that are based on a model for the mechanism that produces the waveform. The parameters of the model that are extracted from the source waveform are quantized, encoded, and transmitted to the receiving end. At the receiving end, based on the received information, the same model is synthesized and used to generate an output similar to the original waveform. These systems are used mainly for speech coding, and in this section we will briefly treat one such system known as *linear predictive coding* or *LPC*.

Speech is produced as a result of excitation of the vocal tract by the vocal cords. This mechanism can be modeled as a time-varying filter (the vocal tract) excited by a signal generator. The vocal tract is a combination of the throat, the mouth, the tongue, the lips, and the nose, that change shape during generation of speech, and, it is modeled therefore, as a time-varying system. The properties of the excitation signal highly depend on the type of speech sounds, either *voiced* or *unvoiced*. For voiced speech, the excitation can be modeled as a periodic sequence of impulses at a frequency f_0, the value of which depends on the speaker. The reciprocal $\frac{1}{f_0}$ is called the *pitch period*. For unvoiced speech the excitation is well modeled as a white noise. This model is shown in Figure 4.27. The vocal tract filter is usually modeled as an all-pole filter described by the difference equation

$$X_n = \sum_{i=1}^{p} a_i X_{n-i} + G w_n \qquad (4.7.1)$$

where w_n denotes the input sequence (white noise or impulses), G is a gain parameter, $\{a_i\}$ are the filter coefficients, and p is the number of poles of the filter. The process w_n, which represents that part of X_n not contained in the previous p samples, is called the *innovation process*.

Speech signals are known to be stationary for short periods of time, of the order of 20–30 msec. This characteristic behavior follows from the observation that the vocal tract cannot change instantaneously. Hence, over 20–30 msec intervals, the all-pole filter coefficients may be assumed to be fixed. At the encoder, we observe a 20–30-msec record of speech from which we estimate the model parameters $\{a_i\}$,

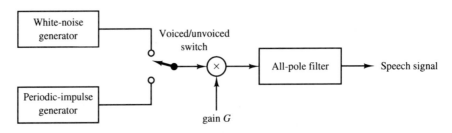

FIGURE 4.27. Model for speech-generation mechanism.

the type of excitation signal (white noise or impulse), the pitch period $\frac{1}{f_0}$ if the speech is voiced, and the gain parameter G.

To elaborate on this process, suppose that the speech signal is filtered to 3 kHz and sampled at a rate of 8000 samples/sec. The signal samples are subdivided into blocks of 160 samples, corresponding to 20-msec intervals. Let $\{x_n, 1 \le n \le 160\}$ be the sequence of samples for a block. The encoder must measure the model parameters to be transmitted to the receiver.

Linear prediction is used to determine the filter coefficients at the encoder. A linear predictor of order p is an all-zero digital filter with input $\{x_n\}$ and output

$$\hat{x}_n = \sum_{k=1}^{p} a_k x_{n-k} \quad \text{for } 1 \le n \le N \tag{4.7.2}$$

where we have assumed that outside the interval of interest $x_n = 0$. Figure 4.28 illustrates the functional block diagram for the prediction process. The difference between the actual speech sample x_n and the predicted value $\hat{x}_n$ constitutes the prediction error e_n, i.e.,

$$e_n = x_n - \hat{x}_n$$

$$= x_n - \sum_{k=1}^{p} a_k x_{n-k} \tag{4.7.3}$$

To extract as much information as possible from the previous values of X_n, we choose the coefficients $\{a_i\}$ so that the average of the squared-error terms, i.e.,

$$\mathcal{E}_p = \frac{1}{N} \sum_{n=1}^{N} e_n^2$$

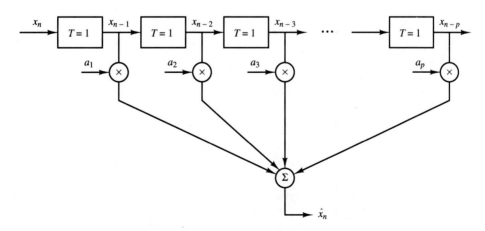

FIGURE 4.28. Functional block diagram of linear prediction.

$$= \frac{1}{N} \sum_{n=1}^{N} \left(x_n - \sum_{k=1}^{p} a_k x_{n-k}\right)^2 \qquad (4.7.4)$$

is minimized. By differentiating $\mathcal{E}_p$ with respect to each of the prediction filter coefficients $\{a_i\}$ and setting the derivative to zero, we obtain a set of linear equations for the filter coefficients, i.e.,

$$\frac{1}{N} \sum_{n=1}^{N} x_n x_{n-i} = \frac{1}{N} \sum_{n=1}^{N} \sum_{k=1}^{p} a_k x_{n-i} x_{n-k} \qquad \text{for } 1 \le i \le p \qquad (4.7.5)$$

Because we have assumed that outside the stationary interval, $1 \le n \le N$, we have $x_n = 0$, we can write the above relation as

$$\frac{1}{N} \sum_{n=-\infty}^{+\infty} x_n x_{n-i} = \frac{1}{N} \sum_{n=-\infty}^{+\infty} \sum_{k=1}^{p} a_k x_{n-i} x_{n-k}$$

$$= \sum_{k=1}^{p} a_k \left[\frac{1}{N} \sum_{n=-\infty}^{+\infty} x_{n-i} x_{n-k} \right] \qquad (4.7.6)$$

Now if we define

$$\hat{R}_i = \frac{1}{N} \sum_{n=-\infty}^{+\infty} x_n x_{n-i} \qquad (4.7.7)$$

we can write (4.7.6) as

$$\hat{R}_i = \sum_{k=1}^{p} a_k \hat{R}_{i-k} \qquad \text{for } 1 \le i \le p \qquad (4.7.8)$$

which is the same as the Yule-Walker equations (4.6.25) derived in Section 4.6. We can further simplify the above equation to the matrix equation

$$\mathbf{r} = \hat{\mathbf{R}} \mathbf{a} \qquad (4.7.9)$$

where $\mathbf{a}$ is the vector of the linear predictor coefficients, $\hat{\mathbf{R}}$ is a $p \times p$ matrix whose $(i, j)^{\text{th}}$ element is $\hat{R}_{i-j}$, and $\mathbf{r}$ is a vector whose components are $\hat{R}_i$'s. It can be easily verified from the definition of $\hat{R}_i$ that

$$\hat{R}_i = \hat{R}_{-i} \qquad (4.7.10)$$

and therefore the matrix $\hat{\mathbf{R}}$ is a symmetric matrix. Also it is obvious that all elements of $\hat{\mathbf{R}}$ that are on a line parallel to the diagonal elements are equal. Such a matrix is called a *Toeplitz matrix* and there exist efficient recursive algorithms for finding its inverse. One such algorithm is the well-known Levinson-Durbin algorithm. The interested reader is referred to the references at the end of this chapter for the details of this algorithm.

For the optimal choice of the predictor coefficients, the squared-error term can be shown to be

$$\mathcal{E}_p^{\min} = \hat{R}_0 - \sum_{k=1}^{p} \hat{R}_k \qquad (4.7.11)$$

According to the speech-production model

$$\mathcal{E}_p^{\min} = \frac{1}{N} \sum_{n=1}^{N} \left[x_n - \sum_{k=1}^{p} a_k x_{n-k} \right]^2$$

$$= G^2 \frac{1}{N} \sum_{n=1}^{N} w_n^2 \qquad (4.7.12)$$

If we normalize the excitation sequence $\{w_n\}$ such that $\frac{1}{N} \sum_{n=1}^{N} w_n^2 = 1$, we obtain the value of the gain predictor as

$$G = \sqrt{\mathcal{E}_p^{\min}} \qquad (4.7.13)$$

The estimation of the type of excitation (impulsive or noise) and the estimate of the pitch period $\frac{1}{f_0}$ when the excitation consists of impulses may be accomplished by various algorithms. One simple approach is to transform the speech data into the frequency domain and look for sharp peaks in the signal spectrum. If the spectrum exhibits peaks at some fundamental frequency f_0, the excitation is taken to be a periodic impulse train with period $\frac{1}{f_0}$. If the spectrum of the speech samples exhibit no sharp peaks, the excitation is taken as white noise.

The prediction filter coefficients, gain, voiced–unvoiced information, and pitch $\frac{1}{f_0}$ are quantized and transmitted to the receiver for each block of sampled speech. The speech signal is synthesized from these parameters using the system model shown in Figure 4.27. Typically the voiced–unvoiced information requires 1 bit, the pitch frequency is represented by 6 bits, the gain parameter can be represented by 5 bits using logarithmic companding, and the prediction coefficients require 8–10 bits/coefficient. Based on linear predictive coding, speech can be compressed to bit rates as low as 2400 bits/sec. One could alternatively use vector quantization when quantizing the LPC parameters. This would further reduce the bit rate. In contrast, PCM applied to speech has a bit rate of 56,000 bits/sec.

LPC is widely used in speech coding to reduce the bandwidth. By vector quantizing the LPC parameters, good quality speech can be achieved at bit rates of about 4800 bits/sec. One version of LPC with vector quantization has been adopted as a standard for speech compression in mobile (cellular) telephone systems. Efficient speech coding is a very active area for research, and we expect to see further reduction in the bit rate of commercial speech encoders over the coming years.

4.8 DIGITAL AUDIO TRANSMISSION AND DIGITAL AUDIO RECORDING

Audio signals constitute a large part of our daily communications. Today thousands of radio stations broadcast audio signals in analog form. The quality of voice-signal broadcasting is generally acceptable as long as the voice signal is intelligible. On the other hand, the quality of music signals that are broadcast over AM radio is relatively low fidelity because the bandwidth of the transmitted signal is restricted through regulation (by the Federal Communication Commission). FM radio broadcasting of analog signals provides higher fidelity by using a significantly larger channel bandwidth for signal transmission. It is conceivable that, in the near future, commercial radio broadcasting of audio signals will convert to digital form.

In the transmission of audio signals on telephone channels, the conversion from analog to digital transmission, which has been taking place over the past three decades, is now nearly complete. We will describe some of the current developments in the digital encoding of audio signals for telephone transmission.

The entertainment industry has experienced the most dramatic changes and benefits in the conversion of analog audio signals to digital form. The development of the compact disc (CD) player and the digital audio tape recorder have rendered the previous analog-recording systems technically obsolete. We shall use the CD player as a case study of the sophisticated source encoding/decoding and channel encoding/decoding methods that have been developed over the past few years for digital audio systems.

4.8.1 Digital Audio in Telephone Transmission Systems

Nearly all of the transmission of speech signals over telephone channels is currently digital. The encoding of speech signals for transmission over telephone channels has been a topic of intense research for over 50 years and continues to be today. A wide variety of methods for speech-source encoding have been developed over the years, many of which are in use today.

The general configuration for a speech signal encoder is shown in Figure 4.29. Because the frequency content of speech signals is limited to below 3400 Hz, the speech signal is first passed through an anti-aliasing lowpass filter and then sampled. To ensure that aliasing is negligible, a sampling rate of 8000 Hz or higher is typically selected. The analog samples are then quantized and represented in digital form for transmission over telephone channels.

PCM and DPCM are widely used waveform-encoding methods for digital speech transmission. Logarithmic $\mu = 255$ compression, given by (4.6.8) is generally used for achieving nonuniform quantization. The typical rate for PCM is 64,000 bits/sec while for DPCM the rate is 32,000 bits/sec.

PCM and DPCM encoding and decoding are performed generally in a telephone central office where telephone lines from subscribers in a common geographical area are connected to the telephone transmission system. The PCM or

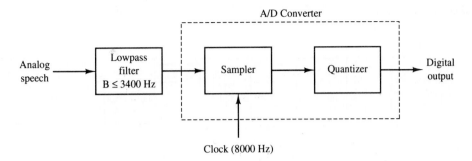

FIGURE 4.29.   Analog-to-digital conversion of speech signals.

DPCM encoded speech signals are transmitted from one telephone central office to another in digital form over so-called trunk lines that are capable of carrying the digitized speech signals of many subscribers. The method for simultaneous transmission of several signals over a common communication channel is called *multiplexing*. In the case of PCM and DPCM transmission, the signals from different subscribers are multiplexed in time, hence the name *time-division multiplexing* (TDM). In TDM, a given time interval T_f is selected as a frame. Each frame is subdivided into N subintervals of duration T_f/N, where N corresponds to the number of users that will use the common communication channel. Then each subscriber who wishes to use the channel for transmission is assigned a subinterval within each frame. In PCM each user transmits one 8-bit sample in each subinterval.

In digital speech transmission over telephone lines via PCM there is a standard TDM hierarchy that has been established for accommodating multiple subscribers. In the first level of the TDM hierarchy, 24 digital subscriber signals are time-division multiplexed into a single high-speed data stream of 1.544 Mbits/sec (24×64 kbits plus a few additional bits for control purposes). The resulting combined TDM signal is usually called a DS-1 channel. In the second level of TDM, four DS-1 channels are multiplexed into a DS-2 channel, having the bit rate of 6.312 Mbits/sec. In a third level of hierarchy, seven DS-2 channels are combined via TDM to produce a DS-3 channel, which has a bit rate of 44.736 Mbits/sec. Beyond DS-3, there are two more levels of TDM hierarchy. Figure 4.30 illustrates the TDM hierarchy for the North American telephone system.

In mobile cellular radio systems (see Section 5.5 for a description) for transmission of speech signals, the available channel bandwidth per user is small and cannot support the high bit rates required by waveform-encoding methods such as PCM and DPCM. For this application, the analysis-synthesis method based on LPC as described in Section 4.7 is used to estimate the set of model parameters from short segments of the speech signal. The speech-model parameters are then transmitted over the channel using vector quantization. Thus, a bit rate in the range of 4800–9600 bits/sec is achieved with LPC.

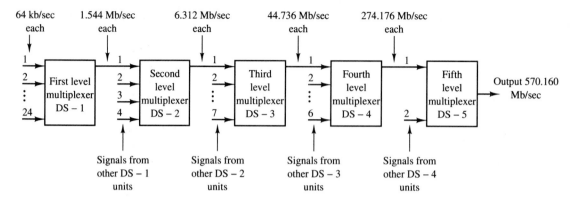

FIGURE 4.30. Digital TDM hierarchy for North American telephone communication system.

In mobile cellular communication system, the base station in each cell serves as the interface to the terrestrial telephone system. LPC speech compression is only required for the radio transmission between the mobile transcriber and the base station in any cell. At the base station interface, the LPC-encoded speech is converted to analog form and resampled and digitized using PCM or DPCM for transmission over the terrestrial telephone system. Hence, we note that a speech signal transmitted from a mobile subscriber to a fixed subscriber will undergo two different types of analog to digital encoding, whereas in speech-signal communication between two mobiles serviced by different base stations will undergo four translations between the analog and the digital domain.

4.8.2 Digital Audio Recording

Historically, audio recording became a reality with the invention of the phonograph during the second half of the nineteenth century. The phonograph had a lifetime of approximately 100 years before it was supplanted by the compact disc, which was introduced in 1982. During the 100-year period we witnessed the introduction of a wide variety of records, the most popular of which proved to be the long-playing (LP) record, which was introduced in 1948. LP records provide relatively high-quality analog audio recording.

In spite of their wide acceptance and popularity, analog audio recordings have a number of limitations, including a limited dynamic range (typically about 70 dB) and a relatively low signal-to-noise ratio (typically about 60 dB). By comparison, the dynamic range of orchestral music is in the range of 100–120 dB. This means that, to record the music in analog form, the dynamic range has to be decreased to fit the range that can be accommodated by the recording system. Otherwise, at low music levels, noise will be audible and, if one wishes to prevent this, saturation will occur at high music levels. The compression of the dynamic range of music during recording is usually done by controlling the volume of the input signal.

Digital audio recording and playback allows us to improve the fidelity of recorded music by increasing the dynamic range and the signal-to-noise ratio. Furthermore, digital recordings are generally more durable and do not deteriorate with playing time, as do analog recordings. We describe below a CD system as an example of a commercially successful digital audio system. Table 4.4 provides a comparison of some important specifications of an LP record and a CD system. The advantages of the latter are clearly evident.

From a systems point of view, the CD system embodies most of the elements of a modern digital communications system. These include analog-to-digital (A/D) and digital-to-analog (D/A) conversion, interpolation, modulation/demodulation, and channel coding/decoding. A general block diagram of the elements of a CD digital audio system are illustrated in Figure 4.31. We will describe the main features of the source encoder and decoder.

The two audio signals from the left (L) and right (R) microphones in a recording studio or a concert hall are sampled and digitized by passing them through an A/D converter. Recall that the frequency band of audible sound is limited to approximately 20 kHz. Therefore, the corresponding Nyquist sampling rate is 40 kHz. To allow for some frequency guard band and to prevent aliasing, the sampling rate in a CD system has been selected to be 44.1 kHz. This frequency is compatible with video recording equipment that is commonly used for digital recording of audio signals on magnetic tape.

The samples of both the L and R signals are quantized using uniform PCM with 16 bits/sample. According to the formula for SQNR given by (4.6.5), 16-bit uniform quantization results in an SQNR of over 90 dB. In addition, the total harmonic distortion achieved is 0.005%. The PCM bytes from the digital recorder are encoded to provide protection against channel errors in the read-back process and passed to the modulator.

At the modulator, digital control and display information is added, including a table of contents of the disc. This information allows for programmability of the

TABLE 4-4   COMPARISON OF LP RECORDS WITH CD SYSTEM

Specification/Feature	LP record	CD system
Frequency response	30 Hz–20 kHZ ± 3 dB	20 Hz–20 kHz $+0.5/-1$dB
Dynamic range	70 dB @ 1 kHz	> 90 dB
Signal-to-noise ratio	60 dB	> 90 dB
Harmonic distortion	1–2%	0.005%
Durability	High-frequency response degrades with playing	Permanent
Stylus life	500–600 hours	5000 hours

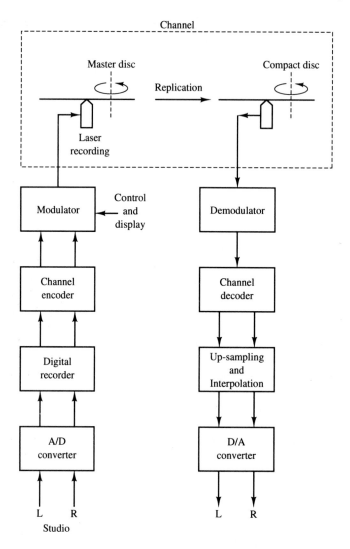

FIGURE 4.31.   CD digital audio system.

CD player. The type of modulation and demodulation and the channel coding and decoding operation will be described in subsequent chapters.

Using a laser, the digital signal from the modulator is optically recorded in the surface of a glass disc that is coated with photoresist. This results in a master disc that is used to produce CD's by a series of processes that ultimately convert the information into tiny pits on the plastic disc. The disc is coated with a reflective aluminum coating and then with a protective layer.

In the CD player, a laser is used to optically scan a track on the disc at a constant velocity of 1.25 m/sec and thus reads the digitally recorded signal. After the L and R signal are demodulated and passed through the channel decoder, the digital audio signal is converted back to an analog audio signal by means of a D/A converter.

The L and R digital audio signals into the D/A converter have a precision of 16 bits. In principle, the digital-to-analog conversion of the two 16-bit signals at the 44.1-kHz sampling rate is relatively simple. However, the practical implementation of a 16-bit D/A converter is very expensive. On the other hand, inexpensive D/A converters with 12-bit (or less) precision are readily available. The problem is to devise a method for D/A conversion that employs a low precision and, hence, low cost D/A converter, while maintaining the 16-bit precision of the digital audio signal.

The practical solution to this problem is to expand the bandwidth of the digital audio signal by oversampling through interpolation and digital filtering prior to analog conversion. The basic approach is shown in the block diagram in Figure 4.32. The 16-bit L and R digital audio signals are up-sampled by some multiple U by inserting $U - 1$ zeros between successive 16-bit signal samples. This process effectively increases the sampling rate to $U \times 44.1$ kHz. The high-rate L and R signals are then filtered by a finite-duration impulse response (FIR) digital filter, which produces a high-rate, high-precision output. The combination of up-sampling and filtering is a practical method for realizing a *digital interpolator*. The FIR filter is designed to have linear phase and a bandwidth of approximately 20 kHz. It serves the purpose of eliminating the spectral images created by the up-sampling process and is sometimes called an *anti-imaging filter*.

If we observe the high sample rate, high-precision L and R digital audio signals of the output of the FIR digital filter, we will find that successive samples are nearly the same, differing only in the low-order bits. Consequently, it is possible to represent successive samples of the digital audio signals by their differences and, thus, to reduce the dynamic range of the signals. If the oversampling factor U is sufficiently large, ΔM may be employed to reduce the quantized output to a precision of 1 bit/sample. Thus, the D/A converter is considerably simplified. An oversampling factor $U = 256$ is normally chosen in practice. This raises the sampling rate to 11.2896 MHz.

Recall that the general configuration for the conventional ΔM system is as shown in Figure 4.33. Suppose we move the integrator from the decoder to the input of the ΔM. This has two effects. First, it preemphasizes the low frequencies in the input signal and, thus, it increases the correlation of the signal into the ΔM. Second, it simplifies the ΔM decoder because the differentiator (the inverse system) required at the decoder is canceled by the integrator. Hence the decoder is reduced to a simple lowpass filter. Furthermore, the two integrators at the encoder can be replaced by a single integrator placed before the quantizer. The resulting system, shown in Figure 4.34, is called a *sigma-delta modulator* (SDM). Figure 4.35

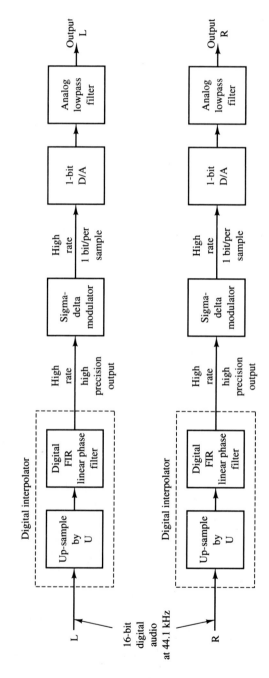

FIGURE 4.32. D/A conversion by oversampling/interpolation.

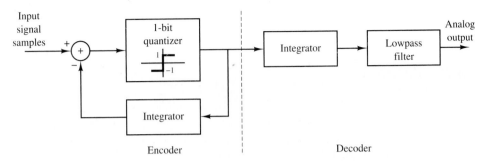

FIGURE 4.33.   Conventional ΔM system.

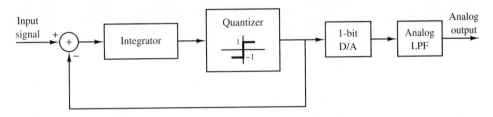

FIGURE 4.34.   Basic configuration of a SDM followed by a 1-bit D/A converter and lowpass smoothing filter.

illustrates a SDM that employs a single-digital integrator (first-order SDM) with a system function

$$H(z) = \frac{z^{-1}}{1 - z^{-1}}$$

Thus, the SDM simplifies the D/A conversion process by requiring only a 1-bit D/A followed by a conventional analog filter (e.g., a Butterworth filter), for providing anti-aliasing protection and signal smoothing. The output analog filters have a pass-band of approximately 20 kHz and thus eliminate any noise above the desired signal band. In modern CD players, the interpolator, the SDM, the 1-bit D/A converter,

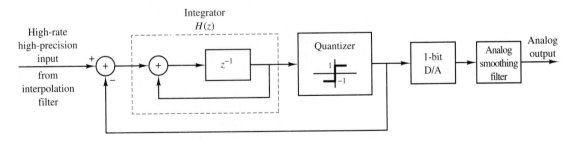

FIGURE 4.35.   SDM D/A converter with first-order integrator.

and the lowpass smoothing filter are generally implemented on a single integrated chip.

4.9 FURTHER READING

Any standard text on information theory covers source-coding theorems and algo-rithms in detail. Gallager (1968), Blahut (1987), and particularly Cover and Thomas (1991) provide nice and readable treatments of the subject. Our treatment of the Lempel-Ziv algorithm follows that of Cover and Thomas (1991). Berger (1971) is devoted entirely to rate-distortion theory, Jayant and Noll (1984) and Gersho and Gray (1992) examine various quantization and waveform-coding techniques in detail. Gersho and Gray (1992) include detailed treatment of vector quantization. Analysis-synthesis techniques and linear predictive coding are treated in books on speech coding including Markel and Gray (1975), Rabiner and Schafer (1978) and Deller, Proakis, and Hansen (1993).

Among the original works contributing to the material covered in this chapter are the classic papers by Shannon (1948a, 1959). The paper by Huffman (1952) describes the Huffman coding algorithm. The Lempel-Ziv algorithm is described in the paper by Lempel and Ziv (1978). The papers by Lloyd (1957) and Max (1960) treat scalar quantization. Vector quantization is treated in the paper by Linde, Buzo, and Gray (1980).

PROBLEMS

4.1 A source has an alphabet $\{a_1, a_2, a_3, a_4, a_5, a_6\}$ with corresponding probabil-ities $\{0.1, 0.2, 0.3, 0.05, 0.15, 0.2\}$. Find the entropy of this source. Compare this entropy with the entropy of a uniformly distributed source with the same alphabet.

4.2 Let the random variable X be the output of the source that is uniformly distributed with size N. Find its entropy.

4.3 Show that $H(X) \geq 0$ with equality holding if and only if X is deterministic.

4.4 Let X be a geometrically distributed random variable, that is,

$$p(X = k) = p(1 - p)^{k-1} \qquad k = 1, 2, 3, \ldots$$

    1. Find the entropy of X.

    2. Knowing that $X > K$ where K is a positive integer, what is the entropy of X?

4.5 Let $Y = g(X)$, where g denotes a deterministic function. Show that in general $H(Y) \leq H(X)$. When does equality hold?

4.6 An information source can be modeled as a bandlimited process with a band-width of 6000 Hz. This process is sampled at a rate higher than the Nyquist rate to provide a guard-band of 2000 Hz. It is observed that the resulting samples take values in the set $\mathcal{A} = \{-4, -3, -1, 2, 4, 7\}$ with probabilities 0.2, 0.1, 0.15, 0.05, 0.3, 0.2. What is the entropy of the discrete-time source in bits/output (sample)? What is the entropy in bits/sec?

4.7 Let X denote a random variable distributed on the set $\mathcal{A} = \{a_1, a_2, \ldots, a_N\}$ with corresponding probabilities $\{p_1, p_2, \ldots, p_N\}$. Let Y be another random variable defined on the same set but distributed uniformly. Show that

$$H(X) \leq H(Y)$$

with equality if and only if X is also uniformly distributed. (Hint: First prove the inequality $\ln x \leq x - 1$ with equality for $x = 1$, then apply this inequality to $\sum_{n=1}^{N} p_n \ln \left(\frac{\frac{1}{N}}{p_n} \right)$.)

4.8 A random variable X is distributed on the set of all positive integers $1, 2, 3, \ldots$ with corresponding probabilities $p_1, p_2, p_3, \ldots$. We also know that the expected value of this random variable is given to be m, i.e.,

$$\sum_{i=1}^{\infty} i p_i = m$$

Show that among all random variables that satisfy the above condition, the geometric random variable, which is defined by

$$p_i = \frac{1}{m} \left(1 - \frac{1}{m} \right)^{i-1} \qquad i = 1, 2, 3, \ldots$$

has the highest entropy. (Hint: Define two distributions on the source, the first one being the geometric distribution given above and the second one an arbitrary distribution denoted by q_i, and then apply the approach of Problem 4.7.)

4.9 Two binary random variables X and Y are distributed according to the joint distribution $p(X = Y = 0) = p(X = 0, Y = 1) = p(X = Y = 1) = \frac{1}{3}$. Compute $H(X)$, $H(Y)$, $H(X|Y)$, $H(Y|X)$, and $H(X, Y)$.

4.10 Show that if $Y = g(X)$ where g denotes a deterministic function, then $H(Y|X) = 0$.

4.11 A memoryless source has the alphabet $\mathcal{A} = \{-5, -3, -1, 0, 1, 3, 5\}$ with corresponding probabilities $\{0.05, 0.1, 0.1, 0.15, 0.05, 0.25, 0.3\}$.

1. Find the entropy of the source.
2. Assume that the source is quantized according to the quantization rule

$$\begin{cases} q(-5) = g(-3) = 4 \\ q(-1) = q(0) = q(1) = 0 \\ q(3) = q(5) = -4 \end{cases}$$

Find the entropy of the quantized source.

4.12 Using both definitions of the entropy rate of a process, prove that for a d.m.s. the entropy rate and the entropy are equal.

4.13 A Markov process is a process with one step memory, i.e., a process such that

$$p(x_n|x_{n-1}, x_{n-2}, x_{n-3}, \ldots) = p(x_n|x_{n-1})$$

for all n. Show that for a stationary Markov process the entropy rate is given by $H(X_n|X_{n-1})$.

4.14 Show that

$$H(X|Y) = \sum_y p(y)H(X|Y = y)$$

4.15 Let X and Y denote two jointly distributed discrete valued random variables

1. Show that

$$H(X) = -\sum_{x,y} p(x, y) \log p(x)$$

and

$$H(Y) = -\sum_{x,y} p(x, y) \log p(y)$$

2. Use the above result to show that

$$H(X, Y) \leq H(X) + H(Y)$$

When does the equality hold? (Hint: Consider the two distributions $p(x, y)$ and $p(x)p(y)$ on the product set $X \times Y$, and apply the inequality proved in Problem 4.7 to $\sum_{x,y} p(x, y) \log \frac{p(x)p(y)}{p(x,y)}$.)

4.16 Use the result of Problem 4.15 to show that

$$H(X|Y) \leq H(X)$$

with equality if and only if X and Y are independent.

4.17 Show that $H(X)$ is a concave function of the probability distribution on X, i.e., for any two probability distribution vectors $\mathbf{p}_1$ and $\mathbf{p}_2$ on X, and any $0 \leq \lambda \leq 1$ we have

$$\lambda H(\mathbf{p}_1) + \bar{\lambda} H(\mathbf{p}_2) \leq H(\lambda \mathbf{p}_1 + \bar{\lambda} \mathbf{p}_2)$$

where $\bar{\lambda} = 1 - \lambda$. (Note: You have to first show that $\lambda \mathbf{p}_1 + \bar{\lambda} \mathbf{p}_2$ is a legitimate probability vector.)

4.18 Show that in general

$$H(X_1, X_2, \ldots, X_n) \le \sum_{i=1}^{n} H(X_i)$$

When does the equality hold?

4.19 Assume that a b.s.s. generates a sequence of n outputs.

1. What is the probability that this sequence consists of all zeros?
2. What is the probability that this sequence consists of all ones?
3. What is the probability that in this sequence the first k symbols are ones and the next $n - k$ symbols are zeros?
4. What is the probability that this sequence has k ones and $n - k$ zeros?
5. How would your answers change if instead of a b.s.s. we were dealing with a general binary d.m.s. with $p(X_i = 1) = p$?

4.20 Give an estimate of the number of binary sequences of length 10,000 with 3000 zeros and 7000 ones.

4.21 A memoryless ternary source with output alphabet a_1, a_2, and a_3 and corresponding probabilities 0.2, 0.3, 0.5 produces sequences of length 1000.

1. Approximately what is the number of typical sequences in the source output?
2. What is the ratio of typical sequences to nontypical sequences?
3. What is the probability of a typical sequence?
4. What is the number of bits required to represent only the typical output sequences?
5. What is the most probable sequence and what is its probability?
6. Is the most probable sequence a typical sequence?

4.22 A source has an alphabet $\{a_1, a_2, a_3, a_4\}$ with corresponding probabilities $\{0.1, 0.2, 0.3, 0.4\}$.

1. Find the entropy of the source.
2. What is the minimum required average codeword length to represent this source for error-free reconstruction?
3. Design a Huffman code for the source and compare the average length of the Huffman code with the entropy of the source.
4. Design a Huffman code for the second extension of the source (take two letters at a time). What is the average codeword length? What is the average required binary letters per each source output letter?

5. Which one is a more efficient coding scheme: Huffman coding of the original source or Huffman coding of the second extension of the source?

4.23 Design a Huffman code for a source with n output letters and corresponding probabilities $\{\frac{1}{2}, \frac{1}{4}, \frac{1}{8}, \ldots, \frac{1}{2^{n-1}}, \frac{1}{2^{n-1}}\}$. Show that the average codeword length for such a source is equal to the source entropy.

4.24 Show that $\{01, 100, 101, 1110, 1111, 0011, 0001\}$ cannot be a Huffman code for any source probability distribution.

4.25 Design a ternary Huffman code, using 0, 1, 2 as letters, for a source with output alphabet probabilities given by $\{0.05, 0.1, 0.15, 0.17, 0.18, 0.22, 0.13\}$. What is the resulting average codeword length? Compare the average codeword length with the entropy of the source. (In what base would you compute the logarithms in the expression for the entropy for a meaningful comparison?)

4.26 Design a ternary Huffman code for a source with output alphabet probabilities given by $\{0.05, 0.1, 0.15, 0.17, 0.13, 0.4\}$. (Hint: You can add a dummy source output with zero probability.)

4.27 Find the Lempel-Ziv source code for the binary source sequence

0001001000000110000100000001000000101000010000001101000000001100

Recover the original sequence back from the Lempel-Ziv source code. (Hint: You require two passes of the binary sequence to decide on the size of the dictionary.)

4.28 Using the definition of $H(X)$ and $H(X|Y)$ show that

$$I(X; Y) = \sum_{x,y} p(x, y) \log \frac{p(x, y)}{p(x)p(y)}$$

Now by using the approach of Problem 4.7 show that $I(X; Y) \geq 0$ with equality if and only if X and Y are independent.

4.29 Show that

1. $I(X; Y) \leq \min\{H(X), H(Y)\}$
2. If $|X|$ and $|\mathcal{Y}|$ represent the size of sets X and $\mathcal{Y}$, respectively, then $I(X; Y) \leq \min\{\log |X|, \log |\mathcal{Y}|\}$.

4.30 Show that $I(X; Y) = H(X) + H(Y) - H(X, Y) = H(Y) - H(Y|X) = I(Y; X)$.

4.31 Let X denote a binary random variable with $p(X = 0) = 1 - p(X = 1) = p$ and let Y be a binary random variable that depends on X through $p(Y = 1|X = 0) = p(Y = 0|X = 1) = \epsilon$.

1. Find $H(X)$, $H(Y)$, $H(Y|X)$, $H(X, Y)$, $H(X|Y)$, and $I(X; Y)$.

2. For a fixed ϵ, which p maximizes $I(X; Y)$?
3. For a fixed p, which ϵ minimizes $I(X; Y)$?

4.32 Show that

$$I(X; YZW) = I(X; Y) + I(X; Z|Y) + I(X; W|ZY)$$

Can you interpret this relation?

4.33 Let X, Y, and Z be three discrete random variables.

1. Show that if $p(x, y, z) = p(z)p(x|z)p(y|x)$ we have

$$I(X; Y|Z) \leq I(X; Y)$$

2. Show that if $p(x, y, z) = p(x)p(y)p(z|x, y)$, then

$$I(Z; Y) \leq I(Z; Y|X)$$

3. In each case give an example where strict inequality holds.

4.34 Let X and Y denote finite sets. We denote the probability vectors on X by $\mathbf{p}$ and the conditional probability matrices on Y given X by $\mathbf{Q}$. Then $I(X; Y)$ can be represented as a function of the probability distribution on X and the conditional probability distribution on Y given X as $I(\mathbf{p}; \mathbf{Q})$.

1. Show that $I(\mathbf{p}; \mathbf{Q})$ is a concave function in $\mathbf{p}$, i.e., for any $\mathbf{Q}$ any $0 \leq \lambda \leq 1$, and for any two probability vectors $\mathbf{p}_1$ and $\mathbf{p}_2$ on X, we have

$$\lambda I(\mathbf{p}_1; \mathbf{Q}) + \bar{\lambda} I(\mathbf{p}_2; \mathbf{Q}) \leq I(\lambda \mathbf{p}_1 + \bar{\lambda} \mathbf{p}_2; \mathbf{Q})$$

where $\bar{\lambda} = 1 - \lambda$.

2. Show that $I(\mathbf{p}; \mathbf{Q})$ is a convex function in $\mathbf{Q}$, i.e., for any $\mathbf{p}$ any $0 \leq \lambda \leq 1$, and any two conditional probabilities $\mathbf{Q}_1$ and $\mathbf{Q}_2$, we have

$$I(\mathbf{p}; \lambda \mathbf{Q}_1 + \bar{\lambda} \mathbf{Q}_2) \leq \lambda I(\mathbf{p}; \mathbf{Q}_1) + \bar{\lambda} I(\mathbf{p}; \mathbf{Q}_2)$$

(Note: You have to first show that $\lambda \mathbf{p}_1 + \bar{\lambda} \mathbf{p}_2$ and $\lambda \mathbf{Q}_1 + \bar{\lambda} \mathbf{Q}_2$ are a legitimate probability vector and conditional probability matrix, respectively.)

4.35 Let the random variable X be continuous with p.d.f. $f_X(x)$ and let $Y = aX$ where a is a nonzero constant.

1. Show that $h(Y) = \log |a| + h(X)$.
2. Does a similar relation hold if X is a discrete random variable?

4.36 Find the differential entropy of the continuous random variable X in the following cases.

1. X is an exponential random variable with parameter $\lambda > 0$, i.e.,

$$f_X(x) = \begin{cases} \dfrac{1}{\lambda} \exp(-x/\lambda) & x > 0 \\ 0, & \text{otherwise} \end{cases}$$

2. X is a Laplacian random variable with parameter $\lambda > 0$, i.e.,

$$f_X(x) = \frac{1}{2\lambda} \exp(-|x|/\lambda)$$

3. X is a triangular random variable with parameter $\lambda > 0$, i.e.,

$$f_X(x) = \begin{cases} \dfrac{x + \lambda}{\lambda^2} & -\lambda \le x \le 0 \\ \dfrac{-x + \lambda}{\lambda^2} & 0 < x \le \lambda \\ 0, & \text{otherwise} \end{cases}$$

4.37 Generalize the technique developed in Problem 4.7 to continuous random variables and show that for continuous X and Y

1. $h(X|Y) \le h(X)$ with equality if and only if X and Y are independent.
2. $I(X; Y) \ge 0$ with equality if and only if X and Y are independent.

4.38 Using an approach similar to Problem 4.8, show that among all continuous random variables distributed on the positive real line and having a given mean m, the exponential random variable has the highest differential entropy.

4.39 Using the method of Problem 4.8, show that among all continuous random variables with a given variance σ^2, the Gaussian random variable has the highest differential entropy.

4.40 A memoryless source emits 2000 binary symbols/sec and each symbol has a probability of .25 to be equal to 1 and .75 to be equal to 0.

1. What is the minimum number of bits/sec required for error-free transmission of this source?
2. What is the minimum number of bits/sec required for reproduction of this source with an error probability not exceeding .1?
3. What is the minimum number of bits/sec required for reproduction of this source with an error not exceeding .25? What is the best decoding strategy in this case?

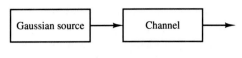

FIGURE P-4.41

4.41 A Gaussian source with mean zero and variance one is to be transmitted via a channel that can provide a transmission capacity of 1 bit/each source output (see Figure P-4.41).

  1. What is the minimum mean-squared error achievable?
  2. If the maximum tolerable distortion is 0.01, what is the required transmission capacity per source output?

4.42 It can be shown that the rate distortion function for a Laplacian source $f_X(x) = \frac{1}{2\lambda} \exp(-|x|/\lambda)$ with an absolute value of error-distortion measure $d(x, \hat{x}) = |x - \hat{x}|$ is given by

$$R(D) = \begin{cases} \log \dfrac{\lambda}{D} & 0 \le D \le \lambda \\ 0 & D > \lambda \end{cases}$$

(see Berger [1971]).

  1. How many bits per sample are required to represent the outputs of this source with an average distortion not exceeding $\frac{\lambda}{2}$?
  2. Plot $R(D)$ for three different values of λ and discuss the effect of changes in λ on these plots.

4.43 It can be shown that if X is a zero-mean continuous random variable with variance σ^2, its rate distortion function, subject to squared-error distortion measure, satisfies the lower and upper bounds given by the inequalities

$$h(X) - \frac{1}{2} \log(2\pi e D) \le R(D) \le \frac{1}{2} \log \frac{\sigma^2}{D}$$

where $h(X)$ denotes the differential entropy of the random variable X (see Cover and Thomas [1991]).

  1. Show that for a Gaussian random variable, the lower and upper bounds coincide.
  2. Plot the lower and upper bounds for a Laplacian source with $\sigma = 1$.
  3. Plot the lower and upper bounds for a triangular source with $\sigma = 1$.

(Hint: Use the results of Problem 4.36.)

4.44 With the assumptions of Example 4.4.6, if we want to reduce the distortion by a factor of 1000, how many extra bits/source symbol have to be introduced?

4.45 Let $X(t)$ be a strictly stationary random process.

1. Show that $Q(X(t))$ is also strictly stationary for any function Q.
2. From above conclude that $X(t) - Q(X(t))$ is also strictly stationary and, therefore, in any quantizer

$$\text{SNR} = \frac{E[X^2]}{E[X - Q(X)]^2} = \frac{P_X}{P_{\tilde{X}}} = \frac{R_X(0)}{R_{\tilde{X}}(0)}$$

where $\tilde{X}(t) = X(t) - Q(X(t))$

4.46 Let $X(t)$ denote a WSS Gaussian process with $P_X = 10$.

1. Using Table 4.2, design a 16-level optimal uniform quantizer for this source.
2. What is the resulting distortion if the quantizer in part 1 is employed?
3. What is the minimum number of bits per source symbol required to represent the quantized source?
4. Compare this result with the result obtained from the rate-distortion bound that achieves the same amount of distortion.
5. What is the amount of improvement in SQNR (in dB) that results from doubling the number of quantization levels from 8 to 16?

4.47 Using Table 4.2, design an optimal quantizer for the source given in Example 4.5.1. Compare the distortion of this quantizer to the distortion obtained there. What is the entropy of the quantized source in this case?

4.48 Solve Problem 4.46 using Table 4.3 instead of 4.2 to design an optimal nonuniform quantizer for the Gaussian source.

4.49 Consider the encoding of the two random variables X and Y, which are uniformly distributed on the region between the two squares as shown in Figure P-4.49.

1. Find $f_X(x)$ and $f_Y(y)$.
2. Assume each of the random variables X and Y are quantized using 4-level uniform quantizers. What is the resulting distortion? What is the resulting number of bits/pair (X, Y)?
3. Now assume that instead of scalar quantizers for X and Y we employ a vector quantizer to achieve the same level of distortion as in part 2. What is the resulting number of bits/source output pair (X, Y)?

4.50 Two random variables X and Y are uniformly distributed on the square shown in Figure P-4.50.

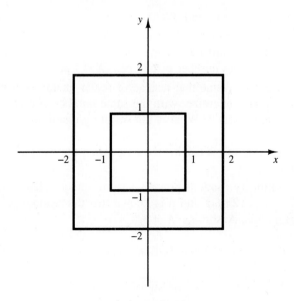

FIGURE P-4.49

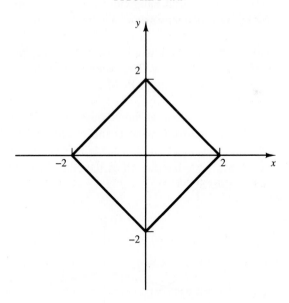

FIGURE P-4.50

1. Find $f_X(x)$ and $f_Y(y)$.

2. Assume that each of the random variables X and Y are quantized using 4-level uniform quantizers. What is the resulting distortion? What is the resulting number of bits per (X, Y) pair?

3. Now assume that instead of scalar quantizers for X and Y we employ a vector quantizer with the same number of bits/source output pair (X, Y) as in part 2. What is the resulting distortion for this vector quantizer?

4.51 Solve Example 4.6.1 for the case when the samples are uniformly distributed on $[-2, 2]$.

4.52 A stationary random process has an autocorrelation function given by $R_X = \frac{A^2}{2} e^{-|\tau|} \cos 2\pi f_0 \tau$ and it is known that the random process never exceeds 6 in magnitude. Assuming $A = 6$,

1. How many quantization levels are required to guarantee a SQNR of at least 60 dB?

2. Assuming that the signal is quantized to satisfy the condition of part 1 and assuming the approximate bandwidth of the signal is W, what is the minimum required bandwidth for transmission of a binary PCM signal based on this quantization scheme?

4.53 A signal can be modeled as a lowpass stationary process $X(t)$ whose p.d.f. at any time t_0 is given below. The bandwidth of this process is 5 kHz, and it is desired to transmit it using a PCM system.

1. If sampling is done at the Nyquist rate and a uniform quantizer with 32 levels is employed, what is the resulting SQNR? What is the resulting bit rate?

2. If the available bandwidth of the channel is 40 kHz, what is the highest achievable SQNR?

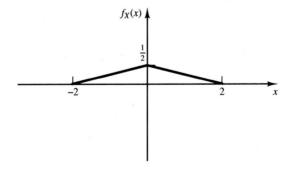

FIGURE P-4.53

3. If instead of sampling at the Nyquist rate we require a guard band of at least 2 kHz, and the bandwidth of the channel is 40 kHz again, what is the highest achievable SQNR?

4.54 A stationary source is distributed according a triangular p.d.f., $f_X(x) = \frac{1}{2}\Lambda\left(\frac{x}{2}\right)$. This source is quantized using a four level uniform quantizer described below

$$
Q(x) = \begin{cases}
1.5 & 1 < x \le 2 \\
0.5 & 0 < x \le 1 \\
-0.5 & -1 < x \le 0 \\
-1.5 & -2 \le x \le -1
\end{cases}
$$

1. Find the entropy of the quantized source.
2. Determine the p.d.f. of the random variable representing the quantization error, i.e., $\tilde{X} = X - Q(X)$.

4.55 The random process $X(t)$ is defined by $X(t) = Y\cos(2\pi f_0 t + \Theta)$ where Y and Θ are two independent random variables, Y uniform on $[-3, 3]$ and Θ uniform on $[0, 2\pi)$.

1. Find the autocorrelation function of $X(t)$ and its power-spectral density.
2. If $X(t)$ is to be transmitted to maintain a SQNR of at least 40 dB using a uniform PCM system, what is the required number of bits/sample and the least bandwidth requirement (in terms of f_0)?
3. If the SQNR is to be increased by 24 dB, how many more bits/sample have to be introduced and what is the new minimum bandwidth requirement in this case?

4.56 In our analysis of PCM systems, it was always assumed that the transmitted bits are received with no errors. However, practical channels cause errors. Let us assume that the output of a PCM system is transmitted via a channel whose error probability is denoted by p_b. It is further assumed that p_b is small enough such that, in transmission of the ν bits resulting from encoding of each quantized sample, either no error occurs or at most one error occurs. This means that the probability of each transmitted bit being in error is p_b and the probability of no error in transmission of ν bits is roughly $1 - \nu p_b$. We also assume that for the binary representation of each quantized value natural binary coding (NBC) is employed, i.e., the lowest quantized level is mapped into a sequence of zeros and the largest level is mapped into a sequence of all ones and all the other levels are mapped according to their relative value.

1. Show that: if an error occurs in the least significant bit, its effect on the quantized value is equivalent to Δ, the spacing between the levels; if an error occurs in the next bit, its effect on the quantized value is 2Δ, ..., if an error occurs in the most significant bit, its effect on the quantized value is $2^{\nu-1}\Delta$.

2. From the above show that the mean-squared error resulting from channel errors is given by

$$D_{\text{channel}} = p_b \Delta^2 \frac{4^\nu - 1}{3}$$

where $\Delta = \frac{2x_{\max}}{N} = \frac{x_{\max}}{2^{\nu-1}}$ is the spacing between adjacent levels.

3. From the above conclude that the total distortion, which is the sum of the quantization distortion and the transmission distortion due to channel errors, can be expressed as

$$D_{\text{total}} = \frac{x_{\max}^2}{3 \times N^2} \left(1 + 4p_b(N^2 - 1)\right) = \frac{x_{\max}^2}{3 \times 4^\nu} \left(1 + 4p_b(4^\nu - 1)\right)$$

4. Finally show that the SNR defined as the ratio of the signal power to the total noise power is given by

$$\text{SNR} = \frac{3N^2 \overline{\check{X}^2}}{1 + 4p_b(N^2 - 1)} = \frac{3 \times 4^\nu \overline{\check{X}^2}}{1 + 4p_b(4^\nu - 1)}$$

where $\check{X} = \frac{X}{x_{\max}}$.

4.57 In this problem we study the performance of nonuniform PCM systems. Let the signal samples be distributed according to the p.d.f. $f_X(x)$.

1. Use (4.6.15) to show that for a μ-law compander

$$D \approx \frac{[\ln(1 + \mu)]^2 x_{\max}^2}{3\mu^2 N^2} \left(\mu^2 E(\check{X}^2) + 2\mu E(|\check{X}|) + 1\right)$$

$$= \frac{[\ln(1 + \mu)]^2 x_{\max}^2}{3\mu^2 4^\nu} \left(\mu^2 E(\check{X}^2) + 2\mu E(|\check{X}|) + 1\right)$$

where $\check{X}$ represents the normalized random variable $\frac{X}{x_{\max}}$.

2. Show that

$$\text{SQNR} = \frac{3\mu^2 N^2}{\left(\ln(1 + \mu)\right)^2} \frac{E(\check{X}^2)}{\mu^2 E(\check{X}^2) + 2\mu E|\check{X}| + 1}$$

$$= \frac{3\mu^2 4^\nu}{\left(\ln(1 + \mu)\right)^2} \frac{E(\check{X}^2)}{\mu^2 E(\check{X}^2) + 2\mu E|\check{X}| + 1}$$

3. Compare the above result with the SQNR for a uniform quantizer and conclude that

$$\text{SQNR}_{\mu-\text{law}} = \text{SQNR}_{\text{uniform}} G(\mu, \check{X})$$

and determine $G(\mu, \check{X})$.

4. Now assume that X is a truncated zero-mean Gaussian random variable truncated to $[-4\sigma_x, 4\sigma_x]$. Plot both the $\text{SQNR}_{\mu-\text{law}}$ and the $\text{SQNR}_{\text{uniform}}$ in dB as a function of $E(\check{X}^2)$ (in dB) for $\mu = 255$ and $\nu = 8$. Compare the results and note the relative insensitivity of the μ-law scheme to the dynamic range of the input signal.

4.58 Design an optimal compander for a source with a triangular p.d.f. given by

$$f_X(x) = \begin{cases} x+1 & -1 \leq x \leq 0 \\ -x+1 & 0 \leq x \leq 1 \\ 0 & \text{otherwise} \end{cases}$$

Plot the resulting $g(x)$ and determine the SQNR.

4.59 In a CD player the sampling rate is 44.1 kHz and the samples are quantized using a 16 bits/sample quantizer. Determine the resulting number of bits for a piece of music with a duration of 50 minutes.

5

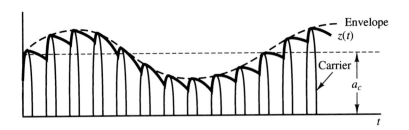

Envelope
$z(t)$

Carrier

a_c

t

Analog Signal Transmission and Reception

In the preceding chapter, we considered the characterization and modeling of discrete-time and continuous-time information sources. We also treated the problem of efficient encoding of signals emitted by sources. We observed that many information sources produce analog output signals, i.e., signals that are continuous in time and in amplitude. Common examples of analog information-bearing signals are speech signals and video signals.

We recall that analog signals may be converted into digital signals by sampling and quantizing the signal amplitude. Once an analog signal has been converted to digital form, the digital signal may be encoded and transmitted through a communication channel. At the receiving terminal, the received digital signal is converted back to an analog signal, if the intended user desires to have the signal in this form. In some applications, the received signal is stored on magnetic or optical media in digital form for future recall and use.

In spite of the general trend toward digital transmission of analog signals, there is a significant amount of analog signal transmission, especially in audio and video broadcasting. In this chapter we treat the transmission of analog signals by carrier modulation, the treatment of the performance of these systems in the presence of noise being deferred to Chapter 6. We consider the transmission of an analog signal by impressing it on either the amplitude, the phase, or the frequency of a sinusoidal carrier. Methods for demodulation of the carrier-modulated signal to recover the analog information signal are also described.

5.1 INTRODUCTION TO MODULATION

The source-output signal to be transmitted is denoted as $m(t)$. In general, $m(t)$ is an information-bearing signal; hence, it is characterized as a sample function of a random process $M(t)$. The random process is assumed to be wide-sense stationary (WSS) with autocorrelation function $R_M(\tau)$ and power-spectral density $S_M(f)$. Without loss of generality, we assume that the analog signal is lowpass, zero mean, and its spectral content is limited to W Hz., i.e., $S_M(f) = 0$ for $|f| > W$. A signal whose frequency content is in the vicinity of $f = 0(dc)$ is usually called a *base-band signal*. An important parameter of the information-bearing signal is its average power, which is defined as

$$P_M = R_M(0) = \int_{-W}^{W} S_M(f)\,df \qquad (5.1.1)$$

The message signal $m(t)$ is transmitted through the communication channel by impressing it on a carrier signal of the form

$$c(t) = A_c \cos(2\pi f_c t + \phi_c) \qquad (5.1.2)$$

where A_c is the amplitude, f_c is the carrier frequency, and ϕ_c is the carrier phase. We say that the message signal $m(t)$ modulates the carrier signal $c(t)$ in either amplitude, frequency, or phase. In effect, modulation converts the message signal $m(t)$ from lowpass to bandpass, in the neighborhood of the center frequency f_c.

Modulation of the carrier $c(t)$ by the message signal $m(t)$ is performed to achieve one or more of the following objectives: (1) the lowpass signal is translated in frequency to the passband of the channel so that the spectrum of the transmitted bandpass signal will match the passband characteristics of the channel; (2) to accommodate for simultaneous transmission of signals from several message sources, by means of frequency division multiplexing; and (3) to expand the bandwidth of the transmitted signal to increase its noise immunity in transmission over a noisy channel (as we will see in our discussion of angle-modulation noise performance in Chapter 6). We will see that objectives (1) and (2) are met by all of the modulation methods described below. Objective (3) is met by employing angle modulation to spread the signal $m(t)$ over a larger bandwidth.

The choice of modulation depends on a number of factors. The important factors generally are (1) the characteristics of the information-bearing signal, such as its bandwidth; (2) the channel characteristics, such as the type of transmission medium (wireline, free space, optical fiber, etc.) and the type of channel disturbances (additive noise, interference from other users, etc.); (3) performance requirements, such as received signal fidelity; and (4) economic factors in a practical implementation of the modulator and the demodulator.

In the following sections of this chapter we consider the transmission and reception of analog signals by carrier-amplitude modulation (AM), carrier-frequency modulation (FM), and carrier-phase modulation (PM). Comparisons will be made among these modulation methods on the basis of their bandwidth requirements and their implementation complexity. Their performance in the presence of additive noise disturbances will be treated in Chapter 6.

5.2 AMPLITUDE MODULATION

In amplitude modulation (AM), the message signal $m(t)$ is impressed on the amplitude of the carrier signal $c(t)$. There are several different ways of amplitude modulating the carrier signal by $m(t)$, each of which results in different spectral characteristics for the transmitted signal. We will describe these methods, which are called (1) double-sideband, suppressed carrier AM, (2) conventional double-sideband AM, (3) single-sideband AM, and (4) vestigial-sideband AM.

5.2.1 Double-Sideband Suppressed Carrier AM

A double-sideband, suppressed carrier (DSB-SC) AM signal is obtained by multiplying the message signal $m(t)$ with the carrier signal $c(t)$. Thus, we have the amplitude-modulated signal

$$u(t) = m(t)c(t)$$
$$= A_c m(t) \cos(2\pi f_c t + \phi_c) \tag{5.2.1}$$

Below we determine the spectral characteristics of the amplitude-modulated signal $u(t)$. Although, in practice, the message signal $m(t)$ is random, it is instructive to consider first the case in which $m(t)$ is a deterministic signal that possesses a Fourier transform $M(f)$.

Deterministic modulating signal. If $m(t)$ is deterministic with Fourier transform (spectrum) $M(f)$, the spectrum of the amplitude-modulated signal $u(t)$ is the convolution of the spectrum of $m(t)$ with the spectrum of the carrier signal. Hence,

$$U(f) = \mathcal{F}[m(t)] \star \mathcal{F}[A_c \cos(2\pi f_c t + \phi_c)]$$
$$= M(f) \star \frac{A_c}{2}\left[e^{j\phi_c}\delta(f - f_c) + e^{-j\phi_c}\delta(f + f_c)\right]$$
$$= \frac{A_c}{2}\left[M(f - f_c)e^{j\phi_c} + M(f + f_c)e^{-j\phi_c}\right] \tag{5.2.2}$$

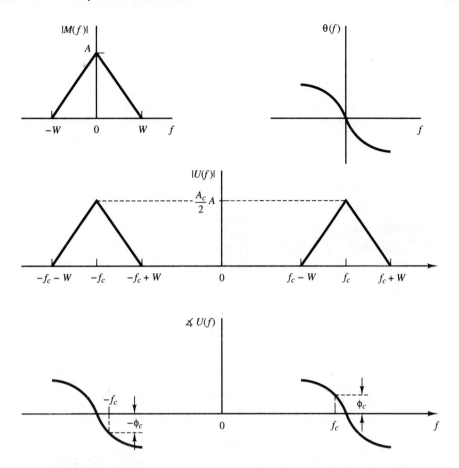

FIGURE 5.1.   Magnitude and phase spectra of the message signal $m(t)$ and the DSB-AM modulated signal $u(t)$.

Figure 5.1 illustrates the magnitude and phase spectra for $M(f)$ and $U(f)$.

We observe that the magnitude of the spectrum of the message signal $m(t)$ has been translated or shifted in frequency by an amount f_c. The phase of the message signal has been translated in frequency and offset by the carrier phase ϕ_c. Furthermore, the bandwidth occupancy of the amplitude-modulated signal is $2W$, whereas the bandwidth of the message signal $m(t)$ is W. Therefore, the channel bandwidth required to transmit the modulated signal $u(t)$ is $B_c = 2W$.

The frequency content of the modulated signal $u(t)$ in the frequency band $|f| > f_c$ is called the *upper sideband* of $U(f)$ and the frequency content in the frequency band $|f| < f_c$ is called the *lower sideband* of $U(f)$. It is important to note that either one of the sidebands of $U(f)$ contains all the frequencies that are in $M(f)$. That is, the frequency content of $U(f)$ for $f > f_c$ corresponds to the frequency content of $M(f)$ for $f > 0$, and the frequency content of $U(f)$ for

$f < -f_c$ corresponds to the frequency content of $M(f)$ for $f < 0$. Hence, the upper sideband of $U(f)$ contains all the frequencies in $M(f)$. A similar statement applies to the lower sideband of $U(f)$. Therefore, the lower sideband of $U(f)$ contains all the frequency content of the message signal $M(f)$. Because $U(f)$ contains both the upper and the lower sidebands, it is called a *double-sideband* (DSB) *AM signal.*

The other characteristic of the modulated signal $u(t)$ is that it does not contain a carrier component. That is, all the transmitted power is contained in the modulating (message) signal $m(t)$. This is evident from observing the spectrum of $U(f)$. We note that, as long as $m(t)$ does not have any DC component, there is no impulse in $U(f)$ at $f = f_c$, which would be the case if a carrier component was contained in the modulated signal $u(t)$. For this reason, $u(t)$ is called a *suppressed-carrier signal.* Therefore, $u(t)$ is a DSB-SC AM signal.

Example 5.2.1

Suppose that the modulating signal $m(t)$ is a sinusoid of the form

$$m(t) = a \cos 2\pi f_m t, \qquad f_m \ll f_c$$

Determine the DSB-SC AM signal and its upper and lower sidebands.

Solution The DSB-SC AM is expressed in the time domain as

$$u(t) = m(t)c(t) = A_c a \cos 2\pi f_m t \cos(2\pi f_c t + \phi_c)$$

$$= \frac{A_c a}{2} \cos[2\pi(f_c - f_m)t + \phi_c]$$

$$+ \frac{A_c a}{2} \cos[2\pi(f_c + f_m)t + \phi_c] \qquad (5.2.3)$$

In the frequency domain, the modulated signal has the form

$$U(f) = \frac{A_c a}{4} \left[e^{j\phi_c} \delta(f - f_c + f_m) + e^{-j\phi_c} \delta(f + f_c - f_m) \right]$$

$$+ \frac{A_c a}{4} \left[e^{j\phi_c} \delta(f - f_c - f_m) + e^{-j\phi_c} \delta(f + f_c + f_m) \right] \qquad (5.2.4)$$

This spectrum is shown in Figure 5.2(a).

The lower sideband of $u(t)$ is the signal

$$u_\ell(t) = \frac{A_c a}{2} \cos[2\pi(f_c - f_m)t + \phi_c]$$

and its spectrum is illustrated in Figure 5.2(b). Finally, the upper sideband of $u(t)$ is the signal

$$u_u(t) = \frac{A_c a}{2} \cos[2\pi(f_c + f_m)t + \phi_c]$$

and its spectrum is illustrated in Figure 5.2(c).

Random modulating signal. In practice the message signal $m(t)$ is a sample function of a random process $M(t)$. Consequently, the amplitude modulated signal $u(t)$ is a sample function of a random process $U(t)$, which is characterized

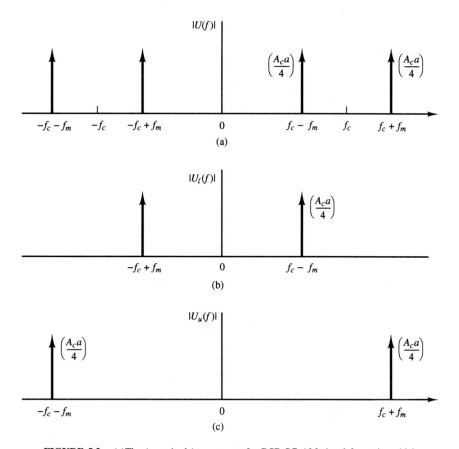

FIGURE 5.2.   (a)The (magnitude) spectrum of a DSB-SC AM signal for a sinusoidal message signal and (b) its lower and (c) upper sidebands.

in the time domain by its autocorrelation function and in the frequency domain by its power-spectral density.

The message process $M(t)$ is assumed to be WSS with zero mean and auto-correlation function $R_M(\tau)$. Consequently, the DSB-SC modulated signal $U(t)$ has zero mean, i.e.,

$$E[U(t)] = A_c E[M(t)] \cos(2\pi f_c t + \phi_c)$$

$$= 0$$

Its autocorrelation function is

$$R_U(t, t+\tau) = E[U(t)U(t+\tau)]$$

$$= A_c^2 E[M(t)M(t+\tau)] \cos(2\pi f_c t + \phi_c) \cos(2\pi f_c(t+\tau) + \phi_c)$$

$$= \frac{1}{2} A_c^2 R_M(\tau)[\cos 2\pi f_c \tau + \cos(2\pi f_c(2t + \tau) + 2\phi_c)] \qquad (5.2.5)$$

Note that $R_U(t, t + \tau)$ is a periodic autocorrelation in the t variable, with period $T_p = 1/2f_c$. Hence, $U(t)$ is a cyclostationary process. We have seen in Chapter 3 that, for such a process, it is convenient to deal with the average autocorrelation function $\bar{R}_U(\tau)$, defined as

$$\bar{R}_U(\tau) = \frac{1}{T_p} \int_{-T_p/2}^{T_p/2} R_U(t, t + \tau) \, dt \qquad (5.2.6)$$

By substituting (5.2.5) in (5.2.6) and performing the integration over the variable t, we obtain

$$\bar{R}_U(\tau) = \frac{1}{2} A_c^2 R_M(\tau) \cos 2\pi f_c \tau \qquad (5.2.7)$$

The average power in the DBS-SC signal $U(t)$ is $\bar{R}_U(0)$, i.e.,

$$\bar{R}_U(0) = \frac{1}{2} A_c^2 R_M(0) \qquad (5.2.8)$$

Since $A_c^2/2$ is the average power of the carrier component and $R_M(0)$ is the power of the message signal, the average power in the DSB-SC signal $U(t)$ is simply the product of the two average powers.

The Fourier transform of $\bar{R}_U(\tau)$ gives us the power-spectral density $S_U(f)$ of the DSB-SC signal. Thus we obtain

$$S_U(f) = \int_{-\infty}^{\infty} \bar{R}_U(\tau)e^{-j2\pi f_c \tau} \, d\tau$$

$$S_U(f) = \frac{A_c^2}{4} \left[S_M(f - f_c) + S_M(f + f_c) \right] \qquad (5.2.9)$$

where $S_M(f)$ is the power-spectral density of the message signal.

Example 5.2.2

Suppose that the message signal $m(t)$ is a sample function of a zero-mean random process with power-spectral density

$$S_M(f) = \begin{cases} 1 - \dfrac{|f|}{F_0}, & |f| \le F_0 < f_c \\ 0, & |f| > F_0 \end{cases} \qquad (5.2.10)$$

Determine the average autocorrelation function $\bar{R}_U(\tau)$ of the DSB-SC AM signal $U(t)$ and its power-spectral density $S_U(f)$.

Solution   Because $S_M(f) = \Lambda\left(\frac{f}{F_0}\right)$, the autocorrelation function of the message signal is

$$R_M(\tau) = \int_{-\infty}^{\infty} S_M(f)e^{j2\pi f \tau} \, df$$

$$= F_0 \, \text{sinc}^2(F_0 \tau) \qquad (5.2.11)$$

Hence, the average autocorrelation function of the DSB-SC AM signal is

$$\bar{R}_U(\tau) = \frac{A_c^2}{2} F_0 \, \text{sinc}^2(F_0 \tau) \cos 2\pi f_c \tau \qquad (5.2.12)$$

and the power-spectral density

$$S_U(f) = \begin{cases} \frac{A_c^2}{4} \left(1 - \frac{|f-f_c|}{F_0}\right), & f_c - F_0 < f < f_c + F_0 \\ \frac{A_c^2}{4} \left(1 - \frac{|f+f_c|}{F_0}\right), & -f_c - F_0 < f < -f_c + F_0 \\ 0, & \text{otherwise} \end{cases} \qquad (5.2.13)$$

Figure 5.3 illustrates the power-spectral density of the modulated signal and its upper and lower sidebands.

Demodulation of DSB-SC AM signals.   In the absence of noise and with the assumption of an ideal channel, the received signal is equal to the modulated signal, i.e.,

$$r(t) = u(t)$$

$$= A_c m(t) \cos(2\pi f_c t + \phi_c) \qquad (5.2.14)$$

Suppose we demodulate the received signal by first multiplying $r(t)$ by a locally generated sinusoid $\cos(2\pi f_c t + \phi)$, where ϕ is the phase of the sinusoid, and then passing the product signal through an ideal lowpass filter having a bandwidth W. The multiplication of $r(t)$ with $\cos(2\pi f_c t + \phi)$ yields

$$r(t) \cos(2\pi f_c t + \phi) = A_c m(t) \cos(2\pi f_c t + \phi_c) \cos(2\pi f_c t + \phi)$$

$$= \frac{1}{2} A_c m(t) \cos(\phi_c - \phi) + \frac{1}{2} A_c m(t) \cos(4\pi f_c t + \phi + \phi_c)$$

$$(5.2.15)$$

The lowpass filter rejects the double-frequency components and passes only the lowpass components. Hence, its output is

$$y_\ell(t) = \frac{1}{2} A_c m(t) \cos(\phi_c - \phi) \qquad (5.2.16)$$

Note that $m(t)$ is multiplied by $\cos(\phi_c - \phi)$. The desired signal thus, is scaled in amplitude by a factor that depends on the phase difference between the phase ϕ_c of the carrier in the received signal and the phase ϕ of the locally generated sinusoid. When $\phi_c \neq \phi$, the amplitude of the desired signal is reduced by the factor $\cos(\phi_c - \phi)$. If $\phi_c - \phi = 45^o$, the amplitude of the desired signal is reduced by $\sqrt{2}$ and the signal power is reduced by a factor of 2. If $\phi_c - \phi = 90^o$, the desired signal component vanishes.

The above discussion demonstrates the need for a *phase-coherent or synchronous demodulator* for recovering the message signal $m(t)$ from the received

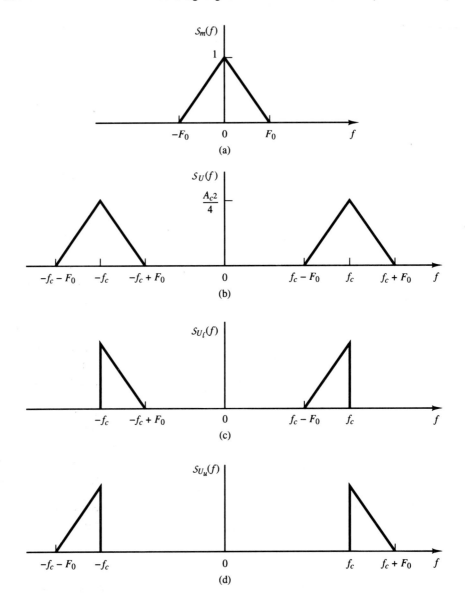

FIGURE 5.3. Power-spectral density of the message signal (a) and modulated signal (b), power spectrum of the lower sideband (c) and the upper sideband (d).

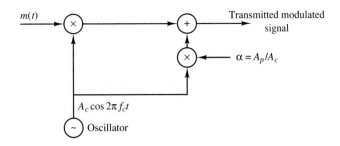

FIGURE 5.4.   Addition of a pilot tone to a DSB AM signal.

signal. That is, the phase ϕ of the locally generated sinusoid should ideally be equal to the phase ϕ_c of the received carrier signal.

A sinusoid that is phase-locked to the phase of the received carrier can be generated at the receiver in one of two ways. One method is to add a carrier component into the transmitted signal as illustrated in Figure 5.4.

We call such a carrier component "a pilot tone." Its amplitude A_p and its power $A_p^2/2$ is selected to be significantly smaller than that of the modulated signal $u(t)$. Thus the transmitted signal is double-sideband, but it is no longer a suppressed carrier signal. At the receiver, a narrowband filter tuned to frequency f_c is used to filter out the pilot signal component, and its output is used to multiply the received signal as shown in Figure 5.5.

The reader may show that the presence of the pilot signal results in a *dc* component in the demodulated signal, which must be subtracted out to recover $m(t)$.

The addition of a pilot tone to the transmitted signal has the disadvantage of requiring that a certain portion of the transmitted signal power must be allocated to the transmission of the pilot. As an alternative, we may generate a phase-locked sinusoidal carrier from the received signal $r(t)$ without the need of a pilot signal. This can be accomplished by using a *phase-locked loop* as described in Section 6.2.

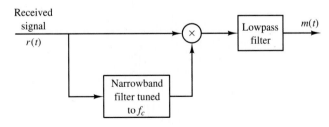

FIGURE 5.5.   Use of a pilot tone to demodulate a DSB AM signal.

5.2.2 Conventional Amplitude Modulation

A conventional AM signal consists of a large carrier component in addition to the DSBD AM modulated signal. The transmitted signal is expressed mathematically as

$$u(t) = A_c[1 + m(t)]\cos(2\pi f_c t + \phi_c) \qquad (5.2.17)$$

where the message waveform is constrained to satisfy the condition that $|m(t)| \leq 1$. We observe that $A_c m(t)\cos(2\pi f_c t + \phi_c)$ is a DSBD AM signal and $A_c\cos(2\pi f_c t + \phi_c)$ is the carrier component. Figure 5.6 illustrates an AM signal in the time domain.

As long as $|m(t)| \leq 1$, the amplitude $A_c[1 + m(t)]$ is always positive. This is the desired condition for conventional DSB AM that makes it easy to demodulate (described below). On the other hand, if $m(t) < -1$ for some t, the AM signal is said to be *overmodulated* and its demodulation is rendered more complex. In practice, $m(t)$ is scaled so that its magnitude is always less than unity.

It is sometimes convenient to express $m(t)$ as

$$m(t) = a m_n(t)$$

where $m_n(t)$ is normalized such that its minimum value is -1. This can be done, for example, by defining

$$m_n(t) = \frac{m(t)}{\max|m(t)|}$$

The scale factor a is called the *modulation index*. Then the modulated signal can be expressed as

$$u(t) = A_c\big[1 + a m_n(t)\big]\cos 2\pi f_c t \qquad (5.2.18)$$

Let us consider the spectral characteristics of the transmitted signal—first, for the case in which $m(t)$ is a deterministic signal and second for the case in which $m(t)$ is a random process.

Deterministic modulating signal.   If $m(t)$ is a deterministic signal with Fourier transform (spectrum) $M(f)$, the spectrum of the AM signal $u(t)$ is

$$U(f) = \mathcal{F}[a m_n(t)] \star \mathcal{F}[A_c\cos(2\pi f_c t + \phi_c)] + \mathcal{F}[A_c\cos(2\pi f_c t + \phi_c)]$$

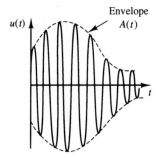

FIGURE 5.6.   A conventional AM signal in the time domain.

$$= aM_n(f) \star \frac{A_c}{2} \left[e^{j\phi_c} \delta(f - f_c) + e^{-j\phi_c} \delta(f + f_c) \right]$$

$$+ \frac{A_c}{2} \left[e^{j\phi_c} \delta(f - f_c) + e^{-j\phi_c} \delta(f + f_c) \right]$$

$$U(f) = \frac{A_c}{2} [e^{j\phi_c} aM_n(f - f_c) + e^{j\phi_c} \delta(f - f_c)$$

$$+ e^{-j\phi_c} aM_n(f + f_c) + e^{-j\phi_c} \delta(f + f_c)] \tag{5.2.19}$$

Obviously the spectrum of a conventional AM signal occupies a bandwidth twice the bandwidth of the message signal.

Example 5.2.3

Suppose that the modulating signal $m(t)$ is a sinusoid of the form

$$m(t) = \cos 2\pi f_m t \qquad f_m \ll f_c$$

Determine the DSB AM signal, its upper and lower sidebands, and its spectrum assuming a modulation index of a.

Solution   From (5.2.18) the DSB AM signal is expressed as

$$u(t) = A_c[1 + a\cos 2\pi f_m t]\cos(2\pi f_c t + \phi_c)$$

$$= A_c \cos(2\pi f_c t + \phi_c) + \frac{A_c a}{2} \cos[2\pi (f_c - f_m)t + \phi_c]$$

$$+ \frac{A_c a}{2} \cos[2\pi (f_c + f_m)t + \phi_c] \tag{5.2.20}$$

The lower-sideband component is

$$u_\ell(t) = \frac{A_c a}{2} \cos[2\pi (f_c - f_m)t + \phi_c]$$

while the upper-sideband component is

$$u_u(t) = \frac{A_c a}{2} \cos[2\pi (f_c + f_m)t + \phi_c]$$

The spectrum of the DSB AM signal $u(t)$ is

$$U(f) = \frac{A_c}{2} \left[e^{j\phi_c} \delta(f - f_c) + e^{-j\phi_c} \delta(f + f_c) \right]$$

$$+ \frac{A_c a}{4} \left[e^{j\phi_c} \delta(f - f_c + f_m) + e^{-j\phi_c} \delta(f + f_c - f_m) \right]$$

$$+ \frac{A_c a}{4} \left[e^{j\phi_c} \delta(f - f_c - f_m) + e^{-j\phi_c} \delta(f + f_c + f_m) \right] \tag{5.2.21}$$

The spectrum $|U(f)|$ is shown in Figure 5.7. It is interesting that the power of the carrier component, which is $A_c^2/2$, exceeds the total power $\left(A_c^2 a^2/2\right)$ of the two sidebands.

Random modulating signal.   When the message signal $m(t)$ is a sample function of a WSS random process with zero mean and autocorrelation $R_M(\tau)$, the

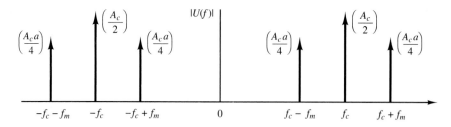

FIGURE 5.7.   Spectrum of a DSB AM signal in Example 5.2.3.

AM signal $U(t)$ has a mean of

$$E[U(t)] = A_c \cos(2\pi f_c t + \phi_c) \tag{5.2.22}$$

Its autocorrelation function is

$$
\begin{aligned}
R_U(t, t+\tau) &= E[U(t)U(t+\tau)] \\
&= A_c^2 E[1 + M(t)][1 + M(t+\tau)] \\
&\quad \times \cos(2\pi f_c t + \phi_c) \cos(2\pi f_c(t+\tau) + \phi_c) \\
&= \frac{A_c^2}{2}[1 + R_M(\tau)][\cos 2\pi f_c \tau + \cos(2\pi f_c(2t+\tau) + 2\phi_c)] \tag{5.2.23}
\end{aligned}
$$

As in the case of the DSB-SC AM signal, $U(t)$ is a cyclostationary process with period $T_p = 1/f_c$ (the common period of mean and autocorrelation). To characterize this process in the frequency domain, we first determine the average autocorrelation function $\bar{R}_U(\tau)$. Thus,

$$
\begin{aligned}
\bar{R}_U(\tau) &= \frac{1}{T_p} \int_{-T_p/2}^{T_p/2} R_U(t, t+\tau)dt \\
&= \frac{A_c^2}{2}[1 + R_M(\tau)] \cos 2\pi f_c \tau \tag{5.2.24}
\end{aligned}
$$

Note that the average power in the transmitted signal $\bar{R}_U(0)$ is

$$
\begin{aligned}
\bar{R}_U(0) &= \frac{A_c^2}{2}[1 + R_M(0)] \\
&= \frac{A_c^2}{2}[1 + a^2 R_{M_n}(0)] \tag{5.2.25}
\end{aligned}
$$

Because $|m_n(t)| \le 1$, it follows that $R_{M_n}(0) \le 1$ and, as a consequence, at least one-half of the transmitted power is contained in the carrier component.

The *modulation efficiency* of a conventional AM signal is defined as the ratio of the power in the modulated signal that contributes to the transmission of the message to the total power in the modulated signal. The modulation efficiency is given by

$$
\begin{aligned}
\nu &= \frac{\frac{A_c^2}{2} a^2 R_{M_n}(0)}{\frac{A_c^2}{2} [1 + a^2 R_{M_n}(0)]} \\[2mm]
&= \frac{a^2 R_{M_n}(0)}{1 + a^2 R_{M_n}(0)} \\[2mm]
&= \frac{a^2 P_{M_n}}{1 + a^2 P_{M_n}}
\end{aligned} \tag{5.2.26}
$$

The power-spectral density of the AM signal is simply determined by taking the Fourier transform of (5.2.24). Thus, we obtain

$$
S_U(f) = \frac{A_c^2}{4} [S_M(f - f_c) + S_M(f + f_c) + \delta(f - f_c) + \delta(f + f_c)] \tag{5.2.27}
$$

Example 5.2.4

Suppose that the message signal $m(t)$ is a sample function of a zero-mean random process with power-density spectrum

$$
S_M(f) = \Lambda\left(\frac{f}{F_0}\right)
$$

Determine the average autocorrelation function $\bar{R}_U(\tau)$ of the DSB AM signal $U(t)$ and its power-spectral density $S_U(f)$.

Solution   The average autocorrelation function $\bar{R}_U(\tau)$ is given by (5.2.24) where the autocorrelation of the message signal is given by (5.2.11). Hence

$$
\bar{R}_U(\tau) = \frac{A_c^2}{2} \left[1 + F_0 \operatorname{sinc}^2(F_0 \tau)\right] \cos 2\pi f_c \tau \tag{5.2.28}
$$

The power-spectral density

$$
S_U(f) = \begin{cases}
\frac{A_c^2}{4}\left[\delta(f - f_c) + \left(1 - \frac{|f - f_c|}{F_0}\right)\right], & |f - f_c| \le F_0 \\[2mm]
\frac{A_c^2}{4}\left[\delta(f + f_c) + \left(1 - \frac{|f + f_c|}{F_0}\right)\right], & |f + f_c| \le F_0 \\[2mm]
0, & \text{otherwise}
\end{cases} \tag{5.2.29}
$$

Figure 5.8 illustrates $S_U(f)$.

Demodulation of conventional DSB AM signals.   The major advantage of conventional AM signal transmission is the ease with which the signal can be demodulated. There is no need for a synchronous demodulator. Because the message signal $m(t)$ satisfies the condition $|m(t)| < 1$, the envelope (amplitude) $1 + m(t) > 0$. If we rectify the received signal, we eliminate the negative values without affecting the message signal as shown in Figure 5.9. The rectified signal is equal to $u(t)$ when $u(t) > 0$ and zero when $u(t) < 0$. The message signal is recovered by passing the rectified signal through a lowpass filter whose bandwidth matches that of the

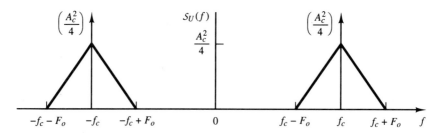

FIGURE 5.8.   Power-density spectrum of a DSB AM signal in Example 5.2.4.

message signal. The combination of the rectifier and the lowpass filter is called *envelope detector.*

Ideally, the output of the envelope detector is of the form

$$d(t) = g_1 + g_2 m(t) \qquad (5.2.30)$$

where g_1 represents a *dc* component and g_2 is a gain factor due to the signal demod-ulator. The *dc* component can be eliminated by passing $d(t)$ through a transformer, whose output is $g_2 m(t)$.

The simplicity of the demodulator has made conventional DSB AM a practical choice for AM radio broadcasting. Because there are literally billions of radio receivers, an inexpensive implementation of the demodulator is extremely important. The power inefficiency of conventional AM is justified by the fact that there are few broadcast transmitters relative to the number of receivers. Consequently it is cost effective to construct powerful transmitters and sacrifice power efficiency to simplify the signal demodulation at the receivers.

5.2.3 Single-Sideband AM

In Section 5.2.1 we showed that a DSB-SC AM signal required a channel bandwidth of $B_c = 2W$ Hz for transmission, where W is the bandwidth of the baseband signal. However, the two sidebands are redundant. Below we demonstrate that the transmission of either sideband is sufficient to reconstruct the message signal $m(t)$ at the receiver. Thus we reduce the bandwidth of the transmitted to that of the baseband signal.

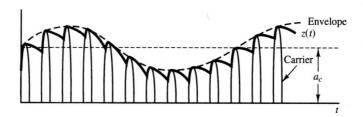

FIGURE 5.9.   Envelope detection of conventional AM signal.

First, we demonstrate that a single-sideband (SSB) AM signal is represented mathematically as

$$u(t) = A_c m(t) \cos 2\pi f_c t \mp A_c \hat{m}(t) \sin 2\pi f_c t \qquad (5.2.31)$$

where $\hat{m}(t)$ is the Hilbert transform of $m(t)$ that was introduced in Section 2.6, and the plus-or-minus sign determines which sideband we obtain. Recall that the Hilbert transform may be viewed as a linear filter with impulse response $h(t) = 1/\pi t$ and frequency response

$$H(f) = \begin{cases} -j, & f > 0 \\ j, & f < 0 \\ 0, & f = 0 \end{cases} \qquad (5.2.32)$$

Therefore, the SSB AM signal $u(t)$ may be generated by using the system configuration shown in Figure 5.10.

The method shown in Figure 5.10 for generating a SSB AM signal is one that employs a Hilbert transform filter. Another method, illustrated in Figure 5.11, generates a DSB-SC AM signal and then employs a filter that selects either the upper sideband or the lower sideband of the DSB AM signal.

To demonstrate that the modulated signal contains only a single sideband, we first treat $m(t)$ as a deterministic signal and then as a random signal.

Deterministic modulating signal.   Let $m(t)$ be a deterministic signal with Fourier transform (spectrum) $M(f)$. An upper single-sideband amplitude-modulated signal (USSB AM) can be obtained by eliminating the lower sidebands of a DSB AM signal. To eliminate the lower sidebands of the DSB AM signal, $u_{\text{DSB}}(t) = 2A_c m(t) \cos 2\pi f_c t$, it is passed through a highpass filter whose transfer function is given by

$$H(f) = \begin{cases} 1 & |f| > f_c \\ 0 & \text{otherwise} \end{cases}$$

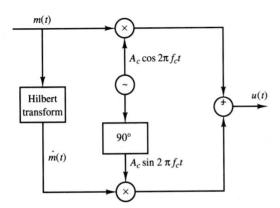

FIGURE 5.10.   Generation of a SSB AM signal.

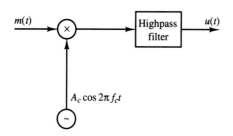

FIGURE 5.11. Generation of a SSB AM signal by filtering one of the sidebands of a DSB-SC AM signal.

as shown in Figure 5.11. Obviously $H(f)$ can be written as

$$H(f) = u_{-1}(f - f_c) + u_{-1}(-f - f_c)$$

where $u_{-1}(\cdot)$ represents the unit step function. Therefore the spectrum of the USSB AM signal is given by

$$U_u(f) = A_c M(f - f_c)u_{-1}(f - f_c) + A_c M(f + f_c)u_{-1}(-f - f_c)$$

or equivalently

$$U_u(f) = A_c M(f)u_{-1}(f)|_{f=f-f_c} + A_c M(f)u_{-1}(-f)|_{f=f+f_c} \tag{5.2.33}$$

Taking the inverse Fourier transform of both sides of (5.2.33) and using the modulation theorem of the Fourier transform, we obtain

$$u_u(t) = A_c m(t) \star \mathcal{F}^{-1}\left[u_{-1}(f)\right]e^{j2\pi f_c t} +$$
$$A_c m(t) \star \mathcal{F}^{-1}\left[u_{-1}(-f)\right]e^{-j2\pi f_c t} \tag{5.2.34}$$

By noting that

$$\mathcal{F}\left[\frac{1}{2}\delta(t) - \frac{j}{2\pi t}\right] = u_{-1}(f)$$

$$\mathcal{F}\left[\frac{1}{2}\delta(t) + \frac{j}{2\pi t}\right] = u_{-1}(-f) \tag{5.2.35}$$

and substituting (5.2.35) in (5.2.34), we obtain

$$u_u(t) = A_c m(t) \star \left[\frac{1}{2}\delta(t) - \frac{j}{2\pi t}\right]e^{j2\pi f_c t} + A_c m(t) \star \left[\frac{1}{2}\delta(t) + \frac{j}{2\pi t}\right]e^{-j2\pi f_c t}$$

$$= \frac{A_c}{2}\left[m(t) - j\hat{m}(t)\right]e^{j2\pi f_c t} +$$

$$\frac{A_c}{2}\left[m(t) + j\hat{m}(t)\right]e^{-j2\pi f_c t} \tag{5.2.36}$$

where we have used the identities

$$m(t) \star \delta(t) = m(t)$$

$$m(t) \star \frac{1}{\pi t} = \hat{m}(t)$$

Using Euler's relations in (5.2.36), we obtain

$$u_u(t) = A_c m(t) \cos 2\pi f_c t - A_c \hat{m}(t) \sin 2\pi f_c t \qquad (5.2.37)$$

which is the time-domain representation of an USSB AM signal. The expression for the lower single sideband (LSSB) AM signal can be derived by noting that

$$u_u(t) + u_\ell(t) = u_{\mathrm{DSB}}(t)$$

or

$$A_c m(t) \cos 2\pi f_c t - A_c \hat{m}(t) \sin 2\pi f_c t + u_\ell(t) = 2A_c m(t) \cos 2\pi f_c t$$

and therefore

$$u_\ell(t) = A_c m(t) \cos 2\pi f_c t + A_c \hat{m}(t) \sin 2\pi f_c t \qquad (5.2.38)$$

Thus the time-domain representation of a SSB AM signal can in general be expressed as

$$u_{\mathrm{SSB}}(t) = A_c m(t) \cos 2\pi f_c t \mp A_c \hat{m}(t) \sin 2\pi f_c t \qquad (5.2.39)$$

where the minus sign corresponds to the USSB AM signal and the plus sign corresponds to the LSSB AM signal.

Example 5.2.5

Suppose that the modulating signal is a sinusoid of the form

$$m(t) = \cos 2\pi f_m t, \qquad f_m \ll f_c$$

Determine the two possible SSB AM signals.

Solution   The Hilbert transform of $m(t)$ is

$$\hat{m}(t) = \sin 2\pi f_m t \qquad (5.2.40)$$

Hence,

$$u(t) = A_c \cos 2\pi f_m t \, \cos 2\pi f_c t \mp A_c \sin 2\pi f_m t \sin 2\pi f_c t \qquad (5.2.41)$$

If we take the upper $(-)$ sign, we obtain the USSB signal

$$u_u(t) = A_c \cos 2\pi (f_c + f_m)t$$

On the other hand, if we take the lower $(+)$ sign in (5.2.41), we obtain the lower sideband signal

$$u_\ell(t) = A_c \cos 2\pi (f_c - f_m)t$$

The spectra of $u_u(t)$ and $u_\ell(t)$ were previously given in Figure 5.2.

Random modulating signal.   Let us express the SSB AM signal given by (5.2.31) as

$$U(t) = A_c \, \mathrm{Re} \left\{ [M(t) \pm j\hat{M}(t)] e^{j2\pi f_c t} \right\} \qquad (5.2.42)$$

where the upper sign $(+)$ corresponds to the upper sideband signal and the lower sign $(-)$ corresponds to the lower sideband signal. As in our preceding discussion, we assume that the message signal $M(t)$ is WSS with zero mean and autocorrelation function $R_M(\tau)$.

The autocorrelation function for the SSB AM signal is

$$R_U(t, t + \tau) = E[U(t)U(t + \tau)]$$

$$= A_c^2 E\left[\text{Re}\left\{[M(t) \pm j\hat{M}(t)]e^{j2\pi f_c t}\right\}\right.$$

$$\left. \times \text{Re}\left\{[M(t + \tau) \pm j\hat{M}(t + \tau)]e^{j2\pi f_c(t+\tau)}\right\}\right]$$

It is convenient to express the real part of a complex-valued function as

$$Re(C) = \frac{1}{2}(C + C^*)$$

Then,

$$R_U(t, t + \tau) = \frac{A_c^2}{4} E\left[\left\{[M(t) \pm j\hat{M}(t)]e^{+j2\pi f_c t} + [M(t) \mp j\hat{M}(t)]e^{-j2\pi f_c t}\right\}\right.$$

$$\times \left\{[M(t + \tau) \pm j\hat{M}(t + \tau)]e^{j2\pi f_c(t+\tau)}\right.$$

$$\left.\left. + [M(t + \tau) \mp j\hat{M}(t + \tau)]e^{-j2\pi f_c(t+\tau)}\right\}\right]$$

By making use of the relations (see Example 3.2.20)

$$E[M(t)M(t + \tau)] = E[\hat{M}(t)\hat{M}(t + \tau)] = R_M(\tau)$$

$$E[\hat{M}(t)M(t + \tau)] = -\hat{R}_M(\tau) \qquad (5.2.43)$$

we find that the autocorrelation function of $U(t)$ reduces to

$$R_U(t, t + \tau) = R_U(\tau) = A_c^2[R_M(\tau)\cos 2\pi f_c \tau \mp \hat{R}_M(\tau)\sin 2\pi f_c \tau] \qquad (5.2.44)$$

Therefore, $U(t)$ is a stationary random process. It is interesting to note from the expression for $R_U(\tau)$ that $R_U(\tau)$ is a single sideband modulation of $R_M(\tau)$ at carrier frequency f_c.

The power-spectral density of the SSB signal $U(t)$ is simply found by taking the Fourier transform of (5.2.44). Thus, with the upper sign $(-)$ in (5.2.44) we obtain

$$S_U(f) = A_c^2 \begin{cases} 0, & |f| < f_c \\ S_M(f - f_c) + S_M(f + f_c), & |f| > f_c \end{cases} \qquad (5.2.45)$$

and with the lower sign $(+)$ in (5.2.44), we obtain

$$S_U(f) = A_c^2 \begin{cases} S_M(f - f_c) + S_M(f + f_c), & |f| < f_c \\ 0, & |f| > f_c \end{cases} \qquad (5.2.46)$$

Integrating the power-spectral density yields the power content of the modulated signals as

$$P_u = P_\ell = A_c^2 P_M$$

Example 5.2.6

Suppose that the message signal $m(t)$ is a sample function of a zero-mean random process with power-spectral density

$$S_M(f) = \begin{cases} 1 - \dfrac{|f|}{F_0}, & |f| \leq F_0 < f_c \\ 0, & |f| > F_0 \end{cases}$$

Determine the power-spectral density of the SSB AM signal.

Solution   From (5.2.45) we obtain the power-spectral density of the upper sideband signal, i.e.,

$$S_U(f) = \begin{cases} 1 - \dfrac{|f-f_c|}{F_0}, & f_c < f < f_c + F_0 \\ 1 - \dfrac{|f+f_c|}{F_0}, & -f_c - F_0 < f < -f_c \\ 0, & \text{otherwise} \end{cases}$$

From (5.2.46) we obtain the power-spectral density of the lower sideband signal, i.e.,

$$S_U(f) = \begin{cases} 1 - \dfrac{|f-f_c|}{F_0}, & f_c - F_0 < f < f_c \\ 1 - \dfrac{|f+f_c|}{F_0}, & -f_c < f < -f_c + F_0 \\ 0, & \text{otherwise} \end{cases}$$

Figure 5.12 illustrates the spectral characteristics of the two SSB AM signals.

Demodulation of SSB AM signals.   To recover the message signal $m(t)$ in the received SSB AM signal, we require a phase-coherent or synchronous demodulator, as was the case for DSB-SC AM signals. Thus, for the USSB signal given by (5.2.39), we have

$$u(t) \cos(2\pi f_c t + \phi) = \frac{1}{2} A_c m(t) \cos\phi + \frac{1}{2} A_c \hat{m}(t) \sin\phi$$

$$+ \text{double frequency terms} \qquad (5.2.47)$$

By passing the product signal in (5.2.47) through an ideal lowpass filter, the double-frequency components are eliminated, leaving us with

$$y_\ell(t) = \frac{1}{2} A_c m(t) \cos\phi + \frac{1}{2} A_c \hat{m}(t) \sin\phi \qquad (5.2.48)$$

Note that the effect of the phase offset is not only to reduce the amplitude of the desired signal $m(t)$ by $\cos\phi$, but it also results in an undesirable sideband signal due to the presence of $\hat{m}(t)$ in $y_\ell(t)$. The latter component was not present

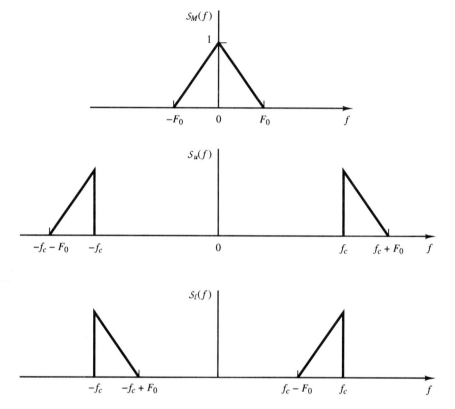

FIGURE 5.12.   Power spectra of SSB AM signal.

in a DSB-SC signal and, hence, it was not a factor. However, it is an important element that contributes to the distortion of the demodulated SSB signal.

The transmission of a pilot tone at the carrier frequency is a very effective method for providing a phase-coherent reference signal for performing synchronous demodulation at the receiver. Thus the undesirable sideband signal component is eliminated. However, this means that a portion of the transmitted power must be allocated to the transmission of the carrier.

The spectral efficiency of SSB AM makes this modulation method very attractive for use in voice communications over telephone channels (wire lines and cables). In this application, a pilot tone is transmitted for synchronous demodulation and shared among several channels.

The filter method shown in Figure 5.11 for selecting one of the two signal sidebands for transmission is particularly difficult to implement when the message signal $m(t)$ has a large power concentrated in the vicinity of $f = 0$. In such a case, the sideband filter must have an extremely sharp cutoff in the vicinity of the

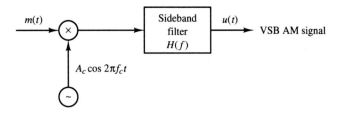

FIGURE 5.13.   Generation of VSB AM signal.

carrier to reject the second sideband. Such filter characteristics are very difficult to implement in practice.

5.2.4 Vestigial-Sideband AM

The stringent frequency-response requirements on the sideband filter in a SSB AM system can be relaxed by allowing a part, called a *vestige,* of the unwanted sideband to appear at the output of the modulator. Thus, we simplify the design of the sideband filter at the cost of a modest increase in the channel bandwidth required to transmit the signal. The resulting signal is called *vestigial-sideband* (VSB) AM.

To generate a VSB AM signal, we begin by generating a DSB-SC AM signal and passing it through a sideband filter with frequency response $H(f)$ as shown in Figure 5.13. In the time domain the VSB signal may be expressed as

$$u(t) = [A_c m(t) \cos 2\pi f_c t] \star h(t) \qquad (5.2.49)$$

where $h(t)$ is the impulse response of the VSB filter. In the frequency domain, the corresponding expression is

$$U(f) = \frac{A_c}{2}[M(f - f_c) + M(f + f_c)]H(f) \qquad (5.2.50)$$

To determine the frequency-response characteristics of the filter, let us consider the demodulation of the VSB signal $u(t)$. We multiply $u(t)$ by the carrier component $\cos 2\pi f_c t$ and pass the result through an ideal lowpass filter, as shown in Figure 5.14. Thus, the product signal is

$$v(t) = u(t) \cos 2\pi f_c t$$

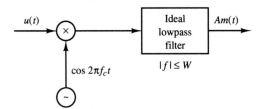

FIGURE 5.14.   Demodulation of VSB signal.

or, equivalently,

$$V(f) = \frac{1}{2}[U(f - f_c) + U(f + f_c)] \qquad (5.2.51)$$

If we substitute for $U(f)$ from (5.2.50) into (5.2.51), we obtain

$$V(f) = \frac{A_c}{4}[M(f - 2f_c) + M(f)]H(f - f_c)$$

$$+ \frac{A_c}{4}[M(f) + M(f + 2f_c)]H(f + f_c) \qquad (5.2.52)$$

The lowpass filter rejects the double-frequency terms and passes only the components in the frequency range $|f| \leq W$. Hence, the signal spectrum at the output of the ideal lowpass filter is

$$V_\ell(f) = \frac{A_c}{4}M(f)[H(f - f_c) + H(f + f_c)] \qquad (5.2.53)$$

We require that the message signal at the output of the lowpass filter be undistorted. Therefore the VSB filter characteristic must satisfy the condition

$$H(f - f_c) + H(f + f_c) = \text{constant}, \qquad |f| \leq W \qquad (5.2.54)$$

This condition is satisfied by a filter that has the frequency response characteristic shown in Figure 5.15. We note that $H(f)$ selects the upper sideband and a vestige of the lower sideband. It has an odd symmetry about the carrier frequency f_c, in the frequency range $f_c - f_a < f < f_c + f_a$, where f_a is a conveniently selected frequency that is some small fraction of W, i.e., $f_a \ll W$. Thus, we obtain an

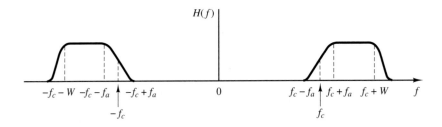

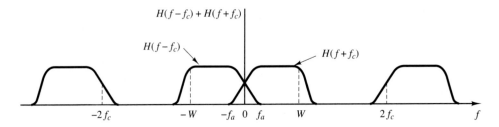

FIGURE 5.15. VSB filter characteristics.

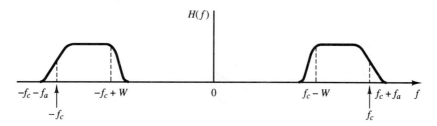

FIGURE 5.16.   Frequency response of VSB filter for selecting the lower sideband of the message signals.

undistorted version of the transmitted signal. Figure 5.16 illustrates the frequency response of a VSB filter that selects the lower sideband and a vestige of the upper sideband.

In practice, the VSB filter is designed to have some specified phase characteristic. To avoid distortion of the message signal, the VSB filter should be designed to have linear phase over its passband $f_c - f_a \leq |f| \leq f_c + W$.

Example 5.2.7

Suppose that the message signal is given as

$$m(t) = 10 + 4\cos 2\pi t + 8\cos 4\pi t + 10\cos 20\pi t$$

Specify the frequency-response characteristic of a VSB filter that passes the upper sideband and the first frequency component of the lower sideband.

Solution   The spectrum of the DSB-SC AM signal $u(t) = m(t)\cos 2\pi f_c t$ is

$$U(f) = 5[\delta(f - f_c) + \delta(f + f_c)] + 2[\delta(f - f_c - 1) + \delta(f + f_c + 1)]$$
$$+ 4[\delta(f - f_c - 2) + \delta(f + f_c + 2)] + 5[\delta(f - f_c - 10) + \delta(f + f_c + 10)]$$

The VSB filter can be designed to have unity gain in the range $2 \leq |f - f_c| \leq 10$, a gain of 1/2 at $f = f_c$, a gain of $1/2 + \alpha$ at $f = f_c + 1$ and a gain of $1/2 - \alpha$ at $f = f_c - 1$, where α is some conveniently selected parameter that satisfies the condition $0 < \alpha < 1/2$. Figure 5.17 illustrates the frequency-response characteristic of the VSB filter.

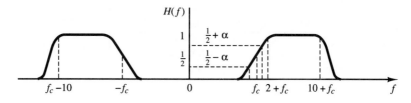

FIGURE 5.17.   Frequency response characteristics of VSB filter in Example 5.2.7.

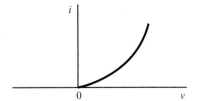

FIGURE 5.18. Voltage-current characteristic of P-N diode.

5.2.5 Implementation of AM Modulators and Demodulators

There are several different methods for generating AM modulated signals. We shall describe the methods most commonly used in practice. Because the process of modulation involves the generation of new frequency components, modulators are generally characterized as nonlinear and/or time-variant systems.

Power-law modulation.   Let us consider the use of a nonlinear device such as a P-N diode, which has a voltage-current characteristic as shown in Figure 5.18. Suppose that the voltage input to such a device is the sum of the message signal $m(t)$ and the carrier $A_c \cos 2\pi f_c t$, as illustrated in Figure 5.19. The nonlinearity will generate a product of the message $m(t)$ with the carrier, plus additional terms. The desired modulated signal can be filtered out by passing the output of the nonlinear device through a bandpass filter.

To elaborate on this method, suppose that the nonlinear device has an input–output (square-law) characteristic of the form

$$v_0(t) = a_1 v_i(t) + a_2 v_i^2(t) \tag{5.2.55}$$

where $v_i(t)$ is the input signal, $v_0(t)$ is the output signal, and the parameters (a_1, a_2) are constants. Then, if the input to the nonlinear device is

$$v_i(t) = m(t) + A_c \cos 2\pi f_c t, \tag{5.2.56}$$

its output is

$$v_o(t) = a_1[m(t) + A_c \cos 2\pi f_c t]$$
$$+ a_2[m(t) + A_c \cos \cos 2\pi f_c t]^2$$
$$= a_1 m(t) + a_2 m^2(t) + a_2 A_c^2 \cos^2 2\pi f_c t$$

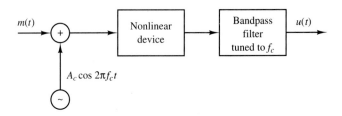

FIGURE 5.19. Block diagram of power-law AM modulator.

$$+ A_c a_1 \left[1 + \frac{2a_2}{a_1} m(t) \right] \cos 2\pi f_c t \qquad (5.2.57)$$

The output of the bandpass filter with bandwidth $2W$ centered at $f = f_c$ yields

$$u(t) = A_c a_1 \left[1 + \frac{2a_2}{a_1} m(t) \right] \cos 2\pi f_c t \qquad (5.2.58)$$

where $2a_2 |m(t)|/a_1 < 1$ by design. Thus, the signal generated by this method is a conventional DSB-AM signal.

Switching modulator.   Another method for generating an AM modulated signal is by means of a switching modulator. Such a modulator can be implemented by the system illustrated in Figure 5.20(a). The sum of the message signal and the

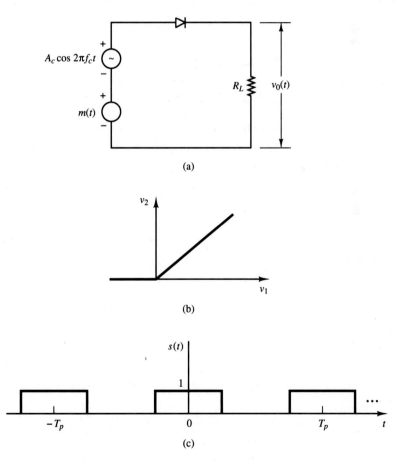

FIGURE 5.20.   Switching modulator (a), diode with input–output voltage (b), and periodic switching signal (c).

carrier, i.e., $v_i(t)$ given by (5.2.56), are applied to a diode that has the input–output voltage characteristic shown in Figure 5.20(b), where $A_c \gg m(t)$. The output across the load resistor is simply

$$v_0(t) = \begin{cases} v_i(t), & c(t) > 0 \\ 0, & c(t) < 0 \end{cases} \qquad (5.2.59)$$

This switching operation may be viewed mathematically as a multiplication of the input $v_i(t)$ with the switching function $s(t)$, i.e.,

$$v_0(t) = [m(t) + A_c \cos 2\pi f_c t]s(t) \qquad (5.2.60)$$

where $s(t)$ is shown in Figure 5.20(c).

Because $s(t)$ is a periodic function, it is represented in the Fourier series as

$$s(t) = \frac{1}{2} + \frac{2}{\pi}\sum_{n=1}^{\infty}\frac{(-1)^{n-1}}{2n-1}\cos[2\pi f_c t(2n-1)] \qquad (5.2.61)$$

Hence,

$$\begin{aligned} v_0(t) &= [m(t) + A_c \cos 2\pi f_c t]s(t) \\ &= \frac{A_c}{2}\left[1 + \frac{4}{\pi A_c}m(t)\right]\cos 2\pi f_c t + \text{other terms} \end{aligned} \qquad (5.2.62)$$

The desired AM modulated signal is obtained by passing $v_0(t)$ through a bandpass filter with center frequency $f = f_c$ and bandwidth $2W$. At its output, we have the desired conventional DSB AM signal

$$u(t) = \frac{A_c}{2}\left[1 + \frac{4}{\pi A_c}m(t)\right]\cos 2\pi f_c t \qquad (5.2.63)$$

Balanced modulator.   A relatively simple method to generate a DSB-SC AM signal is to use two conventional AM modulators arranged in the configuration illustrated in Figure 5.21. For example, we may use two square-law AM modulators

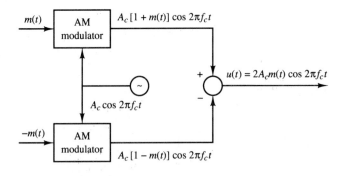

FIGURE 5.21.   Block diagram of a balanced modulator.

as described above. Care must be taken to select modulators with approximately identical characteristics so that the carrier component cancels out at the summing junction.

Ring modulator.   Another type of modulator for generating a DSB-SC AM signal is the *ring modulator* illustrated in Figure 5.22. The switching of the diodes is controlled by a square wave of frequency f_c, denoted as $c(t)$, which is applied to the center taps of the two transformers. When $c(t) > 0$, the top and bottom diodes conduct, while the two diodes in the crossarms are off. In this case, the message signal $m(t)$ is multiplied by $+1$. When $c(t) < 0$, the diodes in the crossarms of the ring conduct, while the other two are switched off. In this case, the message signal $m(t)$ is multiplied by -1. Consequently, the operation of the ring modulator may be described mathematically as a multiplier of $m(t)$ by the square-wave carrier $c(t)$, i.e.,

$$v_0(t) = m(t)c(t) \tag{5.2.64}$$

as shown in Figure 5.22.

Because $c(t)$ is a periodic function, it is represented by the Fourier series

$$c(t) = \frac{4}{\pi} \sum_{n=1}^{\infty} \frac{(-1)^{n-1}}{2n - 1} \cos[2\pi f_c(2n - 1)t] \tag{5.2.65}$$

Hence, the desired DSB-SC AM signal $u(t)$ is obtained by passing $v_0(t)$ through a bandpass filter with center frequency f_c and bandwidth $2W$.

From the discussion above, we observe that the balanced modulator and the ring modulator systems, in effect, multiply the message signal $m(t)$ with the carrier to produce a DSB-SC AM signal. The multiplication of $m(t)$ with $A_c \cos w_c t$ is called a mixing operation. Hence, a *mixer* is basically a balanced modulator.

The method shown in Figure 5.10 for generating a SSB signal requires two mixers, i.e., two balanced modulators, in addition to the Hilbert transformer. On the

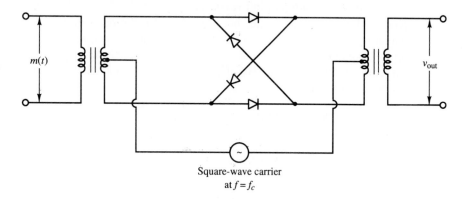

FIGURE 5.22.   Ring modulator for generating DSB-SC AM signal.

other hand, the filter method illustrated in Figure 5.11 for generating a SSB signal requires a single balanced modulator and a sideband filter.

Let us now consider the demodulation of AM signals. We begin with a description of the envelope detector.

Envelope detector.   As previously indicated, conventional DSB AM signals are easily demodulated by means of an envelope detector. A circuit diagram for an envelope detector is shown in Figure 5.23. It consists of a diode and an RC circuit, which is basically a simple lowpass filter.

During the positive half-cycle of the input signal, the diode is conducting and the capacitor charges up to the peak value of the input signal. When the input falls below the voltage on the capacitor, the diode becomes reverse-biased and the input becomes disconnected from the output. During this period the capacitor discharges slowly through the load resistor R. On the next cycle of the carrier, the diode conducts again when the input signal exceeds the voltage across the capacitor. The capacitor charges up again to the peak value of the input signal and the process is repeated again.

The time constant RC must be selected to follow the variations in the envelope of the carrier-modulated signal. In effect,

$$\frac{1}{f_c} \ll RC \ll \frac{1}{W}$$

In such a case, the capacitor discharges slowly through the resistor and, thus, the output of the envelope detector, which we denote as $\hat{m}(t)$, closely follows the message signal.

Demodulation of DSB-SC AM signals.   As previously indicated, the demodulation of a DSB-SC AM signal requires a synchronous demodulator. That is, the demodulator must use a coherent phase reference, which is usually generated by means of a phase-locked loop (PLL), to demodulate the received signal.

The general configuration is shown in Figure 5.24. A PLL is used to generate a phase-coherent carrier signal that is mixed with the received signal in a balanced modulator. The output of the balanced modulator is passed through a lowpass filter of bandwidth W that passes the desired signal and rejects all signal and noise components above W Hz. The characteristics and operation of the PLL are described in Section 6.2

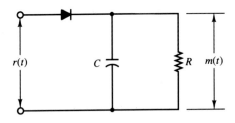

FIGURE 5.23.   An envelope detector.

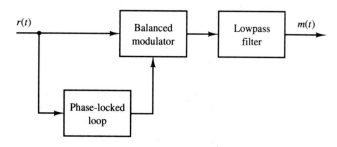

FIGURE 5.24.   Demodulator for DSB-SC signal.

Demodulation of SSB signals.   The demodulation of SSB AM signals also requires the use of a phase-coherent reference. In the case of signals such as speech, that have relatively little or no power content at dc, it is straightforward to generate the SSB signal, as shown in Figure 5.11, and then to insert a small carrier component that is transmitted along with the message. In such a case we may use the configuration shown in Figure 5.25 to demodulate the SSB signal. We observe that a balanced modulator is used for the purpose of frequency conversion of the bandpass signal to lowpass or baseband.

Demodulation of VSB signals.   In VSB a carrier component is generally transmitted along with the message sidebands. The existence of the carrier component makes it possible to extract a phase coherent reference for demodulation in a balanced modulator, as shown in Figure 5.25.

In some applications such as TV broadcasting, a large carrier component is transmitted along with the message in the VSB signal. In such a case, it is possible to recover the message by passing the received VSB signal through an envelope detector.

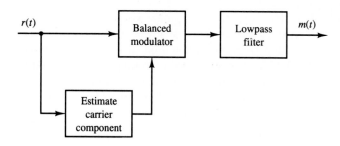

FIGURE 5.25.   Demodulation of SSB AM signal with a carrier component.

5.2.6 Signal Multiplexing

We have seen that amplitude modulation of a sinusoidal carrier by a message signal $m(t)$ translates the message signal in frequency by an amount equal to the carrier frequency f_c. If we have two or more message signals to transmit simultaneously over the communications channel, it is possible to have each message signal modulate a carrier of a different frequency where the minimum separation between two adjacent carriers is either $2W$ (for DSB AM) or W (for SSB AM), where W is the bandwidth of each of the message signals. Thus, the various message signals occupy separate frequency bands of the channel and do not interfere with one another in transmission over the channel.

The process of combining a number of separate message signals into a composite signal for transmission over a common channel is called *multiplexing*. There are two commonly used methods for signal multiplexing: (1) time-division multiplexing, and (2) frequency-division multiplexing. Time-division multiplexing is usually used in the transmission of digital information and was described in Section 4.8.1. Frequency-division multiplexing (FDM) may be used with either analog or digital signal transmission.

In FDM the message signals are separated in frequency as described above. A typical configuration of an FDM system is shown in Figure 5.26. This figure illustrates the FDM of K message signals at the transmitter and their demodulation at the receiver. The lowpass filters at the transmitter are used to ensure that the bandwidth of the message signals is limited to W Hz. Each signal modulates a separate carrier; hence, K modulators are required. Then the signals from the K modulators are summed and transmitted over the channel. For SSB and VSB modulation, the modulator outputs are filtered prior to summing the modulated signals.

At the receiver of an FDM system, the signals are usually separated by passing through a parallel bank of bandpass filters where each filter is tuned to one of the carrier frequencies and has a bandwidth that is sufficiently wide to pass the desired signal. The output of each bandpass filter is demodulated, and each demodulated signal is fed to a lowpass filter that passes the baseband message signal and eliminates the double-frequency components.

FDM is widely used in radio and telephone communications. For example, in telephone communications, each voice-message signal occupies a nominal bandwidth of 3 kHz. The message signal is single-sideband modulated for bandwidth efficient transmission. In the first level of multiplexing, 12 signals are stacked in frequency with a frequency separation of 4 kHz between adjacent carriers. Thus, a composite 48-kHz channel, called a group channel, is used to transmit the 12 voiceband signals simultaneously. In the next level of FDM, a number of group channels (typically five or six) are stacked together in frequency to form a supergroup channel, and the composite signal is transmitted over the channel. Higher-order multiplexing is obtained by combining several supergroup channels. Thus, an FDM hierarchy is employed in telephone communication systems.

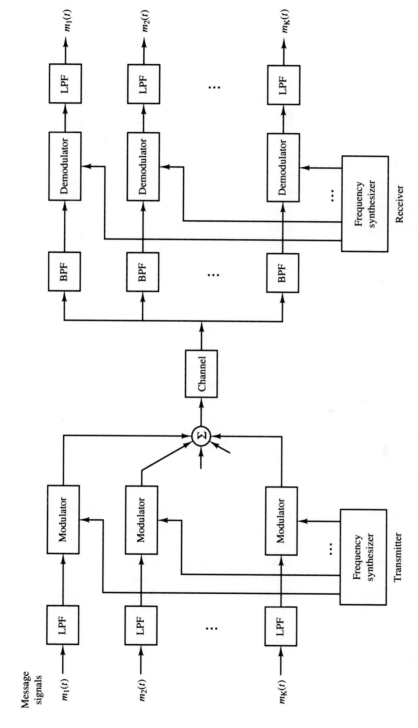

FIGURE 5.26. Frequency-division multiplexing of multiple signals.

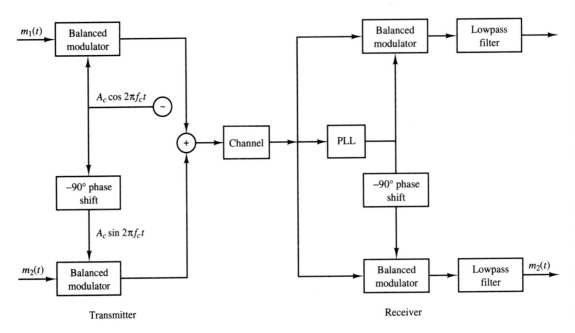

FIGURE 5.27. Quadrative-carrier modulation and demodulation.

Quadrature-carrier multiplexing.   A totally different type of multiplexing allows us to transmit two message signals on the same carrier frequency using two quadrature carriers $A_c \cos 2\pi f_c t$ and $A_c \sin 2\pi f_c t$. To elaborate, suppose that $m_1(t)$ and $m_2(t)$ are two separate message signals to be transmitted over the channel. The signal $m_1(t)$ amplitude modulates the carrier $A_c \cos 2\pi f_c t$ and the signal $m_2(t)$ amplitude modulates the quadrature carrier $A_c \sin 2\pi f_c t$. The two signals are added and transmitted over the channel. Hence, the transmitted signal is

$$u(t) = A_c m_1(t) \cos 2\pi f_c t + A_c m_2(t) \sin 2\pi f_c t \qquad (5.2.66)$$

Therefore, each message signal is transmitted by DSB-SC AM. This type of signal multiplexing is called *quadrature-carrier multiplexing.*

Figure 5.27 illustrates the modulation and demodulation of the quadrature-carrier multiplexed signals. As shown, a synchronous demodulator is required at the receiver to separate and recover the quadrature-carrier modulated signals.

Quadrature-carrier multiplexing results in a bandwidth efficient communication system that is comparable in bandwidth efficiency to SSB AM.

5.3 ANGLE MODULATION

In the previous section we considered amplitude modulation of the carrier as a means for transmitting the message signal. Amplitude modulation methods are also called *linear modulation methods,* although conventional AM is not linear in the strict sense.

Another class of modulation methods are frequency and phase modulation, which are described in this section. In frequency modulation (FM) systems the frequency of the carrier f_c is changed by the message signal, and in phase-modulation (PM) systems the phase of the carrier is changed according to the variations in the message signal. Frequency and phase modulation are obviously quite nonlinear, and very often they are jointly referred to as *angle-modulation methods.* As our analysis in the following sections will show, angle modulation, due to its inherent nonlinearity, is more complex to implement and much more difficult to analyze. In many cases only an approximate analysis can be done. Another property of angle modulation is its bandwidth expansion property. Frequency and phase-modulation systems generally expand the bandwidth such that the effective bandwidth of the modulated signal is usually many times the bandwidth of the message signal[†] With a higher implementation complexity and a higher bandwidth occupancy, one would naturally question the usefulness of these systems. As our analysis in Chapter 6 will show, the major benefit of these systems is their high degree of noise immunity. In fact these systems trade-off bandwidth for high-noise immunity. That is the reason that FM systems are used widely in high-fidelity music broadcasting and point-to-point communication systems where the transmitter power is quite limited.

5.3.1 Representation of FM and PM Signals

An angle-modulated signal can be written as

$$u(t) = A_c \cos(\theta(t))$$

$\theta(t)$ is the phase of the signal, and its instantaneous frequency $f_i(t)$ is given by

$$f_i(t) = \frac{1}{2\pi} \frac{d}{dt} \theta(t) \tag{5.3.1}$$

Because $u(t)$ is a bandpass signal, it can be represented as

$$u(t) = A_c \cos(2\pi f_c t + \phi(t)) \tag{5.3.2}$$

and therefore

$$f_i(t) = f_c + \frac{1}{2\pi} \frac{d}{dt} \phi(t) \tag{5.3.3}$$

If $m(t)$ is the message signal, then in a PM system we have

$$\phi(t) = k_p m(t) \tag{5.3.4}$$

[†]Strictly speaking, the bandwidth of the modulated signal, as it will be shown later, is infinite. That is why we talk about the *effective bandwidth.*

and in an FM system we have

$$f_i(t) - f_c = k_f m(t) = \frac{1}{2\pi} \frac{d}{dt} \phi(t) \tag{5.3.5}$$

where k_p and k_f are phase and frequency *deviation constants*. From the above relationships we have

$$\phi(t) = \begin{cases} k_p m(t) & \text{PM} \\ 2\pi k_f \displaystyle\int_{-\infty}^{t} m(\tau)d\tau & \text{FM} \end{cases} \tag{5.3.6}$$

The above expression shows the close and interesting relation between FM and PM systems. This close relationship makes it possible to analyze these systems in parallel and only emphasize their main differences. The first interesting result observed from the above is that, if we phase modulate the integral of a message, it is equivalent to the frequency modulation of the original message. On the other hand, the above relation can be expressed as

$$\frac{d}{dt}\phi(t) = \begin{cases} k_p \dfrac{d}{dt} m(t) & \text{PM} \\ 2\pi k_f m(t) & \text{FM} \end{cases} \tag{5.3.7}$$

which shows that if we frequency modulate the derivative of a message, the result is equivalent to phase modulation of the message itself. Figure 5.28 shows the above relation between FM and PM. Figure 5.29 illustrates a square-wave signal and its integral, a sawtooth signal, and their corresponding FM and PM signals.

The demodulation of an FM signal involves finding the instantaneous frequency of the modulated signal and then subtracting the carrier frequency from it. In the demodulation of PM, the demodulation process is done by finding the phase of the signal and then recovering $m(t)$. The maximum-phase deviation in a PM system is given by

$$\Delta\phi_{\max} = k_p \max[|m(t)|] \tag{5.3.8}$$

and the maximum-frequency deviation in an FM system is given by

$$\Delta f_{\max} = k_f \max[|m(t)|] \tag{5.3.9}$$

Example 5.3.1

The message signal

$$m(t) = a\cos(2\pi f_m t)$$

is used to either frequency modulate or phase modulate the carrier $A_c\cos(2\pi f_c t)$. Find the modulated signal in each case.

Solution   In PM we have

$$\phi(t) = k_p m(t) = k_p a\cos(2\pi f_m t) \tag{5.3.10}$$

FM Modulator

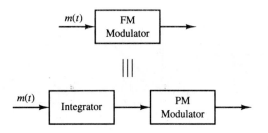

PM Modulator

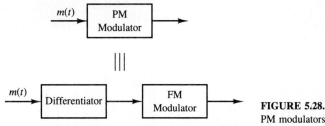

FIGURE 5.28. A comparison of FM and PM modulators.

and in FM we have

$$\phi(t) = 2\pi k_f \int_{-\infty}^{t} m(\tau)\, d\tau = \frac{k_f a}{f_m} \sin(2\pi f_m t) \qquad (5.3.11)$$

Therefore, the modulated signals will be

$$u(t) = \begin{cases} A_c \cos\left(2\pi f_c t + k_p a \cos(2\pi f_m t)\right) & \text{PM} \\[2mm] A_c \cos\left(2\pi f_c t + \dfrac{k_f a}{f_m}\, \sin(2\pi f_m t)\right) & \text{FM} \end{cases} \qquad (5.3.12)$$

By defining

$$\beta_p = k_p a \qquad (5.3.13)$$

$$\beta_f = \frac{k_f a}{f_m} \qquad (5.3.14)$$

we have

$$u(t) = \begin{cases} A_c \cos\left(2\pi f_c t + \beta_p \cos(2\pi f_m t)\right) & \text{PM} \\[2mm] A_c \cos\left(2\pi f_c t + \beta_f \sin(2\pi f_m t)\right) & \text{FM} \end{cases} \qquad (5.3.15)$$

The parameters β_p and β_f are called the *modulation indices* of the PM and FM systems respectively.

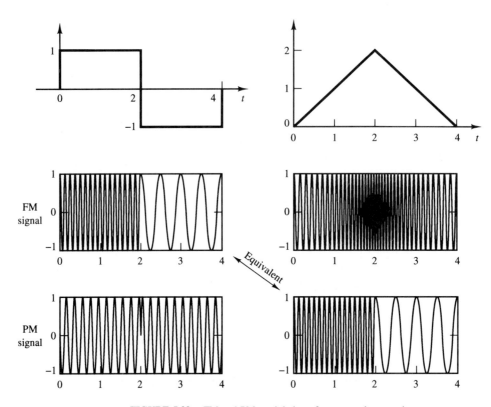

FIGURE 5.29.   FM and PM modulation of square and sawtooth waves.

We can extend the definition of the modulation index for a general nonsinusoidal signal $m(t)$ as

$$\beta_p = k_p \max[|m(t)|] \tag{5.3.16}$$

$$\beta_f = \frac{k_f \max[|m(t)|]}{W} \tag{5.3.17}$$

where W denotes the bandwidth of the message signal $m(t)$. In terms of the maximum phase and frequency deviation $\Delta\phi_{max}$ and Δf_{max}, we have

$$\beta_p = \Delta\phi_{max} \tag{5.3.18}$$

$$\beta_f = \frac{\Delta f_{max}}{W} \tag{5.3.19}$$

Narrowband angle modulation[†].   If in an angle-modulation system the deviation constants k_p and k_f and the message signal $m(t)$ are such that for all t we

---

[†]Also known as *low-index angle modulation*.

have $\phi(t) \ll 1$, then we can use a simple approximation to expand $u(t)$ as

$$u(t) = A_c \cos 2\pi f_c t \cos \phi(t) - A_c \sin 2\pi f_c t \sin \phi(t)$$

$$\approx A_c \cos 2\pi f_c t - A_c \phi(t) \sin 2\pi f_c t \qquad (5.3.20)$$

This last equation shows that in this case the modulated signal is very similar to a conventional AM signal. The only difference is that the message signal $m(t)$ is modulated on a sine carrier rather than a cosine carrier. The bandwidth of this signal is similar to the bandwidth of a conventional AM signal, which is twice the bandwidth of the message signal. Of course this bandwidth is only an approximation to the real bandwidth of the FM signal. A phasor diagram for this signal and the comparable conventional AM signal are given in Figure 5.30. Note that, compared to conventional AM, the narrowband angle modulation scheme has far less amplitude variation. Of course, the angle-modulation system has constant amplitude, and, hence, there should be no amplitude variations in the phasor-diagram representation of the system. The slight variations here are due to the first-order approximation that we have used for the expansions of $\sin(\phi(t))$ and $\cos(\phi(t))$. As we will see later, the narrowband angle modulation method does not provide any better noise immunity compared to a conventional AM system. Therefore narrowband angle modulation is seldom used in practice for communication purposes. However, these systems can be used as an intermediate stage for generation of wideband angle modulated signals as we will discuss in Section 5.3.3.

5.3.2 Spectral Characteristics of Angle Modulated Signals

Due to the inherent nonlinearity of angle-modulation systems, the precise characterization of their spectral properties, even for simple message signals, is mathematically intractable. Therefore, the derivation of the spectral characteristics of these signals usually involves the study of very simple modulating signals and certain approximations. Then the results are generalized to the more complicated

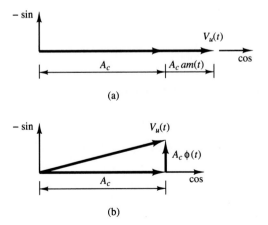

(a)

(b)

FIGURE 5.30.   Phasor diagram for the conventional AM (a) and narrowband angle (b) modulation.

messages. We will study the spectral characteristics of an angle-modulated signal in three cases: when the message signal is a sinusoidal signal, a general periodic signal, and a stationary zero-mean Gaussian process.

Angle modulation by a sinusoidal signal. Let us begin with the case where the message signal is a sinusoidal signal. As we have seen, in this case for both FM and PM we have

$$u(t) = A_c \cos(2\pi f_c t + \beta \cos 2\pi f_m t) \tag{5.3.21}$$

where β is the modulation index that can be either β_p or β_f. Therefore, the modulated signal can be written as

$$u(t) = \text{Re}\left(A_c e^{j2\pi f_c t} e^{j\beta \cos 2\pi f_m t} \right) \tag{5.3.22}$$

Because $\cos 2\pi f_m t$ is periodic with period $T_m = \frac{1}{f_m}$, the same is true for the complex exponential signal

$$e^{j\beta \cos 2\pi f_m t}$$

Therefore, it can be expanded in a Fourier series representation. The Fourier series coefficients are obtained from the integral

$$
\begin{aligned}
c_n &= f_m \int_0^{\frac{1}{f_m}} e^{j\beta \cos 2\pi f_m t} e^{-jn2\pi f_m t}\, dt \\
&\overset{u = 2\pi f_m t}{=} \frac{1}{2\pi} \int_0^{2\pi} e^{j(\beta \cos u - nu)}\, du
\end{aligned} \tag{5.3.23}
$$

This latter integral is a well-known integral known as the *Bessel function of the first kind of order n* and is denoted by $J_n(\beta)$. Therefore, we have the Fourier series for the complex exponential as

$$e^{j\beta \cos 2\pi f_m t} = \sum_{n=-\infty}^{+\infty} J_n(\beta) e^{j2\pi n f_m t} \tag{5.3.24}$$

By substituting (5.3.24) in (5.3.22), we obtain

$$
\begin{aligned}
u(t) &= \text{Re}\left(A_c \sum_{n=-\infty}^{+\infty} J_n(\beta) e^{j2\pi n f_m t} e^{j2\pi f_c t} \right) \\
&= \sum_{n=-\infty}^{+\infty} A_c J_n(\beta) \cos\left(2\pi (f_c + n f_m) t \right)
\end{aligned} \tag{5.3.25}
$$

The above relation shows that even in this very simple case, where the modulating signal is a sinusoid of frequency f_m, the angle-modulated signal contains all frequencies of the form $f_c + n f_m$ for $n = 0, \pm 1, \pm 2, \ldots$. Therefore, the actual bandwidth of the modulated signal is infinite. However, the amplitude of the sinusoidal components of frequencies $f_c \pm nf_m$ for large n is minimal. Hence we can define a finite *effective bandwidth* for the modulated signal. A series expansion for the Bessel function is given by

$$J_n(\beta) = \sum_{k=0}^{\infty} \frac{(-1)^k (\frac{\beta}{2})^{n+2k}}{k!(k+n)!} \tag{5.3.26}$$

The above expansion shows that for small β, we can use the approximation

$$J_n(\beta) \approx \frac{\beta^n}{2^n n!} \tag{5.3.27}$$

Thus for a small modulation index β, only the first sideband corresponding to $n = 1$ is of importance. Also, using the above expansion, it is easy to verify the following symmetry properties of the Bessel function.

$$J_{-n}(\beta) = \begin{cases} J_n(\beta) & n \text{ even} \\ -J_n(\beta) & n \text{ odd} \end{cases} \tag{5.3.28}$$

Plots of $J_n(\beta)$ for various values of n are given in Figure 5.31, and a table of the values of the Bessel function is given in Table 5.1.

Example 5.3.2

Let the carrier be given by $c(t) = 10\cos(2\pi f_c t)$ and let the message signal be $\cos(20\pi t)$. Further assume that the message is used to frequency modulate the carrier

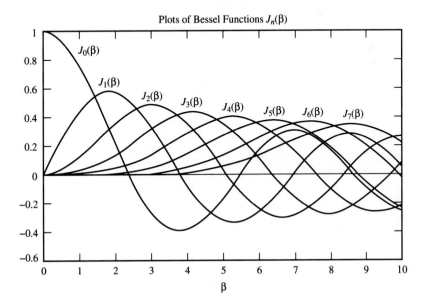

FIGURE 5.31. Bessel functions for various values of n.

TABLE 5-1 VALUES OF THE BESSEL FUNCTION $J_n(\beta)$.

n	$\beta = 0.1$	$\beta = 0.2$	$\beta = 0.5$	$\beta = 1$	$\beta = 2$	$\beta = 5$	$\beta = 8$	$\beta = 10$
0	0.997	0.990	0.938	0.765	0.224	−0.178	0.172	−0.246
1	0.050	0.100	0.242	0.440	0.577	−0.328	0.235	0.043
2	0.001	0.005	0.031	0.115	0.353	0.047	−0.113	0.255
3				0.020	0.129	0.365	−0.291	0.058
4				0.002	0.034	0.391	−0.105	−0.220
5					0.007	0.261	0.186	−0.234
6					0.001	0.131	0.338	−0.014
7						0.053	0.321	0.217
8						0.018	0.223	0.318
9						0.006	0.126	0.292
10						0.001	0.061	0.207
11							0.026	0.123
12							0.010	0.063
13							0.003	0.029
14							0.001	0.012
15								0.004
16								0.001

From Ziemer and Tranter (1990) © Houghton Mifflin, reprinted by persmission.

with $k_f = 50$. Find the expression for the modulated signal and determine how many harmonics should be selected to contain 99% of the modulated signal power.

Solution The power content of the carrier signal is given by

$$P_c = \frac{A_c^2}{2} = \frac{100}{2} = 50 \tag{5.3.29}$$

The modulated signal is represented by

$$u(t) = 10\cos(2\pi f_c t + 2\pi k_f \int_{-\infty}^{t} \cos(20\pi\tau)\,d\tau)$$

$$= 10\cos(2\pi f_c t + \frac{50}{10}\sin(20\pi t))$$

$$= 10\cos(2\pi f_c t + 5\sin(20\pi t)) \tag{5.3.30}$$

The modulation index is given by

$$\beta = k_f \frac{\max[|m(t)|]}{f_m} = 5 \tag{5.3.31}$$

and, therefore, the FM-modulated signal is

$$u(t) = \sum_{n=-\infty}^{+\infty} A_c J_n(\beta)\cos\left(2\pi(f_c + nf_m)t\right)$$

$$= \sum_{n=-\infty}^{+\infty} 10 J_n(5)\cos\left(2\pi(f_c + 10n)t\right) \tag{5.3.32}$$

It is seen that the frequency content of the modulated signal is concentrated at frequencies of the form $f_c + 10n$ for various n. To make sure that at least 99% of the total power is within the effective bandwidth, we have to choose k large enough such that

$$\sum_{n=-k}^{k} \frac{100 J_n^2(5)}{2} \geq 0.99 \times 50 \tag{5.3.33}$$

This is a nonlinear equation and its solution (for k) can be found by trial and error and by using tables of the Bessel functions. Of course, in finding the solution to this equation, we have to employ the symmetry properties of the Bessel function given in (5.3.28). Using these properties, we have

$$50 \left[J_0^2(5) + 2 \sum_{n=1}^{k} J_n^2(5) \right] \geq 49.5 \tag{5.3.34}$$

Starting with small values of k and increasing it, we see that the smallest value of k for which the left-hand side exceeds the right-hand side is $k = 6$. Therefore, taking frequencies $f_c \pm 10k$ for $0 \leq k \leq 6$ guarantees that 99% of the power of the modulated signal has been included and only 1% has been left out. This means that, if the modulated signal is passed through an ideal bandpass filter centered at f_c with a bandwidth of at least 120 Hz, only 1% of the signal power will be eliminated. This gives us a practical way to define the *effective bandwidth* of the angle-modulated signal as being 120 Hz. Figure 5.32 shows the frequencies present in the effective bandwidth of the modulated signal.

In general the effective bandwidth of an angle-modulated signal, which contains at least 98% of the signal power, is given by the relation

$$B_c = 2(\beta + 1) f_m \tag{5.3.35}$$

where β is the modulation index and f_m is the frequency of the sinusoidal message signal. It is instructive to study the effect of the amplitude and frequency of the sinusoidal message signal on the bandwidth and the number of harmonics in the modulated signal. Let the message signal be given by

$$m(t) = a \cos(2\pi f_m t) \tag{5.3.36}$$

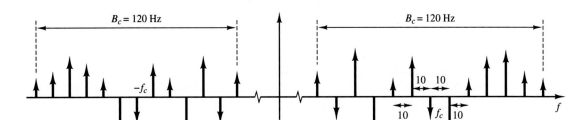

FIGURE 5.32.   The harmonics present inside the effective bandwidth of Example 5.3.2.

The bandwidth† of the modulated signal is given by

$$B_c = 2(\beta + 1)f_m = \begin{cases} 2(k_p a + 1)f_m & \text{PM} \\ 2\left(\dfrac{k_f a}{f_m} + 1\right)f_m & \text{FM} \end{cases} \qquad (5.3.37)$$

or

$$B_c = \begin{cases} 2(k_p a + 1)f_m & \text{PM} \\ 2(k_f a + f_m) & \text{FM} \end{cases} \qquad (5.3.38)$$

The above relation shows that increasing a, the amplitude of the modulating signal, in PM and FM has almost the same effect on increasing the bandwidth B_c. On the other hand, increasing f_m, the frequency of the message signal, has a more profound effect in increasing the bandwidth of a PM signal as compared to an FM signal. In both PM and FM the bandwidth B_c increases by increasing f_m, but in PM this increase is a proportional increase and in FM this is only an additive increase, which in most cases of interest (for large β) is not substantial. Now if we look at the number of harmonics in the bandwidth (including the carrier) and denote it with M_c, we have

$$M_c = 2\lfloor \beta \rfloor + 3 = \begin{cases} 2\lfloor k_p a \rfloor + 3 & \text{PM} \\ 2\lfloor \dfrac{k_f a}{f_m} \rfloor + 3 & \text{FM} \end{cases} \qquad (5.3.39)$$

Increasing the amplitude a increases the number of harmonics in the bandwidth of the modulated signal in both cases. However, increasing f_m, has no effect on the number of harmonics in the bandwidth of the PM signal and decreases the number of harmonics in the FM signal almost linearly. This explains the relative insensitivity of the bandwidth of the FM signal to the message frequency. On the one hand, increasing f_m decreases the number of harmonics in the bandwidth and, at the same time, it increases the spacing between the harmonics. The net effect is a slight increase in the bandwidth. In PM, however, the number of harmonics remains constant and only the spacing between them increases. Therefore, the net effect is a linear increase in bandwidth. Figure 5.33 shows the effect of increasing the frequency of the message in both FM and PM.

Angle modulation by a periodic message signal.   To generalize the above results, we now consider angle modulation by an arbitrary periodic message signal $m(t)$. Let us consider a PM-modulated signal where

$$u(t) = A_c \cos(2\pi f_c t + \beta m(t)) \qquad (5.3.40)$$

---

†From now on, by bandwidth we mean effective bandwidth unless otherwise stated.

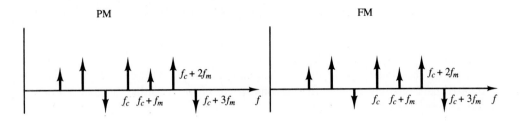

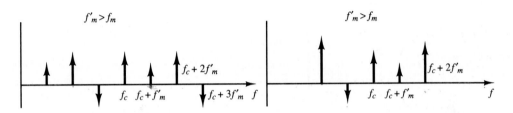

FIGURE 5.33.   The effect of increasing bandwidth of the message in FM and PM.

We can write this as

$$u(t) = A_c \text{Re}\left[e^{j2\pi f_c t} e^{j\beta m(t)} \right] \tag{5.3.41}$$

We are assuming that $m(t)$ is periodic with period $T_m = \frac{1}{f_m}$. Therefore $e^{j\beta m(t)}$ will be a periodic signal with the same period, and we can find its Fourier series expansion as

$$e^{j\beta m(t)} = \sum_{n=-\infty}^{+\infty} c_n e^{j2\pi n f_m t} \tag{5.3.42}$$

where

$$
c_n = \frac{1}{T_m} \int_0^{T_m} e^{j\beta m(t)} e^{-j2\pi n f_m t} \, dt
$$

$$
\overset{u=2\pi f_m t}{=} \frac{1}{2\pi} \int_0^{2\pi} e^{j\left[\beta m\left(\frac{u}{2\pi f_m} \right) - nu \right]} \, du \tag{5.3.43}
$$

and

$$
u(t) = A_c \text{Re}\left[\sum_{n=-\infty}^{+\infty} c_n e^{j2\pi f_c t} e^{j2\pi n f_m t} \right]
$$

$$
= A_c \sum_{n=-\infty}^{+\infty} |c_n| \cos\left(2\pi (f_c + n f_m)t + \angle c_n \right) \tag{5.3.44}
$$

It is seen again that the modulated signal contains all frequencies of the form $f_c + nf_m$.

The detailed treatment of the spectral characteristics of an angle-modulated signal for a general nonperiodic deterministic message signal $m(t)$ is quite involved due to the nonlinear nature of the modulation process. However, there exists an approximate relation for the effective bandwidth of the modulated signal, known as the *Carson's rule,* and given by

$$B_c = 2(\beta + 1)W \tag{5.3.45}$$

where β is the modulation index defined as

$$\beta = \begin{cases} k_p \max[|m(t)|] & \text{PM} \\[2mm] \dfrac{k_f \max[|m(t)|]}{W} & \text{FM} \end{cases} \tag{5.3.46}$$

and W is the bandwidth of the message signal $m(t)$. Because in wideband FM the value of β is usually around 5 or more, it is seen that the bandwidth of an angle-modulated signal is much greater than the bandwidth of various amplitude-modulation schemes, which is either W (in SSB) or $2W$ (in DSB or conventional AM).

Angle modulation by a Gaussian process. In this case we assume that the message signal $m(t)$ is a sample function of a stationary zero-mean Gaussian process $M(t)$, whose autocorrelation function is $R_M(\tau)$. The angle-modulated signal, now a random process, is as usual given by

$$U(t) = A_c \cos(2\pi f_c t + \Phi(t))$$

$$= \begin{cases} A_c \cos(2\pi f_c t + k_p M(t)) & \text{PM} \\[2mm] A_c \cos(2\pi f_c t + 2\pi k_f \displaystyle\int_{-\infty}^{t} M(\eta)\, d\eta) & \text{FM} \end{cases} \tag{5.3.47}$$

To determine the autocorrelation function of $U(t)$, we have

$$\begin{aligned} R_U(t, t + \tau) &= E[U(t + \tau)U(t)] \\ &= A_c^2 E[\cos(2\pi f_c t + \Phi(t)) \cos(2\pi f_c(t + \tau) + \\ &\quad + \Phi(t + \tau))] \end{aligned} \tag{5.3.48}$$

Obviously the process $U(t)$ is not stationary. To find its power spectrum we have to utilize the generalized form of the Wiener-Khinchin theorem as proved in Chapter 3. To this end, we have to first determine the average autocorrelation function

$$\bar{R}_U(\tau) = \lim_{T \to \infty} \frac{1}{T} \int_{-T/2}^{T/2} R_U(t, t + \tau)\, dt \tag{5.3.49}$$

This gives

$$\bar{R}_U(\tau) = A_c^2 \lim_{T\to\infty} \frac{1}{T} E\left[\int_{-T/2}^{T/2} \cos(2\pi f_c t + \Phi(t)) \right.$$

$$\left. \times \cos(2\pi f_c(t+\tau) + \Phi(t+\tau))\, dt \right]$$

$$= \frac{A_c^2}{2} \lim_{T\to\infty} \frac{1}{T} E\left[\int_{-T/2}^{T/2} \left[\cos(4\pi f_c t + 2\pi f_c \tau + \Phi(t) + \Phi(t+\tau)) \right.\right.$$

$$\left.\left. + \cos(2\pi f_c \tau + \Phi(t+\tau) - \Phi(t)) \right] dt \right]$$

$$\stackrel{a}{=} \frac{A_c^2}{2} \lim_{T\to\infty} \frac{1}{T} \int_{-T/2}^{T/2} E\left[\cos(2\pi f_c \tau + \Phi(t+\tau) - \Phi(t)) \right] dt$$

$$= \frac{A_c^2}{2} \lim_{T\to\infty} \frac{1}{T} \int_{-T/2}^{T/2} \mathrm{Re}\left[e^{j2\pi f_c \tau} E\left[e^{j(\Phi(t+\tau)-\Phi(t))} \right] \right] dt \qquad (5.3.50)$$

where the equality in a follows from the fact that $\cos(4\pi f_c t + 2\pi f_c \tau + \Phi(t) + \Phi(t+\tau))$ is a bandpass signal centered at $2f_c$ and its dc value (as demonstrated by the integral) is zero. Now it remains to find

$$E\left[e^{j(\Phi(t+\tau)-\Phi(t))} \right]$$

Because $M(t)$ is a Gaussian process, the process $\Phi(t)$ will also be Gaussian. This is because in PM, $\Phi(t)$ is simply $k_p M(t)$, which is obviously Gaussian; and in FM, $\Phi(t) = 2\pi k_f \int_{-\infty}^{t} M(\eta)\, d\eta$, which is the result of passing $M(t)$ through an LTI system. Therefore, $\Phi(t)$ is a zero-mean stationary Gaussian process with the autocorrelation function denoted by $R_\Phi(\tau)$. We also conclude that for fixed t and τ, the random variable $Z(t,\tau) = \Phi(t+\tau) - \Phi(t)$ is a zero-mean Gaussian random variable because it is a linear combination of two jointly Gaussian random variables. The variance of this random variable is easily computed to be

$$\sigma_Z^2 = 2R_\Phi(0) - 2R_\Phi(\tau) \qquad (5.3.51)$$

Now we have

$$E\left[e^{j(\Phi(t+\tau)-\Phi(t))} \right] = E\left[e^{jZ(t,\tau)} \right]$$

$$= e^{-\frac{1}{2}\sigma_Z^2}$$

$$= e^{-(R_\Phi(0)-R_\Phi(\tau))} \qquad (5.3.52)$$

where we have used the fact that the characteristic function of a zero-mean Gaussian random variable is given by

$$E\left[e^{j\omega X} \right] = e^{-\frac{1}{2}\omega^2 \sigma_X^2} \qquad (5.3.53)$$

Substituting (5.3.52) in (5.3.50), we obtain

$$\bar{R}_U(\tau) = \frac{A_c^2}{2} \lim_{T \to \infty} \frac{1}{T} \int_{-T/2}^{T/2} \mathrm{Re}\left[e^{j2\pi f_c \tau} e^{-(R_\Phi(0) - R_\Phi(\tau))} \right] dt$$

$$= \frac{A_c^2}{2} \cos(2\pi f_c \tau) e^{-(R_\Phi(0) - R_\Phi(\tau))}$$

$$= \frac{A_c^2}{2} \cos(2\pi f_c \tau) g(\tau) \tag{5.3.54}$$

where, by definition,

$$g(\tau) = e^{-(R_\Phi(0) - R_\Phi(\tau))} \tag{5.3.55}$$

From here we observe that the power content of the modulated process is $\bar{R}_U(\tau)|_{\tau=0} = \frac{A_c^2}{2}$. This is of course obvious from the beginning, because an angle-modulated signal is essentially a sinusoidal signal with constant amplitude A_c. Now we can obtain the power-spectral density of the modulated process $U(t)$ by taking the Fourier transform of $\bar{R}_U(\tau)$.

$$S_U(f) = \mathcal{F}\left[\frac{A_c^2}{2} \cos(2\pi f_c \tau) g(\tau) \right]$$

$$= \frac{A_c^2}{4} \left[G(f - f_c) + G(f + f_c) \right] \tag{5.3.56}$$

where

$$G(f) = e^{-R_\Phi(0)} \mathcal{F}[e^{R_\Phi(\tau)}] \tag{5.3.57}$$

From this point we cannot go further without making approximations. For example, we can find the Taylor series expansion of $e^{R_\Phi(\tau)}$ as

$$e^{R_\Phi(\tau)} = 1 + R_\Phi(\tau) + \frac{1}{2!} R_\Phi^2(\tau) + \frac{1}{3!} R_\Phi^3(\tau) + \cdots \tag{5.3.58}$$

and obtain $G(f)$ as

$$G(f) = e^{-R_\Phi(0)}\left[\delta(f) + S_\Phi(f) + \frac{1}{2} S_\Phi(f) \star S_\Phi(f) + \cdots \right] \tag{5.3.59}$$

Since

$$\Phi(t) = \begin{cases} k_p M(t) & \text{PM} \\[2mm] 2\pi k_f \displaystyle\int_{-\infty}^{t} M(\eta)\, d\eta & \text{FM} \end{cases} \tag{5.3.60}$$

we have

$$S_\Phi(f) = \begin{cases} k_p^2 S_M(f) & \text{PM} \\ \dfrac{k_f^2}{f^2} S_M(f) & \text{FM} \end{cases} \tag{5.3.61}$$

Due to the multifold convolutions involved in (5.3.59), the bandwidth of the modulated signal is not finite and the effective bandwidth in general is quite high. It is convenient to define the mean-square bandwidth of $G(f)$, denoted by W_g, as

$$W_g = \sqrt{\frac{\int_{-\infty}^{+\infty} f^2 G(f)\,df}{\int_{-\infty}^{+\infty} G(f)\,df}} \tag{5.3.62}$$

Then it can be shown that

$$W_g = \begin{cases} k_p\sqrt{P_M}\,W_M & \text{PM} \\ k_f\sqrt{P_M} & \text{FM} \end{cases} \tag{5.3.63}$$

where P_M and W_M denote the power content and the mean-square bandwidth of the message process. The derivation of the above relations is treated in a problem at the end of this chapter (see Problem 5.41). It is instructive to note the similarity between (5.3.38) and (5.3.63).

5.3.3 Implementation of Angle Modulators and Demodulators

Any modulation and demodulation process involves the generation of new frequencies that are not present in the input signal. This is true for both amplitude and angle-modulation systems. This means that if we interpret the modulator as a system with the message signal $m(t)$ as the input and with the modulated signal $u(t)$ as the output, this system has frequencies in its output that are not present in the input. Therefore, a modulator (and demodulator) cannot be modeled as a linear time-invariant system because a linear time-invariant system cannot produce any frequency components in the output that are not present in the input signal.

Angle modulators are in general time-varying and nonlinear systems. One method for generating an FM signal directly is to design an oscillator whose frequency changes with the input voltage. When the input voltage is zero, the oscillator generates a sinusoid with frequency f_0, and when the input voltage changes, this frequency changes accordingly. There are two approaches to design such an oscillator, usually called a VCO or *voltage controlled oscillator.* One approach is to use a *varactor diode.* A varactor diode is a capacitor whose capacitance changes with the applied voltage. Therefore, if this capacitor is used in the tuned circuit of the oscillator and the message signal is applied to it, the frequency of the tuned circuit and the oscillator will change in accordance with the message signal. Let us assume that the inductance of the inductor in the tuned circuit of Figure 5.34 is L_0 and the capacitance of the varactor diode is given by

$$C(t) = C_0 + k_0 m(t) \tag{5.3.64}$$

When $m(t) = 0$, the frequency of the tuned circuit is given by $f_c = 1/(2\pi\sqrt{L_0 C_0})$. In general, for nonzero $m(t)$, we have

$$
\begin{aligned}
f_i(t) &= \frac{1}{2\pi\sqrt{L_0(C_0 + k_0 m(t))}} \\
&= \frac{1}{2\pi\sqrt{L_0 C_0}}\, \frac{1}{\sqrt{1 + \frac{k_0}{C_0} m(t)}} \\
&= f_c \frac{1}{\sqrt{1 + \frac{k_0}{C_0} m(t)}} \tag{5.3.65}
\end{aligned}
$$

Assuming that

$$\epsilon = \frac{k_0}{C_0} m(t) \ll 1$$

and using the approximations

$$\sqrt{1 + \epsilon} \approx 1 + \frac{\epsilon}{2} \tag{5.3.66}$$

$$\frac{1}{1 + \epsilon} \approx 1 - \epsilon \tag{5.3.67}$$

we obtain

$$f_i(t) \approx f_c\left(1 - \frac{k_0}{2C_0} m(t)\right) \tag{5.3.68}$$

which is the relation for a frequency-modulated signal.

A second approach for generating an FM signal is through a *reactance tube*. In the reactance-tube implementation, an inductor whose inductance varies with the applied voltage is employed, and the analysis is very similar to the analysis presented for the varactor diode. It should be noted that although we described

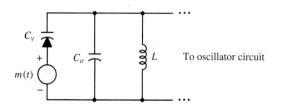

FIGURE 5.34. Varactor diode implementation of an angle modulator.

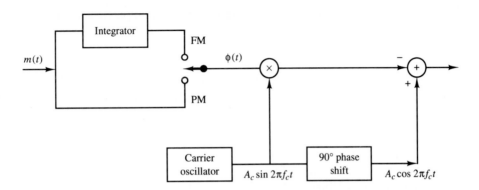

FIGURE 5.35.   Generation of narrowband angle-modulated signal.

these methods for generation of FM signals, due to the close relation between FM and PM signals basically the same methods can be applied for generation of PM signals (see Figure 5.28).

Another approach for generating an angle-modulated signal is to first generate a narrowband angle modulated signal and then change it to a wideband signal. This method is usually known as the *indirect method* for generation of FM and PM signals. Due to the similarity of conventional AM signals, generation of narrowband angle-modulated signals is straightforward. In fact, any modulator for conventional AM generation can be easily modified to generate a narrowband angle-modulated signal. Figure 5.35 shows the block diagram of a narrowband angle modulator. The next step is to use the narrowband angle-modulated signal to generate a wideband angle-modulated signal. Figure 5.36 shows the block diagram of a system that generates wideband angle-modulated signals from narrowband angle-modulated signals. The first stage of such a system is of course a narrowband angle modulator such as the one shown in Figure 5.35. The narrowband angle-modulated signal enters a frequency multiplier that multiplies the instantaneous frequency of the input by some constant n. This is usually done by applying the input signal to a nonlinear

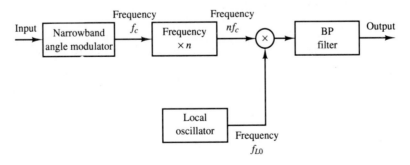

FIGURE 5.36.   Indirect generation of angle-modulated signals.

element and then passing its output through a bandpass filter tuned to the desired central frequency. If the narrowband modulated signal is represented by

$$u_n(t) = A_c \cos(2\pi f_c t + \phi(t)) \tag{5.3.69}$$

the output of the frequency multiplier (output of the bandpass filter) is given by

$$y(t) = A_c \cos(2\pi n f_c t + n\phi(t)) \tag{5.3.70}$$

In general this is, of course, a wideband angle-modulated signal. However, there is no guarantee that the carrier frequency of this signal, nf_c, will be the desired carrier frequency. The last stage of the modulator performs an up/down conversion to shift the modulated signal to the desired center frequency. This stage consists of a mixer and a bandpass filter. If the frequency of the local oscillator of the mixer is f_{LO} and we are using a down converter, the final wideband angle-modulated signal is given by

$$u(t) = A_c \cos\left(2\pi(nf_c - f_{LO})t + n\phi(t)\right) \tag{5.3.71}$$

Because we can freely choose n and f_{LO}, we can generate by this method any modulation index at any desired carrier frequency.

FM demodulators are implemented by generating an AM signal whose amplitude is proportional to the instantaneous frequency of the FM signal, and then using an AM demodulator to recover the message signal. To implement the first step, i.e., transforming the FM signal into an AM signal, it is enough to pass the FM signal through an LTI system whose frequency response is approximately a straight line in the frequency band of the FM signal. If the frequency response of such a system is given by

$$|H(f)| = V_0 + k(f - f_c) \quad \text{for } |f - f_c| < \frac{B_c}{2} \tag{5.3.72}$$

and if the input to the system is

$$u(t) = A_c \cos\left(2\pi f_c t + 2\pi k_f \int_{-\infty}^{t} m(\tau)\,d\tau\right) \tag{5.3.73}$$

then, the output will be the signal

$$v_o(t) = A_c(V_0 + kk_f m(t)) \cos\left(2\pi f_c t + 2\pi k_f \int_{-\infty}^{t} m(\tau)\,d\tau\right) \tag{5.3.74}$$

The next step is to demodulate this signal to obtain $A_c(V_0 + kk_f m(t))$, from which the message $m(t)$ can be recovered. Figure 5.37 shows a block diagram of these two steps.

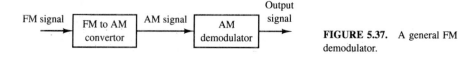

FIGURE 5.37. A general FM demodulator.

There exist many circuits that can be used to implement the first stage of an FM demodulator, i.e., FM to AM conversion. One such candidate is a simple differentiator with

$$|H(f)| = 2\pi f \tag{5.3.75}$$

Another candidate is the rising half of the frequency characteristics of a tuned circuit as shown in Figure 5.38. Such a circuit can be implemented easily but usually the linear region of the frequency characteristic may not be wide enough. To obtain a linear characteristics over a wider range of frequencies, usually two circuits tuned at two frequencies f_1 and f_2 are connected in a configuration which is known as a *balanced discriminator.* A balanced discriminator with the corresponding frequency characteristics is shown in Figure 5.39.

The FM demodulation methods described above that transform the FM signal into an AM signal have a bandwidth equal to the channel bandwidth B_c occupied by the FM signal. Consequently, the noise that is passed by the demodulator is the noise contained within B_c.

A totally different approach to FM signal demodulation is to use feedback in the FM demodulator to narrow the bandwidth of the FM detector and, as will be seen in Chapter 6, to reduce the noise power at the output of the demodulator. Figure 5.40 illustrates a system in which the FM discrimination is placed in the feedback branch of a feedback system that employs a voltage-controlled oscillator (VCO) path. The bandwidth of the discriminator and the subsequent lowpass filter is designed to match the bandwidth of the message signal $m(t)$. The output of the lowpass filter is the desired message signal. This type of FM demodulator is called an FM demodulator with feedback (FMFB). An alternative to the FMFB demodulator is the use of a PLL, as shown in Figure 5.41. The input to the PLL is the angle-modulated signal (we neglect the presence of noise in this discussion)

$$u(t) = A_c \cos[2\pi f_c t + \phi(t)] \tag{5.3.76}$$

where, for FM,

$$\phi(t) = 2\pi k_f \int_{-\infty}^{t} m(\tau)\, d\tau \tag{5.3.77}$$

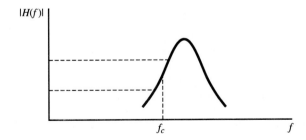

FIGURE 5.38. A tuned circuit used in an FM demodulator.

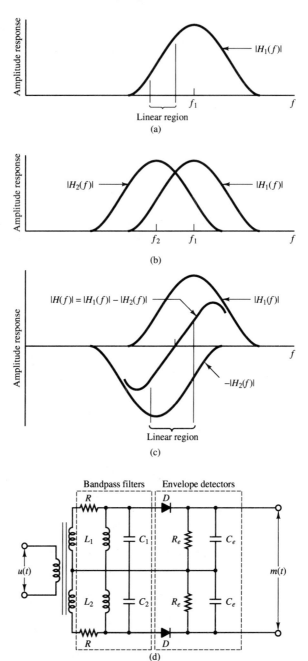

FIGURE 5.39.   A balanced discriminator and the corresponding frequency response.

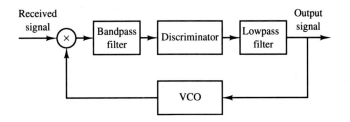

FIGURE 5.40. Block diagram of FMFB demodulator.

The VCO generates a sinusoid of a fixed frequency, in this case the carrier frequency f_c, in the absence of an input control voltage.

Now suppose that the control voltage to the VCO is the output of the loop filter, denoted as $v(t)$. Then the instantaneous frequency of the VCO is

$$f_v(t) = f_c + k_v v(t) \tag{5.3.78}$$

where k_v is a deviation constant with units of Hz/volt. Consequently, the VCO output may be expressed as

$$y_v(t) = A_v \sin[2\pi f_c t + \phi_v(t)] \tag{5.3.79}$$

where

$$\phi_v(t) = 2\pi k_v \int_0^t v(\tau)\, d\tau \tag{5.3.80}$$

The phase comparator is basically a multiplier and a filter that rejects the signal component centered at $2f_c$. Hence its output may be expressed as

$$e(t) = \frac{1}{2} A_v A_c \sin[\phi(t) - \phi_v(t)] \tag{5.3.81}$$

where the difference $\phi(t) - \phi_v(t) \equiv \phi_e(t)$ constitutes the phase error. The signal $e(t)$ is the input to the loop filter.

Let us assume that the PLL is in lock, so that the phase error is small. Then,

$$\sin[\phi(t) - \phi_v(t)] \approx \phi(t) - \phi_v(t) = \phi_e(t) \tag{5.3.82}$$

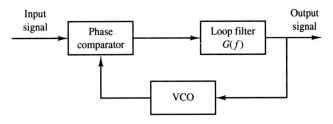

FIGURE 5.41. Block diagram of PLL-FM demodulator.

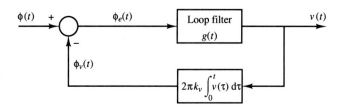

FIGURE 5.42. Linearized PLL.

Under this condition, we may deal with the linearized model of the PLL, shown in Figure 5.42. We may express the phase error as

$$\phi_e(t) = \phi(t) - 2\pi k_v \int_0^t v(\tau)\, d\tau \qquad (5.3.83)$$

or, equivalently, either as

$$\frac{d}{dt}\phi_e(t) + 2\pi k_v v(t) = \frac{d}{dt}\phi(t) \qquad (5.3.84)$$

or

$$\frac{d}{dt}\phi_e(t) + 2\pi k_v \int_{-\infty}^{+\infty} \phi_e(\tau) g(t-\tau)\, d\tau = \frac{d}{dt}\phi(t) \qquad (5.3.85)$$

The Fourier transform of the integro-differential equation in (5.3.85) is

$$(j2\pi f)\Phi_e(f) + 2\pi k_v \Phi_e(f) G(f) = (j2\pi f)\Phi(f) \qquad (5.3.86)$$

and, hence,

$$\Phi_e(f) = \frac{1}{1 + (\frac{k_v}{jf})G(f)} \Phi(f) \qquad (5.3.87)$$

The corresponding equation for the control voltage to the VCO is

$$V(f) = \Phi_e(f)G(f)$$
$$= \frac{G(f)}{1 + (\frac{k_v}{jf})G(f)} \Phi(f) \qquad (5.3.88)$$

Now, suppose that we design $G(f)$ such that

$$\left| k_v \frac{G(f)}{jf} \right| \gg 1 \qquad (5.3.89)$$

in the frequency band $|f| < W$ of the message signal. Then from (5.3.88), we have

$$V(f) = \frac{j2\pi f}{2\pi k_v} \Phi(f) \qquad (5.3.90)$$

or, equivalently,

$$v(t) = \frac{1}{2\pi k_v} \frac{d}{dt} \phi(t)$$

$$= \frac{k_f}{k_v} m(t) \qquad (5.3.91)$$

Because the control voltage of the VCO is proportional to the message signal, $v(t)$ is the demodulated signal.

We observe that the output of the loop filter with frequency response $G(f)$ is the desired message signal. Hence, the bandwidth of $G(f)$ should be the same as the bandwidth W of the message signal. Consequently the noise at the output of the loop filter is also limited to the bandwidth W. On the other hand, the output from the VCO is a wideband FM signal with an instantaneous frequency that follows the instantaneous frequency of the received FM signal.

The major benefit of using feedback in FM signal demodulation is to reduce the threshold effect that occurs when the input signal-to-noise-ratio to the FM demodulator drops below a critical value. The threshold effect is treated in Chapter 6.

5.4 RADIO AND TELEVISION BROADCASTING

Radio and television broadcasting are the most familiar forms of communication via analog signal transmission. Below we describe three types of broadcasting, namely, AM radio, FM radio, and television.

5.4.1 AM Radio Broadcasting

Commercial AM radio broadcasting utilizes the frequency band 535–1605 kHz for transmission of voice and music. The carrier frequency allocations range from 540 kHz to 1600 kHz with 10 kHz spacing.

Radio stations employ conventional AM for signal transmission. The baseband message signal $m(t)$ is limited to a bandwidth of approximately 5 kHz. Because there are billions of receivers and relatively few radio transmitters, the use of conventional AM for broadcasting is justified from an economic standpoint. The major objective is to reduce the cost of implementing the receiver.

The receiver most commonly used in AM radio broadcasting is the *superheterodyne receiver,* shown in Figure 5.43. It consists of a radio frequency (RF) tuned amplifier, a mixer, a local oscillator, an intermediate frequency (IF) amplifier, an envelope detector, an audio frequency amplifier, and a loudspeaker. Tuning for the desired radio frequency is provided by a variable capacitor, which simultaneously tunes the RF amplifier and the frequency of the local oscillator.

In the superheterodyne receiver, every AM radio signal is converted to a common IF frequency of $f_{IF} = 455$ kHz. This conversion allows the use of a single

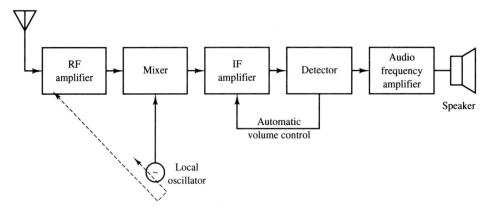

FIGURE 5.43.   Superheterodyne AM receiver.

tuned IF amplifier for signals from any radio station in the frequency band. The IF amplifier is designed to have a bandwidth of 10 kHz, which matches the bandwidth of the transmitted signal.

The frequency conversion to IF is performed by the combination of the RF amplifier and the mixer. The frequency of the local oscillator is

$$f_{\text{LO}} = f_c + f_{\text{IF}}$$

where f_c is the carrier frequency of the desired AM radio signal. The tuning range of the local oscillator is 955–2055 kHz. By tuning the RF amplifier to the frequency f_c and mixing its output with the local oscillator frequency $f_{\text{LO}} = f_c + f_{\text{IF}}$, we obtain two signal components, one centered at the difference frequency f_{IF} and the other centered at the sum frequency $2f_c + f_{IF}$. Only the first component is passed by the IF amplifier.

At the input to the RF amplifier we have signals picked up by the antenna from all radio stations. By limiting the bandwidth of the RF amplifier to the range $B_c < B_{\text{RF}} < 2f_{\text{IF}}$ where B_c is the bandwidth of the AM radio signal (10 kHz), we can reject the radio signal transmitted at the so-called *image frequency*, $f_c' = f_{\text{LO}} + f_{\text{IF}}$. Note that when we mix the local oscillator output, $\cos 2\pi f_{\text{LO}} t$, with the received signals

$$r_1(t) = A_c[1 + m_1(t)]\cos 2\pi f_c t$$

$$r_2(t) = A_c[1 + m_2(t)]\cos 2\pi f_c' t \qquad (5.4.1)$$

where $f_c = f_{\text{LO}} - f_{\text{IF}}$ and $f_c' = f_{\text{LO}} + f_{\text{IF}}$, the mixer output consists of the two signals

$$y_1(t) = A_c[1 + m_1(t)]\cos 2\pi f_{\text{IF}} t + \text{double frequency term}$$

$$y_2(t) = A_c[1 + m_2(t)]\cos 2\pi f_{\text{IF}} t + \text{double frequency term} \qquad (5.4.2)$$

where $m_1(t)$ represents the desired signal and $m_2(t)$ is the signal transmitted by the radio station transmitting at the carrier frequency $f_c' = f_{\text{LO}} + f_{\text{IF}}$. To prevent the signal $r_2(t)$ from interfering with the demodulation of the desired signal $r_1(t)$, the RF amplifier bandwidth is designed to be sufficiently narrow so that the image frequency signal is rejected. Hence, $B_{RF} < 2f_{IF}$ is the upper limit on the bandwidth of the RF amplifier. In spite of this constraint, the bandwidth of the RF amplifier is still considerably wider than the bandwidth of the IF amplifier. Thus, the IF amplifier, with its narrow bandwidth, provides signal rejection from adjacent channels and the RF amplifier provides signal rejection from image channels. Figure 5.44 illustrates the bandwidths of the RF and IF amplifiers and the requirement for rejecting the image frequency signal.

The output of the IF amplifier is passed through an envelope detector that produces the desired audio-band message signal $m(t)$. Finally, the output of the envelope detector is amplified and the amplified signal drives a loudspeaker. Automatic volume control (AVC) is provided by a feedback control loop that adjusts the gain of the IF amplifier based on the power level of the signal at the envelope detector.

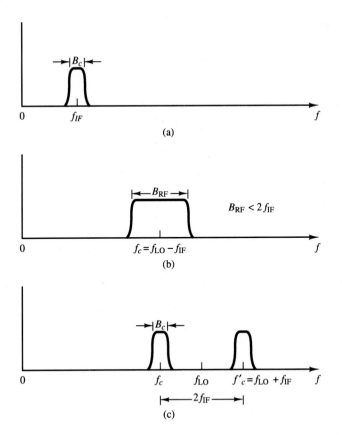

FIGURE 5.44.   Frequency response characteristics of IF and RF amplifiers. Bandpass characteristics of IF (a) and RF (b) amplifiers. (c) Desired signal and IF signal.

5.4.2 FM Radio Broadcasting

Commercial FM radio broadcasting utilizes the frequency band 88–108 MHz for transmission of voice and music signals. The carrier frequencies are separated by 200 kHz and the peak-frequency deviation is fixed at 75 kHz. Pre-emphasis is generally used to improve the demodulator performance at the receiver.

The receiver most commonly used in FM radio broadcast is a superheterodyne type. The block diagram of such a receiver is shown in Figure 5.45. As in AM radio reception, common tuning between the RF amplifier and the local oscillator allows the mixer to bring all FM radio signals to a common IF bandwidth of 200 kHz, centered at $f_{IF} = 10.7$ MHz. Since the message signal $m(t)$ is embedded in the frequency of the carrier, any amplitude variations in the received signal are a result of additive noise and interference. The amplitude limiter removes any amplitude variations in the received signal at the output of the IF amplifier by band-limiting the signal. A bandpass filter centered at $f_{IF} = 10.7$ MHz with a bandwidth of 200 kHz is included in the limiter to remove higher order frequency components introduced by the nonlinearity inherent in the hard limiter.

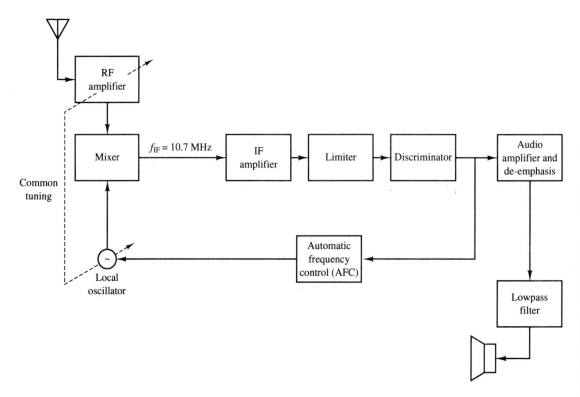

FIGURE 5.45.   Block diagram of a superheterodyne FM radio receiver.

A balanced frequency discriminator is used for frequency demodulation. The resulting message signal is then passed to the audio frequency amplifier which performs the functions of de-emphasis and amplification. The output of the audio amplifier is further filtered by a lowpass filter to remove out-of-band noise and its output is used to drive a loudspeaker.

FM stereo broadcasting.   Many FM radio stations transmit music programs in stereo by using the outputs of two microphones placed in two different parts of the stage. Figure 5.46 shows a block diagram of an FM stereo transmitter. The signals from the left and right microphones, $m_\ell(t)$ and $m_r(t)$, are added and subtracted as shown. The sum signal $m_\ell(t) + m_r(t)$ is left as is and occupies the frequency band 0–15 kHz. The difference signal $m_\ell(t) - m_r(t)$ is used to AM modulate (DSB-SC) a 38-kHz carrier that is generated from a 19-kHz oscillator. A pilot tone at the frequency of 19 kHz is added to the signal for the purpose of demodulating the DSB-SC AM signal. The reason for placing the pilot tone at 19 kHz instead of 38 kHz is that the pilot is more easily separated from the composite signal at the receiver. The combined signal is used to frequency modulate a carrier.

By configuring the baseband signal as an FDM signal, a monophonic FM receiver can recover the sum signal $m_\ell(t) + m_r(t)$ by use of a conventional FM demodulator. Hence, FM stereo broadcasting is compatible with conventional FM. The second requirement is that the resulting FM signal does not exceed the allocated 200-kHz bandwidth.

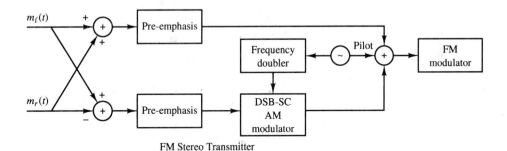

FM Stereo Transmitter

Baseband signal spectrum

FIGURE 5.46.   FM stereo transmitter and signal spacing.

The FM demodulator for FM stereo is basically the same as a conventional FM demodulator down to the limiter/discriminator. Thus, the received signal is converted to baseband. Following the discriminator, the baseband message signal is separated into the two signals $m_\ell(t) + m_r(t)$ and $m_\ell(t) - m_r(t)$ and passed through de-emphasis filters, as shown in Figure 5.47. The difference signal is obtained from the DSB-SC signal by means of a synchronous demodulator using the pilot tone. By taking the sum and the difference of the two composite signals, we recover the two signals $m_\ell(t)$ and $m_r(t)$. These audio signals are amplified by audio-band amplifiers and the two outputs drive dual loudspeakers. As indicated above, an FM receiver that is not configured to receive the FM stereo sees only the baseband signal $m_\ell(t) + m_r(t)$ in the frequency range 0–15 kHz. Thus, it produces a monophonic output signal that consists of the sum of the signals at the two microphones.

5.4.3 Television Broadcasting

Commercial TV broadcasting began as black-and-white picture transmission in London in 1936 by the British Broadcasting Corporation (BBC). Color television was demonstrated a few years later but the move of commercial TV stations to color TV signal transmission was slow in developing. To a large extent, this was due to the high cost of color TV receivers. With the development of the transistor and microelectronics components, the cost of color TV receivers decreased significantly, so that by the middle 1960's color TV broadcasting was used widely by the industry.

The frequencies allocated for TV broadcasting fall in the VHF and UHF frequency bands. Table 5.2 lists the TV channel allocations in the United States.

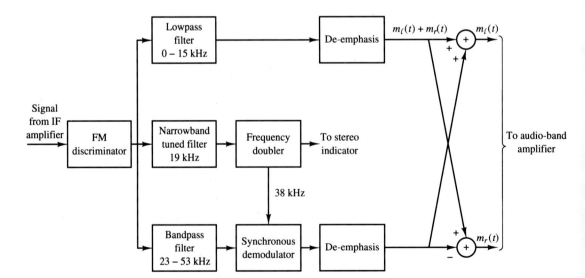

FIGURE 5.47. FM stereo receiver.

TABLE 5-2 VHF AND UHF FREQUENCY ALLOCATIONS FOR COMMERCIAL TELEVISION.

Channel	Frequency band (6-MHz bandwidth/station)
VHF 2–4	54–72 MHz
VHF 5, 6	76–88 MHz
VHF 7–13	174–216 MHz
UHF 14–69	470–896 MHz

We observe that the channel bandwidth allocated for transmission of TV signals is 6 MHz.

In contrast to radio broadcasting, standards for television signal transmission vary from country to country. The U.S. standard, which we describe below, was set by the National Television Systems Committee (NTSC).

Black-and-white TV signals. The first step in TV signal transmission is to convert a visual image into an electrical signal. The two-dimensional image or picture is converted to a one-dimensional electrical signal by sequentially scanning the image and producing an electrical signal that is proportional to the brightness level of the image. The scanning is performed in a TV camera, which optically focuses the image on a photo cathode tube that consists of a photosensitive surface.

The scanning of the image is performed by an electron beam that produces an output current or voltage that is proportional to the brightness of the image. The resulting electrical signal is called a *video signal.*

The scanning of the electron beam is controlled by two voltages applied across the horizontal and vertical deflection plates. These two voltages are shown in Figure 5.48. In this scanning method the image is divided into 525 lines that define a *frame,* as illustrated in Figure 5.49. The resulting signal is transmitted in 1/30 of a second. The number of lines determines the picture resolution and, in combination with the rate of transmission, determine the channel bandwidth required for transmission of the image.

The time interval of 1/30 sec to transmit a complete image is generally not fast enough to avoid the flickering that is annoying to the eyes of the average viewer. To overcome flickering, the scanning of the image is performed in an *interlaced pattern* as shown in Figure 5.49. The interlaced pattern consists of two fields, each consisting of 262.5 lines. Each field is transmitted in 1/60 of a second, which exceeds the flicker rate that is observed by the average eye. The first field begins at point "a" and terminates at point "b." The second field begins at point "c" and terminates at point "d."

A horizontal line is scanned in 53.5 μsec as indicated by the sawtooth signal waveform applied to the horizontal deflection plates. The beam has 10 μsec to move to the next line. During this interval, a blanking pulse is inserted to avoid

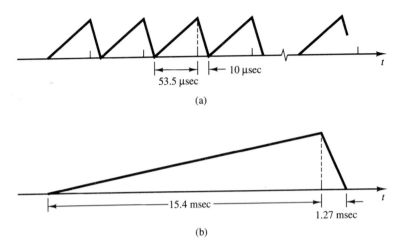

FIGURE 5.48.   Signal waveforms applied to horizontal (a) and vertical (b) deflection plates.

the appearance of retrace lines across the TV receiver. A 5-μsec pulse is added to the blanking pulse to provide synchronization for the horizontal sweep circuit at the receiver. A typical video signal is shown in Figure 5.50.

After the transmission of one interlaced field, the vertical sawtooth signal waveform applied to the vertical deflection plates is reset to zero. The retrace interval of 1.27 msec, corresponding to 20 line scans, allows the beam to move from the bottom to the top of the picture. A vertical blanking pulse is inserted during the interval to avoid the appearance of retrace lines at the TV receiver. When we allow for the vertical retrace (twice per frame), the actual number of horizontal lines in the image is 485.

The bandwidth of the video signal can be estimated by viewing the image as a rectangular array of 485 rows by (485)(4/3) columns, where 4/3 is the *aspect ratio* (the ratio of the width to height of the image). Thus, we have 313,633 picture elements (*pixels*) per frame, which are transmitted in 1/30 of a second. This is equivalent to a sampling rate of 10.5 MHz, which is sufficient to represent a signal as large as 5.25 MHz. However, the light intensity of adjacent pixels in an image is highly correlated. Hence, the bandwidth of the video signal is less than 5.25 MHz. In commercial TV broadcasting, the bandwidth of the video signal is limited to $W = 4.2$ MHz.

Because the allocated channel bandwidth for commercial TV is 6 MHz, it is clear that DSB transmission is not possible. The large low-frequency content of the video signal also rules out SSB as a practical modulation method. Hence, VSB is the only viable alternative. By transmitting a large carrier component, the received VSB signal can be simply demodulated by means of an envelope detector. This type of detection significantly simplifies the implementation of the receiver.

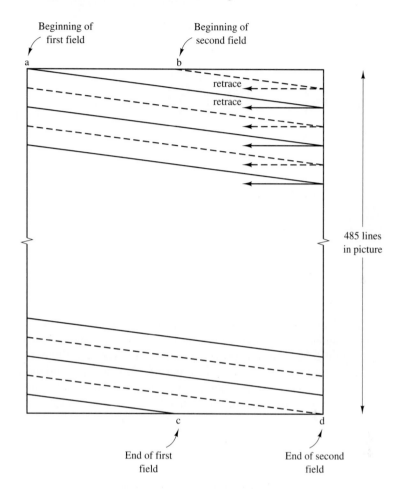

Beginning of first field

Beginning of second field

retrace

retrace

485 lines in picture

End of first field

End of second field

FIGURE 5.49. Interlaced scanning pattern.

The range of frequencies occupied by the transmitted video signal is shown in Figure 5.51. We note that the full upper sideband (4.2 MHz) of the video signal is transmitted along with a portion (1.25 MHz) of the lower sideband. Unlike the conventional VSB spectral shaping described in Section 5.2.4, the lower sideband signal in the frequency ranges f_c and $f_c - 0.75$ MHz is transmitted without attenuation. The frequencies in the range $f_c - 1.25$ MHz to $f_c - 0.75$ MHz are attenuated as shown in Figure 5.51 and all frequency components below $f_c - 1.25$ MHz are blocked. VSB spectral shaping is performed at the IF amplifier of the receiver.

In addition to the video signal, the audio portion of the TV signal is transmitted by frequency modulating a carrier at $f_c + 4.5$ MHz. The audio-signal bandwidth is limited to $W = 10$ kHz. The frequency deviation in the FM modulated signal

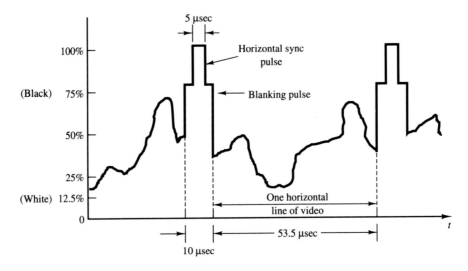

FIGURE 5.50. Typical video signal for one horizontal sweep.

is selected as 25 kHz and the FM signal bandwidth is 70 kHz. Hence, the total channel bandwidth required to transmit the video and audio signals is 5.785 MHz.

Figure 5.52 shows a block diagram of a black-and-white TV transmitter. The corresponding receiver is shown in Figure 5.53; it is a heterodyne receiver. We note that there are two separate tuners, one for the UHF band and one for the VHF band. The TV signals in the UHF band are brought down to the VHF band by

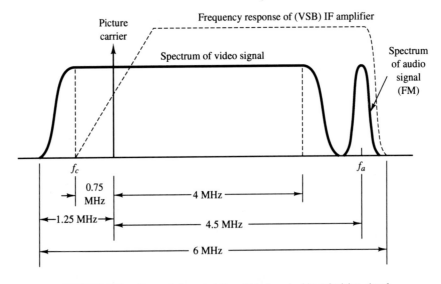

FIGURE 5.51. Spectral characteristics of black-and-white television signal.

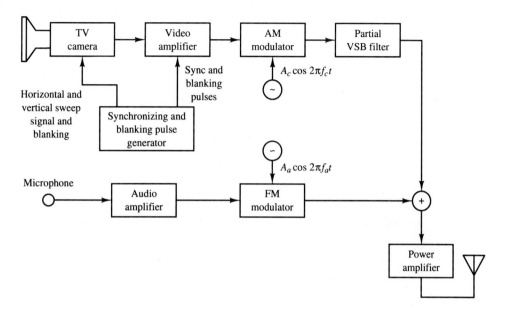

FIGURE 5.52.   Block diagram of a black-and-white TV transmitter.

a UHF mixer. This frequency conversion makes it possible to use a common RF amplifier for the two frequency bands. Then the video signal selected by the tuner is translated to a common IF frequency band of 41–47 MHz. The IF amplifier also provides the VSB shaping required prior to signal detection. The output of the IF amplifier is envelope detected to produce the baseband signal.

The audio portion of the signal centered at 4.5 MHz is filtered out by means of an IF filter amplifier and passed to the FM demodulator. The demodulated audio-band signal is then amplified by an audio amplifier and its output drives the speaker.

The video component of the baseband signal is passed through a video amplifier that passes frequency components in the range 0–4.2 MHz. Its output is passed to the DC restorer that clamps the blanking pulses and sets the correct dc level. The dc-restored video signal is then fed to the picture tube. The synchronizing pulses contained in the received video signal are separated and applied to the horizontal and vertical sweep generators.

Compatible color television.   The transmission of color information contained in an image can be accomplished by decomposing the colors of pixels into primary colors and transmitting the electrical signals corresponding to these colors. In general, all natural colors are well approximated by appropriate mixing of three primary colors: blue, green, and red. Consequently, if we employ three cameras, one with a blue filter, one with a green filter, and one with a red filter, and transmit the electrical signals $m_b(t)$, $m_g(t)$, and $m_r(t)$, generated by the three color cameras that

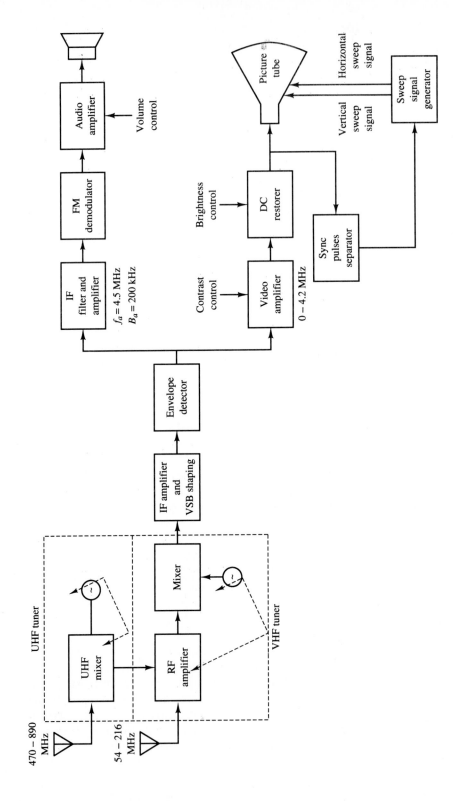

FIGURE 5.53. Block diagram of a black-and-white TV receiver.

view the color image, the received signals can be combined to produce a replica of the original color image.

Such a transmission scheme has two major disadvantages. First of all, it requires three times the channel bandwidth of black-and-white television. Second, the transmitted color TV signal cannot be received by a black-and-white (monochrome) TV receiver.

The NTSC standard adopted in 1953 in the United States avoids these two problems by transmitting a mixture of the three primary-color signals. Specifically, the three signals transmitted in the standard color TV system are the following three linearly independent combinations:

$$m_L(t) = 0.11m_b(t) + 0.59m_g(t) + 0.30m_r(t)$$

$$m_I(t) = -0.32m_b(t) - 0.28m_g(t) + 0.60m_r(t) \qquad (5.4.3)$$

$$m_Q(t) = 0.31m_b(t) - 0.52m_g(t) + 0.21m_r(t)$$

The transformation matrix

$$\mathbf{M} = \begin{bmatrix} 0.11 & 0.59 & 0.30 \\ -0.32 & -0.28 & 0.60 \\ 0.31 & -0.52 & 0.21 \end{bmatrix} \qquad (5.4.4)$$

that is used to construct the new transmitted signals $m_L(t), m_I(t)$, and $m_Q(t)$ is nonsingular and is inverted at the receiver to recover the primary-color signals $m_b(t), m_g(t)$, and $m_r(t)$ from $m_L(t), m_I(t)$, and $m_Q(t)$.

The signal $m_L(t)$ is called the *luminance signal*. It is assigned a bandwidth of 4.2 MHz and is transmitted via VSB-AM as in monochrome TV transmission. When this signal is received by a monochrome receiver, the result is a conventional black-and-white version of the color image. Thus, compatibility with monochrome TV broadcasting is achieved by transmitting $m_L(t)$. There remains the problem of transmitting the additional color information that can be used by a color TV receiver to reconstruct the color image. It is remarkable that the two composite color signals $m_I(t)$ and $m_Q(t)$ can be transmitted in the same bandwidth as $m_L(t)$, without interfering with $m_L(t)$.

The signals $m_I(t)$ and $m_Q(t)$ are called *chrominance signals* and are related to hue and saturation of colors. It has been determined experimentally, through subjective tests, that human vision cannot discriminate changes in $m_I(t)$ and $m_Q(t)$ over short time intervals and, hence, over small areas of the image. This implies that the high-frequency content in the signals $m_I(t)$ and $m_Q(t)$ can be eliminated without significantly compromising the quality of the reconstructed image. The end result is that $m_I(t)$ is limited in bandwidth to 1.6 MHz and $m_Q(t)$ is limited to 0.6 MHz prior to transmission. These two signals are quadrature-carrier multiplexed on a subcarrier frequency $f_{sc} = f_c + 3.579545$ MHz, as illustrated in Figure 5.54. The signal $m_I(t)$ is passed through a VSB filter that removes a part of the upper sideband, above 4.2 MHz. The signal $m_Q(t)$ is transmitted by DSB-SC

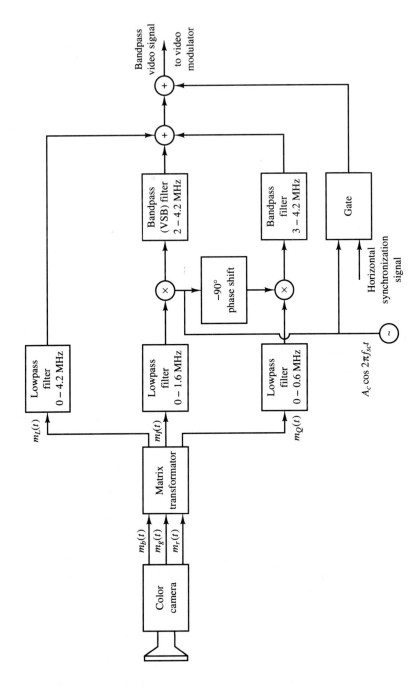

FIGURE 5.54. Transmission of primary-color signals and multiplexing of chrominance and luminance signals.

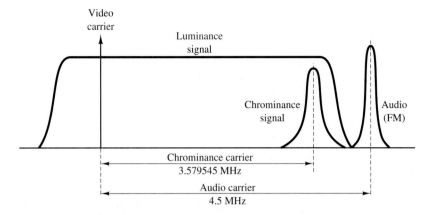

FIGURE 5.55.   Spectral characteristics of color TV signal.

amplitude modulation. Therefore, the composite video signal may be expressed as

$$m(t) = m_L(t) + m_Q(t) \sin 2\pi f_{sc}t + m_I(t) \cos 2\pi f_{sc}t$$
$$+ \hat{m}'_I(t) \sin 2\pi f_{sc}t \qquad\qquad (5.4.5)$$

The last two terms in (5.4.5) involving $m_I(t)$ and $\hat{m}'_I(t)$, constitute the VSB-AM signal for the chrominance $m_I(t)$. The composite signal $m(t)$ is transmitted by VSB plus carrier in a 6 MHz bandwidth, as shown in Figure 5.55.

The spectrum of the luminance signal $m_L(t)$ has periodic gaps between harmonics of the horizontal sweep frequency f_h, which in color TV is 4.5 MHz/286. The subcarrier frequency $f_{sc} = 3.579545$ for transmission of the chrominance signals was chosen because it corresponds to one of these gaps in the spectrum of $m_L(t)$. Specifically, it falls between the 227 and 228 harmonics of f_h. Thus the chrominance signals are interlaced in the frequency domain with the luminance signal as illustrated in Figure 5.56. As a consequence the effect of the chrominance signal on the luminance signal $m_L(t)$ is not perceived by the human eye, due to the persistence of human vision. Therefore the chrominance signals do not interfere

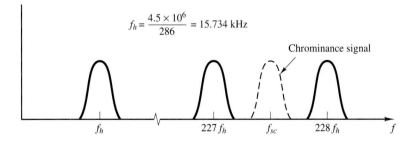

FIGURE 5.56.   Interlacing of chrominance signals with luminance signals.

with the demodulation of the luminance signal in both a monochrome TV receiver and a color TV receiver.

Horizontal and vertical synchronization pulses are added to $m(t)$ at the transmitter. In addition, eight cycles of the color subcarrier $A_c \cos 2\pi f_{sc} t$, called a *color burst,* are superimposed on the trailing edge of the blanking pulses, as shown in Figure 5.57, for the purpose of providing a signal for subcarrier-phase synchronization at the receiver.

The front end of the color TV receiver is basically the same as that of a monochrome receiver, down to the envelope detector that converts the 6-MHz VSB signal to baseband. The remaining demultiplexing operations in the color TV receiver are shown in Figure 5.58. We note that a lowpass filter with bandwidth 4.2 MHz is used to recover the luminance signal $m_L(t)$. The chrominance signals are stripped off by bandpass filtering and demodulated by the quadrature carrier demodulator using the output of a VCO that is phase locked to the received color-carrier frequency burst transmitted in each horizontal sweep. The demodulated chrominance signals are lowpass filtered and, along with the luminance signal, are passed to the "inverse matrix" converter that reconstructs the three color signals $m_b(t), m_g(t)$ and $m_r(t)$, i.e.,

$$\begin{bmatrix} m_b(t) \\ m_g(t) \\ m_r(t) \end{bmatrix} = \begin{bmatrix} 1.00 & -1.10 & 1.70 \\ 1.00 & -0.28 & -0.64 \\ 1.00 & -0.96 & 0.62 \end{bmatrix} \begin{bmatrix} m_L(t) \\ m_I(t) \\ m_Q(t) \end{bmatrix} \qquad (5.4.6)$$

The resulting color signals control the three electron guns that strike corresponding blue, green, and red picture elements in a color picture tube. Although color picture tubes are constructed in many different ways, the color mask tube is commonly used in practice. The face of the picture tube contains a matrix of dots of phosphor of the three primary colors with three such dots in each group. Behind each dot color group there is a mask with holes, one hole for each group. The three electron guns are aligned so that each gun can excite one of the three types of color dots. Thus, the three types of color dots are excited simultaneously in different intensities to generate color images.

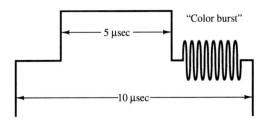

FIGURE 5.57.   Blanking pulse with color subcarrier.

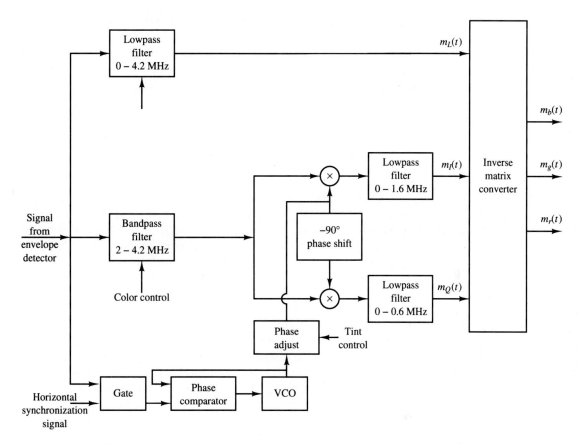

FIGURE 5.58. Demultiplexing and demodulation of luminance and chrominance signals in a color TV receiver.

5.5 MOBILE RADIO SYSTEMS

The demand to provide telephone service for people traveling in automobiles, buses, trains, and airplanes has been steadily increasing over the past three to four decades. To meet this demand, radio transmission systems have been developed that link the mobile telephone user to the terrestrial telephone network. Today, radio-based systems make it possible for people to communicate via telephone while traveling on airplanes and motor vehicles. In this section we will briefly describe the cellular telephone system that provides telephone service to people with hand-held portable telephones and automobile telephones.

A major problem with the establishment of any radio communication system is the availability of a portion of the radio spectrum. In the case of radio telephone service, the Federal Communications Commission (FCC) in the United States has assigned parts of the UHF frequency band in the 806–890 MHz range for this use.

Similar frequency assignments in the UHF band have been made in Europe and Japan.

The cellular radio concept was adopted as a method for efficient utilization of the available frequency spectrum, especially in highly populated metropolitan areas where the demand for mobile telephone services is the greatest. A geographic area is subdivided into cells, each of which contains a *base station*, as illustrated in Figure 5.59. Each base station is connected via telephone lines to a *mobile telephone switching office* (MTSO), which in turn is connected via telephone lines to a telephone central office of the terrestrial telephone network.

A mobile user communicates via radio with the base station within the cell. The base station routes the call through the MTSO to another base station if the called party is located in another cell or to the central office of the terrestrial telephone network if the called party is not mobile. Each mobile telephone is identified by its telephone number and the telephone serial number assigned by the manufacturer. These numbers are automatically transmitted to the MTSO during the initialization of the call for purposes of authentication and billing.

A mobile user initiates a telephone call in the usual manner by keying in the desired telephone number and pressing the "send" button. The MTSO checks the authentication of the mobile user and assigns an available frequency channel for radio transmission of the voice signal from the mobile telephone to the base station. The frequency assignment is sent to the mobile telephone via a supervisory control channel. A second frequency is assigned for radio transmission from the base station to the mobile user. The simultaneous transmission between the two parties is called *full-duplex operation*. The MTSO interfaces with the central office of the telephone network to complete the connection to the called party. All telephone communications between the MTSO and the telephone network are by means of wideband trunk lines that carry speech signals from many users. Upon completion of the telephone call, when the two parties hang up, the radio channel becomes available for another user.

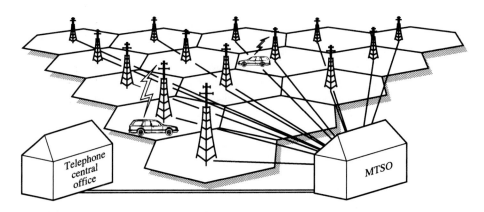

FIGURE 5.59.   Mobile radio system.

During the phone call, the MTSO monitors the signal strength of the radio transmission from the mobile user to the base station and, if the signal strength drops below a preset threshold, the MTSO views this as an indication that the mobile user is moving out of the initial cell into a neighboring cell. By communicating with the base stations of neighboring cells, the MTSO finds a neighboring cell that receives a stronger signal and automatically switches or hands-off the mobile user to the base station of the adjacent cell. The switching is performed in a fraction of a second and is generally transparent to the two parties. When a mobile user is outside of the assigned service area, the mobile telephone may be placed in a "roam" mode, which allows the mobile user to initiate and receive telephone calls.

In analog transmission of voice-band audio signals via radio, between the base station and the mobile user, the 3 kHz wide audio signal is transmitted via FM using a channel bandwidth of 30 kHz. This represents a bandwidth expansion of approximately a factor of 10. Such a large bandwidth expansion is necessary to obtain a sufficiently large signal-to-noise ratio at the output of the FM demodulator. However, the use of FM is highly wasteful of the radio frequency spectrum. The next generation of cellular telephone systems will use digital transmission of digitized compressed speech (at rates below 10,000 bits/sec) based on LPC encoding and vector quantization of the speech-model parameters. With this development, it is envisioned that the cellular telephone system will accommodate a fourfold to tenfold increase in the number of simultaneous users with the same available channel bandwidth.

The cellular radio telephone system is designed so that the transmitter powers of the base station and the mobile users are sufficiently small, so that signals do not propagate beyond immediately adjacent cells. This allows for frequencies to be reused in other cells outside of the immediately adjacent cells. Consequently, by making the cells smaller and reducing the radiated power, it is possible to increase frequency reuse and, thus, to increase the bandwidth efficiency and the number of mobile users. Current cellular systems employ cells with a radius in the range of 5–18 km. The base station normally transmits at a power level of 35 W or less and the mobile users transmit at a power level of approximately 3 W.

The cellular radio concept is being extended to different types of personal communication services using low power, hand-held radio transmitter and receivers. These emerging communication services are made possible by rapid advances in the fabrication of small and powerful integrated circuits that consume very little power and are relatively inexpensive. As a consequence, we will continue to experience exciting new developments in the telecommunications industry, well into the twenty-first century.

5.6 FURTHER READING

Analog communication systems are treated in numerous books on basic communication theory, including Sakrison (1968), Shanmugam (1979), Carlson (1986),

Stremler (1990), Ziemer and Tranter (1990), Couch (1993), and Gibson (1993). Implementation of analog communications systems are dealt with in depth in Clarke and Hess (1971).

PROBLEMS

5.1 The message signal $m(t) = 2\cos 400t + 4\sin(500t + \frac{\pi}{3})$ modulates the carrier signal $c(t) = A\cos(8000\pi t)$, using DSB amplitude modulation. Find the time domain and frequency domain representation of the modulated signal and plot the spectrum (Fourier transform) of the modulated signal. What is the power content of the modulated signal?

5.2 In a DSB system the carrier is $c(t) = A\cos 2\pi f_c t$ and the message signal is given by $m(t) = \text{sinc}(t) + \text{sinc}^2(t)$. Find the frequency-domain representation and the bandwidth of the modulated signal.

5.3 The two signals (a) and (b) shown in Figure P-5.3 DSB modulate a carrier signal $c(t) = A\cos 2\pi f_0 t$. Precisely plot the resulting modulated signals as a function of time and discuss their differences and similarities.

5.4 Show that in a DSB-modulated signal, the envelope of the resulting bandpass signal is proportional to the *absolute value* of the message signal. This means that an envelope detector can be employed as a DSB demodulator if we know that the message signal is always positive.

5.5 Suppose the signal $x(t) = m(t) + \cos 2\pi f_c t$ is applied to a nonlinear system whose output is $y(t) = x(t) + \frac{1}{2}x^2(t)$. Determine and sketch the spectrum of $y(t)$ when $M(f)$ is as shown in Figure P-5.5 and $W \ll f_c$.

5.6 The modulating signal

$$m(t) = 2\cos 4000\pi t + 5\cos 6000\pi t$$

is multiplied by the carrier

$$c(t) = 100\cos 2\pi f_c t$$

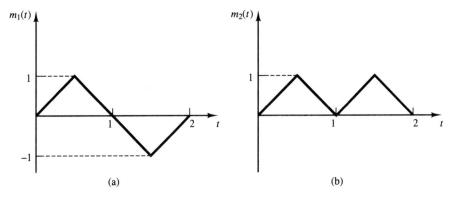

(a) (b)

FIGURE P-5.3

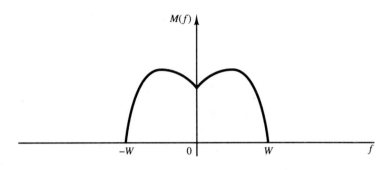

FIGURE P-5.5

where $f_c = 50$ kHz. Determine and sketch the power-spectral density of the DSB signal.

5.7 A DSB-modulated signal $u(t) = Am(t) \cos 2\pi f_c t$ is mixed (multiplied) with a local carrier $x_L(t) = \cos(2\pi f_c t + \theta)$ and the output is passed through a LPF with a bandwidth equal to the bandwidth of the message $m(t)$. Denoting the power of the signal at the output of the lowpass filter by P_{out} and the power of the modulated signal by P_U, plot $\frac{P_{\text{out}}}{P_U}$ as a function of θ for $0 \leq \theta \leq \pi$.

5.8 An AM signal has the form

$$u(t) = \left[20 + 2\cos 3000\pi t + 10\cos 6000\pi t\right] \cos 2\pi f_c t$$

where $f_c = 10^5$ Hz.

1. Sketch the (voltage) spectrum of $u(t)$.
2. Determine the power in each of the frequency components.
3. Determine the modulation index.
4. Determine the power in the sidebands, the total power, and the ratio of the sidebands power to the total power.

5.9 A message signal $m(t) = \cos 2000\pi t + 2\cos 4000\pi t$ modulates the carrier $c(t) = 100 \cos 2\pi f_c t$ where $f_c = 1$ MHz to produce the DSB signal $m(t)c(t)$.

1. Determine the expression for the upper sideband (USB) signal.
2. Determine and sketch the spectrum of the USB signal.

5.10 A DSB-SC signal is generated by multiplying the message signal $m(t)$ with the periodic rectangular waveform shown in Figure P-5.10 and filtering the product with a bandpass filter tuned to the reciprocal of the period T_p, with bandwidth $2W$, where W is the bandwidth of the message signal. Demonstrate that the output $u(t)$ of the BPF is the desired DSB-SC AM signal

$$u(t) = m(t) \sin 2\pi f_c t$$

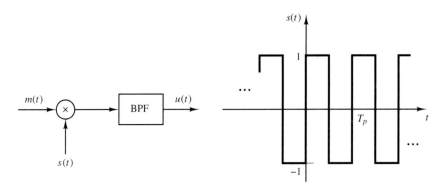

FIGURE P-5.10

where $f_c = 1/T_p$.

5.11 Show that in generating a DSB-SC signal as in Problem 5.10, it is not necessary that the periodic signal be rectangular. This means that any periodic signal with period T_p can substitute for the rectangular signal in Figure P-5.10.

5.12 The message signal $m(t)$ has a Fourier transform shown in Figure P-5.12(a). This signal is applied to the system shown in Figure P-5.12(b) to generate the signal $y(t)$.

   1. Plot $Y(f)$, the Fourier transform of $y(t)$.

   2. Show that if $y(t)$ is transmitted, the receiver can pass it through a replica of the system shown in Figure 5.12(b) to obtain $m(t)$ back. This means that this system can be used as a simple scrambler to enhance communication privacy.

5.13 An AM signal is generated by modulating the carrier $f_c = 800$ kHz by the signal

$$m(t) = \sin 2000\pi t + 5\cos 4000\pi tt$$

The AM signal

$$u(t) = 100\left[1 + m(t)\right]\cos 2\pi f_c t$$

is fed to a 50 Ω load.

   1. Determine and sketch the spectrum of the AM signal.

   2. Determine the average power in the carrier and in the sidebands.

   3. What is the modulation index?

   4. What is the peak power delivered to the load?

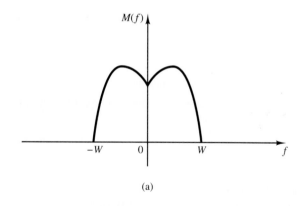

(a)

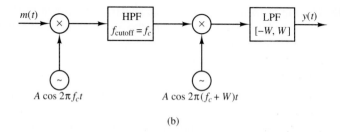

(b)

FIGURE P-5.12

5.14 The output signal from an AM modulator is

$$u(t) = 5\cos 1800\pi t + 20\cos 2000\pi t + 5\cos 2200\pi t$$

1. Determine the modulating signal $m(t)$ and the carrier $c(t)$.
2. Determine the modulation index.
3. Determine the ratio of the power in the sidebands to the power in the carrier.

5.15 A DSB-SC AM signal is modulated by the signal

$$m(t) = 2\cos 2000\pi t + \cos 6000\pi t$$

The modulated signal is

$$u(t) = 100m(t)\cos 2\pi f_c t$$

where $f_c = 1$ MHz.

1. Determine and sketch the spectrum of the AM signal.
2. Determine the average power in the frequency components.

5.16 A SSB-AM signal is generated by modulating an 800-kHz carrier by the signal $m(t) = \cos 2000\pi t + 2\sin 2000\pi t$. The amplitude of the carrier is $A_c = 100$.

    1. Determine the signal $\hat{m}(t)$.

    2. Determine the (time domain) expression for the lower sideband of the SSB-AM signal.

    3. Determine the magnitude spectrum of the lower sideband SSB signal.

5.17 Weaver's SSB modulator is illustrated in Figure P-5.17. By taking the input signal as $m(t) = \cos 2\pi f_m t$, where $f_m < W$, demonstrate that by proper choice of f_1 and f_2 the output is a SSB signal.

5.18 The message signal $m(t)$ whose spectrum is shown in Figure P-5.18 is passed through the system shown in the same figure. The bandpass filter has a bandwidth of $2W$ centered at f_0 and the lowpass filter has a bandwidth of W. Plot the spectra of the signals $x(t)$, $y_1(t)$, $y_2(t)$, $y_3(t)$ and $y_4(t)$. What are the bandwidths of these signals?

5.19 The system shown in Figure P-5.19 is used to generate an AM signal. The modulating signal $m(t)$ has zero mean and its maximum (absolute) value is $A_m = \max |m(t)|$. The nonlinear device has an input–output characteristic

$$y(t) = ax(t) + bx^2(t)$$

    1. Express $y(t)$ in terms of the modulating signal $m(t)$ and the carrier $c(t) = \cos 2\pi f_c t$.

    2. Specify the filter characteristics that yield an AM signal at its output.

    3. What is the modulation index?

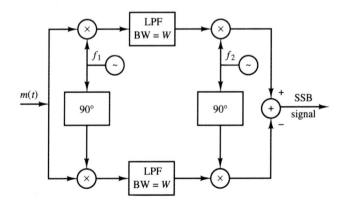

FIGURE P-5.17

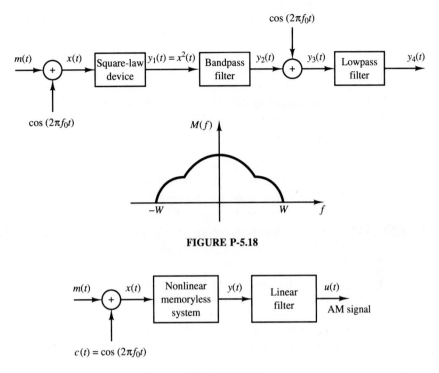

FIGURE P-5.18

FIGURE P-5.19

5.20 The signal $m(t)$ whose Fourier transform $M(f)$ is shown in Figure P-5.20 is to be transmitted from point A to point B. It is known that the signal is normalized, meaning that $-1 \le m(t) \le 1$.

   1. If USSB is employed, what is the bandwidth of the modulated signal?
   2. If DSB is employed, what is the bandwidth of the modulated signal?

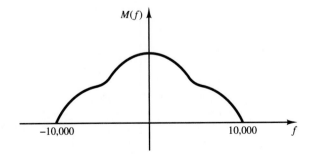

FIGURE P-5.20

3. If an AM modulation scheme with $a = 0.8$ is used, what is the bandwidth of the modulated signal?

4. If an FM signal with $k_f = 60$ kHz is used, what is the bandwidth of the modulated signal?

5.21 A VSB modulation system is shown in Figure P-5.21. The bandwidth of the message signal $m(t)$ is W and the transfer function of the bandpass filter is shown in the figure.

    1. Determine $h_l(t)$ the lowpass equivalent of $h(t)$, where $h(t)$ represents the impulse response of the bandpass filter.

    2. Derive an expression for the modulated signal $u(t)$.

5.22 Find expressions for the in-phase and quadrature components, $x_c(t)$ and $x_s(t)$, and envelope and phase, $V(t)$ and $\Theta(t)$, for DSB, SSB, conventional AM, USSB, LSSB, FM, and PM.

5.23 The normalized signal $m_n(t)$ has a bandwidth of 10,000 Hz and its power content is 0.5 W. The carrier $A \cos 2\pi f_0 t$ has a power content of 200 W.

    1. If $m_n(t)$ modulates the carrier using SSB amplitude modulation, what will be the bandwidth and the power content of the modulated signal?

    2. If the modulation scheme is DSB-SC, what will be the answer to part 1?

    3. If modulation scheme is AM with modulation index of 0.6, what will be the answer to part 1?

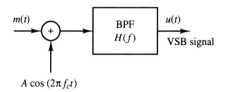

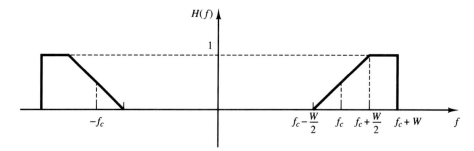

FIGURE P-5.21

4. If modulation is FM with $k_f = 50{,}000$, what will be the answer to part 1?

5.24 The message signal $m(t) = 10\,\text{sinc}(400t)$ frequency modulates the carrier $c(t) = 100\cos 2\pi f_c t$. The modulation index is 6.

    1. Write an expression for the modulated signal $u(t)$.
    2. What is the maximum frequency deviation of the modulated signal?
    3. What is the power content of the modulated signal?
    4. Find the bandwidth of the modulated signal.

5.25 Signal $m(t)$ is shown in Figure P-5.25. This signal is used once to frequency modulate a carrier and once to phase modulate the same carrier.

    1. Find a relation between k_p and k_f such that the maximum phase of the modulated signals in both cases are equal.
    2. If $k_p = k_f = 1$, what is the maximum instantaneous frequency in each case?

5.26 An angle modulated signal has the form

$$u(t) = 100\cos\left[2\pi f_c t + 4\sin 2000\pi t\right]$$

where $f_c = 10$ MHz.

    1. Determine the average transmitted power.
    2. Determine the peak-phase deviation.
    3. Determine the peak-frequency deviation.
    4. Is this an FM or a PM signal? Explain.

5.27 Find the smallest value of the modulation index in an FM system that guarantees that all the modulated signal power is contained in the sidebands and no power is transmitted at the carrier frequency.

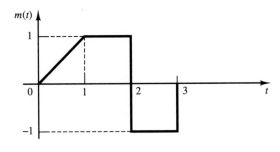

FIGURE P-5.25

5.28 Wideband FM can be generated by first generating a narrowband FM signal and then using frequency multiplication to spread the signal bandwidth. Figure P-5.28 illustrates such a scheme, which is called an Armstrong-type FM modulator. The narrowband FM signal has a maximum angular deviation of 0.10 radians in order to keep distortion under control.

1. If the message signal has a bandwidth of 15 kHz and the output frequency from the oscillator is 100 kHz, determine the frequency multiplication that is necessary to generate an FM signal at a carrier frequency of $f_c = 104$ MHz and a frequency deviation of $f = 75$ kHz.

2. If the carrier frequency for the wideband FM signal is to be within ± 2 Hz, determine the maximum allowable drift of the 100 kHz oscillator.

5.29 Determine the amplitude and phase of various frequency components of a PM signal with $k_p = 1$ and $m(t)$, a periodic signal given by

$$m(t) = \begin{cases} 1 & 0 \le t \le \frac{T_m}{2} \\ -1 & \frac{T_m}{2} \le t \le T_m \end{cases} \tag{5.6.1}$$

in one period.

5.30 An FM signal is given as

$$u(t) = 100\cos\left[2\pi f_c t + 100 \int_{-\infty}^{t} m(\tau)\,d\tau \right]$$

where $m(t)$ is shown in Figure P-5.30.

1. Sketch the instantaneous frequency as a function of time.

2. Determine the peak-frequency deviation.

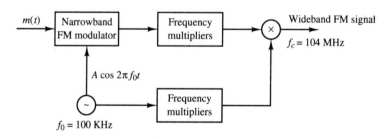

FIGURE P-5.28. Armstrong-type FM modulator.

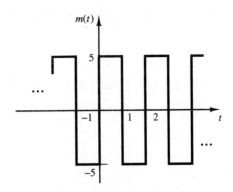

FIGURE P-5.30

5.31 The carrier $c(t) = 100 \cos 2\pi f_c t$ is frequency modulated by the signal $m(t) = 5 \cos 20000\pi t$, where $f_c = 10^8$ Hz. The peak frequency deviation is 20 kHz.

1. Determine the amplitude and frequency of all signal components that have a power level of at least 10% of the power of the unmodulated carrier component.
2. From Carson's rule, determine the approximate bandwidth of the FM signal.

5.32 The carrier $c(t) = A \cos 2\pi 10^6 t$ is angle modulated (PM or FM) by the sinusoid signal $m(t) = 2 \cos 2000\pi t$. The deviation constants are $k_p = 1.5$ rad/V and $k_f = 3000$ Hz/V.

1. Determine β_f and β_p.
2. Determine the bandwidth in each case using Carson's rule.
3. Plot the spectrum of the modulated signal in each case (plot only those frequency components that lie within the bandwidth derived in part 2).
4. If the amplitude of $m(t)$ is decreased by a factor of 2, how would your answers to parts 1–3 change?
5. If the frequency of $m(t)$ is increased by a factor of 2, how would your answers to parts 1–3 change?

5.33 The carrier $c(t) = 100 \cos 2\pi f_c t$ is phase modulated by the signal $m(t) = 5 \cos 2000\pi t$. The PM signal has a peak-phase deviation of $\pi/2$. The carrier frequency is $f_c = 10^8$ Hz.

1. Determine the magnitude spectrum of the sinusoidal components and sketch the results.

2. Using Carson's rule, determine the approximate bandwidth of the PM signal and compare the results with the analytical result in part 1.

5.34 An angle-modulated signal has the form

$$u(t) = 100\cos\left[2\pi f_c t + 4\sin 2\pi f_m t\right]$$

where $f_c = 10$ MHz and $f_m = 1000$ Hz.

1. Assuming that this is an FM signal, determine the modulation index and the transmitted signal bandwidth.
2. Repeat part 1 if f_m is doubled.
3. Assuming that this is a PM signal, determine the modulation index and the transmitted signal bandwidth.
4. Repeat part 3 if f_m is doubled.

5.35 It is easy to demonstrate that amplitude modulation satisfies the superposition principle, whereas angle modulation does not. To be specific, let $m_1(t)$ and $m_2(t)$ be two message signals and let $u_1(t)$ and $u_2(t)$ be the corresponding modulated versions.

1. Show that when the combined message signal $m_1(t) + m_2(t)$ DSB modulates a carrier $A_c \cos 2\pi f_c t$, the result is the sum of the two DSB amplitude-modulated signals $u_1(t) + u_2(t)$.
2. Show that if $m_1(t) + m_2(t)$ frequency modulates a carrier, the modulated signal is not equal to $u_1(t) + u_2(t)$.

5.36 An FM discriminator is shown in Figure P-5.36. The envelope detector is assumed to be ideal and has an infinite input impedance. Select the values for L and C if the discriminator is to be used to demodulate an FM signal with a carrier $f_c = 80$ MHz and a peak frequency deviation of 6 MHz.

5.37 An angle-modulated signal is given as

$$u(t) = 100\cos\left[2000\pi t + \phi(t)\right]$$

where (a) $\phi(t) = 5\sin 20\pi t$ and (b) $\phi(t) = 5\cos 20\pi t$. Determine and sketch the amplitude and phase spectra for (a) and (b) and compare the results.

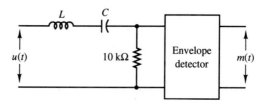

FIGURE P-5.36

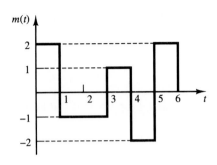

FIGURE P-5.38

5.38 The message signal $m(t)$ into an FM modulator with peak-frequency deviation $k_f = 25$ Hz/V is shown in Figure P-5.38. Plot the frequency deviation in Hz and the phase deviation in radians.

5.39 A message signal $m(t)$ has a bandwidth of 10 kHz and a peak magnitude $|m(t)|$ of 1 volt. Estimate the bandwidth of the signal $u(t)$ obtained when $m(t)$ frequency modulates a carrier with a peak-frequency deviation of (a) $f_d = 10$ Hz/V, (b) 100 Hz/V, and (c) 1000 Hz/V.

5.40 The modulating signal into an FM modulator is

$$m(t) = 10\cos 16\pi t$$

The output of the FM modulator is

$$u(t) = 10\cos\left[4000\pi t + 2\pi k_f \int_{-\infty}^{t} m(\tau)\,d\tau\right]$$

where $k_f = 10$. If the output of the FM modulator is passed through an ideal BPF centered at $f_c = 2000$ with a bandwidth of 62 Hz, determine the power of the frequency components at the output of the filter. What percentage of the transmitter power appears at the output of the BPF?

5.41 In this problem we derive an expression for the mean-square bandwidth of an FM modulated stationary Gaussian process as expressed in (5.3.63). The

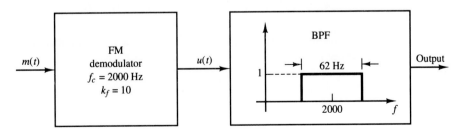

FIGURE P-5.40

mean-square bandwidth of $G(f)$ is defined in (5.3.62) as

$$W_g = \sqrt{\frac{\int_{-\infty}^{+\infty} f^2 G(f)\,df}{\int_{-\infty}^{+\infty} G(f)\,df}}$$

1. Using the basic properties of the Fourier transform, show that

$$W_g^2 = -\frac{\left[\frac{d^2}{d\tau^2}g(\tau)\right]_{\tau=0}}{4\pi^2 g(0)}$$

where

$$g(\tau) = \mathcal{F}^{-1}\left[G(f)\right]$$

2. Using the definition of $g(\tau)$

$$g(\tau) = e^{-(R_\Phi(0)-R_\Phi(\tau))}$$

and the result of part 1 show that in PM, if $R'_\Phi(\tau)$ is differentiable at $\tau = 0$, then

$$W_g^2 = -\left(\frac{k_p}{2\pi}\right)^2 \frac{d^2}{d\tau^2} R_M(\tau)|_{\tau=0}$$

$$= k_p^2 R_M(0) W_M^2$$

$$= k_p^2 P_M W_M^2$$

or

$$W_g = k_p \sqrt{P_M}\, W_M$$

where

$$W_M^2 = \frac{\int_{-\infty}^{+\infty} f^2 S_M(f)\,df}{\int_{-\infty}^{+\infty} S_M(f)\,df}$$

3. In FM

$$\Phi(t) = 2\pi k_f \int_{-\infty}^{t} M(\eta)\,d\eta$$

Using this relation and Leibnitz's rule for differentiation of an integral, show that when $\lim_{\tau\to\infty} R_M(\tau) = 0$, we have

$$\frac{d^2 \Phi(\tau)}{d\tau^2}\bigg|_{\tau=\infty} = 4\pi^2 k_f^2 R_M(0)$$

Use this relation to show that in FM

$$W_g^2 = k_f^2 R_M(0)$$

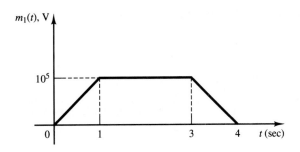

FIGURE P-5.42

$$= k_f^2 P_M$$

and therefore

$$W_g = k_f \sqrt{P_M}$$

5.42 The message signal $m_1(t)$ is shown in Figure P-5.42. and the message signal $m_2(t) = \text{sinc}(2 \times 10^4 t)$, in volts.

1. If $m_1(t)$ is frequency modulated on a carrier with frequency 10^6 Hz with a frequency deviation constant (k_f) equal to 5 Hz/V, what is the maximum instantaneous frequency of the modulated signal?

2. If $m_1(t)$ is phase modulated with phase-deviation constant $k_p = 3$ rad/V, what is the maximum instantaneous frequency of the modulated signal? What is the minimum instantaneous frequency of the modulated signal?

3. If $m_2(t)$ is frequency modulated with $k_f = 10^3$ Hz/V, what is the maximum instantaneous frequency of the modulated signal? What is the bandwidth of the modulated signal?

5.43 We wish to transmit 60 voice-band signals by SSB (upper sideband) modulation and frequency-division multiplexing (FDM). Each of the 60 signals has a spectrum as shown in Figure P-5.43. Note that the voice-band signal is

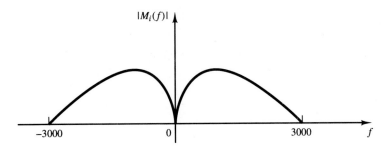

FIGURE P-5.43

band limited to 3 kHz. If each signal in frequency translated separately, we would require a frequency synthesizer that produces 60 carrier frequencies to perform the frequency division multiplexing. On the other hand, if we subdivide the channels into L groups of K subchannels each, such that $LK = 60$, we may reduce the number of frequencies from the synthesizer to $L + K$.

1. Illustrate the spectrum of the SSB signals in a group of K subchannels. Assume that a 1 kHz guard-band separates the signals in adjacent frequency subchannels and that the carrier frequencies are $f_{c_1} = 10$ kHz, $f_{c_2} = 14$ kHz, $\ldots$, etc.
2. Sketch L and K such that $LK = 60$ and $L + K$ is a minimum.
3. Determine the frequencies of the carriers if the 60 FDM signals occupy the frequency band 300–540 kHz, and each group of K signals occupies the band 10 kHz to $(10 + 4K)$ kHz.

5.44 A superheterodyne FM receiver operates in the frequency range of 88–108 MHz. The IF and local oscillator frequencies are chosen such that $f_{\text{IF}} < f_{\text{LO}}$. We require that the image frequency f_c' fall outside of the 88–108 MHz region. Determine the minimum required f_{IF} and the range of variations in f_{LO}.

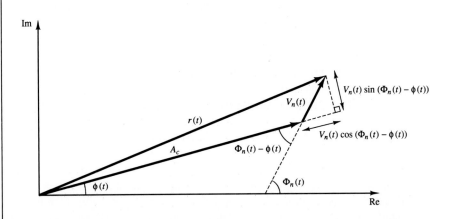

Effect of Noise on Analog Communication Systems

In the preceding chapter we studied the important characteristics of analog communication systems. These characteristics included time-domain and frequency-domain representations of the modulated signal, bandwidth requirements, and power content of the modulated signal, and, finally, modulator and demodulator implementation of various analog communication systems.

In this chapter the effect of noise on various analog communication systems will be analyzed. As we will see, angle-modulation systems and particularly FM can provide a high degree of noise immunity and therefore are desirable in cases of severe noise and/or low signal power. This noise immunity is obtained at the price of sacrificing channel bandwidth because, as we have seen in Chapter 5, the bandwidth requirements of angle-modulation systems is considerably higher than amplitude-modulation systems.

This chapter starts with an analysis of the performance of linear-modulation systems in the presence of noise. Then, the performance of phase-locked loops (PLL) that are used for carrier synchronization is studied. The effect of noise on angle-modulation systems is treated next, and, finally, the effects of transmission loss and noise on analog communication systems in general are analyzed.

6.1 EFFECT OF NOISE ON LINEAR MODULATION SYSTEMS

In this section we determine the signal-to-noise ratio (SNR) of the output of the receiver that demodulates the amplitude-modulated signals. In evaluating the effect of noise on the various types of analog-modulated signals, it is also interesting to compare the result with the effect of noise on an equivalent baseband communication system. We begin the evaluation of the effect of noise on a baseband system.

6.1.1 Effect of Noise on a Baseband System

Because baseband systems serve as a basis for comparison of various modulation systems, we begin with a noise analysis of a baseband system. In this case there exists no demodulator, and the receiver consists only of a lowpass filter with bandwidth W. The noise power at the output of the receiver is therefore

$$P_{n_o} = \int_{-W}^{+W} \frac{N_0}{2} \, df$$

$$= N_0 W \tag{6.1.1}$$

If we denote the received power by P_R, the baseband SNR is given by

$$\left(\frac{S}{N}\right)_b = \frac{P_R}{N_0 W} \tag{6.1.2}$$

6.1.2 Effect of Noise on DSB-SC AM

In DSB we have

$$u(t) = A_c m(t) \cos(2\pi f_c t + \phi_c) \tag{6.1.3}$$

and therefore the received signal at the output of the receiver noise-limiting filter is

$$r(t) = u(t) + n(t)$$
$$= A_c m(t) \cos(2\pi f_c t + \phi_c)$$
$$+ n_c(t) \cos 2\pi f_c t - n_s(t) \sin 2\pi f_c t \tag{6.1.4}$$

Suppose we demodulate the received signal by first multiplying $r(t)$ by a locally generated sinusoid $\cos(2\pi f_c t + \phi)$, where ϕ is the phase of the sinusoid, and then passing the product signal through an ideal lowpass filter having a bandwidth W. The multiplication of $r(t)$ with $\cos(2\pi f_c t + \phi)$ yields

$$r(t) \cos(2\pi f_c t + \phi) = A_c m(t) \cos(2\pi f_c t + \phi_c) \cos(2\pi f_c t + \phi)$$
$$+ n(t) \cos(2\pi f_c t + \phi)$$
$$= \frac{1}{2} A_c m(t) \cos(\phi_c - \phi) + \frac{1}{2} A_c m(t) \cos(4\pi f_c t + \phi + \phi_c)$$

$$+ \frac{1}{2}[n_c(t)\cos\phi + n_s(t)\sin\phi]$$

$$+ \frac{1}{2}\left[n_c(t)\cos(4\pi f_c t + \phi) - n_s(t)\sin(4\pi f_c t + \phi)\right] \quad (6.1.5)$$

The lowpass filter rejects the double-frequency components and passes only the lowpass components. Hence, its output is

$$y(t) = \frac{1}{2}A_c m(t)\cos(\phi_c - \phi) + \frac{1}{2}\left[n_c(t)\cos\phi + n_s(t)\sin\phi\right] \quad (6.1.6)$$

As it was discussed in Chapter 5, the effect of phase incoherence between the transmitter and the receiver is a drop equal to $\cos^2(\phi_c - \phi)$ in the received signal power. This can be avoided by employing a PLL as described in Section 6.2. The effect of a PLL is to generate a sinusoid at the receiver with the same frequency and phase as the carrier. If a PLL is employed, then $\phi = \phi_c$, and the demodulator is called a *coherent* or *synchronous demodulator*. In our analysis we assume that we are employing a coherent demodulator. With this assumption and without loss of generality, we can assume $\phi = \phi_c = 0$ and

$$y(t) = \frac{1}{2}\left[A_c m(t) + n_c(t)\right] \quad (6.1.7)$$

Therefore, at the receiver output the message signal and the noise components are additive and we are able to define a meaningful SNR. The message signal power is given by

$$P_o = \frac{A_c^2}{4}E[M^2(t)] = \frac{A_c^2}{4}R_M(0) = \frac{A_c^2}{4}P_M \quad (6.1.8)$$

where $P_M = R_M(0)$ is the power content of the message process. The noise power is given by

$$P_{n_o} = \frac{1}{4}P_{n_c}$$

$$= \frac{1}{4}P_n \quad (6.1.9)$$

where we have used the fact that the power contents of $n_c(t)$ and $n(t)$ are equal. This was shown in Section 3.6, in (3.6.12). The power content of $n(t)$ can be found by noting that it is the result of passing $n_w(t)$ through a filter with bandwidth B_c. Therefore, the power-spectral density of $n(t)$ is given by

$$S_n(f) = \begin{cases} \dfrac{N_0}{2}, & |f - f_c| < W \\ 0, & \text{otherwise} \end{cases} \quad (6.1.10)$$

The noise power is

$$P_n = \int_{-\infty}^{+\infty} S_n(f)\,df$$

$$= \frac{N_0}{2} \times 4W$$

$$= 2W N_0 \tag{6.1.11}$$

Now we can find the output SNR as

$$\left(\frac{S}{N}\right)_o = \frac{P_o}{P_{n_o}}$$

$$= \frac{\frac{A_c^2}{4} P_M}{\frac{1}{4} 2W N_0}$$

$$= \frac{A_c^2 P_M}{2W N_0} \tag{6.1.12}$$

In this case, the received signal power is $P_R = \frac{A_c^2 P_M}{2}$. Therefore, by using (6.1.2), we obtain

$$\left(\frac{S}{N}\right)_{o_{DSB}} = \frac{P_R}{N_0 W}$$

$$= \left(\frac{S}{N}\right)_b \tag{6.1.13}$$

It is seen that in DSB-SC AM the output SNR is the same as the SNR for a baseband system. Therefore, DSB-SC AM does not provide any SNR improvement over a simple baseband communication system.

6.1.3 Effect of Noise on SSB AM

In this case

$$u(t) = A_c m(t) \cos 2\pi f_c t \pm A_c \hat{m}(t) \sin 2\pi f_c t \tag{6.1.14}$$

Therefore, the input to the demodulator is

$$r(t) = A_c m(t) \cos 2\pi f_c t \pm A_c \hat{m}(t) \sin 2\pi f_c t + n(t)$$

$$= (A_c m(t) + n_c(t)) \cos 2\pi f_c t +$$

$$+ (\pm A_c \hat{m}(t) - n_s(t)) \sin 2\pi f_c t \tag{6.1.15}$$

Here we assume that demodulation occurs with an ideal-phase reference. Hence, the output of the lowpass filter is the in-phase component (with a coefficient of $\frac{1}{2}$) of the above signal; that is,

$$y(t) = \frac{A_c}{2} m(t) + \frac{1}{2} n_c(t) \qquad (6.1.16)$$

It is observed that in this case again the signal and the noise components are additive and a meaningful SNR at the receiver output can be defined. Parallel to our discussion of DSB we have

$$P_o = \frac{A_c^2}{4} P_M \qquad (6.1.17)$$

and

$$P_{n_o} = \frac{1}{4} P_{n_c} = \frac{1}{4} P_n \qquad (6.1.18)$$

where

$$P_n = \int_{-\infty}^{+\infty} S_n(f)\,df = \frac{N_0}{2} \times 2W = W N_0 \qquad (6.1.19)$$

Therefore

$$\left(\frac{S}{N}\right)_{o_{SSB}} = \frac{P_o}{P_{n_o}} = \frac{A_c^2 P_M}{W N_0} \qquad (6.1.20)$$

But in this case

$$P_R = P_U = A_c^2 P_M \qquad (6.1.21)$$

and therefore

$$\left(\frac{S}{N}\right)_{o_{SSB}} = \frac{P_R}{W N_0} = \left(\frac{S}{N}\right)_b \qquad (6.1.22)$$

Therefore the SNR in a SSB system is is equivalent to that of a DSB system.

6.1.4 Effect of Noise on Conventional AM

In conventional DSB AM the modulated signal is

$$u(t) = A_c \left[1 + a m_n(t)\right] \cos 2\pi f_c t \qquad (6.1.23)$$

Therefore, the received signal at the input to the demodulator is

$$r(t) = \left\{A_c \left[1 + a m_n(t)\right] + n_c(t)\right\} \cos 2\pi f_c t - n_s(t) \sin 2\pi f_c t \qquad (6.1.24)$$

where a is the modulation index and $m_n(t)$ is normalized so that its minimum value is -1. If a synchronous demodulator is employed, the situation is basically similar to the DSB case, except that we have $1 + a m_n(t)$ instead of $m(t)$. Therefore, in this case, after mixing and lowpass filtering we have

$$y_1(t) = \frac{1}{2} \left\{A_c \left[1 + a m_n(t)\right] + n_c(t)\right\} \qquad (6.1.25)$$

However, in this case, the desired signal is $m(t)$, not $1 + a m_n(t)$. The dc component in the demodulated waveform is removed by a dc block and, hence, the lowpass filter output is

$$y(t) = \frac{1}{2} A_c a m_n(t) + \frac{n_c(t)}{2} \qquad (6.1.26)$$

In this case, the received signal power P_R is given by

$$P_R = \frac{A_c^2}{2} \left[1 + a^2 P_{M_n} \right] \qquad (6.1.27)$$

where we have assumed that the message process is a zero-mean process. Now we can derive the output SNR as

$$
\begin{aligned}
\left(\frac{S}{N} \right)_{o_{AM}} &= \frac{\frac{1}{4} A_c^2 a^2 P_{M_n}}{\frac{1}{4} P_{n_c}} \\
&= \frac{A_c^2 a^2 P_{M_n}}{2 N_0 W} \\
&= \frac{a^2 P_{M_n}}{1 + a^2 P_{M_n}} \frac{\frac{A_c^2}{2} \left[1 + a^2 P_{M_n} \right]}{N_0 W} \\
&= \frac{a^2 P_{M_n}}{1 + a^2 P_{M_n}} \frac{P_R}{N_0 W} \\
&= \frac{a^2 P_{M_n}}{1 + a^2 P_{M_n}} \left(\frac{S}{N} \right)_b \\
&= \eta \left(\frac{S}{N} \right)_b \qquad (6.1.28)
\end{aligned}
$$

where we have used (6.1.2) and η denotes the modulation efficiency.

From above it is seen that, because $a^2 P_{M_n} < 1 + a^2 P_{M_n}$, the SNR in conventional AM is always smaller than the SNR in a baseband system. In practical applications the modulation index a is in the range of 0.8–0.9. The power content of the normalized message process depends a lot on the message source. For speech signals that usually have a large dynamic range, P_{M_n} is in the neighborhood of 0.1. This means that the overall loss in SNR compared to a baseband system is a factor of 0.075 or equivalent to a loss of 11 dB. The reason for this loss is that a large part of the transmitter power is used to send the carrier component of the modulated signal and not the desired signal.

To analyze the envelope-detector performance in the presence of noise, we have to use certain approximations. This is due mainly to the nonlinear structure of an envelope detector that makes an exact analysis difficult. In this case the demodulator detects the envelope of the received signal and the noise process. The input to the envelope detector is

$$r(t) = \left\{ A_c \left[1 + am_n(t) \right] + n_c(t) \right\} \cos 2\pi f_c t - n_s(t) \sin 2\pi f_c t \qquad (6.1.29)$$

and, therefore, the envelope of $r(t)$ is given by

$$V_r(t) = \sqrt{\left[A_c \left[1 + am_n(t) \right] + n_c(t) \right]^2 + n_s^2(t)} \qquad (6.1.30)$$

Now we assume that the signal component in $r(t)$ is much stronger than the noise component. With this assumption we have

$$p \left(n_s(t) \ll \left[A_c \left[1 + am_n(t) \right] \right] \right) \approx 1 \qquad (6.1.31)$$

and, therefore, with high probability

$$V_r(t) \approx \left[A_c \left[1 + am_n(t) \right] \right] + n_c(t) \qquad (6.1.32)$$

After removing the dc component, we obtain

$$y(t) \approx A_c a m_n(t) + n_c(t) \qquad (6.1.33)$$

which is basically the same as $y(t)$ for the synchronous demodulation without the $\frac{1}{2}$ coefficient. This coefficient, of course, has no effect on the final SNR, and therefore we conclude that, under the assumption of high SNR at the receiver input, the performance of synchronous and envelope demodulators is the same. However, if the above assumption is not true, for synchronous demodulation we still have additive signal and noise at the receiver output but with envelope demodulation the signal and noise become intermingled. To see this, let us assume that at the receiver input the noise power[†] is much stronger than the signal power. This means that

$$V_r(t) = \sqrt{\left[A_c \left[1 + am_n(t) \right] + n_c(t) \right]^2 + n_s^2(t)}$$

$$= \sqrt{A_c^2 (1 + am_n(t))^2 + n_c^2(t) + n_s^2(t) + 2 A_c n_c(t)(1 + am_n(t))}$$

$$\overset{a}{\approx} \sqrt{(n_c^2(t) + n_s^2(t)) \left[1 + \frac{2 A_c n_c(t)}{n_c^2(t) + n_s^2(t)} (1 + am_n(t)) \right]}$$

$$\overset{b}{\approx} V_n(t) \left[1 + \frac{A_c n_c(t)}{V_n^2(t)} (1 + am_n(t)) \right]$$

$$= V_n(t) + \frac{A_c n_c(t)}{V_n(t)} (1 + am_n(t)) \qquad (6.1.34)$$

where in (a) we have used the fact that $A_c^2 (1 + am_n^2(t))^2$ is small compared to the other components and in (b) we have denoted $\sqrt{n_c^2(t) + n_s^2(t)}$ by $V_n(t)$, the

---

[†]By noise power at the receiver input we mean the power of the noise within the bandwidth of the modulated signal, or equivalently, the noise power at the output of the noise-limiting filter.

envelope of the noise process, and have used the approximation $\sqrt{1+\epsilon} \approx 1 + \frac{\epsilon}{2}$, for small ϵ, where

$$\epsilon = \frac{2A_c n_c(t)}{n_c^2(t) + n_s^2(t)} (1 + a m_n(t)) \tag{6.1.35}$$

From (6.1.34) it is observed that at the demodulator output the signal and the noise components are no longer additive and in fact the signal component is multiplied by the noise and is no longer distinguishable. In this case no meaningful SNR can be defined. It is said that this system *is operating below the threshold*. The subject of threshold and its effect on the performance of a communication system will be covered in more detail when we discuss the noise performance in angle modulation.

Example 6.1.1

The message process $M(t)$ is a stationary process with the autocorrelation function

$$R_M(\tau) = 16 \operatorname{sinc}^2(10{,}000\tau)$$

It is also known that all the realizations of the message process satisfy the condition $\max |m(t)| = 6$. It is desirable to transmit this message to a destination via a channel with 80-dB attenuation and additive white noise with power-spectral density $S_n(f) = \frac{N_0}{2} = 10^{-12}$ W/Hz, and achieve a SNR at the modulator output of at least 50 dB. What is the required transmitter power and channel bandwidth if the following modulation schemes are employed?

1. DSB AM.
2. SSB AM.
3. Conventional AM with modulation index equal to 0.8.

Solution   First we determine the bandwidth of the message process. To do this we obtain the power-spectral density of the message process.

$$S_M(f) = \mathcal{F}[R_M(\tau)] = \frac{16}{10{,}000} \Lambda \left(\frac{f}{10{,}000} \right)$$

which is nonzero for $-10{,}000 < f < 10{,}000$, and therefore $W = 10{,}000$ Hz. Now we can determine $\left(\frac{S}{N} \right)_b$ as a basis of comparison.

$$\left(\frac{S}{N} \right)_b = \frac{P_R}{N_0 W} = \frac{P_R}{2 \times 10^{-12} \times 10^4} = \frac{10^8 P_R}{2}$$

Because the channel attenuation is 80 dB,

$$10 \log \frac{P_T}{P_R} = 80$$

and, therefore

$$P_R = 10^{-8} P_T$$

Hence

$$\left(\frac{S}{N}\right)_b = \frac{10^8 \times 10^{-8} P_T}{2} = \frac{P_T}{2}$$

1. For DSB AM, we have

$$\left(\frac{S}{N}\right)_o = \left(\frac{S}{N}\right)_b = \frac{P_T}{2} \sim 50 \text{ dB} = 10^5$$

therefore

$$\frac{P_T}{2} = 10^5 \Longrightarrow P_T = 2 \times 10^5 \text{ W} \sim 200 \text{ kW}$$

and

$$\text{BW} = 2W = 2 \times 10,000 = 20,000 \text{ Hz} \sim 20 \text{ kHz}$$

2. For SSB AM,

$$\left(\frac{S}{N}\right)_o = \left(\frac{S}{N}\right)_b = \frac{P_T}{2} = 10^5 \Longrightarrow P_T = 200 \text{ kW}$$

and

$$\text{BW} = W = 10,000 \text{ Hz} = 10 \text{ kHz}$$

3. For conventional AM, with $a = 0.8$,

$$\left(\frac{S}{N}\right)_o = \eta \left(\frac{S}{N}\right)_b = \eta \frac{P_T}{2}$$

where η is the modulation efficiency given by

$$\eta = \frac{a^2 P_{M_n}}{1 + a^2 P_{M_n}}$$

First we find P_{M_n}, the power content of the normalized message signal. Because $\max |m(t)| = 6$, we have

$$P_{M_n} = \frac{P_M}{\left(\max |m(t)|\right)^2} = \frac{P_M}{36}$$

To determine P_M, we have

$$P_M = R_M(\tau)|_{\tau=0} = 16$$

and, therefore

$$P_{M_n} = \frac{16}{36} = \frac{4}{9}$$

hence,

$$\eta = \frac{0.8^2 \times \frac{4}{9}}{1 + 0.8^2 \times \frac{4}{9}} \approx 0.22$$

Therefore,

$$\left(\frac{S}{N}\right)_o \approx 0.22\frac{P_T}{2} = 0.11 P_T = 10^5$$

or

$$P_T \approx 909 \text{ kW}$$

The bandwidth of conventional AM is equal to the bandwidth of DSB AM, i.e.,

$$\text{BW} = 2W = 20 \text{ kHz}$$

6.2 CARRIER PHASE ESTIMATION WITH A PHASE-LOCKED LOOP

In this section, we describe a method for generating a phase reference for synchronous demodulation of a DSB-SC AM signal. The received noise-corrupted signal at the input to the demodulator is given by (compare with 5.2.14)

$$r(t) = u(t) + n(t)$$

$$= A_c m(t)\cos(2\pi f_c t + \phi_c) + n(t) \qquad (6.2.1)$$

First, we note that the received signal $r(t)$ has a zero mean because the message signal $m(t)$ is zero mean, i.e., $m(t)$ contains no *dc* component. Consequently, the average power at the output of a narrowband filter tuned to the carrier frequency f_c is zero. This fact implies that we cannot extract a carrier-signal component directly from $r(t)$.

If we square $r(t)$, the squared signal contains a spectral component at twice the carrier frequency; that is,

$$r^2(t) = A_c^2 m^2(t)\cos^2(2\pi f_c t + \phi_c) + \text{noise terms}$$

$$= \frac{1}{2}A_c^2 m^2(t) + \frac{1}{2}A_c^2 m^2(t)\cos(4\pi f_c t + 2\phi_c)$$

$$+ \text{noise terms} \qquad (6.2.2)$$

Because $E[M^2(t)] = R_M(0) > 0$, there is signal power at the frequency $2f_c$, which can be used to drive a PLL.

To isolate the desired double-frequency component from the rest of the frequency components, the squared-input signal is passed through a narrowband filter that is tuned to the frequency $2f_c$. The mean value of the output of such a filter is a sinusoid with frequency $2f_c$, phase $2\phi_c$, and amplitude $A_c^2 E[M^2(t)]|H(2f_c)|/2$, where $|H(2f_c)|$ is the gain (attenuation) of the filter at $f = 2f_c$. Thus, squaring the input signal produces a sinusoidal component at twice the carrier frequency, which can be used as the input to a PLL. The general configuration for the carrier-phase-estimation system is illustrated in Figure 6.1.

The PLL consists of a multiplier, a loop filter, and a voltage-controlled oscillator (VCO), as shown in Figure 6.2. If the input to the PLL is the sinusoid

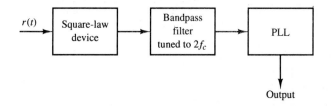

FIGURE 6.1.   System for carrier-phase estimation.

Output

$\cos(4\pi f_c t - 2\phi)$ and the output of the VCO is $\sin(4\pi f_c t - 2\hat{\phi})$, where $\hat{\phi}$ represents the estimate of ϕ, the product of these two signals produces the signal

$$e(t) = \cos(4\pi f_c t - 2\phi)\sin(4\pi f_c t - 2\hat{\phi})$$

$$= \frac{1}{2}\sin 2(\hat{\phi} - \phi) + \frac{1}{2}\sin(8\pi f_c t - 2\hat{\phi} - 2\phi) \tag{6.2.3}$$

Note that $e(t)$ contains a low-frequency term (dc) and a term at four times the carrier.

The loop filter is a lowpass filter that responds only to the low-frequency component $\sin(\hat{\phi} - \phi)$ and removes the component at $4f_c$. This filter is usually selected to have the relatively simple transfer function

$$G(s) = \frac{1 + \tau_2 s}{1 + \tau_1 s} \tag{6.2.4}$$

where the time constants τ_1 and τ_2 are design parameters ($\tau_1 \gg \tau_2$) that control the bandwidth of the loop. A higher-order filter that contains additional poles may be used, if necessary, to obtain a better loop response.

The output of the loop filter, $v(t)$, provides the control voltage for the VCO, whose implementation is described in Section 5.3.3, in the context of FM modulation. The VCO is basically a sinusoidal signal generator with an instantaneous phase given by

$$4\pi f_c t - 2\hat{\phi} = 4\pi f_c t - K_v \int_{-\infty}^{t} v(\tau)d\tau \tag{6.2.5}$$

where K_v is a gain constant in radians/volt-sec. Hence, the carrier-phase estimate at the output of the VCO is

$$2\hat{\phi} = K_v \int_{-\infty}^{t} v(\tau)d\tau \tag{6.2.6}$$

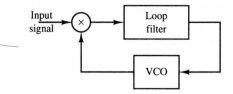

FIGURE 6.2.   Basic elements of a PLL.

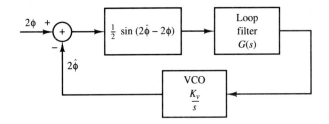

FIGURE 6.3. Model of a PLL.

and its transfer function is K_v/s.

Because the double-frequency term resulting from the multiplication of the input signal to the loop with the output of the VCO is removed by the loop filter, the PLL may be represented by the closed-loop system model shown in Figure 6.3. The sine function of the phase difference $2(\hat{\phi} - \phi)$ makes the system nonlinear and, as a consequence, the analysis of the PLL performance in the presence of noise is somewhat involved although it is mathematically tractable for simple-loop filters.

In steady-state operation, when the loop is tracking the phase of the received carrier, the phase error $\hat{\phi} - \phi$ is small and, hence

$$\frac{1}{2} \sin 2(\hat{\phi} - \phi) \approx \hat{\phi} - \phi \tag{6.2.7}$$

With this approximation, the PLL is represented by the *linear model* shown in Figure 6.4. This linear model has a closed-loop transfer function.

$$H(s) = \frac{KG(s)/s}{1 + KG(s)/s} \tag{6.2.8}$$

where the factor of $1/2$ has been absorbed into the gain parameter K. By substituting for $G(s)$ from (6.2.4) into (6.2.8), we obtain

$$H(s) = \frac{1 + \tau_2 s}{1 + \left(\tau_2 + \frac{1}{K}\right)s + \frac{\tau_1}{K} s^2} \tag{6.2.9}$$

Hence, the closed-loop system function for the linearized PLL is second order when the loop filter has a single pole and a single zero. The parameter τ_2 determines the position of the zero in $H(s)$, while K, τ_1, and τ_2 control the position of the closed-loop system poles.

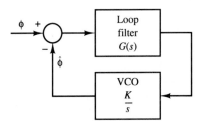

FIGURE 6.4. Linear model of a PLL.

The denominator of $H(s)$ may be expressed in the standard form

$$D(s) = s^2 + 2\zeta\omega_n s + \omega_n^2 \tag{6.2.10}$$

where ζ is called the *loop-damping factor* and ω_n is the *natural frequency* of the loop. In terms of the loop parameters, $\omega_n = \sqrt{K/\tau_1}$ and $\zeta = \omega_n(\tau_2 + 1/K)/2$, the closed-loop transfer function becomes

$$H(s) = \frac{(2\zeta\omega_n - \omega_n^2/K)s + \omega_n^2}{s^2 + 2\zeta\omega_n s + \omega_n^2} \tag{6.2.11}$$

The magnitude response $20\log|H(j\omega)|$ as a function of the normalized frequency ω/ω_n is illustrated in Figure 6.5, with the damping factor as a parameter and $\tau_1 \gg 1$. Note that $\zeta = 1$ results in a critically damped loop response, $\zeta < 1$ produces an underdamped loop response, and $\zeta > 1$ yields an overdamped loop response.

The (one-sided) noise-equivalent bandwidth of the loop is (see Problem 6.6)

$$B_{neq} = \frac{\tau_2^2\left(1/\tau_2^2 + K/\tau_1\right)}{4\left(\tau_2 + \frac{1}{K}\right)} = \frac{1 + (\tau_2\omega_n)^2}{8\zeta/\omega_n} \tag{6.2.12}$$

In practice, the selection of the bandwidth of the PLL involves a trade-off between speed of response and noise in the phase estimate. On the one hand, it is desirable to

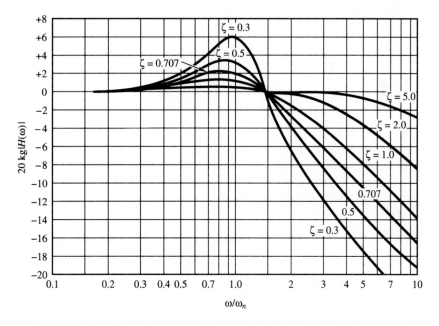

FIGURE 6.5. Frequency response of a second-order loop. (From *Phaselock Techniques,* by Sec. Ed., by F.M. Gardner, © 1979 by John Wiley and Sons, Inc. Reprinted with permission of the publisher).

select the bandwidth of the loop to be sufficiently wide to track any time variations in the phase of the received carrier. On the other hand, a wideband PLL allows more noise to pass into the loop, which corrupts the phase estimate. We assess below the effects of noise in the quality of the phase estimate.

6.2.1 Effect of Additive Noise on Phase Estimation

To evaluate the effects of noise on the estimate of the carrier phase, let us assume that the PLL is tracking a sinusoidal signal of the form

$$s(t) = A_c \cos[2\pi f_c t + \phi(t)] \qquad (6.2.13)$$

which is corrupted by the additive narrowband noise

$$n(t) = n_c(t) \cos 2\pi f_c t - n_s(t) \sin 2\pi f_c t \qquad (6.2.14)$$

The in-phase and quadrature components of the noise are assumed to be statistically independent, stationary Gaussian noise processes with (two-sided) power-spectral density $N_0/2$ W/Hz. By using simple trigonometric identities, the noise term in (6.2.14) can be expressed as

$$n(t) = x_c(t) \cos(2\pi f_c t + \phi(t)) - x_s(t) \sin(2\pi f_c t + \phi(t)) \qquad (6.2.15)$$

where

$$x_c(t) = n_c(t) \cos \phi(t) + n_s(t) \sin \phi(t)$$
$$x_s(t) = -n_c(t) \sin \phi(t) + n_s(t) \cos \phi(t) \qquad (6.2.16)$$

We note that

$$x_c(t) + j x_s(t) = [n_c(t) + j n_s(t)] e^{-j\phi(t)} \qquad (6.2.17)$$

It is easy to verify that a phase shift does not change the first two moments of $n_c(t)$ and $n_s(t)$, so that the quadrature components $x_c(t)$ and $x_s(t)$ have exactly the same statistical characteristics as $n_c(t)$ and $n_s(t)$ (see also Problem 3.29).

Now, if $s(t) + n(t)$ is multiplied by the output of the VCO and the double-frequency terms are neglected, the input to the loop filter is the noise-corrupted signal

$$e(t) = A_c \sin \Delta\phi + x_c(t) \sin \Delta\phi - x_s(t) \cos \Delta\phi \qquad (6.2.18)$$

where, by definition, $\Delta\phi = \hat{\phi} - \phi$ is the phase error. Thus, we have the equivalent model for the PLL with additive noise as shown in Figure 6.6.

When the power $P_c = A_c^2/2$ of the incoming signal is much larger than the noise power, the phase estimate $\hat{\phi} \approx \phi$. Then we may linearize the PLL and thus easily determine the effect of the additive noise on the quality of the estimate $\hat{\phi}$. Under these conditions, the model for the linearized PLL with additive noise is

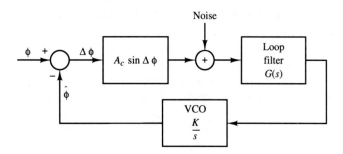

FIGURE 6.6.   Equivalent model of PLL with additive noise.

illustrated in Figure 6.7. Note that the gain parameter A_c may be normalized to unity, provided that the noise term is scaled by $1/A_c$. Thus the noise term becomes

$$n_1(t) = \frac{x_c(t)}{A_c} \sin \Delta\phi - \frac{x_s(t)}{A_c} \cos \Delta\phi \qquad (6.2.19)$$

Because the noise $n_1(t)$ is additive at the input to the loop, the variance of the phase error $\Delta\phi$, which is also the variance of the VCO output phase, is

$$\sigma_{\hat\phi}^2 = \frac{N_0 B_{neq}}{A_c^2} \qquad (6.2.20)$$

where B_{neq} is the (one-sided) noise equivalent bandwidth of the loop, given by (6.2.12). Note that $A_c^2/2$ is the power of the input sinusoid, and $\sigma_{\hat\phi}^2$ is simply proportional to the total noise power within the bandwidth of the PLL divided by the input signal power. Hence,

$$\sigma_{\hat\phi}^2 = \frac{1}{\rho_L} \qquad (6.2.21)$$

where ρ_L is defined as the SNR

$$\rho_L = \frac{A_c^2/2}{B_{neq} N_0/2} \qquad (6.2.22)$$

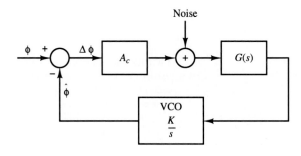

FIGURE 6.7.   Linearized model of PLL with additive noise.

Thus, the variance of $\hat{\phi}$ is inversely proportional to the SNR.

The expression for the variance $\sigma_{\hat{\phi}}^2$ of the VCO-phase error applies to the case where the SNR is sufficiently high so that the linear model for the PLL applies. An exact analysis based on the nonlinear PLL is mathematically tractable when $G(s) = 1$, which results in a first-order loop. In this case, the p.d.f. for the phase error can be derived (see Viterbi, 1966), and has the form

$$f(\Delta\phi) = \frac{\exp(\rho_L \cos \Delta\phi)}{2\pi I_0(\rho_L)} \tag{6.2.23}$$

where ρ_L is the SNR defined in (6.2.22), B_{neq} is the appropriate noise-equivalent bandwidth of the first-order loop, and $I_0(\cdot)$ is the modified Bessel function of order zero.

From the expression for $f(\Delta\phi)$, we may obtain the exact value of the variance $\sigma_{\hat{\phi}}^2$ for the phase error of a first-order PLL. This is plotted in Figure 6.8 as a function of $1/\rho_L$. Also shown for comparison is the result obtained with the linearized PLL model. Note that the variance for the linear model is close to the exact variance for $\rho_L > 3$. Hence, the linear model is adequate for practical purposes.

Approximate analysis of the statistical characteristics of the phase error for the nonlinear PLL have been also performed. Of particular importance is the transient behavior of the nonlinear PLL during initial acquisition. Another important problem

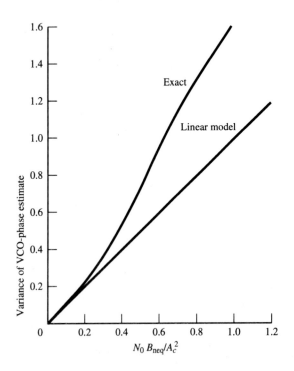

FIGURE 6.8. Comparison of VCO-phase variance for exact and approximate (linear order) first-order PLL. (From *Principles of Coherent Communication*, by A.J. Viterbi; © 1966 by McGraw-Hill Book Co. Reprinted with permission of the publisher).

is the behavior of the PLL at low SNR. It is known, for example, that when the SNR at the input to the PLL drops below a certain value, there is a rapid deterioration in the performance of the PLL. The loop begins to lose lock and an impulsive-type of noise, characterized as clicks, is generated, which degrades the performance of the loop. Results on these topics can be found in the texts by Viterbi (1966), Lindsey (1972), Lindsey and Simon (1973), and Gardner (1979), and in the survey papers by Gupta (1975) and Lindsey and Chie (1981).

Now that we have established the effect of noise on the performance of the PLL, let us return to the problem of carrier synchronization based on the system shown in Figure 6.9. The squaring of the received signal that produces the frequency component at $2f_c$ also enhances the noise power level at the input to the PLL, and thus it increases the variance of the phase error.

To elaborate on this point, let the input to the squarer be $u(t) + n(t)$. The output is

$$y(t) = u^2(t) + 2u(t)n(t) + n^2(t) \qquad (6.2.24)$$

The noise terms are $2u(t)n(t)$ and $n^2(t)$. By computing the autocorrelation and power-spectral density of these two noise components, one can show that both components have spectral power in the frequency band centered at $2f_c$. Consequently, the bandpass filter with bandwidth B_{neq} centered at $2f_c$, which produces the desired sinusoidal signal component that drives the PLL, also passes noise due to these two noise terms.

Let us select the bandwidth of the loop to be significantly smaller than the bandwidth B_{bp} of the bandpass filter, so that the total noise spectrum at the input to the PLL may be approximated by a constant within the loop bandwidth. This

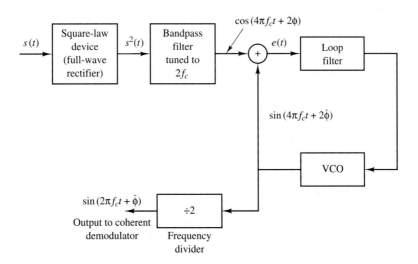

FIGURE 6.9. Carrier recovery using a square law device.

approximation allows us to obtain a simple expression for the variance of the phase error as

$$\sigma_{\hat{\phi}}^2 = \frac{1}{\rho_L S_L} \tag{6.2.25}$$

where S_L is called the squaring loss and is given as

$$S_L = \frac{1}{1 + \frac{B_{bp}/2B_{neq}}{\rho_L}} \tag{6.2.26}$$

Because $S_L < 1$, we have an increase in the variance of the phase error caused by the added noise power that results from the squaring operation. For example, when $\rho_L = B_{bp}/2B_{neq}$, the loss is 3 dB or, equivalently, the variance in the estimate increases by a factor of 2.

Finally, we observe that the output of the VCO from the squaring loop must be frequency divided by a factor of 2 and phase shifted by 90° to generate the carrier signal for demodulating the received signal.

Costas loop.   A second method for generating a properly phased carrier for a DSB-SC AM signal is illustrated by the block diagram in Figure 6.10. The received signal

$$r(t) = A_c m(t) \cos(2\pi f_c t - \phi) + n(t)$$

is multiplied by $\cos(2\pi f_c t - \hat{\phi})$ and $\sin(2\pi f_c t - \hat{\phi})$, which are outputs from the

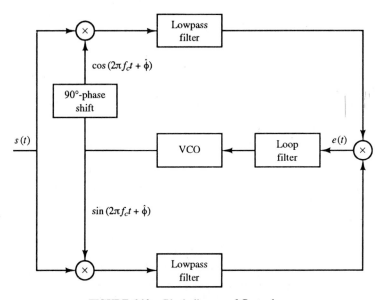

FIGURE 6.10.   Block diagram of Costas loop.

VCO. The two products are

$$y_c(t) = [A_c m(t) \cos(2\pi f_c t - \phi)$$
$$+ n_c(t) \cos 2\pi f_c t - n_s(t) \sin 2\pi f_c t] \cos(2\pi f_c t - \hat{\phi})$$
$$= \frac{A_c}{2} m(t) \cos \Delta\phi + \frac{1}{2}[n_c(t) \cos \hat{\phi} - n_s(t) \sin \hat{\phi}]$$
$$+ \text{double-frequency terms} \qquad (6.2.27)$$

$$y_s(t) = [A_c m(t) \cos(2\pi f_c t - \phi)$$
$$+ n_c(t) \cos 2\pi f_c t - n_s(t) \sin 2\pi f_c t] \sin(2\pi f_c t - \hat{\phi})$$
$$= \frac{A_c}{2} m(t) \sin \Delta\phi + \frac{1}{2}[n_c(t) \sin \hat{\phi} + n_s(t) \cos \hat{\phi}]$$
$$+ \text{double-frequency terms} \qquad (6.2.28)$$

where $\Delta\phi = \hat{\phi} - \phi$. The double-frequency terms are eliminated by the lowpass filters following the multiplications.

An error signal is generated by multiplying the two outputs $y_c'(t)$ and $y_s'(t)$ of the lowpass filters; thus

$$e(t) = y_c'(t) y_s'(t)$$
$$= \frac{A_c^2}{4} m^2(t) \sin 2\Delta\phi$$
$$+ \frac{A_c}{4} m(t)[n_c(t) \cos \hat{\phi} - n_s(t) \sin \hat{\phi}] \sin \Delta\phi$$
$$+ \frac{A_c}{4} m(t)[n_c(t) \sin \hat{\phi} + n_s(t) \cos \hat{\phi}] \cos \Delta\phi$$
$$+ \frac{1}{4}[n_c(t) \cos \hat{\phi} - n_s(t) \sin \hat{\phi}][n_c(t) \sin \hat{\phi} + n_s(t) \cos \hat{\phi}]$$

This error signal is filtered by the loop filter whose output is the control voltage that drives the VCO.

We note that the error signal into the loop filter consists of the desired term $(A_c^2 m^2(t)/4) \sin 2\Delta\phi$, plus terms that involve signal × noise and noise × noise. These terms are similar to the two noise terms at the input of the PLL for the squaring method. In fact, if the loop filter in the Costas loop is identical to that used in the squaring loop, the two loops are equivalent. Under this condition the p.d.f. of the phase error, and the performance of the two loops are identical.

In conclusion, the squaring PLL and the Costas PLL are two practical methods for deriving a carrier-phase estimate for synchronous demodulation of a DSB-SC AM signal.

6.3 EFFECT OF NOISE ON ANGLE MODULATION

In this section we will study the performance of angle-modulated signals when contaminated by additive white Gaussian noise and compare this performance with the performance of amplitude-modulated signals. Recall that in amplitude modulation, the message information is contained in the amplitude of the modulated signal, and because noise is additive the noise is directly added to the signal. However, in a frequency-modulated signal, the noise is added to the amplitude and the message information is contained in the frequency of the modulated signal. Therefore, the message is contaminated by the noise to the extent that the added noise changes the frequency of the modulated signal. The frequency of a signal can be described by its zero crossings. Therefore, the effect of additive noise on the demodulated FM signal can be described by the changes that it produces in the zero crossings of the modulated FM signal. Figure 6.11 shows the effect of additive noise on the zero crossings of two frequency-modulated signals, one with high power and the other with low power. From the above discussion and also from Figure 6.11, it should be clear that the effect of noise in an FM system is less than that for an AM system. It is observed also that the effect of noise in a low-power FM system is more than in a high-power FM system. The analysis that we present in this chapter verifies our intuition based on these observations.

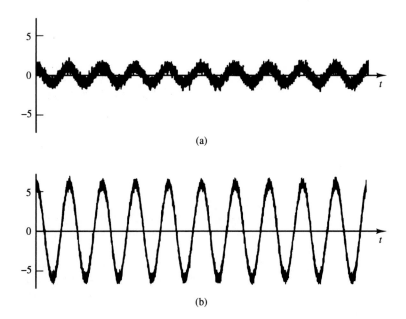

FIGURE 6.11. Effect of noise on the zero crossings of (a) low-power and (b) high-power modulated signals.

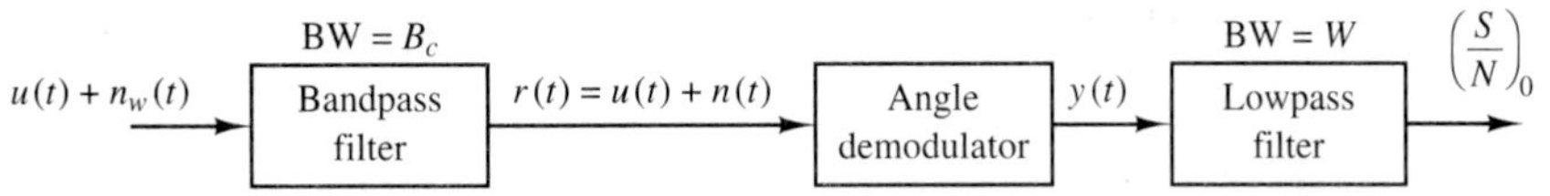

FIGURE 6.12. Block diagram of the receiver for a general angle-modulated signal.

The block diagram of the receiver for a general angle-modulated signal is shown in Figure 6.12. The angle-modulated signal is represented as[†]

$$u(t) = A_c \cos\left(2\pi f_c t + \phi(t)\right)$$

$$= \begin{cases} A_c \cos\left(2\pi f_c t + 2\pi k_f \int_{-\infty}^{t} m(\tau)\, d\tau\right) & \text{FM} \\[4mm] A_c \cos\left(2\pi f_c t + k_p m(t)\right) & \text{PM} \end{cases} \quad (6.3.1)$$

The additive white Gaussian noise $n_w(t)$ is added to $u(t)$ and the result is passed through a noise-limiting filter whose role is to remove the out-of-band noise. The bandwidth of this filter is equal to the bandwidth of the modulated signal and, therefore, it passes the modulated signal without distortion. However, it eliminates the out-of-band noise and, hence, the noise output of the filter is a bandpass Gaussian noise denoted by $n(t)$. The output of this filter is

$$r(t) = u(t) + n(t)$$

$$= u(t) + n_c(t)\cos 2\pi f_c t - n_s(t)\sin 2\pi f_c t \quad (6.3.2)$$

As with conventional AM noise-performance analysis, a precise analysis is quite involved due to the nonlinearity of the demodulation process. Let us make the assumption that the signal power is much higher than the noise power. Then, if the bandpass noise is represented as

$$n(t) = \sqrt{n_c^2(t) + n_s^2(t)}\, \cos\left(2\pi f_c t + \arctan \frac{n_s(t)}{n_c(t)}\right)$$

$$= V_n(t)\cos\left(2\pi f_c t + \Phi_n(t)\right), \quad (6.3.3)$$

where $V_n(t)$ and $\Phi_n(t)$ represent the envelope and the phase of the bandpass noise process, respectively, the assumption that the signal is much larger than the noise means that

$$p(V_n(t) \ll A_c) \approx 1 \quad (6.3.4)$$

Therefore, the phasor diagram of the signal and the noise are as shown in

---

[†]When we refer to the modulated signal, we mean the signal as received by the receiver. Therefore, the signal power is the power in the received signal, not the transmitted power.

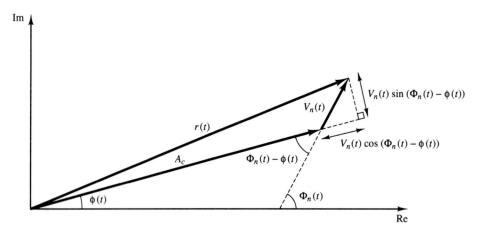

FIGURE 6.13. Phasor diagram of signal and noise in an angle-modulated system.

Figure 6.13. From this figure it is obvious that we can write

$$r(t) \approx \left(A_c + V_n(t) \cos \left(\Phi_n(t) - \phi(t) \right) \right)$$

$$\times \cos \left(2\pi f_c t + \phi(t) \right.$$

$$\left. + \arctan \frac{V_n(t) \sin \left(\Phi_n(t) - \phi(t) \right)}{A_c + V_n(t) \cos \left(\Phi_n(t) - \phi(t) \right)} \right)$$

$$\approx \left(A_c + V_n(t) \cos \left(\Phi_n(t) - \phi(t) \right) \right)$$

$$\times \cos \left(2\pi f_c t + \phi(t) + \frac{V_n(t)}{A_c} \sin \left(\Phi_n(t) - \phi(t) \right) \right)$$

The demodulator processes this signal and, depending whether it is a phase or a frequency demodulator, its output will be the phase or the instantaneous frequency of this signal.[†] Therefore, noting that

$$\phi(t) = \begin{cases} k_p m(t) & \text{PM} \\ 2\pi k_f \displaystyle\int_{-\infty}^{t} m(\tau)\, d\tau & \text{FM} \end{cases} \tag{6.3.5}$$

[†]Of course, in the FM case the demodulator output is the instantaneous frequency deviation of $v(t)$ from the carrier frequency f_c.

the output of the demodulator is given by

$$
y(t) = \begin{cases} k_p m(t) + Y_n(t) & \text{PM} \\ k_f m(t) + \dfrac{1}{2\pi}\dfrac{d}{dt} Y_n(t) & \text{FM} \end{cases}
$$

$$
= \begin{cases} k_p m(t) + \dfrac{V_n(t)}{A_c} \sin\left(\Phi_n(t) - \phi(t)\right) & \text{PM} \\ k_f m(t) + \dfrac{1}{2\pi}\dfrac{d}{dt}\dfrac{V_n(t)}{A_c} \sin\left(\Phi_n(t) - \phi(t)\right) & \text{FM} \end{cases} \tag{6.3.6}
$$

where we have defined

$$
Y_n(t) = \frac{V_n(t)}{A_c} \sin\left(\Phi_n(t) - \phi(t)\right). \tag{6.3.7}
$$

The first term in the above expressions is the desired signal component and the second term is the noise component. From this expression we observe that the noise component is inversely proportional to the signal amplitude A_c. Hence, the higher the signal level, the lower the noise level will be. This is in agreement with the intuitive reasoning presented at the beginning of this section based on Figure 6.11. Note also that this is not the case with amplitude modulation. In AM systems the noise component is independent of the signal component, and a scaling of the signal power does not affect the received noise power.

Let us study the properties of the noise component given by

$$
Y_n(t) = \frac{V_n(t)}{A_c} \sin\left(\Phi_n(t) - \phi(t)\right)
$$

$$
= \frac{1}{A_c}\Big[V_n(t) \sin \Phi_n(t) \cos\phi(t)
$$

$$
- V_n(t) \cos \Phi_n(t) \sin\phi(t) \Big]
$$

$$
= \frac{1}{A_c}\big[n_s(t) \cos\phi(t) - n_c(t) \sin\phi(t) \big] \tag{6.3.8}
$$

The autocorrelation function of this process is given by

$$
E[Y_n(t+\tau)Y_n(t)] = \frac{1}{A_c^2} E\Big[R_{n_s}(\tau)\cos\phi(t)\cos\phi(t+\tau)
$$

$$
+ R_{n_c}(\tau)\sin\phi(t+\tau)\sin\phi(t) \Big]
$$

$$
= \frac{1}{A_c^2} R_{n_c}(\tau) E\big[\cos[\phi(t+\tau) - \phi(t)] \big] \tag{6.3.9}
$$

where we have used the fact that the noise process is stationary and $R_{n_c}(\tau) = R_{n_s}(\tau)$ and $R_{n_c n_s}(\tau) = 0$ (see (3.6.5) and Example 3.6.1). Now we assume that the message $m(t)$ is a sample function of a zero-mean, stationary Gaussian process $M(t)$ with autocorrelation function $R_M(\tau)$. Then, in both PM and FM modulation, $\phi(t)$ will also be a sample function of a zero-mean stationary, Gaussian process $\Phi(t)$. For PM this is obvious because

$$\Phi(t) = k_p M(t) \tag{6.3.10}$$

and in the FM case we have

$$\Phi(t) = 2\pi k_f \int_{-\infty}^{t} M(\tau) \, d\tau \tag{6.3.11}$$

Noting that $\int_{-\infty}^{t}$ represents a linear time-invariant operation, it is seen that in this case $\Phi(t)$ is the output of an LTI system whose input is a zero mean, stationary Gaussian process. Consequently, $\Phi(t)$ will also be a zero-mean, stationary Gaussian process.

The analysis from this point is almost parallel to our analysis of the spectral density of an angle-modulated Gaussian process presented in Section 5.3.2. At any fixed time t, the random variable $Z(t, \tau) = \Phi(t + \tau) - \Phi(t)$ is the difference between two jointly Gaussian random variables. Therefore, it is itself a Gaussian random variable with mean equal to zero and variance

$$\sigma_Z^2 = E[\Phi^2(t + \tau)] + E[\Phi^2(t)] - 2R_\Phi(\tau)$$
$$= 2[R_\Phi(0) - R_\Phi(\tau)] \tag{6.3.12}$$

Now, using this result in (6.3.9) we obtain

$$E[Y_n(t + \tau)Y_n(t)] = \frac{1}{A_c^2} R_{n_c}(\tau) E\left[\cos[\Phi(t + \tau) - \Phi(t)]\right]$$

$$= \frac{1}{A_c^2} R_{n_c}(\tau) \operatorname{Re}[E e^{j(\Phi(t+\tau) - \Phi(t))}]$$

$$= \frac{1}{A_c^2} R_{n_c}(\tau) \operatorname{Re}[E e^{jZ(t,\tau)}]$$

$$= \frac{1}{A_c^2} R_{n_c}(\tau)[e^{-\frac{1}{2}\sigma_Z^2}]$$

$$= \frac{1}{A_c^2} R_{n_c}(\tau)[e^{-(R_\Phi(0) - R_\Phi(\tau))}]$$

$$= \frac{1}{A_c^2} R_{n_c}(\tau) e^{-(R_\Phi(0) - R_\Phi(\tau))} \tag{6.3.13}$$

This result shows that, under the assumption of a stationary Gaussian message, the noise process at the output of the demodulator is also a stationary process whose autocorrelation function is given above and whose power-spectral density is

$$S_Y(f) = \mathcal{F}[R_Y(\tau)]$$

$$= \mathcal{F}\left[\frac{1}{A_c^2} R_{n_c}(\tau) e^{-(R_\Phi(0) - R_\Phi(\tau))}\right]$$

$$= \frac{e^{-R_\Phi(0)}}{A_c^2} \mathcal{F}\left[R_{n_c}(\tau) e^{R_\Phi(\tau)}\right]$$

$$= \frac{e^{-R_\Phi(0)}}{A_c^2} \mathcal{F}\left[R_{n_c}(\tau) g(\tau)\right]$$

$$= \frac{e^{-R_\Phi(0)}}{A_c^2} S_{n_c}(f) \star G(f) \tag{6.3.14}$$

where $g(\tau) = e^{R_\Phi(\tau)}$ and $G(f)$ is its Fourier transform.

The bandwidth of $g(\tau)$ is the bandwidth B_c of the angle-modulated signal, as indicated in (5.3.56), which for high modulation indices is much larger than W, the message bandwidth (see 5.3.59). Because the bandwidth of the angle-modulated signal is defined as the frequencies that contain 98–99% of the signal power, $G(f)$ is very small in the neighborhood of $|f| = B_c/2$ and, of course,

$$S_{n_c}(f) = \begin{cases} N_0, & |f| < \frac{B_c}{2} \\ 0, & \text{otherwise} \end{cases} \tag{6.3.15}$$

A typical example of $G(f)$, $S_{n_c}(f)$ and the result of their convolution is shown in Figure 6.14. A little thought shows that, because $G(f)$ is very small in the neighborhood of $|f| = \frac{B_c}{2}$, the resulting $S_Y(f)$ has almost a flat spectrum for $|f| < W$, the bandwidth of the message. From Figure 6.14 it is obvious that for all $|f| < W$, we have

$$S_Y(f) = \frac{e^{-R_\Phi(0)}}{A_c^2} S_{n_c}(f) \star G(f)$$

$$= \frac{e^{-R_\Phi(0)}}{A_c^2} N_0 \int_{-\frac{B_c}{2}}^{\frac{B_c}{2}} G(f)\, df$$

$$\approx \frac{e^{-R_\Phi(0)}}{A_c^2} N_0 \int_{-\infty}^{+\infty} G(f)\, df$$

$$= \frac{e^{-R_\Phi(0)}}{A_c^2} N_0 g(\tau)\Big|_{\tau=0}$$

$$= \frac{e^{-R_\Phi(0)}}{A_c^2} N_0 e^{R_\Phi(0)}$$

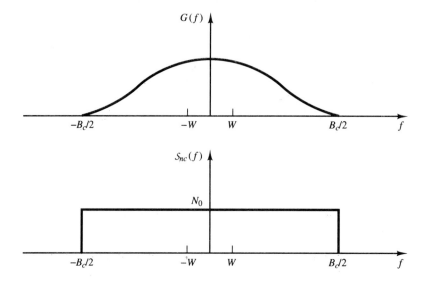

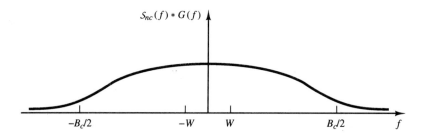

FIGURE 6.14. Typical plots of $G(f)$, $S_{nc}(f)$, and the result of their convolution.

$$= \frac{N_0}{A_c^2} \tag{6.3.16}$$

It should be noted that the above relation is a good approximation only for $|f| < W$. This means that for $|f| < W$, the spectrum of the noise components in the PM and FM case are given by

$$S_{n_o}(f) = \begin{cases} \dfrac{N_0}{A_c^2} & \text{PM} \\[2ex] \dfrac{N_0}{A_c^2} f^2 & \text{FM} \end{cases} \tag{6.3.17}$$

where we have used the fact that in FM the noise component is given by $\frac{1}{2\pi} \frac{d}{dt} Y_n(t)$ as previously indicated in (6.3.6). The power spectrum of the noise component at the output of the demodulator in the frequency interval $|f| < W$ for PM and FM is shown in Figure 6.15. It is interesting to note that PM has a flat noise spectrum and FM has a parabolic noise spectrum. Therefore, the effect of noise in FM for higher-frequency components is much higher than the effect of noise on lower-frequency components. The noise power at the output of the lowpass filter is the noise power in the frequency range $[-W, +W]$. Therefore, it is given by

$$P_{n_o} = \int_{-W}^{+W} S_{n_o}(f)\,df$$

$$= \begin{cases} \displaystyle\int_{-W}^{+W} \frac{N_0}{A_c^2}\,df & \text{PM} \\[4mm] \displaystyle\int_{-W}^{+W} f^2 \frac{N_0}{A_c^2}\,df & \text{FM} \end{cases}$$

$$= \begin{cases} \dfrac{2W N_0}{A_c^2} & \text{PM} \\[4mm] \dfrac{2 N_0 W^3}{3 A_c^2} & \text{FM} \end{cases} \qquad (6.3.18)$$

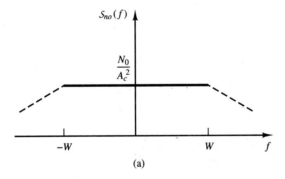

(a)

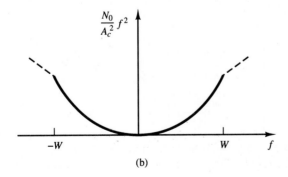

(b)

FIGURE 6.15. Noise-power spectrum at demodulator output in (a) PM and (b) FM.

Now we can use (6.3.6) to determine the output SNR in angle modulation. First, we have the output signal power

$$
P_{S_o} = \begin{cases} k_p^2 P_M & \text{PM} \\ k_f^2 P_M & \text{FM} \end{cases} \tag{6.3.19}
$$

Then, the SNR, defined as

$$
\left(\frac{S}{N}\right)_o = \frac{P_{S_o}}{P_{n_o}}
$$

becomes

$$
\left(\frac{S}{N}\right)_o = \begin{cases} \dfrac{k_p^2 A_c^2}{2} \dfrac{P_M}{N_0 W} & \text{PM} \\[3mm] \dfrac{3k_f^2 A_c^2}{2W^2} \dfrac{P_M}{N_0 W} & \text{FM} \end{cases} \tag{6.3.20}
$$

Noting that $\frac{A_c^2}{2}$ is the received signal power, denoted by P_R, and

$$
\begin{cases} \beta_p = k_p \max |m(t)| & \text{PM} \\[3mm] \beta_f = \dfrac{k_f \max |m(t|}{W} & \text{FM} \end{cases} \tag{6.3.21}
$$

we may express the output SNR as

$$
\left(\frac{S}{N}\right)_o = \begin{cases} P_R \left(\dfrac{\beta_p}{\max |m(t)|}\right)^2 \dfrac{P_M}{N_0 W} & \text{PM} \\[3mm] 3P_R \left(\dfrac{\beta_f}{\max |m(t)|}\right)^2 \dfrac{P_M}{N_0 W} & \text{FM} \end{cases} \tag{6.3.22}
$$

If we denote $\frac{P_R}{N_0 W}$ by $\left(\frac{S}{N}\right)_b$, the SNR of a baseband system with the same received power, we obtain

$$
\left(\frac{S}{N}\right)_o = \begin{cases} \dfrac{P_M \beta_p^2}{(\max |m(t)|)^2} \left(\dfrac{S}{N}\right)_b & \text{PM} \\[3mm] 3\dfrac{P_M \beta_f^2}{(\max |m(t|)^2} \left(\dfrac{S}{N}\right)_b & \text{FM} \end{cases} \tag{6.3.23}
$$

Note that in the above expression $\frac{P_M}{(\max |m(t|)^2}$ is the average-to-peak-power ratio of the message signal (or, equivalently, the power content of the normalized message,

P_{M_n}). Therefore,

$$
\left(\frac{S}{N}\right) = \begin{cases} \beta_p^2 P_{M_n} \left(\dfrac{S}{N}\right)_b, & \text{PM} \\[4mm] 3\beta_f^2 P_{M_n} \left(\dfrac{S}{N}\right)_b, & \text{FM} \end{cases}
\tag{6.3.24}
$$

Now using Carson's rule, $B_c = 2(\beta + 1)W$, we can express the output SNR in terms of the bandwidth expansion factor, which is defined as the ratio of the channel bandwidth to the message bandwidth and denoted by Ω

$$
\Omega = \frac{B_c}{W} = 2(\beta + 1)
\tag{6.3.25}
$$

From this relationship, we have $\beta = \frac{\Omega}{2} - 1$; therefore,

$$
\left(\frac{S}{N}\right)_o = \begin{cases} P_M \left(\dfrac{\frac{\Omega}{2} - 1}{\max |m(t)|} \right)^2 \left(\dfrac{S}{N}\right)_b, & \text{PM} \\[7mm] 3 P_M \left(\dfrac{\frac{\Omega}{2} - 1}{\max |m(t)|} \right)^2 \left(\dfrac{S}{N}\right)_b, & \text{FM} \end{cases}
\tag{6.3.26}
$$

From (6.3.22) and (6.3.26), we observe the following points.

1. In both PM and FM the output SNR is proportional to the square of the modulation index β. Therefore, increasing β increases the output SNR even with low received power. This is in contrast to amplitude modulation where such an increase in the received SNR is not possible.

2. The increase in the received SNR is obtained by increasing the bandwidth. Therefore, angle modulation provides a way to trade-off bandwidth for transmitted power.

3. The relation between the output SNR and the bandwidth expansion factor, Ω, is a quadratic relation. This is far from optimal.[†] We will see in Chapter 10 that the optimal relation between the output SNR and the bandwidth expansion factor is an exponential relation.

4. Although we can increase the output SNR by increasing β, having a large β means having a large B_c (by Carson's rule). Having a large B_c means having a large noise power at the input of the demodulator. This means that the approximation $p(V_n(t) \ll A_c) \approx 1$ will no longer apply and that the

---

[†] By optimal relation we mean the maximum saving in transmitter power for a given expansion in bandwidth. An optimal system achieves the fundamental limits on communication predicted by information theory

above analysis will not hold. In fact, if we increase β such that the above approximation does not hold, a phenomenon known as the *threshold effect* will occur, and the signal will be lost in the noise.

5. A comparison of the above result with the SNR in amplitude modulation shows that in both cases increasing the transmitter power (or the received power) will increase the output SNR but the mechanisms are totally different. In AM, any increase in the received power directly increases the signal power at the output of the receiver. This is due basically to the fact the message is in the amplitude of the transmitted signal, and an increase in the transmitted power directly affects the demodulated signal power. However, in angle modulation the message is in the phase of the modulated signal and, consequently, increasing the transmitter power does not increase the demodulated message power. In angle modulation what increases the output SNR is a *decrease in the received noise power* as seen from (6.3.18) and Figure 6.11.

6. In FM, the effect of noise is higher at higher frequencies. This means that signal components at higher frequencies will suffer more from noise than the lower-frequency components. In some applications, where FM is used to transmit SSB-FDM signals, those channels that are modulated on higher-frequency carriers suffer from more noise. To compensate for this effect such channels must have a higher-signal level. The quadratic characteristics of the demodulated noise spectrum in FM is the basis of pre-emphasis and de-emphasis filtering that will be discussed later in this chapter.

6.3.1 Threshold Effect in Angle Modulation

The noise analysis of angle-demodulation schemes is based on the assumption that the SNR at the demodulator input is high. With this crucial assumption we observed that the signal and noise components at the demodulator output are additive, and we were able to carry out the analysis. This assumption of high SNR is a simplifying assumption that is usually made in analysis of nonlinear-modulation systems. Due to the nonlinear nature of the demodulation process, there is no requirement that the additive signal and noise components at the input of the modulator result in additive signal and noise components at the output of the demodulator. In fact, this assumption is not at all correct in general, and the signal and noise processes at the output of the demodulator are completely mixed in a single process by a complicated nonlinear functional. Only under the high SNR assumption is this highly nonlinear functional approximated as an additive form. Particularly at low SNR's, signal and noise components are so intermingled that one cannot recognize the signal from the noise and, therefore, no meaningful SNR as a measure of performance can be defined. In such cases, the signal is not distinguishable from the noise and a *mutilation* or *threshold effect* is present. There exists a specific SNR at the input of the demodulator known as the *threshold SNR,* beyond which signal mutilation occurs. The existence of the threshold effect places an upper limit on the trade-off between bandwidth and power in an FM system. This limit is a practical limit in the value of the modulation index β_f. The analysis of the threshold effect and derivation of the threshold index β_f is quite involved and beyond the scope of our analysis. The interested reader is referred to the references cited at the end of this chapter for an analytic treatment of the subject. Here we only mention some results on the threshold effect in FM.

It can be shown that at threshold the following approximate relation between $\frac{P_R}{N_0 W} = \left(\frac{S}{N}\right)_b$ and β_f holds in an FM system:

$$\left(\frac{S}{N}\right)_{b,\text{th}} = 20(\beta + 1) \tag{6.3.27}$$

From the above relation, given a received power P_R, we can calculate the maximum allowed β to make sure that the system works above threshold. Also, given a bandwidth allocation of B_c, we can find an appropriate β using Carson's rule, $B_c = 2(\beta + 1)W$. Then, using the threshold relation given above, we determine the required minimum received power to make the whole allocated bandwidth usable.

In general, there are two factors that limit the value of the modulation index β. The first is the limitation on channel bandwidth that affects β through Carson's rule. The second is the limitation on the received power that limits the value of β to less than what is derived from (6.3.27). Figure 6.16 shows plots of the SNR in an FM system as a function of the baseband SNR. The SNR values in these curves are in decibels, and different curves correspond to different values of β as marked. The effect of threshold is apparent from the sudden drops in the output SNR. These plots are drawn for a sinusoidal message for which

$$\frac{P_M}{(\max |m(t)|)^2} = \frac{1}{2} \tag{6.3.28}$$

In such a case,

$$\left(\frac{S}{N}\right)_o = \frac{3}{2}\beta^2 \left(\frac{S}{N}\right)_b \tag{6.3.29}$$

As an example, for $\beta = 5$, the above relation yields

$$\left.\left(\frac{S}{N}\right)_o\right|_{\text{dB}} = 15.7 + \left.\left(\frac{S}{N}\right)_b\right|_{\text{dB}} \tag{6.3.30}$$

$$\left(\frac{S}{N}\right)_{b,\text{th}} = 120 \sim 20.8 \text{ dB} \tag{6.3.31}$$

For $\beta = 2$, we have

$$\left.\left(\frac{S}{N}\right)_o\right|_{\text{dB}} = 7.8 + \left.\left(\frac{S}{N}\right)_b\right|_{\text{dB}} \tag{6.3.32}$$

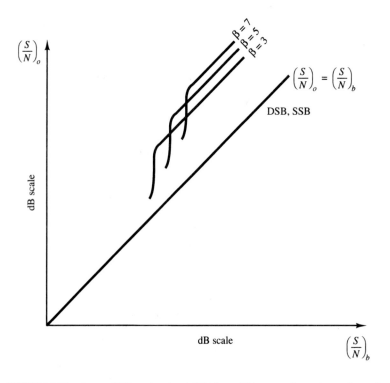

FIGURE 6.16. Output SNR vs. baseband SNR in an FM system for various values of β.

$$\left(\frac{S}{N}\right)_{b,\text{th}} = 60 \sim 17.8 \text{ dB} \tag{6.3.33}$$

From this discussion it is apparent that, if for example $\left(\frac{S}{N}\right)_b = 20$ dB, then, regardless of the available bandwidth, we cannot use $\beta = 5$ for such a system because the system will be operating below threshold. For this case, $\beta = 2$ can be used, which yields an SNR equal to 27.8 dB at the output of the receiver. This is an improvement of 7.8 dB compared to a baseband system.

In general, if we want to employ the maximum available bandwidth, we must choose the largest possible β that guarantees the system operates above threshold. This is the value of β that satisfies

$$\left(\frac{S}{N}\right)_{b,\text{th}} = 20(\beta + 1) \tag{6.3.34}$$

By substituting this value in (6.3.24), we obtain

$$\left(\frac{S}{N}\right)_o = 60\beta^2(\beta+1)P_{M_n} \tag{6.3.35}$$

which relates a desired output SNR to the highest possible β that achieves that SNR.

Example 6.3.1

Design an FM system that achieves an SNR at the receiver equal to 40 dB and requires the minimum amount of transmitter power. The bandwidth of the channel is 120 kHz, the message bandwidth is 10 kHz, the average-to-peak-power ratio for the message, $P_{M_n} = \frac{P_M}{(\max |m(t)|)^2}$ is $\frac{1}{2}$, and the (one-sided) noise power-spectral density is $N_0 = 10^{-8}$ W/Hz. What is the required transmitter power if the signal is attenuated by 40 dB in transmission through the channel?

Solution   First we have to see whether the threshold or the bandwidth impose a more restrictive bound on the modulation index. By Carson's rule,

$$B_c = 2(\beta+1)W$$

$$120,000 = 2(\beta+1) \times 10,000$$

from which we obtain $\beta = 5$. Using the relation

$$\left(\frac{S}{N}\right)_o = 60\beta^2(\beta+1)P_{M_n} \tag{6.3.36}$$

with $\left(\frac{S}{N}\right)_o = 10^4$, we obtain $\beta \approx 6.6$. Because the value of β given by the bandwidth constraint is less than the value of β given by the power constraint, we are limited in bandwidth as opposed to being limited in power. Therefore, we choose $\beta = 5$, which when substituted in the expansion for the output SNR

$$\left(\frac{S}{N}\right)_o = \frac{3}{2}\beta^2\left(\frac{S}{N}\right)_b \tag{6.3.37}$$

yields

$$\left(\frac{S}{N}\right)_b = \frac{800}{3} = 266.6 \sim 24.26 \text{ dB} \tag{6.3.38}$$

Since $\left(\frac{S}{N}\right)_b = \frac{P_R}{N_0 W}$ with $W = 10,000$ and $N_0 = 10^{-8}$, we obtain

$$P_R = \frac{8}{300} = 0.0266 \sim -15.74 \text{ dB} \tag{6.3.39}$$

and

$$P_T = -15.74 + 40 = 24.26 \text{ dB} \sim 266.66 \text{ W} \tag{6.3.40}$$

Had there been no bandwidth constraint, we could have chosen $\beta = 6.6$, which results in $\left(\frac{S}{N}\right)_b \approx 153$. In turn, we have $P_R \approx 0.0153$ and $P_T \approx 153$ W.

6.3.2 Pre-emphasis and De-emphasis Filtering

As observed in Figure 6.15, the noise power-spectral density at the output of the demodulator in PM is flat within the message bandwidth, whereas for FM the noise power spectrum has a parabolic shape. This means that for low-frequency components of the message signal FM performs better and for high-frequency components PM is a better choice. Therefore, if we can design a system that for low-frequency components of the message signal performs frequency modulation and for high-frequency components works as a phase modulator, we have a better overall performance compared to each system alone. This is the idea behind pre-emphasis and de-emphasis filtering techniques.

The objective in pre-emphasis and de-emphasis filtering is to design a system that behaves like an ordinary frequency modulator–demodulator pair in the low-frequency band of the message signal and like a phase modulator–demodulator pair in the high-frequency band of the message signal. A phase modulator is nothing but the cascade connection of a differentiator and a frequency modulator; therefore, we need a filter in cascade with the modulator that at low frequencies does not affect the signal and at high frequencies acts as a differentiator. A simple highpass filter is a very good approximation to such a system. Such a filter has a constant gain for low frequencies, and at higher frequencies it has a frequency characteristic approximated by $K|f|$, which is the frequency characteristic of a differentiator. At the demodulator side, for low frequencies we have a simple FM demodulator and for high-frequency components we have a phase demodulator, which is the cascade of a simple FM demodulator and an integrator. Therefore, at the demodulator we need a filter that at low frequencies has a constant gain and at high frequencies behaves as an integrator. A good approximation to such a filter is a simple lowpass filter. The modulator filter that emphasizes high frequencies is called the *pre-emphasis filter*, and the demodulator filter, which is the inverse of the modulator filter is called the *de-emphasis filter*. Frequency responses of a sample pre-emphasis and de-emphasis filters are given in Figure 6.17.

Another way to look at pre-emphasis and de-emphasis filtering is to note that, due to the high level of noise at high-frequency components of the message in FM, it is desirable to attenuate the high-frequency components of the demodulated signal. This results in a reduction in the noise level, but it causes the higher-frequency components of the message signal to be attenuated also. To compensate for the attenuation of the higher components of the message signal, we can amplify these components at the transmitter before modulation. Therefore, at the transmitter we need a highpass filter and at the receiver we must use a lowpass filter. The net effect of these filters should be a flat-frequency response. Therefore, the receiver filter should be the inverse of the transmitter filter.

The characteristics of the pre-emphasis and de-emphasis filters depend largely on the power-spectral density of the message process. In the commercial FM broadcasting of music and voice, first-order lowpass and highpass RC filters with a time constant of 75 μsec are employed. In this case, the frequency response of the

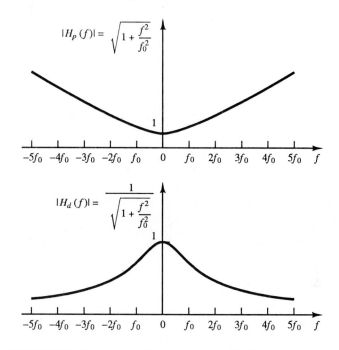

$$|H_p(f)| = \sqrt{1 + \frac{f^2}{f_0^2}}$$

$$|H_d(f)| = \frac{1}{\sqrt{1 + \frac{f^2}{f_0^2}}}$$

FIGURE 6.17. Pre-emphasis (a) and de-emphasis (b) filter characteristics.

receiver (de-emphasis) filter is given by

$$H_d(f) = \frac{1}{1 + j\frac{f}{f_0}} \tag{6.3.41}$$

where $f_0 = \frac{1}{2\pi \times 75 \times 10^{-6}} \approx 2100$ Hz is the 3-dB frequency of the filter.

To analyze the effect of pre-emphasis and de-emphasis filtering on the overall SNR in FM broadcasting, we note that, because the transmitter and the receiver filters cancel the effect of each other, the received power in the message signal remains unchanged and we only have to consider the effect of filtering on the received noise. Of course, the only filter that has an effect on the received noise is the receiver filter that shapes the power-spectral density of the noise within the message bandwidth. The noise component before filtering has a parabolic power spectrum. Therefore, the noise component after de-emphasis filtering has a power spectral density given by

$$S_{n_{PD}}(f) = S_{n_o}(f)|H_d(f)|^2$$

$$= \frac{N_0}{A_c^2} f^2 \frac{1}{1 + \frac{f^2}{f_0^2}} \tag{6.3.42}$$

where we have used (6.3.17). The noise power at the output of the demodulator now can be obtained as

$$
P_{n_{PD}} = \int_{-W}^{+W} S_{n_{PD}}(f)\,df
$$

$$
= \frac{N_0}{A_c^2} \int_{-W}^{+W} \frac{f^2}{1 + \frac{f^2}{f_0^2}}\,df
$$

$$
= \frac{2N_0 f_0^3}{A_c^2} \left[\frac{W}{f_0} - \arctan \frac{W}{f_0} \right] \tag{6.3.43}
$$

Because the demodulated message signal power in this case is equal to that of a simple FM system with no pre-emphasis and de-emphasis filtering, the ratio of the output SNR's in these two cases is inversely proportional to the noise-power ratios, i.e.,

$$
\frac{\left(\frac{S}{N}\right)_{o_{PD}}}{\left(\frac{S}{N}\right)_o} = \frac{P_{n_o}}{P_{n_{PD}}}
$$

$$
= \frac{\dfrac{2N_0 W^3}{3A_c^2}}{\dfrac{2N_0 f_0^3}{A_c^2} \left[\dfrac{W}{f_0} - \arctan \dfrac{W}{f_0} \right]}
$$

$$
= \frac{1}{3} \frac{\left(\dfrac{W}{f_0}\right)^3}{\dfrac{W}{f_0} - \arctan \dfrac{W}{f_0}} \tag{6.3.44}
$$

where we have used (6.3.18). The above equation gives the improvement obtained by employing pre-emphasis and de-emphasis filtering.

Example 6.3.2

In commercial FM broadcasting $W = 15$ kHz, $f_0 = 2100$ Hz, and $\beta = 5$. Assuming that the average-to-peak-power ratio of the message signal is 0.5, find the improvement in output SNR of FM with pre-emphasis and de-emphasis filtering compared to a baseband system.

Solution   From (6.3.24) we have

$$
\left(\frac{S}{N}\right)_o = 3 \times 5^2 \times 0.5 \times \left(\frac{S}{N}\right)_b
$$

$$
= 37.5 \left(\frac{S}{N}\right)_b
$$

$$
\sim 15.7 + \left(\frac{S}{N}\right)_b \Bigg|_{dB} \tag{6.3.45}
$$

Therefore, FM with no pre-emphasis and de-emphasis filtering performs 15.7 dB better than a baseband system. For FM with pre-emphasis and de-emphasis filtering, we have

$$
\begin{aligned}
\left(\frac{S}{N}\right)_{o_{PD}} &= \frac{1}{3} \frac{\left(\frac{W}{f_0}\right)^3}{\frac{W}{f_0} - \arctan \frac{W}{f_0}} \left(\frac{S}{N}\right)_o \\
&= \frac{1}{3} \frac{\left(\frac{15{,}000}{2100}\right)^3}{\frac{15{,}000}{2100} - \arctan \frac{15{,}000}{2100}} \left(\frac{S}{N}\right)_o \\
&= 21.3 \left(\frac{S}{N}\right)_o \\
&\sim 13.3 + \left.\left(\frac{S}{N}\right)_o\right|_{dB} \\
&\sim 13.3 + 15.7 + \left.\left(\frac{S}{N}\right)_b\right|_{dB} \\
&\sim 29 + \left.\left(\frac{S}{N}\right)_b\right|_{dB}
\end{aligned}
\qquad (6.3.46)
$$

The overall improvement compared to a baseband system is, therefore, 29 dB.

6.4 COMPARISON OF ANALOG-MODULATION SYSTEMS

Now we are at a point that we can present an overall comparison of different analog communication systems. The systems that we have studied include linear-modulation systems (DSB-SC, conventional AM, SSB-SC, VSB) and nonlinear systems (FM and PM).

The comparison of these systems can be done from various viewpoints. Here we present a comparison based on three important practical criteria:

1. The bandwidth efficiency of the system.
2. The power efficiency of the system as reflected in its performance in the presence of noise.
3. The ease of implementation of the system (transmitter and receiver).

Bandwidth efficiency.   The most bandwidth-efficient analog communication system is the SSB-SC system with a transmission bandwidth equal to the signal bandwidth. This system is used widely in bandwidth-critical applications such as voice transmission over microwave and satellite links, and some point-to-point communication systems in congested areas. Because SSB-SC cannot effectively transmit

dc, it cannot be used for transmission of signals that have a significant *dc* component, such as image signals. A good compromise is the VSB system, which has a
bandwidth slightly larger than SSB and is capable of transmitting *dc* values. VSB
is used widely in TV broadcasting, and also in some data communication systems.
PM and particularly FM are least-favorable systems when bandwidth is the major
concern, and their use is only justified by their high level of noise immunity.

Power efficiency.   A criterion for comparing power efficiency of various
systems is the comparison of their output SNR at a given received signal power.
We have seen already that angle-modulation schemes and particularly FM provide a
high level of noise immunity and therefore power efficiency. FM is used widely on
power-critical communication links such as point-to-point communication systems
and high-fidelity radio broadcasting. It also is used for transmission of voice (which
has been already SSB/FDM multiplexed) on microwave line-of-sight and satellite
links. Conventional AM and VSB+C are the least power-efficient systems and are
not used when the transmitter power is a major concern. However, their use is
justified by the simplicity of the receiver structure.

Ease of implementation.   The simplest receiver structure is the receiver for
conventional AM, and the structure of the receiver for a VSB+C system is only
slightly more complicated. FM receivers are also easy to implement. These three
systems are used widely for AM, TV, and high-fidelity FM broadcasting (including
FM stereo). The power inefficiency of the AM transmitter is compensated for by
the extremely simple structure of literally hundreds of millions of receivers. DSB-
SC and SSB-SC require synchronous demodulation and, therefore, their receiver
structure is much more complicated. These systems are, therefore, never used for
broadcasting purposes. Because the receiver structure of SSB-SC and DSB-SC have
almost the same complexity and the transmitter of SSB-SC is slightly more complicated compared to DSB-SC, DSB-SC is hardly used due to its relative bandwidth
inefficiency.

6.5 EFFECTS OF TRANSMISSION LOSSES AND NOISE IN ANALOG COMMUNICATION SYSTEMS

In any communication system there are usually two dominant factors that limit
the performance of the system. One important factor is additive noise generated
by electronic devices that are used to filter and amplify the communication signal.
A second factor that affects the performance of a communication system is signal
attenuation. Basically, all physical channels, including wire line and radio channels,
are lossy. Hence the signal is attenuated (reduced in amplitude) as it travels through
the channel. A simple mathematical model of the attenuation may be constructed,
as shown in Figure 6.18, by multiplying the transmitted signal by a factor $\alpha < 1$.

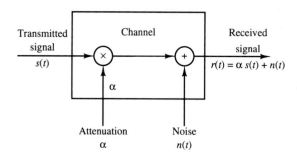

FIGURE 6.18.   Mathematical model of channel with attenuation and additive noise.

Consequently, if the transmitted signal is $s(t)$, the received signal is

$$r(t) = \alpha s(t) + n(t) \tag{6.5.1}$$

Clearly the effect of signal attenuation is to reduce the amplitude of the desired signal $s(t)$ and, thus, to render the communication signal more vulnerable to additive noise.

In many channels, such as wire lines and microwave line-of-sight channels, signal attenuation can be offset by using amplifiers to boost the level of the signal during transmission. However, an amplifier also introduces additive noise in the process of amplification and, thus, corrupts the signal. This additional noise must be taken into consideration in the design of the communication system.

In this section we consider the effects of attenuation encountered in signal transmission through a channel and additive thermal noise generated in electronic amplifiers. We also demonstrate how these two factors influence the design of a communication system.

6.5.1 Characterization of Thermal Noise Sources

Any conductive two-terminal device is characterized generally as lossy and has some resistance, say R ohms. A resistor, which is at a temperature $\mathcal{T}$ above absolute zero, contains free electrons that exhibit random motion and, thus, result in a noise voltage across the terminals of the resistor. Such a noise voltage is called *thermal noise*.

In general, any physical resistor (or lossy device) may be modeled by a noise source in series with a noiseless resistor, as shown in Figure 6.19. The output $n(t)$

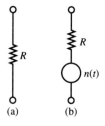

FIGURE 6.19.   A physical resistor (a) is modeled as a noiseless resistor in series with a noise source (b).

of the noise source is characterized as a sample function of a random process. Based on quantum mechanics, the power-spectral density of thermal noise (see Section 3.4) is given as

$$S_R(f) = \frac{2Rh|f|}{\left(e^{\frac{h|f|}{kT}} - 1\right)} \ \text{V}^2/\text{Hz} \tag{6.5.2}$$

where h is Plank's constant, k is Boltzmann's constant, and T is the temperature of the resistor in degree Kelvin, i.e., $T = 273 + C$, where C is in degrees Centigrade. As indicated in Section 3.4, at frequencies below 10^{12} Hz (which includes all conventional communication systems) and at room temperature,

$$e^{\frac{h|f|}{kT}} \approx 1 + \frac{h|f|}{kT} \tag{6.5.3}$$

Consequently, the power-spectral density is well approximated as

$$S_R(f) = 2RkT \ \text{V}^2/\text{Hz} \tag{6.5.4}$$

When connected to a load resistance with value R_L, the noise voltage shown in Figure 6.20 delivers the maximum power when $R = R_L$. In such a case, the load is matched to the source and the maximum power delivered to the load is $E[N^2(t)]/4R_L$. Therefore, the power-spectral density of the noise voltage across the load resistor is

$$S_n(f) = \frac{kT}{2} \ \text{W/Hz} \tag{6.5.5}$$

As previously indicated in Section 3.4.2, kT is usually denoted by N_0. Hence the power-spectral density of thermal noise is generally expressed as

$$S_n(f) = \frac{N_0}{2} \ watts/Hz \tag{6.5.6}$$

For example, at room temperature ($T_0 = 290^o$K), $N_0 = 4 \times 10^{-21}$ W/Hz.

6.5.2 Effective Noise Temperature and Noise Figure

When we employ amplifiers in communication systems to boost the level of a signal, we are also amplifying the noise corrupting the signal. Because any amplifier

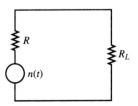

FIGURE 6.20. Noisy resistor connected to a load resistance R_L.

has some finite passband, we may model an amplifier as a filter with frequency-response characteristic $H(f)$. Let us evaluate the effect of the amplifier on an input thermal noise source.

Figure 6.21 illustrates a thermal-noise source connected to a matched two-port network having frequency response $H(f)$. The output of this network is connected to a matched load. First, we recall that the noise power at the output of the network is

$$P_{no} = \int_{-\infty}^{\infty} S_n(f)|H(f)|^2 df = \frac{N_0}{2} \int_{-\infty}^{\infty} |H(f)|^2 \, df \qquad (6.5.7)$$

From Section 3.4.2, we recall that the noise-equivalent bandwidth of the filter is defined as

$$B_{neq} = \frac{1}{2G} \int_{-\infty}^{\infty} |H(f)|^2 \, df \qquad (6.5.8)$$

where, by definition, $G = |H(f)|^2_{\max}$ is the *maximum available power gain* of the amplifier. Consequently, the output noise power from an ideal amplifier that introduces no additional noise may be expressed as

$$P_{no} = G N_0 B_{neq} \qquad (6.5.9)$$

Any practical amplifier introduces additional noise at its output due to internally generated noise. Hence, the noise power at its output may be expressed as

$$P_{no} = G N_0 B_{neq} + P_{ni}$$
$$= G k \, \mathcal{T} B_{neq} + P_{ni} \qquad (6.5.10)$$

where P_{ni} is the power of the amplifier output due to internally generated noise. Therefore,

$$P_{no} = G k B_{neq} \left(\mathcal{T} + \frac{P_{ni}}{G k B_{neq}} \right) \qquad (6.5.11)$$

This leads us to define a quantity

$$\mathcal{T}_e = \frac{P_{ni}}{G k B_{neq}} \qquad (6.5.12)$$

which we call the *effective noise temperature* of the two-port network (amplifier). Then,

$$P_{no} = G k B_{neq} (\mathcal{T} + \mathcal{T}_e) \qquad (6.5.13)$$

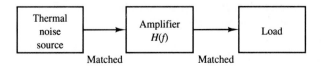

FIGURE 6.21. Thermal noise converted to amplifier and load.

Thus, we interpret the output noise as originating from a thermal-noise source at temperature $T + T_e$.

A signal source with power P_{si} at the input to the amplifier with gain G will produce an output with power

$$P_{so} = G P_{si} \tag{6.5.14}$$

Hence, the output SNR from the two-port network is

$$
\begin{aligned}
\left(\frac{S}{N} \right)_o &= \frac{P_{so}}{P_{no}} = \frac{G P_{si}}{G k T B_{\text{neq}} (1 + T_e/T)} \\
&= \frac{P_{si}}{N_0 B_{\text{neq}} (1 + T_e/T)} \\
&= \frac{1}{1 + T_e/T} \left(\frac{S}{N} \right)_i
\end{aligned}
\tag{6.5.15}
$$

where, by definition, $(S/N)_i$ is the input SNR to the two-port network. We observe that the SNR at the output of the amplifier is degraded (reduced) by the factor $(1 + T_e/T)$. Thus, T_e is a measure of the noisiness of the amplifier. An ideal amplifier is one for which $T_e = 0$.

When T is taken as room temperature $T_0 (290^o \text{K})$ the factor $(1 + T_e/T_0)$ is called the *noise figure* of the amplifier. Specifically, the noise figure of a two-port network is defined as the ratio of the output noise power P_{no} to the output noise power of an ideal (noiseless) two-part network for which the thermal noise source is at room temperature ($T = 290^o \text{K}$). Clearly, the ratio

$$F = \left(1 + \frac{T_e}{T_0} \right) \tag{6.5.16}$$

is the noise figure of the amplifier. Consequently, (6.5.15) may be expressed as

$$\left(\frac{S}{N} \right)_o = \frac{1}{F} \left(\frac{S}{N} \right)_i \tag{6.5.17}$$

By taking the logarithm of both sides of (6.5.17) we obtain

$$10 \log \left(\frac{S}{N} \right)_o = -10 \log F + 10 \log \left(\frac{S}{N} \right)_i \tag{6.5.18}$$

Hence, $10 \log F$ represents the loss in SNR due to the additional noise introduced by the amplifier. The noise figure for many low-noise amplifiers such as traveling wave tubes is below 3 dB. Conventional integrated circuit amplifiers have noise figures of 6–7 dB.

It is easy to show (see Problem 6.18) that the overall noise figure of a cascade of K amplifiers with gains G_k and corresponding noise figures F_k, $1 \leq k \leq K$ is

$$F = F_1 + \frac{F_2 - 1}{G_1} + \frac{F_3 - 1}{G_1 G_2} + \cdots + \frac{F_K - 1}{G_1 G_2 \cdots G_{K-1}} \tag{6.5.19}$$

This expression is known as *Friis' formula*. We observe that the dominant term is F_1, which is the noise figure of the first amplifier stage. Therefore, the front end of a receiver should have a low-noise figure and a high gain. In that case, the remaining terms in the sum will be negligible.

Example 6.5.1

Suppose an amplifier is designed of three identical states, each of which has a gain of $G_i = 5$ and a noise figure $F_i = 6$, $i = 1, 2, 3$. Determine the overall noise figure of the cascade of the three stages.

Solution   From (6.5.19) we obtain

$$F = F_1 + \frac{F_2 - 1}{G_1} + \frac{F_3 - 1}{G_1 G_2}$$

where $F_1 = F_2 = F_3 = 6$ and $G_1 = G_2 = 5$. Hence,

$$F_1 = 6 + 1 + 0.2 = 7.2$$

or, equivalently, $F_{1dB} = 8.57$ dB.

6.5.3 Transmission Losses

As we indicated previously, any physical channel attenuates the signal transmitted through it. The amount of signal attenuation generally depends on the physical medium, the frequency of operation, and the distance between the transmitter and the receiver. We define the loss $\mathcal{L}$ in signal transmission as the ratio of the input (transmitted) power to the output (received) power of the channel, i.e.,

$$\mathcal{L} = \frac{P_T}{P_R} \tag{6.5.20}$$

or, in decibels, as

$$\mathcal{L}_{dB} \equiv 10 \log \mathcal{L} = 10 \log P_T - 10 \log P_R \tag{6.5.21}$$

In wireline channels, the transmission loss is usually given in terms of decibels/unit length, e.g., dB/km. For example, the transmission loss in coaxial cable of 1 cm diameter is about 2 dB/km at a frequency of 1 Mhz. This loss generally increases with an increase in frequency.

Example 6.5.2

Determine the transmission loss for a 10-km and a 20-km coaxial cable if the loss/km is 2 dB at the frequency of operation.

Solution   The loss for the 10-km channel is $\mathcal{L}_{dB} = 20$ dB. Hence, the output (received) power is $P_R = P_T/\mathcal{L} = 10^{-2} P_T$. For the 20-km channel, the loss is $\mathcal{L}_{dB} = 40$ dB; hence, $P_R = 10^{-4} P_T$. Note that increasing the cable length from 10-km to 20-km increased the attenuation by two orders of magnitude.

In line-of-sight radio systems the transmission loss is given as

$$\mathcal{L} = \left(\frac{4\pi d}{\lambda}\right)^2 \tag{6.5.22}$$

where $\lambda = c/f$ is the wavelength of the transmitted signal, c is the speed of light (3×10^8m/sec), f is the frequency of the transmitted signal, and d is the distance between the transmitter and the receiver in meters. In radio transmission, L is called the *free-space path loss.*

Example 6.5.3

Determine the free-space path loss for a signal transmitted at $f = 1$ MHz over distances of 10 km and 20 km.

Solution   The loss given in (6.5.22) for a signal at a wavelength $\lambda = 300$ m is

$$L_{dB} = 20\log_{10}(4\pi \times 10^4/300)$$

$$= 52.44 \text{ dB} \tag{6.5.23}$$

for the 10-km path and

$$L_{dB} = 20\log_{10}(8\pi \times 10^4/300)$$

$$= 58.44 \text{ dB} \tag{6.5.24}$$

for the 20-km path. It is interesting to note that doubling the distance in radio transmission increases the free-space path loss by 6 dB.

Example 6.5.4

A signal is transmitted through a 10-km coaxial line channel that exhibits a loss of 2 dB/km. The transmitted signal power is $P_{T\,dB} = -30$ dBW (-30 dBW means 30 dB below 1 W or, simply, 1 mW). Determine the received signal power and the power at the output of an amplifier that has a gain of $G_{dB} = 15$ dB.

Solution   The transmission loss for the 10-km channel is $L_{dB} = 20$ dB. Hence, the received signal power is

$$P_{R\,dB} = P_{T\,dB} - L_{dB} = -30 - 20 = -50 \text{ dBW} \tag{6.5.25}$$

The amplifier boosts the received signal power by 15 dB. Hence, the power at the output of the amplifier is

$$P_{o\,dB} = P_{R\,dB} + G_{dB}$$

$$= -50 + 15 = -35 \text{ dBW} \tag{6.5.26}$$

6.5.4 Repeaters for Signal Transmission

Analog repeaters are basically amplifiers that are used generally in telephone wireline channels and microwave line-of-sight radio channels to boost the signal level and, thus, to offset the effect of signal attenuation in transmission through the channel.

Figure 6.22 illustrates a system in which a repeater is used to amplify the signal that has been attenuated by the lossy transmission medium. Hence, the signal power at the input to the repeater is

$$P_R = P_T/L \tag{6.5.27}$$

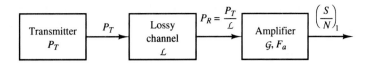

FIGURE 6.22. A communication system employing a repeater to compensate for channel loss.

The output power from the repeater is

$$P_o = G P_R = G P_T / L \tag{6.5.28}$$

We may select the amplifier gain G to offset the transmission loss; hence, $G = L$ and $P_o = P_T$.

Now, the SNR at the output of the repeater is

$$\left(\frac{S}{N}\right)_1 = \frac{1}{F_a}\left(\frac{S}{N}\right)_i$$

$$= \frac{1}{F_a}\left(\frac{P_R}{N_0 B_{\text{neq}}}\right) = \frac{1}{F_a}\left(\frac{P_T}{L N_0 B_{\text{neq}}}\right)$$

$$= \frac{1}{F_a L}\left(\frac{P_T}{N_0 B_{\text{neq}}}\right) \tag{6.5.29}$$

Based on this result, we may view the lossy transmission medium followed by the amplifier as a cascade of two networks, one with a noise figure L and the other with a noise figure F_a. Then, for the cascade connection, the overall noise figure is

$$F = L + \frac{F_a - 1}{G_a} \tag{6.5.30}$$

If we select $G_a = 1/L$, then,

$$F = L + \frac{F_a - 1}{1/L} = L F_a \tag{6.5.31}$$

Hence, the cascade of the lossy transmission medium and the amplifier is equivalent to a single network with noise figure $L F_a$.

Now, suppose that we transmit the signal over K segments of the channel where each segment has its own repeater, as shown in Figure 6.23. Then, if $F_i = L_i F_{ai}$ is the noise figure of the ith section, the overall noise figure for the K sections is

$$F = L_1 F_{a1} + \frac{L_2 F_{a2} - 1}{L_1 / G_{a1}} + \frac{L_3 F_{a3} - 1}{(L_1 / G_{a1})(L_2 / G_{a2})} \cdots$$

$$+ \frac{L_K F_{aK} - 1}{(L_1 / G_{a1})(L_2 / G_{a2}) \dots (L_K / G_{aK})} \tag{6.5.32}$$

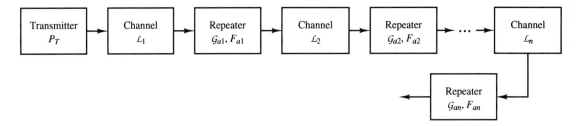

FIGURE 6.23. A communication system employing repeaters.

Therefore, the SNR at the output of the repeater (amplifier) at the receiver is

$$\left(\frac{S}{N}\right)_o = \frac{1}{F}\left(\frac{S}{N}\right)_i$$

$$= \frac{1}{F}\left(\frac{P_T}{N_0 B_{\text{neq}}}\right) \qquad (6.5.33)$$

In the important special case where the K segments are identical, i.e., $L_i = L$ for all i and $F_{ai} = F_a$ for all i, and where the amplifier gains are designed to offset the losses in each segment, i.e., $G_{ai} = L_i$ for all i, then the overall noise figure becomes

$$F = KLF_a - (K - 1) \approx KLF_a \qquad (6.5.34)$$

Hence,

$$\left(\frac{S}{N}\right)_o \approx \frac{1}{KLF_a}\left(\frac{P_T}{N_0 B_{\text{neq}}}\right) \qquad (6.5.35)$$

Therefore, the overall noise figure for the cascade of the K identical segments is simply K times the noise figure of one segment.

Example 6.5.5

A signal with bandwidth 4 kHz is to be transmitted a distance of 200 km over a wireline channel that has an attenuation of 2 dB/km. (1) Determine the transmitter power P_T required to achieve an SNR of $(S/N)_0 = 30$ dB at the output of the receiver amplifier that has a noise figure $F_{a\text{dB}} = 5$ dB. (2) Repeat the calculation when a repeater is inserted every 10 km in the wireline channel, where the repeater has a gain of 20 dB and a noise figure of $F_a = 5$ dB. Assume that the noise-equivalent bandwidth of each repeater is $B_{\text{neq}} = 4$ kHz and that $N_0 = 4 \times 10^{-21}$ W/Hz.

Solution (1) The total loss in the 200-km wire line is 400 dB. From (6.4.30), with $K = 1$, we have

$$10\log(S/N) = -10\log L - 10\log F_a - 10\log(N_0 B_{\text{neq}}) + 10\log P_T$$

Hence,

$$P_{T\,\text{dB}} = (S/N)_{0\,\text{dB}} + F_{a\,\text{dB}} + (N_0 B_{\text{neq}})_{\text{dB}} + 10\log L$$

$$= 30 + 5 + 400 + (N_0 B_{neq})_{dB}$$

But,

$$(N_0 B_{neq})_{dB} = 10 \log(1.6 \times 10^{-17}) = -168 \text{ dBW}$$

where dBW denotes the power level relative to 1 W. Therefore,

$$P_{T\,dB} = 435 - 168 = 267 \text{ dB}$$

$$P_T = 5 \times 10^{26} \text{ W}$$

which is an astronomical figure.

    (2) The use of a repeater every 10 km reduces the per-segment loss to $\mathcal{L}_{dB} = 20$ dB. There are 20 repeaters, and each repeater has a noise figure of 5 dB. Hence, (6.4.30) yields

$$(S/N)_{0\,dB} = -10 \log K - 10 \log \mathcal{L} - 10 \log F_a - 10 \log(N_0 B_{neq}) + 10 \log P_T$$

$$30 = -13 - 20 - 5 + 168 + P_{T\,dB}$$

Therefore,

$$P_{T\,dB} = -100 \text{ dBW}$$

or, equivalently,

$$P_T = 10^{-10} \text{ W} \quad (0.1 \text{ pW})$$

    The above example illustrates clearly the advantage of using analog repeaters in communication channels that span large distances. However, we also observed that analog repeaters add noise to the signal and, consequently, degrade the output SNR. It is clear from (6.5.35) that the transmitted power P_T must be increased linearly with the number K of repeaters to maintain the same $(S/N)_0$ as K increases. Hence, for every factor of two increase in K, the transmitted power P_T must be increased by 3 dB.

6.6 FURTHER READING

Analysis of the effect of noise on analog communication systems can be found in Carlson (1986), Ziemer and Tranter (1990), Couch (1993), and Gibson (1993). The book by Sakrison (1968) provides a detailed analysis of FM in the presence of noise. Phase-locked loops are treated in detail in Viterbi (1966), Lindsey (1972), Lindsey and Simon (1973), and Gardner (1979), and in the survey papers by Gupta (1975) and Lindsey and Chie (1981). Taub and Schilling (1986) provide in-depth treatment of the effect of threshold and various methods for threshold extension in FM.

PROBLEMS

6.1 The received signal $r(t) = s(t) + n(t)$ in a communication system is passed through an ideal LPF with bandwidth W and unity gain. The signal component $s(t)$ has a power-spectral density

$$S_s(f) = \frac{P_0}{1 + (f/B)^2}$$

where B is the 3-dB bandwidth. The noise component $n(t)$ has a power-spectral density $N_0/2$ for all frequencies. Determine and plot the SNR as a function of the ratio W/B. What is the filter bandwidth W that yields a maximum SNR?

6.2 The input to the system shown in Figure P-6.2 is the signal plus noise waveform

$$r(t) = A_c \cos 2\pi f_c t + n(t)$$

where $n(t)$ is a sample function of a white-noise process with spectral density $N_0/2$.

1. Determine and sketch the frequency response of the RC filter.
2. Sketch the frequency response of the overall system.
3. Determine the SNR at the output of the ideal LPF assuming that $W > f_c$. Sketch the SNR as a function of W for fixed values of R and C.

6.3 A DSB AM signal with power-spectral density as shown in Figure P-6.3(a) is corrupted with additive noise that has a power-spectral density $N_0/2$ within the passband of the signal. The received signal-plus-noise is demodulated and lowpass filtered as shown in Figure P-6.3(b). Determine the SNR at the output of the LPF.

6.4 A certain communication channel is characterized by 90-dB attenuation and additive white noise with power-spectral density of $\frac{N_0}{2} = 0.5 \times 10^{-14}$ W/Hz. The bandwidth of the message signal is 1.5 MHz and its amplitude is uniformly distributed in the interval $[-1, 1]$. If we require that the SNR after demodulation be 30 dB, in each of the following cases find the necessary transmitter power.

1. USSB modulation.
2. Conventional AM with a modulation index of 0.5.

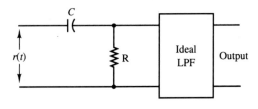

FIGURE P-6.2

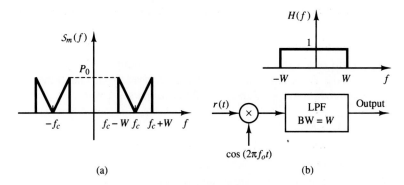

(a)   (b)

FIGURE P-6.3

3. DSB-SC modulation.

6.5 A sinusoidal message signal whose frequency is less than 1000 Hz, modulates the carrier $c(t) = 10^{-3} \cos 2\pi f_c t$. The modulation scheme is conventional AM and the modulation index is 0.5. The channel noise is additive white with power-spectral density of $\frac{N_0}{2} = 10^{-12}$ W/Hz. At the receiver the signal is processed as shown in Figure P-6.5(a). The frequency response of the bandpass noise-limiting filter is shown in Figure P-6.5(b).

1. Find the signal power and the noise power at the output of the noise-limiting filter.
2. Find the output SNR.

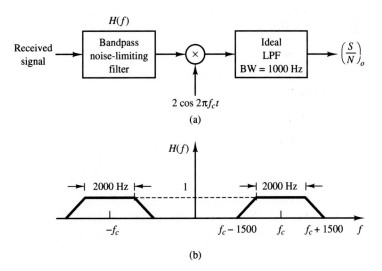

(a)

(b)

FIGURE P-6.5

6.6 Derive the expression for the (one-sided) noise-equivalent bandwidth of the PLL given by (6.2.12).

6.7 In an analog communication system, the *demodulation gain* is defined as the ratio of the SNR at the output of the demodulator to the SNR at the output of the noise-limiting filter at the receiver front end. Find expressions for the demodulation gain in each of the following cases:

1. DSB.
2. SSB.
3. Conventional AM with a modulation index of a. What is the largest possible demodulation gain in this case?
4. FM with modulation index β_f.
5. PM with modulation index β_p.

6.8 In a broadcasting communication system the transmitter power is 40 kW, the channel attenuation is 80 dB, and the noise power-spectral density is 10^{-10} W/Hz. The message signal has a bandwidth of 10^4 Hz.

1. Find the predetection SNR in $r(t) = ku(t) + n(t)$.
2. Find the output SNR if the modulation is DSB.
3. Find the output SNR if the modulation is SSB.
4. Find the output SNR if the modulation is conventional AM with a modulation index of 0.85 and normalized message power of 0.2.

6.9 A communication channel has a bandwidth of 100 kHz. This channel is to be used for transmission of an analog source $m(t)$, where $|m(t)| < 1$, whose bandwidth is $W = 4$ kHz. The power content of the message signal is 0.1 W.

1. Find the ratio of the output SNR of an FM system that utilizes the whole bandwidth to the output SNR of a conventional AM system with a modulation index of $a = 0.85$. What is this ratio in decibels?
2. Show that if an FM system and a PM system are employed and these systems have the same output signal to noise ratio, we have

$$\frac{\mathrm{BW_{PM}}}{\mathrm{BW_{FM}}} = \frac{\sqrt{3}\beta_f + 1}{\beta_f + 1}$$

6.10 The normalized message signal $m_n(t)$ has a bandwidth of 5000 Hz and power of 0.1 W, the channel has a bandwidth of 100 kHz and attenuation of 80 dB. The noise is white with power-spectral density 0.5×10^{-12} W/Hz and the transmitter power is 10 kW.

1. If AM with $a = 0.8$ is employed, what is $\left(\frac{S}{N}\right)_o$?

2. If FM is employed, what is the highest possible $(\frac{S}{N})_o$?

6.11 A normalized message signal has a bandwidth of $W = 8$ kHz and a power of $P_{M_n} = \frac{1}{2}$. It is required to transmit this signal via a channel with an available bandwidth of 60 kHz and attenuation of 40 dB. The channel noise is additive and white with a power-spectral density of $\frac{N_0}{2} = 10^{-12}$ W/Hz. A frequency modulation scheme, with no pre-emphasis/de-emphasis filtering, has been proposed for this purpose.

    1. If it is desirable to have an SNR of at least 40 dB at the receiver output, what is the minimum required transmitter power and the corresponding modulation index?

    2. If the minimum required SNR is increased to 60 dB, how would your answer change?

    3. If, in part 2, we are allowed to employ pre-emphasis/de-emphasis filters with a time constant of $\tau = 75$ μsec, how would the answer to part 2 change?

6.12 In the transmission of telephone signals over line-of-sight microwave links, a combination of FDM-SSB and FM is often employed. A block diagram of such a system is shown in Figure P-6.12. Each of the signals $m_i(t)$ is bandlimited to W Hz, and these signals are USSB modulated on carriers $c_i(t) = A_i \cos 2\pi f_{ci} t$ where $f_{ci} = (i-1)W$, $1 \leq i \leq K$ and $m(t)$ is the sum of all USSB modulated signals. This signal is FM modulated on a carrier with frequency f_c with a modulation index of β.

    1. Plot a typical spectrum of the USSB modulated signal $m(t)$.

    2. Determine the bandwidth of $m(t)$.

    3. At the receiver side, the received signal $r(t) = u(t) + n_w(t)$ is first FM demodulated and then passed through a bank of USSB demodulators. Show that the noise power entering these demodulators depends on i.

    4. Determine an expression for the ratio of the noise power entering the demodulator whose carrier frequency is f_i to the noise power entering the demodulator with carrier frequency f_j, $1 \leq i, j \leq K$.

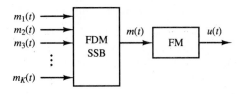

FIGURE P-6.12

5. How should the carrier amplitudes A_i be chosen to guarantee that, after USSB demodulation, the SNR for all channels is the same?

6.13 A power meter that measures average power is connected to the output of a transmitter as shown in Figure P-6.13. The meter reading is 20 W when it is connected to a 50 Ω load. Determine:

1. the voltage across the load resistance.
2. the current through the load resistance.
3. the power level in dBm units.

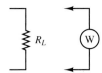

FIGURE P-6.13

6.14 A twisted-pair telephone wireline channel with characteristic impedance $Z_c = 300$ Ω is terminated with a load of $Z_L = 300$ Ω. The telephone line is 200-km long and has a loss of 2 dB/km.

1. If the average transmitted power $P_T = 10$ dBm, determine the received power P_R if the line contains no repeaters.
2. If repeaters with a gain of 20 dB are used to boost the signal on the channel and if each repeater requires an input signal level of 10 dBm, determine the number of repeaters and their spacing. The noise figure of each repeater is 6 dB.

6.15 A radio antenna pointed in a direction of the sky has a noise temperatuture of 50° K. The antenna feeds the received signal to the preamplifier, which has a gain of 35 dB over a bandwidth of 10 MHz and a noise figure of 2 dB.

1. Determine the effective noise temperature at the input to the preamplifier.
2. Determine the noise power at the output of the preamplifier.

6.16 An amplifier has a noise equivalent bandwith $B_{neq} = 25$ kHz and a maximum available power gain of $G = 30$ dB. Its output noise power is $10^8 k T_0$. Determine the effective noise temperature and the noise figure.

6.17 Prove that the effective noise temperature of k two-port networks in cascade is

$$T_e = T_{e1} + \frac{T_{e2}}{G_1} + \frac{T_{e3}}{G_1 G_2} + \cdots + \frac{T_{ek}}{G_1 G_2 \cdots G_k}$$

7

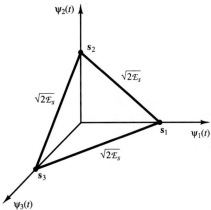

Digital Transmission Through an Additive White Gaussian Noise Channel

In Chapter 4, we described various methods for converting the output of a signal source into a sequence of binary digits. In this chapter, we consider the transmission of the digital information sequence over communication channels that are characterized as additive white Gaussian noise (AWGN) channels. The AWGN channel is one of the simplest mathematical models for various physical communication channels, including wire lines and some radio channels.

Our treatment focuses on digital modulation and demodulation methods that are appropriate for *baseband channels,* i.e., channels having frequency passbands that usually include zero frequency ($f = 0$). Examples of baseband channels are wire line channels such as coaxial cable and twisted-pair telephone channels. For such channels, the digital information can be transmitted directly by baseband pulse modulation techniques, i.e., by modulation methods that do not require a sinusoidal carrier to perform frequency translation of the transmitted signal spectrum. Later, in Chapter 9, we will see how these digital modulation methods can be applied to *bandpass channels,* i.e., channels having frequency passbands far removed from $f = 0$.

We begin by developing a geometric representation for several different types of pulse modulation signals. Optimum demodulation and detection of these signals is then described. Finally, we evaluate the probability of error as a performance measure for the different types of modulation signals on an AWGN channel. The various modulation methods are compared on the basis of their performance characteristics, their bandwidth requirements, and their implementation complexity.

Initially, we will not impose bandwidth constraints in the design of signals for digital modulation. However, because the channel bandwidth is an important parameter that influences the design of most communication systems, the design of the modulator and demodulator for bandlimited channels is treated in depth in Chapter 8.

7.1 PULSE MODULATION SIGNALS AND THEIR GEOMETRIC REPRESENTATION

In this section, we introduce several types of pulse modulation signals that are used for the transmission of digital information and develop a geometric representation of such signals. The pulse modulation signals considered include (1) pulse amplitude modulated signals, (2) pulse position modulated (orthogonal) signals, (3) biorthogonal signals, (4) simplex signals, and (5) signals generated from binary code sequences. First we describe how the digital information is conveyed with these types of signals.

7.1.1 Pulse Modulation Signals

In *pulse amplitude modulation* (PAM), the information is conveyed by the amplitude of the pulse. For example, in binary PAM, the information bit 1 is represented by a pulse of amplitude A and the information bit 0 is represented by a pulse of amplitude $-A$ as shown in Figure 7.1. Pulses are transmitted at a bit rate $R_b = 1/T_b$ bits per second, where T_b is called the bit interval. Although the pulses are shown as rectangular, in practical systems, the rise time and decay time are nonzero and the pulses are generally smoother. The effect of the pulse shape on the spectral characteristics of the transmitted signal is considered in Chapter 8.

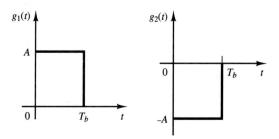

FIGURE 7.1.   Example of signals for binary PAM.

In *pulse position modulation* (PPM), the information is conveyed by the time interval in which the pulse is transmitted. For example, in binary PPM, the bit interval is divided into two time slots of $T_b/2$ seconds each. The information bit 1 is represented by a pulse of amplitude A in the first time slot and the information bit 0 is represented by a pulse of amplitude A in the second time slot, as shown in Figure 7.2. The pulse shape within a time slot need not be rectangular.

Another pulse modulation method for transmitting binary information is called on-off keying (OOK). The transmitted signal for transmitting a 1 is a pulse of duration T_b. If a 0 is to be transmitted, no pulse is transmitted in the signal interval of duration T_b.

The generalization of PAM and PPM to nonbinary (M-ary) pulse transmission is relatively straightforward. Instead of transmitting one bit at a time, the binary information sequence is subdivided into blocks of k bits, called *symbols,* and each block, or symbol, is represented by one of $M = 2^k$ pulse amplitude values for PAM and pulse position values for PPM. Thus, with $k = 2$, we have $M = 4$ pulse amplitude values, or pulse position values. Figure 7.3 illustrates the PAM and PPM signals for $k = 2$, $M = 4$. Note that when the bit rate R_b is fixed, the *symbol interval* is

$$T = \frac{k}{R_b} = kT_b \qquad (7.1.1)$$

as shown in Figure 7.4.

It is interesting to characterize the PAM and PPM signals in terms of their basic properties. For example, the M-ary PAM signal waveforms may be expressed as

$$s_m(t) = A_m g_T(t), \quad m = 1, 2, \dots, M$$

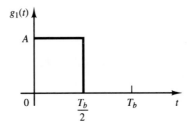

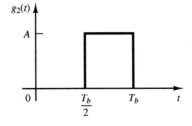

FIGURE 7.2.   Example of signals for binary PPM.

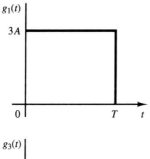

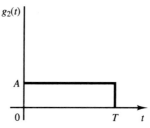

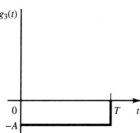

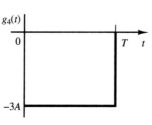

(a) $M = 4$ PAM signals

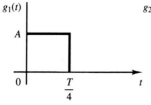

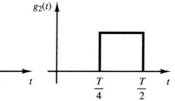

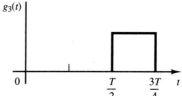

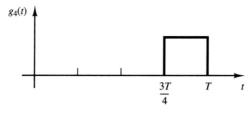

(b) $M = 4$ PPM signals

FIGURE 7.3.   Examples of $M = 4$ PAM and PPM signals.

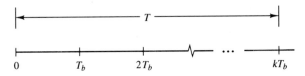

T_b = bit interval

T = symbol interval

FIGURE 7.4. Relationship between the symbol interval and the bit interval.

$$0 \leq t \leq T \qquad (7.1.2)$$

where $g_T(t)$ is a pulse of some arbitrary shape as shown in Figure 7.5(a). We observe that the distinguishing feature among the M signals is the signal amplitude. All the M signals have the same pulse shape. Another important feature of these signals is their energies. We note that the signals have different energies, i.e.,

$$\mathcal{E}_m = \int_0^T s_m^2(t)dt = A_m^2 \int_0^T g_T^2(t)dt, \quad m = 1, 2, \ldots, M \qquad (7.1.3)$$

In the case of PPM, the signal waveforms may be expressed as

$$s_m(t) = Ag_T(t - (m-1)T/M), \quad m = 1, 2, \ldots, M$$

$$(m-1)T/M \leq t \leq mT/M \qquad (7.1.4)$$

where $g_T(t)$ is a pulse of duration T/M and of arbitrary shape, as shown for example in Figure 7.5(b). A major distinguishing characteristic of these waveforms is that they are nonoverlapping. Consequently,

$$\int_0^T s_m(t)s_n(t)dt = 0, \quad m \neq n \qquad (7.1.5)$$

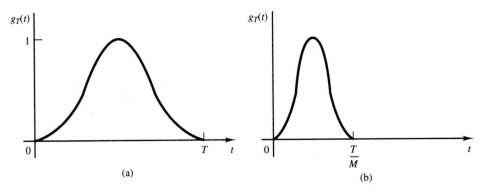

FIGURE 7.5. Signal pulses for (a) PAM and (b) PPM.

Such signal waveforms are said to be *orthogonal*. A second distinguishing feature of PPM signal waveforms is that all the waveforms have equal energy, i.e.,

$$\int_0^T s_m^2(t)dt = A^2 \int_{(m-1)T/M}^{mT/M} g_T^2(t-(m-1)T/M)\, dt$$

$$= A^2 \int_0^{T/M} g_T^2(t)dt = \mathcal{E}_s, \quad \text{all } m \qquad (7.1.6)$$

A major difference between a set of M PAM signal waveforms and a set of M PPM signal waveforms is the channel bandwidth required for their transmission. We observe that the channel bandwidth required to transmit the PAM signals is determined by the frequency characteristics of the pulse $g_T(t)$. This basic pulse has a duration T. On the other hand, the basic pulse $g_T(t)$ in a set of M PPM signal waveforms has a duration T/M. Whatever pulse shape we choose for PAM and PPM and whatever definition of bandwidth that we employ, it is clear that the spectrum of the PPM pulse is M times wider than the spectrum of a PAM pulse. Consequently, the PAM signals do not require an expansion of the channel bandwidth as M increases. On the other hand, the PPM signal waveforms require an increase in the channel bandwidth as M increases.

PAM and PPM are two examples of a variety of different types of signal sets that can be constructed for transmission of digital information over baseband channels. For example, if we take a set of $M/2$ PPM signals and construct the $M/2$ negative signal pulses, the combined set of M signal waveforms constitute a set of M *biorthogonal signals*. An example for $M = 4$ is shown in Figure 7.6. It is easy to see that all the M signals have equal energy. Furthermore, the channel bandwidth required to transmit the M signals is just one-half of that required to transmit M PPM signals.

As another example, we demonstrate that from any set of M orthogonal signal waveforms, we can construct another set of M signal waveforms that are known as *simplex signal waveforms*. From the M orthogonal signals we subtract the average of the M signals. Thus,

$$s_m'(t) = s_m(t) - \frac{1}{M}\sum_{k=1}^{M} s_k(t) \qquad (7.1.7)$$

Then, it follows that (see Problem 7.7) the energy of these signals $s_m'(t)$ is

$$\mathcal{E}_s' = \int_0^T \left[s_m'(t)\right]^2 dt = \left(1 - \frac{1}{M}\right)\mathcal{E}_s \qquad (7.1.8)$$

and

$$\int_0^T s_m'(t)s_n'(t)dt = -\frac{1}{M-1}\mathcal{E}_s, \quad m \neq n \qquad (7.1.9)$$

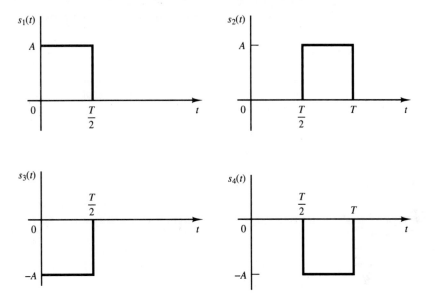

FIGURE 7.6.   A set of $M = 4$ biorthogonal signal waveforms.

where $\mathcal{E}_s$ is the energy of each of the orthogonal signals and $\mathcal{E}'_s$ is the energy of each of the signals in the simplex signal set. Note that the waveforms in the simplex set have smaller energy than the waveforms in the orthogonal signal set. Second, we note that simplex signal waveforms are not orthogonal. Instead, they have a negative correlation, which is equal for all pairs of signal waveforms. It has been conjectured for several decades that among all the possible M-ary signal waveforms of equal energy $\mathcal{E}_s$, the simplex signal set results in the smallest probability of error when used to transmit information on an additive white Gaussian noise channel. However, this conjecture has not been proved.

As a final example of the construction of M signal waveforms, let us consider a set of M binary code words of the form

$$\mathbf{c}_m = (c_{m1}, c_{m2}, \ldots, c_{mN}), \quad m = 1, 2, \ldots, M \qquad (7.1.10)$$

where $c_{mj} = 0$ or 1 for all m and j. N is called the *block length* or dimension of the code words. Given M code words, we can construct M signal waveforms by mapping a code bit $c_{mj} = 1$ into a pulse $g_T(t)$ of duration T/N and a code bit $c_{mj} = 0$ into the negative pulse $-g_T(t)$.

Example 7.1.1

Given the code words

$$
\begin{aligned}
\mathbf{c}_1 &= [1 \quad 1 \quad 1 \quad 1] \\
\mathbf{c}_2 &= [1 \quad 1 \quad 0 \quad 0] \\
\mathbf{c}_3 &= [1 \quad 0 \quad 1 \quad 0] \\
\mathbf{c}_4 &= [0 \quad 1 \quad 0 \quad 1]
\end{aligned}
$$

construct a set of $M = 4$ signal waveforms, as described above, using a rectangular pulse $g_T(t)$.

Solution   As indicated above, a code bit 1 is mapped into the rectangular pulse $g_T(t)$ of duration $T/4$ and a code bit 0 is mapped into the rectangular pulse $-g_T(t)$. Thus, we construct the four waveforms shown in Figure 7.7 that correspond to the four code words. It is interesting to note that the first three signal waveforms in Figure 7.7 are mutually orthogonal, but the fourth waveform is the negative of the third.

There are numerous binary block codes that have been constructed over the past several decades. Several of these codes are described in Chapter 10. Any of these codes can be used to construct signal waveforms for M-ary signal transmission over a baseband channel.

In addition to the pulse modulated signals described above, in Section 7.1.4, we describe other types of baseband signals that are synthesized from a set of code words.

7.1.2 Geometric Representation of Signal Waveforms

We recall from Section 2.2 that the Gram-Schmidt orthogonalization procedure may be used to construct an orthonormal basis for a set of signals. In this section, we develop a geometric representation of signal waveforms as points in a signal space. Such a representation provides a compact characterization of signal sets for transmitting information over a channel and simplifies the analysis of their performance.

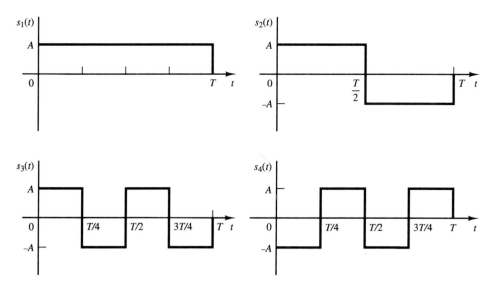

FIGURE 7.7.   A set of $M = 4$ signal waveforms constructed from the code words in Example 7.1.1.

Suppose we have a set of M signal waveforms $s_m(t)$, $1 \le m \le M$ which are to be used for transmitting information over a communication channel. From the set of M waveforms we first construct a set of $N \le M$ orthonormal waveforms, where N is the dimension of the signal space. For this purpose, we use the Gram-Schmidt orthogonalization procedure, which was introduced in Section 2.2.

Gram-Schmidt orthogonalization procedure.   We begin with the first waveform $s_1(t)$, which is assumed to have energy $\mathcal{E}_1$. The first waveform of the orthonormal set is constructed simply as

$$\psi_1(t) = \frac{s_1(t)}{\sqrt{\mathcal{E}_1}} \qquad (7.1.11)$$

Thus, $\psi_1(t)$ is simply $s_1(t)$ normalized to unit energy.

The second waveform is constructed from $s_2(t)$ by first computing the projection of $s_2(t)$ onto $\psi_1(t)$, which is

$$c_{12} = \int_{-\infty}^{\infty} s_2(t)\psi_1(t)\,dt \qquad (7.1.12)$$

Then, $c_{12}\psi_1(t)$ is subtracted from $s_2(t)$ to yield

$$d_2(t) = s_2(t) - c_{12}\psi_1(t) \qquad (7.1.13)$$

Now, $d_2(t)$ is orthogonal to $\psi_1(t)$, but it does not possess unit energy. If $\mathcal{E}_2$ denotes the energy in $d_2(t)$, then the energy normalized waveform that is orthogonal to $\psi_1(t)$ is

$$\psi_2(t) = \frac{d_2(t)}{\sqrt{\mathcal{E}_2}} \qquad (7.1.14)$$

In general, the orthogonalization of the kth function leads to

$$\psi_k(t) = \frac{d_k(t)}{\sqrt{\mathcal{E}_k}} \qquad (7.1.15)$$

where

$$d_k(t) = s_k(t) - \sum_{i=1}^{k-1} c_{ik}\psi_i(t) \qquad (7.1.16)$$

and

$$c_{ik} = \int_{-\infty}^{\infty} s_k(t)\psi_i(t)dt, \quad i = 1, 2, \ldots, k-1 \qquad (7.1.17)$$

Thus, the orthogonalization process is continued until all the M signal waveforms $\{s_m(t)\}$ have been exhausted and $N \le M$ orthonormal waveforms have been constructed. The N orthonormal waveforms $\{\psi_n(t)\}$ form a *basis* in the N-dimensional signal space. The dimensionality N of the signal space will be equal to M if all the

M signal waveforms are linearly independent, i.e., if none of the signal waveforms is a linear combination of the other signal waveforms.

Example 7.1.2

Let us apply the Gram-Schmidt procedure to the set of four waveforms illustrated in Figure 7.8(a). The waveform $s_1(t)$ has energy $\mathcal{E}_1 = 2$, so that $\psi_1(t) = s_1(t)/\sqrt{2}$. Next we observe that $c_{12} = 0$, so that $\psi_1(t)$ and $s_2(t)$ are orthogonal. Therefore, $\psi_2(t) = s_2(t)/\sqrt{\mathcal{E}_2} = s_2(t)/\sqrt{2}$. To obtain $\psi_3(t)$, we compute c_{13} and c_{23}, which are $c_{13} = 0$ and $c_{23} = -\sqrt{2}$. Hence,

$$d_3(t) = s_3(t) + \sqrt{2}\psi_2(t)$$

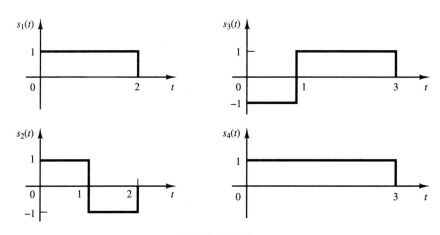

(a) Original signal set.

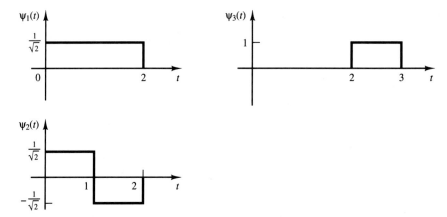

(b) Orthonormal waveforms.

FIGURE 7.8. Application of Gram-Schmidt orthogonalizaton procedure to signals $\{s_i(t)\}$.

Since $d_3(t)$ has unit energy, it follows that $\psi_3(t) = d_3(t)$. Finally, we find that $c_{14} = \sqrt{2}$, $c_{24} = 0$, $c_{34} = 1$. Hence,

$$d_4(t) = s_4(t) - \sqrt{2}\,\psi_1(t) - \psi_3(t) = 0$$

Thus, $s_4(t)$ is a linear combination of $\psi_1(t)$ and $\psi_3(t)$, and consequently, the dimensionality of the signal set is $N = 3$. The functions $\psi_1(t)$, $\psi_2(t)$, and $\psi_3(t)$ are shown in Figure 7.8(b).

Once we have constructed the set of orthogonal waveforms $\{\psi_n(t)\}$, we can express the M signals $\{s_m(t)\}$ as exact linear combinations of the $\{\psi_n(t)\}$. Hence, we may write

$$s_m(t) = \sum_{n=1}^{N} s_{mn}\psi_n(t), \quad m = 1, 2, \ldots, M \tag{7.1.18}$$

where

$$s_{mn} = \int_0^T s_m(t)\psi_n(t)dt$$

and

$$\mathcal{E}_m = \int_{-\infty}^{\infty} s_m^2(t)dt = \sum_{n=1}^{N} s_{mn}^2 \tag{7.1.19}$$

Based on the expression in (7.1.18), each signal waveform may be represented by the vector

$$\mathbf{s}_m = (s_{m1}, s_{m2}, \ldots, s_{mN}) \tag{7.1.20}$$

or equivalently, as a point in N-dimensional signal space with coordinates $\{s_{mi}, i = 1, 2, \ldots, N\}$. The energy of the mth signal waveform is simply the square of the length of the vector or, equivalently, the square of the Euclidean distance from the origin to the point in the N-dimensional space. Thus, any N-dimensional signal can be represented geometrically as a point in the signal space spanned by the N orthonormal functions $\{\psi_n(t)\}$.

Example 7.1.3

Let us determine the vector representation of the four signals shown in Figure 7.8(a) by using the orthonormal set of functions in Figure 7.8(b). Since the dimensionality of the signal space is $N = 3$, each signal is described by three components, which are obtained by projecting each of the four signal waveforms on the three orthonormal basis functions $\psi_1(t)$, $\psi_2(t)$, $\psi_3(t)$. Thus, we obtain $\mathbf{s}_1 = \left(\sqrt{2}, 0, 0\right)$, $\mathbf{s}_2 = \left(0, \sqrt{2}, 0\right)$, $\mathbf{s}_3 = \left(0, -\sqrt{2}, 1\right)$, $\mathbf{s}_4 = \left(\sqrt{2}, 0, 1\right)$. These signal vectors are shown in Figure 7.9.

Finally, we should observe that the set of basis functions $\{\psi_n(t)\}$ obtained by the Gram-Schmidt procedure is not unique. For example, another set of basis

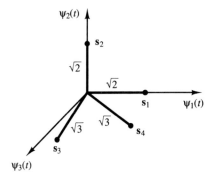

$\psi_2(t)$

$\bullet\, s_2$

$\sqrt{2}$

$\sqrt{2}$

$s_1 \quad \psi_1(t)$

$\sqrt{3} \quad \sqrt{3}$

s_4

s_3

$\psi_3(t)$

FIGURE 7.9. Signal vectors corresponding to the signals $s_i(t)$, $i = 1, 2, 3, 4$.

functions that span the three-dimensional space is shown in Figure 7.10. For this basis, the signal vectors are $s_1 = (1, 1, 0)$, $s_2 = (1, -1, 0)$, $s_3 = (-1, 1, 1)$, and $s_4 = (1, 1, 1)$. The reader should note that the change in the basis functions has not changed the lengths (energies) of the signal vectors.

7.1.3 Geometric Representation of M-ary Pulse Modulation Signals

Let us begin by representing a PAM signal set in geometric form. The signal waveforms for M-ary PAM may be expressed as

$$s_m(t) = A_m g_T(t), \quad 0 \le t \le T, \quad m = 1, 2, \dots, M \quad (7.1.21)$$

where $M = 2^k$ and $g_T(t)$ is a pulse of unit amplitude as previously illustrated in Figure 7.5(a).

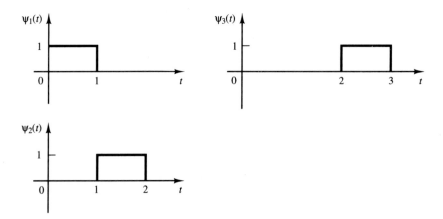

FIGURE 7.10. Alternate set of basis functions.

The M-ary PAM waveforms are one-dimensional signals, which may be expressed as

$$s_m(t) = s_m \psi(t), \quad m = 1, 2, \ldots, M \tag{7.1.22}$$

where the basis function $\psi(t)$ is defined as

$$\psi(t) = \frac{1}{\sqrt{\mathcal{E}_g}} g_T(t), \quad 0 \le t \le T \tag{7.1.23}$$

$\mathcal{E}_g$ is the energy of the signal pulse $g_T(t)$, and the signal coefficients (one-dimensional vectors) are simply

$$s_m = \sqrt{\mathcal{E}_g} A_m, \quad m = 1, 2, \ldots, M \tag{7.1.24}$$

The Euclidean distance between two signal points is

$$d_{mn} = \sqrt{|s_m - s_n|^2} = \sqrt{\mathcal{E}_g (A_m - A_n)^2} \tag{7.1.25}$$

If we select the signal amplitudes $\{A_m\}$ to be symmetrically spaced about zero and equally distant between adjacent signal amplitudes, we obtain the signal points for symmetric PAM, as shown in Figure 7.11.

We observe that the PAM signals have different energies. In particular, the energy of the mth signal is

$$\mathcal{E}_m = s_m^2 = \mathcal{E}_g A_m^2 \tag{7.1.26}$$

For equally probable signals, the average energy is

$$\mathcal{E}_{av} = \frac{1}{M} \sum_{m=1}^{M} \mathcal{E}_m = \frac{\mathcal{E}_g}{M} \sum_{m=1}^{M} A_m^2 \tag{7.1.27}$$

Next, we consider a set of M PPM signals which are represented as in (7.1.4), where $g(t)$ is a pulse of duration T/M, as previously illustrated in Figure 7.5(b). We recall that these M waveforms are orthogonal and have equal energy $\mathcal{E}_s$. Therefore, the representation of these waveforms requires M basis functions. We define the basis functions as

$$\psi_m(t) = \begin{cases} \frac{1}{\sqrt{\mathcal{E}_s}} g(t - (m-1)T/M), & (m-1)T/M \le t \le mT/M \\ 0, & \text{otherwise} \end{cases} \tag{7.1.28}$$

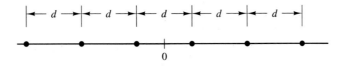

FIGURE 7.11. Signal points (constellation) for symmetric PAM.

for $m = 1, 2, \ldots, M$. Hence, M-ary PPM signal waveforms are represented geometrically by the M-dimensional vectors.

$$\mathbf{s}_1 = \left(\sqrt{\mathcal{E}_s}, 0, 0, \ldots, 0\right)$$

$$\mathbf{s}_2 = \left(0, \sqrt{\mathcal{E}_s}, 0, \ldots, 0\right)$$

$$\vdots \qquad \vdots \qquad\qquad (7.1.29)$$

$$\mathbf{s}_M = \left(0, 0, 0, \ldots, \sqrt{\mathcal{E}_s}\right)$$

Clearly, these vectors are orthogonal, i.e., $\mathbf{s}_i \cdot \mathbf{s}_j = 0$ when $i \neq j$. It is also interesting to note that the M signal vectors are mutually equidistant, i.e.,

$$d_{mn} = \sqrt{|\mathbf{s}_m - \mathbf{s}_n|^2} = \sqrt{2\mathcal{E}_s}, \quad \text{for all } m \neq n \qquad (7.1.30)$$

Hence, the minimum distance between signal points is $\sqrt{2\mathcal{E}_s}$. Figure 7.12 shows an example of $M = 3$ orthogonal signals.

The geometric representation of M-ary biorthogonal signals $\{s_m(t), \quad 1 \leq m \leq M/2, -s_m(t), \quad M/2+1 \leq m \leq M\}$ is straightforward. We begin with $M/2$ orthogonal vectors in $N = M/2$ dimensions and then append their negatives. Thus, we obtain

$$
\begin{aligned}
\mathbf{s}_1 &= \ (\sqrt{\mathcal{E}_s}, 0, 0, \ldots, 0) \\
\mathbf{s}_2 &= \ (0, \sqrt{\mathcal{E}_s}, 0, \ldots, 0) \\
&\ \ \vdots \qquad\qquad \vdots \\
\mathbf{s}_{M/2} &= \ (0, 0, 0, \ldots, \sqrt{\mathcal{E}_s}) \\
\mathbf{s}_{\frac{M}{2}+1} &= (-\sqrt{\mathcal{E}_s}, 0, 0, \ldots, 0) \\
&\ \ \vdots \qquad\qquad \vdots \\
\mathbf{s}_M &= (0, 0, 0, \ldots, -\sqrt{\mathcal{E}_s})
\end{aligned}
\qquad (7.1.31)
$$

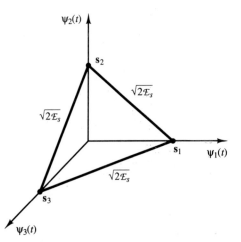

FIGURE 7.12. Orthogonal signals for $M = N = 3$.

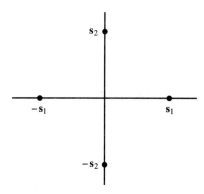

FIGURE 7.13. Signal constellation for $M = 4$ biorthogonal signals.

Figure 7.13 illustrates the biorthogonal signal vectors for $M = 4$.

We note that the distance between any pair of signal vectors is either $2\sqrt{\mathcal{E}_s}$ or $\sqrt{2\mathcal{E}_s}$. Hence, the minimum distance between pairs of signal vectors is $d_{\min} = \sqrt{2\mathcal{E}_s}$.

The geometric representation of a set of M simplex signals is obtained by subtracting the mean signal vector from a set of M orthogonal vectors. Thus, we have

$$\mathbf{s}'_m = \mathbf{s}_m - \frac{1}{M}\sum_{k=1}^{M}\mathbf{s}_k, \quad m = 1, 2, \ldots, M \tag{7.1.32}$$

The effect of subtracting the mean signal

$$\bar{\mathbf{s}} = \frac{1}{M}\sum_{k=1}^{M}\mathbf{s}_k \tag{7.1.33}$$

from each orthogonal vector is to translate the origin of the M orthogonal signals to the point $\bar{\mathbf{s}}$ and to minimize the energy in the signal set $\{\mathbf{s}'_m\}$.

If the energy per signal for the orthogonal signals is $\mathcal{E}_s = |\mathbf{s}_m|^2$, then the energy for the simplex signals is

$$\mathcal{E}'_s = \left|\mathbf{s}'_m\right|^2 = |\mathbf{s}_m - \bar{\mathbf{s}}|^2$$

$$= \left(1 - \frac{1}{M}\right)\mathcal{E}_s \tag{7.1.34}$$

The distance between any two signal points is not changed by the translation of the origin, i.e., the distance between signal points remains at $d = \sqrt{2\mathcal{E}_s}$. Finally, the M simplex signals are correlated. The *crosscorrelation coefficient* (normalized crosscorrelation) between the mth and n^{th} signals is

$$\gamma_{mn} = \frac{\mathbf{s}'_m \cdot \mathbf{s}'_n}{|\mathbf{s}'_m||\mathbf{s}'_n|}$$

$$= \frac{-1/M}{(1 - 1/M)} = -\frac{1}{M - 1} \tag{7.1.35}$$

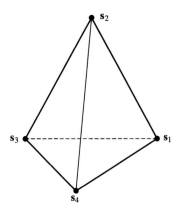

FIGURE 7.14. Signal constellation for $M = 4$ simplex signals.

Hence, all the signals have the same pair-wise correlation. Figure 7.14 illustrates a set of $M = 4$ simplex signals.

Finally, we consider the geometric representation of a set of M signal waveforms generated from a set of M binary words of the form

$$\mathbf{c}_m = (c_{m1}, c_{m2}, \ldots, c_{mN}), \quad m = 1, 2, \ldots, M \tag{7.1.36}$$

where $c_{mj} = 0$ or 1 for all m and j, as described in Section 7.1.1. The M signal waveforms are of dimension $N \leq M$ and are represented geometrically in vector form as

$$\mathbf{s}_m = (s_{m1}, s_{m2}, \ldots, s_{mN}), \quad m = 1, 2, \ldots, M \tag{7.1.37}$$

where $s_{mj} = \pm\sqrt{\mathcal{E}_s/N}$ for all m and j.

We observe that there are 2^N possible signals that can be constructed from the 2^N possible binary code words. We may select a subset of $M < 2^N$ signals for transmission of the information. We also observe that the 2^N possible signal points correspond to the vertices of an N-dimensional hypercube with its center at the origin. Figure 7.15 illustrates the signal points in $N = 2$ and $N = 3$ dimensions.

The M signals constructed in this manner have equal energy $\mathcal{E}_s$. The cross-correlation coefficient between any pair of signals depends on how we select the M signals from the 2^N possible signals. This topic is treated in Chapter 10. It is apparent that any adjacent signal points have a crosscorrelation coefficient of (see Problem 7.2)

$$\gamma = \frac{N - 2}{N} \tag{7.1.38}$$

and a corresponding Euclidean distance

$$d = 2\sqrt{\mathcal{E}_s/N} \tag{7.1.39}$$

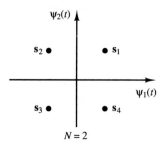

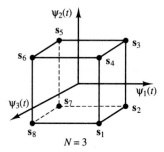

FIGURE 7.15.   Signal-space diagrams for signals generated from binary codes.

7.1.4 Modulation Codes and Modulation Signals with Memory

Coding is generally used in a digital communication system to achieve several objectives. We have already described source coding as a means for compressing the output of a source and representing the compressed output by a sequence of binary digits. Coding is also used for introducing redundancy into the data from the source encoder to correct errors that may occur as a result of noise and other interference in the transmission of the data through the communications channel. This type of coding is called *channel coding* and is treated in Chapter 10.

Coding may also be used in a digital communication system to shape the spectrum of the transmitted signal so that it matches the spectral characteristics of the baseband channel. Codes that are used for spectrum shaping are generally called either *modulation codes,* or *line codes,* or *data translation codes.* Such codes generally place restrictions on the sequence of bits into the modulator and, thus, introduce memory into the transmitted signal. It is this type of coding that is treated in this section. We shall consider the spectral-shaping properties of modulation codes in the next chapter, where we treat the problem of signal design for bandlimited channels.

Modulation codes are usually employed in magnetic recording, in optical recording, and in digital communications over cable systems to achieve spectral shaping of the modulated signal that matches the passband characteristics of the channel. Let us consider magnetic recording as an example.

In magnetic recording, we encounter two basic problems. One problem is concerned with the packing density that is used to write the data onto the magnetic

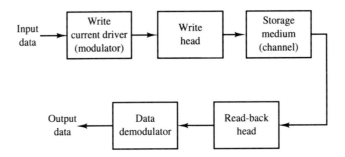

FIGURE 7.16. Block diagram of magnetic storage read/write system.

medium (disk or tape). Of course, we would like to write as many bits as possible on a single track. However, there is a limit as to how close successive bits in a sequence are stored, and this limit is imposed by the medium. Let us explore this problem further.

Figure 7.16 illustrates a block diagram of the magnetic recording system. The binary data sequence to be stored is used to generate a write current. This current may be viewed as the output of the "modulator." The two most commonly used methods to map the data sequence into the write current waveform are the so-called NRZ (non-return-to-zero) and NRZI (non-return-to-zero-inverse) methods. These two waveforms are illustrated in Figure 7.17. We note that NRZ is identical to binary PAM in which the information bit 1 is represented by a rectangular pulse of amplitude A and the information bit 0 is represented by a rectangular pulse of amplitude $-A$. In contrast, the NRZI signal waveform is different from NRZ in

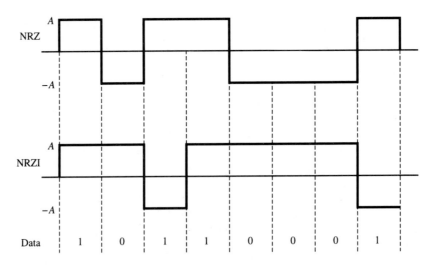

FIGURE 7.17. NRZ and NRZI signals.

that transitions from one amplitude level to another, (A to $-A$ or $-A$ to A), occur only when the information bit is a 1. No transition occurs when the information bit is a 0, i.e., the amplitude level remains the same as the previous signal level. The positive amplitude pulse results in magnetizing the medium on one (direction) polarity and the negative pulse magnetizes the medium in the opposite (direction) polarity.

Since the input data sequence is basically random with equally probable 1's and 0's, whether we use NRZ or NRZI, we will encounter level transitions for A to $-A$ or $-A$ to A with probability $1/2$ for every data bit. The readback signal for a positive transition ($-A$ to A) is a pulse that is well modeled mathematically as

$$p(t) = \frac{1}{1 + (2t/T_{50})^2} \qquad (7.1.40)$$

where T_{50} is defined as the width of the pulse at its 50% amplitude level, as shown in Figure 7.18. Similarly, the readback signal for a negative transition (A to $-A$) is the pulse $-p(t)$. The value of T_{50} is determined by the characteristics of the medium, the read/write heads, and the distance of the head to the medium.

Now, suppose we write a positive transition followed by a negative transition. Let's vary the time interval between the two transitions, which we denote as T_b (the bit time interval). Figure 7.19 illustrates the readback signal pulses, which are obtained by a superposition of $p(t)$ with $-p(t - T_b)$. The parameter $\Delta = T_{50}/T_b$ is defined as the *normalized density*. The closer the bit transitions (T_b small), the larger will be the value of the normalized density and, hence, the larger will be the packing density. We notice that as Δ is increased, the peak amplitudes of the readback signal are reduced and are also shifted in time from the desired time instants. In other words, the pulses interfere with one another, thus limiting the density with which we can write. This problem serves as a motivation to design modulation codes that take the original data sequence and transform (encode) it into another sequence that results in a write waveform in which amplitude transitions

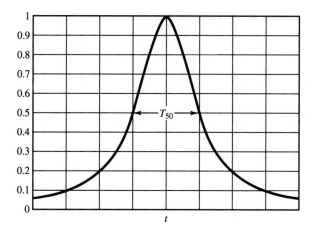

FIGURE 7.18.   Read-back pulse in magnetic recording system.

Pulse response

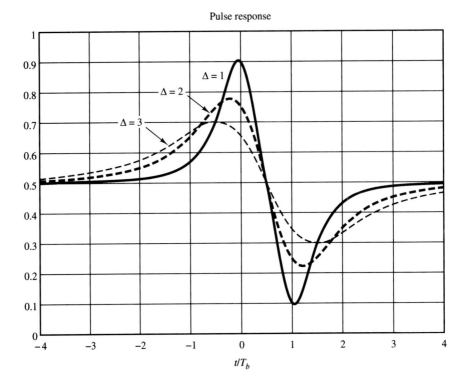

FIGURE 7.19. Read-back signal response to a pulse.

are spaced farther apart. For example, if we use NRZI, the encoded sequence into the modulator must contain one or more 0's between 1's.

The second problem encountered in magnetic recording is the need to avoid (or minimize) having a *dc* content in the modulated signal (the write current) due to the frequency response characteristics of the readback system and associated electronics. This requirement also arises in digital communication over cable channels. This problem can be overcome by altering (encoding) the data sequence into the modulator. A class of codes that satisfy these objectives are the modulation codes described below.

Runlength-limited codes.   Codes that have a restriction on the number of consecutive 1's or 0's in a sequence are generally called *runlength-limited codes*. These codes are generally described by two parameters, say, d and κ, where d denotes the minimum number of 0's between two 1's in a sequence, and κ denotes the maximum number of 0's between two 1's in a sequence. When used with NRZI modulation, the effect of placing d zeros between successive 1's is to spread the transitions farther apart, thus reducing the overlap in the channel response due to successive transitions. Setting an upper limit κ on the runlength of 0's ensures that transitions occur frequently enough so that symbol timing information can be recovered from the received modulated signal. Runlength-limited codes are usually called (d, κ) codes.[†]

The (d, κ) code sequence constraints may be represented by a finite-state sequential machine with $\kappa + 1$ states, denoted as σ_i, $1 \le i \le \kappa + 1$, as shown in Figure 7.20. We observe that an output data bit 0 takes the sequence from state σ_i to σ_{i+1}, $i \le \kappa$. The output data bit 1 takes the sequence to state σ_1. The output bit from the encoder may be a 1 only when the sequence is in state σ_i, $d + 1 \le i \le \kappa + 1$. When the sequence is in state $\sigma_{\kappa+1}$, the output bit is always 1.

The finite-state sequential machine may also be represented by a *state transition matrix,* denoted as $\mathbf{D}$, which is a square $(\kappa + 1) \times (\kappa + 1)$ with elements d_{ij}, where

$$d_{i1} = 1, \quad i \ge d + 1$$

$$d_{ij} = 1, \quad j = i + 1$$

$$d_{ij} = 0, \quad \text{otherwise} \tag{7.1.41}$$

Example 7.1.4

Determine the state transition matrix for a $(d, \kappa) = (1, 3)$ code.

Solution   The $(1, 3)$ code has four states. From Figure 7.20 we obtain its state transition matrix which is

$$\mathbf{D} = \begin{bmatrix} 0 & 1 & 0 & 0 \\ 1 & 0 & 1 & 0 \\ 1 & 0 & 0 & 1 \\ 1 & 0 & 0 & 0 \end{bmatrix} \tag{7.1.42}$$

An important parameter of any (d, κ) code is the number of sequences of a certain length, say n, which satisfy the (d, κ) constraints. As n is allowed to increase, the number of sequences $N(n)$ that satisfy the (d, κ) constraint also increases. The number of information bits that can be uniquely represented with $N(n)$

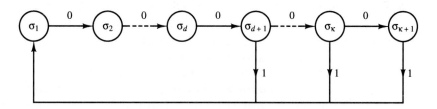

FIGURE 7.20.   Finite-state sequential machine for a (d, κ)-coded sequence.

[†]Runlength-limited codes are usually called (d, k) codes, where k is the maximum runlength of zeros. We have substituted the Greek letter κ for k, to avoid confusion with our previous use of k.

code sequences is

$$k = \lfloor \log_2 N(n) \rfloor$$

where $\lfloor x \rfloor$ denotes the largest integer contained in x. The maximum code rate is then $R_c = k/n$.

The capacity of a (d, κ) code is defined as

$$C(d, \kappa) = \lim_{n \to \infty} \frac{1}{n} \log_2 N(n) \qquad (7.1.43)$$

Clearly, $C(d, \kappa)$ is the maximum possible rate that can be achieved with the (d, κ) constraints. Shannon (1948) showed that the capacity is given as

$$C(d, \kappa) = \log_2 \lambda_{\max} \qquad (7.1.44)$$

where $\lambda_{\max}$ is the largest real eigenvalue of the state transition matrix $\mathbf{D}$.

Example 7.1.5

Determine the capacity of a $(d, \kappa) = (1, 3)$ code.

Solution Using the state-transition matrix given in Example 7.1.4 for the $(1, 3)$ code, we have

$$\det(\mathbf{D} - \lambda\mathbf{I}) = \det \begin{bmatrix} -\lambda & 1 & 0 & 0 \\ 1 & -\lambda & 1 & 0 \\ 1 & 0 & -\lambda & 1 \\ 1 & 0 & 0 & -\lambda \end{bmatrix} \qquad (7.1.45)$$

$$= \lambda^4 - \lambda^2 - \lambda - 1 = 0$$

The maximum real root of this polynomial is found to be $\lambda_{\max} = 1.4656$. Therefore, the capacity $C(1, 3) = \log_2 \lambda_{\max} = 0.5515$.

The capacities of (d, κ) for $0 \le d \le 6$ and $2 \le k \le 15$ are given in Table 7.1. We observe that $C(d, \kappa) < 1/2$ for $d \ge 3$ and any value of κ. The most commonly used codes for magnetic recording employ a $d \le 2$, hence, their rate R_c is at least $1/2$.

Now let us turn our attention to the construction of some runlength-limited codes. In general, (d, κ) codes can be constructed either as fixed-length codes or as variable-length codes. In a fixed-length code, each bit or block of k bits is encoded into a block of $n > k$ bits.

In principle, the construction of a fixed-length code is straightforward. For a given block length n, we may select the subset of the 2^n code words that satisfy the specified runlength constraints. From this subset, we eliminate code words that do not satisfy the runlength constraints when concatenated. Thus, we obtain a set of code words that satisfy the constraints and can be used in the mapping of the input data bits to the encoder. The encoding and decoding operations can be performed by use of a look-up table.

TABLE 7-1   CAPACITY $C(d, \kappa)$ VERSUS RUNLENGTH PARAMETERS d AND κ.

k	$d = 0$	$d = 1$	$d = 2$	$d = 3$	$d = 4$	$d = 5$	$d = 6$
2	.8791	.4057					
3	.9468	.5515	.2878				
4	.9752	.6174	.4057	.2232			
5	.9881	.6509	.4650	.3218	.1823		
6	.9942	.6690	.4979	.3746	.2269	.1542	
7	.9971	.6793	.5174	.4057	.3142	.2281	.1335
8	.9986	.6853	.5293	.4251	.3432	.2709	.1993
9	.9993	.6888	.5369	.4376	.3620	.2979	.2382
10	.9996	.6909	.5418	.4460	.3746	.3158	.2633
11	.9998	.6922	.5450	.4516	.3833	.3285	.2804
12	.9999	.6930	.5471	.4555	.3894	.3369	.2924
13	.9999	.6935	.5485	.4583	.3937	.3432	.3011
14	.9999	.6938	.5495	.4602	.3968	.3478	.3074
15	.9999	.6939	.5501	.4615	.3991	.3513	.3122
∞	1.000	.6942	.5515	.4650	.4057	.3620	.3282

Example 7.1.6

Construct a $d = 0, \kappa = 2$ code of length $n = 3$, and determine its efficiency.

Solution   By listing all the code words, we find that the following five code words satisfy the $(0, 2)$ constraint: $(0\ 1\ 0)$, $(0\ 1\ 1)$, $(1\ 0\ 1)$, $(1\ 1\ 0)$, $(1\ 1\ 1)$. We may select any four of these code words and use them to encode the pairs of data bits $(00, 01, 10, 11)$. Thus, we have a rate $k/n = 2/3$ code that satisfies the $(0, 2)$ constraint.

The fixed-length code in this example is not very efficient. The capacity is $C(0, 2) = 0.8791$, so that this code has an *efficiency* of

$$\text{efficiency} = \frac{R_c}{C(d, \kappa)} = \frac{2/3}{0.8791} = 0.76$$

Surely, better $(0, 2)$ codes can be constructed by increasing the block length n.

In the following example, we place no restriction on the maximum runlength zeros.

Example 7.1.7

Construct a $d = 1$, $\kappa = \infty$ code of length $n = 5$.

Solution   In this case, we are placing no constraint on the number of consecutive zeros. To construct the code, we select from the set of 32 possible code words, those that satisfy the $d = 1$ constraint. There are eight such code words, which implies that we can encode three information bits with each code word. The code is given in Table 7.2. Note that the first bit of each code word is a 0, whereas the last bit may be either 0 or 1. Consequently, the $d = 1$ constraint is satisfied when these code words are concatenated. This code has a rate $R_c = 3/5$. When compared with the capacity $C(1, \infty) = 0.6942$ obtained from Table 7.1, the code efficiency is 0.864, which is quite acceptable.

TABLE 7-2 FIXED LENGTH $d = 1$, $\kappa = \infty$ CODE.

Input data bits	Output coded sequence
0 0 0	0 0 0 0 0
0 0 1	0 0 0 0 1
0 1 0	0 0 0 1 0
0 1 1	0 0 1 0 0
1 0 0	0 0 1 0 1
1 0 1	0 1 0 0 0
1 1 0	0 1 0 0 1
1 1 1	0 1 0 1 0

The code construction method described in the two examples above produces fixed-length (d, κ) codes that are *state independent*. By state independent, we mean that fixed-length code words can be concatenated without violating the (d, κ) constraints. In general, fixed-length state-independent (d, κ) codes require large block lengths, except in cases such as those in the examples above where d is small. Simpler (shorter-length) codes are generally possible by allowing for state-dependence and for variable length code words. Below, we consider codes for which both input blocks to the encoder and the output blocks may have variable length. For the code words to be uniquely decodable at the receiver, the variable length code should satisfy the prefix condition, previously described in Chapter 4.

Example 7.1.8

Construct a simple variable-length $d = 0, \kappa = 2$ code.

Solution A very simple uniquely decodable $(0, 2)$ code is the following:

$$
\begin{array}{ccc}
0 & \rightarrow & 01 \\
10 & \rightarrow & 10 \\
11 & \rightarrow & 11
\end{array}
$$

The code in the above example has a fixed output block size but a variable input block size. In general, both the input and output blocks may be variable. The following example illustrates the latter case.

Example 7.1.9

Construct a $(2, 7)$ variable block size code.

Solution The solution to this code construction is certainly not unique nor is it trivial. We picked this example because the $(2, 7)$ code has been widely used by IBM in many of its disk storage systems. The code is listed in Table 7.3. We observe that the input data blocks of 2, 3, and 4 bits are mapped into output data blocks of 4, 6, and 8 bits, respectively. Hence, the code rate is $R_c = 1/2$. Since this is the code rate for all code words, the code is called a *fixed-rate* code. This code has an efficiency of $0.5/0.5174 = 0.966$. Note that this code satisfies the prefix condition.

TABLE 7-3   CODE BOOK FOR
VARIABLE-LENGTH $(2, 7)$ CODE.

Input data bits	Output coded sequence
1 0	1 0 0 0
1 1	0 1 0 0
0 1 1	0 0 0 1 0 0
0 1 0	0 0 1 0 0 0
0 0 0	1 0 0 1 0 0
0 0 1 1	0 0 1 0 0 1 0 0
0 0 1 0	0 0 0 0 1 0 0 0

TABLE 7-4   ENCODER FOR $(1, 3)$
MILLER CODE.

Input data bits	Output coded sequence
0	x 0
1	0 1

$x = 0$, if preceding input bit is 1
$x = 1$, if preceding input bit is 0

Another code that has been widely used in magnetic recording is the rate $1/2$, $(d, \kappa) = (1, 3)$ code given in Table 7.4. We observe that when the information bit is a 0, the first output bit is 1 if the previous input bit was 0, or a 0 if the previous input bit was a 1. When the information bit is a 1, the encoder output is 01. Decoding of this code is simple. The first bit of the two-bit block is redundant and may be discarded. The second bit is the information bit. This code is usually called the *Miller code*. We observe that this is a state-dependent code, which is described by the state diagram shown in Figure 7.21. There are two states labeled S_1 and S_2 with transitions as shown in the figure. When the encoder is a state S_1, an input bit 1 results in the encoder staying in state S_1 and outputs 01. This is denoted as $1/01$. If the input bit is a 0, the encoder enters state S_2 and outputs 00. This is denoted as $0/00$. Similarly, if the encoder is in state S_2, an input bit 0 causes no transition and the encoder output is 10. On the other hand, if the input bit is a 1, the encoder enters state S_1 and outputs 01.

Trellis representation of state-dependent (d, κ) **codes.**   The state diagram provides a relatively compact representation of a state-dependent code. Another way

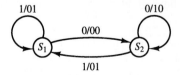

FIGURE 7.21.   State diagrams for $d = 1$, $\kappa = 3$ (Miller) code.

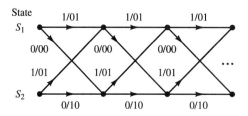

FIGURE 7.22. Trellis for $d = 1, \kappa = 3$ (Miller) code.

to describe such codes which have memory is by means of a graph called a *trellis*. A trellis is a graph that illustrates the state transitions as a function of time. It consists of a set of nodes representing the states that characterize the memory in the code at different instants in time and interconnections between pairs of nodes that indicate the transitions between successive instants of time. For example, Figure 7.22 shows the trellis for the $d = 1, \kappa = 3$ Miller code whose state diagram is shown in Figure 7.21.

The mapping of coded bits into signal waveforms.   The output sequence from a (d, κ) encoder is mapped by the modulator into signal waveforms for transmission over the channel. If the binary digit 1 is mapped into a rectangular pulse of amplitude A and the binary digit 0 is mapped into a rectangular pulse of amplitude $-A$, the result is a (d, κ) coded NRZ modulated signal. We note that the duration of the rectangular pulses is $T_c = R_c/R_b = R_c T_b$, where R_b is the information (bit) rate into the encoder, T_b is the corresponding (uncoded) bit interval, and R_c is the code rate for the (d, κ) code.

When the (d, κ) code is a state-independent fixed-length code with code rate $R_c = k/n$, we may consider each n-bit block as generating one signal waveform of duration nT_c. Thus, we have $M = 2^k$ signal waveforms, one for each of the 2^k possible k-bit data blocks. These coded waveforms have the general form given by (7.1.36) and (7.1.37). In this case, there is no dependence between the transmission of successive waveforms.

In contrast to the situation considered above, where the (d, κ) code is state independent and NRZ modulation is used for transmission of the coded bits, the modulation signal is no longer memoryless when NRZI is used and/or the (d, κ) code is state dependent. Let us first consider the effect of mapping the coded bits into an NRZI signal waveform.

An NRZI modulated signal is itself state dependent. The signal amplitude level is changed from its current value $(\pm A)$ only when the bit to be transmitted is a 1. It is interesting to note that the NRZI signal may be viewed as an NRZ signal preceded by another encoding operation, called *precoding,* of the binary sequence, as shown in Figure 7.23. The precoding operation is described mathematically by

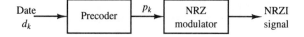

FIGURE 7.23. Method for generating an NRZI signal using precoding.

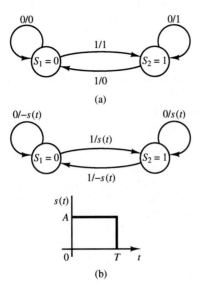

(a)

(b)

FIGURE 7.24.   State diagram for NRZI signal.

the relation

$$p_k = d_k \oplus p_{k-1}$$

where $\{d_k\}$ is the binary sequence into the precoder, $\{p_k\}$ is the output binary sequence from the precoder, and $\oplus$ denotes modulo-2 addition. This type of encoding is called *differential encoding,* and ıs characterized by the state diagram shown in Figure 7.24(a). Then, the sequence $\{p_k\}$ is transmitted by NRZ. Thus, when $p_k = 1$, the modulator output is a rectangular pulse of amplitude A, and when $p_k = 0$ the modulator output is a rectangular pulse of amplitude $-A$. When the signal waveforms are superimposed on the state diagram of Figure 7.24(a), we obtain the corresponding state diagram shown in Figure 7.24(b).

It is apparent from the state diagram that differential encoding or precoding as described above introduces memory in the modulated signal. As in the case of state-dependent (d, κ) codes, a trellis diagram may be used to illustrate the time dependence of the modulated signal. The trellis diagram for the NRZI signal is shown in Figure 7.25. When the output of a state-dependent (d, κ) encoder is followed by an NRZI modulator, we may simply combine the two-state diagrams into a single-state diagram for the (d, κ) code with precoding. A similar combination

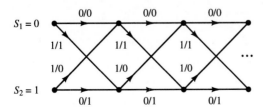

FIGURE 7.25.   The trellis for an NRZI signal.

can be performed with the corresponding trellises. The following example illustrates the approach for the $(1, 3)$ Miller code followed by NRZI modulation.

Example 7.1.10

Determine the state diagram of the combined $(1, 3)$ Miller code followed by the precoding inherent in NRZI modulation.

Solution   Since the $(1, 3)$ Miller code has two states and the precoder has two states, the state diagram for the combined encoder has four states, which we denote as $(S_M, S_N) = (\sigma_1, s_1), (\sigma_1, s_2), (\sigma_2, s_1), (\sigma_2, s_2)$, where $S_M = \{\sigma_1, \sigma_2\}$ represents the two states of the Miller code and $S_N = \{s_1, s_2\}$ represents the two states of the precoder for NRZI. For each data input bit into the Miller encoder, we obtain two output bits which are then precoded to yield two precoded output bits. The resulting state diagram is shown in Figure 7.26, where the first bit denotes the information bit into the Miller encoder and the next two bits represent the corresponding output of the precoder.

The trellis diagram for the Miller precoded sequence may be obtained directly from the combined state diagram or from a combination of the trellises of the two codes. The result of this combination is the four-state trellis, one stage of which is shown in Figure 7.27.

It is interesting to note that the four signal waveforms obtained by mapping each pair of bits of the Miller-precoded sequence into an NRZ signal are biorthogonal. In particular, the pair of bits 11 map into the waveform $s_1(t)$ and the bits 01 map into the waveform $s_2(t)$, shown in Figure 7.28. Then, the encoded bits 00 map into $-s_1(t)$ and the bits 10 map into $-s_2(t)$. Since $s_1(t)$ and $s_2(t)$ are orthogonal, the set of four waveforms constitute a biorthogonal set of $M = 4$ waveforms. Hence, in the state diagram for the Miller precoded sequence, we may substitute the signal waveforms from the modulator in place of the encoder output bits. This state diagram is illustrated in Figure 7.29, where the four states are simply designated as S_i, $1 \le i \le 4$. The resulting modulated signal has also been called *delay modulation*.

Modulated signals with memory such as NRZI and Miller coded/NRZI (delay modulation) are generally characterized by a K-state Markov chain with *stationary*

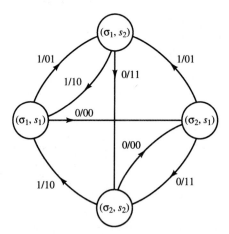

FIGURE 7.26.   State diagram of the Miller code followed by the precoder.

State (S_M, S_N)

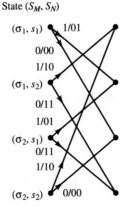

FIGURE 7.27.   One stage of trellis diagram for the Miller code followed by the precoder.

state probabilities $\{p_i, i = 1, 2, \ldots, K\}$. Associated with each transition is a signal waveform $s_j(t)$, $j = 1, 2, \ldots, K$. Thus the transition probability p_{ij} denotes the probability that signal waveform $s_j(t)$ is transmitted in a given signaling interval after the transmission of the signal waveform $s_i(t)$ in the previous signaling interval. The transition probabilities may be arranged in matrix form as

$$\mathbf{P} = \begin{bmatrix} p_{11} & p_{12} & \cdots & p_{1K} \\ p_{21} & p_{22} & \cdots & p_{2K} \\ \vdots & & & \\ p_{K1} & p_{K2} & \cdots & p_{KK} \end{bmatrix} \tag{7.1.46}$$

where $\mathbf{P}$ is the *transition probability matrix*. The transition probability matrix is easily obtained from the state diagram and the corresponding probabilities of occurrence of the input bits (or equivalently, the stationary state probabilities $\{p_i\}$).

For the NRZI signal with equal state probabilities $p_1 = p_2 = 1/2$ and the state diagram shown in Figure 7.24, the transition probability matrix is

$$\mathbf{P} = \begin{bmatrix} \frac{1}{2} & \frac{1}{2} \\ \frac{1}{2} & \frac{1}{2} \end{bmatrix} \tag{7.1.47}$$

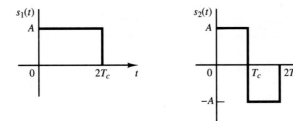

FIGURE 7.28.   Signal waveforms for Miller-precoded pairs of bits.

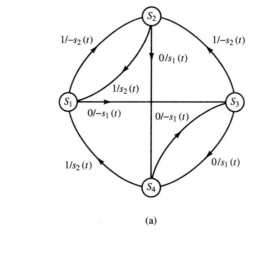

(a)

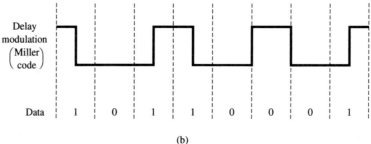

(b)

FIGURE 7.29. State diagram for the Miller-precoded signal and sample waveform.

Similarly, the transition probability matrix for the Miller-coded NRZI modulated signal with equally likely symbols ($p_1 = p_2 = p_3 = p_4 = 1/4$) is

$$\mathbf{P} = \begin{bmatrix} 0 & \frac{1}{2} & 0 & \frac{1}{2} \\ 0 & 0 & \frac{1}{2} & \frac{1}{2} \\ \frac{1}{2} & \frac{1}{2} & 0 & 0 \\ \frac{1}{2} & 0 & \frac{1}{2} & 0 \end{bmatrix} \qquad (7.1.48)$$

As we shall see later, in Chapter 8, the transition probability matrix is useful in determining the power-spectral density of a modulated signal with memory.

Modulation code for the compact disc system.   In the design of a modulation code for the compact disc system, several factors had to be considered. One constraint is that the maximum runlength of zeros must be sufficiently small to allow the system to synchronize from the readback signal. To satisfy this constraint, $\kappa = 10$ was chosen. A second factor is the frequency content of the modulated signal below 20 kHz. In a CD system, the servo-control systems that keep the laser beam on a specified track and the focusing circuits are controlled by signals in the frequency range 0–20 kHz. To avoid interference with these control signals, the modulated information signal must fall above this frequency range. The runlength-limited code that was implemented for this purpose is a $(d, \kappa) = (2, 10)$ code that results in a coded information-bearing signal that occupies the frequency range 20 kHz to 1.5 MHz.

The $d = 2, \kappa = 10$ code selected is a fixed-length code that encodes eight information bits into 14 coded bits and is called EFM (eight-to-fourteen modulation). Since each audio signal sample is quantized to 16 bits, a 16-bit sample is divided into two 8-bit bytes and encoded. By enumerating all the 14-bit code words, one can show that there are 267 distinct code words that satisfy the (d, κ) constraints. Of these, 256 code words are selected to form the code. However, the (d, κ) constraints are not satisfied when the code words are concatenated (merged in a sequence). To remedy this problem, three additional bits are added to the 14, called "merging" bits. The three merging bits serve two purposes. First, if the d-constraint is not satisfied in concatenation, we choose 0's for the merging bits. On the other hand, if the κ-constraint is being violated due to concatenation, we select one of the bits as 1. Since two bits are sufficient to accomplish these goals, the third merging bit may be viewed as an added degree of freedom. This added degree of freedom allows us to use the merging bits to minimize the low-frequency content of the modulated signal. Since the merging bits carry no audio signal information, they are discarded in the readback process prior to decoding.

A measure of the low-frequency content of a digital modulated signal is the *running digital sum* (RDS), which is the difference between the total zeros and the total ones in the coded sequence accumulated from the beginning of the disc. In addition to satisfying the (d, κ) constraints, the three merging bits are selected so as to bring the RDS as close to zero as possible. Thus, the three merging bits are instrumental in reducing the low-frequency content below 20 kHz by an additional factor of 10 (in power).

In addition to the coded information bits and merging bits, additional bits are added for control and display (C&D), synchronization bits, and parity bits. The data bits, control bits, parity bits, and synchronization bits are arranged in a frame structure, consisting of 588 bits per frame. This frame structure is illustrated in Figure 7.30.

7.2 OPTIMUM RECEIVER FOR PULSE-MODULATED SIGNALS IN ADDITIVE WHITE GAUSSIAN NOISE

Let us consider a digital communication system that transmits digital information by use of any one of the M-ary signal waveforms described in the preceding section. Thus, the input sequence to the modulator is subdivided into k-bit blocks or symbols and each of the $M = 2^k$ symbols is associated with a corresponding waveform from

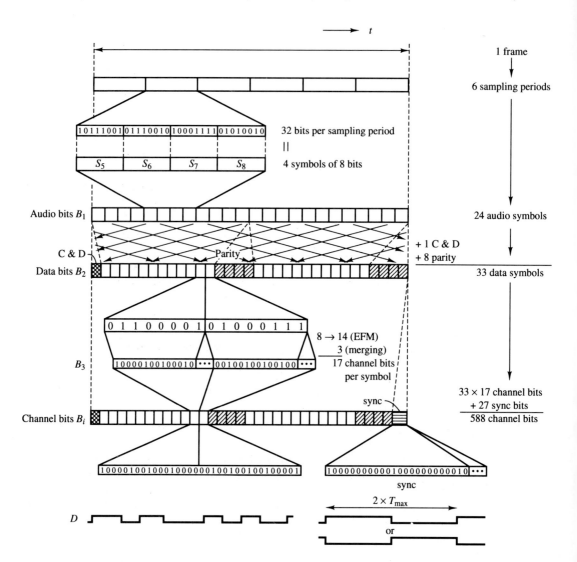

FIGURE 7.30. Frame structure of data in a compact disc. The information is divided into frames; the figure gives one frame of the successive bit streams. There are six sampling periods for one frame, each sampling period giving 32 bits (16 for each of the two audio channels). These 32 bits are divided to make four symbols in the "audio bit stream" B_1. In the "data bit stream," B_2 eight parity symbols and one C&D symbol have been added to the 24 audio symbols. To scatter possible errors, the symbols of different frames in B_1 are interleaved, so that the audio signals in one frame of B_2 originate from different frames in B_1. The modulation translated the eight data bits of a symbol of B_2 into fourteen channel bits, to which three 'merging bits' are added (B_3). The frames are marked with a synchronization signal of the form illustrated (bottom right); the final result is the "channel bit stream" (B_i) used for writing on the master disc, in such a way that each 1 indicates a pit edge (D). [From paper by Heemskerk and Schouhamer Immink (1982).]

the set $\{s_m(t), m = 1, 2, \ldots, M\}$. Each waveform is transmitted within the symbol (signaling) interval or time slot of duration T. To be specific, we consider the transmission of information over the interval $0 \leq t \leq T$.

The channel is assumed to corrupt the signal by the addition of white Gaussian noise as shown in Figure 7.31. Thus, the received signal in the interval $0 \leq t \leq T$ may be expressed as

$$r(t) = s_m(t) + n(t), \quad 0 \leq t \leq T \tag{7.2.1}$$

where $n(t)$ denotes the sample function of the additive white Gaussian noise (AWGN) process with power-spectral density $S_n(f) = \frac{N_0}{2}$ W/Hz. Based on the observation of $r(t)$ over the signal interval, we wish to design a receiver that is optimum in the sense that it minimizes the probability of making an error.

It is convenient to subdivide the receiver into two parts, the signal demodulator and the detector. The function of the signal demodulator is to convert the received waveform $r(t)$ into an N-dimensional vector $\mathbf{r} = (r_1, r_2, \ldots, r_N)$, where N is the dimension of the transmitted signal waveforms. The function of the detector is to decide which of the M possible signal waveforms was transmitted based on observation of the vector $\mathbf{r}$.

Two realizations of the signal demodulator are described in the next two sections. One is based on the use of signal correlators. The second is based on the use of matched filters. The optimum detector that follows the signal demodulator is designed to minimize the probability of error.

7.2.1 Correlation-Type Demodulator

In this section, we describe a correlation-type demodulator that decomposes the received signal and the noise into N-dimensional vectors. In other words, the signal and the noise are expanded into a series of linearly weighted orthonormal basis functions $\{\psi_n(t)\}$. It is assumed that the N basis functions $\{\psi_n(t)\}$ span the signal space, so that every one of the possible transmitted signals of the set $\{s_m(t), 1 \leq m \leq M\}$ can be represented as a weighted linear combination of $\{\psi_n(t)\}$. In the

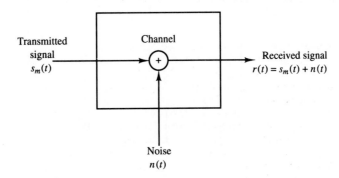

FIGURE 7.31. Model for received signal passed through an AWGN channel.

case of the noise, the functions $\{\psi_n(t)\}$ do not span the noise space. However, we show below that the noise terms that fall outside the signal space are irrelevant to the detection of the signal.

Suppose the received signal $r(t)$ is passed through a parallel bank of N crosscorrelators which basically compute the projection of $r(t)$ onto the N basis functions $\{\psi_n(t)\}$, as illustrated in Figure 7.32. Thus, we have

$$\int_0^T r(t)\psi_k(t)dt = \int_0^T [s_m(t) + n(t)]\psi_k(t)\,dt$$

$$r_k = s_{mk} + n_k, \quad k = 1, 2, \ldots, N \tag{7.2.2}$$

where

$$s_{mk} = \int_0^T s_m(t)\psi_k(t)\,dt, \quad k = 1, 2, \ldots, N$$

$$n_k = \int_0^T n(t)\psi_k(t)\,dt, \quad k = 1, 2, \ldots, N \tag{7.2.3}$$

The signal is now represented by the vector s_m with components s_{mk}, $k = 1, 2, \ldots, N$. Their values depend on which of the M signals was transmitted. The components $\{n_k\}$ are random variables that arise from the presence of the additive noise.

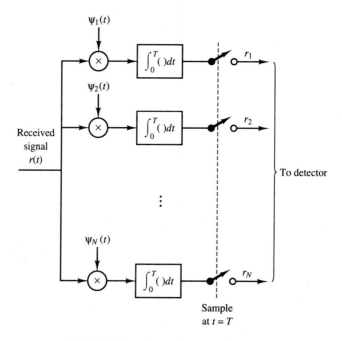

FIGURE 7.32. Correlation-type demodulator.

In fact, we can express the received signal $r(t)$ in the interval $0 \leq t \leq T$ as

$$r(t) = \sum_{k=1}^{N} s_{mk} \psi_k(t) + \sum_{k=1}^{N} n_k \psi_k(t) + n'(t)$$

$$= \sum_{k=1}^{N} r_k \psi_k(t) + n'(t) \qquad (7.2.4)$$

The term $n'(t)$, defined as

$$n'(t) = n(t) - \sum_{k=1}^{N} n_k \psi_k(t) \qquad (7.2.5)$$

is a zero-mean Gaussian noise process that represents the difference between the original noise process $n(t)$ and that part which corresponds to the projection of $n(t)$ onto the basis functions $\{\psi_k(t)\}$. We will show below that $n'(t)$ is irrelevant to the decision as to which signal was transmitted. Consequently, the decision may be based entirely on the correlator output signal and noise components $r_k = s_{mk} + n_k$, $k = 1, 2, \ldots, N$.

Since the signals $\{s_m(t)\}$ are deterministic, the signal components are deterministic. The noise components $\{n_k\}$ are Gaussian. Their mean values are

$$E[n_k] = \int_0^T E[n(t)]\psi_k(t)\,dt = 0 \qquad (7.2.6)$$

for all k. Their covariances are

$$E[n_k n_m] = \int_0^T \int_0^T E[n(t)n(\tau)]\psi_k(t)\psi_m(\tau)\,dt\,d\tau$$

$$= \frac{N_0}{2} \int_0^T \int_0^T \delta(t - \tau)\psi_k(t)\psi_m(\tau)\,dt\,d\tau$$

$$= \frac{N_0}{2} \int_0^T \psi_k(t)\psi_m(t)\,dt$$

$$= \frac{N_0}{2} \delta_{mk} \qquad (7.2.7)$$

where $\delta_{mk} = 1$ when $m = k$ and zero otherwise. Therefore, the N noise components $\{n_k\}$ are zero-mean uncorrelated Gaussian random variables with a common variance $\sigma_n^2 = N_0/2$.

From the above development, it follows that the correlator outputs $\{r_k\}$ conditioned on the mth signal being transmitted are Gaussian random variables with mean

$$E[r_k] = E[s_{mk} + n_k] = s_{mk} \qquad (7.2.8)$$

and equal variance

$$\sigma_r^2 = \sigma_n^2 = N_0/2 \qquad (7.2.9)$$

Since the noise components $\{n_k\}$ are uncorrelated Gaussian random variables, they are also statistically independent. As a consequence, the correlator outputs $\{r_k\}$ conditioned on the mth signal being transmitted are statistically independent Gaussian variables. Hence, the conditional probability density functions (p.d.f.'s) of the random variables $(r_1, r_2, \ldots, r_N) = \mathbf{r}$ are simply

$$f(\mathbf{r}|\mathbf{s}_m) = \prod_{k=1}^{N} f(r_k|s_{mk}), \qquad m = 1, 2, \ldots, M \qquad (7.2.10)$$

where

$$f(r_k|s_{mk}) = \frac{1}{\sqrt{\pi N_0}} e^{-(r_k - s_{mk})^2/N_0} \qquad k = 1, 2, \ldots, N \qquad (7.2.11)$$

By substituting (7.2.11) into (7.2.10), we obtain the joint conditional p.d.f.'s as

$$f(\mathbf{r}|\mathbf{s}_m) = \frac{1}{(\pi N_0)^{N/2}} \exp\left[-\sum_{k=1}^{N} (r_k - s_{mk})^2/N_0 \right], \qquad m = 1, 2, \ldots, M \qquad (7.2.12)$$

As a final point, we wish to show that the correlator outputs $(r_1, r_2, \ldots, r_N)$ are *sufficient statistics* for reaching a decision on which of the M signals was transmitted, i.e., that no additional relevant information can be extracted from the remaining noise process $n'(t)$. Indeed, $n'(t)$ is uncorrelated with the N correlator outputs $\{r_k\}$, i.e.,

$$E[n'(t)r_k] = E[n'(t)]s_{mk} + E[n'(t)n_k]$$

$$= E[n'(t)n_k]$$

$$= E\left\{ \left[n(t) - \sum_{j=1}^{N} n_j \psi_j(t) \right] n_k \right\}$$

$$= \int_0^T E[n(t)n(\tau)]\psi_k(\tau)d\tau - \sum_{j=1}^{N} E[n_j n_k]\psi_j(t)$$

$$= \frac{N_0}{2} \psi_k(t) - \frac{N_0}{2} \psi_k(t) = 0 \qquad (7.2.13)$$

Since $n'(t)$ and $\{r_k\}$ are Gaussian and uncorrelated, they are also statistically independent. Consequently, $n'(t)$ does not contain any information that is relevant to the decision as to which signal waveform was transmitted. All the relevant information is contained in the correlator outputs $\{r_k\}$. Hence, $n'(t)$ may be ignored.

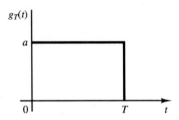

FIGURE 7.33. Signal pulse for
Example 7.2.1.

Example 7.2.1

Consider an M-ary PAM signal set in which the basic pulse shape $g_T(t)$ is rectangular as shown in Figure 7.33. The additive noise is a zero-mean white Gaussian noise process. Determine the basis function $\psi(t)$ and the output of the correlation-type demodulator.

Solution   The energy in the rectangular pulse is

$$\mathcal{E}_g = \int_0^T g_T^2(t)dt = \int_0^T a^2 dt = a^2 T$$

Since the PAM signal set has a dimension $N = 1$, there is only one basis function $\psi(t)$. This is obtained from (7.1.23) and given as

$$\psi(t) = \frac{1}{\sqrt{a^2 T}} g_T(t)$$

$$= \begin{cases} \frac{1}{\sqrt{T}}, & 0 \leq t \leq T \\ 0, & \text{otherwise} \end{cases}$$

The output of the correlation-type demodulator is

$$r = \int_0^T r(t)\psi(t)dt = \frac{1}{\sqrt{T}} \int_0^T r(t)\,dt$$

It is interesting to note that the correlator becomes a simple integrator when $\psi(t)$ is rectangular. If we substitute for $r(t)$ we obtain

$$r = \frac{1}{\sqrt{T}} \left[\int_0^T [s_m(t) + n(t)] \right] dt$$

$$= \frac{1}{\sqrt{T}} \left[\int_0^T s_m \psi(t)\,dt + \int_0^T n(t)\,dt \right]$$

$$r = s_m + n$$

where the noise term $E[n] = 0$ and

$$\sigma_n^2 = E\left[\frac{1}{T} \int_0^T \int_0^T n(t)n(\tau)\,dt\,d\tau \right]$$

$$= \frac{1}{T} \int_0^T \int_0^T E[n(t)n(\tau)] \, dt \, d\tau$$

$$= \frac{N_0}{2T} \int_0^T \int_0^T \delta(t - \tau) \, dt \, d\tau = \frac{N_0}{2}$$

The probability density function for the sampled output is

$$f(r|s_m) = \frac{1}{\sqrt{\pi N_0}} e^{-(r - s_m)^2/N_0}$$

7.2.2 Matched-Filter-Type Demodulator

Instead of using a bank of N correlators to generate the variables $\{r_k\}$, we may use a bank of N linear filters. To be specific, let us suppose that the impulse responses of the N filters are

$$h_k(t) = \psi_k(T - t), \quad 0 \le t \le T \tag{7.2.14}$$

where $\{\psi_k(t)\}$ are the N basis functions and $h_k(t) = 0$ outside of the interval $0 \le t \le T$. The outputs of these filters are

$$y_k(t) = \int_0^t r(\tau)h_k(t - \tau) \, d\tau$$

$$= \int_0^t r(\tau)\psi_k(T - t + \tau) \, d\tau, \quad k = 1, 2, \dots, N \tag{7.2.15}$$

Now, if we sample the outputs of the filters at $t = T$, we obtain

$$y_k(T) = \int_0^T r(\tau)\psi_k(\tau) \, d\tau = r_k, \quad k = 1, 2, \dots, N \tag{7.2.16}$$

Hence, the sampled outputs of the filters at time $t = T$ are exactly the set of values $\{r_k\}$ obtained from the N linear correlators.

A filter whose impulse response $h(t) = s(T - t)$, where $s(t)$ is assumed to be confined to the time interval $0 \le t \le T$, is called the *matched filter* to the signal $s(t)$. An example of a signal and its matched filter are shown in Figure 7.34. The response of $h(t) = s(T - t)$ to the signal $s(t)$ is

$$y(t) = \int_0^t s(\tau)s(T - t + \tau) \, d\tau$$

which is basically the time-autocorrelation function of the signal $s(t)$. Figure 7.35 illustrates $y(t)$ for the triangular signal pulse shown in Figure 7.34. We note that the autocorrrelation function $y(t)$ is an even function of t, which attains a peak at $\tau = T$.

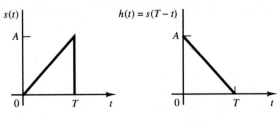

(a) Signal $s(t)$

(b) Impulse response
of filter matched to $s(t)$

FIGURE 7.34. Signal $s(t)$ and filter
matched to $s(t)$.

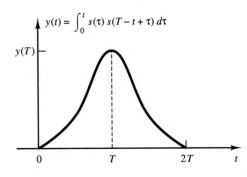

FIGURE 7.35. Matched filter output is
the autocorrelation function of $s(t)$.

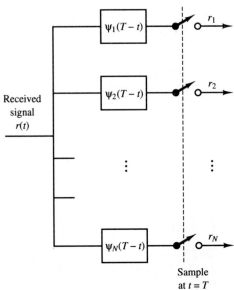

FIGURE 7.36. Matched filter-type
demodulator.

In the case of the demodulator described above, the N matched filters are matched to the basis functions $\{\psi_k(t)\}$. Figure 7.36 illustrates the matched filter-type demodulator that generates the observed variables $\{r_k\}$.

Properties of the matched filter. A matched filter has some interesting properties. Let us prove the most important property, which may be stated as follows: If a signal $s(t)$ is corrupted by AWGN, the filter with impulse response matched to $s(t)$ maximizes the output signal-to-noise ratio (SNR).

To prove this property, let us assume that the received signal $r(t)$ consists of the signal $s(t)$ and AWGN $n(t)$ which has zero-mean and power-spectral density $S_n(f) = N_0/2$ W/Hz. Suppose the signal $r(t)$ is passed through a filter with impulse response $h(t), 0 \leq t \leq T$, and its output is sampled at time $t = T$. The filter response to the signal and noise components is

$$y(t) = \int_0^t r(\tau)h(t - \tau)d\tau$$

$$= \int_0^t s(\tau)h(t - \tau)d\tau + \int_0^t n(\tau)h(t - \tau)d\tau \qquad (7.2.17)$$

At the sampling instant $t = T$, the signal and noise components are

$$y(T) = \int_0^T s(\tau)h(T - \tau)d\tau + \int_0^T n(\tau)h(T - \tau)d\tau$$

$$= y_s(T) + y_n(T) \qquad (7.2.18)$$

where $y_s(T)$ represents the signal component and $y_n(T)$ represents the noise component. The problem is to select the filter impulse response that maximizes the output signal-to-noise ratio (SNR) defined as

$$\left(\frac{S}{N}\right)_o = \frac{y_s^2(T)}{E\left[y_n^2(T)\right]} \qquad (7.2.19)$$

The denominator in (7.2.19) is simply the variance of the noise term at the output of the filter. Let us evaluate $E\left[y_n^2(T)\right]$. We have

$$E\left[y_n^2(T)\right] = \int_0^T \int_0^T E[n(\tau)n(t)]h(T - \tau)h(T - t)\,dt\,d\tau$$

$$= \frac{N_0}{2} \int_0^T \int_0^T \delta(t - \tau)h(T - \tau)h(T - t)\,dt\,d\tau$$

$$= \frac{N_0}{2} \int_0^T h^2(T - t)\,dt \qquad (7.2.20)$$

Note that the variance depends on the power-spectral density of the noise and the energy in the impulse response $h(t)$.

By substituting for $y_s(T)$ and $E\left[y_n^2(T)\right]$ into (7.2.19), we obtain the expression for the output SNR as

$$\left(\frac{S}{N}\right)_0 = \frac{\left[\int_0^T s(\tau)h(T-\tau)d\tau\right]^2}{\frac{N_0}{2}\int_0^T h^2(T-t)dt} = \frac{\left[\int_0^T h(\tau)s(T-\tau)d\tau\right]^2}{\frac{N_0}{2}\int_0^T h^2(T-t)dt} \tag{7.2.21}$$

Since the denominator of the SNR depends on the energy in $h(t)$, the maximum output SNR over $h(t)$ is obtained by maximizing the numerator of $(S/N)_0$ subject to the constraint that the denominator is held constant. The maximization of the numerator is most easily performed by use of the Cauchy-Schwarz inequality (see Section 2.2.1), which states, in general, that if $g_1(t)$ and $g_2(t)$ are finite-energy signals, then

$$\left[\int_{-\infty}^{\infty} g_1(t)g_2(t)dt\right]^2 \leq \int_{-\infty}^{\infty} g_1^2(t)dt \int_{-\infty}^{\infty} g_2^2(t)\,dt \tag{7.2.22}$$

where equality holds when $g_1(t) = Cg_2(t)$ for any arbitrary constant C. If we set $g_1(t) = h(t)$ and $g_2(t) = s(T-t)$, it is clear that the $(S/N)_0$ is maximized when $h(t) = Cs(T-t)$, i.e., $h(t)$ is matched to the signal $s(t)$. The scale factor C^2 drops out of the expression for $(S/N)_0$, since it appears in both the numerator and the denominator.

The output (maximum) SNR obtained with the matched filter is

$$\left(\frac{S}{N}\right)_o = \frac{2}{N_0}\int_0^T s^2(t)\,dt$$

$$= \frac{2\mathcal{E}_s}{N_0} \tag{7.2.23}$$

Note that the output SNR from the matched filter depends on the energy of the waveform $s(t)$ but not on the detailed characteristics of $s(t)$. This is another interesting property of the matched filter.

Frequency domain interpretation of the matched filter.   The matched filter has an interesting frequency domain interpretation. Since $h(t) = s(T-t)$, the Fourier transform of this relationship is

$$H(f) = \int_0^T s(T-t)e^{-j2\pi ft}\,dt$$

$$= \left[\int_0^T s(\tau)e^{j2\pi f\tau}\,d\tau\right]e^{-j2\pi fT}$$

$$= S^*(f)e^{-j2\pi fT} \tag{7.2.24}$$

We observe that the matched filter has a frequency response which is the complex conjugate of the transmitted signal spectrum multiplied by the phase factor $e^{-j2\pi fT}$, which represents the sampling delay of T. In other words, $|H(f)| = |S(f)|$, so that the magnitude response of the matched filter is identical to the transmitted signal spectrum. On the other hand, the phase of $H(f)$ is the negative of the phase of $S(f)$.

Now, if the signal $s(t)$ with spectrum $S(f)$ is passed through the matched filter, the filter output has a spectrum $Y(f) = |S(f)|^2 e^{-j2\pi fT}$. Hence, the output waveform is

$$y_s(t) = \int_{-\infty}^{\infty} Y(f) e^{j2\pi ft} \, df$$

$$= \int_{-\infty}^{\infty} |S(f)|^2 e^{-j2\pi fT} e^{j2\pi ft} \, df \tag{7.2.25}$$

By sampling the output of the matched filter at $t = T$, we obtain

$$y_s(T) = \int_{-\infty}^{\infty} |S(f)|^2 \, df = \int_0^T s^2(t) \, dt = \mathcal{E}_s \tag{7.2.26}$$

where the last step follows from Parseval's relation.

The noise of the output of the matched filter has a power-spectral density

$$S_o(f) = |H(f)|^2 N_0/2 \tag{7.2.27}$$

Hence, the total noise power at the output of the matched filter is

$$P_n = \int_{-\infty}^{\infty} S_o(f) \, df$$

$$= \frac{N_0}{2} \int_{-\infty}^{\infty} |H(f)|^2 \, df = \frac{N_0}{2} \int_{-\infty}^{\infty} |S(f)|^2 \, df = \frac{\mathcal{E}_s N_0}{2} \tag{7.2.28}$$

The output SNR is simply the ratio of the signal power P_s, given by

$$P_s = y_s^2(T) \tag{7.2.29}$$

to the noise power P_n. Hence,

$$\left(\frac{S}{N}\right)_o = \frac{P_s}{P_n} = \frac{\mathcal{E}_s^2}{\mathcal{E}_s N_0/2} = \frac{2\mathcal{E}_s}{N_0} \tag{7.2.30}$$

which agrees with the result given by (7.2.23).

Example 7.2.2

Consider the $M = 4$ biorthogonal signals shown in Figure 7.6 for transmitting information over an AWGN channel. The noise is assumed to have zero mean and power-spectral density $N_0/2$. Determine the basis functions for this signal set, the impulse response of the matched-filter demodulators, and the output waveforms of the matched-filter demodulators when the transmitted signal is $s_1(t)$.

Solution   The $M = 4$ biorthogonal signals have dimension $N = 2$. Hence, two basis functions are needed to represent the signals. From Figure 7.6, we choose $\psi_1(t)$ and $\psi_2(t)$ as

$$\psi_1(t) = \begin{cases} \sqrt{\frac{2}{T}}, & 0 \le t \le \frac{T}{2} \\ 0, & \text{otherwise} \end{cases}$$

$$\psi_2(t) = \begin{cases} \sqrt{\frac{2}{T}}, & \frac{T}{2} \le t \le T \\ 0, & \text{otherwise} \end{cases} \qquad (7.2.31)$$

These waveforms are illustrated in Figure 7.37(a). The impulse responses of the two matched filters are

$$h_1(t) = \psi_1(T - t) = \begin{cases} \sqrt{\frac{2}{T}}, & \frac{T}{2} \le t \le T \\ 0, & \text{otherwise} \end{cases}$$

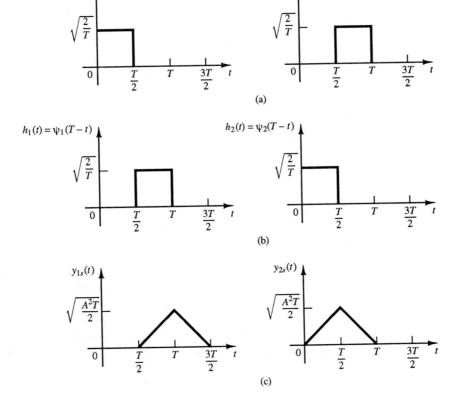

FIGURE 7.37.   Basis functions and matched filter responses for Example 7.2.2.

$$h_2(t) = \psi_2(T - t) = \begin{cases} \sqrt{\frac{2}{T}}, & 0 \le t \le T/2 \\ 0, & \text{otherwise} \end{cases} \tag{7.2.32}$$

and are illustrated in Figure 7.37(b).

If $s_1(t)$ is transmitted, the (noise-free) responses of the two matched filters are shown in Figure 7.37(c). Since $y_1(t)$ and $y_2(t)$ are sampled at $t = T$, we observe that $y_{1s}(T) = \sqrt{\frac{A^2 T}{2}}$ and $y_{2s}(T) = 0$. Note that $A^2 T/2 = \mathcal{E}_s$, the signal energy. Hence, the received vector formed from the two matched filter outputs at the sampling instant $t = T$ is

$$\mathbf{r} = (r_1, r_2) = \left(\sqrt{\mathcal{E}_s} + n_1, n_2 \right) \tag{7.2.33}$$

where $n_1 = y_{1n}(T)$ and $n_2 = y_{2n}(T)$ are the noise components at the outputs of the matched filters, given by

$$y_{kn}(T) = \int_0^T n(t)\psi_k(t)\,dt, \quad k = 1, 2 \tag{7.2.34}$$

Clearly, $E[n_k] = E[y_{kn}(T)] = 0$. Their variance is

$$\sigma_n^2 = E\left[y_{kn}^2(T) \right] = \int_0^T \int_0^T E[n(t)n(\tau)]\psi_k(t)\psi_k(\tau)\,dt\,d\tau$$

$$= \frac{N_0}{2} \int_0^T \int_0^T \delta(t - \tau)\psi_k(\tau)\psi_k(t)\,dt\,d\tau$$

$$= \frac{N_0}{2} \int_0^T \psi_k^2(t)\,dt = \frac{N_0}{2} \tag{7.2.35}$$

Observe that the $(S/N)_0$ for the first matched filter is

$$\left(\frac{S}{N} \right)_o = \frac{\left(\sqrt{\mathcal{E}_s} \right)^2}{N_0/2} = \frac{2\mathcal{E}_s}{N_0}$$

which agrees with our previous result. Also note that the four possible outputs of the two matched filters, corresponding to the four possible transmitted signals in Figure 7.6 are $(r_1, r_2) = \left(\sqrt{\mathcal{E}_s} + n_1, n_2 \right), \left(n_1, \sqrt{\mathcal{E}_s} + n_2 \right), (-\sqrt{\mathcal{E}_s} + n_1, n_2)$, and $(n_1, -\sqrt{\mathcal{E}_s} + n_2)$.

7.2.3 The Optimum Detector

In Sections 7.2.1 and 7.2.2 we demonstrated that for a signal transmitted over an AWGN channel either a correlation-type demodulator or a matched filter-type demodulator produces the vector $\mathbf{r} = (r_1, r_2, \ldots, r_N)$ which contains all the relevant information in the received signal waveform. In this section, we describe the optimum decision rule based on the observation vector $\mathbf{r}$. For this development, we assume that there is no memory in signals transmitted in successive signal intervals.

We wish to design a signal detector that makes a decision on the transmitted signal in each signal interval based on the observation of the vector $\mathbf{r}$ in each interval such that the probability of a correct decision is maximized. With this goal in mind, we consider a decision rule based on the computation of the *posterior probabilities* defined as

$$P(\text{signal } \mathbf{s}_m \text{ was transmitted}|\mathbf{r}), \quad m = 1, 2, \ldots, M$$

which we abbreviate as $P(\mathbf{s}_m|\mathbf{r})$. The decision criterion is based on selecting the signal corresponding to the maximum of the set of posterior probabilities $\{P(\mathbf{s}_m|\mathbf{r})\}$. Later we show that this criterion maximizes the probability of a correct decision and, hence, minimizes the probability of error. This decision criterion is called the *maximum a posteriori probability* (MAP) criterion.

Using Bayes rule, the posterior probabilities may be expressed as

$$P(\mathbf{s}_m|\mathbf{r}) = \frac{f(\mathbf{r}|\mathbf{s}_m)P(\mathbf{s}_m)}{f(\mathbf{r})} \tag{7.2.36}$$

where $f(\mathbf{r}|\mathbf{s}_m)$ is the conditional p.d.f. of the observed vector given $\mathbf{s}_m$, and $P(\mathbf{s}_m)$ is the *a priori probability* of the mth signal being transmitted. The denominator of (7.2.36) may be expressed as

$$f(\mathbf{r}) = \sum_{m=1}^{M} f(\mathbf{r}|\mathbf{s}_m)P(\mathbf{s}_m) \tag{7.2.37}$$

From (7.2.36) and (7.2.37) we observe that the computation of the posterior probabilities $P(\mathbf{s}_m|\mathbf{r})$ requires knowledge of the *a priori* probabilities $P(\mathbf{s}_m)$ and the conditional p.d.f.'s $f(\mathbf{r}|\mathbf{s}_m)$ for $m = 1, 2, \ldots, M$.

Some simplification occurs in the MAP criterion when the M signals are equally probable a priori, i.e., $P(\mathbf{s}_m) = 1/M$ for all M. Furthermore, we note that the denominator in (7.2.36) is independent of which signal is transmitted. Consequently, the decision rule based on finding the signal that maximizes $P(\mathbf{s}_m|\mathbf{r})$ is equivalent to finding the signal that maximizes $f(\mathbf{r}|\mathbf{s}_m)$.

The conditional p.d.f. $f(\mathbf{r}|\mathbf{s}_m)$ or any monotonic function of it is usually called the *likelihood function*. The decision criterion based on the maximum of $f(\mathbf{r}|\mathbf{s}_m)$ over the M signals is called the *maximum-likelihood* (ML) *criterion*. We observe that a detector based on the MAP criterion and one that is based on the ML criterion make the same decisions as long as the a priori probabilities $P(\mathbf{s}_m)$ are all equal, i.e., the signals $\{\mathbf{s}_m\}$ are equiprobable.

In the case of an AWGN channel, the likelihood function $f(\mathbf{r}|\mathbf{s}_m)$ is given by (7.2.11). To simplify the computations, we may work with the natural logarithm of $f(\mathbf{r}|\mathbf{s}_m)$, which is a monotonic function. Thus,

$$\ln f(\mathbf{r}|\mathbf{s}_m) = \frac{-N}{2} \ln(\pi N_0) - \frac{1}{N_0} \sum_{k=1}^{N} (r_k - s_{mk})^2 \tag{7.2.38}$$

The maximum of $\ln f(\mathbf{r}|\mathbf{s}_m)$ over $\mathbf{s}_m$ is equivalent to finding the signal $\mathbf{s}_m$ that minimizes the Euclidean distance

$$D(\mathbf{r}, \mathbf{s}_m) = \sum_{k=1}^{N} (r_k - s_{mk})^2 \qquad (7.2.39)$$

We call $D(\mathbf{r}, \mathbf{s}_m)$ $m = 1, 2, \ldots, M$, the *distance metrics*. Hence, for the AWGN channel, the decision rule based on the ML criterion reduces to finding the signal $\mathbf{s}_m$ that is closest in distance to the received signal vector $\mathbf{r}$. We will refer to this decision rule as *minimum distance detection*.

Another interpretation of the optimum decision rule based on the ML criterion is obtained by expanding the distance metrics in (7.2.39) as

$$D(\mathbf{r}, \mathbf{s}_m) = \sum_{n=1}^{N} r_n^2 - 2\sum_{n=1}^{N} r_n s_{mn} + \sum_{n=1}^{N} s_{mn}^2$$

$$= |\mathbf{r}|^2 - 2\mathbf{r} \cdot \mathbf{s}_m + |\mathbf{s}_m|^2, \qquad m = 1, 2, \ldots, M \qquad (7.2.40)$$

The term $|\mathbf{r}|^2$ is common to all decision metrics, and hence, it may be ignored in the computations of the metrics. The result is a set of modified distance metrics

$$D'(\mathbf{r}, \mathbf{s}_m) = -2\mathbf{r} \cdot \mathbf{s}_m + |\mathbf{s}_m|^2 \qquad (7.2.41)$$

Note that selecting the signal $\mathbf{s}_m$ that minimizes $D'(\mathbf{r}, \mathbf{s}_m)$ is equivalent to selecting the signal that maximizes the metric $C(\mathbf{r}, \mathbf{s}_m) = -D'(\mathbf{r}, \mathbf{s}_m)$, i.e.,

$$C(\mathbf{r}, \mathbf{s}_m) = 2\mathbf{r} \cdot \mathbf{s}_m - |\mathbf{s}_m|^2 \qquad (7.2.42)$$

The term $\mathbf{r} \cdot \mathbf{s}_m$ represents the projection of the received signal vector onto each of the M possible transmitted signal vectors. The value of each of these projections is a measure of the correlation between the received vector and the mth signal. For this reason, we call $C(\mathbf{r}, \mathbf{s}_m)$, $m = 1, 2, \ldots, M$, the *correlation metrics* for deciding which of the M signals was transmitted. Finally, the terms $|\mathbf{s}_m|^2 = \mathcal{E}_m$, $m = 1, 2, \ldots, M$, may be viewed as bias terms that serve as compensation for signal sets that have unequal energies, such as PAM. If all signals have the same energy, $|\mathbf{s}_m|^2$ may also be ignored in the computation of the correlation metrics $C(\mathbf{r}, \mathbf{s}_m)$ and the distance metrics $D(\mathbf{r}, \mathbf{s}_m)$ or $D'(\mathbf{r}, \mathbf{s}_m)$.

In summary, we have demonstrated that the optimum ML detector computes a set of M distances $D(\mathbf{r}, \mathbf{s}_m)$ or $D'(\mathbf{r}, \mathbf{s}_m)$ and selects the signal corresponding to the smallest (distance) metric. Equivalently, the optimum ML detector computes a set of M correlation metrics $C(\mathbf{r}, \mathbf{s}_m)$ and selects the signal corresponding to the largest correlation metric.

The above development for the optimum detector treated the important case in which all signals are equally probable. In this case, the MAP criterion is equivalent to the ML criterion. However, when the signals are not equally probable, the optimum MAP detector bases its decision on the probabilities $P(\mathbf{s}_m|\mathbf{r})$, $m =$

$1, 2, \ldots, M$, given by (7.2.36) or, equivalently, on the *metrics,*

$$PM(\mathbf{r}, \mathbf{s}_m) = f(\mathbf{r}|\mathbf{s}_m)P(\mathbf{s}_m) \tag{7.2.43}$$

The following example illustrates this computation for binary PAM signals.

Example 7.2.3

Consider the case of binary PAM signals in which the two possible signal points are $s_1 = -s_2 = \sqrt{\mathcal{E}_b}$, where $\mathcal{E}_b$ is the energy per bit. The prior probabilities are $P(s_1) = p$ and $P(s_2) = 1 - p$. Determine the metrics for the optimum MAP detector when the transmitted signal is corrupted with AWGN.

Solution   The received signal vector (one dimensional) for binary PAM is

$$r = \pm\sqrt{\mathcal{E}_b} + y_n(T) \tag{7.2.44}$$

where $y_n(T)$ is a zero-mean Gaussian random variable with variance $\sigma_n^2 = N_0/2$. Consequently, the conditional p.d.f.'s $f(r|s_m)$ for the two signals are

$$f(r|s_1) = \frac{1}{\sqrt{2\pi}\,\sigma_n}\, e^{-(r-\sqrt{\mathcal{E}_b})^2/2\sigma_n^2} \tag{7.2.45}$$

$$f(r|s_2) = \frac{1}{\sqrt{2\pi}\,\sigma_n}\, e^{-(r+\sqrt{\mathcal{E}_b})^2/2\sigma_n^2} \tag{7.2.46}$$

Then the metrics $PM(\mathbf{r}, \mathbf{s}_1)$ and $PM(\mathbf{r}, \mathbf{s}_2)$ defined by (7.2.43) are

$$PM(\mathbf{r}, \mathbf{s}_1) = pf(r|s_1)$$

$$= \frac{p}{\sqrt{2\pi}\,\sigma_n}\, e^{-(r-\sqrt{\mathcal{E}_b})^2/2\sigma_n^2} \tag{7.2.47}$$

$$PM(\mathbf{r}, \mathbf{s}_2) = \frac{1-p}{\sqrt{2\pi}\,\sigma_n}\, e^{-(r+\sqrt{\mathcal{E}_b})^2/2\sigma_n^2} \tag{7.2.48}$$

If $PM(\mathbf{r}, \mathbf{s}_1) > PM(\mathbf{r}, \mathbf{s}_2)$, we select s_1 as the transmitted signal; otherwise, we select s_2. This decision rule may be expressed as

$$\frac{PM(\mathbf{r}, \mathbf{s}_1)}{PM(\mathbf{r}, \mathbf{s}_2)} \underset{s_2}{\overset{s_1}{\gtrless}} 1 \tag{7.2.49}$$

But

$$\frac{PM(\mathbf{r}, \mathbf{s}_1)}{PM(\mathbf{r}, \mathbf{s}_2)} = \frac{p}{1-p}\, e^{\left[(r+\sqrt{\mathcal{E}_b})^2 - (r-\sqrt{\mathcal{E}_b})^2\right]/2\sigma_n^2} \tag{7.2.50}$$

so that (7.2.50) may be expressed as

$$\frac{\left(r + \sqrt{\mathcal{E}_b}\right)^2 - \left(r - \sqrt{\mathcal{E}_b}\right)^2}{2\sigma_n^2} \underset{s_2}{\overset{s_1}{\gtrless}} \ln\frac{1-p}{p} \tag{7.2.51}$$

or equivalently,

$$\sqrt{\mathcal{E}_b}\, r \underset{s_2}{\overset{s_1}{\gtrless}} \frac{\sigma_n^2}{2}\ln\frac{1-p}{p} = \frac{N_0}{4}\ln\frac{1-p}{p} \tag{7.2.52}$$

This is the final form for the optimum detector. It computes the correlation metric $C(\mathbf{r}, \mathbf{s}_1) = r\sqrt{\mathcal{E}_b}$ and compares it with the threshold $(N_0/4)\ln(1-p)/p$.

It is interesting to note that in the case of unequal prior probabilities, it is necessary to know not only the values of the prior probabilities but also the value of the power-spectral density N_0 in order to compute the threshold. When $p = 1/2$, the threshold is zero, and knowledge of N_0 is not required by the detector.

We conclude this section with the proof that the decision rule based on the maximum-likelihood criterion minimizes the probability of error when the M signals are equally probable a priori. Let us denote by R_m the region in the N-dimensional space for which we decide that signal $s_m(t)$ was transmitted when the vector $\mathbf{r} = (r_1, r_2, \ldots, r_N)$ is received. The probability of a decision error given that $s_m(t)$ was transmitted is

$$P(e|\mathbf{s}_m) = \int_{R_m^c} f(\mathbf{r}|\mathbf{s}_m) \, d\mathbf{r} \tag{7.2.53}$$

where R_m^c is the complement of R_m. The average probability of error is

$$P(e) = \sum_{m=1}^{M} \frac{1}{M} P(e|\mathbf{s}_m)$$

$$= \sum_{m=1}^{M} \frac{1}{M} \int_{R_m^c} f(\mathbf{r}|\mathbf{s}_m) \, d\mathbf{r}$$

$$= \sum_{m=1}^{M} \frac{1}{M} \left[1 - \int_{R_m} f(\mathbf{r}|\mathbf{s}_m) \, d\mathbf{r} \right] \tag{7.2.54}$$

We note that $P(e)$ is minimized by selecting the signal $\mathbf{s}_m$ if $f(\mathbf{r}|\mathbf{s}_m)$ is larger than $f(\mathbf{r}|\mathbf{s}_k)$ for all $m \neq k$.

When the M signals are not equally probable, the proof given above can be generalized to show that the MAP criterion minimizes the average probability of error.

7.2.4 The Maximum-Likelihood Sequence Detector

When the signal has no memory, the symbol-by-symbol detector described in the preceding section is optimum in the sense of minimizing the probability of a symbol error. On the other hand, when the transmitted signal has memory, i.e., the signals transmitted in successive symbol intervals are interdependent, the optimum detector is a maximum-likelihood sequence detector which bases its decisions on observation of a sequence of received signals over successive signal intervals.

Consider as an example the NRZI signal described in Section 7.1.4. Its memory is characterized by the trellis shown in Figure 7.25. The signal transmitted in each signal interval is binary PAM. Hence, there are two possible transmitted signals corresponding to the signal points $s_2 = -s_1 = \sqrt{\mathcal{E}_b}$, where $\mathcal{E}_b$ is the energy per bit. As shown in Example 7.2.3, the output of the matched-filter or correlation-type demodulator for binary PAM in the kth signal interval may be expressed as (see (7.2.44))

$$r_k = \pm\sqrt{\mathcal{E}_b} + n_k \tag{7.2.55}$$

where n_k is a zero-mean Gaussian random variable with variance $\sigma_n^2 = N_0/2$. Consequently, the conditional p.d.f.'s for the two possible transmitted signals are

$$f(r_k|s_1) = \frac{1}{\sqrt{2\pi}\,\sigma_n} e^{-(r_k - \sqrt{\mathcal{E}_b})^2/2\sigma_n^2}$$

$$f(r_k|s_2) = \frac{1}{\sqrt{2\pi}\,\sigma_n} e^{-(r_k + \sqrt{\mathcal{E}_b})^2/2\sigma_n^2} \tag{7.2.56}$$

Now, suppose we observe the sequence of matched-filter outputs $r_1, r_2, \ldots, r_K$. Since the channel noise is assumed to be white and Gaussian, and $\psi(t - iT)$, $\psi(t - jT)$ for $i \neq j$ are orthogonal, it follows that $E(n_i n_j) = 0, i \neq j$. Hence, the noise sequence $n_1, n_2, \ldots, n_k$ is also white. Consequently, for any given transmitted sequence $\mathbf{s}^{(m)}$, the joint p.d.f. of $r_1, r_2, \ldots, r_K$ may be expressed as a product of k marginal p.d.f.'s, i.e.,

$$p(r_1, r_2, \ldots, r_K|\mathbf{s}^{(m)}) = \prod_{k=1}^{K} p(r_k|s_k^{(m)})$$

$$= \prod_{k=1}^{K} \frac{1}{\sqrt{2\pi}\,\sigma_n} e^{-(r_k - s_k^{(m)})^2/2\sigma_n^2}$$

$$= \left(\frac{1}{\sqrt{2\pi}\,\sigma_n}\right)^K \exp\left[-\sum_{k=1}^{K}(r_k - s_k^{(m)})^2/2\sigma_n^2\right] \tag{7.2.57}$$

where either $s_k = \sqrt{\mathcal{E}_b}$ or $s_k = -\sqrt{\mathcal{E}_b}$. Then, given the received sequence $r_1, r_2, \ldots, r_K$ at the output of the matched-filter or correlation-type demodulation, the detector determines the sequence $\mathbf{s}^{(m)} = \{s_1^{(m)}, s_2^{(m)}, \ldots, s_K^{(m)}\}$ that maximizes the conditional p.d.f. $p(r_1, r_2, \ldots, r_K|\mathbf{s}^{(m)})$. Such a detector is called the *maximum-likelihood* (ML) *sequence detector*.

By taking the logarithm of (7.2.57) and neglecting the terms that are independent of $(r_1, r_2, \ldots, r_K)$, we find that an equivalent ML sequence detector selects the sequence $\mathbf{s}^{(m)}$ that minimizes the *Euclidean distance metric*

$$D(\mathbf{r}, \mathbf{s}^{(m)}) = \sum_{k=1}^{K} \left(r_k - s_k^{(m)}\right)^2 \tag{7.2.58}$$

In searching through the trellis for the sequence that minimizes the Euclidean distance $D(\mathbf{r}, \mathbf{s}^{(m)})$, it may appear that we must compute the distance $D(\mathbf{r}, \mathbf{s}^{(m)})$ for every possible path (sequence). For the NRZI example given above which employs binary modulation, the total number of paths is 2^K, where K is the number of outputs obtained from the demodulator. However, this is not the case. We may reduce the number of sequences in the trellis search by using the *Viterbi algorithm* to eliminate sequences as new data is received from the demodulator.

The Viterbi algorithm is a sequential trellis search algorithm for performing ML sequence detection. It is described in Chapter 10 as a decoding algorithm for channel coded systems. We describe it below in the context of the NRZI signal. We assume that the search process begins initially at state S_1. The corresponding trellis is shown in Figure 7.38.

At time $t = T$, we receive $r_1 = s_1^{(m)} + n$ from the demodulator, and at $t = 2T$ we receive $r_2 = s_2^{(m)} + n_2$. Since the signal memory is one bit, which we denote as $L = 1$, we observe that the trellis reaches its regular (steady-state) form after the first transition. Thus, upon receipt of r_2 at $t = 2T$ (and thereafter), we observe that there are two signal paths entering each of the two nodes and two signal paths leaving each node. The two paths entering node S_1 at $t = 2T$ correspond to the information bits $(0, 0)$ and $(1, 1)$ or, equivalently, to the signal points $(-\sqrt{\mathcal{E}_b}, -\sqrt{\mathcal{E}_b})$ and $(\sqrt{\mathcal{E}_b}, -\sqrt{\mathcal{E}_b})$, respectively. The two paths entering node S_2 at $t = 2T$ correspond to the information bits $(0, 1)$ and $(1, 0)$ or, equivalently, to the signal points $(-\sqrt{\mathcal{E}_b}, \sqrt{\mathcal{E}_b})$ and $(\sqrt{\mathcal{E}_b}, \sqrt{\mathcal{E}_b})$, respectively.

For the two paths entering node S_1, we compute the two Euclidean distance metrics

$$\mu_2(0, 0) = (r_1 + \sqrt{\mathcal{E}_b})^2 + (r_2 + \sqrt{\mathcal{E}_b})^2$$
$$\mu_2(1, 1) = (r_1 - \sqrt{\mathcal{E}_b})^2 + (r_2 + \sqrt{\mathcal{E}_b})^2 \qquad (7.2.59)$$

by using the outputs r_1 and r_2 from the demodulator. The Viterbi algorithm compares these two metrics and discards the path having the larger (greater distance) metric. The other path with the lower metric is saved and is called the *survivor* at $t = 2T$. The elimination of one of the two paths may be done without compromising the optimality of the trellis search, because any extension of the path with the larger distance beyond $t = 2T$ will always have a larger metric than the survivor that is extended along the same path beyond $t = 2T$.

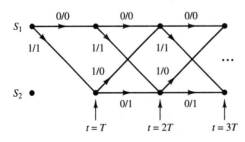

FIGURE 7.38. Trellis for NRZI signal with initial state s_1.

Similarly, for the two paths entering node S_2 at $t = 2T$, we compute the two Euclidean distance metrics

$$\mu_2(0, 1) = (r_1 + \sqrt{E_b})^2 + (r_2 - \sqrt{E_b})^2$$
$$\mu_2(1, 0) = (r_1 - \sqrt{E_b})^2 + (r_2 - \sqrt{E_b})^2 \qquad (7.2.60)$$

by using the outputs r_1 and r_2 from the demodulator. The two metrics are compared and the signal path with the larger metric is eliminated. Thus, at $t = 2T$, we are left with two survivor paths, one at node S_1 and the other at node S_2, and their corresponding metrics. The signal paths at nodes S_1 and S_2 are then extended along the two survivor paths.

Upon receipt of r_3 at $t = 3T$, we compute the metrics of the two paths entering state S_1. Suppose the survivors at $t = 2T$ are the paths $(0, 0)$ at S_1 and $(0, 1)$ at S_2. Then, the two metrics for the paths entering S_1 at $t = 3T$ are

$$\mu_3(0, 0, 0) = \mu_2(0, 0) + (r_3 + \sqrt{E_b})^2$$
$$\mu_3(0, 1, 1) = \mu_2(0, 1) + (r_3 + \sqrt{E_b})^2 \qquad (7.2.61)$$

These two metrics are compared and the path with the larger (distance) metric is eliminated. Similarly, the metrics for the two paths entering S_2 at $t = 3T$ are

$$\mu_3(0, 0, 1) = \mu_2(0, 0) + (r_3 - \sqrt{E_b})^2$$
$$\mu_3(0, 1, 0) = \mu_2(0, 1) + (r_3 - \sqrt{E_b})^2 \qquad (7.2.62)$$

These two metrics are compared and the path with the larger (distance) metric is eliminated.

This process is continued as each new signal sample is received from the demodulator. Thus, the Viterbi algorithm computes two metrics for the two signal paths entering a node at each stage of the trellis search and eliminates one of the two paths at each node. The two survivor paths are then extended forward to the next state. Therefore, the number of paths searched in the trellis is reduced by a factor of two at each stage.

It is relatively easy to generalize the trellis search performed by the Viterbi algorithm for M-ary modulation. For example, delay modulation employs $M = 4$ signals and is characterized by the four-state trellis shown in Figure 7.27. We observe that each state has two signal paths entering and two signal paths leaving each node. The memory of the signal is $L = 1$. Hence, the Viterbi algorithm will have four survivors at each stage and their corresponding metrics. Two metrics corresponding to the two entering paths are computed at each node, and one of the two signal paths entering the node is eliminated at each state of the trellis. Thus, the Viterbi algorithm minimizes the number of trellis paths searched in performing ML sequence detection.

From the description of the Viterbi algorithm given above, it is unclear as to how decisions are made on the individual detected information symbols given the surviving sequences. If we have advanced to some stage, say K, where $K \gg L$ in the trellis, and we compare the surviving sequences, we will find that with probability approaching one all surviving sequences will be identical in bit (or symbol) positions $K - 5L$ and less. In a practical implementation of the Viterbi algorithm, decisions on each information bit (or symbol) are forced after a delay of $5L$ bits (or symbols), and hence, the surviving sequences are truncated to the $5L$ most recent bits (or symbols). Thus, a variable delay in bit or symbol detection is avoided. The loss in performance resulting from the suboptimum detection procedure is negligible if the delay is at least $5L$.

Example 7.2.4

Describe the decision rule for detecting the data sequence in an NRZI signal with a Viterbi algorithm having a delay of $5L$ bits.

Solution   The trellis for the NRZI signal is shown in Figure 7.25. In this case, $L = 1$, hence the delay in bit detection is set to 5 bits. Hence, at $t = 6T$, we will have two surviving sequences, one for each of the two states and the corresponding metrics $\mu_6(b_1, b_2, b_3, b_4, b_5, b_6)$ and $\mu_6(b'_1, b'_2, b'_3, b'_4, b'_5, b'_6)$. At this stage, with probability nearly equal to one, the bit b_1 will be the same as b'_1; that is, both surviving sequences will have a common first branch. If $b_1 \neq b'_1$, we may select the bit (b_1 or b'_1) corresponding to the smaller of the two metrics. Then the first bit is dropped from the two surviving sequences. At $t = 7T$, the two metrics $\mu_7(b_2, b_3, b_4, b_5, b_6, b_7)$ and $\mu_7(b'_2, b'_3, b'_4, b'_5, b'_6, b'_7)$ will be used to determine the decision on bit b_2. This process continues at each stage of the search through the trellis for the minimum distance sequence. Thus the detection delay is fixed at 5 bits.

7.3 PROBABILITY OF ERROR FOR SIGNALS IN ADDITIVE WHITE GAUSSIAN NOISE

In this section, we evaluate the probability of error for several of the modulation signals described in Section 7.1. First, we consider binary modulation signals and then M-ary signals.

7.3.1 Probability of Error for Binary Modulation

Let us consider binary PAM signals where the two signal waveforms are $s_1(t) = g_T(t)$ and $s_2(t) = -g_T(t)$, and $g_T(t)$ is an arbitrary pulse which is nonzero in the interval $0 \leq t \leq T_b$ and zero elsewhere.

Since $s_1(t) = -s_2(t)$, these signals are said to be *antipodal*. The energy in the pulse $g_T(t)$ is $\mathcal{E}_b$. As indicated in Section 7.1.3, PAM signals are one-dimensional, and hence, their geometric representation is simply the one-dimensional vector $s_1 = \sqrt{\mathcal{E}_b}$, $s_2 = -\sqrt{\mathcal{E}_b}$. Figure 7.39 illustrates the two signal points.

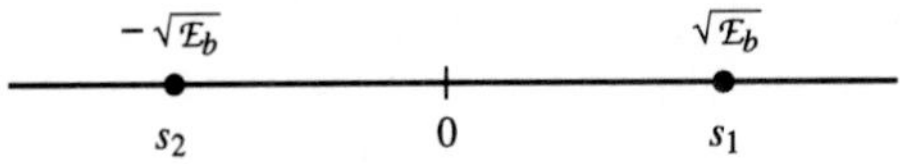

FIGURE 7.39. Signal points for binary antipodal signals.

Let us assume that the two signals are equally likely and that signal $s_1(t)$ was transmitted. Then, the received signal from the (matched filter or correlation-type) demodulator is

$$r = s_1 + n = \sqrt{\mathcal{E}_b} + n \qquad (7.3.1)$$

where n represents the additive Gaussian noise component which has zero mean and variance $\sigma_n^2 = N_0/2$. In this case, the decision rule based on the correlation metric given by (7.2.42) compares r with the threshold zero. If $r > 0$, the decision is made in favor of $s_1(t)$ and if $r < 0$, the decision is made that $s_2(t)$ was transmitted. Clearly, the two conditional p.d.f.'s of r are

$$f(r|s_1) = \frac{1}{\sqrt{\pi N_0}} e^{-(r-\sqrt{\mathcal{E}_b})^2/N_0} \qquad (7.3.2)$$

$$f(r|s_2) = \frac{1}{\sqrt{\pi N_0}} e^{-(r+\sqrt{\mathcal{E}_b})^2/N_0} \qquad (7.3.3)$$

These two conditional p.d.f.'s are shown in Figure 7.40.

Given that $s_1(t)$ was transmitted, the probability of error is simply the probability that $r < 0$, i.e.,

$$P(e|s_1) = \int_{-\infty}^{0} f(r|s_1)\, dr$$

$$= \frac{1}{\sqrt{\pi N_0}} \int_{-\infty}^{0} e^{-(r-\sqrt{\mathcal{E}_b})^2/N_0}\, dr$$

$$= \frac{1}{\sqrt{2\pi}} \int_{-\infty}^{-\sqrt{2\mathcal{E}_b/N_0}} e^{-x^2/2}\, dx$$

$$= \frac{1}{\sqrt{2\pi}} \int_{\sqrt{2\mathcal{E}_b/N_0}}^{\infty} e^{-x^2/2}\, dx$$

$$= Q\left[\sqrt{\frac{2\mathcal{E}_b}{N_0}}\right] \qquad (7.3.4)$$

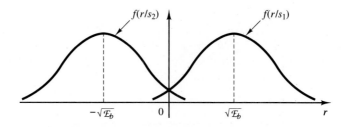

FIGURE 7.40. Conditional p.d.f.'s of two signals.

where $Q(x)$ is the Q-function defined previously in Section 3.1. Similarly, if we assume that $s_2(t)$ was transmitted, $r = -\sqrt{\mathcal{E}_b} + n$ and the probability that $r > 0$ is also $P(e|s_2) = Q\left(\sqrt{\frac{2\mathcal{E}_b}{N_0}}\right)$. Since the signals $s_1(t)$ and $s_2(t)$ are equally likely to be transmitted, the average probability of error is

$$P_b = \frac{1}{2}P(e|s_1) + \frac{1}{2}P(e|s_2)$$

$$= Q\left(\sqrt{\frac{2\mathcal{E}_b}{N_0}}\right) \qquad (7.3.5)$$

We should observe two important characteristics of this performance measure. First, we note that the probability of error depends only on the ratio $\mathcal{E}_b/N_0$ and not on any other detailed characteristics of the signals and the noise. Secondly, we note that $2\mathcal{E}_b/N_0$ is also the output SNR from the matched-filter (and correlation-type) demodulator. The ratio $\mathcal{E}_b/N_0$ is usually called the *signal-to-noise ratio.*

We also observe that the probability of error may be expressed in terms of the distance between the two signals s_1 and s_2. From Figure 7.39, we observe that the two signals are separated by the distance $d_{12} = 2\sqrt{\mathcal{E}_b}$. By substituting $\mathcal{E}_b = d_{12}^2/4$ in (7.2.54) we obtain

$$P_b = Q\left(\sqrt{\frac{d_{12}^2}{2N_0}}\right) \qquad (7.3.6)$$

This expression illustrates the dependence of the error probability on the distance between the two signal points.

Next, let us evaluate the error probability for binary orthogonal signals. Recall the binary PPM is an example of binary orthogonal signaling. In this case, the signal vectors $\mathbf{s}_1$ and $\mathbf{s}_2$ are two-dimensional, as shown in Figure 7.41, and may be

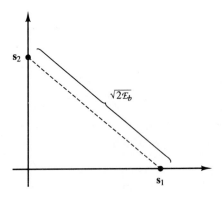

FIGURE 7.41. Signal points for binary orthogonal signals.

expressed, according to (7.1.29), as

$$\mathbf{s}_1 = \left(\sqrt{\mathcal{E}_b}, 0\right)$$

$$\mathbf{s}_2 = \left(0, \sqrt{\mathcal{E}_b}\right) \tag{7.3.7}$$

where $\mathcal{E}_b$ denotes the energy for each of the waveforms. Note that the distance between these signal points is $d_{12} = \sqrt{2\mathcal{E}_b}$.

To evaluate the probability of error, let us assume that $\mathbf{s}_1$ was transmitted. Then, the received vector at the output of the demodulator is

$$\mathbf{r} = \left[\sqrt{\mathcal{E}_b} + n_1, n_2\right] \tag{7.3.8}$$

We can now substitute for $\mathbf{r}$ into the correlation metrics given by (7.2.42) to obtain $C(\mathbf{r}, \mathbf{s}_1)$ and $C(\mathbf{r}, \mathbf{s}_2)$. Then the probability of error is the probability that $C(\mathbf{r}, \mathbf{s}_2) > C(\mathbf{r}, \mathbf{s}_1)$. Thus,

$$P(e|\mathbf{s}_1) = P[C(\mathbf{r}, \mathbf{s}_2) > C(\mathbf{r}_1, \mathbf{s}_1)] = P\left[n_2 - n_1 > \sqrt{\mathcal{E}_b}\right] \tag{7.3.9}$$

Since n_1 and n_2 are zero-mean statistically independent Gaussian random variables each with variance $N_0/2$, the random variable $x = n_2 - n_1$ is zero-mean Gaussian with variance N_0. Hence,

$$P\left(n_2 - n_1 > \sqrt{\mathcal{E}_b}\right) = \frac{1}{\sqrt{2\pi N_0}} \int_{\sqrt{\mathcal{E}_b}}^{\infty} e^{-x^2/2N_0} \, dx$$

$$= \frac{1}{\sqrt{2\pi}} \int_{\sqrt{\mathcal{E}_b/N_0}}^{\infty} e^{-x^2/2} \, dx$$

$$= Q\left[\sqrt{\frac{\mathcal{E}_b}{N_0}}\right] \tag{7.3.10}$$

Due to symmetry, the same error probability is obtained when we assume that $\mathbf{s}_2$ is transmitted. Consequently, the average error probability for binary orthogonal signals is

$$P_b = Q\left[\sqrt{\frac{\mathcal{E}_b}{N_0}}\right] \tag{7.3.11}$$

If we compare the probability of error for binary antipodal signals with that for binary orthogonal signals, we find that orthogonal signals require a factor of two increase in energy to achieve the same error probability as antipodal signals. Since $10\log_{10} 2 = 3$ dB, we say that orthogonal signals are 3 dB poorer than antipodal signals. The difference of 3 dB is simply due to the distance between the two signal points, which is $d_{12}^2 = 2\mathcal{E}_b$ for orthogonal signals, whereas $d_{12}^2 = 4\mathcal{E}_b$ for antipodal signals.

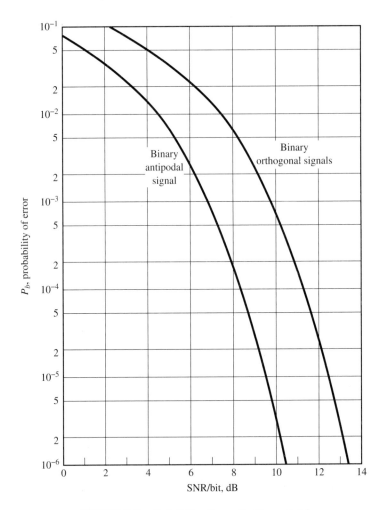

FIGURE 7.42. Probability of error for binary signals.

The error probability versus $10 \log_{10} \mathcal{E}_b/N_0$ for these two types of signals is shown in Figure 7.42. As observed from this figure, at any given error probability, the $\mathcal{E}_b/N_0$ required for orthogonal signals is 3 dB more than that for antipodal signals.

7.3.2 Probability of Error for M-ary Modulation

In this section, we derive the probability of error for M-ary signal modulation transmitted over an AWGN channel.

M-ary PAM signals. Recall that M-ary PAM signals are represented geometrically as M one-dimensional signal points with values

$$s_m = \sqrt{\mathcal{E}_g} A_m, \quad m = 1, 2, \ldots, M \tag{7.3.12}$$

where $\mathcal{E}_g$ is the energy of the basic signal pulse $g_T(t)$. The amplitude values may be expressed as

$$A_m = (2m - 1 - M), \quad m = 1, 2, \ldots, M \tag{7.3.13}$$

where the distance between adjacent signal points is $2\sqrt{\mathcal{E}_g}$.

As previously indicated in Section 7.1, the PAM signals have different energies, denoted as $\{\mathcal{E}_m\}$. The average energy is

$$\mathcal{E}_{av} = \frac{1}{M} \sum_{m=1}^{M} \mathcal{E}_m = \frac{\mathcal{E}_g}{M} \sum_{m=1}^{M} (2m - 1 - M)^2$$

$$= \frac{\mathcal{E}_g}{M} \frac{M(M^2 - 1)}{3} = \left(\frac{M^2 - 1}{3} \right) \mathcal{E}_g \tag{7.3.14}$$

Equivalently, we may characterize these signals in terms of their average power, which is

$$P_{av} = \frac{\mathcal{E}_{av}}{T} = \left(\frac{M^2 - 1}{3} \right) \frac{\mathcal{E}_g}{T} \tag{7.3.15}$$

The average probability of error for M-ary PAM can be determined from the decision rule that maximizes the correlation metrics given by (7.2.42). Equivalently, the detector compares the demodulator output r with a set of $M - 1$ thresholds, which are placed at the midpoints of successive amplitude levels, as shown in Figure 7.43. Thus, a decision is made in favor of the amplitude level that is closest to r.

The placing of the thresholds as shown in Figure 7.43 helps in evaluating the probability of error. We note that if the mth amplitude level is transmitted, the demodulator output is

$$r = s_m + n = \sqrt{\mathcal{E}_g} A_m + n \tag{7.3.16}$$

where the noise variable n has zero mean and variance $\sigma_n^2 = N_0/2$. On the basis that all amplitude levels are equally likely a priori, the average probability of a symbol

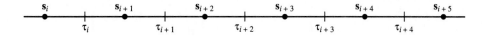

s_i – signal point
τ_i – thresholds

FIGURE 7.43.   Placement of thresholds at midpoints of successive amplitude levels.

error is simply the probability that the noise variable n exceeds in magnitude one-half of the distance between levels. However, when either one of the two outside levels $\pm(M-1)$ is transmitted, an error can occur in one direction only. Thus, we have

$$
\begin{aligned}
P_M &= \frac{M-1}{M} P\left(|r - s_m| > \sqrt{\mathcal{E}_g}\right) \\
&= \frac{M-1}{M} \frac{2}{\sqrt{\pi N_0}} \int_{\sqrt{\mathcal{E}_g}}^{\infty} e^{-x^2/N_0} \, dx \\
&= \frac{M-1}{M} \frac{2}{\sqrt{2\pi}} \int_{\sqrt{2\mathcal{E}_g/N_0}}^{\infty} e^{-x^2/2} \, dx \\
&= \frac{2(M-1)}{M} Q\left(\sqrt{\frac{2\mathcal{E}_g}{N_0}}\right)
\end{aligned}
\tag{7.3.17}
$$

The error probability in (7.3.17) can also be expressed in terms of the average transmitted power. From (7.3.15) we note that

$$
\mathcal{E}_g = \frac{3}{M^2 - 1} P_{av} T
\tag{7.3.18}
$$

By substituting for $\mathcal{E}_g$ in (7.3.17) using (7.3.18), we obtain the average probability of a symbol error for PAM in terms of the average power as

$$
P_M = \frac{2(M-1)}{M} Q\left(\sqrt{\frac{6 P_{av} T}{(M^2 - 1) N_0}}\right)
\tag{7.3.19}
$$

or equivalently,

$$
P_M = \frac{2(M-1)}{M} Q\left(\sqrt{\frac{6 \mathcal{E}_{av}}{(M^2 - 1) N_0}}\right)
\tag{7.3.20}
$$

where $\mathcal{E}_{av} = P_{av} T$ is the average energy.

In plotting the probability of a symbol error for M-ary signals such as M-ary PAM, it is customary to use the SNR per bit as the basic parameter. Since $T = k T_b$ and $k = \log_2 M$, (7.3.20) may be expressed as

$$
P_M = \frac{2(M-1)}{M} Q\left(\sqrt{\frac{6 (\log_2 M) \mathcal{E}_{bav}}{(M^2 - 1) N_0}}\right)
\tag{7.3.21}
$$

where $\mathcal{E}_{bav} = P_{av} T_b$ is the average bit energy and $\mathcal{E}_{bav}/N_0$ is the average SNR per bit. Figure 7.44 illustrates the probability of a symbol error as a function of $10 \log_{10} \mathcal{E}_{bav}/N_0$ with M as a parameter. Note that the case $M = 2$ corresponds to

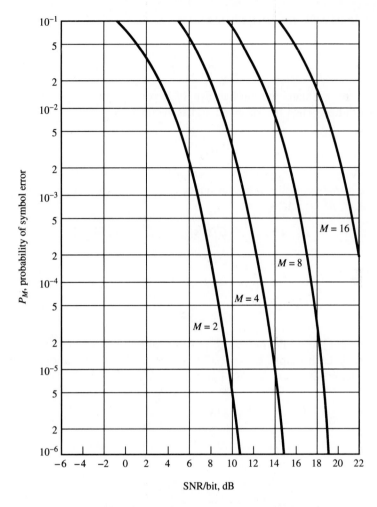

FIGURE 7.44.   Probability of a symbol error for PAM.

the error probability for binary antipodal signals. We also observe that the SNR/bit increases by over 4 dB for every factor of two increase in M. For large M, the additional SNR/bit required to increase M by a factor of two approaches 6 dB.

M-ary orthogonal signals.   As shown in Section 7.1.1, PPM is a modulation method that results in orthogonal signals whose vector space representation is given by (7.1.29). Note that all signals have equal energy, denoted as $\mathcal{E}_s$.

For equal energy orthogonal signals, the optimum detector selects the signal resulting in the largest crosscorrelation between the received vector $\mathbf{r}$ and each of the M possible transmitted signal vectors $\{s_m\}$, i.e.,

$$C(\mathbf{r}, \mathbf{s}_m) = \mathbf{r} \cdot \mathbf{s}_m = \sum_{k=1}^{M} r_k s_{mk}, \quad m = 1, 2, \ldots, M \qquad (7.3.22)$$

To evaluate the probability of error, let us suppose that the signal $\mathbf{s}_1$ is transmitted. Then the received signal vector is

$$\mathbf{r} = \left(\sqrt{\mathcal{E}_s} + n_1, n_2, n_3, \ldots, n_M \right) \qquad (7.3.23)$$

where $r_1, r_2, \ldots, r_M$ are zero-mean, mutually statistically independent Gaussian random variables with equal variance $\sigma_n^2 = N_0/2$. In this case, the outputs from the bank of M correlators are

$$C(\mathbf{r}, \mathbf{s}_1) = \sqrt{\mathcal{E}_s} \left(\sqrt{\mathcal{E}_s} + n_1 \right)$$

$$C(\mathbf{r}, \mathbf{s}_2) = \sqrt{\mathcal{E}_s} n_2$$

$$\vdots \qquad \vdots$$

$$C(\mathbf{r}, \mathbf{s}_M) = \sqrt{\mathcal{E}_s} n_M \qquad (7.3.24)$$

Note that the scale factor $\mathcal{E}_s$ may be eliminated from the correlator outputs by dividing each output by $\sqrt{\mathcal{E}_s}$. Then, with this normalization, the p.d.f. of the first correlator output $(r_1 = \sqrt{\mathcal{E}_s} + n_1)$ is

$$f_{r_1}(x_1) = \frac{1}{\sqrt{\pi N_0}} e^{-(x_1 - \sqrt{\mathcal{E}_s})^2 / N_0} \qquad (7.3.25)$$

and the p.d.f.'s of the other $M - 1$ correlator outputs are

$$f_{r_m}(x_m) = \frac{1}{\sqrt{\pi N_0}} e^{-x_m^2 / N_0}, \quad m = 2, 3, \ldots, M \qquad (7.3.26)$$

It is mathematically convenient to first derive the probability that the detector makes a correct decision. This is the probability that r_1 is larger than each of the other $M - 1$ correlator outputs $n_2, n_3, \ldots, n_M$. This probability may be expressed as

$$P_c = \int_{-\infty}^{\infty} P(n_2 < r_1, n_3 < r_1, \ldots, n_M < r_1 | r_1) f(r_1) dr_1 \qquad (7.3.27)$$

where $P(n_2 < r_1, n_3 < r_1, \ldots, n_M < r_1 | r_1)$ denotes the joint probability that $n_2, n_3, \ldots, n_M$ are all less than r_1, conditioned on any given r_1. Then this joint probability is averaged over all r_1. Since the $\{r_m\}$ are statistically independent, the joint probability factors into a product of $M - 1$ marginal probabilities of the form

$$P(n_m < r_1 | r_1) = \int_{-\infty}^{r_1} f_{r_m}(x_m) dx_m, \quad m = 2, 3, \ldots, M$$

$$= \frac{1}{\sqrt{2\pi}} \int_{-\infty}^{\sqrt{2r_1^2/N_0}} e^{-x^2/2}\, dx$$

$$= 1 - Q\left(\sqrt{\frac{2r_1^2}{N_0}}\right) \tag{7.3.28}$$

These probabilities are identical for $m = 2, 3, \ldots, M$, and hence, the joint probability under consideration is simply the result in (7.3.28) raised to the $(M - 1)$ power. Thus, the probability of a correct decision is

$$P_c = \int_{-\infty}^{\infty} \left[1 - Q\left(\sqrt{\frac{2r_1^2}{N_0}}\right)\right]^{M-1} f(r_1)\, dr_1 \tag{7.3.29}$$

and the probability of a (k-bit) symbol error is

$$P_M = 1 - P_c \tag{7.3.30}$$

where

$$P_c = \frac{1}{\sqrt{2\pi}} \int_{-\infty}^{\infty} \left\{1 - \left[Q\left(\sqrt{x^2}\right)\right]^{M-1}\right\} e^{-\left(x - \sqrt{2\mathcal{E}_s/N_0}\right)^2/2}\, dx \tag{7.3.31}$$

The same expression for the probability of error is obtained when any one of the other $M - 1$ signals is transmitted. Since all the M signals are equally likely, the expression for P_M given in (7.3.31) is the average probability of a symbol error. This expression can be evaluated numerically.

In comparing the performance of various digital modulation methods, it is desirable to have the probability of error expressed in terms of the SNR per bit, $\mathcal{E}_b/N_0$, instead of the SNR per symbol $\mathcal{E}_s/N_0$. With $M = 2^k$, each symbol conveys k bits of information, and hence, $\mathcal{E}_s = k\mathcal{E}_b$. Thus, (7.3.31) may be expressed in terms of $\mathcal{E}_b/N_0$ by substituting for $\mathcal{E}_s$.

Sometimes, it is also desirable to convert the probability of a symbol error into an equivalent probability of a binary digit error. For equiprobable orthogonal signals, all symbol errors are equiprobable and occur with probability

$$\frac{P_M}{M - 1} = \frac{P_M}{2^k - 1} \tag{7.3.32}$$

Furthermore, there are $\binom{k}{n}$ ways in which n bits out of k may be in error. Hence, the average number of bit errors per k-bit symbol is

$$\sum_{n=1}^{k} n \binom{k}{n} \frac{P_M}{2^k - 1} = k \frac{2^{k-1}}{2^k - 1} P_M \tag{7.3.33}$$

and the average bit error probability is just the result in (7.3.33) divided by k, the number of bits per symbol. Thus,

$$P_b = \frac{2^{k-1}}{2^k - 1} P_M \approx \frac{P_M}{2} \quad k \gg 1 \tag{7.3.34}$$

The graphs of the probability of a binary digit error as a function of the SNR per bit, $\mathcal{E}_b/N_0$, are shown in Figure 7.45 for $M = 2, 4, 8, 16, 32, 64$. This figure illustrates that by increasing the number M of waveforms, one can reduce the SNR per bit required to achieve a given probability of a bit error. For example, to achieve a $P_b = 10^{-5}$, the required SNR per bit is a little more than 12 dB for $M = 2$, but if M is increased to 64 signal waveforms ($k = 6$ bits/symbol), the required SNR per

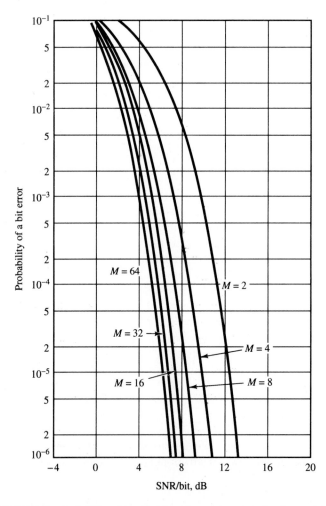

FIGURE 7.45. Probability of bit error for coherent detection of orthogonal signals.

bit is approximately 6 dB. Thus, a savings of over 6 dB (a factor of four reduction) is realized in transmitter power (or energy) required to achieve a $P_b = 10^{-5}$ by increasing M from $M = 2$ to $M = 64$.

What is the minimum required $\mathcal{E}_b/N_0$ to achieve an arbitrarily small probability of error as $M \to \infty$? This question is answered below.

A union bound on the probability of error.   Let us investigate the effect of increasing M on the probability of error for orthogonal signals. To simplify the mathematical development, we first derive an upper bound on the probability of a symbol error which is much simpler than the exact form given in (7.3.31).

Recall that the probability of error for binary orthogonal signals is given by (7.3.11). Now, if we view the detector for M orthogonal signals as one that makes $M - 1$ binary decisions between the correlator output $C(\mathbf{r}, \mathbf{s}_1)$ that contains the signal and the other $M - 1$ correlator outputs $C(\mathbf{r}, \mathbf{s}_m)$, $m = 2, 3, \ldots, M$, the probability of error is upper bounded by the *union bound* of the $M - 1$ events. That is, if E_i represents the event that $C(\mathbf{r}, \mathbf{s}_m) > C(\mathbf{r}, \mathbf{s}_1)$ for $m \neq 1$, then we have $P_M = P\left(\cup_{i=1}^n E_i\right) \leq \sum_{i=1}^n P(E_i)$. Hence,

$$P_M \leq (M-1)P_2 = (M-1)Q\left(\sqrt{\mathcal{E}_s/N_0}\right) < MQ\left(\sqrt{\mathcal{E}_s/N_0}\right) \qquad (7.3.35)$$

This bound can be simplified further by upper-bounding $Q\left(\sqrt{\mathcal{E}_s/N_0}\right)$ (see (3.1.7)). We have

$$Q\left(\sqrt{\mathcal{E}_s/N_0}\right) < e^{-\mathcal{E}_s/2N_0} \qquad (7.3.36)$$

Thus,

$$P_M < Me^{-\mathcal{E}_s/2N_0} = 2^k e^{-k\mathcal{E}_b/2N_0}$$

$$P_M < e^{-k(\mathcal{E}_b/N_0 - 2\ln 2)/2} \qquad (7.3.37)$$

As $k \to \infty$, or equivalently, as $M \to \infty$, the probability of error approaches zero exponentially, provided that $\mathcal{E}_b/N_0$ is greater than $2\ln 2$, i.e.,

$$\frac{\mathcal{E}_b}{N_0} > 2\ln 2 = 1.39 \ (1.42 \text{ dB}) \qquad (7.3.38)$$

The simple upper bound on the probability of error given by (7.3.37) implies that as the long as SNR > 1.42 dB, we can achieve an arbitrarily low P_M. However, this union bound is not a very tight upper bound at a sufficiently low SNR due to the fact that the upper bound for the Q-function in (7.3.36) is loose. In fact, by more elaborate bounding techniques, it is shown in Section 10.4 that the upper bound in (7.3.37) is sufficiently tight for $\mathcal{E}_b/N_0 > 4\ln 2$. For $\mathcal{E}_b/N_0 < 4\ln 2$, a tighter upper bound on P_M is

$$P_M < 2e^{-k\left(\sqrt{\mathcal{E}_b/N_0} - \sqrt{\ln 2}\right)^2} \qquad (7.3.39)$$

Consequently, $P_M \to 0$ as $k \to \infty$, provided that

$$\frac{\mathcal{E}_b}{N_0} > \ln 2 = 0.693 \ (-1.6 \text{ dB}) \tag{7.3.40}$$

Hence -1.6 dB is the minimum required SNR per bit to achieve an arbitrarily small probability of error in the limit as $k \to \infty (M \to \infty)$. This minimum SNR per bit (-1.6 dB) is called the *Shannon limit* for an additive white Gaussian noise channel.

Biorthogonal signals. As previously indicated in Section 7.1, a set of $M = 2^k$ biorthogonal signals are constructed from $M/2$ orthogonal signals by including the negatives of the orthogonal signals. Thus, we achieve a reduction in the complexity of the demodulator for the biorthogonal signals relative to that for the orthogonal signals, since the former is implemented with $M/2$ crosscorrelators or matched filters, whereas the latter requires M matched filters or crosscorrelators.

To evaluate the probability of error for the optimum detector, let us assume that the signal $s_1(t)$ corresponding to the vector $\mathbf{s}_1 = \left(\sqrt{\mathcal{E}_s}, 0, 0, \ldots, 0\right)$ was transmitted. Then, the received signal vector is

$$\mathbf{r} = \left(\sqrt{\mathcal{E}_s} + n_1, n_2, \ldots, n_{M/2}\right) \tag{7.3.41}$$

where the $\{n_m\}$ are zero-mean, mutually statistically independent and identically distributed Gaussian random variables with variance $\sigma_n^2 = N_0/2$. The optimum detector decides in favor of the signal corresponding to the largest in magnitude of the crosscorrrelators

$$C(\mathbf{r}, \mathbf{s}_m) = \mathbf{r} \cdot \mathbf{s}_m = \sum_{k=1}^{M/2} r_k s_{mk}, \quad m = 1, 2, \ldots, M/2 \tag{7.3.42}$$

while the sign of this largest term is used to decide whether $s_m(t)$ or $-s_m(t)$ was transmitted. According to this decision rule, the probability of a correct decision is equal to the probability that $r_1 = \sqrt{\mathcal{E}_s} + n_1 > 0$ and r_1 exceeds $|r_m| = |n_m|$ for $m = 2, 3, \ldots, M/2$. But

$$P(|n_m| < r_1 | r_1 > 0) = \frac{1}{\sqrt{\pi N_0}} \int_{-r_1}^{r_1} e^{-x^2/N_0} dx = \frac{1}{\sqrt{2\pi}} \int_{-r_1/\sqrt{N_0/2}}^{r_1/\sqrt{N_0/2}} e^{-x^2/2} dx \tag{7.3.43}$$

Then, the probability of a correct decision is

$$P_c = \int_0^\infty \left[\frac{1}{\sqrt{2\pi}} \int_{-r_1/\sqrt{N_0/2}}^{r_1/\sqrt{N_0/2}} e^{-x^2/2} dx \right]^{\frac{M}{2}-1} f(r_1) \, dr_1$$

which, upon substitution for $f(r_1)$, we obtain

$$P_c = \int_{-\sqrt{2\mathcal{E}_s/N_0}}^\infty \left[\frac{1}{\sqrt{2\pi}} \int_{-\left(v+\sqrt{2\mathcal{E}_s/N_0}\right)}^{v+\sqrt{2\mathcal{E}_s/N_0}} e^{-x^2/2} dx \right]^{\frac{M}{2}-1} e^{-v^2/2} dv \tag{7.3.44}$$

where we have used the p.d.f. of r_1 given in (7.3.25). Finally, the probability of a symbol error $P_M = 1 - P_c$.

P_c, and hence, P_M may be evaluated numerically for different values of M from (7.3.44). The graph shown in Figure 7.46 illustrates P_M as a function of $\mathcal{E}_b/N_0$, where $\mathcal{E}_s = k\mathcal{E}_b$, for $M = 2, 4, 8, 16$, and 32. We observe that this graph is similar to that for orthogonal signals (see Figure 7.45). However, in this case, the probability of error for $M = 4$ is greater than that for $M = 2$. This is due to the fact that we have plotted the symbol error probability P_M in Figure 7.46. If we plot the equivalent bit error probability we would find that the graphs for $M = 2$ and $M = 4$ coincide. As in the case of orthogonal signals, as $M \to \infty$ (or $k \to \infty$),

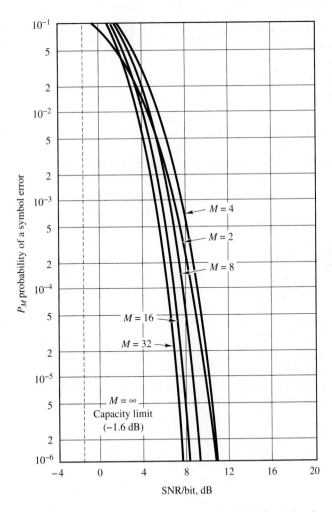

FIGURE 7.46.   Probability of symbol error for biorthogonal signals.

the minimum required $\mathcal{E}_b/N_0$ to achieve an arbitrarily small probability of error is -1.6 dB, the Shannon limit.

Simplex signals.   Next we consider the probability of error for M simplex signals. Recall from Section 7.1 that simplex signals are a set of M equally cor-related signals with mutual crosscorrelation coefficient $\gamma_{mn} = -1/(M-1)$. These signals have the same minimum separation of $\sqrt{2\mathcal{E}_s}$ between adjacent signal points in M-dimensional space as orthogonal signals. They achieve this mutual separation with a transmitted energy of $\mathcal{E}_s(M-1)/M$, which is less than that required for orthogonal signals by a factor of $(M-1)/M$. Consequently, the probability of error for simplex signals is identical to the probability of error for orthogonal signals, but this performance is achieved with a savings of

$$10\log(1 - \gamma_{mn}) = 10\log\frac{M}{M-1}\ \text{dB} \qquad (7.3.45)$$

in SNR.

This concludes our derivations of the probability of error performance for the various pulse modulation methods. Below, we compare the performance character-istics of these modulations for the AWGN channel. However, we first consider the probability of error for ML sequence detection.

7.3.3 Probability of Error for ML Sequence Detection

As we have observed from our discussion of ML sequence detection in Sec-tion 7.2.4, sequence detection involves the comparison of multi-branch signal paths that merge at each node in the trellis. When the additive noise is large enough to cause errors in the detection process, the errors are generally signal path errors involving two or more information bits or symbols. The selection of the wrong signal path at a node is called an *error event*. In general, an error event begins at the point where the surviving path diverges from the correct path and ends when the surviving path remerges with the correct path.

Figure 7.47 illustrates two paths through a four-state trellis. The upper path represents the correct path, which we assume to be the all-zero path, and the lower path represents the path selected by the ML sequence detector. This figure depicts two error events, one of length two and the other of length five. We should note that within an error event, some bits or symbols may be correct while others may be wrong. Nevertheless, the probability of (bit or symbol) error may be obtained by

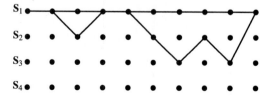

FIGURE 7.47.   Illustration of two error events in a trellis search.

properly weighting each event error probability with the corresponding fraction of incorrect symbols.

In general, there are many (possibly infinite) numbers of distinct error events, each of which has its own probability of occurrence. For this reason, it is extremely difficult to obtain an exact expression for the probability of error. Instead, we usually resort to upper (union) bounds on error performance.

The simplest approximation to the probability of error is obtained by considering distinct error events and their corresponding Euclidean distances. Then, the probability of error is well approximated (upper-bounded) by the weighted sum of the error probabilities of the distinct error events, where the weighting factors are the number of distinct error events at the corresponding Euclidean distances, i.e.,

$$P_e \leq \sum_i N_{d_i} Q\left(\sqrt{\frac{d_i^2}{2N_0}}\right) \tag{7.3.46}$$

where d_i^2 is the squared Euclidean distance of the ith path relative to the correct path, and N_{di} is the number of paths at distance d_i.

When the distances $\{d_i\}$ of the error events vary widely with i, we may neglect those terms with large distances in favor of the terms with smaller distances. Of course, the simplest approximation to the probability of error is to use only the dominant term in the sum of (7.3.46), which is the term at the minimum Euclidean distance, $d_{\min}$, i.e.,

$$P_e \approx N_{d_{\min}} Q\left(\sqrt{\frac{d_{\min}^2}{2N_0}}\right) \tag{7.3.47}$$

where $N_{d_{\min}}$ is the number of error events with distance $d_{\min}$.

To demonstrate this computation, let us consider the trellis for the NRZI signal. Figure 7.48 illustrates the trellis and indicates two distinct error events. The top path represents the correct path. Let us evaluate the probability of the two error events. The probability of the length two error event is simply the probability that $\mu_2(1, 1) < \mu_2(0, 0)$, where $\mu_2(0, 0)$ and $\mu_2(1, 1)$ are given by (7.2.59) and $r_1 = -\sqrt{\mathcal{E}_b} + n_1$ and $r_2 = -\sqrt{\mathcal{E}_b} + n_2$. Hence,

$$P[\mu_2(1, 1) < \mu_2(0, 0)] = P(n_1 > \sqrt{\mathcal{E}_b})$$

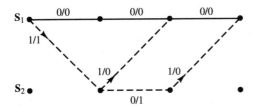

FIGURE 7.48.   Two error events in a trellis for an NRZI signal.

$$= \frac{1}{\sqrt{2\pi}\,\sigma_n} \int_{\sqrt{\mathcal{E}_b}}^{\infty} e^{-x^2/2\sigma_n^2}\, dx$$

$$= Q\left(\sqrt{\frac{2\mathcal{E}_b}{N_0}}\right) \tag{7.3.48}$$

where $\sigma_n^2 = N_0/2$. It is interesting to note that this is exactly the same probability of error as that of a symbol detector that ignores the memory inherent in the NRZI signal. Thus, the NRZI signal is an example of a signal with memory in which the Euclidean distance for a length two path is not increased relative to the Euclidean distance of a length one path. As a consequence, there is no gain in performance that is obtained by performing ML sequence detection compared with symbol detection.

The probability of the length three error event is the probability that $\mu_2(1,0,1) < \mu_2(0,0,0)$, where

$$\mu_3(0,0,0) = (r_1 + \sqrt{\mathcal{E}_b})^2 + (r_2 + \sqrt{\mathcal{E}_b})^2 + (r_3 + \sqrt{\mathcal{E}_b})^2$$

$$\mu_3(1,0,1) = (r_1 - \sqrt{\mathcal{E}_b})^2 + (r_2 - \sqrt{\mathcal{E}_b})^2 + (r_3 + \sqrt{\mathcal{E}_b})^2$$

where $r_i = -\sqrt{\mathcal{E}_b} + n_i$, $i = 1, 2, 3$. Hence,

$$P[\mu_3(1,0,1) < \mu_3(0,0,0)] = P(n_1 + n_2 > 2\sqrt{\mathcal{E}_b})$$

The random variable $x = n_1 + n_2$ is zero-mean Gaussian with variance N_0. Therefore,

$$P\left(n_1 + n_2 > 2\sqrt{\mathcal{E}_b}\right) = \frac{1}{\sqrt{2\pi}} \int_{\sqrt{4\mathcal{E}_b/N_0}}^{\infty} e^{-x^2/2}\, dx$$

$$= Q\left(\sqrt{\frac{4\mathcal{E}_b}{N_0}}\right) \tag{7.3.49}$$

We observe that this error event occurs with a significantly lower error probability than the length two error event.

In general, longer error events will have larger Euclidean distance between them and the correct path, and hence, their probability of occurrence will be significantly smaller. Therefore, it is apparent that, in general, the probability of error for the ML detector will be dominated by the error events at the minimum Euclidean distance $d_{\min}$.

7.3.4 Comparison of Modulation Methods

The digital modulation methods described in this chapter can be compared in a number of ways. For example, one can compare them on the basis of the SNR required to achieve a specified probability of error. However, such a comparison would not be very meaningful, unless it were made on the basis of some constraint, such as a fixed data rate of transmission.

Suppose that the bit rate R_b is fixed, and let us consider the channel bandwidth required to transmit the various signals. If we employ M-ary PAM, where $M = 2^k$, the symbol interval is $T = k/R_b$. A signal pulse of duration T has a spectral characteristic of width $1/2T$ (approximately). Hence, the channel bandwidth required to transmit the M-ary PAM signal is

$$W = R_b/2k = R_b/2\log_2 M \text{ Hz} \qquad (7.3.50)$$

In PPM, the symbol interval T is subdivided into M subintervals of duration T/M, and pulses of width T/M are transmitted in the corresponding subintervals. Consequently, the spectrum of each pulse is approximately $M/2T$ wide. Consequently, the channel bandwidth required to transmit the PPM signals is

$$W = \frac{M}{2T} = \frac{M}{2(k/R_b)} = \frac{MR_b}{2\log_2 M} \text{ Hz} \qquad (7.3.51)$$

Biorthogonal and simplex signals result in similar relationships as PPM (orthogonal). In the case of biorthogonal signals, the required bandwidth is one-half of that for orthogonal signals.

A compact and meaningful comparison of these modulation methods is one that is based on the normalized data rate R_b/W (bits per second per hertz of bandwidth) versus the SNR per bit ($\mathcal{E}_b/N_0$) required to achieve a given error probability. For PAM and orthogonal signals, we have

$$\text{PAM}: \qquad \frac{R_b}{W} = 2\log_2 M \qquad (7.3.52)$$

$$\text{PPM (orthogonal)}: \qquad \frac{R_b}{W} = \frac{2\log_2 M}{M} \qquad (7.3.53)$$

Figure 7.49 illustrates the graph of R_b/W versus SNR/bit for PAM and orthogonal signals for the case in which the symbol error probability is $P_M = 10^{-5}$. We observe that in the case of PAM, increasing the number of amplitudes M results in a higher bit rate to bandwidth ratio R_b/W. However, the cost of achieving the higher data rate is an increase in the SNR/bit. Consequently, M-ary PAM modulation is appropriate for communication channels that are bandwidth limited, where we desire a bit rate-to-bandwidth ratio $R_b/W > 1$ and where there is sufficiently high SNR to support multiple amplitude levels. Telephone channels are examples of such bandlimited channels.

In contrast, M-ary orthogonal signals yield a bit rate-to-bandwidth ratio of $R_b/W \leq 1$. As M increases, R_b/W decreases due to an increase in the required channel bandwidth. However, the SNR/bit required to achieve a given error probability (in this case, $P_M = 10^{-5}$) decreases as M increases. Consequently, M-ary orthogonal signals are appropriate for power-limited channels that have sufficiently large bandwidth to accommodate a large number of signals. In this case, as $M \to \infty$,

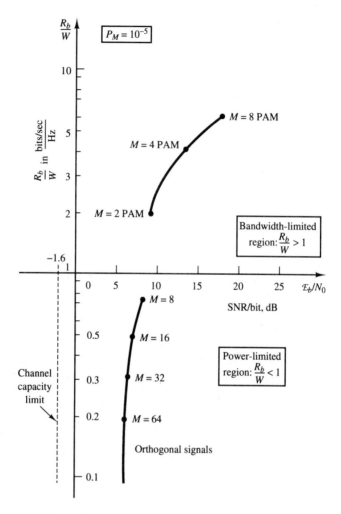

FIGURE 7.49. Comparison of several modulation methods at 10^{-5} symbol error probability.

the error probability can be made as small as desired, provided that $\mathcal{E}_b/N_0 > 0.693$ (-1.6 dB). This is the minimum SNR/bit required to achieve reliable transmission in the limit as the channel bandwidth $W \to \infty$ and the corresponding bit rate-to-bandwidth ratio $R_b/W \to 0$.

7.4 REGENERATIVE REPEATERS AND LINK BUDGET ANALYSIS

In the transmission of digital signals through an AWGN channel, we have observed that the performance of the communication system, measured in terms of the probability of error, depends on the received SNR, $\mathcal{E}_b/N_0$, where $\mathcal{E}_b$ is the transmitted energy per bit and $N_0/2$ is the power-spectral density of the additive noise. Hence, the additive noise ultimately limits the performance of the communication system.

In addition to the additive noise, another factor that affects the performance of a communication system is channel attenuation. As indicated in Chapter 6, all physical channels, including wire lines and radio channels, are lossy. Hence, the signal is attenuated as it travels through the channel. The simple mathematical model for the attenuation shown previously in Figure 6.18 may also be used for the purpose of digital communication. Consequently, if the transmitted signal is $s(t)$, the received signal is

$$r(t) = \alpha s(t) + n(t) \tag{7.4.1}$$

Then, if the energy in the transmitted signal is $\mathcal{E}_b$, the energy in the received signal is $\alpha^2 \mathcal{E}_b$. Consequently, the received signal has an SNR $\alpha^2 \mathcal{E}_b/N_0$. As in the case of analog communication systems, the effect of signal attenuation in digital communication systems is to reduce the energy in the received signal and thus to render the communication system more vulnerable to additive noise.

Recall that in analog communication systems, amplifiers called repeaters are used to periodically boost the signal strength in transmission through the channel. However, each amplifier also boosts the noise in the system. In contrast, digital communication systems allow us to detect and regenerate a clean (noise-free) signal in a transmission channel. Such devices, called *regenerative repeaters,* are frequently used in wireline and fiber optic communication channels.

7.4.1 Regenerative Repeaters

The front end of each regenerative repeater consists of a demodulator/detector that demodulates and detects the transmitted digital information sequence sent by the preceding repeater. Once detected, the sequence is passed to the transmitter side of the repeater which maps the sequence into signal waveforms that are transmitted to the next repeater. This type of repeater is called a regenerative repeater.

Since a noise-free signal is regenerated at each repeater, the additive noise does not accumulate. However, when errors occur in the detector of a repeater, the errors are propagated forward to the following repeaters in the channel. To evaluate the effect of errors on the performance of the overall system, suppose that the modulation is binary PAM, so that the probability of a bit error for one hop (signal transmission from one repeater to the next repeater in the chain) is

$$P_2 = Q\left(\sqrt{\frac{2\mathcal{E}_b}{N_0}}\right)$$

Since errors occur with low probability, we may ignore the probability that any one bit will be detected incorrectly more than once in transmission through a channel with K repeaters. Consequently, the number of errors will increase linearly with the number of regenerative repeaters in the channel, and therefore, the overall probability of error may be approximated as

$$P_b \approx K Q \left(\sqrt{\frac{2\mathcal{E}_b}{N_0}} \right) \tag{7.4.2}$$

In contrast, the use of K analog repeaters in the channel reduces the received SNR by K, and hence, the bit error probability is

$$P_b \approx Q \left(\sqrt{\frac{2\mathcal{E}_b}{K N_0}} \right) \tag{7.4.3}$$

Clearly, for the same probability of error performance, the use of regenerative repeaters results in a significant savings in transmitter power compared with analog repeaters. Hence, in digital communication systems, regenerative repeaters are preferable. However, in wireline telephone channels that are used to transmit both analog and digital signals, analog repeaters are generally employed.

Example 7.4.1

A binary digital communication system transmits data over a wireline channel of length 1000 Km. Repeaters are used every 10 Km to offset the effect of channel attenuation. Determine the $\mathcal{E}_b/N_0$ that is required to achieve a probability of a bit error of 10^{-5} if (a) analog repeaters are employed, and (b) regenerative repeaters are employed.

Solution The number of repeaters used in the system is $K = 100$. If regenerative repeaters are used, the $\mathcal{E}_b/N_0$ obtained from (7.4.2) is

$$10^{-5} = 100 Q \left(\sqrt{\frac{2\mathcal{E}_b}{N_0}} \right)$$

$$10^{-7} = Q \left(\sqrt{\frac{2\mathcal{E}_b}{N_0}} \right)$$

which yields approximately 11.3 dB. If analog repeaters are used, the $\mathcal{E}_b/N_0$ obtained from (7.4.3) is

$$10^{-5} = Q \left(\sqrt{\frac{2\mathcal{E}_b}{100 N_0}} \right)$$

which yields an $\mathcal{E}_b/N_0$ of 29.6 dB. Hence, the difference on the required SNR is about 18.3 dB, or approximately, 70 times the transmitter power of the digital communication system.

7.4.2 Link Budget Analysis for Radio Channels

In the design of radio communication systems[†] that transmit over line-of-sight microwave channels and satellite channels, we must also consider the effect of the antenna characteristics in determining the SNR at the receiver that is required to achieve a given level of performance. The system design procedure is described below.

Suppose that a transmitting antenna radiates isotropically in free space at a power level P_T watts, as shown in Figure 7.50. The power density at a distance d from the antenna is $P_T/4\pi d^2$ w/m^2. If the transmitting antenna has directivity in a particular direction, the power density in that direction is increased by a factor called the *antenna gain* G_T. Then, the power density at a distance d is $P_T G_T/4\pi d^2$ w/m^2. The product $P_T G_T$ is usually called the *effective isotropic radiated power* (EIRP), which is basically the radiated power relative to an isotropic antenna for which $G_T = 1$.

A receiving antenna pointed in the direction of the radiated power gathers a portion of the power that is proportional to its cross-sectional area. Hence, the received power extracted by the antenna is expressed as

$$P_R = \frac{P_T G_T A_R}{4\pi d^2} \tag{7.4.4}$$

where A_R is the *effective area of the antenna*. The basic relationship between the antenna gain G_R and its effective area, obtained from basic electromagnetic theory, is

$$A_R = \frac{G_R \lambda^2}{4\pi} \text{ m}^2 \tag{7.4.5}$$

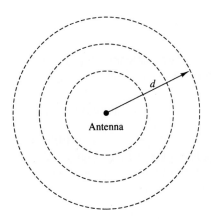

FIGURE 7.50.   Antenna that radiates isotropically in free space.

[†]Radio channels are not usually considered to be baseband channels. Information-bearing signals are usually transmitted over radio channels by means of carrier modulation, as described in Chapter 9. Nevertheless, it is appropriate at this point to describe the factors that are involved in a communication link budget analysis for digital transmission.

where λ is the wavelength of the transmitted signal.

If we substitute for A_R from (7.4.5) into (7.4.4), we obtain the expression for the received power as

$$P_R = \frac{P_T G_T G_R}{(4\pi d/\lambda)^2} \tag{7.4.6}$$

The factor $(4\pi d/\lambda)^2 = \mathcal{L}_s$ was previously defined as the free-space path loss. Other losses, such as atmospheric losses, that may be encountered in the transmission of the signal are accounted for by introducing an additional loss factor $\mathcal{L}_a$. Therefore, the received power may be expressed as

$$P_R = \frac{P_T G_T G_R}{\mathcal{L}_s \mathcal{L}_a} \tag{7.4.7}$$

or equivalently,

$$P_R\Big|_{\text{dB}} = P_T\Big|_{\text{dB}} + G_T\Big|_{\text{dB}} + G_R\Big|_{\text{dB}} - \mathcal{L}_s\Big|_{\text{dB}} - \mathcal{L}_a\Big|_{\text{dB}} \tag{7.4.8}$$

The effective area for an antenna generally depends on the wavelength λ of the radiated power and the physical dimension of the antenna. For example, a parabolic (dish) antenna of diameter D has an effective area

$$A_R = \frac{\pi D^2}{4}\eta \tag{7.4.9}$$

where $\pi D^2/4$ is the physical area and η is the *illumination efficiency factor,* which is typically in the range $0.5 \leq \eta \leq 0.6$. Hence, the antenna gain for a parabolic antenna of diameter D is

$$G_R = \eta\left(\frac{\pi D}{\lambda}\right)^2, \quad \text{parabolic antenna} \tag{7.4.10}$$

As a second example, a horn antenna of physical area A has an efficiency factor of 0.8, an effective area of $A_R = 0.8A$, and a gain of

$$G_R = \frac{10A}{\lambda^2}, \quad \text{horn antenna} \tag{7.4.11}$$

Another parameter that is related to the gain (directivity) of an antenna is its *beamwidth,* denoted as Θ_B and illustrated in Figure 7.51. Usually, the beamwidth is measured as the -3 dB width of the antenna pattern. For example, the -3 dB beamwidth of a parabolic antenna is approximately

$$\Theta_b \approx 70\lambda/D \text{ deg}$$

so that G_T is inversely proportional to Θ_B^2. Hence, a decrease of the beamwidth by a factor of two, which is obtained by doubling the diameter, increases the antenna gain by a factor of four (6 dB).

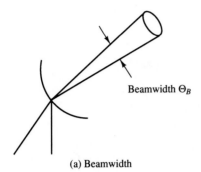

Beamwidth Θ_B

(a) Beamwidth

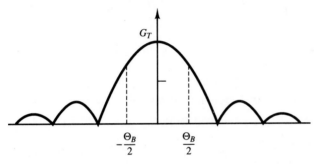

(b) Antenna pattern

FIGURE 7.51. A narrow beam antenna and its radiation pattern.

Example 7.4.2

A satellite in geosynchronous orbit (36,000 km above the earth's surface) radiates 100 watts of power (20 dBW). The transmitting antenna has a gain of 18 dB, so that the EIRP = 38 dBW. The earth station employs a 3 meter parabolic antenna and the downlink is transmitting at a frequency of 4 GHz. Determine the received power.

Solution   The wavelength $\lambda = 0.075$ m. Hence, the free-space path loss is

$$L_s\Big|_{dB} = 20\log\left(\frac{4\pi d}{\lambda}\right) = 195.6 \text{ dB}$$

Assuming $\eta = 0.5$, the antenna gain is 39 dB. Since no other losses are assumed,

$$P_R\Big|_{dB} = 20 + 18 + 39 - 195.6$$

$$= -118.6 \text{ dBW}$$

or equivalently,

$$P_R = 2.5 \times 10^{-11} \text{ watts}$$

We may carry the computation one step further by relating the $\mathcal{E}_b/N_0$ required to achieve a specified level of performance to P_R. Since

$$\frac{\mathcal{E}_b}{N_0} = \frac{T_b P_R}{N_0} = \frac{1}{R_b} \frac{P_R}{N_0} \tag{7.4.12}$$

it follows that

$$\frac{P_R}{N_0} = R_b \left(\frac{\mathcal{E}_b}{N_0} \right)_{\text{req}} \tag{7.4.13}$$

where $(\mathcal{E}_b/N_0)_{\text{req}}$ is the required SNR per bit to achieve the desired performance. The relation in (7.4.13) allows us to determine the bit rate R_b. We have

$$10 \log_{10} R_b = \left(\frac{P_R}{N_0} \right)_{\text{dB}} - \log_{10} \left(\frac{\mathcal{E}_b}{N_0} \right)_{\text{req}} \tag{7.4.14}$$

Example 7.4.3

If $(\mathcal{E}_b/N_0)_{\text{req}} = 10$ dB, determine the bit rate for the satellite communication system in Example 7.4.2. Assume that the receiver front end has a noise temperature of 300° K, which is typical for a receiver in the 4 GHz range.

Solution   Since $\mathcal{T}_0 = 290^\circ$ K and $\mathcal{T}_e = 10^\circ$ K, it follows that

$$N_0 = k\mathcal{T} = 4.1 \times 10^{-21} \text{ W/Hz}$$

or equivalently, -203.9 dBW/Hz. Then,

$$\left(\frac{P_R}{N_0} \right)_{\text{dB}} = -118.6 + 203.9$$

$$= 85.3 \text{ dB-Hz}$$

Therefore, from (7.4.14) we obtain

$$10 \log_{10} R_b = 85.3 - 10$$

$$= 75.3$$

or, equivalently,

$$R_b = 33.9 \times 10^6 \text{ b/s}$$

We conclude that this satellite channel can support a bit rate of 33.9 megabits per second.

7.5 FURTHER READING

The geometrical representation of digital signals as vectors was first used by Kotelnikov (1947) and by Shannon (1948) in his classic papers. This approach was popularized by Wozencraft and Jacobs (1965). Today this approach to signal analysis and design is widely used. Similar treatments to that given in the text may be found in most books on digital communications.

Modulation codes were also first introduced by Shannon (1948). Some of the early work on the construction of runlength-limited codes is found in the papers by Freiman and Wyner (1964), Gabor (1967), Franaszek (1968, 1969, 1970), Tang and Bahl (1970), and Jacoby (1977). More recent works are found in papers by Adler Coppersmith and Hassner (1983), and Karabed and Siegel (1991). The motivation for most of the work on runlength-limited codes was provided by applications to magnetic and optical recording. A well-written tutorial paper on runlength-limited codes has been written by Schouhamer Immink (1990).

The matched filter was introduced by North (1943), who showed that it maximized the signal-to-noise ratio. Maximum-likelihood sequence detection based on the Viterbi algorithm was first described in a paper by Forney (1972). Analysis of various binary and M-ary modulation signals in AWGN were performed in the two decades following Shannon's work. Treatments similar to that given in this chapter may be found in most books on digital communications.

PROBLEMS

7.1 Determine the average energy of a set of M PAM signals of the form

$$s_m(t) = s_m \psi(t), \quad m = 1, 2, \ldots, M$$
$$0 \le t \le T$$

where

$$s_m = \sqrt{\mathcal{E}_g} A_m, \quad m = 1, 2, \ldots, M$$

The signals are equally probable with amplitudes that are symmetric about zero and are uniformly spaced with distance d between adjacent amplitudes as shown in Figure 7.11.

7.2 Show that the correlation coefficient of two adjacent signal points corresponding to the vertices of an N-dimensional hypercube with its center at the origin is given by

$$\gamma = \frac{N-2}{N}$$

and their Euclidean distance is

$$d = 2\sqrt{\mathcal{E}_s/N}$$

7.3 Consider the three waveforms $\psi_n(t)$ shown in Figure P-7.3.

1. Show that these waveforms are orthonormal.
2. Express the waveform $x(t)$ as a weighted linear combination of $\psi_n(t), n = 1, 2, 3$, if

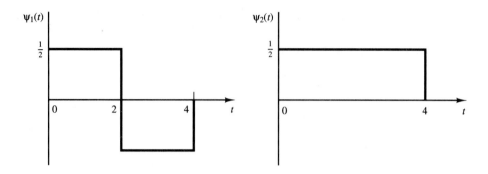

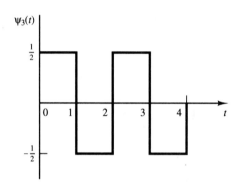

FIGURE P-7.3

$$x(t) = \begin{cases} -1, & 0 \le t \le 1 \\ 1, & 1 \le t \le 3 \\ -1, & 3 \le t \le 4 \end{cases}$$

and determine the weighting coefficients.

7.4 Use the orthonormal waveforms in Problem P-7.3 to approximate the function

$$x(t) = \sin(\pi t/4)$$

over the interval $0 \le t \le 4$ by the linear combination

$$\widehat{x}(t) = \sum_{n=1}^{3} c_n \psi_n(t)$$

1. Determine the expansion coefficients $\{c_n\}$ that minimize the mean-square approximation error

$$E = \int_0^4 \left[x(t) - \hat{x}(t)\right]^2 dt$$

2. Determine the residual mean-square error E_{min}.

7.5 Consider the four waveforms shown in Figure P-7.5.

1. Determine the dimensionality of the waveforms and a set of basis functions.
2. Use the basis functions to represent the four waveforms by vectors s_1, s_2, s_3, s_4.
3. Determine the minimum distance between any pair of vectors.

7.6 Determine a set of orthonormal functions for the four signals shown in Figure P-7.6.

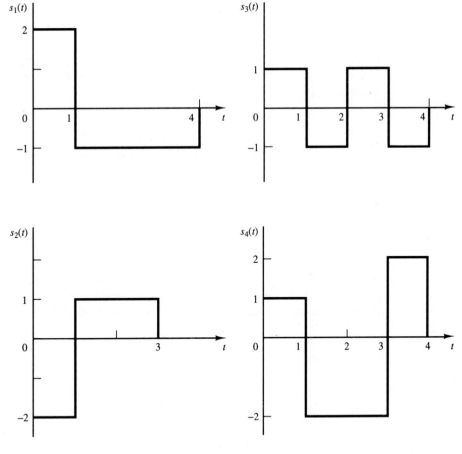

FIGURE P-7.5

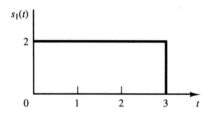
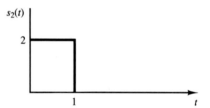
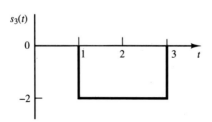
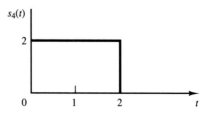

FIGURE P-7.6

7.7 Consider a set of M orthogonal signal waveforms $s_m(t), 1 \le m \le M, \quad 0 \le t \le T$, all of which have the same energy $\mathcal{E}$. Define a new set of M waveforms as

$$s'_m(t) = s_m(t) - \frac{1}{M} \sum_{k=1}^{M} s_k(t), \quad \begin{matrix} 1 \le m \le M \\ 0 \le t \le T \end{matrix}$$

Show that the M signal waveforms $\{s'_m(t)\}$ have equal energy, given by

$$\mathcal{E}' = (M-1)\,\mathcal{E}/M$$

and are equally correlated, with correlation coefficient

$$\gamma_{mn} = \frac{1}{\mathcal{E}'} \int_0^T s'_m(t)\,s'_n(t)\,dt = -\frac{1}{M-1}$$

7.8 As indicated in Section 7.1.4, the running digital sum (RDS) is defined as the difference between the total number of accumulated zeros and the total number of accumulated ones in a sequence. If the RDS is limited to the range

$$-2 \le \text{RDS} \le 2$$

determine the state transition matrix and the capacity of the corresponding code.

7.9 Determine the capacity of a $(0, 1)$ runlength-limited code. Compare its capacity with that of a $(1, \infty)$ code and explain the relationship.

7.10 A ternary signal format is designed for a channel that does not pass dc. The binary input information sequence is transmitted by mapping a 1 into either a positive pulse or a negative pulse, and a zero is transmitted by the absence of a pulse. Hence, for the transmission of 1's, the polarity of the pulses alternate. This is called an AMI (alternate mark inversion) code. Determine the capacity of the code.

7.11 Give an alternative description of the AMI code described in Problem 7.10 using the RDS with the constraint that the RDS can take only the values 0 and +1.

7.12 (*kBnT* codes) From Problem 7.10 we note that the AMI code is a "pseudo ternary" code in that it transmits one bit per symbol using a ternary alphabet, which has the capacity of $\log_2 3 = 1.58$ bits. Such a code does not provide sufficient spectral shaping. Better spectral shaping is achieved by the class of block codes designated as *kBnT*, where *k* denotes the number of information bits and *n* denotes the number of ternary symbols per block. By selecting the largest *k* possible for each *n*, we obtain the table shown below:

k	n	Code
1	1	1B1T
3	2	3B2T
4	3	4B3T
6	4	6B4T

Determine the efficiency of these codes by computing the ratio of the rate of the code in bits/symbol divided by $\log_2 3$. Note that 1B1T is the AMI code.

7.13 This problem deals with the capacity of two (d, κ) codes.

1. Determine the capacity of a (d, κ) code that has the following state transition matrix:

$$D = \begin{bmatrix} 1 & 1 \\ 1 & 0 \end{bmatrix}$$

2. Repeat (step 1) when D is given as

$$D = \begin{bmatrix} 1 & 1 \\ 0 & 1 \end{bmatrix}$$

3. Comment on the differences between parts 1 and 2.

7.14 A simplified model of the telegraph code consists of two symbols (Blahut, 1990). A dot consists of one time unit of line closure followed by one time unit of line open. A dash consists of three units of line closure followed by one time unit of line open.

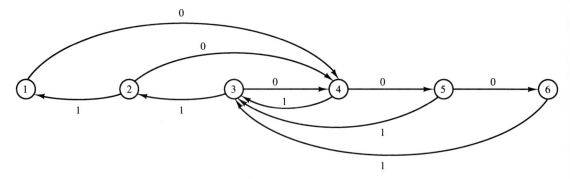

FIGURE P-7.15

1. If we view this code as a constrained code with symbols of equal duration, give the constraints.
2. Determine the state transition matrix.
3. Determine the capacity.

7.15 Determine the state transition matrix for the runlength constrained code described by the state diagram shown in Figure P-7.15. Sketch the corresponding trellis.

7.16 Determine the state transition matrix for the (2,7) runlength-limited code specified by the state diagram shown in Figure P-7.16.

7.17 Suppose that two signal waveforms $s_1(t)$ and $s_2(t)$ are orthogonal over the interval $(0, T)$. A sample function $n(t)$ of a zero-mean white noise process is crosscorrelated with $s_1(t)$ and $s_2(t)$ to yield

$$n_1 = \int_0^T s_1(t)n(t)\,dt$$
$$n_2 = \int_0^T s_2(t)n(t)\,dt$$

Prove that $E[n_1 n_2] = 0$.

7.18 A binary digital communication system employs the signals

$$s_0(t) = 0, \quad 0 \leq t \leq T$$
$$s_1(t) = A, \quad 0 \leq t \leq T$$

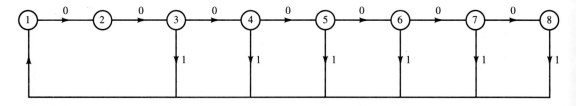

FIGURE P-7.16

for transmitting the information. This is called *on–off signaling*. The demodulator crosscorrelates the received signal $r(t)$ with $s(t)$ and samples the output of the correlator at $t = T$.

1. Determine the optimum detector for an AWGN channel and the optimum threshold, assuming that the signals are equally probable.
2. Determine the probability of error as a function of the SNR. How does on–off signaling compare with antipodal signaling?

7.19 A binary PAM communication system employs rectangular pulses of duration T_b and amplitudes $\pm A$ to transmit digital information at a rate $R = 10^5$ bits/sec. If the power-spectral density of the additive Gaussian noise is $N_0/2$, where $N_0 = 10^{-2}$ W/Hz, determine the value of A that is required to achieve a probability of error $P_2 = 10^{-6}$.

7.20 In a binary PAM system for which the two signals occur with unequal probabilities $(p \text{ and } 1 - p)$, the optimum detector is specified by (7.2.52).

1. Determine the average probability of error as a function of $(\mathcal{E}_b/N_0)$ and p.
2. Evaluate the probability of error for $p = 0.3$ and $p = 0.5$, with $\mathcal{E}_b/N_0 = 10$.

7.21 A binary PAM communication system is used to transmit data over an AWGN channel. The prior probabilities for the bits are $P(a_m = 1) = 1/3$ and $P(a_m = -1) = 2/3$.

1. Determine the optimum threshold at the detector.
2. Determine the average probability of error.

7.22 Binary antipodal signals are used to transmit information over an AWGN channel. The prior probabilities for the two input symbols (bits) are 1/3 and 2/3.

1. Determine the optimum maximum-likelihood decision rule for the detector.
2. Determine the average probability of error as a function of $\mathcal{E}_b/N_0$.

7.23 The received signal in a binary communication system that employs antipodal signals is

$$r(t) = s(t) + n(t)$$

where $s(t)$ is shown in Figure P-7.23 and $n(t)$ is AWGN with power-spectral density $N_0/2$ W/Hz.

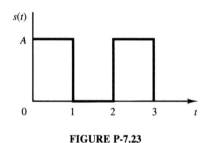

FIGURE P-7.23

1. Sketch the impulse response of the filter matched to $s(t)$.
2. Sketch the output of the matched filter to the input $s(t)$.
3. Determine the variance of the noise of the output of the matched filter at $t = 3$.
4. Determine the probability of error as a function of A and N_0.

7.24 A matched filter has the frequency response

$$H(f) = \frac{1 - e^{-j2\pi fT}}{j2\pi f}$$

1. Determine the impulse response $h(t)$ corresponding to $H(f)$.
2. Determine the signal waveform to which the filter characteristic is matched.

7.25 Prove that when a sinc pulse $g_T(t)$ is passed through its matched filter, the output is the same sinc pulse.

7.26 The demodulation of the binary antipodal signals

$$s_1(t) = -s_2(t) = \begin{cases} \sqrt{\dfrac{E_b}{T}} & 0 \le t \le T \\ 0, & \text{otherwise} \end{cases}$$

can be accomplished by use of a single integrator, as shown in Figure P-7.26, which is sampled periodically at $t = kT$, $k = 0, \pm 1, \pm 2, \ldots$. The additive noise is zero-mean Gaussian with power-spectral density of $\frac{N_0}{2}$ W/Hz.

1. Determine the output SNR of the demodulator at $t = T$.
2. If the ideal integrator is replaced by the RC filter shown in Figure P-7.26, determine the output SNR as a function of the time constant RC.
3. Determine the value of RC that maximizes the output SNR.

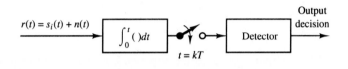

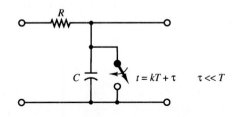

FIGURE P-7.26

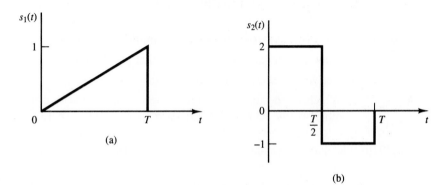

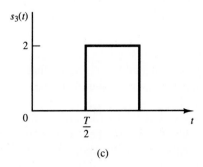

FIGURE P-7.27

7.27 Sketch the impulse response of the filter matched to the pulses shown in Figure P-7.27. Also determine and sketch the outputs of each of the matched filters.

7.28 Three equally probable messages m_1, m_2, and m_3 are to be transmitted over an AWGN channel with noise power-spectral density $\frac{N_0}{2}$. The messages are

$$s_1(t) = \begin{cases} 1 & 0 \leq t \leq T \\ 0 & \text{otherwise} \end{cases}$$

$$s_2(t) = -s_3(t) = \begin{cases} 1 & 0 \leq t \leq \frac{T}{2} \\ -1 & \frac{T}{2} \leq 0 \leq T \\ 0 & \text{otherwise} \end{cases}$$

1. What is the dimensionality of the signal space?
2. Find an appropriate basis for the signal space (Hint: You can find the basis without using the Gram-Schmidt procedure).
3. Draw the signal constellation for this problem.
4. Derive and sketch the optimal decision regions R_1, R_2, and R_3.
5. Which of the three messages is more vulnerable to errors and why? In other words, which of $p(\text{Error}|m_i$ transmitted), $i = 1, 2, 3$ is larger?

7.29 In this chapter, we showed that an optimal demodulator can be realized as:

- a correlation-type demodulator
- a matched-filter-type demodulator

where in both cases $\psi_j(t)$, $1 \leq j \leq N$ were used for correlating $r(t)$ or designing the matched filters. Show that an optimal demodulator for a general M-ary communication system can also be designed based on correlating $r(t)$ with $s_i(t)$, $1 \leq i \leq M$, or designing filters that are matched to $s_i(t)$'s, $1 \leq i \leq M$. Precisely describe the structure of such receivers by giving their block diagram and all relevant design parameters, and compare their complexity with the complexity of the receivers obtained in the text.

7.30 In a binary antipodal signaling scheme, the signals are given by

$$s_1(t) = -s_2(t) = \begin{cases} \dfrac{2At}{T} & 0 \leq t \leq \frac{T}{2} \\ 2A\left(1 - \dfrac{t}{T}\right) & \frac{T}{2} \leq t \leq T \\ 0 & \text{otherwise} \end{cases}$$

The channel is AWGN and $S_n(f) = \frac{N_0}{2}$. The two signals have prior probabilities p_1 and $p_2 = 1 - p_1$.

1. Determine the structure of the optimal receiver.
2. Determine an expression for the error probability.
3. Plot error probability as a function of p_1 for $0 \le p_1 \le 1$.

7.31 In an additive white Gaussian noise channel with a noise power-spectral density of $\frac{N_0}{2}$, two equiprobable messages are transmitted by

$$s_1(t) = \begin{cases} \dfrac{At}{T} & 0 \le t \le T \\ 0 & \text{otherwise} \end{cases}$$

$$s_2(t) = \begin{cases} A - \dfrac{At}{T} & 0 \le t \le T \\ 0 & \text{otherwise} \end{cases}$$

1. Determine the structure of the optimal receiver.
2. Determine the probability of error.

7.32 Consider a signal detector with an input

$$r = \pm A + n$$

where $+A$ and $-A$ occur with equal probability and the noise variable n is characterized by the (Laplacian) p.d.f. shown in Figure P-7.32.

1. Determine the probability of error as a function of the parameters A and σ
2. Determine the "SNR" required to achieve an error probability of 10^{-5}. How does the SNR compare with the result for a Gaussian p.d.f.?

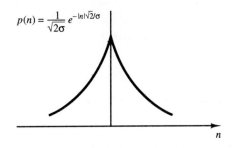

$$p(n) = \frac{1}{\sqrt{2}\sigma} e^{-|n|\sqrt{2}/\sigma}$$

FIGURE P-7.32

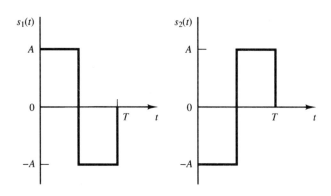

FIGURE P-7.33

7.33 A Manchester encoder maps an information 1 into 10 and a 0 into 01. If the output of the encoder is transmitted by use of NRZ, the signal waveforms corresponding to the Manchester code are shown in Figure P-7.33. Determine the probability of error if the two signals are equally probable.

7.34 Determine the probability of error for the optimum symbol-by-symbol detector of the AMI code described in Problem 7.10, if the additive noise is white Gaussian with power-spectral density N_0 W/Hz.

7.35 A three-level PAM system is used to transmit the output of a memoryless ternary source. The signal constellation is shown in Figure P-7.35. Determine the input to the detector, the optimum threshold that minimizes the average probability of error, and the average probability of error.

7.36 Consider a biorthogonal signal set with $M = 8$ signal points. Determine a union bound for the probability of a symbol error as a function of $\mathcal{E}_b/N_0$. The signal points are equally likely a priori.

7.37 Consider an M-ary digital communication system where $M = 2^N$, and N is the dimension of the signal space. Suppose that the M signal vectors lie on the vertices of a hypercube that is centered at the origin, as illustrated in Figure 7.15. Determine the average probability of a symbol error as a function of $\mathcal{E}_s/N_0$ where $\mathcal{E}_s$ is the energy per symbol, $N_0/2$ is the power-spectral density of the AWGN, and all signal points are equally probable.

7.38 Consider the signal waveform

$$s(t) = \sum_{i=1}^{n} c_i \, p(t - kT_c)$$

FIGURE P-7.35

where $p(t)$ is a rectangular pulse of unit amplitude and duration T_c. The $\{c_i\}$ may be viewed as a code vector $\underline{C} = [c_1, c_2, \ldots, c_n]$, where the elements $c_i = \pm 1$. Show that the filter matched to the waveform $s(t)$ may be realized as a cascade of a filter matched to $p(t)$ followed by a discrete-time filter matched to the vector $\underline{C}$. Determine the value of the output of the matched filter at the sampling instant $t = nT_c$.

7.39 A speech signal is sampled at a rate of 8 kHz, logarithmically compressed and encoded into a PCM format using 8 bits per sample. The PCM data is transmitted through an AWGN baseband channel via M-level PAM. Determine the bandwidth required for transmission when (a) $M = 4$, (b) $M = 8$, and (c) $M = 16$.

7.40 Two equiprobable messages are transmitted via an additive white Gaussian noise channel with noise power-spectral density of $\frac{N_0}{2} = 1$. The messages are transmitted by the following two signals

$$s_1(t) = \begin{cases} 1 & 0 \le t \le 1 \\ 0 & \text{otherwise} \end{cases}$$

and $s_2(t) = s_1(t-1)$. It is intended to implement the receiver using a correlation type structure, but due to imperfections in the design of the correlators, the structure shown in Figure P-7.40 has been implemented. The imperfection appears in the integrator in the upper branch where instead of $\int_0^1$ we have $\int_0^{1.5}$. The decision box, therefore, observes r_1 and r_2 and based on this observation has to decide which message was transmitted. What decision rule should be adopted by the decision box for an optimal decision?

7.41 A Hadamard matrix is defined as a matrix whose elements are ± 1 and its row vectors are pairwise orthogonal. In the case when n is a power of 2, an $n \times n$ Hadamard matrix is constructed by means of the recursion

$$\underline{H}_2 = \begin{bmatrix} 1 & 1 \\ 1 & -1 \end{bmatrix} \qquad H_{2n} = \begin{bmatrix} \underline{H}_n & \underline{H}_n \\ \underline{H}_n & -\underline{H}_n \end{bmatrix}$$

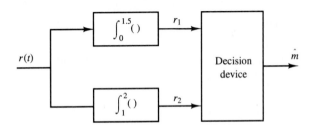

FIGURE P-7.40

1. Let $\underline{C}_i$ denote the ith row of an $n \times n$ Hadamard matrix as defined above. Show that the waveforms constructed as

$$s_i(t) = \sum_{k=1}^{n} c_{ik} p(t - kT_c), \quad i = 1, 2, \dots, n$$

are orthogonal, where $p(t)$ is an arbitrary pulse confined to the time interval $0 \le t \le T_c$.

2. Show that the matched filters (or crosscorrelators) for the n waveforms $\{s_i(t)\}$ can be realized by a single filter (or correlator) matched to the pulse $p(t)$ followed by a set of n crosscorrelators using the code words $\{\underline{C}_i\}$.

7.42 The discrete sequence

$$r_k = \sqrt{\mathcal{E}_b} c_k + n_k, \quad k = 1, 2, \dots, n$$

represents the output sequence of samples from a demodulator, where $c_k = \pm 1$ are elements of one of two possible code words, $\underline{C}_1 = [1, 1, \dots, 1]$ and $\underline{C}_2 = [1, 1, \dots, 1, -1, \dots, -1]$. The code word $\underline{C}_2$ has w elements which are $+1$ and $n - w$ elements which are -1, where w is some positive integer. The noise sequence $\{n_k\}$ is white Gaussian with variance σ.

1. What is the optimum maximum likelihood detector for the two possible transmitted signals?
2. Determine the probability error as a function of the parameters $\left(\sigma^2, \mathcal{E}_b, w\right)$.
3. What is the value of w that minimizes the error probability?

7.43 A baseband digital communication system employs the signals shown in Figure P-7.43(b) for transmission of two equiprobable messages. It is assumed the the communication problem studied here is a "one shot" communication problem; that is, the above messages are transmitted just once and no transmission takes place afterward. The channel has no attenuation ($\alpha = 1$), and the noise is AWG with power-spectral density $\frac{N_0}{2}$.

1. Find an appropriate orthonormal basis for the representation of the signals.
2. In a block diagram, give the precise specifications of the optimal receiver using matched filters. Label the block diagram carefully.
3. Find the error probability of the optimal receiver.
4. Show that the optimal receiver can be implemented by using just *one* filter (see block diagram shown in Figure P-7.43(b)). What are the characteristics of the matched filter and the sampler and decision device?

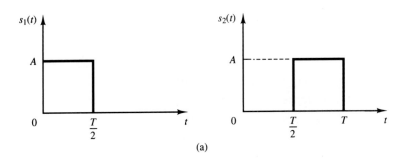

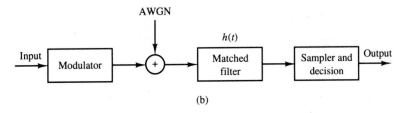

FIGURE P-7.43

5. Now assume the channel is not ideal but has an impulse response of $c(t) = \delta(t) + \frac{1}{2}\delta(t - \frac{T}{2})$. Using the same matched filter you used in the previous part, design an optimal receiver.

6. Assuming that the channel impulse response is $c(t) = \delta(t) + a\delta(t - \frac{T}{2})$, where a is a random variable uniformly distributed on $[0, 1]$, and using the same matched filter, design the optimal receiver.

7.44 Consider a transmission line channel that employs $n - 1$ regenerative repeaters plus the terminal receiver in the transmission of binary information. We assume that the probability or error at the detector of each receiver is p and that errors among repeaters are statistically independent.

1. Show that the binary error probability at the terminal receiver is

$$P_n = \frac{1}{2}\left[1 - (1 - 2p)^n\right]$$

2. If $p = 10^{-6}$ and $n = 100$, determine an approximate value of P_n.

7.45 A digital communication system consists of a transmission line with 100 digital (regenerative) repeaters. Binary antipodal signals are used for transmitting the information. If the overall end-to-end error probability is 10^{-6}, determine the probability of error for each repeater and the required E_b/N_0 to achieve this performance in AWGN.

7.46 A radio transmitter has a power output of $P_T = 1$ watt at a frequency of 10^9 Hz (1 GHz). The transmitting and receiving antennas are parabolic dishes with diameter $D = 3$ feet.

   1. Determine the antenna gains.
   2. Determine the EIRP for the transmitter.
   3. The distance (free space) between the transmitting and receiving antennas is 20 km. Determine the signal power at the output of the receiving antenna in dBm.

7.47 A radio communication system transmits at a power level of 0.1 watt at 1 GHz. The transmitting and receiving antennas are parabolic, each having a diameter of one meter. The receiver is located 30 km from the transmitter.

   1. Determine the gains of the transmitting and receiving antennas.
   2. Determine the EIRP of the transmitted signal.
   3. Determine the signal power from the receiving antenna.

7.48 A satellite in synchronous orbit is used to communicate with an earth station at a distance of 4×10^7 meters. The satellite has an antenna with a gain of 15 dB and a transmitter power of 3 watts. The earth station uses a 10 meter parabolic antenna with an efficiency of 0.6. The frequency band is at $f = 10$ GHz. Determine the received power level at the output of the receiver antenna.

7.49 A spacecraft in space located 10^8 meters from the earth is sending data at a rate of R bps. The frequency band is centered at 2 GHz and the transmitted power is 10 watts. The earth station uses a parabolic antenna, 50 meters in diameter, and the spacecraft has an antenna with a gain of 10 dB. The noise temperature of the receiver front end is $T = 300°$K.

   1. Determine the received power level.
   2. If the desired $\mathcal{E}_b/N_0 = 10$ dB, determine the maximum bit rate that the spacecraft can transmit.

7.50 Consider the front end of the receiver shown in the block diagram in Figure P-7.50. The received signal power at the input to the first amplifier is -113 dBm, and the received noise power-spectral density is -175 dBm/Hz. The bandpass filter has a bandwidth of 10 MHz, and gains and noise figures are as shown. Determine the signal to-noise ratio P_s/P_n at the input to the demodulator.

7.51 A satellite in geosynchronous orbit is used as a regenerative repeater in a digital communication system. Let us consider the satellite-to-earth link in which the satellite antenna has a gain of 6 dB and the earth station antenna

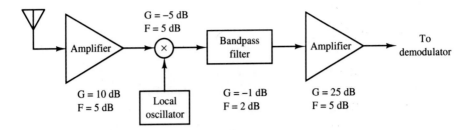

FIGURE P-7.50

has a gain of 50 dB. The downlink is operated at a center frequency of 4 GHz, and the signal bandwidth is 1 MHz. If the required $(\mathcal{E}_b/N_0)$ for reliable communication is 15 dB, determine the transmitted power for the satellite downlink. It is assumed that $N_0 = 4.1 \times 10^{-21}$ W/Hz.

7.52 One of the Mariner spacecrafts that traveled to the planet Mercury sent its data to earth through a distance of 1.6×10^{11} meters. The transmitting antenna had a gain of 27 dB and operated at a frequency $f = 2.3$ GHz. The transmitter power was 17 watts. The earth station employed parabolic antenna with a 64-meter diameter and an efficiency of 0.55. The receiver had an effective noise temperature of $T_e = 15°$K. If the desired SNR per bit $(\mathcal{E}_b/N_0)$ was 6 dB, determine the data rate that could have been supported by the communication link.

8

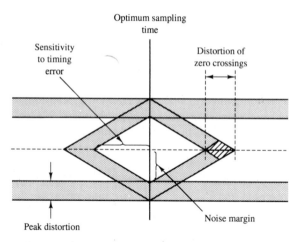

Optimum sampling time

Sensitivity to timing error

Distortion of zero crossings

Peak distortion

Noise margin

Digital PAM Transmission Through Bandlimited AWGN Channels

In the preceding chapter we considered digital communication over an AWGN channel and evaluated the probability of error performance of the optimum receiver for several different types of pulse modulation methods. In this chapter we treat digital communication over a baseband channel that is modeled as a linear filter with a bandwidth limitation. Bandlimited channels most frequently encountered in practice are telephone channels, microwave line-of-sight radio channels, satellite channels, and underwater acoustic channels.

In general, a linear filter channel imposes more stringent requirements on the design of modulation signals. Specifically, the transmitted signals must be designed to satisfy the bandwidth constraint imposed by the channel. The bandwidth constraint generally precludes the use of rectangular pulses at the output of the modulator. Instead, the transmitted signals must be shaped to restrict their bandwidth to that available on the channel. The design of bandlimited signals is one of the topics treated in this chapter.

We will see that a linear filter channel distorts the transmitted signal. The channel distortion results in intersymbol interference at the output of the demodulator and leads to an increase in the probability of error at the detector. Devices or methods for correcting or undoing the channel distortion, called *channel equalizers*, are then described.

The basic modulation method considered in this chapter is digital PAM. This type of modulation is particularly suitable for bandlimited channels due to its high bandwidth efficiency. Additional bandwidth efficient modulation methods are described in the next chapter in the context of carrier modulation methods.

8.1   DIGITAL TRANSMISSION THROUGH BANDLIMITED BASEBAND CHANNELS

A bandlimited channel such as a telephone wireline is characterized as a linear filter with impulse response $c(t)$ and frequency response $C(f)$, where

$$C(f) = \int_{-\infty}^{\infty} c(t)e^{-j2\pi ft}\, dt \qquad (8.1.1)$$

If the channel is bandlimited to B_c Hz, then $C(f) = 0$ for $|f| > B_c$. Any frequency components at the input to the channel that are higher than B_c Hz will not be passed by the channel. For this reason, we consider the design of signals for transmission through the channel that are bandlimited to $W = B_c$ Hz, as shown in Figure 8.1. Henceforth, W will denote the bandwidth limitation of the signal and the channel.

Now, suppose that the input to a bandlimited channel is a signal waveform $g_T(t)$. Then, the response of the channel is the convolution of $g_T(t)$ with $c(t)$, i.e.,

$$h(t) = \int_{-\infty}^{\infty} c(\tau)g_T(t-\tau)d\tau = c(t) \star g_T(t) \qquad (8.1.2)$$

or, when expressed in the frequency domain, we have

$$H(f) = C(f)G_T(f) \qquad (8.1.3)$$

where $G_T(f)$ is the spectrum (Fourier transform) of the signal $g_T(t)$ and $H(f)$ is the spectrum of $h(t)$. Thus, the channel alters or distorts the transmitted signal $g_T(t)$.

Let us assume that the signal at the output of the channel is corrupted by AWGN. Then the signal at the input to the demodulator is of the form $h(t) + n(t)$, where $n(t)$ denotes the AWGN. Recall from the preceding chapter that in the presence of AWGN a demodulator that employs a filter which is matched to the signal $h(t)$ maximizes the signal-to-noise ratio (SNR) at its output. Therefore, let us pass the received signal $h(t) + n(t)$ through a filter that has a frequency response

$$G_R(f) = H^\star(f)e^{-j2\pi ft_0} \qquad (8.1.4)$$

where t_0 is some nominal time delay at which we sample the filter output.

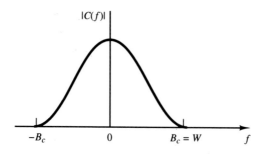

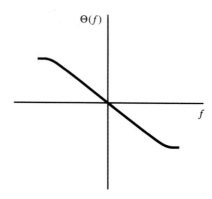

FIGURE 8.1. Magnitude and phase responses of bandlimited channel.

The signal component at the output of the matched filter at the sampling instant $t = t_0$ is

$$y_s(t_0) = \int_{-\infty}^{\infty} |H(f)|^2 df = \mathcal{E}_h \tag{8.1.5}$$

which is the energy in the channel output $h(t)$. The noise component at the output of the matched filter has a zero mean and a power spectral density

$$S_n(f) = \frac{N_0}{2} |H(f)|^2 \tag{8.1.6}$$

Hence, the noise power at the output of the matched filter has a variance

$$\sigma_n^2 = \int_{-\infty}^{\infty} S_n(f)\,df = \frac{N_0}{2} \int_{-\infty}^{\infty} |H(f)|^2 df = \frac{N_0 \mathcal{E}_h}{2} \tag{8.1.7}$$

The SNR at the output of the matched filter is

$$\left(\frac{S}{N}\right)_0 = \frac{\mathcal{E}_h^2}{N_0 \mathcal{E}_h / 2} = \frac{2\mathcal{E}_h}{N_0} \tag{8.1.8}$$

This is the same result as for the SNR at the output of the matched filter that was obtained in the preceding chapter except that the received signal energy $\mathcal{E}_h$ has replaced the transmitted signal energy $\mathcal{E}_s$. Compared to the previous result, the major difference in this development is that the filter impulse response is matched to the received signal $h(t)$ instead of the transmitted signal. Note that the implementation of the matched filter at the receiver requires that $h(t)$, or equivalently, the channel impulse response $c(t)$, must be known to the receiver.

Example 8.1.1

The signal pulse $g_T(t)$, defined as

$$g_T(t) = \frac{1}{2}\left[1 + \cos\frac{2\pi}{T}\left(t - \frac{T}{2}\right)\right], \quad 0 \le t \le T$$

is transmitted through a channel with frequency response characteristic as shown in Figure 8.2(a). The signal pulse is illustrated in Figure 8.2(b). The channel output is corrupted by AWGN with power spectral density $N_0/2$. Determine the matched filter to the received signal and the output SNR.

Solution This problem is most easily solved in the frequency domain. First, the spectrum of the signal pulse is

$$G_T(f) = \frac{T}{2}\frac{\sin \pi f T}{\pi f T(1 - f^2 T^2)}e^{-j\pi f T}$$

$$= \frac{T}{2}\frac{\text{sinc}(fT)}{1 - f^2 T^2}e^{-j\pi f T}$$

The spectrum $|G_T(f)|^2$ is shown in Figure 8.2(c). Hence,

$$H(f) = C(f)G_T(f)$$

$$= \begin{cases} G_T(f), & |f| \le W \\ 0, & \text{otherwise} \end{cases}$$

Then, the signal energy component at the output of the filter matched to $H(f)$ is

$$\mathcal{E}_h = \int_{-W}^{W} |G_T(f)|^2 \, df$$

$$= \frac{1}{(2\pi)^2}\int_{-W}^{W}\frac{(\sin \pi f T)^2}{f^2(1 - f^2 T^2)^2}\,df$$

$$= \frac{T}{(2\pi)^2}\int_{-WT}^{WT}\frac{\sin^2 \pi \alpha}{\alpha^2(1 - \alpha^2)^2}\,d\alpha$$

The variance of the noise component is

$$\sigma_n^2 = \frac{N_0}{2}\int_{-W}^{W}|G_T(f)|^2 df = \frac{N_0 \mathcal{E}_h}{2}$$

Hence, the output SNR is

$$\left(\frac{S}{N}\right)_0 = \frac{2\mathcal{E}_h}{N_0}$$

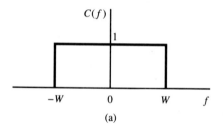

(a)

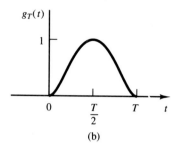

(b)

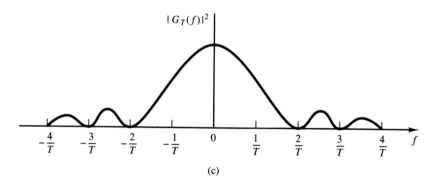

(c)

FIGURE 8.2. The signal pulse in (b) is transmitted through the ideal bandlimited channel shown in (a). The spectrum of $g_T(t)$ is shown in (c).

In this example we observe that the signal at the input to the channel is not bandlimited. Hence, only a part of the transmitted signal energy is received. The amount of signal energy at the output of the matched filter depends on the value of the channel bandwidth W when the signal pulse duration is fixed (see Problem 8.1). The maximum value of $\mathcal{E}_h$, obtained as $W \to \infty$, is

$$\max \mathcal{E}_h = \int_{-\infty}^{\infty} |G_T(f)|^2 \, df = \int_0^T g_T^2(t) \, dt$$

In the above development, we considered the transmission and reception of only a single signal waveform $g_T(t)$ through a bandlimited channel with impulse response $c(t)$. We observed that the performance of the system is determined by $\mathcal{E}_h$, the energy in the received signal $h(t)$. To maximize the received energy, we have to make sure that the power spectral density of the transmitted signal matches the frequency band of the channel. To this end we must study the power spectral density of the input signal. This will be done in Section 8.1.2. The impact of the channel bandwidth limitation is felt when we consider the transmission of a sequence of signal waveforms. This problem is treated in the following section.

8.1.1 Digital PAM Transmission Through Bandlimited Baseband Channels

Let us consider the baseband PAM communication system illustrated by the functional block diagram in Figure 8.3. The system consists of a transmitting filter having an impulse response $g_T(t)$, the linear filter channel with AWGN, a receiving filter with impulse response $g_R(t)$, a sampler that periodically samples the output of the receiving filter, and a symbol detector. The sampler requires the extraction of a timing signal from the received signal. This timing signal serves as a clock that specifies the appropriate time instants for sampling the output of the receiving filter. The extraction of a timing signal is considered in Section 8.5.

In general, we consider digital communications by means of M-ary PAM. Hence, the input binary data sequence is subdivided into k-bit symbols, and each symbol is mapped into a corresponding amplitude level that modulates the output of the transmitting filter. The signal at the output of the transmitting filter (the input to the channel) may be expressed as

$$v(t) = \sum_{n=-\infty}^{\infty} a_n g_T(t - nT) \tag{8.1.9}$$

where $T = k/R_b$ is the symbol interval ($1/T = R_b/k$ is the symbol rate), R_b is the bit rate, and $\{a_n\}$ is a sequence of amplitude levels corresponding to the sequence of k-bit blocks of information bits.

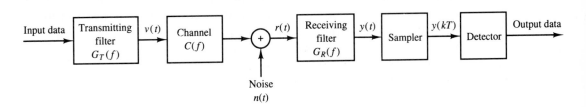

FIGURE 8.3.   Block diagram of digital PAM system.

The channel output, which is the received signal at the demodulator, may be expressed as

$$r(t) = \sum_{n=-\infty}^{\infty} a_n h(t - nT) + n(t) \qquad (8.1.10)$$

where $h(t)$ is the impulse response of the cascade of the transmitting filter and the channel, i.e., $h(t) = c(t) \star g_T(t)$, $c(t)$ is the impulse response of the channel, and $n(t)$ represents the AWGN.

The received signal is passed through a linear receiving filter with impulse response $g_R(t)$ and frequency response $G_R(f)$. If $g_R(t)$ is matched to $h(t)$, then its output SNR is a maximum at the proper sampling instant. The output of the receiving filter may be expressed as

$$y(t) = \sum_{n=-\infty}^{\infty} a_n x(t - nT) + v(t) \qquad (8.1.11)$$

where $x(t) = h(t) \star g_R(t) = g_T(t) \star c(t) \star g_R(t)$ and $v(t) = n(t) \star g_R(t)$ denotes the additive noise at the output of the receiving filter.

To recover the information symbols $\{a_n\}$, the output of the receiving filter is sampled periodically, every T seconds. Thus, the sampler produces

$$y(mT) = \sum_{n=-\infty}^{\infty} a_n x(mT - nT) + v(mT) \qquad (8.1.12)$$

or, equivalently,

$$y_m = \sum_{n=-\infty}^{\infty} a_n x_{m-n} + v_m$$

$$= x_0 a_m + \sum_{n \neq m} a_n x_{m-n} + v_m \qquad (8.1.13)$$

where $x_m = x(mT)$, $v_m = v(mT)$, and $m = 0, \pm 1, \pm 2, \dots$. A timing signal extracted from the received signal as described in Section 8.5 is used as a clock for sampling the received signal.

The first term on the right-hand side (RHS) of (8.1.13) is the desired symbol a_m, scaled by the gain parameter x_0. When the receiving filter is matched to the received signal $h(t)$, the scale factor is

$$x_0 = \int_{-\infty}^{\infty} h^2(t) dt = \int_{-\infty}^{\infty} |H(f)|^2 df$$

$$= \int_{-W}^{W} |G_T(f)|^2 |C(f)|^2 df \equiv \mathcal{E}_h \qquad (8.1.14)$$

as indicated by the development of (8.1.4) and (8.1.5). The second term on the RHS of (8.1.13) represents the effect of the other symbols at the sampling instant

$t = mT$, called the *intersymbol interference* (ISI). This is an undesirable term that degrades the performance of the digital PAM communication system. Finally, the third term, v_m, that represents the additive noise, is a zero-mean Gaussian random variable with variance $\sigma_v^2 = N_0 \mathcal{E}_h / 2$, previously given by (8.1.7).

By appropriate design of the transmitting and receiving filters it is possible to satisfy the condition $x_n = 0$ for $n \neq 0$, so that the ISI term vanishes. In this case, the only term that can cause errors in the received digital sequence is the additive noise.

In Section 8.2 we consider the design of bandlimited transmitting and receiving filters that either eliminate or control the ISI. First, however, we will determine the power spectral density of the transmitted digital PAM signal. Thus, we will establish the relationship between the spectral characteristics of the PAM signal and the channel bandwidth requirements for transmitting the signal.

8.1.2 The Power Spectrum of a Digital PAM Signal

The transmitted signal for a digital PAM signal is represented in the general form as

$$v(t) = \sum_{n=-\infty}^{\infty} a_n g_T(t - nT) \tag{8.1.15}$$

where $\{a_n\}$ is the sequence of amplitudes corresponding to the information symbols from the source and $g_T(t)$ is the impulse response of the transmitting filter. Since the information sequence $\{a_n\}$ is random, $v(t)$ is a sample function of a random process $V(t)$. In this section we evaluate the power-density spectrum of $V(t)$. Our approach is to derive the autocorrelation function of $V(t)$ and then to determine its Fourier transform.

First, the mean value of $v(t)$ is

$$E[V(t)] = \sum_{n=-\infty}^{\infty} E[a_n] g_T(t - nT)$$

$$= m_a \sum_{n=-\infty}^{\infty} g_T(t - nT) \tag{8.1.16}$$

where m_a is the mean amplitude value of the random sequence $\{a_n\}$. Note that although m_a is a constant, the term $\sum_n g_T(t - nT)$ is a periodic function with period T. Hence, the mean value of $V(t)$ is periodic with period T.

The autocorrelation function of $V(t)$ is

$$R_V(t + \tau, t) = E[V(t)V(t + \tau)] = \sum_{n=-\infty}^{\infty} \sum_{l=-\infty}^{\infty} E(a_n a_l) g_T(t - nT) g_T(t + \tau - lT) \tag{8.1.17}$$

In general, we assume that the information sequence $\{a_n\}$ is WSS with autocorrelation sequence

$$R_a(n) = E[a_m a_{n+m}] \tag{8.1.18}$$

Hence, (8.1.17) may be expressed as

$$R_V(t + \tau, t) = \sum_{n=-\infty}^{\infty} \sum_{l=-\infty}^{\infty} R_a(l - n) g_T(t - nT) g_T(t + \tau - lT)$$

$$= \sum_{m=-\infty}^{\infty} R_a(m) \sum_{n=-\infty}^{\infty} g_T(t - nT) g_T(t + \tau - nT - mT) \tag{8.1.19}$$

We observe that the second summation in (8.1.19), namely

$$\sum_{n=-\infty}^{\infty} g_T(t - nT) g_T(t + \tau - nT - mT) \tag{8.1.20}$$

is periodic with period T. Consequently, the autocorrelation function $R_V(t + \tau, t)$ is periodic in the variable t, i.e.,

$$R_V(t + T + \tau, t + T) = R_V(t + \tau, t) \tag{8.1.21}$$

Therefore, the random process $V(t)$ has a periodic mean and a periodic autocorrelation. Such a random process is *cyclostationary* (see Definition 3.2.7).

The power-spectral density of a cyclostationary process can be determined by first averaging the autocorrelation function $R_V(t + \tau, t)$ over a single period T and then computing the Fourier transform of the average autocorrelation function (see Corollary to Theorem 3.3.1). Thus we have

$$\bar{R}_V(\tau) = \frac{1}{T} \int_{-T/2}^{T/2} R_V(t + \tau, t) dt$$

$$= \sum_{m=-\infty}^{\infty} R_a(m) \sum_{n=-\infty}^{\infty} \frac{1}{T} \int_{-T/2}^{T/2} g_T(t - nT) g_T(t + \tau - nT - mT) dt$$

$$= \sum_{m=-\infty}^{\infty} R_a(m) \sum_{n=-\infty}^{\infty} \frac{1}{T} \int_{nT-T/2}^{nT+T/2} g_T(t) g_T(t + \tau - mT) dt$$

$$= \frac{1}{T} \sum_{m=-\infty}^{\infty} R_a(m) \int_{-\infty}^{\infty} g_T(t) g_T(t + \tau - mT) dt \tag{8.1.22}$$

We interpret the integral in (8.1.22) as the time-autocorrelation function of $g_T(t)$ and define it as (see (2.4.1))

$$R_g(\tau) = \int_{-\infty}^{\infty} g_T(t) g_T(t + \tau) dt \tag{8.1.23}$$

With this definition, the average autocorrelation function of $V(t)$ becomes

$$\bar{R}_V(\tau) = \frac{1}{T} \sum_{m=-\infty}^{\infty} R_a(m) R_g(\tau - mT) \tag{8.1.24}$$

We observe that the expression for $\bar{R}_V(\tau)$ in (8.1.24) has the form of a convolution sum. Hence the Fourier transform of (8.1.24) becomes

$$\begin{aligned} S_V(f) &= \int_{-\infty}^{\infty} \bar{R}_V(\tau) e^{-j2\pi f\tau} \, d\tau \\ &= \frac{1}{T} \sum_{m=-\infty}^{\infty} R_a(m) \int_{-\infty}^{\infty} R_g(\tau - mT) e^{-j2\pi f\tau} \, d\tau \\ &= \frac{1}{T} S_a(f) |G_T(f)|^2 \end{aligned} \tag{8.1.25}$$

where $S_a(f)$ is the power spectrum of the information sequence $\{a_n\}$, defined as

$$S_a(f) = \sum_{m=-\infty}^{\infty} R_a(m) e^{-j2\pi fmT} \tag{8.1.26}$$

and $G_T(f)$ is the spectrum of the transmitting filter. $|G_T(f)|^2$ is the Fourier transform of $R_g(\tau)$. (See (2.3.45).)

The result in (8.1.25) illustrates the dependence of the power-spectral density $S_V(f)$ of the transmitted signal on (1) the spectral characteristics $G_T(f)$ of the transmitting filter and (2) the spectral characteristics $S_a(f)$ of the information sequence $\{a_n\}$. Both $G_T(f)$ and $S_a(f)$ can be designed to control the shape and form of the power-spectral density of the transmitted signal.

Whereas the dependence of $S_V(f)$ on $G_T(f)$ is easily understood, the effect of the autocorrelation properties of the information sequence $\{a_n\}$ is more subtle. First, we observe that for an arbitrary autocorrelation $R_a(m)$, the corresponding power-spectral density $S_a(f)$ is periodic in frequency with period $1/T$. In fact, we note that $S_a(f)$, given by (8.1.26), has the form of an exponential Fourier series with $\{R_a(m)\}$ as the Fourier coefficients. Consequently, the autocorrelation sequence $\{R_a(m)\}$ is simply

$$R_a(m) = T \int_{-1/2T}^{1/2T} S_a(f) e^{j2\pi fmT} \, df \tag{8.1.27}$$

Secondly, let us consider the case in which the information symbols in the sequence $\{a_n\}$ are mutually uncorrelated. Then,

$$R_a(m) = \begin{cases} \sigma_a^2 + m_a^2, & m = 0 \\ m_a^2, & m \neq 0 \end{cases} \tag{8.1.28}$$

where $\sigma_a^2 = E\left[a_n^2\right] - m_a^2$ is the variance of an information symbol. By substituting for $R_a(m)$ into (8.1.26), we obtain the power-spectral density

$$S_a(f) = \sigma_a^2 + m_a^2 \sum_{m=-\infty}^{\infty} e^{-j2\pi f m T} \tag{8.1.29}$$

The term involving the summation on the RHS of (8.1.29) is periodic with period $1/T$. It may be viewed as the exponential Fourier series of a periodic train of impulses where each impulse has an area $1/T$ (see 2.3.26). Therefore, (8.1.29) can be expressed as

$$S_a(f) = \sigma_a^2 + \frac{m_a^2}{T} \sum_{m=-\infty}^{\infty} \delta\left(f - \frac{m}{T}\right) \tag{8.1.30}$$

Substitution of the expression into $S_V(f)$ given by (8.1.25) yields the desired result for the power-spectral density of the transmitted signal $V(t)$ when the sequence of information symbols is uncorrelated. That is,

$$S_V(f) = \frac{\sigma_a^2}{T} |G_T(f)|^2 + \frac{m_a^2}{T^2} \sum_{m=-\infty}^{\infty} \left|G_T\left(\frac{m}{T}\right)\right|^2 \delta\left(f - \frac{m}{T}\right) \tag{8.1.31}$$

The expression for the power-spectral density of the transmitted signal given by (8.1.31) is purposely separated into two terms to emphasize the two different types of spectral components. The first term, $\sigma_a^2|G_T(f)|^2/T$, is the continuous spectrum and its shape depends of $G_T(f)$. The second term in (8.1.31) consists of discrete frequency components spaced $1/T$ apart in frequency. Each spectral line has a power that is proportional to $|G_T(f)|^2$ evaluated at $f = m/T$. We note that the discrete frequency components can be eliminated by selecting the information symbol sequence $\{a_n\}$ to have zero mean. This condition is usually imposed in digital modulation methods because discrete spectral lines are considered to be undesirable. To be specific, the mean m_a in digital PAM is easily forced to be zero by selecting amplitude levels that are equally likely and symmetrically positioned on the real line relative to zero. Under the condition that $m_a = 0$, we have

$$S_V(f) = \frac{\sigma_a^2}{T} |G_T(f)|^2 \tag{8.1.32}$$

Thus, the system designer can control the spectral characteristics of the transmitted digital PAM signal. The following example illustrates the spectral shaping resulting from $g_T(t)$.

Example 8.1.2

Determine the power-spectral density in (8.1.31) when $g_T(t)$ is the rectangular pulse shown in Figure 8.4(a).

Solution   The Fourier transform of $g_T(t)$ is

$$G_T(f) = AT \frac{\sin \pi f T}{\pi f T} e^{-j\pi f T}$$

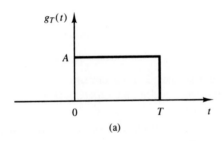

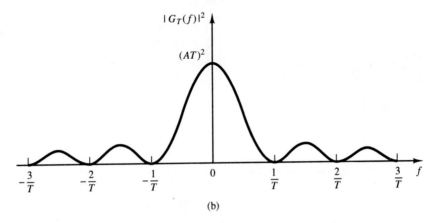

FIGURE 8.4.   A rectangular pulse $g_T(t)$ and its energy density spectrum $|G_T(f)|^2$.

Hence,

$$|G_T(f)|^2 = (AT)^2 \left(\frac{\sin \pi f T}{\pi f T} \right)^2$$

$$= (AT)^2 \operatorname{sinc}^2(fT)$$

This spectrum is illustrated in Figure 8.4(b). We note that it contains nulls at multiples of $1/T$ in frequency and that it decays inversely as the square of the frequency variable. As a consequence of the spectral nulls in $G_T(f)$, all but one of the discrete spectral components in (8.1.31) vanish. Thus, upon substitution for $|G_T(f)|^2$ into (8.1.31), we obtain the result

$$S_V(f) = \sigma_a^2 A^2 T \left(\frac{\sin \pi f T}{\pi f T} \right)^2 + A^2 m_a^2 \delta(f)$$

$$= \sigma_a^2 A^2 T \operatorname{sinc}^2(fT) + A^2 m_a^2 \delta(f)$$

The following example illustrates the spectral shaping that can be achieved by operations performed on the input information sequence.

Example 8.1.3

Consider a binary sequence $\{b_n\}$, from which we form the symbols

$$a_n = b_n + b_{n-1}$$

The $\{b_n\}$ are assumed to be uncorrelated binary valued (± 1) random variables, each having a zero mean and a unit variance. Determine the power-spectral density of the transmitted signal.

Solution   The autocorrelation function of the sequence $\{a_n\}$ is

$$
\begin{aligned}
R_a(m) &= E[a_n a_{n+m}] \\
&= E\left[(b_n + b_{n-1})(b_{n+m} + b_{n+m-1})\right] \\
&= \begin{cases} 2 & m = 0 \\ 1 & m = \pm 1 \\ 0 & \text{otherwise} \end{cases}
\end{aligned}
$$

Hence the power-spectral density of the input sequence is

$$
\begin{aligned}
S_a(f) &= 2(1 + \cos 2\pi f T) \\
&= 4 \cos^2 \pi f T
\end{aligned}
$$

and the corresponding power spectrum for the modulated signal is, from (8.1.25),

$$S_V(f) = \frac{4}{T} |G_T(f)|^2 \cos^2 \pi f T$$

Figure 8.5 illustrates the power-density spectrum $S_a(f)$ of the input sequence, and the corresponding $S_V(f)$ when $G_T(f)$ is the spectrum of the rectangular pulse.

As demonstrated in the example, the transmitted signal spectrum can be shaped by having a correlated sequence $\{a_n\}$ as the input to the modulator. We recall from Section 7.1.4 that state-dependent modulation codes generally produce correlated sequences and thus provide spectral shaping. State-dependent runlength-limited codes are often used for this purpose. In the following section we consider the spectral characteristics of modulated signals produced by such codes.

8.1.3 The Power Spectrum of Digital Signals with Memory

In Section 7.1.4 we demonstrated that state-dependent modulation codes resulted in modulated signals with memory. Such signals were described by Markov chains, which are basically graphs that include the possible "states" of the modulator with corresponding state probabilities $\{p_i\}$, and state transitions with corresponding state transition probabilities $\{p_{ij}\}$.

The power-spectral density of digitally modulated signals that are characterized by Markov chains may be derived by following the basic procedure given in the previous section. Thus, we may determine the autocorrelation function and then evaluate its Fourier transform to obtain the power-spectral density. For signals that

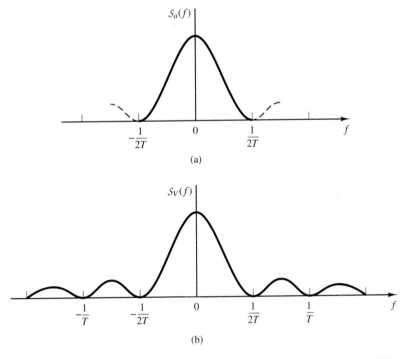

FIGURE 8.5. Power-density spectra for (a) information sequence and (b) PAM modulated signal.

are generated by a Markov chain with transition probability matrix **P**, as generally given by (7.1.46), the power-spectral density of the modulated signal may be expressed in the general form

$$S(f) = \frac{1}{T^2} \sum_{n=-\infty}^{\infty} \left| \sum_{i=1}^{K} p_i S_i \left(\frac{n}{T} \right) \right|^2 \delta \left(f - \frac{n}{T} \right) + \frac{1}{T} \sum_{i=1}^{K} p_i \left| S_i'(f) \right|^2$$

$$+ \frac{2}{T} \mathrm{Re} \left[\sum_{i=1}^{K} \sum_{j=1}^{K} p_i S_i'^*(f) S_j'(f) P_{ij}(f) \right] \qquad (8.1.33)$$

where K is the number of states of the modulator, $S_i'(f)$ is the Fourier transform of the signal waveform $s_i'(t)$, where $s_i'(t) = s_i(t) - \sum_{k=1}^{K} p_k s_k(t)$, and $P_{ij}(f)$ is the Fourier transform of the discrete-time sequence $p_{ij}(n)$, defined as

$$P_{ij}(f) = \sum_{n=1}^{\infty} p_{ij}(n) e^{-j2\pi n f T} \qquad (8.1.34)$$

The term $p_{ij}(n)$ denotes the probability that a signal $s_j(t)$ is transmitted n signaling intervals after the transmission of $s_i(t)$. Hence, $\{p_{ij}(n)\}$ are the transition probabilities in the transition probability matrix $\mathbf{P}^n$. Note that when $n = 1, \mathbf{P}^n \equiv \mathbf{P}$, which is the matrix given by (7.1.46).

Example 8.1.4

Determine the power-spectral density of the NRZI signal.

Solution   The NRZI signal is characterized by the transition probability matrix

$$\mathbf{P} = \begin{bmatrix} \frac{1}{2} & \frac{1}{2} \\ \frac{1}{2} & \frac{1}{2} \end{bmatrix}$$

We note that $\mathbf{P}^n = \mathbf{P}$ for all $n > 1$. Hence, with $K = 2$ states and $g_T(t) = s_1(t) = -s_2(t)$, we obtain

$$S(f) = \frac{(2p-1)^2}{T^2} \sum_{n=-\infty}^{\infty} \left| G_T\left(\frac{n}{T}\right) \right|^2 \delta\left(f - \frac{n}{T}\right)$$

$$+ \frac{4p(1-p)}{T} |G_T(f)|^2 \tag{8.1.35}$$

where for a rectangular pulse of amplitude A,

$$|G_T(f)|^2 = (AT)^2 \left(\frac{\sin \pi f T}{\pi f T}\right)^2$$

$$= (AT)^2 \operatorname{sinc}^2(fT)$$

We observe that when $p = 1/2$ (equally probable signals) the impulse spectrum vanishes and $S(f)$ reduces to

$$S(f) = \frac{1}{T} |G_T(f)|^2$$

We observe that the power spectrum of the NRZI signal for equally probable signals is identical to the expression in (8.1.32), which applies to an uncorrelated sequence $\{a_n\}$ into the modulator. Hence, we conclude that the simple precoding operation in NRZI does not result in a correlated sequence.

Example 8.1.5

Determine the power-spectral density of the delay modulated (Miller encoded) signal described in Section 7.1.4.

Solution   The transition probability matrix of a delay modulated signal is (see (7.1.48))

$$\mathbf{P} = \begin{bmatrix} 0 & \frac{1}{2} & 0 & \frac{1}{2} \\ 0 & 0 & \frac{1}{2} & \frac{1}{2} \\ \frac{1}{2} & \frac{1}{2} & 0 & 0 \\ \frac{1}{2} & 0 & \frac{1}{2} & 0 \end{bmatrix}$$

The state probabilities are $p_i = 1/4$, for $i = 1, 2, 3, 4$. Powers of **P** are easily obtained by use of the relation (see Problem 8.6)

$$\mathbf{P}^4 \gamma = -\frac{1}{4}\gamma \tag{8.1.36}$$

where γ is the signal correlation matrix with elements

$$\gamma_{ij} = \frac{1}{T}\int_0^T s_i(t)s_j(t)dt$$

and the four signals $\{s_i(t), i = 1, 2, 3, 4\}$ are shown in Figure 7.28, where $s_3(t) = -s_2(t)$ and $s_4(t) = -s_1(t)$. It is easily seen that

$$\gamma = \begin{bmatrix} 1 & 0 & 0 & -1 \\ 0 & 1 & -1 & 0 \\ 0 & -1 & 1 & 0 \\ -1 & 0 & 0 & 1 \end{bmatrix} \tag{8.1.37}$$

Consequently, powers of **P** can be generated from the relation

$$\mathbf{P}^{k+4}\gamma = -\frac{1}{4}\mathbf{P}^k\gamma, \quad k \geq 1 \tag{8.1.38}$$

With the aid of these relations, the power-spectral density of the delay modulated signal is obtained from (8.1.33). It may be expressed in the form

$$S(f) = \frac{1}{2(\pi fT)^2(17 + 8\cos 8\pi fT)}\big[23 - 2\cos \pi fT - 22\cos 2\pi fT$$
$$-12\cos 3\pi fT + 5\cos 4\pi fT + 12\cos 5\pi fT + 2\cos 6\pi fT$$
$$-8\cos 7\pi fT + 2\cos 8\pi fT\big] \tag{8.1.39}$$

The power-spectral densities of the NRZI and the delay modulated signal are shown in Figure 8.6. We observe that the NRZI signal has a lowpass power spectrum

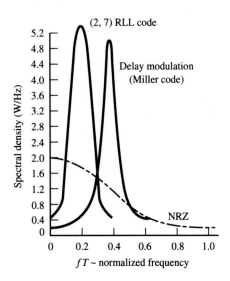

FIGURE 8.6.   Power-spectral density (one-sided) of Miller code (delay modulation), and NRZ/NRZI baseband signals.

with a peak of $f = 0$. On the other hand, the delay modulated signal has very little power in the vicinity of $f = 0$. It also has a relatively narrow power-spectral density. These two characteristics make it particularly suitable for use in magnetic recording storage channels that use flux-sensing heads for writing on and reading off a disk. Also shown for comparison in Figure 8.6 is the power-spectral density of the modulated signal generated by a $(d, \kappa) = (2, 7)$ code followed by NRZI. The $(2, 7)$ runlength-limited code, previously given in Table 7.3, is also widely used in magnetic recording channels.

When there is no memory in the modulation method, the signal waveform transmitted in each signaling interval is independent of the waveforms transmitted in previous signaling intervals. The power-spectral density of the resulting signal may still be expressed in the form of (8.1.33) if the transition probability matrix is replaced by

$$
\mathbf{P} =
\begin{bmatrix}
p_1 & p_2 & \cdots & p_K \\
p_1 & p_2 & \cdots & p_K \\
\vdots & \vdots & & \vdots \\
p_1 & p_2 & \cdots & p_K
\end{bmatrix}
\tag{8.1.40}
$$

and we impose the condition that $\mathbf{P}^n = \mathbf{P}$ for all $n > 1$ as a consequence of the memoryless modulation. Under these conditions, the expression for $S(f)$ given (8.1.33) reduces to the simpler form

$$
S(f) = \frac{1}{T^2} \sum_{n=-\infty}^{\infty} \left| \sum_{i=1}^{K} p_i S_i \left(\frac{n}{T} \right) \right|^2 \delta \left(f - \frac{n}{T} \right)
$$

$$
+ \frac{1}{T} \sum_{i=1}^{K} p_i (1 - p_i) |S_i(f)|^2 - \frac{2}{T} \sum_{i=1}^{K} \sum_{\substack{j=1 \\ i<j}}^{K} p_i p_j \operatorname{Re}[S_i(f) S_j^*(f)]
\tag{8.1.41}
$$

We observe that our previous result for the power-spectral density of memoryless PAM modulation given by (8.1.31) is a special case of the expression in (8.1.41). Specifically, if we have $K = M$ signals, which are amplitude-scaled versions of a basic pulse $g_T(t)$, i.e., $s_m(t) = a_m g_T(t)$ where a_m is the signal amplitude; then $S_m(f) = a_m G_T(f)$ and (8.1.41) becomes

$$
S(f) = \frac{1}{T^2} \left(\sum_{i=1}^{M} a_i p_i \right)^2 \sum_{n=-\infty}^{\infty} \left| G_T \left(\frac{n}{T} \right) \right|^2 \delta \left(f - \frac{n}{T} \right)
$$

$$
+ \frac{1}{T} \left[\sum_{i=1}^{M} p_i (1 - p_i) a_i^2 - 2 \sum_{i=1}^{M} \sum_{\substack{j=1 \\ i<j}}^{M} p_i p_j a_i a_j \right] |G_T(f)|^2
\tag{8.1.42}
$$

If we compare (8.1.31) with (8.1.42), we find that these expressions are identical, where the mean and variance of the information sequence is

$$m_a = \sum_{i=1}^{M} a_i p_i$$

$$\sigma_a^2 = \sum_{i=1}^{M} p_i (1 - p_i) a_i^2 - 2 \sum_{\substack{i=1 \\ i<j}}^{M} \sum_{j=1}^{M} p_i p_j a_i a_j \tag{8.1.43}$$

Therefore (8.1.41) is the more general form for the power-spectral density of memoryless modulated signals, since it applies to signals that may have different pulse shapes.

8.2 SIGNAL DESIGN FOR BANDLIMITED CHANNELS

Recall from Section 8.1.1 that the output of the transmitting filter in a digital PAM communication system may be expressed as

$$v(t) = \sum_{n=-\infty}^{\infty} a_n g_T(t - nT) \tag{8.2.1}$$

and the output of the channel, which is the received signal at the demodulator, may be expressed as

$$r(t) = \sum_{n=-\infty}^{\infty} a_m h(t - nT) + n(t) \tag{8.2.2}$$

where $h(t) = c(t) \star g_T(t)$, $c(t)$ is the impulse response of the channel, $g_T(t)$ is the impulse response of the transmitting filter, and $n(t)$ is a sample function of an additive, white Gaussian noise process.

In this section, we consider the problem of designing a bandlimited transmitting filter. The design will be done first under the condition that there is no channel distortion. Later we consider the problem of filter design when the channel distorts the transmitted signal. Since $H(f) = C(f)G_T(f)$, the condition for distortion-free transmission is that the frequency response characteristic $C(f)$ of the channel has a constant magnitude and a linear phase over the bandwidth of the transmitted signal. That is,

$$C(f) = \begin{cases} C_0 e^{-j2\pi f t_0}, & |f| \leq W \\ 0, & |f| > W \end{cases} \tag{8.2.3}$$

where W is the available channel bandwidth, t_0 represents an arbitrary finite delay, which we set to zero for convenience, and C_0 is a constant gain factor that we set to unity for convenience. Thus, under the condition that the channel is distortion-free, $H(f) = G_T(f)$ for $|f| \leq W$ and zero for $|f| > W$. Consequently, the matched filter has a frequency response $H^\star(f) = G_T^\star(f)$, and its output at the periodic sampling times $t = mT$ has the form

$$y(mT) = x(0)a_m + \sum_{n \neq m} a_n x(mT - nT) + v(mT) \qquad (8.2.4)$$

or, more simply,

$$y_m = x_0 a_m + \sum_{n \neq m} a_n x_{m-n} + v_m \qquad (8.2.5)$$

where $x(t) = g_T(t) \star g_R(t)$ and $v(t)$ is the output response of the matched filter to the input AWGN process $n(t)$.

The middle term on the RHS of (8.2.5) represents the ISI. The amount of ISI and noise that is present in a digital PAM system can be viewed on an oscilloscope. Specifically, we may display the received signal on the vertical input with the horizontal sweep rate set at $1/T$. The resulting oscilloscope display is called an *eye pattern* because of its resemblance to the human eye. Examples of two eye patterns, one for binary PAM and the other for quaternary ($M = 4$) PAM, are illustrated in Figure 8.7(a).

The effect of ISI is to cause the eye to close, thereby reducing the margin for additive noise to cause errors. Figure 8.7(b) illustrates the effect of ISI in reducing the opening of the eye. Note that ISI distorts the position of the zero crossings and causes a reduction in the eye opening. As a consequence, the system is more sensitive to a synchronization error and exhibits a smaller margin against additive noise.

Below we consider the problem of signal design under two conditions, namely, (1) that there is no ISI at the sampling instants, and (2) that a controlled amount of ISI is allowed.

8.2.1 Design of Bandlimited Signals for Zero ISI—The Nyquist Criterion

As indicated above, in a general digital communication system that transmits through a bandlimited channel, the Fourier transform of the signal at the output of the receiving filter is given by $X(f) = G_T(f)C(f)G_R(f)$ where $G_T(f)$ and $G_R(f)$ denote the transmitter and receiver filters frequency responses and $C(f)$ denotes the frequency response of the channel. We have also seen that the output of the receiving filter, sampled at $t = mT$, is given by

$$y_m = x(0)a_m + \sum_{\substack{n=-\infty \\ n \neq m}}^{\infty} x(mT - nT)a_n + v(mT) \qquad (8.2.6)$$

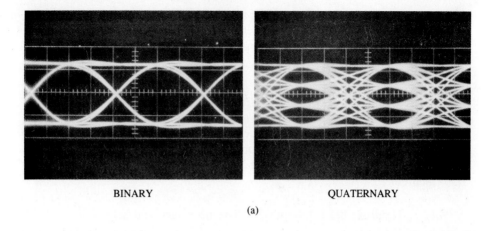

BINARY   QUATERNARY

(a)

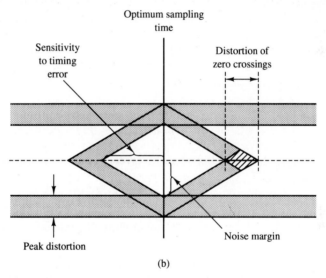

(b)

FIGURE 8.7.   Eye patterns. (a) Examples of eye pattersn for binary and quaternary amplitude-shift keying (or PAM). (b) Effect of ISI on eye opening.

To remove the effect of ISI, it is necessary and sufficient that $x(mT - nT) = 0$ for $n \neq m$ and $x(0) \neq 0$, where without loss of generality we can assume $x(0) = 1$.[†] This means that the overall communication system has to be designed such that

$$x(nT) = \begin{cases} 1 & n = 0 \\ 0 & n \neq 0 \end{cases} \tag{8.2.7}$$

In this section we derive the necessary and sufficient condition for $X(f)$ in order for $x(t)$ to satisfy the above relation. This condition is known as the *Nyquist pulse-shaping criterion* or *Nyquist condition for zero ISI* and is stated in the following theorem.

Theorem 8.2.1 [Nyquist].   The necessary and sufficient condition for $x(t)$ to satisfy

$$x(nT) = \begin{cases} 1 & n = 0 \\ 0 & n \neq 0 \end{cases} \tag{8.2.8}$$

is that its Fourier transform $X(f)$ satisfy

$$\sum_{m=-\infty}^{\infty} X\left(f + \frac{m}{T}\right) = T \tag{8.2.9}$$

Proof.   In general, $x(t)$ is the inverse Fourier transform of $X(f)$. Hence,

$$x(t) = \int_{-\infty}^{\infty} X(f)e^{j2\pi ft}\,df \tag{8.2.10}$$

At the sampling instants $t = nT$, this relation becomes

$$x(nT) = \int_{-\infty}^{\infty} X(f)e^{j2\pi fnT}\,df \tag{8.2.11}$$

Let us break up the integral in (8.2.11) into integrals covering the finite range of $1/T$. Thus we obtain

$$x(nT) = \sum_{m=-\infty}^{\infty} \int_{(2m-1)/2T}^{(2m+1)/2T} X(f)e^{j2\pi fnT}\,df$$

$$= \sum_{m=-\infty}^{\infty} \int_{-1/2T}^{1/2T} X\left(f + \frac{m}{T}\right) e^{j2\pi fnT}\,dt$$

[†]The choice of $x(0)$ is equivalent to the choice of a constant gain factor in the receiving filter. This constant gain factor has no effect on the overall system performance since it enhances both the signal and the noise.

$$= \int_{-1/2T}^{1/2T} \left[\sum_{m=-\infty}^{\infty} X\left(f + \frac{m}{T} \right) \right] e^{j2\pi fnT} \, df$$

$$= \int_{-1/2T}^{1/2T} Z(f) e^{j2\pi fnT} \, df \tag{8.2.12}$$

where we have defined $Z(f)$ by

$$Z(f) = \sum_{m=-\infty}^{\infty} X\left(f + \frac{m}{T} \right) \tag{8.2.13}$$

Obviously $Z(f)$ is a periodic function with period $\frac{1}{T}$, and therefore it can be expanded in terms of its Fourier series coefficients $\{z_n\}$ as

$$Z(f) = \sum_{n=-\infty}^{\infty} z_n e^{j2\pi n f T} \tag{8.2.14}$$

where

$$z_n = T \int_{-\frac{1}{2T}}^{\frac{1}{2T}} Z(f) e^{-j2\pi n f T} \, df \tag{8.2.15}$$

Comparing (8.2.15) and (8.2.12) we obtain

$$z_n = Tx(-nT) \tag{8.2.16}$$

Therefore, the necessary and sufficient conditions for (8.2.8) to be satisfied is that

$$z_n = \begin{cases} T & n = 0 \\ 0 & n \neq 0 \end{cases} \tag{8.2.17}$$

which, when substituted into (8.2.14), yields

$$Z(f) = T \tag{8.2.18}$$

or, equivalently,

$$\sum_{m=-\infty}^{\infty} X\left(f + \frac{m}{T} \right) = T \tag{8.2.19}$$

This concludes the proof of the theorem.

Now suppose that the channel has a bandwidth of W. Then $C(f) \equiv 0$ for $|f| > W$ and consequently, $X(f) = 0$ for $|f| > W$. We distinguish three cases:

1. $T < \frac{1}{2W}$, or equivalently $\frac{1}{T} > 2W$. Since $Z(f) = \sum_{n=-\infty}^{+\infty} X\left(f + \frac{n}{T} \right)$ consists of nonoverlapping replicas of $X(f)$, separated by $1/T$ as shown in Figure 8.8, there is no choice for $X(f)$ to ensure $Z(f) \equiv T$ in this case and there is no way that we can design a system with no ISI.

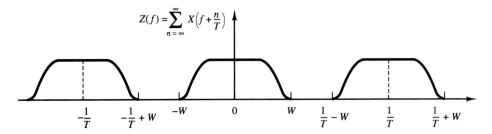

$$Z(f) = \sum_{n=\infty}^{\infty} X\left(f + \frac{n}{T}\right)$$

FIGURE 8.8.   Plot of $Z(f)$ for the case $T < \frac{1}{2W}$.

2. $T = \frac{1}{2W}$, or equivalently $\frac{1}{T} = 2W$ (the Nyquist rate). In this case the replications of $X(f)$, separated by $\frac{1}{T}$, are about to overlap as shown in Figure 8.9. It is clear that in this case there exists only one $X(f)$ that results in $Z(f) = T$, namely,

$$X(f) = \begin{cases} T & |f| < W \\ 0 & \text{otherwise} \end{cases} \tag{8.2.20}$$

or, $X(f) = T \Pi \left(\frac{f}{2W}\right)$, which results in

$$x(t) = \text{sinc}\left(\frac{t}{T}\right) \tag{8.2.21}$$

This means that the smallest value of T for which transmission with zero ISI is possible is $T = \frac{1}{2W}$ and for this value, $x(t)$ has to be a sinc function. The difficulty with this choice of $x(t)$ is that it is noncausal and therefore nonrealizable. To make it realizable, usually a delayed version of it, i.e., $\text{sinc}\left(\frac{t-t_0}{T}\right)$ is used and t_0 is chosen such that for $t < 0$, we have $\text{sinc}\left(\frac{t-t_0}{T}\right) \approx 0$. Of course with this choice of $x(t)$, the sampling time must also be shifted to $mT + t_0$. A second difficulty with this pulse shape is that its rate of convergence to zero is slow. The tails of $x(t)$ decay as $1/t$, consequently, a small mistiming

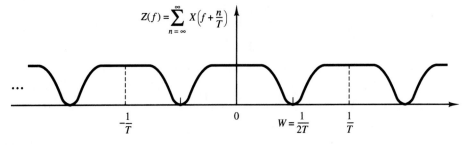

$$Z(f) = \sum_{n=\infty}^{\infty} X\left(f + \frac{n}{T}\right)$$

FIGURE 8.9.   Plot of $Z(f)$ for the case $T = \frac{1}{2W}$.

error in sampling the output of the matched filter at the demodulator results in an infinite series of ISI components. Such a series is not absolutely summable because of the $1/t$ rate of decay of the pulse and, hence, the sum of the resulting ISI does not converge.

3. For $T > \frac{1}{2W}$, $Z(f)$ consists of overlapping replications of $X(f)$ separated by $\frac{1}{T}$, as shown in Figure 8.10. In this case, there exist numerous choices for $X(f)$ such that $Z(f) \equiv T$.

A particular pulse spectrum, for the $T > \frac{1}{2W}$ case, that has desirable spectral properties and has been widely used in practice is the raised cosine spectrum. The raised cosine frequency characteristic is given as (see Problem 8.9)

$$X_{rc}(f) = \begin{cases} T, & 0 \le |f| \le (1-\alpha)/2T \\ \frac{T}{2}\left[1 + \cos\frac{\pi T}{\alpha}\left(|f| - \frac{1-\alpha}{2T}\right)\right], & \frac{1-\alpha}{2T} \le |f| \le \frac{1+\alpha}{2T} \\ 0, & |f| > \frac{1+\alpha}{2T} \end{cases} \qquad (8.2.22)$$

where α is called the *rolloff factor*, which takes values in the range $0 \le \alpha \le 1$. The bandwidth occupied by the signal beyond the Nyquist frequency $\frac{1}{2T}$ is called the *excess bandwidth* and is usually expressed as a percentage of the Nyquist frequency. For example, when $\alpha = \frac{1}{2}$, the excess bandwidth is 50% and when $\alpha = 1$ the excess bandwidth is 100%. The pulse $x(t)$, having the raised cosine spectrum, is

$$x(t) = \frac{\sin \pi t/T}{\pi t/T} \frac{\cos(\pi\alpha t/T)}{1 - 4\alpha^2 t^2/T^2}$$

$$= \mathrm{sinc}(t/T)\frac{\cos(\pi\alpha t/T)}{1 - 4\alpha^2 t^2/T^2} \qquad (8.2.23)$$

Note that $x(t)$ is normalized so that $x(0) = 1$. Figure 8.11 illustrates the raised cosine spectral characteristics and the corresponding pulses for $\alpha = 0, 1/2, 1$. We note that for $\alpha = 0$, the pulse reduces to $x(t) = \mathrm{sinc}(t/T)$, and the symbol rate $1/T = 2W$. When $\alpha = 1$, the symbol rate is $1/T = W$. In general, the tails of $x(t)$

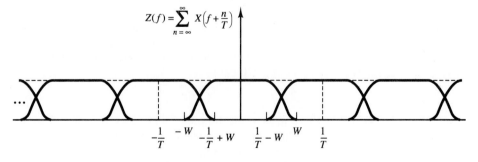

FIGURE 8.10.   Plot of $Z(f)$ for the case $T > 1/2W$.

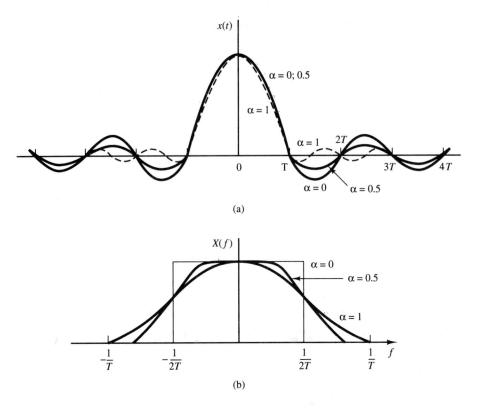

FIGURE 8.11.   Pulses having a raised cosine spectrum.

decay as $1/t^3$ for $\alpha > 0$. Consequently, a mistiming error in sampling leads to a series of ISI components that converges to a finite value.

Due to the smooth characteristics of the raised cosine spectrum, it is possible to design practical filters for the transmitter and the receiver that approximate the overall desired frequency response. In the special case where the channel is ideal with $C(f) = \Pi\left(\frac{f}{2W}\right)$, we have

$$X_{rc}(f) = G_T(f)G_R(f) \tag{8.2.24}$$

In this case, if the receiver filter is matched to the transmitter filter, we have $X_{rc}(f) = G_T(f)G_R(f) = |G_T(f)|^2$. Ideally,

$$G_T(f) = \sqrt{|X_{rc}(f)|}e^{-j2\pi f t_0} \tag{8.2.25}$$

and $G_R(f) = G_T^*(f)$, where t_0 is some nominal delay that is required to ensure physical realizability of the filter. Thus, the overall raised cosine spectral characteristic is split evenly between the transmitting filter and the receiving filter. We should also note that an additional delay is necessary to ensure the physical realizability of the receiving filter.

8.2.2 Design of Bandlimited Signals with Controlled ISI—Partial Response Signals

As we have observed from our discussion of signal design for zero ISI, it is necessary to reduce the symbol rate $1/T$ below the Nyquist rate of $2W$ symbols per second to realize practical transmitting and receiving filters. On the other hand, suppose we choose to relax the condition of zero ISI and, thus, achieve a symbol transmission rate of $2W$ symbols per second. By allowing for a controlled amount of ISI, we can achieve this symbol rate.

We have already seen that the condition of zero ISI is $x(nT) = 0$ for $n \neq 0$. However, suppose that we design the bandlimited signal to have controlled ISI at one time instant. This means that we allow one additional nonzero value in the samples $\{x(nT)\}$. The ISI that we introduce is deterministic or "controlled" and, hence, it can be taken into account at the receiver, as discussed below.

One special case that leads to (approximately) physically realizable transmitting and receiving filters is specified by the samples[†]

$$x(nT) = \begin{cases} 1, & n = 0, 1 \\ 0, & \text{otherwise} \end{cases} \tag{8.2.26}$$

Now, using (8.2.16), we obtain

$$z_n = \begin{cases} T & n = 0, -1 \\ 0 & \text{otherwise} \end{cases} \tag{8.2.27}$$

which when substituted into (8.2.14) yields

$$Z(f) = T + Te^{-j2\pi fT} \tag{8.2.28}$$

As in the preceding section, it is impossible to satisfy the above equation for $T < \frac{1}{2W}$. However, for $T = \frac{1}{2W}$, we obtain

$$X(f) = \begin{cases} \dfrac{1}{2W}\left[1 + e^{-j\frac{\pi f}{W}}\right], & |f| < W \\ 0, & \text{otherwise} \end{cases}$$

$$= \begin{cases} \dfrac{1}{W}e^{-j2\frac{\pi f}{W}}\cos\left(\dfrac{\pi f}{2W}\right), & |f| < W \\ 0, & \text{otherwise} \end{cases} \tag{8.2.29}$$

Therefore, $x(t)$ is given by

$$x(t) = \text{sinc}(2Wt) + \text{sinc}(2Wt - 1) \tag{8.2.30}$$

[†]It is convenient to deal with samples of $x(t)$ that are normalized to unity for $n = 0, 1$.

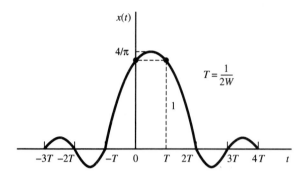

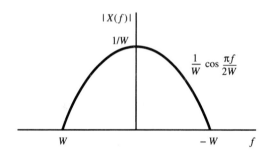

FIGURE 8.12.   Time domain and frequency domain characteristics of a duobionary signal.

This pulse is called a *duobinary signal pulse.* It is illustrated along with its magnitude spectrum in Figure 8.12. We note that the spectrum decays to zero smoothly, which means that physically realizable filters can be designed that approximate this spectrum very closely. Thus, a symbol rate of $2W$ is achieved.

Another special case that leads to (approximately) physically realizable transmitting and receiving filters is specified by the samples

$$x\left(\frac{n}{2W}\right) = x(nT) = \begin{cases} 1, & n = 1 \\ -1, & n = -1 \\ 0, & \text{otherwise} \end{cases} \quad (8.2.31)$$

The corresponding pulse $x(t)$ is given as

$$x(t) = \text{sinc}(t + T)/T - \text{sinc}(t - T)/T \quad (8.2.32)$$

and its spectrum is

$$X(f) = \begin{cases} \frac{1}{2W}\left(e^{j\pi f/W} - e^{-j\pi f/W}\right) = \frac{j}{W}\sin\frac{\pi f}{W}, & |f| \leq W \\ 0, & |f| > W \end{cases} \quad (8.2.33)$$

This pulse and its magnitude spectrum are illustrated in Figure 8.13. It is called a *modified duobinary signal pulse.* It is interesting to note that the spectrum of this

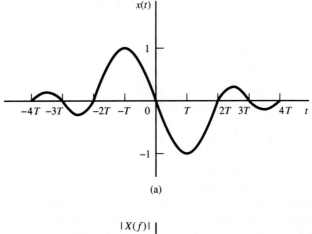

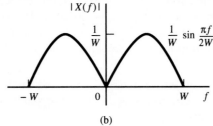

FIGURE 8.13. Time domain and frequency domain characteristics of a modified duobinary signal.

signal has a zero at $f = 0$, making it suitable for transmission over a channel that does not pass dc.

One can obtain other interesting and physically realizable filter characteristics, as shown by Kretzmer (1966) and Lucky et. al. (1968), by selecting different values for the samples $\{x(n/2W)\}$ and more than two nonzero samples. However, as we select more nonzero samples, the problem of unraveling the controlled ISI becomes more cumbersome and impractical.

In general, the class of bandlimited signals pulses that have the form

$$x(t) = \sum_{n=-\infty}^{\infty} x\left(\frac{n}{2W}\right) \frac{\sin 2\pi W(t - n/2W)}{2\pi W(t - n/2W)} \tag{8.2.34}$$

and their corresponding spectra

$$X(f) = \begin{cases} \dfrac{1}{2W} \displaystyle\sum_{n=-\infty}^{\infty} x\left(\dfrac{n}{2W}\right) e^{-jn\pi f/W}, & |f| \le W \\[4mm] 0, & |f| > W \end{cases} \tag{8.2.35}$$

are called *partial response signals* when controlled ISI is purposely introduced by selecting two or more nonzero samples from the set $\{x(n/2W)\}$. The resulting signal pulses allow us to transmit information symbols at the Nyquist rate of $2W$ symbols per second. The detection of the received symbols in the presence of controlled ISI is described below.

8.2.3 Data Detection for Controlled ISI

In this section we describe two methods for detecting the information symbols at the receiver when the received signal contains controlled ISI. One is a symbol-by-symbol detection method that is relatively easy to implement. The second method is based on the maximum-likelihood criterion for detecting a sequence of symbols. The latter method minimizes the probability of error but is a little more complex to implement. In particular, we consider the detection of the duobinary and the modified duobinary partial response signals. In both cases, we assume that the desired spectral characteristic $X(f)$ for the partial response signal is split evenly between the transmitting and receiving filters, i.e., $|G_T(f)| = |G_R(f)| = |X(f)|^{1/2}$.

Symbol-by-symbol detection.   For the duobinary signal pulse, $x(nT) = 1$, for $n = 0$, 1, and zero otherwise. Hence, the samples at the output of the receiving filter have the form

$$y_m = b_m + v_m = a_m + a_{m-1} + v_m \tag{8.2.36}$$

where $\{a_m\}$ is the transmitted sequence of amplitudes and $\{v_m\}$ is a sequence of additive Gaussian noise samples. Let us ignore the noise for the moment and consider the binary case where $a_m = \pm 1$ with equal probability. Then b_m takes on one of three possible values, namely, $b_m = -2, 0, 2$ with corresponding probabilities $1/4, 1/2, 1/4$. If a_{m-1} is the detected symbol from the $(m-1)$st signaling interval, its effect on b_m, the received signal in the mth signaling interval, can be eliminated by subtraction, thus allowing a_m to be detected. This process can be repeated sequentially for every received symbol.

The major problem with this procedure is that errors arising from the additive noise tend to propagate. For example, if a_{m-1} is in error, its effect on b_m is not eliminated but, in fact, it is reinforced by the incorrect subtraction. Consequently, the detection of a_m is also likely to be in error.

Error propagation can be avoided by *precoding* the data at the transmitter instead of eliminating the controlled ISI by subtraction at the receiver. The precoding is performed on the binary data sequence prior to modulation. From the data sequence $\{d_n\}$ of 1's and 0's that is to be transmitted, a new sequence $\{p_n\}$, called the *precoded sequence,* is generated. For the duobinary signal, the precoded sequence is defined as

$$p_m = d_m \ominus p_{m-1}, \quad m = 1, 2, \ldots \tag{8.2.37}$$

where the symbol $\ominus$ denotes modulo-2 subtraction.[†] Then, we set $a_m = -1$ if $p_m = 0$ and $a_m = 1$ if $p_m = 1$, i.e., $a_m = 2p_m - 1$. We note that this precoding operation is identical to that described previously in Section 7.1.4, in context of our discussion of an NRZI signal.

The noise-free samples at the output of the receiving filter are given as

$$b_m = a_m + a_{m-1}$$

$$= (2p_m - 1) + (2p_{m-1} - 1)$$

$$= 2(p_m + p_{m-1} - 1) \tag{8.2.38}$$

Consequently,

$$p_m + p_{m-1} = \frac{b_m}{2} + 1 \tag{8.2.39}$$

Since $d_m = p_m \oplus p_{m-1}$, it follows that the data sequence d_m is obtained from b_m by using the relation

$$d_m = \frac{b_m}{2} - 1 \quad \text{(mod 2)} \tag{8.2.40}$$

Consequently, if $b_m = \pm 2, d_m = 0$, and if $b_m = 0, d_m = 1$. An example that illustrates the precoding and decoding operations is given in Table 8.1. In the presence of additive noise the sampled outputs from the receiving filter are given by (8.2.36). In this case $y_m = b_m + v_m$ is compared with the two thresholds set at $+1$ and -1. The data sequence $\{d_n\}$ is obtained according to the detection rule

$$d_m = \begin{cases} 1, & \text{if } -1 < y_m < 1 \\ 0, & \text{if } |y_m| \geq 1 \end{cases} \tag{8.2.41}$$

TABLE 8-1   BINARY SIGNALING WITH DUOBINARY PULSES

Data sequence d_n		1	1	1	0	1	0	0	1	0	0	0	1	1	0	1
Precoded sequence p_n	0	1	0	1	1	0	0	0	1	1	1	1	0	1	1	0
Transmitted sequence a_n	-1	1	-1	1	1	-1	-1	-1	1	1	1	1	-1	1	1	-1
Received sequence b_n		0	0	0	2	0	-2	-2	0	2	2	2	0	0	2	0
Decoded sequence d_n		1	1	1	0	1	0	0	1	0	0	0	1	1	0	1

[†]Although this is identical to modulo-2 addition, it is convenient to view the precoding operation for duobinary in terms of modulo-2 subtraction.

The extension from binary PAM to multilevel PAM signaling using the duobinary pulses is straightforward. In this case the M-level amplitude sequence $\{a_m\}$ results in a (noise-free) sequence

$$b_m = a_m + a_{m-1}, \quad m = 1, 2, \ldots \tag{8.2.42}$$

which has $2M - 1$ possible equally spaced levels. The amplitude levels are determined from the relation

$$a_m = 2p_m - (M - 1) \tag{8.2.43}$$

where $\{p_m\}$ is the precoded sequence that is obtained from an M-level data sequence $\{d_m\}$ according to the relation

$$p_m = d_m \ominus p_{m-1} \quad (\text{mod } M) \tag{8.2.44}$$

where the possible values of the sequence $\{d_m\}$ are $0, 1, 2, \ldots, M - 1$.

In the absence of noise, the samples at the output of the receiving filter may be expressed as

$$b_m = a_m + a_{m-1}$$
$$= 2[p_m + p_{m-1} - (M - 1)] \tag{8.2.45}$$

Hence,

$$p_m + p_{m-1} = \frac{b_m}{2} + (M - 1) \tag{8.2.46}$$

Since $d_m = p_m + p_{m-1}$ (mod M), it follows that

$$d_m = \frac{b_m}{2} + (M - 1) \quad (\text{mod } M) \tag{8.2.47}$$

An example illustrating multilevel precoding and decoding is given in Table 8.2.

In the presence of noise, the received signal-plus-noise is quantized to the nearest of the possible signal levels and the rule given above is used on the quantized values to recover the data sequence.

TABLE 8-2   FOUR-LEVEL SIGNAL TRANSMISSION WITH DUOBINARY PULSES

Data sequence d_n		0	0	1	3	1	2	0	3	3	2	0	1	0
Precoded sequence p_n	0	0	0	1	2	3	3	1	2	1	1	3	2	2
Transmitted sequence a_n	−3	−3	−3	−1	1	3	3	−1	1	−1	−1	3	1	1
Received sequence b_n		−6	−6	−4	0	4	6	2	0	0	−2	2	4	2
Decoded sequence d_n		0	0	1	3	1	2	0	3	3	2	0	1	0

In the case of the modified duobinary pulse, the controlled ISI is specified by the values $x(n/2W) = -1$, for $n = 1$, $x(n/2W) = 1$ for $n = -1$, and zero otherwise. Consequently, the noise-free sampled output from the receiving filter is given as

$$b_m = a_m - a_{m-2} \qquad (8.2.48)$$

where the M-level sequence $\{a_n\}$ is obtained by mapping a precoded sequence according to the relation (8.2.43) and

$$p_m = d_m \oplus p_{m-2} \quad (\text{mod } M) \qquad (8.2.49)$$

From these relations it is easy to show that the detection rule for recovering the data sequence $\{d_m\}$ from $\{b_m\}$ in the absence of noise is

$$d_m = \frac{b_m}{2} \quad (\text{mod } M) \qquad (8.2.50)$$

As demonstrated above, the precoding of the data at the transmitter makes it possible to detect the received data on a symbol-by-symbol basis without having to look back at previously detected symbols. Thus, error propagation is avoided.

The symbol-by-symbol detection rule described above is not the optimum detection scheme for partial response signals. Nevertheless, symbol-by-symbol detection is relatively simple to implement and is used in many practical applications involving duobinary and modified duobinary pulse signals. Its performance is evaluated in the following section.

Maximum-likelihood sequence detection.   It is clear from the above discussion that partial response waveforms are signal waveforms with memory. This memory is conveniently represented by a trellis. For example, the trellis for the duobinary partial response signal for binary data transmission is illustrated in Figure 8.14. For binary modulation, this trellis contains two states, corresponding to the two possible input values of a_m, i.e., $a_m = \pm 1$. Each branch in the trellis is labeled by two numbers. The first number on the left is the new data bit, i.e., $a_{m+1} = \pm 1$. This number determines the transition to the new state. The number on the right is the received signal level.

The duobinary signal has a memory of length $L = 1$. Hence, for binary modulation the trellis has $S_t = 2^L$ states. In general, for M-ary modulation, the number of trellis states is M^L.

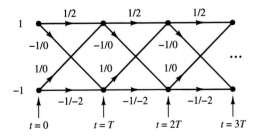

FIGURE 8.14.   Trellis for duobinary partial response signal.

The optimum maximum likelihood (ML) sequence detector selects the most probable path through the trellis upon observing the received data sequence $\{y_m\}$ at the sampling instants $t = mT$, $m = 1, 2, \ldots$. In general, each node in the trellis will have M incoming paths and M corresponding metrics. One out of the M incoming paths is selected as the most probable, based on the values of the metrics and the other $M - 1$ paths and their metrics, are discarded. The surviving path at each node is then extended to M new paths, one for each of the M possible input symbols, and the search process continues. This is basically the Viterbi algorithm for performing the trellis search.

For the class of partial response signals, the received sequence $\{y_m, \ 1 \le m \le N\}$ is generally described statistically by the joint p.d.f. $f(\mathbf{y}_N | \mathbf{a}_N)$, where $\mathbf{y}_N = (y_1, y_2, \ldots, y_N)^t$ and $\mathbf{a}_N = (a_1, a_2, \ldots, a_N)^t$ and $N > L$. When the additive noise is zero-mean Gaussian, $f(\mathbf{y}_N | \mathbf{a}_N)$ is a multivariate Gaussian p.d.f., i.e.,

$$f(\mathbf{y}_N | \mathbf{a}_N) = \frac{1}{[2\pi \det(\mathbf{C})]^{N/2}} \exp\left[-\frac{1}{2}(\mathbf{y}_N - \mathbf{b}_N)^t \mathbf{C}^{-1}(\mathbf{y}_N - \mathbf{b}_N)\right] \quad (8.2.51)$$

where $\mathbf{b}_N = (b_1, b_2, \ldots, b_N)^t$ is the mean of the vector $\mathbf{y}_N$ and $\mathbf{C}$ is the N × N covariance matrix of $\mathbf{y}_N$. Then, the ML sequence detector selects the sequence through the trellis that maximizes the p.d.f. $f(\mathbf{y}_N | \mathbf{a}_N)$.

The computations for finding the most probable sequence through the trellis is simplified by taking the natural logarithms of $f(\mathbf{y}_N | \mathbf{a}_N)$. Thus,

$$\ln f(\mathbf{y}_N | \mathbf{a}_N) = -\frac{N}{2} \ln[2\pi \det(\mathbf{C})] - \frac{1}{2}(\mathbf{y}_N - \mathbf{b}_N)^t \mathbf{C}^{-1}(\mathbf{y}_N - \mathbf{b}_N) \quad (8.2.52)$$

Given the received sequence $\{y_m\}$, the data sequence $\{a_m\}$ that maximizes $\ln f(\mathbf{y}_N | \mathbf{a}_N)$ is identical to the sequence $\{a_m\}$ that minimizes $(\mathbf{y}_N - \mathbf{b}_N)^t \mathbf{C}^{-1}(\mathbf{y}_N - \mathbf{b}_N)$, i.e.,

$$\hat{\mathbf{a}}_N = \arg\min_{\mathbf{a}_N} \left[(\mathbf{y}_N - \mathbf{b}_N)^t \mathbf{C}^{-1}(\mathbf{y}_N - \mathbf{b}_N)\right] \quad (8.2.53)$$

The search through the trellis for the minimum distance path may be performed sequentially by use of the Viterbi algorithm. Let us consider the duobinary signal waveform with binary modulation and suppose that we begin at the initial state with $a_0 = 1$. Then upon receiving $y_1 = a_1 + a_0 + v_1$ at time $t = T$ and $y_2 = a_2 + a_1 + v_2$ at time $t = 2T$, we have four candidate paths, corresponding to $(a_1, a_2) = (1, 1), (-1, 1), (1, -1)$ and $(-1, -1)$. The first two candidate paths merge at state 1 at $t = 2T$. For the two paths merging at state 1, we compute the metrics $\mu_2(1, 1)$ and $\mu_2(-1, 1)$ and select the more probable path. A similar computation is performed at state -1 for the two sequences $(1, -1)$ and $(-1, -1)$. Thus, one of the two sequences at each node is saved and the other is discarded. The trellis search continues upon receipt of the signal sample y_3 at time $t = 3T$, by extending the two surviving paths from time $t = 2T$.

The metric computations are complicated by the correlation of the noise samples at the output of the matched filter for the partial response signal. For example, in the case of the duobinary signal waveform, the correlation of the noise sequence $\{v_m\}$ is over two successive signal samples. Hence, v_m and v_{m+k} are correlated for $k = 1$ and uncorrelated for $k > 1$. In general, a partial response signal waveform with memory L will result in a correlated noise sequence at the output of the matched filter, which satisfies the condition $E[v_m v_{m+k}] = 0$ for $k > L$. Ungerboeck (1974) described a sequential trellis search (Viterbi) algorithm for correlated noise (see Problem 8.20).

Some simplification in the metric computations result if we ignore the noise correlation by assuming that $E[v_m v_{m+k}] = 0$ for $k > 0$. Then, by assumption, the covariance matrix $\mathbf{C} = \sigma_v^2 \mathbf{I}$ where $\sigma_v^2 = E[v_m^2]$ and $\mathbf{I}$ in the $N \times N$ identity matrix. In this case, (8.2.53) simplifies to

$$\hat{\mathbf{a}}_N = \underset{\mathbf{a}_N}{\arg\min} \left[(\mathbf{y}_N - \mathbf{b}_N)^t (\mathbf{y}_N - \mathbf{b}_N) \right]$$

$$= \underset{\mathbf{a}_N}{\arg\min} \left[\sum_{m=1}^{N} \left(y_m - \sum_{k=0}^{L} x_k a_{m-k} \right)^2 \right] \tag{8.2.54}$$

where

$$b_m = \sum_{k=0}^{L} x_k a_{m-k} \tag{8.2.55}$$

and $x_k = x(kT)$ are the sampled values of the partial response signal waveform. In this case, the metric computations at each node of the trellis have the form

$$\mu_m(\mathbf{a}_m) = \mu_{m-1}(\mathbf{a}_{m-1}) + \left(y_m - \sum_{k=0}^{L} x_k a_{m-k} \right)^2 \tag{8.2.56}$$

where $\mu_m(\mathbf{a}_m)$ are the metrics at time $t = mT$, $\mu_{m-1}(\mathbf{a}_{m-1})$ are the metrics at time $t = (m-1)T$ and the second term on the right-hand side of (8.2.56) represents the new increments to the metrics based on the new received sample y_m.

As previously indicated in Section 7.2.4, ML sequence detection introduces a variable delay in detecting each transmitted information symbol. In practice, the variable delay in avoided by truncating the surviving sequences to N_t most recent symbols, where $N_t \gg 5L$, thus achieving a fixed delay. In case the M^L surviving sequences at time $t = mT$ disagree on the symbol a_{m-N_t}, the symbol in the most probable surviving sequence may be chosen. The loss in performance resulting from this truncation is negligible if $N_t > 5L$.

Example 8.2.1

For the duobinary partial response signal, express the metric computations performed at $t = 2T$ and $t = 3T$ based on the received signal samples $y_m = b_m + v_m$ for $m = 1, 2, 3$, where the noise correlation is ignored.

Solution The metrics are generally given by (8.2.56). Upon receipt of y_1 and y_2 and with $a_0 = 1$, the metrics for the two paths merging at state 1 are

$$\mu_2(1, 1) = (y_1 - 2)^2 + (y_2 - 2)^2$$
$$\mu_2(-1, 1) = y_1^2 + y_2^2$$

If $\mu_2(1, 1) < \mu_2(-1, 1)$, we select the path $(1, 1)$ as the more probable and discard $(-1, 1)$. Otherwise, we select the path $(-1, 1)$ and discard the path $(1, 1)$. The path with the smaller metric is called the survivor and the sequence (a_1, a_2) and the corresponding metric are saved.

A similar computation is performed at state -1 for the sequences $(1, -1)$ and $(-1, -1)$. Then, we have

$$\mu_2(1, -1) = (y_1 - 2)^2 + y_2^2$$
$$\mu_2(-1, -1) = y_1^2 + (y_2 + 2)^2$$

We compare the metrics $\mu_2(1, -1)$ and $\mu_2(-1, -1)$ and select the sequence with the smaller metric.

Upon receipt of y_3 at $t = 3T$, we consider the extensions of the survivor paths. Suppose that the two survivor paths are $(a_1, a_2) = (1, 1)$ and $(a_1, a_2) = (1, -1)$. Then, at state 1 (at $t = 3T$) we have the two merging paths $(a_1, a_2, a_3) = (1, 1, 1)$ and $(1, -1, 1)$. Their corresponding metrics are

$$\mu_3(1, 1, 1) = \mu_2(1, 1) + (y_3 - 2)^2$$
$$\mu_3(1, -1, 1) = \mu_2(1, -1) + y_3^2$$

We compare the metrics for these two merging paths and select the path with the smaller metric as the survivor.

Similarly, at state -1 (at $t = 3T$), we have the two merging paths $(1, 1, -1)$ and $(1, -1, -1)$, and their corresponding metrics

$$\mu_3(1, 1, -1) = \mu_2(1, 1) + y_3^2$$
$$\mu_2(1, -1, -1) = \mu_2(1, -1) + (y_3 + 2)^2$$

We select the path with the smaller metric as the survivor at state -1. This process continues upon receipt of additional data at $t = kT$, $k = 4, 5, \ldots$.

8.3 PROBABILITY OF ERROR IN DETECTION OF DIGITAL PAM

In this section we evaluate the performance of the receiver for demodulating and detecting an M-ary PAM signal in the presence of additive, white Gaussian noise at its input. First, we consider the case in which the transmitter and receiver filters $G_T(f)$ and $G_R(f)$ are designed for zero ISI. Then, we consider the case in which $G_T(f)$ and $G_R(f)$ are designed such that $x(t) = g_T(t) \star g_R(t)$ is either a duobinary signal or a modified duobinary signal.

8.3.1 Probability of Error for Detection of Digital PAM with Zero ISI

In the absence of ISI, the received signal sample at the output of the receiving matched filter has the form

$$y_m = x_0 a_m + v_m \tag{8.3.1}$$

where

$$x_0 = \int_{-W}^{W} |G_T(f)|^2 \, df = \mathcal{E}_g \tag{8.3.2}$$

and v_m is the additive Gaussian noise which has zero mean and variance

$$\sigma_v^2 = \mathcal{E}_g N_0 / 2 \tag{8.3.3}$$

In general, a_m takes one of M possible equally spaced amplitude values with equal probability. Given a particular amplitude level, the problem is to determine the probability of error.

The problem of evaluating the probability of error for digital PAM in a band-limited, additive white Gaussian noise channel, in the absence of ISI, is identical to the evaluation of the error probability for M-ary PAM as given in Section 7.3.2. The final result that is obtained from the derivation (see (7.3.17)) is

$$P_M = \frac{2(M - 1)}{M} Q\left[\sqrt{\frac{2\mathcal{E}_g}{N_0}}\right] \tag{8.3.4}$$

But $\mathcal{E}_g = 3\mathcal{E}_{av}/(M^2 - 1)$, $\mathcal{E}_{av} = k\mathcal{E}_{bav}$ is the average energy per symbol and $\mathcal{E}_{bav}$ is the average energy per bit. Hence,

$$P_M = \frac{2(M - 1)}{M} Q\left[\sqrt{\frac{6(\log_2 M)\mathcal{E}_{bav}}{(M^2 - 1)N_0}}\right] \tag{8.3.5}$$

This is exactly the form for the probability of error of M-ary PAM derived previously in Section 7.3.2 (see (7.3.21)). In the treatment of PAM given in this chapter we imposed the additional constraint that the transmitted signal is bandlimited to the bandwidth allocated for the channel. Consequently, the transmitted signal pulses were designed to be bandlimited and to have zero ISI.

In contrast, no bandwidth constraint was imposed on the PAM signals considered in Section 7.3.2. Nevertheless, the receivers (demodulators and detectors) in both cases are optimum (matched filters) for the corresponding transmitted signals. Consequently, no loss in error rate performance results from the bandwidth constraint when the signal pulse is designed for zero ISI and the channel does not distort the transmitted signal.

8.3.2 Probability of Error for Detection of Partial Response Signals

In this section we determine the probability of error for detection of digital M-ary PAM signaling using duobinary and modified duobinary pulses. The channel is assumed to be an ideal bandlimited channel with additive white Gaussian noise. The model for the communications system is shown in Figure 8.15.

We consider two types of detectors. The first is the symbol-by-symbol detector and the second is the optimum ML sequence detector described in the previous section.

Symbol-by-symbol detector.   At the transmitter, the M-level data sequence $\{d_n\}$ is precoded as described previously. The precoder output is mapped into one of M possible amplitude levels. Then the transmitting filter with frequency response $G_T(f)$ has an output

$$v(t) = \sum_{n=-\infty}^{\infty} a_n g_T(t - nT) \tag{8.3.6}$$

The partial response function $X(f)$ is divided equally between the transmitting and receiving filters. Hence, the receiving filter is matched to the transmitted pulse, and the cascade of the two filters results in the frequency characteristic

$$|G_T(f)G_R(f)| = |X(f)| \tag{8.3.7}$$

The matched filter output is sampled at $t = nT = n/2W$ and the samples are fed to the decoder. For the duobinary signal, the output of the matched filter at the sampling instant may be expressed as

$$y_m = a_m + a_{m-1} + v_m$$
$$= b_m + v_m \tag{8.3.8}$$

where v_m is the additive noise component. Similarly, the output of the matched filter for the modified duobinary signal is

$$y_m = a_m - a_{m-2} + v_m$$
$$= b_m + v_m \tag{8.3.9}$$

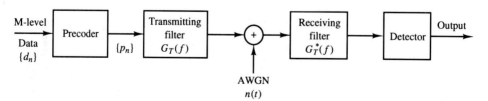

FIGURE 8.15.   Block diagram of modulator and demodulator for partial response signals.

For binary transmission, let $a_m = \pm d$, where $2d$ is the distance between signal levels. Then, the corresponding values of b_m are $(2d, 0, -2d)$. For M-ary PAM signal transmission, where $a_m = \pm d, \pm 3d, \ldots, \pm(M-1)d$, the received signal levels are $b_m = 0, \pm 2d, \pm 4d, \ldots, \pm 2(M-1)d$. Hence, the number of received levels is $2M - 1$, and the scale factor d is equivalent to $x_0 = \mathcal{E}_g$.

The input transmitted symbols $\{a_m\}$ are assumed to be equally probable. Then, for duobinary and modified duobinary signals, it is easily demonstrated that, in the absence of noise, the received output levels have a (triangular) probability mass function of the form

$$P(b = 2md) = \frac{M - |m|}{M^2}, \quad m = 0, \pm 1, \pm 2, \ldots, \pm(M-1) \qquad (8.3.10)$$

where b denotes the noise-free received level and $2d$ is the distance between any two adjacent received signal levels.

The channel corrupts the signal transmitted through it by the addition of white Gaussian noise with zero mean and power-spectral density $N_0/2$.

We assume that a symbol error occurs whenever the magnitude of the additive noise exceeds the distance d. This assumption neglects the rare event that a large noise component with magnitude exceeding d may result in a received signal level that yields a correct symbol decision. The noise component v_m is zero mean, Gaussian with variance

$$\sigma_v^2 = \frac{N_0}{2} \int_{-W}^{W} |G_R(f)|^2 \, df$$

$$= \frac{N_0}{2} \int_{-W}^{W} |X(f)| df = 2N_0/\pi \qquad (8.3.11)$$

for both the duobinary and the modified duobinary signals. Hence, an upper bound on the symbol probability of error is

$$P_M < \sum_{m=-(M-2)}^{M-2} P(|y - 2md| > d|b = 2md) P(b = 2md)$$

$$+ 2P(y + 2(M-1)d > d|b = -2(M-1)d) P(b = -2(M-1)d)$$

$$= P(|y| > d|b = 0) \left[2 \sum_{m=0}^{M-1} P(b = 2md) - P(b = 0) - P(b = -2(M-1)d) \right]$$

$$= \left(1 - \frac{1}{M^2} \right) P(|y| > d|b = 0) \qquad (8.3.12)$$

But

$$P(|y| > d|b = 0) = \frac{2}{\sqrt{2\pi}\sigma_v} \int_{d}^{\infty} e^{-x^2/2\sigma_v^2} \, dx$$

$$= 2Q\left(\sqrt{\frac{\pi d^2}{2N_0}}\right) \tag{8.3.13}$$

Therefore, the average probability of a symbol error is upper bounded as

$$P_M < 2\left(1 - \frac{1}{M^2}\right) Q\left(\sqrt{\frac{\pi d^2}{2N_0}}\right) \tag{8.3.14}$$

The scale factor d in (8.3.14) can be eliminated by expressing d in terms of the average power transmitted into the channel. For the M-ary PAM signal in which the transmitted levels are equally probable, the average power at the output of the transmitting filter is

$$P_{av} = \frac{E[a_m^2]}{T} \int_{-W}^{W} |G_T(f)|^2 \, df$$

$$= \frac{E[a_m^2]}{T} \int_{-W}^{W} |X(f)| \, df = \frac{4}{\pi T} E[a_m^2] \tag{8.3.15}$$

where $E\left[a_m^2\right]$ is the mean square value of the M signal levels, which is

$$E\left[a_m^2\right] = \frac{d^2(M^2 - 1)}{3} \tag{8.3.16}$$

Therefore,

$$d^2 = \frac{3\pi P_{av} T}{4(M^2 - 1)} \tag{8.3.17}$$

By substituting the value of d^2 from (8.3.17) into (8.3.14), we obtain the upper bound for the symbol error probability as

$$P_M < 2\left(1 - \frac{1}{M^2}\right) Q\left(\sqrt{\left(\frac{\pi}{4}\right)^2 \frac{6}{M^2 - 1} \frac{\mathcal{E}_{av}}{N_0}}\right) \tag{8.3.18}$$

where $\mathcal{E}_{av}$ is the average energy per transmitted symbol, which can be also expressed in terms of the average bit energy as $\mathcal{E}_{av} = k\mathcal{E}_{bav} = (\log_2 M)\mathcal{E}_{bav}$.

The expression in (8.3.18) for the probability of error of M-ary PAM holds for both a duobinary and a modified duobinary partial response signal. If we compare this result with the error probability of M-ary PAM with zero ISI, which can be obtained by using a signal pulse with a raised cosine spectrum, we note that the performance of partial response duobinary or modified duobinary has a loss of $(\pi/4)^2$ or 2.1 dB. This loss in SNR is due to the fact that the detector for the partial response signals makes decisions on a symbol-by symbol basis, thus, ignoring the inherent memory contained in the received signal at the input to the detector.

To observe the memory in the received sequence, let us look at the noise-free received sequence for binary transmission given in Table 8.1. The sequence $\{b_m\}$ is $0, -2, 0, 2, 0, -2, 0, 2, 2, \ldots$. We note that it is not possible to have a transition from -2 to $+2$ or from $+2$ to -2 in one symbol interval. For example, if the signal level at the input to the detector is -2 the next signal level can be either -2 or 0. Similarly, if the signal level at a given sampling instant is 2, the signal level in the following time instant can be either 2 or 0. In other words, it is not possible to encounter a transition from -2 to 2 or vice-versa between two successive received samples from the matched filter. However, a symbol-by-symbol detector does not exploit this constraint or inherent memory in the received sequence. Below, we derive the performance of the ML sequence detector that exploits the inherent memory in the modulation and, consequently, regains a large part of the 2.1 dB loss suffered by the symbol-by-symbol detector.

Maximum-likelihood sequence detector. The ML sequence detector searches through the trellis for the most probable transmitted sequence $\{a_m\}$, as previously described in Section 8.2.3. At each stage of the search process the detector compares the metrics of paths that merge at each of the nodes and selects the path that is most probable at each node. The performance of the detector may be evaluated by determining the probability that a wrong path through the trellis has a Euclidean distance metric that is smaller than the metric for the correct path.

In general, the computation of the exact probability of error is extremely difficult. Instead, we shall determine an approximation to the probability of error, which is based on comparing the metrics of two paths that merge at a node and are separated by the smallest Euclidean distance of all other paths. Our derivation is performed for the duobinary partial response signal waveform.

Let us consider the trellis for the duobinary partial response signal shown in Figure 8.14. We assume that we start in state 1 at $t = 0$ and that the first two transmitted symbols are $a_1 = 1$ and $a_2 = 1$. Then, at $t = T$ we receive $y_1 = 2d + v_1$ and at $t = 2T$ we receive $y_2 = 2d + v_2$. An error is made at state 1 if the path $(a_1, a_2) = (-1, 1)$ is more probable than the path $(a_1, a_2) = (1, 1)$, given the received values of y_1 and y_2. This path error event is the dominant path error event and, hence, it serves as a good approximation to the probability of error for the ML sequence detector.

From our discussion in Section 8.2.3, we recall that the metric for the path $(a_1, a_2) = (1, 1)$ is

$$\mu_2(1, 1) = [y_1 - 2d \quad y_2 - 2d]\mathbf{C}^{-1}\begin{bmatrix} y_1 - 2d \\ y_2 - 2d \end{bmatrix} \qquad (8.3.19)$$

where the covariance matrix $\mathbf{C}$ is given by (see Problem 8.24)

$$\mathbf{C} = \frac{2N_0}{\pi}\begin{bmatrix} 1 & \frac{1}{3} \\ \frac{1}{3} & 1 \end{bmatrix} \qquad (8.3.20)$$

For the path $(a_1, a_2) = (-1, 1)$, the corresponding metric is

$$\mu_2(-1, 1) = [y_1, y_2]\mathbf{C}^{-1} \begin{bmatrix} y_1 \\ y_2 \end{bmatrix} \tag{8.3.21}$$

The probability of a path error event is simply the probability that the metric $\mu_2(-1, 1)$ is smaller than the metric $\mu_2(1, 1)$, i.e.,

$$P_2 = P[\mu_2(-1, 1) < \mu_2(1, 1)] \tag{8.3.22}$$

By substituting $y_1 = 2d + v_1$ and $y_2 = 2d + v_2$ into (8.3.19) and (8.3.21), we find that

$$P_2 = P(v_1 + v_2 < -2d) \tag{8.3.23}$$

Since v_1 and v_2 are zero-mean (correlated) Gaussian variables, their sum is also zero-mean Gaussian. The variance of the sum $z = v_1 + v_2$ is simply $\sigma_z^2 = 16N_0/3\pi$. Therefore,

$$P_2 = P(z < -2d) = Q\left(\frac{2d}{\sigma_z}\right) = Q\left(\sqrt{\frac{4d^2}{\sigma_z^2}}\right) \tag{8.3.24}$$

From (8.3.17) we have (with $M = 2$) the expression for d^2 as

$$d^2 = \frac{\pi P_{av} T}{4} = \frac{\pi \mathcal{E}_b}{4} \tag{8.3.25}$$

Hence, the probability of the path error event is

$$P_2 = Q\left(\sqrt{\frac{1.5\pi^2}{16}\left(\frac{2\mathcal{E}_b}{N_0}\right)}\right) \tag{8.3.26}$$

First, we note that this path error event results in one bit error in the sequence of two bits. Hence, the bit error probability is $P_2/2$. Second, there is a reduction in SNR of $10\log(1.5\pi^2/16) = -0.34$ dB relative to the case of no ISI. This small SNR degradation is apparently the penalty incurred in exchange for the bandwidth efficiency of the partial response signal. Finally, we observe that the ML sequence detector has gained back 1.76 dB of the 2.1 dB degradation inherent in the symbol-by-symbol detector.

8.4 SYSTEM DESIGN IN THE PRESENCE OF CHANNEL DISTORTION

In Section 8.2.1 we described a signal design criterion that results in zero ISI at the output of the receiving filter. Recall that a signal pulse $x(t)$ will satisfy the condition of zero ISI at the sampling instants $t = nT, n = \pm 1, \pm 2, \ldots$, if its spectrum $X(f)$ satisfies the condition given by (8.2.9). From this condition we conclude that for ISI-free transmission over a channel, the transmitter-receiver filters and the channel transfer function must satisfy

$$G_T(f)C(f)G_R(f) = X_{rc}(f) \qquad (8.4.1)$$

where $X_{rc}(f)$ denotes the Fourier transform of an appropriate raised cosine pulse whose parameters depend on the channel bandwidth W and the transmission interval T. Obviously there are an infinite number of transmitter-receiver filter pairs that satisfy the above condition.

In this section we consider the design of the transmitter-receiver filter pairs that maximize the SNR at the output of the receiving filter. Therefore, in this section we are concerned with the design of a digital communication system with zero ISI and minimum error probability for a channel with distortion. We first present a brief coverage of various types of channel distortion and then derive the equations for transmitter-receiver filters.

We distinguish two types of distortion. *Amplitude distortion* results when the amplitude characteristic $|C(f)|$ is not constant for $|f| \leq W$. The second type of distortion, called *phase distortion*, results when the phase characteristic $\Theta_c(f)$ is nonlinear in frequency.

Another view of phase distortion is obtained by considering the derivative of $\Theta_c(f)$. Thus, we define the *envelope delay* characteristic as (see Problems 2.76 and 8.25)

$$\tau(f) = -\frac{1}{2\pi}\frac{d\Theta_c(f)}{df} \qquad (8.4.2)$$

When $\Theta_c(f)$ is linear in f, the envelope delay is constant for all frequencies. In this case, all frequencies in the transmitted signal pass through the channel with the same fixed-time delay. In such a case, there is no phase distortion. However, when $\Theta_c(f)$ is nonlinear, the envelope delay $\tau(f)$ varies with frequency, and the various frequency components in the input signal undergo different delays in passing through the channel. In such a case we say that the transmitted signal has suffered from *delay distortion*.

Both amplitude and delay distortion cause ISI in the received signal. For example, let us assume that we have designed a pulse with a raised cosine spectrum that has zero ISI at the sampling instants. An example of such a pulse is illustrated in Figure 8.16(a). When the pulse is passed through a channel filter with constant amplitude $|C(f)| = 1$ for $|f| < W$ and a quadratic phase characteristic (linear envelope delay), the received pulse at the output of the channel is shown in Figure 8.16(b). Note that the periodic zero crossings have been shifted by the delay distortion so that the resulting pulse suffers from ISI. Consequently, a sequence of successive pulses would be smeared into one another and the peaks of the pulses would no longer be distinguishable due to the ISI.

Below we consider two problems. First, we consider the design of optimum transmitting and receiving filters in the presence of channel distortion when the channel characteristics are known. Second, we consider the design of special filters, called *channel equalizers,* that automatically and adaptively correct for the channel distortion when the channel characteristics, i.e., $|C(f)|$ and $\angle\Theta_c(f)$, are unknown.

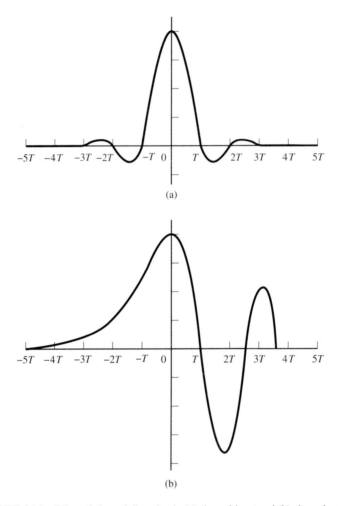

FIGURE 8.16. Effect of channel distortion in (a) channel input and (b) channel output.

8.4.1 Design of Optimum Transmitting and Receiving Filters

In this section we assume that the channel frequency response characteristic $C(f)$ is known and consider the problem of designing a transmitting filter and a receiving filter that maximize the SNR at the output of the receiving filter and results in zero ISI. Figure 8.17 illustrates the overall system under consideration.

For the signal component, we must satisfy the condition

$$G_T(f)C(f)G_R(f) = X_{rc}(f)e^{-j2\pi f t_0}, \quad |f| \le W \qquad (8.4.3)$$

where $X_{rc}(f)$ is the desired raised cosine spectrum that yields zero ISI at the sampling instants, and t_0 is a time delay that is necessary to ensure the physical realizability of the transmitter and receiver filters.

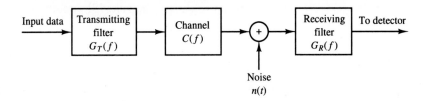

FIGURE 8.17.   System configuration for design of $G_T(f)$ and $G_R(f)$.

The noise at the output of the receiving filter may be expressed as

$$v(t) = \int_{-\infty}^{\infty} n(t - \tau)g_R(\tau)d\tau \qquad (8.4.4)$$

where $n(t)$ is the input to the filter. The noise $n(t)$ is assumed to be zero mean Gaussian. Hence, $v(t)$ is zero-mean Gaussian, with a power-spectral density

$$S_v(f) = S_n(f)|G_R(f)|^2 \qquad (8.4.5)$$

where $S_n(f)$ is the spectral density of the noise process $n(t)$.

For simplicity, we consider binary PAM transmission. Then, the sampled output of the matched filter is

$$y_m = x_0 a_m + v_m = a_m + v_m \qquad (8.4.6)$$

where x_0 is normalized[†] to unity, $a_m = \pm d$, and v_m represents the noise term, which is zero-mean Gaussian with variance

$$\sigma_v^2 = \int_{-\infty}^{\infty} S_n(f)|G_R(f)|^2 \, df \qquad (8.4.7)$$

Consequently, the probability of error is

$$P_2 = \frac{1}{\sqrt{2\pi}} \int_{d/\sigma_v}^{\infty} e^{-y^2/2} \, dy = Q\left(\sqrt{\frac{d^2}{\sigma_v^2}}\right) \qquad (8.4.8)$$

The probability of error is minimized by maximizing the SNR $= d^2/\sigma_v^2$, or, equivalently, by minimizing the noise-to-signal ratio σ_v^2/d^2. But d^2 is related to the transmitted signal power as follows:

$$P_{av} = \frac{E(a_m^2)}{T} \int_{-\infty}^{\infty} g_T^2(t) \, dt = \frac{d^2}{T} \int_{-\infty}^{\infty} g_T^2(t) dt$$

$$\frac{1}{d^2} = \frac{1}{P_{av}T} \int_{-\infty}^{\infty} |G_T(f)|^2 \, df \qquad (8.4.9)$$

[†]By setting $x_0 = 1$ and $a_m = \pm d$, the scaling by x_0 is incorporated into the parameter d.

However, $G_T(f)$ must be chosen to satisfy the zero ISI condition given by (8.4.3). Consequently,

$$|G_T(f)| = \frac{|X_{rc}(f)|}{|C(f)|\,|G_R(f)|}, \quad |f| \le W \tag{8.4.10}$$

and $G_T(f) = 0$ for $|f| \ge W$. Hence,

$$\frac{1}{d^2} = \frac{1}{P_{av}T} \int_{-W}^{W} \frac{|X_{rc}(f)|^2}{|C(f)|^2|G_R(f)|^2}\,df \tag{8.4.11}$$

Therefore, the noise-to-signal ratio that must be minimized with respect to $|G_R(f)|$ for $|f| \le W$ is

$$\frac{\sigma_\nu^2}{d^2} = \frac{1}{P_{av}T} \int_{-W}^{W} S_n(f)|G_R(f)|^2 df \int_{-W}^{W} \frac{|X_{rc}(f)|^2}{|C(f)|^2|G_R(f)|^2}\,df \tag{8.4.12}$$

The optimum $|G_R(f)|$ can be found by applying the Cauchy-Schwartz inequality,

$$\int_{-\infty}^{\infty} |U_1(f)|^2 df \int_{-\infty}^{\infty} |U_2(f)|^2 df \ge \left[\int_{-\infty}^{\infty} |U_1(f)|\,|U_2(f)| df \right]^2 \tag{8.4.13}$$

where $|U_1(f)|$ and $|U_2(f)|$ are defined as

$$|U_1(f)| = |\sqrt{S_n(f)}|\,|G_R(f)|$$

$$|U_2(f)| = \frac{|X_{rc}(f)|}{|C(f)|\,|G_R(f)|} \tag{8.4.14}$$

The minimum value of (8.4.12) is obtained when $|U_1(f)|$ is proportional to $|U_2(f)|$, or, equivalently, when

$$|G_R(f)| = K \frac{|X_{rc}(f)|^{1/2}}{[S_n(f)]^{1/4}|C(f)|^{1/2}}, \quad |f| \le W \tag{8.4.15}$$

where K is a constant. The corresponding transmitting filter has a magnitude characteristic

$$|G_T(f)| = \frac{1}{K} \frac{|X_{rc}(f)|^{1/2}[S_n(f)]^{1/4}}{|C(f)|^{1/2}}, \quad |f| \le W \tag{8.4.16}$$

Finally, the maximum SNR achieved by these optimum transmitting and receiving filters is

$$\frac{d^2}{\sigma_\nu^2} = \frac{P_{av}T}{\left[\int_{-W}^{W} \frac{|X_{rc}(f)|[S_n(f)]^{1/2}}{|C(f)|}\,df \right]^2} \tag{8.4.17}$$

We note that the optimum transmitting and receiving filters are specified in magnitude only. The phase characteristics for $G_T(f)$ and $G_R(f)$ may be selected so as to satisfy the condition in (8.4.3), i.e.,

$$\Theta_T(f) + \Theta_c(f) + \Theta_R(f) = 2\pi f t_0 \tag{8.4.18}$$

where $\Theta_T(f)$, $\Theta_c(f)$, and $\Theta_R(f)$ are the phase characteristics of the transmitting filter, the channel, and the receiving filter, respectively.

In the special case where the additive noise at the input to the demodulator is white Gaussian with spectral density $N_0/2$, the optimum filter characteristics specified by (8.4.15) and (8.4.16) reduce to

$$|G_R(f)| = K_1 \frac{|X_{rc}(f)|^{1/2}}{|C(f)|^{1/2}}, \quad |f| \le W$$

$$|G_T(f)| = K_2 \frac{|X_{rc}(f)|^{1/2}}{|C(f)|^{1/2}}, \quad |f| \le W \tag{8.4.19}$$

where K_1 and K_2 are arbitrary scale factors. Note that, in this case, $|G_R(f)|$ is the matched filter to $|G_T(f)|$. The corresponding SNR at the detector, given by (8.4.17) reduces to

$$\frac{d^2}{\sigma_v^2} = \frac{2P_{av}T}{N_0} \left[\int_{-W}^{W} \frac{|X_{rc}(f)|}{|C(f)|} df \right]^{-2} \tag{8.4.20}$$

Example 8.4.1

Determine the optimum transmitting and receiving filters for a binary communication system that transmits data at a rate of 4800 bits/sec over a channel with frequency (magnitude) response

$$|C(f)| = \frac{1}{\sqrt{1 + \left(\frac{f}{W}\right)^2}}, \quad |f| \le W \tag{8.4.21}$$

where $W = 4800$ Hz. The additive noise is zero mean, white, Gaussian with spectral density $N_0/2 = 10^{-15}$ W/Hz.

Solution Since $W = 1/T = 4800$, we use a signal pulse with a raised cosine spectrum and $\alpha = 1$. Thus,

$$X_{rc}(f) = \frac{T}{2}[1 + \cos(\pi T|f|)]$$

$$= T\cos^2\left(\frac{\pi|f|}{9600}\right) \tag{8.4.22}$$

Then,

$$|G_T(f)| = |G_R(f)| = \left[1 + \left(\frac{f}{4800}\right)^2\right]^{1/4} \cos\left(\frac{\pi|f|}{9600}\right), \quad |f| \le 4800 \tag{8.4.23}$$

and $|G_T(f)| = |G_R(f)| = 0$, otherwise. Figure 8.18 illustrates the filter characteristic $G_T(f)$.

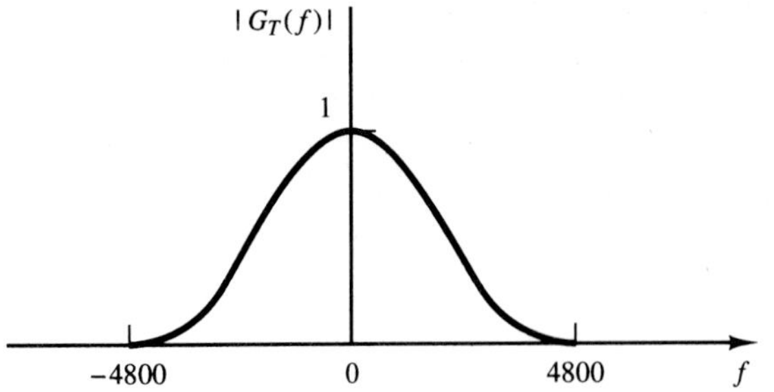

FIGURE 8.18.   Frequency response of optimum transmitter filter.

One can now use these optimum filters to determine the amount of transmitted energy $\mathcal{E}$ required to achieve a specified error probability. This problem is left as an exercise for the reader.

The derivation of the optimum transmitting and receiving filters for transmission of M-ary PAM through a nonideal channel $C(f)$ is identical to that for binary PAM. The only modifications are in the expressions for the average transmitted power, which for M-ary PAM is

$$P_{av} = \frac{E[a_m^2]}{T} \int_{-W}^{W} |G_T(f)|^2 \, df$$

$$= \frac{(M^2 - 1)d^2}{3T} \int_{-W}^{W} |G_T(f)|^2 \, df \tag{8.4.24}$$

and in the corresponding expression for the probability of error, which is

$$P_M = \frac{2(M-1)}{M} Q\left(\sqrt{\frac{d^2}{\sigma_v^2}}\right) \tag{8.4.25}$$

For AWGN, (8.4.25) may be expressed as

$$P_M = \frac{2(M-1)}{M} Q\left(\sqrt{\frac{6\mathcal{E}_{av}}{(M^2-1)N_0} \left[\int_{-W}^{W} \frac{|X_{rc}(f)|}{|C(f)|} \, df\right]^{-2}}\right) \tag{8.4.26}$$

Finally, we observe that the loss due to channel distortion is

$$20 \log_{10} \left[\int_{-W}^{W} \frac{|X_{rc}(f)|}{|C(f)|} \, df\right] \tag{8.4.27}$$

Note that when $C(f) = 1$ for $|f| \le W$, the channel is ideal and

$$\int_{-W}^{W} X_{rc}(f) \, df = 1$$

so that no loss is incurred. On the other hand, when there is amplitude distortion, $|C(f)| < 1$ for some range of frequencies in the band $|f| \leq W$ and, hence, there is a loss in SNR incurred, as given by (8.4.27). This loss is independent of channel phase distortion, because phase distortion has been perfectly compensated, as implied by (8.4.18). The loss given by (8.4.27) is due entirely to amplitude distortion and is a measure of the noise enhancement resulting from the receiving filter, which compensates for the channel distortion.

8.4.2 Channel Equalization

In the preceding section we described the design of transmitting and receiving filters for digital PAM transmission when the frequency response characteristics of the channel are known. Our objectives were to design these filters for zero ISI and maximum SNR at the sampling instants. This design methodology is appropriate when the channel is precisely known and its characteristics do not change with time.

In practice we often encounter channels whose frequency response characteristics are either unknown or change with time. For example, in data transmission over the dial-up telephone network, the communication channel will be different every time we dial a number because the channel route will be different. Once a connection is made, however, the channel will be time-invariant for a relatively long period of time. This is an example of a channel whose characteristics are unknown a priori. Examples of time-varying channels are radio channels, such as ionospheric propagation channels. These channels are characterized by time-varying frequency response characteristics. These types of channels are examples where the optimization of the transmitting and receiving filters, as described in the preceding section, is not possible.

Under these circumstances, we may design the transmitting filter to have a square-root raised cosine frequency response, i.e.,

$$G_T(f) = \begin{cases} \sqrt{X_{rc}(f)}e^{-j2\pi f t_0}, & |f| \leq W \\ 0, & |f| > W \end{cases}$$

and the receiving filter, with frequency response $G_R(f)$, to be matched to $G_T(f)$. Therefore,

$$|G_T(f)||G_R(f)| = X_{rc}(f) \tag{8.4.28}$$

Then, due to channel distortion, the output of the receiving filter is

$$y(t) = \sum_{n=-\infty}^{+\infty} a_n x(t - nT) + v(t) \tag{8.4.29}$$

where $x(t) = g_T(t) \star c(t) \star g_R(t)$. The filter output may be sampled periodically to produce the sequence

$$y_m = \sum_{n=-\infty}^{+\infty} a_n x_{m-n} + v_m$$

$$= x_0 a_m + \sum_{n \neq m} a_n x_{m-n} + v_m \tag{8.4.30}$$

where $x_n = x(nT)$, $n = 0, \pm 1, \pm 2, \ldots$. The middle term on the right-hand side of (8.4.30) represents the ISI.

In any practical system it is reasonable to assume that the ISI affects a finite number of symbols. Hence, we may assume that $x_n = 0$ for $n < -L_1$ and $n > L_2$, where L_1 and L_2 are finite, positive integers. Consequently, the ISI observed at the output of the receiving filter may be viewed as being generated by passing the data sequence $\{a_m\}$ through an FIR filter with coefficients $\{x_n, -L_1 \leq n \leq L_2\}$, as shown in Figure 8.19. This filter is called the *equivalent discrete-time channel filter*. Since its input is the discrete information sequence (binary or M-ary), the output of the discrete-time channel filter may be characterized as the output of a finite-state machine corrupted by additive Gaussian noise. Hence, the noise-free output of the filter is described by a trellis having M^L states where $L = L_1 + L_2$.

Maximum likelihood sequence detection.   The optimum detector for the information sequence $\{a_m\}$ based on the observation of the received sequence $\{y_m\}$, given by (8.4.30), is a ML sequence detector. The detector is akin to the ML sequence detector described in the context of detecting partial response signals that

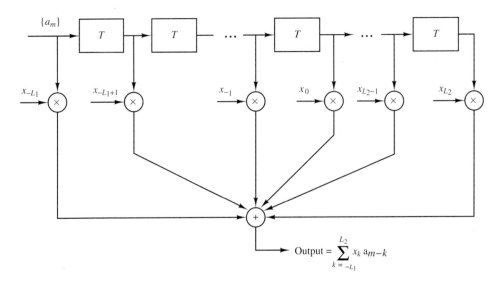

FIGURE 8.19. Equivalent discrete-time channel filter.

have controlled ISI. The Viterbi algorithm provides a method for searching through the trellis for the ML signal path. To accomplish this search, the equivalent channel filter coefficients $\{x_n\}$ must be known or measured by some method. At each stage of the trellis search, there are M^L surviving sequences with M^L corresponding Euclidean distance path metrics.

Due to the exponential increase in the computational complexity of the Viterbi algorithm with the span (length L) of the ISI, this type of detection is practical only when M and L are small. For example, in mobile cellular telephone systems, which employ digital transmission of speech signals, M is usually selected to be small, e.g., $M = 4$, and $L = 2$. In this case, the ML sequence detector may be implemented with reasonable complexity. However, when M and L are large, the ML sequence detector becomes impractical. In such a case other more practical but suboptimum methods are used to detect the information sequence $\{a_m\}$ in the presence of ISI. Nevertheless, the performance of the ML sequence detector for a channel with ISI serves as a benchmark for comparing its performance with that of suboptimum methods. Two suboptimum methods are described below.

Linear equalizers.   To compensate for the channel distortion, we may employ a linear filter with adjustable parameters. The filter parameters are adjusted on the basis of measurements of the channel characteristics. These adjustable filters are called *channel equalizers* or, simply, *equalizers.*

On channels whose frequency response characteristics are unknown, but time-invariant, we may measure the channel characteristics, adjust the parameters of the equalizer, and once adjusted, the parameters remain fixed during the transmission of data. Such equalizers are called *preset equalizers.* On the other hand, *adaptive equalizers* update their parameters on a periodic basis during the transmission of data.

First, we consider the design characteristics for a linear equalizer from a frequency domain viewpoint. Figure 8.20 shows a block diagram of a system that employs a linear filter as a channel equalizer.

The demodulator consists of a receiving filter with frequency response $G_R(f)$ in cascade with a channel equalizing filter that has a frequency response $G_E(f)$. Since $G_R(f)$ is matched to $G_T(f)$ and they are designed so that their product satisfies (8.4.28), $|G_E(f)|$ must compensate for the channel distortion. Hence, the

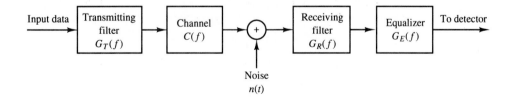

FIGURE 8.20.   Block diagram of a system with an equalizer.

equalizer frequency response must equal the inverse of the channel response, i.e.,

$$G_E(f) = \frac{1}{C(f)} = \frac{1}{|C(f)|} e^{-j\Theta_c(f)} \qquad |f| \le W \qquad (8.4.31)$$

where $|G_E(f)| = 1/|C(f)|$ and the equalizer phase characteristic $\Theta_E(f) = -\Theta_c(f)$. In this case, the equalizer is said to be the *inverse channel filter* to the channel response.

We note that the inverse channel filter completely eliminates ISI caused by the channel. Since it forces the ISI to be zero at the sampling times $t = nT$, the equalizer is called a *zero-forcing equalizer*. Hence, the input to the detector is of the form

$$y_m = a_m + v_m$$

where v_m is the noise component, which is zero-mean Gaussian with a variance

$$\sigma_v^2 = \int_{-\infty}^{\infty} S_n(f)|G_R(f)|^2 |G_E(f)|^2 df$$

$$= \int_{-W}^{W} \frac{S_n(f)|X_{rc}(f)|}{|C(f)|^2} df \qquad (8.4.32)$$

where $S_n(f)$ is the power-spectral density of the noise. When the noise is white, $S_n(f) = N_0/2$, and the variance becomes

$$\sigma_v^2 = \frac{N_0}{2} \int_{-W}^{W} \frac{|X_{rc}(f)|}{|C(f)|^2} df \qquad (8.4.33)$$

Note that the noise variance at the output of the zero-forcing equalizer is, in general, higher than the noise variance at the output of the optimum receiving filter $|G_R(f)|$ given by (8.4.7) for the case in which the channel is known.

Example 8.4.2

The channel given in Example 8.4.1 is equalized by a zero-forcing equalizer. Assuming that the transmitting and receiving filters satisfy (8.4.28), determine the value of the noise variance at the sampling instants and the probability of error.

Solution   When the noise is white, the variance of the noise at the output of the zero-forcing equalizer (input to the detector) is given by (8.4.33). Hence,

$$\sigma_v^2 = \frac{N_0}{2} \int_{-W}^{W} \frac{|X_{rc}(f)|}{|C(f)|^2} df$$

$$= \frac{T N_0}{2} \int_{-W}^{W} \left[1 + \left(\frac{f}{W}\right)^2\right] \cos^2 \frac{\pi|f|}{2W} df$$

$$= N_0 \int_0^1 (1 + x^2) \cos^2 \frac{\pi x}{2} dx$$

$$= \left(\frac{2}{3} - \frac{1}{\pi^2}\right) N_0$$

From (8.4.24) we obtain the average transmitted power as

$$P_{av} = \frac{(M^2 - 1)d^2}{3T} \int_{-W}^{W} |G_T(f)|^2 \, df$$

$$= \frac{(M^2 - 1)d^2}{3T} \int_{-W}^{W} |X_{rc}(f)| \, df$$

$$= \frac{(M^2 - 1)d^2}{3T}$$

The general expression for the probability of error is given by (8.4.25). Therefore, we have

$$P_M = \frac{2(M - 1)}{M} Q \left(\sqrt{\frac{3P_{av}T}{(M^2 - 1)(2/3 - 1/\pi^2)N_0}} \right)$$

If the channel were ideal, the argument of the Q-function would be $6P_{av}T/(M^2-1)N_0$. Hence, the loss in performance due to the equalized nonideal channel is given by the factor $2(2/3 - \frac{1}{\pi^2}) = 1.133$ or 0.54 dB.

Let us now consider the design of a linear equalizer from a time-domain viewpoint. We noted previously that in real channels, the ISI is limited to a finite number of samples, say L samples. As a consequence, in practice the channel equalizer is approximated by a finite duration impulse response (FIR) filter, or transversal filter, with adjustable tap coefficients $\{c_n\}$, as illustrated in Figure 8.21. The time delay τ between adjacent taps may be selected as large as T, the symbol interval, in which case the FIR equalizer is called a *symbol-spaced*

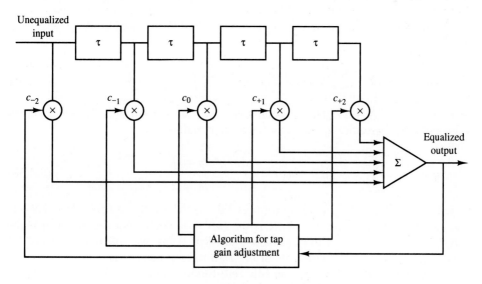

FIGURE 8.21.   Linear transversal filter.

equalizer. In this case the input to the equalizer is the sampled sequence given by (8.4.30). However, we note that when $1/T < 2W$, frequencies in the received signal above the folding frequency $1/T$ are aliased into frequencies below $1/T$. In this case, the equalizer compensates for the aliased channel-distorted signal.

On the other hand, when the time delay τ between adjacent taps is selected such that $1/\tau \geq 2W > 1/T$, no aliasing occurs and, hence, the inverse channel equalizer compensates for the true channel distortion. Since $\tau < T$ the channel equalizer is said to have *fractionally spaced taps* and it is called a *fractionally spaced equalizer.* In practice, τ is often selected as $\tau = T/2$. Notice that, in this case, the sampling rate at the output of the filter $G_R(f)$ is $\frac{2}{T}$.

The impulse response of the FIR equalizer is

$$g_E(t) = \sum_{n=-N}^{N} c_n \delta(t - n\tau) \tag{8.4.34}$$

and the corresponding frequency response is

$$G_E(f) = \sum_{n=-N}^{N} c_n e^{-j2\pi f n\tau} \tag{8.4.35}$$

where $\{c_n\}$ are the $(2N + 1)$ equalizer coefficients, and N is chosen sufficiently large so that the equalizer spans the length of the ISI, i.e., $2N + 1 \geq L$. Since $X(f) = G_T(f)C(f)G_R(f)$ and $x(t)$ is the signal pulse corresponding to $X(f)$, then the equalized output signal pulse is

$$q(t) = \sum_{n=-N}^{N} c_n x(t - n\tau) \tag{8.4.36}$$

The zero-forcing condition can now be applied to the samples of $q(t)$ taken at times $t = mT$. These samples are

$$q(mT) = \sum_{n=-N}^{N} c_n x(mT - n\tau), \quad m = 0, \pm 1, \ldots, \pm N \tag{8.4.37}$$

Since there are $2N + 1$ equalizer coefficients, we can control only $2N + 1$ sampled values of $q(t)$. Specifically, we may force the conditions

$$q(mT) = \sum_{n=-N}^{N} c_n x(mT - n\tau) = \begin{cases} 1, & m = 0 \\ 0, & m = \pm 1, \pm 2, \ldots, \pm N \end{cases} \tag{8.4.38}$$

which may be expressed in matrix form as $\mathbf{Xc} = \mathbf{q}$, where $\mathbf{X}$ is a $(2N+1) \times (2N+1)$ matrix with elements $x(mT - n\tau)$, $\mathbf{c}$ is the $(2N + 1)$ coefficient vector and $\mathbf{q}$ is the $(2N + 1)$ column vector with one nonzero element. Thus, we obtain a set of $2N + 1$ linear equations for the coefficients of the zero-forcing equalizer.

We should emphasize that the FIR zero-forcing equalizer does not completely eliminate ISI because it has a finite length. However, as N is increased, the residual ISI can be reduced and in the limit as $N \to \infty$, the ISI is completely eliminated.

Example 8.4.3

Consider a channel distorted pulse $x(t)$, at the input to the equalizer, given by the expression

$$x(t) = \frac{1}{1 + \left(\frac{2t}{T}\right)^2}$$

where $1/T$ is the symbol rate. The pulse is sampled at the rate $2/T$ and equalized by a zero-forcing equalizer. Determine the coefficients of a five-tap zero-forcing equalizer.

Solution   According to (8.4.38), the zero-forcing equalizer must satisfy the equations

$$q(mT) = \sum_{n=-2}^{2} c_n x(mT - nT/2) = \begin{cases} 1, & m = 0 \\ 0, & m = \pm 1, \pm 2 \end{cases}$$

The matrix $\mathbf{X}$ with elements $x(mT - nT/2)$ is given as

$$\mathbf{X} = \begin{bmatrix} \frac{1}{5} & \frac{1}{10} & \frac{1}{17} & \frac{1}{26} & \frac{1}{37} \\ 1 & \frac{1}{2} & \frac{1}{5} & \frac{1}{10} & \frac{1}{17} \\ \frac{1}{5} & \frac{1}{2} & 1 & \frac{1}{2} & \frac{1}{5} \\ \frac{1}{17} & \frac{1}{10} & \frac{1}{5} & \frac{1}{2} & 1 \\ \frac{1}{37} & \frac{1}{26} & \frac{1}{17} & \frac{1}{10} & \frac{1}{5} \end{bmatrix} \quad (8.4.39)$$

The coefficient vector $\mathbf{C}$ and the vector $\mathbf{q}$ are given as

$$\mathbf{c} = \begin{bmatrix} c_{-2} \\ c_{-1} \\ c_0 \\ c_1 \\ c_2 \end{bmatrix} \quad \mathbf{q} = \begin{bmatrix} 0 \\ 0 \\ 1 \\ 0 \\ 0 \end{bmatrix} \quad (8.4.40)$$

Then, the linear equations $\mathbf{Xc} = \mathbf{q}$ can be solved by inverting the matrix $\mathbf{X}$. Thus, we obtain

$$\mathbf{c}_{\text{opt}} = \mathbf{X}^{-1}\mathbf{q} = \begin{bmatrix} -2.2 \\ 4.9 \\ -3 \\ 4.9 \\ -2.2 \end{bmatrix} \quad (8.4.41)$$

One drawback to the zero-forcing equalizer is that it ignores the presence of additive noise. As a consequence, its use may result in significant noise enhancement. This is easily seen by noting that in a frequency range where $C(f)$ is small, the channel equalizer $G_E(f) = 1/C(f)$ compensates by placing a large gain in that frequency range. Consequently, the noise in that frequency range is greatly enhanced. An alternative is to relax the zero-ISI condition and select the channel equalizer characteristic such that the combined power in the residual ISI and the additive noise at the output of the equalizer is minimized. A channel equalizer that is optimized based on the minimum mean-square error (MMSE) criterion accomplishes the desired goal.

To elaborate, let us consider the noise corrupted output of the FIR equalizer, which is

$$z(t) = \sum_{n=-N}^{N} c_n y(t - n\tau) \tag{8.4.42}$$

where $y(t)$ is the input to the equalizer, given by (8.4.29). The output is sampled at times $t = mT$. Thus, we obtain

$$z(mT) = \sum_{n=-N}^{N} c_n y(mT - n\tau) \tag{8.4.43}$$

The desired response samples at the output of the equalizer at $t = mT$ is the transmitted symbol a_m. The error is defined as the difference between a_m and $z(mT)$. Then, the mean-square error (MSE) between the actual output sample $z(mT)$ and the desired values a_m is

$$\text{MSE} = E[z(mT) - a_m]^2$$

$$= E\left[\sum_{n=-N}^{N} c_n y(mT - n\tau) - a_m \right]^2$$

$$= \sum_{n=-N}^{N} \sum_{k=-N}^{N} c_n c_k R_Y(n - k) -$$

$$2 \sum_{k=-N}^{N} c_k R_{AY}(k) + E[a_m^2] \tag{8.4.44}$$

where the correlations are defined as

$$R_Y(n - k) = E[y(mT - n\tau)y(mT - k\tau)]$$
$$R_{AY}(k) = E[y(mT - k\tau)a_m] \tag{8.4.45}$$

and the expectation is taken with respect to the random information sequence $\{a_m\}$ and the additive noise.

The MMSE solution is obtained by differentiating (8.4.44) with respect to the equalizer coefficients $\{c_n\}$. Thus, we obtain the necessary conditions for the MMSE as

$$\sum_{n=-N}^{N} c_n R_Y(n - k) = R_{AY}(k), \quad k = 0, \pm 1, 2, \ldots, \pm N \tag{8.4.46}$$

These are $(2N + 1)$ linear equations for the equalizer coefficients. In contrast to the zero-forcing solution described previously, these equations depend on the statistical properties (the autocorrelation) of the noise as well as the ISI through the autocorrelation $R_Y(n)$.

In practice, we would not normally know the autocorrelation $R_Y(n)$ and the crosscorrelation $R_{AY}(n)$. However, these correlation sequences can be estimated by transmitting a test signal over the channel and using the time-average estimates

$$\hat{R}_Y(n) = \frac{1}{K} \sum_{k=1}^{K} y(kT - n\tau)y(kT)$$

$$\hat{R}_{AY}(n) = \frac{1}{K} \sum_{k=1}^{K} y(kT - n\tau)a_k \qquad (8.4.47)$$

in place of the ensemble averages to solve for the equalizer coefficients given by (8.4.46).

Adaptive equalizers. We have shown that the tap coefficients of a linear equalizer can be determined by solving a set of linear equations. In the zero-forcing optimization criterion, the linear equations are given by (8.4.38). On the other hand, if the optimization criterion is based on minimizing the MSE, the optimum equalizer coefficients are determined by solving the set of linear equations given by (8.4.46).

In both cases, we may express the set of linear equations in the general matrix form

$$\mathbf{Bc} = \mathbf{d} \qquad (8.4.48)$$

where $\mathbf{B}$ is a $(2N + 1) \times (2N + 1)$ matrix, $\mathbf{c}$ is a column vector representing the $2N + 1$ equalizer coefficients, and $\mathbf{d}$ is a $(2N + 1)$ dimensional column vector. The solution of (8.4.48) yields

$$\mathbf{c}_{\text{opt}} = \mathbf{B}^{-1}\mathbf{d} \qquad (8.4.49)$$

In practical implementations of equalizers, the solution of (8.4.48) for the optimum coefficient vector is usually obtained by an iterative procedure that avoids the explicit computation of the inverse of the matrix $\mathbf{B}$. The simplest iterative procedure is the *method of steepest descent,* in which one begins by choosing arbitrarily the coefficient vector $\mathbf{c}$, say $\mathbf{c}_0$. This initial choice of coefficients corresponds to a point on the criterion function that is being optimized. For example, in the case of the MSE criterion, the initial guess $\mathbf{c}_0$ corresponds to a point on the quadratic MSE surface in the $(2N + 1)$ dimensional space of coefficients. The gradient vector, defined as $\mathbf{g}_0$, which is the derivative of the MSE with respect to the $2N + 1$ filter coefficients, is then computed at this point on the criterion surface and each tap coefficient is changed in the direction opposite to its corresponding gradient component. The change in the jth tap coefficient is proportional to the size of the jth gradient component.

For example, the gradient vector, denoted as $\mathbf{g}_k$, for the MSE criterion, found by taking the derivatives of the MSE with respect to each of the $2N + 1$ coefficients, is

$$\mathbf{g}_k = \mathbf{Bc}_k - \mathbf{d}, \quad k = 0, 1, 2, \ldots \tag{8.4.50}$$

Then the coefficient vector $\mathbf{c}_k$ is updated according to the relation

$$\mathbf{c}_{k+1} = \mathbf{c}_k - \Delta \mathbf{g}_k \tag{8.4.51}$$

where Δ is the *step-size parameter* for the iterative procedure. To ensure convergence of the iterative procedure, Δ is chosen to be a small, positive number. In such a case, the gradient vector $\mathbf{g}_k$ converges toward zero, i.e., $\mathbf{g}_k \to \mathbf{0}$ as $k \to \infty$, and the coefficient vector $\mathbf{c}_k \to \mathbf{c}_{\text{opt}}$ as illustrated in Figure 8.22 based on two-dimensional optimization. In general, convergence of the equalizer tap coefficients to $\mathbf{c}_{\text{opt}}$ cannot be attained in a finite number of iterations with the steepest-descent method. However, the optimum solution $\mathbf{c}_{\text{opt}}$ can be approached as closely as desired in a few hundred iterations. In digital communication systems that employ channel equalizers, each iteration corresponds to a time interval for sending one symbol and, hence, a few hundred iterations to achieve convergence to $\mathbf{c}_{\text{opt}}$ corresponds to a fraction of a second.

Adaptive channel equalization is required for channels whose characteristics change with time. In such a case, the ISI varies with time. The channel equalizer must track such time variations in the channel response and adapt its coefficients to reduce the ISI. In the context of the above discussion, the optimum coefficient vector $\mathbf{c}_{\text{opt}}$ varies with time due to time variations in the matrix $\mathbf{B}$ and, for the case of the MSE criterion, time variations in the vector $\mathbf{d}$. Under these conditions, the iterative method described above can be modified to use estimates of the gradient components. Thus, the algorithm for adjusting the equalizer tap coefficients may be expressed as

$$\hat{\mathbf{c}}_{k+1} = \hat{\mathbf{c}}_k - \Delta \hat{\mathbf{g}}_k \tag{8.4.52}$$

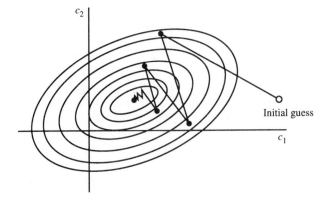

FIGURE 8.22. Example of convergence characteristics of a gradient algorithm. [From *Introduction to Adaptive Arrays*, by R.A. Monzigo and T.W. Miller; © 1980 by John Wiley & Sons, Inc. Reprented with permission of the publisher.]

where $\hat{\mathbf{g}}_k$ denotes an estimate of the gradient vector $\mathbf{g}_k$, and $\hat{\mathbf{c}}_k$ denotes the estimate of the tap coefficient vector.

In the case of the MSE criterion, the gradient vector $\mathbf{g}_k$ given by (8.4.50) may also be expressed as (see Problem 8.31).

$$\mathbf{g}_k = -E\left[e_k \mathbf{y}_k\right]$$

An estimate $\hat{\mathbf{g}}_k$ of the gradient vector at the kth iteration is computed as

$$\hat{\mathbf{g}}_k = -e_k \mathbf{y}_k \tag{8.4.53}$$

where e_k denotes the difference between the desired output from the equalizer at the kth time instant and the actual output $z(kT)$, and $\mathbf{y}_k$ denotes the column vector of $2N + 1$ received signal values contained in the equalizer at time instant k. The *error signal* e_k is expressed as

$$e_k = a_k - z_k \tag{8.4.54}$$

where $z_k = z(kT)$ is the equalizer output given by (8.4.42), and a_k is the desired symbol. Hence, by substituting (8.4.53) into (8.4.52), we obtain the adaptive algorithm for optimizing the tap coefficients (based on the MSE criterion) as

$$\hat{\mathbf{c}}_{k+1} = \hat{\mathbf{c}}_k + \Delta e_k \mathbf{y}_k \tag{8.4.55}$$

Since an estimate of the gradient vector is used in (8.4.55), the algorithm is called a *stochastic gradient algorithm*. It is also known as the *LMS algorithm*.

A block diagram of an adaptive equalizer that adapts its tap coefficients according to (8.4.55) is illustrated in Figure 8.23. Note that the difference between the desired output a_k and the actual output z_k from the equalizer is used to form the error signal e_k. This error is scaled by the step-size parameter Δ, and the scaled-error signal Δe_k multiplies the received signal values $\{y(kT - n\tau)\}$ at the $2N + 1$ taps. The products $\Delta e_k y(kT - n\tau)$ at the $(2N + 1)$ taps are then added to the previous values of the tap coefficients to obtain the updated tap coefficients, according to (8.4.55). This computation is repeated as each new signal sample is received. Thus, the equalizer coefficients are updated at the symbol rate.

Initially, the adaptive equalizer is trained by the transmission of a known pseudorandom sequence $\{a_m\}$ over the channel. At the demodulator, the equalizer employs the known sequence to adjust its coefficients. Upon initial adjustment, the adaptive equalizer switches from a *training mode* to a *decision-directed mode,* in which case the decisions at the output of the detector are sufficiently reliable so that the error signal is formed by computing the difference between the detector output and the equalizer output, i.e.,

$$e_k = \tilde{a}_k - z_k \tag{8.4.56}$$

where $\tilde{a}_k$ is the output of the detector. In general, decision errors at the output of the detector occur infrequently and, consequently, such errors have little effect on the performance of the tracking algorithm given by (8.4.55).

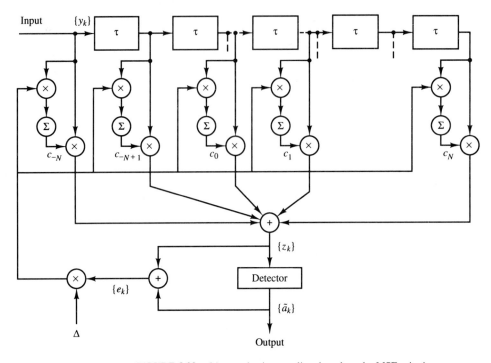

FIGURE 8.23. Linear adaptive equalizer based on the MSE criterion.

A rule of thumb for selecting the step-size parameter so as to ensure convergence and good tracking capabilities in slowly varying channels is

$$\Delta = \frac{1}{5(2N+1)P_R} \qquad (8.4.57)$$

where P_R denotes the received signal-plus-noise power, which can be estimated from the received signal.

The convergence characteristics of the stochastic gradient algorithm in (8.4.55) is illustrated in Figure 8.24. These graphs were obtained from a computer simulation of an 11-tap adaptive equalizer operating a channel with a rather modest amount of ISI. The input signal-plus-noise power P_R was normalized to unity. The rule of thumb given in (8.4.57) for selecting the step size gives $\Delta = 0.018$. The effect of making Δ too large is illustrated by the large jumps in MSE as shown for $\Delta = 0.115$. As Δ is decreased, the convergence is slowed somewhat, but a lower MSE is achieved, indicating that the estimated coefficients are closer to $\mathbf{c}_{opt}$.

Although we have described in some detail the operation of an adaptive equalizer which is optimized on the basis of the MSE criterion, the operation of an adaptive equalizer based on the zero-forcing method is very similar. The major difference lies in the method for estimating the gradient vectors $\mathbf{g}_k$ at each iteration. A block diagram of an adaptive zero-forcing equalizer is shown in Figure 8.25. For

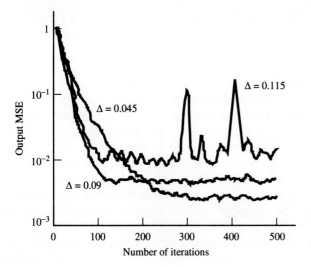

FIGURE 8.24.   Initial convergence characteristics of the LMS algorithm with different step sizes. (From *Digital Signal Processing* by J. G. Proakis and D. G. Manolakis, ©1988, Macmillan Publishing Company. Reprinted with permission of the publisher.)

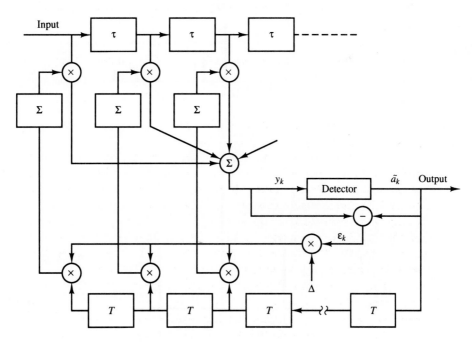

FIGURE 8.25.   An adaptive zero-forcing equalizer.

more details on the tap coefficient update method for a zero-forcing equalizer, the reader is referred to the papers by Lucky (1965, 1966), and the texts by Lucky, Salz, and Weldon (1968) and Proakis (1989).

Decision-feedback equalizer.   The linear filter equalizers described above are very effective on channels, such as wireline telephone channels, where the ISI is not severe. The severity of the ISI is directly related to the spectral characteristics and not necessarily to the time span of the ISI. For example, consider the ISI resulting from two channels, which are illustrated in Figure 8.26. The time span for the ISI in Channel A is 5 symbol intervals on each side of the desired signal component, which has a value of 0.72. On the other hand, the time span for the ISI in Channel B is one symbol interval on each side of the desired signal component, which has a value of 0.815. The energy of the total response is normalized to unity for both channels.

In spite of the shorter ISI span, Channel B results in more severe ISI. This is evidenced in the frequency response characteristics of these channels, which are shown in Figure 8.27. We observe that Channel B has a spectral null (the frequency response $C(f) = 0$ for some frequencies in the band $|f| \leq W$) at $f = 1/2T$, whereas this does not occur in the case of Channel A. Consequently, a linear equalizer will introduce a large gain in its frequency response to compensate for

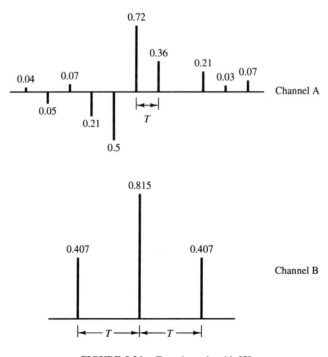

FIGURE 8.26.   Two channels with ISI.

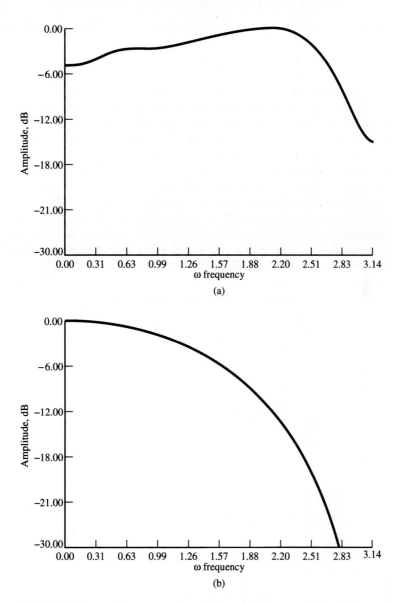

FIGURE 8.27.   Amplitude spectra for (a) channel A shown in Figure 8.26(a) and (b) channel B shown in Figure 8.26(b).

the channel null. Thus, the noise in Channel B will be enhanced much more than in Channel A. This implies that the performance of the linear equalizer for Channel B will be sufficiently poorer than that for Channel A. This fact is borne out by the computer simulation results for the performance of that two linear equalizers shown in Figure 8.28. Hence, the basic limitation of a linear equalizer is that it performs poorly on channels having spectral nulls. Such channels are often encountered in radio communications, such as ionospheric transmission at frequencies below 30 MHz and mobile radio channels, such as those used for cellular radio communications.

A *decision-feedback equalizer* (DFE) is a nonlinear equalizer that employs previous decisions to eliminate the ISI caused by previously detected symbols on the current symbol to be detected. A simple block diagram for a DFE is shown in Figure 8.29. The DFE consists of two filters. The first filter is called a *feedforward filter,* and it is generally a fractionally spaced FIR filter with adjustable tap coefficients. This filter is identical in form to the linear equalizer described above. Its input is the received filtered signal $y(t)$. The second filter is a *feedback filter.* It is implemented as an FIR filter with symbol-spaced taps having adjustable coefficients. Its input is the set of previously detected symbols. The output of the feedback filter is subtracted from the output of the feedforward filter to form the

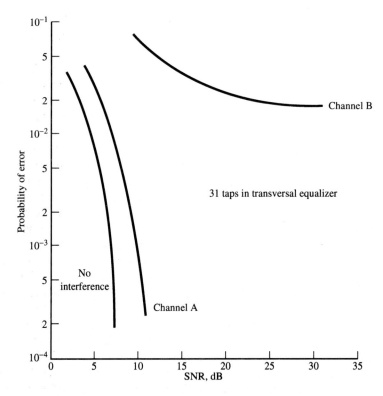

FIGURE 8.28.   Error rate performance of linear MSE equalizer.

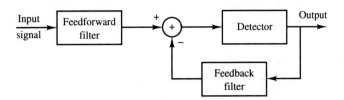

FIGURE 8.29.   Block diagram of DFE.

input to the detector. Thus, we have

$$z_m = \sum_{n=1}^{N_1} c_n y(mT - n\tau) - \sum_{n=1}^{N_2} b_n \tilde{a}_{m-n} \qquad (8.4.58)$$

where $\{c_n\}$ and $\{b_n\}$ are the adjustable coefficients of the feedforward and feedback filters, respectively, $\tilde{a}_{m-n}, n = 1, 2, \ldots, N_2$ are the previously detected symbols, N_1 is the length of the feedforward filter, and N_2 is the length of the feedback filter. Based on the input z_m, the detector determines which of the possible transmitted symbols is closest in distance to the input signal z_m. Thus, it makes its decision and outputs $\tilde{a}_m$. What makes the DFE nonlinear is the nonlinear characteristic of the detector, which provides the input to the feedback filter.

The tap coefficients of the feedforward and feedback filters are selected to optimize some desired performance measure. For mathematical simplicity, the MSE criterion is usually applied and a stochastic gradient algorithm is commonly used to implement an adaptive DFE. Figure 8.30 illustrates the block diagram of an adaptive DFE whose tap coefficients are adjusted by means of the LMS stochastic gradient algorithm. Figure 8.31 illustrates the probability of error performance of the DFE, obtained by computer simulation, for binary PAM transmission over Channel B. The gain in performance relative to that of a linear equalizer is clearly evident.

We should mention that decision errors from the detector that are fed to the feedback filter have a small effect on the performance of the DFE. In general, a small loss in performance of 1–2 dB is possible at error rates below 10^{-2}, but the decision errors in the feedback filters are not catastrophic.

Although the DFE outperforms a linear equalizer, it is not the optimum equalizer from the viewpoint of minimizing the probability of error. As indicated previously, the optimum detector in a digital communication system in the presence of ISI is a ML symbol sequence detector. It is particularly appropriate for channels with severe ISI, when the ISI spans only a few signals. For example Figure 8.32 illustrates the error probability performance of the Viterbi algorithm for a binary PAM signal transmitted through Channel B (see Figure 8.26). For purposes of comparison, we also illustrate the probability of error for a DFE. Both results were obtained by computer simulation. We observe that the performance of the ML sequence detector is about 4.5 dB better than that of the DFE at an error probability

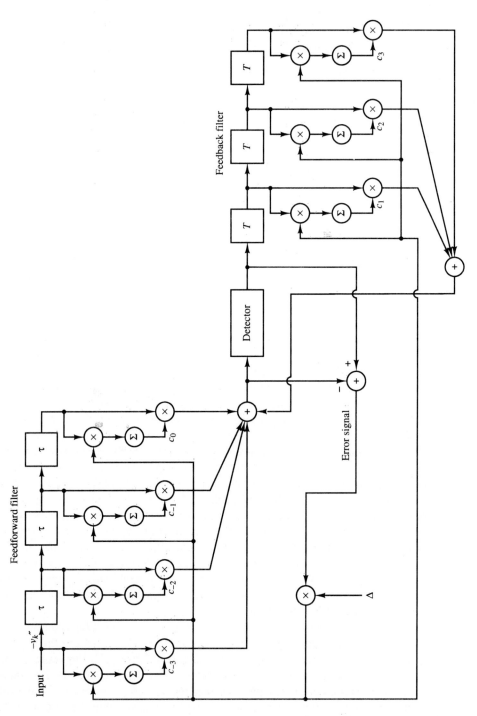

FIGURE 8.30.   Adaptive DFE.

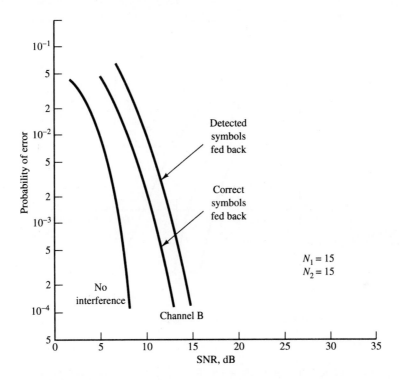

FIGURE 8.31.   Performance of DFE with and without error propagation.

of 10^{-4}. Hence, this is one example where the ML sequence detector provides a significant performance gain on a channel with a relatively short ISI span.

In conclusion, we mention that adaptive equalizers are used widely in high-speed digital communication systems for telephone channels. High-speed telephone line modems (at bit rates above 2400 bits/sec) generally include an adaptive equalizer that is implemented as an FIR filter with coefficients that are adjusted based on the minimum mean squared error criterion. Depending on the data speed, the equalizer typically spans between 20 and 70 symbols. The LMS algorithm given by (8.4.55) is usually employed for the adjustment of the equalizer coefficients adaptively.

8.5 SYMBOL SYNCHRONIZATION

In a digital communication system, the output of the receiving filter $y(t)$ must be sampled periodically at the symbol rate, at the precise sampling time instants $t_m = mT + \tau_0$, where T is the symbol interval and τ_0 is a nominal time delay that accounts for the propagation time of the signal from the transmitter to the receiver. To perform this periodic sampling, we require a clock signal at the receiver. The

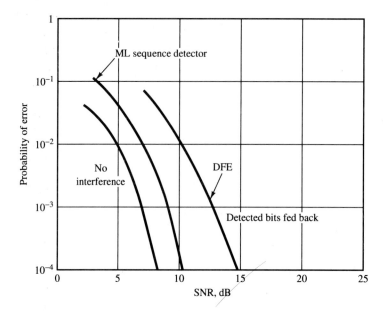

FIGURE 8.32.   Performance of Viterbi detector and DFE for channel B.

process of extracting such a clock signal at the receiver is usually called *symbol synchronization* or *timing recovery.*

Timing recovery is one of the most critical functions that is performed at the receiver of a synchronous digital communication system. We should note that the receiver must know not only the frequency $(1/T)$ at which the outputs of the matched filters or correlators are sampled, but also where to take the samples within each symbol interval. The choice of sampling instant within the symbol interval of duration T is called the *timing phase.*

Viewed in relation to the eye patterns shown in Figure 8.7, the best timing phase corresponds to the time instant within the symbol interval where the eye opening is a maximum. In a practical communication system, the receiver clock must be continuously adjusted in frequency $(1/T)$ and in timing phase τ_0 to compensate for frequency drifts between the oscillators used in the transmitter and receiver clocks and, thus, to optimize the sampling time instants of the matched filter or correlator outputs.

Symbol synchronization can be accomplished in one of several ways. In some communication systems the transmitter and receiver clocks are synchronized to a master clock, which provides a very precise timing signal. In this case, the receiver must estimate and compensate for the relative time delay between the transmitted and received signals. Such may be the case for radio communication systems that operate in the very low frequency (VLF) band (below 30 kHz), where precise clock signals are transmitted from a master radio station.

Another method for achieving symbol synchronization is for the transmitter to simultaneously transmit the clock frequency $1/T$ or a multiple of $1/T$ along with the information signal. The receiver may simply employ a narrowband filter tuned to the transmitted clock frequency and, thus, extract the clock signal for sampling. This approach has the advantage of being simple to implement. There are several disadvantages, however. One is that the transmitter must allocate some of its available power to the transmission of the clock signal. Another is that some small fraction of the available channel bandwidth must be allocated for the transmission of the clock signal. In spite of these disadvantages, this method is frequently used in telephone transmission systems that employ large bandwidths to transmit the signals of many users. In such a case, the transmission of a clock signal is shared in the demodulation of the signals among the many users. Through this shared use of the clock signal, the penalty in transmitter power and in bandwidth allocation is reduced proportionally by the number of users.

A clock signal can also be extracted from the received data signal. There are a number of different methods that can be used at the receiver to achieve self-synchronization. Below we consider four approaches to the problem of achieving symbol synchronization from the received signal. We assume that the received signal at the output of the receiving filter $g_R(t)$ has the general form

$$y(t) = \sum_{n=-\infty}^{\infty} a_n x(t - nT - \tau_0) + v(t) \qquad (8.5.1)$$

where $x(t) = g_T(t) \star c(t) \star g_R(t)$, $\{a_n\}$ is the sequence of information symbols (binary or M-ary), $v(t)$ represents the noise at the output of the receiving filter, and τ_0 represents the timing phase. We wish to derive a clock signal that has a frequency equal to the symbol rate $1/T$ and a timing phase τ_0. The data sequence $\{a_n\}$ is assumed to be a zero-mean, stationary sequence with statistically i.i.d. elements. Therefore, the resulting PAM signal

$$v(t) = \sum_{n=-\infty}^{\infty} a_n x(t - nT - \tau_0) \qquad (8.5.2)$$

is a zero-mean, cyclostationary random process.

8.5.1 Spectral-Line Methods

Since the signal component $v(t)$ in the received signal $y(t)$ is periodic with period T, we can recover a clock signal with frequency $1/T$ by filtering out a signal component at $f = 1/T$. We observe, however, that $E[v(t)] = 0$ because $E(a_n) = 0$. Therefore, $v(t)$ cannot be used directly to generate a frequency at $f = 1/T$. On the other hand, we may perform a nonlinear operation on $v(t)$ to generate power at $f = 1/T$ and its harmonics.

Let us consider a square-law nonlinearity. If we square the signal $v(t)$ and take the expected value with respect to the data sequence $\{a_n\}$, we obtain

$$E[v^2(t)] = E\left[\sum_n \sum_m a_n a_m x(t - mT - \tau_0) x(t - nT - \tau_0)\right]$$

$$= \sigma_a^2 \sum_{n=-\infty}^{\infty} x^2(t - nT - \tau_0) \qquad (8.5.3)$$

where $\sigma_a^2 = E\left[a_n^2\right]$. Since $E[v^2(t)] > 0$, we may use $v^2(t)$ to generate the desired frequency component.

Let us use the Poisson Sum Formula (see Problem 2.51) to express (8.5.3) in the form of a Fourier series. Hence,

$$E[v^2(t)] = \sigma_a^2 \sum_n x^2(t - nT - \tau_0)$$

$$= \frac{\sigma_a^2}{T} \sum_m c_m e^{j2\pi m(t-\tau_0)/T} \qquad (8.5.4)$$

where

$$c_m = \int_{-\infty}^{\infty} X(f) X\left(\frac{m}{T} - f\right) df \qquad (8.5.5)$$

By design, the transmitted signal spectrum is confined to frequencies below $1/T$. Hence, $X(f) = 0$ for $|f| > 1/T$ and, consequently, there are only three nonzero terms ($m = 0, \pm 1$) in (8.5.4). Therefore, the signal $v^2(t)$ contains a *dc* component and a component at the frequency $1/T$.

The above development suggests that we square the signal $y(t)$ at the output of the receiving filter and filter $y^2(t)$ with a narrowband filter $B(f)$ tuned to the symbol rate $1/T$. If we set the filter response $B(1/T) = 1$, then

$$E[v^2(t)] = \frac{\sigma_a^2}{T} \text{Re}\left[c_1 e^{j2\pi(t-\tau_0)/T}\right]$$

$$= \frac{\sigma_a^2}{T} c_1 \cos \frac{2\pi}{T}(t - \tau_0) \qquad (8.5.6)$$

so that the timing signal is a sinusoid with a phase of $-2\pi\tau_0/T$, assuming that $X(f)$ is real. We may use alternate zero crossings of the timing signal as an indication of the correct sampling times. However, the alternate zero crossings of the signal given by (8.5.6) occur at

$$\frac{2\pi}{T}(t - \tau_0) = (4k + 1)\frac{\pi}{2} \qquad (8.5.7)$$

or, equivalently, at

$$t = kT + \tau_0 + \frac{T}{4} \qquad (8.5.8)$$

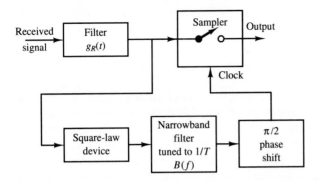

FIGURE 8.33. Symbol timing based on spectral-line method.

which is offset in time by $T/4$ relative to the desired zero crossings. In a practical system the timing offset can be easily compensated either by relatively simple clock circuitry or by designing the bandpass filter $B(f)$ to have a $\pi/2$ phase shift at $f = 1/T$. Figure 8.33 illustrates this method for generating a timing signal at the receiver.

The additive noise that corrupts the signal will generally cause fluctuations in the zero crossings of the desired signal. The effect of the fluctuations will depend on the amplitude c_1 of the mean timing sinusoidal signal given by (8.5.6). We note that the signal amplitude c_1 is proportional to the slope of the timing signal in the vicinity of the zero crossing as shown in Figure 8.34. Therefore, the larger the amplitude c_1, the larger will be the slope and, consequently, the timing errors due to the noise will be smaller. From (8.5.5) we observe that c_1 depends on the amount of spectral overlap of $X(f)$ and $X(1/T - f)$. Thus, c_1 depends on the amount by which the bandwidth of $X(f)$ exceeds the Nyquist bandwidth $1/2T$, i.e., c_1 depends on the excess bandwidth. If the excess bandwidth is zero, i.e., $X(f) = 0$, for $|f| > 1/2T$, then $c_1 = 0$, and this method fails to provide a timing signal. If the

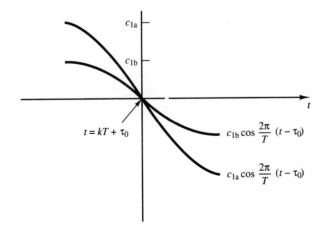

FIGURE 8.34. Illustration of the slope of the sinusoid at the zero crossing as a function of the amplitude.

excess bandwidth is large, say 50% to 100%, the timing signal amplitude will be sufficiently large to yield relatively accurate symbol timing estimates.

8.5.2 Early–Late Gate Synchronizers

There is another method for generating a symbol timing signal at the receiver that exploits the symmetry properties of the signal at the output of the matched filter or correlator. To describe this method, let us consider the rectangular pulse $s(t), 0 \leq t \leq T$, shown in Figure 8.35(a). The output of the filter matched to $s(t)$ attains its maximum value at time $t = T$, as shown in Figure 8.35(b). Thus, the output of the matched filter is the time autocorrelation function of the pulse $s(t)$. Of course, this statement holds for any arbitrary pulse shape, so the approach that we describe applies in general to any signal pulse. Clearly, the proper time to sample the output of the matched filter for a maximum output is at $t = T$, i.e., at the peak of the correlation function.

In the presence of noise, the identification of the peak value of the signal is generally difficult. Instead of sampling the signal at the peak, suppose we sample early, at $t = T - \delta T$ and late at $t = T + \delta T$. The absolute values of the early samples $|y[m(t - \delta T)]|$ and the late samples $|y[m(T + \delta T)]|$ will be smaller (on the average in the presence of noise) than the samples of the peak value $|y(mT)|$. Since the autocorrelation function is even with respect to the optimum sampling time $t = T$, the absolute values of the correlation function at $t = T - \delta T$ and $t = T + \delta T$ are equal. Under this condition, the proper sampling time is the midpoint between

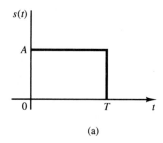

(a)

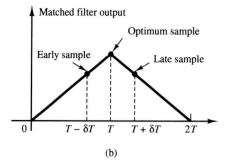

(b)

FIGURE 8.35.   (a) Rectangular signal pulse and (b) its matched filter output.

$t = T - \delta T$ and $t = T + \delta T$. This condition forms the basis for the *early–late gate symbol synchronizer*.

Figure 8.36 illustrates the block diagram of an early–late gate synchronizer. In this figure, correlators are used in place of the equivalent matched filters. The two correlators integrate over the symbol interval T, but one correlator starts integrating δT early relative to the estimated optimum sampling time and the other integrator starts integrating δT late relative to the estimated optimum sampling time. An error signal is formed by taking the difference between the absolute values of the two correlator outputs. To smooth the noise corrupting the signal samples, the error signal is passed through a lowpass filter. If the timing is off relative to the optimum sampling time, the average error signal at the output of the lowpass filter is nonzero, and the clock signal is either retarded or advanced, depending on the sign of the error. Thus, the smoothed error signal is used to drive a voltage-controlled oscillator (VCO), whose output is the desired clock signal that is used for sampling. The output of the VCO is also used as a clock signal for a symbol waveform generator that puts out the same basic pulse waveform as that of the transmitting filter. This pulse waveform is advanced and delayed and then fed to the two correlators, as

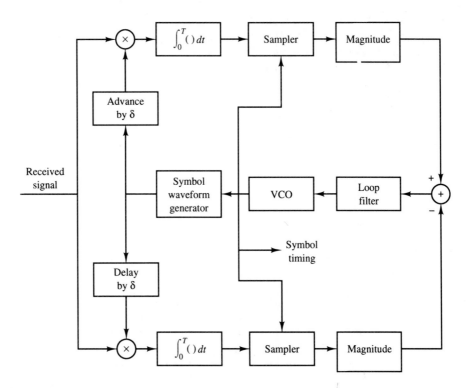

FIGURE 8.36.   Block diagram of early–late gate synchronizer.

shown in Figure 8.36. Note that if the signal pulses are rectangular, there is no need for a signal pulse generator within the tracking loop.

We observe that the early–late gate synchronizer is basically a closed-loop control system whose bandwidth is relatively narrow compared to the symbol rate $1/T$. The bandwidth of the loop determines the quality of the timing estimate. A narrowband loop provides more averaging over the additive noise and, thus, improves the quality of the estimated sampling instants, provided that the channel propagation delay is constant and the clock oscillator at the transmitter is not drifting with time (or drifting very slowly with time). On the other hand, if the channel propagation delay is changing with time and/or the transmitter clock is also drifting with time, then the bandwidth of the loop must be increased to provide for faster tracking of time variations in symbol timing.

In the tracking mode, the two corrrelators are affected by adjacent symbols. However, if the sequence of information symbols has zero mean, as is the case for PAM and some other signal modulations, the contribution to the output of the corrrelators from adjacent symbols averages out to zero in the lowpass filter.

An equivalent realization of the early–late gate synchronizer that is somewhat easier to implement is shown in Figure 8.37. In this case the clock from the VCO

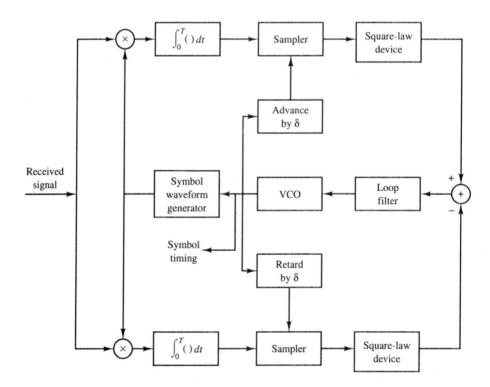

FIGURE 8.37. Block diagram of early–late gate synchronizer—an alternative form.

is advanced and delayed by δT, and these clock signals are used to sample the outputs of the two correlators.

8.5.3 Minimum Mean-Square-Error Method

Another approach to the problem of timing recovery from the received signal is based on the minimization of the MSE between the samples at the output of the receiving filter and the desired symbols. In practice, the MSE

$$\text{MSE} = E\{[y_m(\tau_0) - a_m]^2\} \tag{8.5.9}$$

is approximated by the time-average-square error

$$\langle \text{MSE} \rangle = \sum_m \left[y_m(\tau_0) - \hat{a}_m \right]^2 \tag{8.5.10}$$

where

$$y_m(\tau_0) = \sum_n a_n x(mT - nT - \tau_0) + v(mT) \tag{8.5.11}$$

and $\hat{a}_m$ is the output symbol from the detector.

The minimum of $\langle \text{MSE} \rangle$ with respect to the timing phase τ_0 is found by differentiating (8.5.10) with respect to τ_0. Thus, we obtain the necessary condition

$$\sum_m \left[y_m(\tau_0) - \hat{a}_m \right] \frac{dy_m(\tau_0)}{d\tau_0} = 0 \tag{8.5.12}$$

An interpretation of the necessary condition in (8.5.12) is that the optimum sampling time corresponds to the condition that the error signal $\left[y_m(\tau_0) - \hat{a}_m \right]$ is uncorrelated with the derivative $dy_m(\tau_0)/d\tau_0$. Since the detector output is used in the formation of the error signal $\left[y_m(\tau_0) - \hat{a}_m \right]$, this timing phase estimation method is said to be *decision-directed*.

Figure 8.38 illustrates an implementation of the system that is based on the condition given in (8.5.12). Note that the summation operation is implemented as a lowpass filter, which averages a number of symbols. The averaging time is roughly equal to the reciprocal of the bandwidth of the filter. The filter output drives the voltage-controlled oscillator (VCO), which provides the best MSE estimate of the timing phase τ_0.

8.5.4 Maximum-Likelihood Methods

In the ML criterion, the optimum symbol timing is obtained by maximizing the *likelihood function*

$$\Lambda(\tau_0) = \sum_m a_m y_m(\tau_0) \tag{8.5.13}$$

where $y_m(\tau_0)$ is the sampled output of the receiving filter given by (8.5.11). From a mathematical viewpoint, the likelihood function can be shown to be proportional to

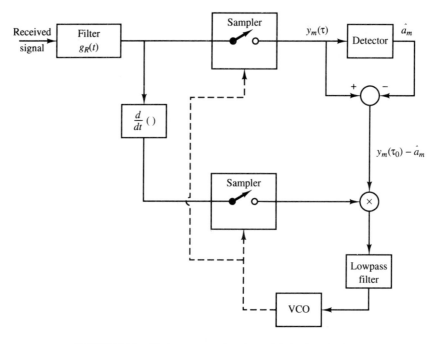

FIGURE 8.38.   Timing recovery based on minimization of MSE.

the probability of the received signal (vector) conditioned on a known transmitted signal. Physically, $\Lambda(\tau_0)$ is simply the output of the matched filter or correlator at the receiver averaged over a number of symbols.

A necessary condition for τ_0 to be the ML estimate is that

$$\frac{d\Lambda(\tau_0)}{d\tau_0} = \sum_m a_m \frac{dy_m(\tau_0)}{d\tau_0} = 0 \qquad (8.5.14)$$

This result suggests the implementation of the tracking loop shown in Figure 8.39. We observe that the product of the detector output $\hat{a}_m$ with the input $dy_m(\tau_0)/d\tau_0$ is averaged by a lowpass filter that drives the VCO. Since the detector output is used in the estimation method, the estimate $\hat{\tau}$ is decision-directed.

As an alternative to the use of the output symbols from the detector, we may use a *non-decision-directed method* that does not require knowledge of the information symbols. This method is based on averaging over the statistics of the symbols. For example, we may square the output of the receiving filter and maximize the function

$$\Lambda_2(\tau_0) = \sum_m y_m^2(\tau_0) \qquad (8.5.15)$$

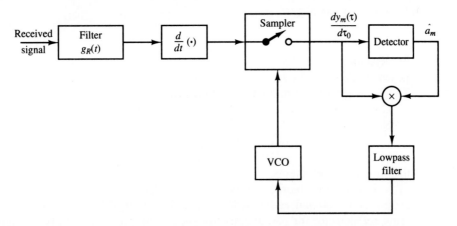

FIGURE 8.39.   Decision-directed ML timing recovery method for baseband PAM.

with respect to τ_0. Thus, we obtain

$$\frac{d\Lambda_2(\tau_0)}{d\tau_0} = 2\sum_m y_m(\tau_0)\frac{dy_m(\tau_0)}{d\tau_0} = 0 \qquad (8.5.16)$$

The condition for the optimum τ_0 given (8.5.16) may be satisfied by the implementation shown in Figure 8.40. In this case, there is no need to know the data sequence $\{a_m\}$. Hence, the method is non-decision-directed.

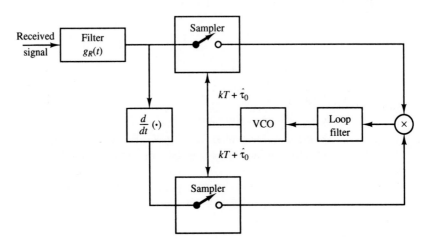

FIGURE 8.40.   Non-decision-directed estimation of timing for baseband PAM.

8.6 FURTHER READING

The pioneering work on signal design for bandwidth-constrained channels was done by Nyquist (1928). The use of binary partial response signals was originally proposed in the paper by Lender (1963) and was later generalized by Kretzmer (1966). The problem of optimum transmitter and receiver filter design was investigated by Gerst and Diamond (1961), Tufts (1965), Smith (1965), and Berger and Tufts (1967).

Adaptive equalization for digital communication was introduced by Lucky (1965, 1966). Widrow (1966) devised the LMS algorithm for adaptively adjusting the equalizer coefficients.

The Viterbi algorithm was devised by Viterbi (1967) for the purpose of decoding convolutional codes, which are described in Chapter 10. Its use as the ML sequence detector for partial response signals and, more generally, for symbols corrupted by ISI was proposed and analyzed by Forney (1972) and Omura (1971). A comprehensive treatment of adaptive equalization algorithms is given in the book by Proakis (1989).

A number of books and tutorial papers have been published on the topic of time synchronization. Books that cover both carrier phase recovery and time synchronization have been written by Stiffler (1971), Lindsey (1972), and Lindsey and Simon (1973), and Meyr and Ascheid (1992). The tutorial paper by Franks (1980) presents a very readable introduction to this topic.

PROBLEMS

8.1 In Example 8.1.1, the ideal channel of bandwidth W limits the transmitted signal energy that passes through the channel. The received signal energy as a function of the channel bandwidth is

$$\mathcal{E}_h(W) = \frac{T}{(2\pi)^2} \int_{-WT}^{WT} \frac{\sin^2 \pi\alpha}{\alpha^2(1-\alpha^2)^2}\, d\alpha$$

where $\alpha = fT$.

1. Evaluate (numerically) $\mathcal{E}_h(W)$ for $W = \frac{1}{2T},\ \frac{1}{T},\ \frac{1.5}{T},\ \frac{2}{T},\ \frac{2.5}{T},\ \frac{3}{T}$, and plot $\frac{\mathcal{E}_h(W)}{T}$ as a function of W.

2. Determine the value of $\mathcal{E}_h(W)$ in the limit as $W \to \infty$. For the computation you may use the time-domain relation

$$\lim_{W \to \infty} \mathcal{E}_h(W) = \int_{-\infty}^{+\infty} g_T^2(t)\, dt$$

8.2 In a binary PAM system, the input to the detector is

$$y_m = a_m + n_m + i_m$$

where $a_m = \pm 1$ is the desired signal, n_m is a zero-mean Gaussian random variable with variance σ_n^2, and i_m represents the ISI due to channel distortion. The ISI term is a random variable that takes the value, $-\frac{1}{2}, 0, \frac{1}{2}$ with probabilities $\frac{1}{4}, \frac{1}{2}, \frac{1}{4}$, respectively. Determine the average probability of error as a function of σ_n^2.

8.3 In a binary PAM system, the clock that specifies the sampling of the correlator output is offset from the optimum sampling time by 10%.

1. If the signal pulse used is rectangular, determine the loss in SNR due to the mistiming.

2. Determine the amount of ISI introduced by the mistiming and determine its effect on performance.

8.4 The elements of the sequence $\{a_n\}_{n=-\infty}^{+\infty}$ are independent binary random variables taking values of ± 1 with equal probability. This data sequence is used to modulate the basic pulse $g(t)$ shown in Figure P-8.4(a). The modulated signal is

$$X(t) = \sum_{n=-\infty}^{+\infty} a_n g(t - nT)$$

1. Find the power-spectral density of $X(t)$.

2. If $g_1(t)$ (shown in Figure P-8.4(b)) is used instead of $g(t)$, how would the power spectrum in part 1 change?

3. In part 2, assume we want to have a null in the spectrum at $f = \frac{1}{3T}$. This is done by a precoding of the form $b_n = a_n + \alpha a_{n-3}$. Find α that provides the desired null.

4. Is it possible to employ a precoding of the form $b_n = a_n + \sum_{i=1}^{N} \alpha_i a_{n-i}$ for some finite N such that the final power spectrum will be identical to zero for $\frac{1}{3T} \le |f| \le \frac{1}{2T}$? If yes, how? If no, why? (Hint: Use properties of analytic functions.)

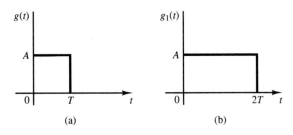

(a)   (b)

FIGURE P-8.4

8.5 The information sequence $\{a_n\}_{n=-\infty}^{\infty}$ is a sequence of i.i.d. random variables, each taking values $+1$ and -1 with equal probability. This sequence is to be transmitted at baseband by a biphase coding scheme, described by

$$s(t) = \sum_{n=-\infty}^{\infty} a_n g(t - nT)$$

where $g(t)$ is shown in Figure P-8.5.

1. Find the power-spectral density of $s(t)$.
2. Assume that it is desirable to have a null in the power spectrum at $f = \frac{1}{T}$. To this end we use a precoding scheme by introducing $b_n = a_n + k a_{n-1}$, where k is some constant, and then transmit the $\{b_n\}$ sequence using the same $g(t)$. Is it possible to choose k to produce a frequency null at $f = \frac{1}{T}$? If yes, what are the appropriate value and the resulting power spectrum?
3. Now assume we want to have nulls at all multiples of $f_0 = \frac{1}{4T}$. Is it possible to have these nulls with an appropriate choice of k in the previous part? If not what kind of precoding do you suggest to result in the desired nulls?

8.6 Starting with the definition of the transition probability matrix for delay modulation given in Example 8.1.5, demonstrate that the relation

$$\mathbf{P}^4 \gamma = -\frac{1}{4} \gamma$$

holds, and, hence,

$$\mathbf{P}^{k+4} \gamma = -\frac{1}{4} \mathbf{P}^k \gamma, \quad k \geq 1$$

8.7 The frequency response characteristic of a lowpass channel can be approximated by

$$H(f) = \begin{cases} 1 + \alpha \cos 2\pi f t_0, & |\alpha| < 1, \quad |f| \leq W \\ 0, & \text{otherwise} \end{cases}$$

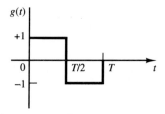

FIGURE P-8.5

where W is channel bandwidth. An input signal $s(t)$, whose spectrum is bandlimited to W Hz, is passed through the channel.

1. Show that

$$y(t) = s(t) + \frac{\alpha}{2} \left[s(t - t_0) + s(t + t_0) \right]$$

Thus, the channel produces a pair of echoes.

2. Suppose the received signal $y(t)$ is passed through a filter matched to $s(t)$. Determine the output of the matched filter at $t = kT$, $k = 0, \pm 1, \pm 2, \ldots$, where T is the symbol duration.

3. What is the ISI pattern resulting from the channel if $t_0 = T$?

8.8 A wireline channel of length 1000 km is used to transmit data by means of binary PAM. Regenerative repeaters are spaced 50 km apart along the system. Each segment of the channel has an ideal (constant) frequency response over the frequency band $0 \le f \le 1200$, and an attenuation of 1 dB/km. The channel noise is AWGN.

1. What is the highest bit rate that can be transmitted without ISI?

2. Determine the required $\mathcal{E}_b/N_0$ to achieve a bit error of $P_2 = 10^{-7}$ for each repeater.

3. Determine the transmitted power at each repeater to achieve the desired $\mathcal{E}_b/N_0$, where $N_0 = 4.1 \times 10^{-21}$ W/Hz.

8.9 Show that a pulse having the raised cosine spectrum given by (8.2.22) satisfies the Nyquist criterion given by (8.2.9) for any value of the roll-off factor α.

8.10 Show that for any value of α the raised cosine spectrum given by (8.2.22) satisfies

$$\int_{-\infty}^{+\infty} X_{rc}(f)\, df = 1$$

(Hint: Use the fact that $X_{rc}(f)$ satisfies the Nyquist criterion given by (8.2.9)).

8.11 Theorem 8.2.1 gives the necessary and sufficient condition for the spectrum $X(f)$ of the pulse $x(t)$ that yields zero ISI. Prove that for any pulse that is bandlimited to $|f| < 1/T$, the zero ISI condition is satisfied if $\text{Re}\left[X(f)\right]$, for $f > 0$, consists of a rectangular function plus an arbitrary odd function about $f = 1/2T$ and $\text{Im}\left[X(f)\right]$ is any arbitrary even function about $f = 1/2T$.

8.12 A voice-band telephone channel has a passband characteristic in the frequency range $300 < f < 3000$ Hz.

1. Select a symbol rate and a power efficient constellation size to achieve 9600 bits/sec signal transmission.

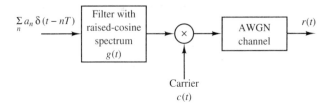

FIGURE P-8.14

FIGURE P-8.14

2. If a square-root raised cosine pulse is used for the transmitter pulse $g_T(t)$, select the roll-off factor. Assume that the channel has an ideal frequency response characteristic.

8.13 Design an M-ary PAM system that transmits digital information over an ideal channel with bandwidth $W = 2400$ Hz. The bit rate is 14,400 bit/sec. Specify the number of transmitted points, the number of received signal points using a duobinary signal pulse, and the required $\mathcal{E}_b$ to achieve an error probability of 10^{-6}. The additive noise is zero-mean Gaussian with a power-spectral density 10^{-4} W/Hz.

8.14 A binary PAM signal is generated by exciting a raised cosine roll-off filter with a 50% rolloff factor and is then DSB-SC amplitude modulated on a sinusoidal carrier as illustrated in Figure P-8.14. The bit rate is 2400 bit/sec.

1. Determine the spectrum of the modulated binary PAM signal and sketch it.

2. Draw the block diagram illustrating the optimum demodulator/detector for the received signal which is equal to the transmitted signal plus additive white Gaussian noise.

8.15 When the additive noise at the input to the modulator is colored, the filter matched to the signal no longer maximizes the output SNR. In such a case we may consider the use of a prefilter that "whitens" the colored noise. The prefilter is followed by a filter matched to the prefiltered signal. Towards this end, consider the configuration shown in Figure P-8.15.

1. Determine the frequency response characteristic of the prefilter that whitens the noise.

FIGURE P-8.15

2. Determine the frequency response characteristic of the filter matched to $\tilde{s}(t)$.

3. Consider the prefilter and the matched filter as a single "generalized matched filter." What is the frequency response characteristic of this filter?

4. Determine the SNR at the input to the detector.

8.16 Consider the transmission of data via PAM over a voice-band telephone channel that has a bandwidth of 3000 Hz. Show how the symbol rate varies as a function of the excess bandwidth. In particular, determine the symbol rate for excess bandwidths of 25%, 33%, 50%, 67%, 75%, and 100%.

8.17 The binary sequence 10010110010 is the input to a precoder whose output is used to modulate a duobinary transmitting filter. Construct a table as in Table 8.1 showing the precoded sequence, the transmitted amplitude levels, the received signal levels and the decoded sequence.

8.18 Repeat the Problem 8.17 for a modified duobinary signal pulse.

8.19 A precoder for a partial response signal fails to work if the desired partial response at $n = 0$ is zero modulo M. For example, consider the desired response for $M = 2$:

$$x(nT) = \begin{cases} 2, & n = 0 \\ 1, & n = 1 \\ -1, & n = 2 \\ 0, & \text{otherwise} \end{cases}$$

Show why this response cannot be precoded.

8.20 In the case of correlated noise, the relation in the Viterbi algorithm may be expressed in general as (Ungerboeck, (1974))

$$\mu(\mathbf{a}) = 2 \sum_{n} a_n r_n - \sum_{n} \sum_{m} a_n a_m x_{n-m}$$

where $x_n = x(nT)$ is the sampled signal output of the matched filter, $\{a_n\}$ is the data sequence, and $\{r_n\}$ is the received signal sequence at the output of the matched filter. Determine the metric for the duobinary signal.

8.21 Sketch and label the trellis for a duobinary signal waveform used in conjunction with the precoding given by (8.2.37). Repeat this for the modified duobinary signal waveform with the precoder given by (8.2.49). Comment on any similarities and differences.

8.22 Consider the use of a (square-root) raised cosine signal pulse with a roll-off factor of unity for transmission of binary PAM over an ideal bandlimited channel that passes the pulse without distortion. Thus, the transmitted signal is

$$v(t) = \sum_{k=-\infty}^{\infty} a_k g_T(t - kT_b)$$

where the signal interval $T_b = T/2$. Thus, the symbol rate is double of that for no ISI.

1. Determine the ISI values at the output of a matched filter demodulator.
2. Sketch the trellis for the maximum likelihood sequence detector and label the states.

8.23 A binary antipodal signal is transmitted over a nonideal bandlimited channel, which introduces ISI over two adjacent symbols. For an isolated transmitted signal pulse $s(t)$, the (noise-free) output of the demodulator is $\sqrt{E_b}$ at $t = T$, $\sqrt{E_b}/4$ at $t = 2T$, at zero for $t = kT$, $k > 2$, where E_b is the signal energy and T is the signaling interval.

1. Determine the average probability of error assuming that the two signals are equally probable and the additive noise is white and Gaussian.
2. By plotting the error probability obtained in part 1 and that for the case of no ISI, determine the relative difference in SNR of the error probability of 10^{-6}.

8.24 Show that the covariance matrix $\mathbf{C}$ for the noise at the output of the matched filter for the duobinary pulse is given by (8.3.20).

8.25 Determine the frequency response characteristics for the RC circuit shown in Figure P-8.25. Also determine the expression for the envelope delay.

8.26 Consider the RC lowpass filter shown in Figure P-8.25 where $\tau = RC = 10^{-6}$.

1. Determine and sketch the envelope (group) delay of the filter as a function of frequency (see Problem 2.76).
2. Suppose that the input to the filter is a lowpass signal of bandwidth $\Delta f = 1$ kHz. Determine the effect of the RC filter on this signal.

8.27 A microwave radio channel has a frequency response

$$C(f) = 1 + 0.3 \cos 2\pi f T$$

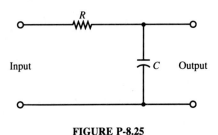

FIGURE P-8.25

Determine the frequency response characteristic for the optimum transmitting and receiving filters that yield zero ISI at a rate of 1/T symbols/sec and have a 50%-excess bandwidth. Assume that the additive noise spectrum is flat.

8.28 $M = 4$ PAM modulation is used for transmitting at a bit rate of 9600 bits/sec on a channel having a frequency response

$$C(f) = \frac{1}{1 + j\frac{f}{2400}},$$

$|f| \leq 2400$, and $C(f) = 0$, otherwise. The additive noise is zero-mean, white Gaussian with power-spectral density $\frac{N_0}{2}$ W/Hz. Determine the (magnitude) frequency response characteristic of the optimum transmitting and receiving filters.

8.29 Binary PAM is used to transmit information over an unequalized linear filter channel. When $a = 1$ is transmitted the noise-free output of the demodulator is

$$x_m = \begin{cases} 0.3, & m = 1 \\ 0.9, & m = 0 \\ 0.3, & m = -1 \\ 0, & \text{otherwise} \end{cases}$$

1. Design a three-tap zero forcing linear equalizer so that the output is

$$q_m = \begin{cases} 1, & m = 0 \\ 0, & m = \pm 1 \end{cases}$$

2. Determine q_m for $m = \pm 2, \pm 3$, by convolving the impulse response of the equalizer with the channel response.

8.30 The transmission of a signal pulse with a raised cosine spectrum through a channel results in the following (noise-free) sampled output from the demodulator:

$$x_k = \begin{cases} -0.5, & k = -2 \\ 0.1, & k = -1 \\ 1, & k = 0 \\ -0.2, & k = 1 \\ 0.05, & k = 2 \\ 0, & \text{otherwise} \end{cases}$$

1. Determine the tap coefficients of a three-tap linear equalizer based on the zero-forcing criterion.

2. For the coefficients determined in part 1, determine the output of the equalizer for the case of the isolated pulse. Thus, determine the residual ISI and its span in time.

8.31 Show that the gradient vector in the minimization of the MSE may be expressed as

$$\mathbf{g}_k = -E[e_k \mathbf{y}_k]$$

where the error $e_k = a_k - z_k$, and the estimate of $\mathbf{g}_k$, i.e.,

$$\hat{\mathbf{g}}_k = -e_k \mathbf{y}_k$$

satisfies the condition that $E[\hat{\mathbf{g}}_k] = \mathbf{g}_k$.

8.32 A nonideal bandlimited channel introduces ISI over three successive symbols. The (noise-free) response of the matched filter demodulator sampled at the sampling time $t = kT$ is

$$\int_{-\infty}^{\infty} s(t)s(t - kT)\,dt = \begin{cases} \mathcal{E}_b, & k = 0 \\ 0.9\mathcal{E}_b, & k = \pm 1 \\ 0.1\mathcal{E}_b, & k = \pm 2 \\ 0, & \text{otherwise} \end{cases}$$

1. Determine the tap coefficients of a three-tap linear equalizer that equalizes the channel (received signal) response to an equivalent partial response (duobinary) signal

$$y_k = \begin{cases} \mathcal{E}_b, & k = 0, 1 \\ 0, & \text{otherwise} \end{cases}$$

2. Suppose that the linear equalizer in part 1 is followed by a Viterbi sequence detector for the partial signal. Give an estimate of the error probability if the additive noise is white and Gaussian, with power-spectral density $N_0/2$ W/Hz.

8.33 Determine the tap weight coefficients of a three-tap zero-forcing equalizer if the ISI spans three symbols and is characterized by the values $x(0) = 1, x(-1) = 0.3, x(1) = 0.2$. Also determine the residual ISI at the output of the equalizer for the optimum tap coefficients.

8.34 In line-of-sight microwave radio transmission, the signal arrives at the receiver via two propagation paths, the direct path and a delayed path that occurs due to signal reflection from surrounding terrain. Suppose that the received signal has the form

$$r(t) = s(t) + \alpha s(t - T) + n(t)$$

where $s(t)$ is the transmitted signal, α is the attenuation $(\alpha < 1)$ of the secondary path and $n(t)$ is AWGN.

1. Determine the output of the demodulator at $t = T$ and $t = 2T$ that employs a filter matched to $s(t)$.

2. Determine the probability of error for a symbol-by-symbol detector if the transmitted signal is binary antipodal and the detector ignores the ISI.

3. What is the error-rate performance of a simple (one-tap) (DFE) that estimates α and removes the ISI? Sketch the detector structure that employs a DFE.

8.35 Repeat Problem 8.29 using the MMSE as the criterion for optimizing the tap coefficients. Assume that the noise power spectrum is 0.1 W/Hz.

8.36 In a magnetic recording channel, where the readback pulse resulting from a positive transition in the write current has the form (see Figure 7.18)

$$p(t) = \frac{1}{1 + \left(\frac{2t}{T_{50}}\right)^2},$$

a linear equalizer is used to equalize the pulse to a partial response. The parameter T_{50} is defined as the width of the pulse at the 50%-amplitude level. The bit rate is $1/T_b$ and the ratio of $T_{50}/T_b = \Delta$ is the normalized density of the recording. Suppose the pulse is equalized to the partial response values

$$x(nT) = \begin{cases} 1, & n = -1, 1 \\ 2, & n = 0 \\ 0, & \text{otherwise} \end{cases}$$

where $x(t)$ represents the equalized pulse shape.

1. Determine the spectrum $X(f)$ of the bandlimited equalized pulse.

2. Determine the possible output levels at the detector, assuming that successive transition can occur at the rate $1/T_b$.

3. Determine the error rate performance of the symbol-by-symbol detector for this signal assuming that the additive noise is zero-mean Gaussian with variance σ^2.

8.37 Sketch the trellis for the Viterbi detector of the equalized signal in Problem 8.36 and label all the states. Also, determine the minimum Euclidean distance between merging paths.

8.38 Show that the early–late gate synchronizer illustrated in Figure 8.37 is a close approximation to the timing recovery system illustrated in Figure P-8.38.

8.39 Based on a ML criterion, determine a carrier phase estimation method for binary on–off keying modulation.

8.40 A communication system for a voice-band (3 kHz) channel is designed for a received SNR at the detector of 30 dB when the transmitter power is $P_S = -3$ dBw. Determine the value of P_S if it's desired to expand the bandwidth of the system to 10 kHz, while maintaining the same SNR at the detector.

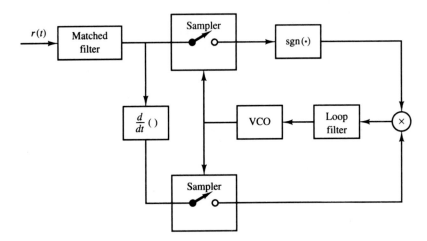

FIGURE P-8.38

9

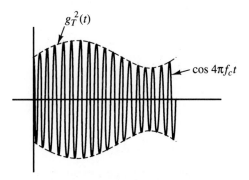

$g_T^2(t)$

$\cos 4\pi f_c t$

Digital Transmission via Carrier Modulation

In the two preceding chapters we described digital modulation methods for baseband channels, i.e., channels with lowpass frequency-response characteristics. Digitally modulated signals with lowpass spectral characteristics can be transmitted directly through such channels without the need for frequency translation of the signal.

There are many communication channels, including telephone channels, radio channels, and satellite channels, that pass signals within a band of frequencies that is far removed from *dc*. Such channels are called *bandpass* channels. As in the case of analog signal transmission treated in Chapter 5, digital information may be transmitted through such channels by using a sinusoidal carrier that is modulated by the information sequence in either amplitude, phase, or frequency, or some combination of amplitude and phase. The effect of impressing the information signal on one or more of the sinusoidal parameters is to shift the frequency content of the transmitted signal to the appropriate frequency band that is passed by the channel. Thus, the signal is transmitted by carrier modulation.

In this chapter we treat several carrier-modulation methods, including amplitude modulation, phase modulation, combined amplitude and phase modulation, and frequency modulation. We will consider the spectral characteristics of these carrier-modulation signals, describe their optimum demodulation when corrupted by AGWN, and evaluate the probability of error performance. We begin our discussion with digital PAM, which is transmitted by modulating the amplitude of the carrier.

9.1 CARRIER-AMPLITUDE MODULATION

From Section 7.1 we recall that in baseband digital PAM, the signal waveforms are of the form

$$s_m(t) = A_m g_T(t), \quad m = 1, 2, \ldots, M \tag{9.1.1}$$

where $g_T(t)$ is the transmitting filter impulse response whose shape determines the spectral characteristics of the transmitted signal and A_m is the signal amplitude that takes the discrete values

$$A_m = (2m - 1 - M), \quad m = 1, 2, \ldots, M \tag{9.1.2}$$

As shown in Section 8.1.2, the spectrum of the baseband signals is contained in the frequency band $|f| \leq W$, where W is the bandwidth of $|G_T(f)|^2$.

To transmit the digital signal waveforms through a bandpass channel by amplitude modulation, the baseband signal waveforms $s_m(t)$, $m = 1, 2, \ldots, M$ are multiplied by a sinusoidal carrier of the form $\cos 2\pi f_c t$, as shown in Figure 9.1, where f_c is the carrier frequency ($f_c > W$) and corresponds to the center frequency in the passband of the channel. Thus, the transmitted signal waveforms may be expressed as

$$u_m(t) = A_m g_T(t) \cos 2\pi f_c t, \quad m = 1, 2, \ldots, M \tag{9.1.3}$$

As previously described in Section 5.2, amplitude modulation of the carrier $\cos 2\pi f_c t$ by the baseband signal waveforms $s_m(t) = A_m g_T(t)$ shifts the spectrum of the baseband signal by an amount f_c and, thus, places the signal into the passband of the channel. Recall that the Fourier transform of the carrier is $[\delta(f - f_c) + \delta(f + f_c)]/2$. Because multiplication of two signals in the time domain corresponds to the convolution of their spectra in the frequency domain, the spectrum of the amplitude-modulated signal given by (9.1.3) is

$$U_m(f) = \frac{A_m}{2} \left[G_T(f - f_c) + G_T(f + f_c) \right] \tag{9.1.4}$$

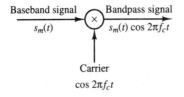

FIGURE 9.1. Amplitude modulation of a sinusoidal carrier by the baseband PAM signal.

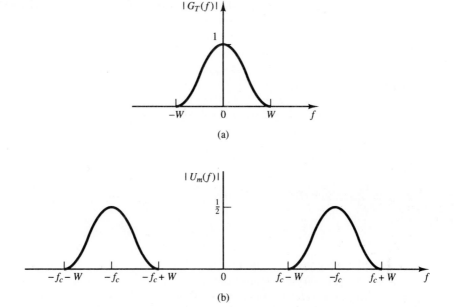

FIGURE 9.2.   Spectra of (a) baseband and (b) amplitude-modulated signals.

Thus the spectrum of the baseband signal $s_m(t) = A_m g_T(t)$ is shifted in frequency by the carrier frequency f_c. The result is a DSB-SC AM signal, as illustrated in Figure 9.2.

The above illustration of the spectral characteristics of the amplitude-modulated signal carry over to the power-spectral density of a sequence of transmitted symbols. Thus, if

$$v(t) = \sum_{n=-\infty}^{\infty} a_n g_T(t - nT) \tag{9.1.5}$$

is the baseband PAM signal and $\{a_n\}$ is the sequence of amplitude values, the power-spectral density of the amplitude-modulated signal

$$u(t) = v(t) \cos 2\pi f_c t \tag{9.1.6}$$

is

$$S_U(f) = \frac{1}{4} \left[S_V(f - f_c) + S_V(f + f_c) \right] \tag{9.1.7}$$

where $S_V(f)$ is the power-spectral density of the baseband signal previously derived in Section 8.1.2, and $S_U(f)$ is the power-spectral density of the bandpass signal $u(t)$.

The modulated signal $u(t)$ given by (9.1.6) is a DSB-SC amplitude-modulated signal. The upper sideband of the signal $u(t)$ is comprised of the frequency content of $u(t)$ for $|f| > f_c$, i.e., for $f_c < |f| \le f_c + W$. The lower sideband of $u(t)$ is comprised of the frequency content for $|f| < f_c$, i.e., for $f_c - W \le |f| < f_c$. Hence, the DSB-SC amplitude-modulated signal occupies a channel bandwidth of $2W$, which is twice the bandwidth required to transmit the baseband signal.

The energy of the passband signal waveforms $u_m(t)$, $\quad m = 1, 2, \ldots, M$, given by (9.1.3) is defined as

$$\mathcal{E}_m = \int_{-\infty}^{\infty} u_m^2(t)dt = \int_{-\infty}^{\infty} A_m^2 g_T^2(t) \cos^2 2\pi f_c t \, dt$$

$$= \frac{A_m^2}{2} \int_{-\infty}^{\infty} g_T^2(t) \, dt + \frac{A_m^2}{2} \int_{-\infty}^{\infty} g_T^2(t) \cos 4\pi f_c t \, dt \qquad (9.1.8)$$

We note that when $f_c \gg W$, the term

$$\int_{-\infty}^{\infty} g_T^2(t) \cos 4\pi f_c t \, dt \qquad (9.1.9)$$

involves the integration of the product of a slowly varying function, namely $g_T^2(t)$, with a rapidly varying sinusoidal term, namely $\cos 4\pi f_c t$ as shown in Figure 9.3. Because $g_T(t)$ is slowly varying relative to $\cos 4\pi f_c t$, the integral in (9.1.9) over a single cycle of $\cos 4\pi f_c t$ is zero, and, hence, the integral over an arbitrary number of cycles is also zero. Consequently,

$$\mathcal{E}_m = \frac{A_m^2}{2} \int_{-\infty}^{\infty} g_T^2(t) \, dt = \frac{A_m^2}{2} \mathcal{E}_g \qquad (9.1.10)$$

where $\mathcal{E}_g$ is the energy in the signal pulse $g_T(t)$. Thus, we have shown that the energy in the passband signal is one-half of the energy in the baseband signal. The scale factor of $\frac{1}{2}$ is due to the carrier component $\cos 2\pi f_c t$, which has an average power of $\frac{1}{2}$.

When the transmitted pulse shape $g_T(t)$ is rectangular, i.e.,

$$g_T(t) = \begin{cases} \sqrt{\dfrac{\mathcal{E}_g}{T}} & 0 \le t \le T \\ 0 & \text{otherwise} \end{cases} \qquad (9.1.11)$$

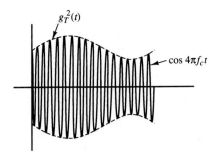

FIGURE 9.3. The signal $g_T^2(t) \cos 4\pi f_c t$.

the amplitude-modulated carrier signal is usually called *amplitude-shift keying* (ASK).

Finally, we note that impressing the baseband signals $s_m(t)$ onto the amplitude of the carrier signal $\cos 2\pi f_c t$ does not change the basic geometric representation of the digital PAM signal waveforms. The signal waveforms $u_m(t)$ may be expressed as

$$u_m(t) = s_m \psi(t) \tag{9.1.12}$$

where the basic signal waveform $\psi(t)$ is defined as

$$\psi(t) = \sqrt{\frac{2}{\mathcal{E}_g}} g_T(t) \cos 2\pi f_c t \tag{9.1.13}$$

and

$$s_m = \sqrt{\frac{\mathcal{E}_g}{2}} A_m, \quad m = 1, 2, \ldots, M \tag{9.1.14}$$

Note that the only change in the geometric representation, compared to baseband signals, is the scale factor $\sqrt{2}$, which appears in (9.1.13) and (9.1.14).

9.1.1 Amplitude Demodulation and Detection

The demodulation of a bandpass digital PAM signal may be accomplished in one of several ways by means of correlation or matched filtering. For illustrative purposes, we consider a correlation-type demodulator.

Suppose the transmitted signal is

$$u_m(t) = A_m g_T(t) \cos 2\pi f_c t, \quad 0 \le t \le T \tag{9.1.15}$$

The received signal may be expressed as

$$r(t) = A_m g_T(t) \cos 2\pi f_c t + n(t), \quad 0 \le t \le T \tag{9.1.16}$$

where $n(t)$ is a bandpass noise process, which is represented as

$$n(t) = n_c(t) \cos 2\pi f_c t - n_s(t) \sin 2\pi f_c t \tag{9.1.17}$$

By crosscorrelating the received signal $r(t)$ with the basic function $\psi(t)$ as shown in Figure 9.4, we obtain the output

$$\int_{-\infty}^{\infty} r(t)\psi(t)\,dt = A_m \sqrt{\frac{2}{\mathcal{E}_g}} \int_{-\infty}^{\infty} g_T^2(t) \cos^2 2\pi f_c t\, dt + \int_{-\infty}^{\infty} n(t)\psi(t)\,dt$$

$$= A_m \sqrt{\mathcal{E}_g/2} + n \tag{9.1.18}$$

where n represents the additive noise component at the output of the correlator.

Carrier phase recovery.   In the above development we assumed that the function $\psi(t)$ is perfectly synchronized with the signal component in $r(t)$ in both

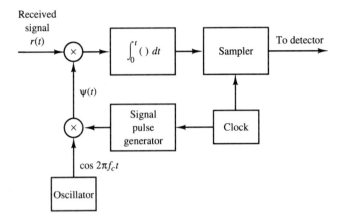

FIGURE 9.4. Demodulation of bandpass digital PAM signal.

time and carrier phase. In practice, however, these ideal conditions do not hold because of the propagation delay encountered in transmitting a signal through the channel and because of frequency and phase drifts that occur in any practical oscillator that generates the carrier signal $\cos 2\pi f_c t$. As a consequence, it is necessary to generate a phase coherent carrier at the receiver to perform the demodulation of the received signal.

Because the message signal, given by (9.1.5) is generally zero mean, the DSB-SC amplitude-modulated signal $u(t)$ given by (9.1.6) has zero-average power at $f = f_c$. Consequently, it is not possible to estimate the carrier phase directly from $r(t)$. However, if we square $r(t)$, we generate a frequency component at $f = 2f_c$ that has nonzero-average power. This component can be filtered out by a narrowband filter tuned to $2f_c$, which can be used to drive a PLL, as previously described in Section 6.2. A functional block diagram of the receiver that employs a PLL for estimating the carrier phase is shown in Figure 9.5.

The Costas loop, also described in Section 6.2, serves as an alternative method for estimating the carrier phase from the received signal $r(t)$. Recall from our

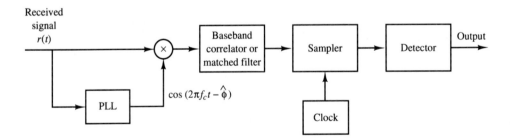

FIGURE 9.5. Demodulation of carrier amplitude-modulated signal.

discussion in Section 6.2 that the PLL and the Costas loop yield phase estimates that are comparable in quality in the presence of additive channel noise.

   Demodulation and detection.   As an alternative to baseband demodulation, we may perform crosscorrelation or matched filtering either at passband or at some convenient intermediate frequency. In particular, a bandpass correlator may be used to multiply the received signal $r(t)$ by the amplitude-modulated carrier $g_T(t) \cos(2\pi f_c t - \hat{\phi})$, where $\cos(2\pi f_c t - \hat{\phi})$ is the output of the PLL. The product signal is integrated over the signaling interval T, the output of the integrator is sampled at $t = T$, and the sample is passed to the detector. If a matched filter instead of a correlator is used, the filter impulse response is $g_T(T - t) \cos[2\pi f_c(T - t) - \hat{\phi}]$. The functional block diagram for these demodulators are shown in Figure 9.6.

   To describe the operation of the optimum detector for the case of an AWGN channel, let us assume that the transmitted signal is $u_m(t) = s_m \psi(t)$, where $\psi(t)$ and s_m are given by (9.1.13) and (9.1.14), respectively. Furthermore, we assume that the carrier-phase estimate $\hat{\phi} = \phi$ where ϕ is the carrier phase. Then, the input

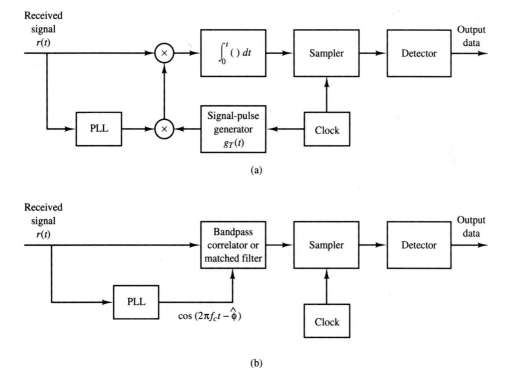

FIGURE 9.6. Bandpass demodulation of digital PAM signal via (a) bandpass correlation and (b) bandpass matched filtering.

to the detector is

$$r = \int_{-\infty}^{\infty} r(t)\psi(t)\,dt$$

$$= s_m \int_{-\infty}^{\infty} \psi^2(t)\,dt + \int_{-\infty}^{\infty} n(t)\psi(t)\,dt$$

$$= A_m\sqrt{\mathcal{E}_g/2} + n \qquad (9.1.19)$$

where $n(t)$ is a sample function of the bandpass noise process, which is given by (9.1.17) in terms of its quadrature components.

For equiprobable messages, the optimum detector bases its decision on the distance metrics

$$D(r, s_m) = (r - s_m)^2, \quad m = 1, 2, \ldots, M$$

or, equivalently, on the correlation metrics

$$C(r, s_m) = 2rs_m - s_m^2$$

as previously shown in Section 7.2.3.

Symbol-timing recovery.   Several methods for symbol-timing recovery were described in Section 8.5. These methods generally apply to carrier-modulated systems as well as to baseband systems. Because any carrier-modulated signal can be converted to a baseband signal by a simple frequency translation, symbol timing can be recovered from the received signal after frequency conversion to baseband.

In many modern communication systems, the received signal is processed (demodulated) digitally after it has been sampled at the Nyquist rate or faster. In such a case, symbol timing and carrier phase are recovered by signal-processing operations performed on the signal samples. Thus, a PLL for carrier recovery is implemented as a digital PLL and a clock recovery loop of a type described in Section 8.5 is also implemented as a digital loop. Timing recovery methods based on sampled signals have been described and analyzed by Mueller and Muller (1976).

9.1.2 Probability of Error for PAM in an AWGN Channel

The error-rate performance of the demodulators described above for an AWGN channel is exactly the same as the error-rate performance of baseband PAM transmitted through an AWGN channel and demodulated by either a correlator or a matched filter. We demonstrate this point by computing the SNR at the output of the correlator shown in Figure 9.5. For this computation we assume that the input to the detector is given by (9.1.19). The transmitted pulse $g_T(t)$ is assumed to be bandlimited to W Hz, i.e., $G(f) = 0$, for $|f| > W$.

Because the noise process is zero-mean Gaussian, we have

$$E(n) = \int_{-\infty}^{\infty} E[n(t)]\psi(t)\,dt = 0 \qquad (9.1.20)$$

The variance of the noise sample is

$$\sigma_n^2 = E\left\{\int_{-\infty}^{\infty}\int_{-\infty}^{\infty} n(t)n(\tau)\psi(t)\psi(\tau)dtd\tau\right\}$$

$$= \int_{-\infty}^{\infty}\int_{-\infty}^{\infty} E[n(t)n(\tau)]\psi(t)\psi(\tau)dtd\tau$$

$$= \int_{-\infty}^{\infty}\int_{-\infty}^{\infty} R_n(t-\tau)\psi(\tau)\psi(t)\,dt\,d\tau$$

$$= \int_{-\infty}^{\infty} \psi(t)dt\left[\int_{-\infty}^{\infty} \Psi(f)S_n(f)e^{j2\pi ft}\,df\right]$$

$$= \int_{-\infty}^{\infty} |\Psi(f)|^2 S_n(f)\,df \tag{9.1.21}$$

where $\Psi(f)$ is the Fourier transform of $\psi(t)$ and $S_n(f)$ is the power-spectral density of the additive noise. From (9.1.13) we obtain

$$\Psi(f) = \frac{1}{\sqrt{2\mathcal{E}_g}}\left[G_T(f-f_c) + G_T(f+f_c)\right] \tag{9.1.22}$$

Furthermore, the power-spectral density of the bandlimited noise process is

$$S_n(f) = \begin{cases} \frac{N_0}{2}, & |f - f_c| \le W \\ 0, & \text{otherwise} \end{cases} \tag{9.1.23}$$

By substituting (9.1.22) and (9.1.23) into (9.1.21), we obtain

$$\sigma_n^2 = \frac{N_0}{4\mathcal{E}_g}\left[\int_{f_c-W}^{f_c+W} |G_T(f-f_c)|^2 df + \int_{-f_c-W}^{-f_c+W} |G_T(f+f_c)|^2 df\right]$$

$$= \frac{N_0}{2\mathcal{E}_g}\int_{-W}^{W} |G_T(f)|^2 df = \frac{N_0}{2} \tag{9.1.24}$$

Following the development of Section 7.3.2, the probability of a symbol error is

$$P_M = \frac{M-1}{M} P\left(|r - s_m| > \sqrt{\mathcal{E}_g/2}\right)$$

$$= \frac{2(M-1)}{M} Q\left(\sqrt{\frac{\mathcal{E}_g}{N_0}}\right) \tag{9.1.25}$$

Using (9.1.10) we can relate the average transmitted energy to $\mathcal{E}_g$. We have

$$\mathcal{E}_{av} = P_{av}T = \sum_{m=1}^{M}\mathcal{E}_m = \frac{\mathcal{E}_g}{2}\sum_{m=1}^{M}(2m-1-M)^2$$

$$= \frac{M^2 - 1}{6} \, \mathcal{E}_g \tag{9.1.26}$$

Hence,

$$\mathcal{E}_g = \frac{6 P_{av} T}{M^2 - 1} \tag{9.1.27}$$

Substitution of (9.1.27) into (9.1.25) yields the desired form for the probability of error for M-ary PAM, i.e.,

$$P_M = \frac{2(M-1)}{M} \, Q\left(\sqrt{\frac{6 P_{av} T}{(M^2 - 1) N_0}}\right) \tag{9.1.28}$$

This is exactly the same result as the probability of a symbol error for M-ary PAM transmission over a baseband AWGN channel (see Section 7.3.2).

9.1.3 Signal Demodulation in the Presence of Channel Distortion

In most practical communication systems, the channel distorts the transmitted signal. The most common type of channel distortion encountered in practice is linear distortion (amplitude and phase distortion), previously described in Section 8.4. This type of distortion results in intersymbol interference (ISI) that degrades the performance of the communication system.

The ISI may be reduced by proper design of the transmitting and receiving filters. The methodology described in Section 8.4.1 for designing optimum transmitting and receiving filters for a channel with linear distortion may be applied directly by converting the bandpass channel into an equivalent baseband channel. To be specific, suppose that the bandpass channel is characterized by the frequency response $C_{bp}(f)$, and let f_c denote the center of the frequency band. Let $C(f)$ denote the frequency response of an equivalent baseband (lowpass) channel, where $C(f)$ is defined as

$$C(f - f_c) = \begin{cases} 2 C_{bp}(f), & f > f_c \\ 0, & f < f_c \end{cases} \tag{9.1.29}$$

Because $C_{bp}^\star(-f) = C_{bp}(f)$, it follows that

$$C^\star(-f - f_c) = \begin{cases} 0, & f > -f_c \\ 2 C_{bp}^\star(-f), & f < -f_c \end{cases} \tag{9.1.30}$$

Therefore, the frequency response $C_{bp}(f)$ of the bandpass channel may be expressed as (see (2.6.19))

$$C_{bp}(f) = \frac{1}{2} [C(f - f_c) + C^\star(-f - f_c)] \tag{9.1.31}$$

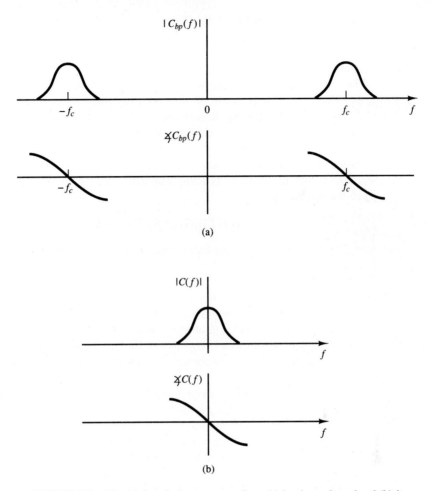

FIGURE 9.7. Magnitude and phase response for a (a) bandpass channel and (b) its equivalent baseband channel.

where $C(f)$ is the frequency response of the equivalent baseband channel. This relationship is illustrated in Figure 9.7. The corresponding relationship between the impulse responses of the bandpass and baseband channels is easily established from (9.1.31). If $c_{bp}(t)$ denotes the impulse response of the bandpass channel and $c(t)$ denotes the impulse response of the baseband channel, then[†]

$$c_{bp}(t) = \text{Re}[c(t)e^{j2\pi f_c t}] \qquad (9.1.32)$$

[†]In general, $c_{bp}(t)$ is real, whereas the impulse response $c(t)$ of the baseband channel may be complex-valued. If the magnitude $|C_{bp}(f)|$ is even about $f = f_c$ and the phase $\Theta(f) = \angle C_{bp}(f)$ is odd about $f = f_c$, then $C_{bp}^*(-f) = C_{bp}(f)$ and $c(t)$ will be real-valued. Otherwise, it is complex-valued.

Once we have determined the frequency response $C(f)$ of the equivalent baseband channel as described above, the optimum transmitting and receiving filters can be designed following the procedure outlined in Section 8.4.1, which applies to the baseband channel.

When the channel frequency-response characteristic $C_{bp}(f)$ is unknown a priori, as is the case, for example, in dial-up telephone channels, the methodology described in Section 8.4.1 is no longer applicable. In such a case, we resort to the use of an adaptive equalizer to compensate for the channel distortion. The basic approach in this case is to design the transmitting and receiving filters $G_T(f)$ and $G_R(f)$ either for zero ISI or for controlled ISI (partial-response signals), based on the assumption that the channel is ideal. For example, we might select $G_T(f)$ and $G_R(f)$ to satisfy the condition

$$|G_T(f)||G_R(f)| = X_{rc}(f) \qquad (9.1.33)$$

which results in zero ISI when the equivalent baseband channel is ideal. Then, if we transmit an isolated pulse $g_T(t)$, the signal component at the output of the receiving filter has the frequency-response characteristic $G_T(f)G_R(f)C(f)$, where $C(f)$ represents the frequency response of the nonideal equivalent baseband channel. The ISI resulting from the nonideal channel may be compensated by passing the received signal through a linear adaptive equalizer, with frequency response $G_E(f)$. Thus, $G_E(f)$ serves as the system that compensates for the nonideal channel characteristic $C(f)$. This is basically the approach described in Section 8.4.2 for adaptively equalizing baseband channels.

In a system that employs carrier modulation, the adaptive equalizer may be implemented either at bandpass or at baseband. Figure 9.8 illustrates the two configurations. These two implementations are functionally equivalent.

Finally, we should say that either a decision-feedback equalizer (DFE) or a maximum-likelihood sequence detector (Viterbi algorithm) may be used as an alternative to a linear equalizer for combating ISI due to channel distortion. These two nonlinear equalizers yield a significantly better performance compared to a linear equalizer on channels with severe distortion.

9.2 CARRIER-PHASE MODULATION

In carrier-phase modulation the information that is transmitted over a communication channel is impressed on the phase of the carrier. Because the range of the carrier phase is $0 \le \theta < 2\pi$, the range of the carrier phases used to transmit digital information via digital-phase modulation are $\theta_k = 2\pi k/M$, $k = 0, 1, \ldots, M - 1$. Thus, for binary phase modulation ($M = 2$), the two carrier phases are $\theta = 0$ and $\theta_1 = \pi$ radians. For M-ary-phase modulation, $M = 2^k$ where k is the number of information bits per transmitted symbol.

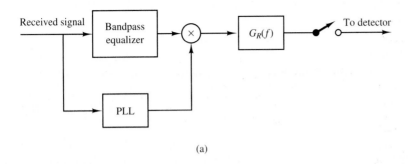

(a)

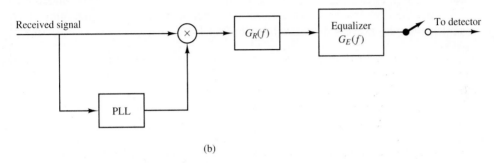

(b)

FIGURE 9.8.   Channel equalization at (a) bandpass and (b) baseband.

The general representation of a set of M carrier-phase-modulated signal waveforms is

$$u_m(t) = g_T(t) \cos\left(2\pi f_c t + \frac{2\pi m}{M}\right), \quad m = 0, 1, \ldots, M-1 \qquad (9.2.1)$$

where $g_T(t)$ is the transmitting filter pulse shape, which determines the spectral characteristics of the transmitted signal. These signals have identical energy, i.e.,

$$\mathcal{E}_m = \int_{-\infty}^{\infty} u_m^2(t)dt = \int_{-\infty}^{\infty} g_T^2(t) \cos^2\left(2\pi f_c t + \frac{2\pi m}{M}\right) dt$$

$$= \frac{1}{2}\int_{-\infty}^{\infty} g_T^2(t)dt + \frac{1}{2}\int_{-\infty}^{\infty} g_T^2(t)\cos\left(4\pi f_c t + \frac{4\pi m}{M}\right) dt$$

$$= \frac{1}{2}\mathcal{E}_g \equiv \mathcal{E}_s, \qquad \text{for all } m \qquad (9.2.2)$$

where $\mathcal{E}_g$ is the energy of the pulse $g_T(t)$ and $\mathcal{E}_s$ denotes the energy per transmitted symbol. The term involving the double-frequency component in (9.2.2) averages out to zero when $f_c \gg W$, where W is the bandwidth of $g_T(t)$. Hence, all waveforms possess the same energy. Note that when $g_T(t)$ is a rectangular pulse, it is defined as

$$g_T(t) = \sqrt{\frac{2\mathcal{E}_s}{T}}, \qquad 0 \le t \le T \tag{9.2.3}$$

When a rectangular pulse is employed, the transmitted signal waveforms

$$u_m(t) = \sqrt{\frac{2\mathcal{E}_s}{T}} \cos\left(2\pi f_c t + \frac{2\pi m}{M}\right), \qquad m = 0, 1, \ldots, M-1 \tag{9.2.4}$$

have a constant envelope and the carrier phase changes abruptly at the beginning of each signal interval. This type of digital-phase modulation is called *phase-shift keying* (PSK). Figure 9.9 illustrates a four-phase ($M = 4$) PSK signal waveform.

By viewing the angle of the cosine function in (9.2.4) as the sum of two angles, we may express the waveforms in (9.2.1) as

$$u_m(t) = g_T(t) A_{mc} \cos 2\pi f_c t - g_T(t) A_{ms} \sin 2\pi f_c t \tag{9.2.5}$$

where

$$A_{mc} = \cos 2\pi m/M, \quad m = 0, 1, \ldots, M-1$$

$$A_{ms} = \sin 2\pi m/M, \quad m = 0, 1, \ldots, M-1 \tag{9.2.6}$$

Thus, a phase-modulated signal may be viewed as two quadrature carriers with amplitudes $g_T(t) A_{mc}$ and $g_T(t) A_{ms}$ as shown in Figure 9.10, which depend on the transmitted phase in each signal interval.

It follows from (9.2.5) that digital-phase-modulated signals can be represented geometrically as two-dimensional vectors with components $\sqrt{\mathcal{E}_s} \cos 2\pi m/M$ and $\sqrt{\mathcal{E}_s} \sin 2\pi m/M$, i.e.,

$$\mathbf{s}_m = \left(\sqrt{\mathcal{E}_s} \cos 2\pi m/M, \quad \sqrt{\mathcal{E}_s} \sin 2\pi m/M\right) \tag{9.2.7}$$

Note that the orthogonal basis functions are $\psi_1(t) = \sqrt{\frac{2}{\mathcal{E}_g}} g_T(t) \cos 2\pi f_c t$ and $\psi_2(t) = -\sqrt{\frac{2}{\mathcal{E}_g}} g_T(t) \sin 2\pi f_c t$. Signal point constellations for $M = 2, 4, 8$ are illus-

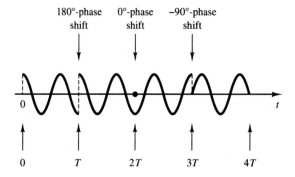

FIGURE 9.9. Example of a four-phase PSK signal.

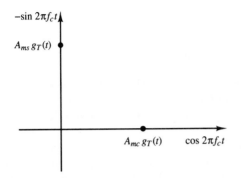

FIGURE 9.10. Digital-phase modulation viewed as two amplitude-modulated quadrature carriers.

trated in Figure 9.11. We observe that binary-phase modulation is identical to binary PAM.

The mapping or assignment of k information bits into the $M = 2^k$ possible phases may be done in a number of ways. The preferred assignment is to use *Gray encoding,* in which adjacent phases differ by one binary digit as illustrated in Figure 9.11. Because the most likely errors caused by noise involve the erroneous selection of an adjacent phase to the transmitted phase, only a single bit error occurs in the k-bit sequence with Gray encoding.

The spectral characteristics of a digital-phase-modulated signal are similar to the spectral characteristics of a PAM signal. From (9.2.1) we observe that the band-pass signal waveforms $u_m(t)$ in the signaling interval $0 \leq t \leq T$ may be expressed as

$$u_m(t) = \text{Re}\left[g_T(t) e^{j2\pi m/M} e^{j2\pi f_c t} \right], \quad m = 0, 1, \ldots, M - 1 \tag{9.2.8}$$

The corresponding transmitted bandpass signal for a sequence of information symbols may then be expressed as

$$u(t) = \text{Re}\left[v(t) e^{j2\pi f_c t} \right] \tag{9.2.9}$$

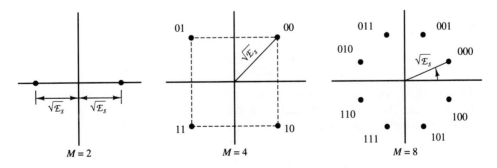

FIGURE 9.11. PSK signal constellations.

where, by definition,

$$v(t) = \sum_{n=-\infty}^{\infty} e^{j\theta_n} g_T(t - nT) \qquad (9.2.10)$$

The transmitted sequence of carrier phases $\{\theta_n\}$ is obtained by mapping the sequence of k-bit symbols into the carrier phases selected from the set of M phases $\{2\pi m/M, m = 0, 1, \ldots, M - 1\}$. By defining a complex-valued amplitude A_n as

$$A_n = e^{j\theta_n} = A_{nc} + jA_{ns} \qquad (9.2.11)$$

we obtain an expression for the lowpass signal $v(t)$, which is identical to the expression for a PAM signal (see Section 8.1.2), except that now the sequence $\{A_n\}$ is complex-valued. This difference requires that we define the autocorrelation function for the sequence $\{A_n\}$ as

$$R_A(n) = E\left[A_k^\star A_{n+k}\right]$$
$$= E\left[e^{-j\theta_k} e^{j\theta_{n+k}}\right] = E\left[e^{j(\theta_{n+k} - \theta_k)}\right] \qquad (9.2.12)$$

Aside from this minor difference, the derivation for the power-density spectrum of the transmitted signal $u(t)$ given by (9.2.9) follows the procedure given in Section 8.1.2. Thus, we obtain the average autocorrelation function of the transmitted signal as

$$\bar{R}_U(\tau) = \frac{1}{2} \sum_{n=-\infty}^{\infty} R_A(m) R_g(\tau - mT) \cos 2\pi f_c \tau \qquad (9.2.13)$$

where $R_g(\tau)$ is the time autocorrelation function of the pulse $g_T(f)$, defined by (8.1.23). The Fourier transform of $\bar{R}_U(\tau)$ yields the power-spectral density as

$$S_U(f) = \frac{1}{4}\left[S_V(f - f_c) + S_V(-f - f_c)\right] \qquad (9.2.14)$$

where the power spectrum of the lowpass signal $v(t)$ is

$$S_V(f) = \frac{1}{T} S_A(f) |G_T(f)|^2 \qquad (9.2.15)$$

and

$$S_A(f) = \sum_{n=-\infty}^{\infty} R_A(n) e^{-j2\pi f nT} \qquad (9.2.16)$$

This is the same form for the average power spectrum as that obtained for a PAM signal. The only difference is that the autocorrelation function $R_A(n)$ in the phase-modulated carrier signal is now complex-valued.

Example 9.2.1

Determine the power-spectral density of a constant envelope M-ary PSK signal when the information sequence is white.

Solution   The pulse shape $g_T(t)$ for PSK is rectangular. Hence,

$$|G_T(f)|^2 = 2\mathcal{E}_s T \left[\frac{\sin \pi f T}{\pi f T} \right]^2 = 2\mathcal{E}_s T \, \mathrm{sinc}^2(fT)$$

The M possible transmitted phases are symmetric and assumed to be equiprobable. Hence,

$$R_A(m) = \begin{cases} \sigma_A^2, & m = 0 \\ 0, & m \neq 0 \end{cases}$$

where $\sigma_A^2 = 1$. Then, the power spectrum of the baseband signal process $V(t)$ is

$$S_V(f) = 2\mathcal{E}_s \left[\frac{\sin \pi f T}{\pi f T} \right]^2 = 2\mathcal{E}_s \, \mathrm{sinc}^2(fT)$$

and the power spectrum of the bandpass signal is

$$S_U(f) = \frac{\mathcal{E}_s}{2} \left[\frac{\sin \pi (f - f_c) T}{\pi (f - f_c) T} \right]^2 + \frac{\mathcal{E}_s}{2} \left[\frac{\sin \pi (f + f_c) T}{\pi (f + f_c) T} \right]^2$$

A sketch of $S_U(f)$ is shown in Figure 9.12, plotted as a function of the normalized frequency fT.

9.2.1 Phase Demodulation and Detection

The received bandpass signal in a signaling interval from an AWGN channel may be expressed as

$$r(t) = u_m(t) + n(t)$$

$$= A_{mc} g_T(t) \cos 2\pi f_c t - A_{ms} g_T(t) \sin 2\pi f_c t$$

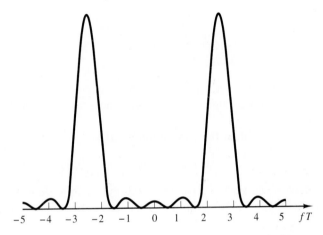

FIGURE 9.12.   The power-spectral density for M-ary PSK signal with uncorrelated information symbols.

$$+ n_c(t) \cos 2\pi f_c t - n_s(t) \sin 2\pi f_c t \qquad (9.2.17)$$

where A_{mc} and A_{ms} are defined by (9.2.6). The received signal may be correlated with $\psi_1(t) = \sqrt{\frac{2}{\mathcal{E}_g}} g_T(t) \cos 2\pi f_c t$ and $\psi_2(t) = \sqrt{\frac{2}{\mathcal{E}_g}} g_T(t) \sin 2\pi f_c t$. The outputs of the two correlators yield the two noise-corrupted signal components, which may be expressed as

$$\mathbf{r} = \mathbf{s}_m + \mathbf{n}$$

$$= \left(\sqrt{\mathcal{E}_s} \cos 2\pi m/M + n_c, \quad \sqrt{\mathcal{E}_s} \sin 2\pi m/M + n_s \right) \qquad (9.2.18)$$

where, by definition

$$n_c = \frac{1}{\sqrt{2\mathcal{E}_g}} \int_{-\infty}^{\infty} g_T(t) n_c(t) dt$$

$$n_s = \frac{1}{\sqrt{2\mathcal{E}_g}} \int_{-\infty}^{\infty} n_s(t) g_T(t) dt \qquad (9.2.19)$$

Because the quadrature noise components $n_c(t)$ and $n_s(t)$ are zero mean and uncorrelated (see the argument following (3.6.10)), it follows that $E[n_c] = E[n_s] = 0$ and $E[n_c n_s] = 0$.

The variance of the noise components is

$$E\left[n_c^2\right] = E\left[n_s^2\right] = \frac{1}{2\mathcal{E}_g} \int_{-\infty}^{\infty} \int_{-\infty}^{\infty} g_T(t) g_T(\tau) E[n_c(t) n_c(\tau)]$$

$$= \frac{1}{2\mathcal{E}_g} \int_{-\infty}^{\infty} \int_{-\infty}^{\infty} g_T(t) g_T(\tau) R_{n_c}(t - \tau) dt \, d\tau$$

$$= \frac{N_0}{2\mathcal{E}_g} \int_{-W}^{W} |G_T(f)|^2 df$$

$$= N_0/2 \qquad (9.2.20)$$

The optimum detector projects the received signal vector onto each of the M possible transmitted signal vectors $\{\mathbf{s}_m\}$ and selects the vector corresponding to the largest projection.

Thus, we obtain the correlation metrics

$$C(\mathbf{r}, \mathbf{s}_m) = \mathbf{r} \cdot \mathbf{s}_m, \quad m = 0, 1, \dots, M - 1 \qquad (9.2.21)$$

Because all signals have equal energy, an equivalent detector metric for digital-phase modulation is to compute the phase of the received signal vector $\mathbf{r} = (r_1, r_2)$ as

$$\Theta_r = \tan^{-1} \frac{r_2}{r_1} \qquad (9.2.22)$$

and select the signal from the set $\{\mathbf{s}_m\}$ whose phase is closest to Θ_r. In the following section, we evaluate the probability of error based on the phase metric given by (9.2.22).

We should mention the problem associated with a carrier-phase offset of the received signal. In general, the oscillators employed at the transmitter and the receiver are not phase-locked. As a consequence, the received signal will be of the form

$$r(t) = A_{mc} g_T(t) \cos(2\pi f_c t + \phi) - A_{ms} g_T(t) \sin(2\pi f_c t + \phi) + n(t) \qquad (9.2.23)$$

where ϕ is the carrier-phase offset. This phase offset must be estimated at the demodulator as described in Section 9.2.3, and the phase estimate must be used in the correlation operation. Hence, the received signal must be correlated with the two orthogonal functions $\psi_1(t) = \sqrt{\frac{2}{\mathcal{E}_g}} g_T(t) \cos(2\pi f_c t + \hat{\phi})$ and $\psi_2(t) = -\sqrt{\frac{2}{\mathcal{E}_g}} g_T(t) \sin(2\pi f_c t + \hat{\phi})$, where $\hat{\phi}$ is the estimate of the carrier phase, as shown in Figure 9.13, for the case in which $g_T(t)$ is a rectanglar pulse.

Symbol timing for a digital-phase-modulated signal can be recovered from the received signal by any of the methods described in Section 8.5, which are appropriate for symbol synchronization from the analog-received signal. Also, as indicated in Section 9.1.1, symbol timing can be obtained by processing a sampled version of the received signal. In such a case, the timing recovery loop is implemented as a digital loop. The paper by Mueller and Muller (1976) is a good reference for timing recovery methods based on sampled signals.

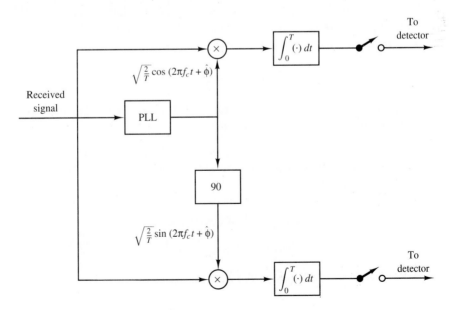

FIGURE 9.13.   Demodulator for PSK signals.

Channel distortion is another problem that is likely to be encountered in a practical communication system employing phase modulation. In such a case, we may design optimum transmitting and receiving filters when the channel frequency response characteristic is known, using the design methodology described in Section 9.1.3. On the other hand, if the channel frequency response is not known a priori, an adaptive equalizer may be used to compensate for the channel distortion.

9.2.2 Probability of Error for Phase Modulation in an AWGN Channel

In this section we shall evaluate the probability of error for M-ary-phase modulation in AWGN with the optimum demodulator and detector. The optimum detector based on the phase metric given by (9.2.22) will be used in this computation. It is assumed that a perfect estimate of the received carrier phase is available. Consequently, the performance that we derive below is for ideal phase modulation.

Let us consider the case in which the transmitted signal phase is $\theta = 0$, corresponding to the signal $u_0(t)$. Hence, the transmitted signal vector is

$$\mathbf{s}_0 = \left(\sqrt{\mathcal{E}_s}, 0\right) \tag{9.2.24}$$

and the received signal vector has components

$$r_1 = \sqrt{\mathcal{E}_s} + n_c$$

$$r_2 = n_s \tag{9.2.25}$$

Because n_c and n_s are jointly Gaussian random variables, it follows that r_1 and r_2 are jointly Gaussian random variables with $E[r_1] = \sqrt{\mathcal{E}_s}$, $E[r_2] = 0$ and $\sigma_{r_1}^2 = \sigma_{r_2}^2 = N_0/2 = \sigma_r^2$. Consequently,

$$f_{\mathbf{r}}(r_1, r_2) = \frac{1}{2\pi\sigma_r^2} e^{-\left[(r_1 - \sqrt{\mathcal{E}_s})^2 + r_2^2\right]/2\sigma_r^2} \tag{9.2.26}$$

The detector metric is the phase $\Theta_r = \tan^{-1} r_2/r_1$. The p.d.f. of Θ_r is obtained by a change in variables from (r_1, r_2) to

$$V = \sqrt{r_1^2 + r_2^2}$$

$$\Theta_r = \tan^{-1} \frac{r_2}{r_1} \tag{9.2.27}$$

This change in variables yields the joint p.d.f.

$$f_{V,\Theta_r}(v, \theta_r) = \frac{v}{2\pi\sigma_r^2} e^{-(v^2 + \mathcal{E}_s - 2\sqrt{\mathcal{E}_s}v\cos\theta_r)/2\sigma_r^2} \tag{9.2.28}$$

Integration of $f_{V,\Theta_r}(v, \theta_r)$ over the range of v yields $f_{\Theta_r}(\theta_r)$. That is,

$$f_{\Theta_r}(\theta_r) = \int_0^\infty f_{V,\Theta_r}(v, \theta_r)\, dv$$

$$= \frac{1}{2\pi} e^{-2\rho_s \sin^2 \theta_r} \int_0^\infty v e^{-\left(v - \sqrt{4\rho_s} \cos\theta_r\right)^2 / 2} \, dv \qquad (9.2.29)$$

where for convenience, we have defined the symbol SNR as $\rho_s = \mathcal{E}_s / N_0$. Figure 9.14 illustrates $f_{\Theta_r}(\theta_r)$ for several values of the SNR parameter ρ_s when the transmitted phase is zero. Note that $f_{\Theta_r}(\theta_r)$ becomes narrower and more peaked about $\theta_r = 0$ as the SNR ρ_s increases.

When $u_0(t)$ is transmitted, a decision error is made if the noise causes the phase to fall outside the range $-\pi/M \leq \Theta_r \leq \pi/M$. Hence, the probability of a symbol error is

$$P_M = 1 - \int_{-\pi/M}^{\pi/M} f_{\Theta_r}(\theta) d\theta \qquad (9.2.30)$$

In general, the integral of $f_{\Theta_r}(\theta)$ does not reduce to a simple form and must be evaluated numerically, except for $M = 2$ and $M = 4$.

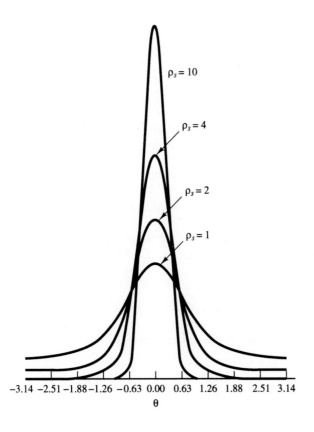

$\rho_s = 10$

$\rho_s = 4$

$\rho_s = 2$

$\rho_s = 1$

−3.14 −2.51 −1.88 −1.26 −0.63 0.00 0.63 1.26 1.88 2.51 3.14

θ

FIGURE 9.14. Probability density function $p(\theta)$ for $\rho_s = 1, 2, 4, 10$.

For binary-phase modulation, the two signals $u_0(t)$ and $u_1(t)$ are antipodal and, hence, the error probability is

$$P_2 = Q\left(\sqrt{\frac{2\mathcal{E}_b}{N_0}}\right) \tag{9.2.31}$$

When $M = 4$, we have in effect two binary-phase-modulation signals in phase quadrature. With a perfect estimate of the carrier phase, there is no crosstalk or interference between the signals on the two quadrature carriers and, hence, the bit error probability is identical to that in (9.2.31). On the other hand, the symbol error probability for $M = 4$ is determined by noting that

$$P_c = (1 - P_2)^2 = \left[1 - Q\left(\sqrt{\frac{2\mathcal{E}_b}{N_0}}\right)\right]^2 \tag{9.2.32}$$

where P_c is the probability of a correct decision for the 2-bit symbol. The result in (9.2.32) follows from the statistical independence of the noise on the quadrature carriers. Therefore, the symbol error probability for $M = 4$ is

$$P_4 = 1 - P_c$$

$$= 2Q\left(\sqrt{\frac{2\mathcal{E}_b}{N_0}}\right)\left[1 - \frac{1}{2}Q\left(\sqrt{\frac{2\mathcal{E}_b}{N_0}}\right)\right] \tag{9.2.33}$$

For $M > 4$, the symbol error probability P_M is obtained by numerically integrating (9.2.30). Figure 9.15 illustrates this error probability as a function of the SNR per bit for $M = 2, 4, 8, 16$, and 32. The graphs clearly illustrate the penalty in SNR per bit as M increases beyond $M = 4$. For example, at $P_M = 10^{-5}$, the difference between $M = 4$ and $M = 8$ is approximately 4 dB, and the difference between $M = 8$ and $M = 16$ is approximately 5 dB. For large values of M, doubling the number of phases requires an additional 6 dB per bit to achieve the same performance.

An approximation to the error probability for large values of M and for large SNR may be obtained by first approximating $f_{\Theta_r}(\theta)$. For $\mathcal{E}_s/N_0 \gg 1$ and $|\Theta_r| \leq \pi/2$, $f_{\Theta_r}(\theta_r)$ is well approximated as

$$f_{\Theta_r}(\theta_r) \approx \sqrt{\frac{2\rho_s}{\pi}} \cos\theta_r e^{-2\rho_s \sin^2\theta_r} \tag{9.2.34}$$

By substituting for $f_{\Theta_r}(\theta_r)$ in (9.2.30) and performing the change in variable from θ_r to $u = 2\sqrt{\rho_s}\sin\theta_r$, we find that

$$P_M \approx 1 - \int_{-\pi/M}^{\pi/M} \sqrt{\frac{2\rho_s}{\pi}} \cos\theta_r e^{-2\rho_s \sin^2\theta_r}\, d\theta_r$$

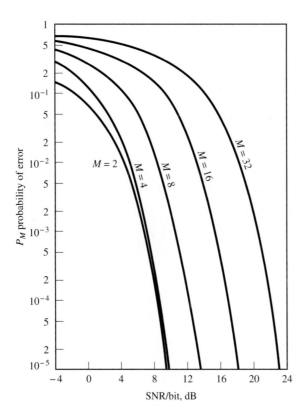

FIGURE 9.15. Probability of a symbol error for PSK signals.

$$\approx \frac{2}{\sqrt{2\pi}} \int_{\sqrt{2\rho_s}\,\sin\pi/M}^{\infty} e^{-u^2/2}\,du$$

$$= 2Q\left(\sqrt{2\rho_s}\,\sin\frac{\pi}{M}\right) = 2Q\left(\sqrt{2k\rho_b}\,\sin\frac{\pi}{M}\right) \qquad (9.2.35)$$

where $k = \log_2 M$ and $\rho_s = k\rho_b$. We note that this approximation to the error probability is good for all values of M. For example, when $M = 2$ and $M = 4$, we have $P_2 = P_4 = 2Q\left(\sqrt{2\rho_s}\right)$, which compares favorably (a factor of 2 difference) with the exact error probability given by (9.2.31).

The equivalent bit error probability for M-ary-phase modulation is rather tedious to derive due to its dependence on the mapping of k-bit symbols into the corresponding signal phases. When a Gray code is used in the mapping, two k-bit symbols corresponding to adjacent signal phases differ in only a single bit. Because the most probable errors due to noise result in the erroneous selection of an adjacent phase to the true phase, most k-bit symbol errors contain only a single bit error. Hence, the equivalent bit-error probability for M-ary-phase modulation is well approximated as

$$P_b \approx \frac{1}{k} P_M \qquad (9.2.36)$$

9.2.3 Carrier Phase Estimation

When the digital information is transmitted via M-phase modulation of a carrier, a PLL may be used to estimate the carrier phase offset. For $M = 2$, the squaring PLL and the Costas loop described in Section 6.2 are directly applicable.

For $M > 2$, the received signal may first be raised to the Mth power as shown in Figure 9.16. Thus, if the received signal $r(t)$ has the form

$$r(t) = s_m(t) + n(t)$$
$$= g_T(t) \cos\left(2\pi f_c t + \phi + \frac{2\pi m}{M}\right) + n(t) \qquad (9.2.37)$$

and we pass $r(t)$ through an Mth power device, the output signal will contain harmonics of the carrier f_c. The harmonic that we wish to select is $\cos\left(2\pi M f_c t + M\theta\right)$ for driving the PLL. We note that

$$M\theta = M\left(\frac{2\pi m}{M}\right) = 2\pi m = 0 \ (\text{mod } 2\pi), \quad m = 1, 2, \ldots, M \qquad (9.2.38)$$

Thus, the information is removed from the Mth harmonic. The bandpass filter tuned to the frequency $M f_c$ produces the desired frequency component $\cos(2\pi M f_c t + M\phi)$ driving the PLL. The VCO output is $\sin(2\pi M f_c t + M\hat{\phi})$, so this output is divided

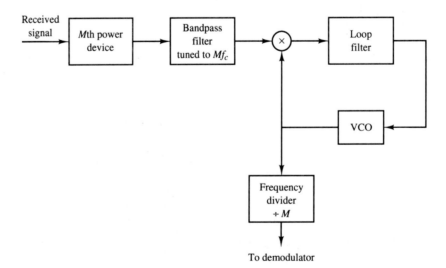

FIGURE 9.16. Carrier-phase estimation for M-ary PSK signals.

in frequency by M to yield $\sin(2\pi f_c t + \hat{\phi})$ and phase-shifted by $\pi/2$ to yield $\cos(2\pi f_c t + \hat{\phi})$. The two quadrature-carrier components are then passed to the demodulator.

We should note that the quadrature-phase carrier components generated as described above contain phase ambiguities of multiples of $2\pi/M$ that result from multiplying the carrier phase ϕ by M. Because $M\phi \pmod{2\pi}$ is less than 2π, dividing the resulting angle by M yields a phase estimate of $|\hat{\phi}| < 2\pi/M$, when, in fact, the true carrier phase may exceed this estimate by multiples of $2\pi/M$, i.e., by $2\pi k/M$, for $k = 1, 2, \ldots, M - 1$. Such phase ambiguities can be overcome by differentially encoding the data at the transmitter and differentially decoding at the detector, as described in Section 9.2.4.

Just as in the case of the squaring PLL, the Mth power PLL operates in the presence of noise that has been enhanced by the Mth power-law device. The variance of the phase error in the PLL resulting from the additive noise may be expressed in the simple form

$$\sigma_{\hat{\phi}}^2 = \frac{1}{S_{ML}\rho_L} \tag{9.2.39}$$

where ρ_L is the loop SNR and S_{ML} is the *M phase power loss*. S_{ML} has been evaluated by Lindsey and Simon (1973) for $M = 4$ and $M = 8$.

Another method for extracting a carrier-phase estimate $\hat{\phi}$ from the received signal for M-ary-phase modulation is the decision-feedback PLL (DFPLL), which is shown in Figure 9.17. The received signal is demodulated by using two quadrature phase-locked carriers to yield $\mathbf{r} = (r_1, r_2)$ at the sampling instants. The phase estimate $\hat{\Theta}_r = \tan^{-1} r_2/r_1$ is computed at the detector and quantized to the nearest of the M possible transmitted phase, which we denote as θ_m. The two outputs of the quadrature multipliers are delayed by one symbol interval T and multiplied by $\cos\theta_m$ and $-\sin\theta_m$. Thus, we obtain

$$-r(t)\cos(2\pi f_c t + \hat{\phi})\sin\theta_m = -\frac{1}{2}\left[g_T(t)\cos\theta_m + n_c(t)\right]\sin\theta_m\cos\left(\phi - \hat{\phi}\right)$$

$$+ \frac{1}{2}\left[g_T(t)\sin\theta_m + n_s(t)\right]\sin\theta_m\sin(\phi - \hat{\phi})$$

$$+ \text{double-frequency terms}$$

$$-r(t)\sin\left(2\pi f_c t + \hat{\phi}\right)\cos\theta_m = \frac{1}{2}\left[g_T(t)\cos\theta_m + n_c(t)\right]\cos\theta_m\sin\left(\phi - \hat{\phi}\right)$$

$$+ \frac{1}{2}\left[g_T(t)\sin\theta_m + n_s(t)\right]\cos\theta_m\cos\left(\phi - \hat{\phi}\right)$$

$$+ \text{double-frequency terms}$$

These two signals are added together to generate the error signal

$$e(t) = g_T(t)\sin(\phi - \hat{\phi}) + \frac{1}{2}n_c(t)\sin\left(\phi - \hat{\phi} - \theta_m\right)$$

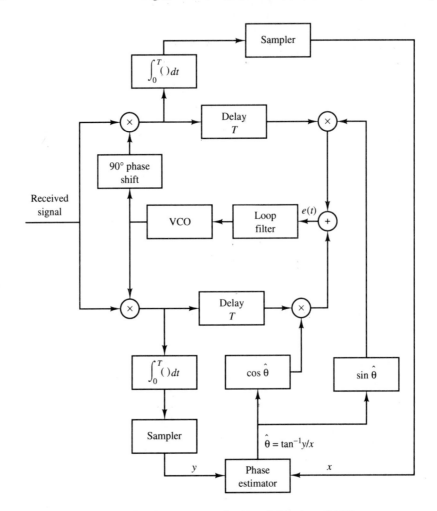

FIGURE 9.17. Carrier recovery for M-ary PSK using a DFPLL.

$$+ \frac{1}{2} n_s(t) \cos\left(\phi - \hat{\phi} - \theta_m\right)$$

$+$ double-frequency terms   (9.2.40)

This error signal is the input to the loop filter that provides the control signal for the VCO.

We observe that the two quadrature noise components in (9.2.40) appear as additive terms and no term involves a product of two noise components as in the output of the Mth power-law device. Consequently, there is no power loss resulting from nonlinear operations on the received signal in the DFPLL.

The M-phase decision-feedback tracking loop also has a phase ambiguity of $2\pi/M$, necessitating the need for differentially encoding the information sequence prior to transmission and differentially decoding the received sequence at the detector to recover the information.

9.2.4 Differential-Phase Modulation and Demodulation

The performance of ideal, coherent phase modulation/demodulation is closely attained in communication systems that transmit a carrier-signal along with the information signal. The carrier signal component may be filtered from the received signal and used to perform phase-coherent demodulation. However, when no separate carrier signal is transmitted, the receiver must estimate the carrier phase from the received signal. As indicated in the preceding section, the phase at the output of a PLL has ambiguities of multiples of $2\pi/M$, necessitating the need to differentially encode the data prior to modulation. This differential encoding allows us to decode the received data at the detector in the presence of the phase ambiguities.

In differential encoding, the information is conveyed by phase shifts relative to the previous signal interval. For example, in binary phase modulation the information bit 1 may be transmitted by shifting the phase of the carrier by $180°$ relative to the previous carrier phase, while the information bit 0 is transmitted by a zero-phase shift relative to the phase in the preceding signaling interval. In four-phase modulation, the relative phase shifts between successive intervals are $0°, 90°, 180°$, and $270°$, corresponding to the information bits $00, 01, 11, 10$, respectively. The generalization of differential encoding for $M > 4$ is straightforward. The phase-modulated signals resulting from this encoding process are called *differentially encoded*. The encoding is performed by a relatively simple logic circuit preceding the modulator.

Demodulation and detection of the differentially encoded phase-modulated signal may be performed as described in the preceding section using the output of a PLL to perform the demodulation. The received signal phase $\Theta_r = \tan^{-1} r_2/r_1$ at the detector is mapped into one of the M possible transmitted signal phases $\{\theta_m\}$ that is closest to Θ_r. Following the detector is a relatively simple phase comparator that compares the phases of the detected signal over two consecutive intervals to extract the transmitted information. Thus, phase ambiguities of $2\pi/M$ are rendered irrelevant.

Coherent demodulation of differentially encoded phase-modulated signals results in a higher probability of error than the error probability derived for absolute-phase encoding. With differentially encoded signals, an error in the detected phase due to noise will frequently result in decoding errors over two consecutive signaling intervals. This is especially the case for error probabilities below 10^{-1}. Therefore, the probability of error for differentially encoded M-ary-phase-modulated signals is approximately twice the probability of error for M-ary-phase modulation with absolute phase encoding. However, a factor-of-2 increase in the error probability translates into a relatively small loss in SNR, as can be seen from Figure 9.15.

A differentially encoded phase-modulated signal also allows another type of demodulation that does not require the estimation of the carrier phase. Instead, the received signal in any given signaling interval is compared to the phase of the received signal from the preceding signaling interval. To elaborate, suppose that we demodulate the differentially encoded signal by multiplying $r(t)$ with $\cos 2\pi f_c t$ and $\sin 2\pi f_c t$ and integrating the two products over the interval T. At the kth signaling interval, the demodulater output is

$$r_k = \sqrt{\mathcal{E}_s} e^{j(\theta_k - \phi)} + n_k \tag{9.2.41}$$

where θ_k is the phase angle of the transmitted signal at the kth signaling interval, ϕ is the carrier phase, and $n_k = n_{kc} + jn_{ks}$ is the noise vector. Similarly, the received signal vector at the output of the demodulator in the preceding signaling interval is

$$r_{k-1} = \sqrt{\mathcal{E}_s} e^{j(\theta_{k-1} - \phi)} + n_{k-1} \tag{9.2.42}$$

The decision variable for the phase detector is the phase difference between these two complex numbers. Equivalently, we can project r_k onto r_{k-1} and use the phase of the resulting complex number; that is,

$$r_k r_{k-1}^\star = \mathcal{E}_s e^{j(\theta_k - \theta_{k-1})} + \sqrt{\mathcal{E}_s} e^{j(\theta_k - \phi)} n_{k-1}$$
$$+ \sqrt{\mathcal{E}_s} e^{-j(\theta_{k-1} - \phi)} n_k + n_k n_{k-1}^\star \tag{9.2.43}$$

which, in the absence of noise, yields the phase difference $\theta_k - \theta_{k-1}$. Thus, the mean value of $r_k r_{k-1}^\star$ is independent of the carrier phase. Differentially encoded PSK signaling that is demodulated and detected as described above is called *differential PSK* (DPSK).

The demodulation and detection of DSPK using matched filters is illustrated in Figure 9.18. If the pulse $g_T(t)$ is rectangular, the matched filters may be replaced by integrate-and-dump filters.

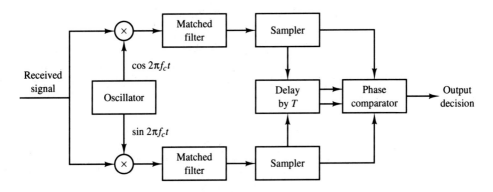

FIGURE 9.18.   Block diagram of DPSK demodulator.

9.2.5 Probability of Error for DPSK in an AWGN Channel

Let us now consider the evaluation of the error probability performance of a DPSK demodulator and detector. The derivation of the exact value of the probability of error for M-ary DPSK is extremely difficult, except for $M = 2$. The major difficulty is encountered in the determination of the p.d.f. for the phase of the random variable $r_k r_{k-1}^\star$, given by (9.2.43). However, an approximation to the performance of DPSK is easily obtained, as we now demonstrate.

Without loss of generality, suppose the phase difference $\theta_k - \theta_{k-1} = 0$. Furthermore, the exponential factors $e^{-j(\theta_{k-1}-\phi)}$ and $e^{j(\theta_k-\phi)}$ in (9.2.43) can be absorbed into the Gaussian noise components n_{k-1} and n_k, (see Problem 3.29) without changing their statistical properties. Therefore, $r_n r_{n-1}^\star$ in (9.2.43) can be expressed as

$$r_k r_{k-1}^\star = \mathcal{E}_s + \sqrt{\mathcal{E}_s}(n_k + n_{k-1}^\star) + n_k n_{k-1}^\star \qquad (9.2.44)$$

The complication in determining the p.d.f. of the phase is the term $n_k n_{k-1}^\star$. However, at SNR's of practical interest, the term $n_k n_{k-1}^\star$ is small relative to the dominant noise term $\sqrt{\mathcal{E}_s}(n_k + n_{k-1}^\star)$. If we neglect the term $n_k n_{k-1}^\star$ and we also normalize $r_k r_{k-1}^\star$ by dividing through by $\sqrt{\mathcal{E}_s}$, the new set of decision metrics become

$$x = \sqrt{\mathcal{E}_s} + \mathrm{Re}(n_k + n_{k-1}^\star)$$
$$y = \mathrm{Im}(n_k + n_{k-1}^\star) \qquad (9.2.45)$$

The variables x and y are uncorrelated Gaussian random variables with identical variances $\sigma_n^2 = N_0$. The phase is

$$\Theta_r = \tan^{-1}\frac{y}{x} \qquad (9.2.46)$$

At this stage we have a problem that is identical to the one we solved previously for phase-coherent demodulation. The only difference is that the noise variance is now twice as large as in the case of PSK. Thus we conclude that the performance of DPSK is 3 dB poorer than that for PSK. This result is relatively good for $M \geq 4$, but it is pessimistic for $M = 2$ in the sense that the loss in binary DPSK relative to binary PSK is less than 3 dB at large SNR. This is demonstrated below.

In binary DPSK, the two possible transmitted phase differences are zero and π radians. As a consequence, only the real part of $r_k r_{k-1}^\star$ is needed for recovering the information. Using (9.2.45), we express the real part as

$$\mathrm{Re}(r_k r_{k-1}^\star) = \frac{1}{2}(r_k r_{k-1}^\star + r_k^\star r_{k-1})$$

Because the phase difference between the two successive signaling intervals is zero, an error is made if $\mathrm{Re}(r_k r_{k-1}^\star)$ is less than zero. The probability that $r_k r_{k-1}^\star + r_k^\star r_{k-1} < 0$ is a special case of a derivation, given in Appendix A, concerned with the probability that a general quadratic form in complex-valued Gaussian random

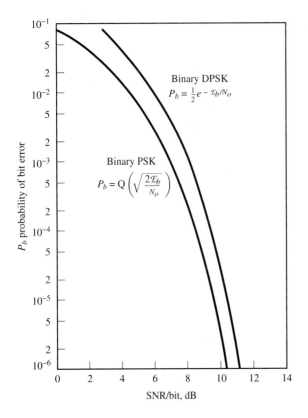

FIGURE 9.19. Probability of error for binary PSK and DPSK.

variables is less than zero. The result for the error probability of binary DPSK is

$$P_2 = \frac{1}{2} e^{-\rho_b} \qquad (9.2.47)$$

where $\rho_b = \mathcal{E}_b/N_0$ is the SNR per bit.

The graph of (9.2.47) is shown in Figure 9.19. Also shown in this figure is the probability of error for binary PSK. We observe that at error probabilities below 10^{-4}, the difference in SNR between binary PSK and binary DPSK is less than 1 dB.

9.3　QUADRATURE AMPLITUDE MODULATION

A quadrature amplitude-modulated (QAM) signal employs two quadrature carriers $\cos 2\pi f_c t$ and $\sin 2\pi f_c t$, each of which is amplitude modulated in accordance with the sequence of information bits. We may view this method of signal transmission as a form of quadrature-carrier multiplexing, previously described in Section 5.2.6.

The transmitted signal waveforms have the form

$$u_m(t) = A_{mc}g_T(t)\cos 2\pi f_c t + A_{ms}g_T(t)\sin 2\pi f_c t, \quad m = 1, 2, \ldots, M \quad (9.3.1)$$

where $\{A_{mc}\}$ and $\{A_{ms}\}$ are the sets of amplitude levels that are obtained by mapping k-bit sequences into signal amplitudes. For example, Figure 9.20 illustrates a 16-QAM signal constellation that is obtained by amplitude modulating each quadrature carrier by $M = 4$ PAM. In general, rectangular signal constellations result when two quadrature carriers are each modulated by PAM.

More generally, QAM may be viewed as a form of combined digital amplitude and digital-phase modulation. Thus, the transmitted QAM signal waveforms may be expressed as

$$u_{mn}(t) = A_m g_T(t)\cos\left(2\pi f_c t + \theta_n\right), \quad m = 1, 2, \ldots, M_1$$
$$n = 1, 2, \ldots, M_2 \quad (9.3.2)$$

If $M_1 = 2^{k_1}$, and $M_2 = 2^{k_2}$, the combined amplitude and phase-modulation method results in the simultaneous transmission of $k_1 + k_2 = \log_2 M_1 M_2$ binary digits occurring at a symbol rate $R_b/(k_1 + k_2)$. Figure 9.21 illustrates the functional block diagram of a QAM modulator.

It is clear that the geometric signal representation of the signals given by (9.3.1) and (9.3.2) is in terms of two-dimensional signal vectors of the form

$$\mathbf{s}_m = \left(\sqrt{\mathcal{E}_s}A_{mc}, \sqrt{\mathcal{E}_s}A_{ms}\right) \quad m = 1, 2, \ldots, M \quad (9.3.3)$$

Examples of signal space constellations for QAM are shown in Figure 9.22.

The power-spectral density of a QAM signal is identical in form to that for a digital-phase-modulation signal given by (9.2.14)–(9.2.16). The transmitted bandpass signal $u(t)$ for a sequence of information symbols may be expressed as in (9.2.9), where the baseband information-bearing signal is

$$v(t) = \sum_{n=-\infty}^{\infty} A_n e^{j\theta_n} g_T(t - nT) \quad (9.3.4)$$

In (9.3.4) A_n is the signal amplitude and θ_n is the signal phase of the transmitted carrier. Hence, the autocorrelation of the transmitted sequence of amplitudes and

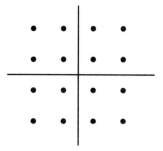

FIGURE 9.20. $M = 16$-QAM signal constellation.

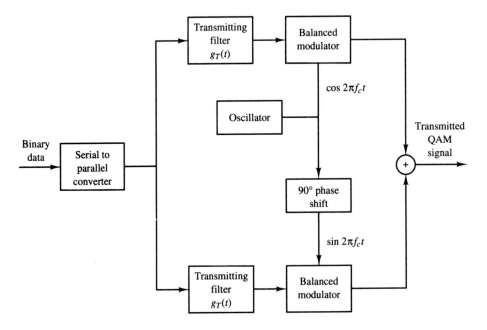

FIGURE 9.21.   Functional block diagram of modulator for QAM.

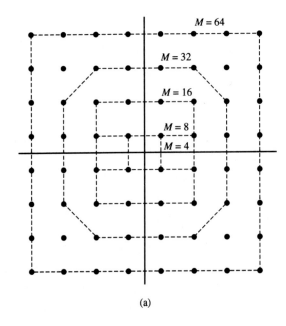

(a)

FIGURE 9.22.   (a) Rectangular
signal-space constellations for QAM.

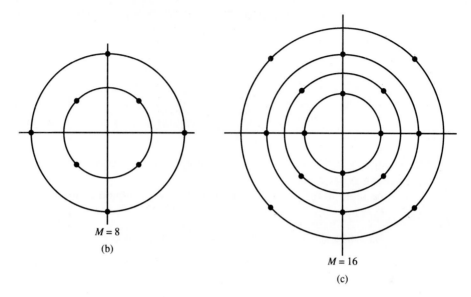

$M = 8$

(b)

$M = 16$

(c)

FIGURE 9.22. (*continued*) (b,c) Examples of combined PAM-PSK signal-space diagrams.

phases is

$$R_A(n) = E\left[A_k e^{-j\theta_k} A_{k+n} e^{j\theta_{k+n}}\right]$$

$$= E\left[A_k A_{k+n} e^{j(\theta_{k+n}-\theta_k)}\right] \tag{9.3.5}$$

In the special case in which the symbols in the transmitted sequence are uncorrelated and zero mean, the autocorrelation in (9.3.5) reduces to

$$R_A(n) = \begin{cases} \sigma_A^2, & n = 0 \\ 0, & n \neq 0 \end{cases} \tag{9.3.6}$$

where $\sigma_A^2 = E[A_k^2]$. In this case the power-spectral density of the input sequence is white and, hence, the average power-spectral density of the transmitted signal depends solely on the spectral characteristics of the transmitted pulse, i.e., on $|G_T(f)|^2$.

9.3.1 Demodulation and Detection of QAM

Let us assume that a carrier-phase offset is introduced in the transmission of the signal through the channel. In addition, the received signal is corrupted by additive Gaussian noise. Hence, $r(t)$ may be expressed as

$$r(t) = A_{mc}g_T(t)\cos(2\pi f_c t + \phi) + A_{ms}g_T(t)\sin(2\pi f_c t + \phi) + n(t) \tag{9.3.7}$$

Suppose that an estimate $\hat{\phi}$ of the carrier phase is available at the demodulator. Then, the received signal may be correlated with the two basis functions

$$\psi_1(t) = \sqrt{\frac{2}{\mathcal{E}_g}}\, g_T(t) \cos(2\pi f_c t + \hat{\phi})$$

$$\psi_2(t) = \sqrt{\frac{2}{\mathcal{E}_g}}\, g_T(t) \sin(2\pi f_c t + \hat{\phi}) \tag{9.3.8}$$

as illustrated in Figure 9.23, and the outputs of the correlators are sampled and passed to the detector.

The input to the detector consists of the two sampled components r_1, r_2, where $(\mathcal{E}_s \equiv \mathcal{E}_g/2)$

$$r_1 = A_{mc}\sqrt{\mathcal{E}_s}\cos(\phi - \hat{\phi}) + A_{ms}\sqrt{\mathcal{E}_s}\sin(\phi - \hat{\phi})$$
$$+ n_c \sin\hat{\phi} - n_s \cos\hat{\phi} \tag{9.3.9}$$

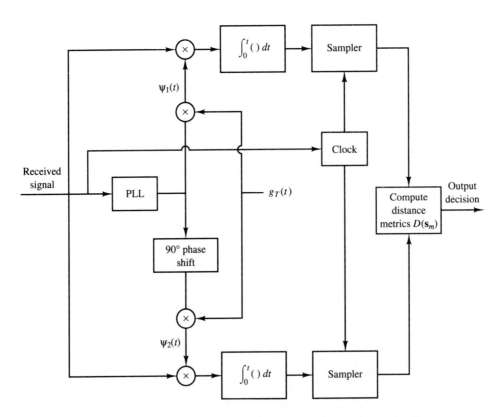

FIGURE 9.23. Demodulation and detection of QAM signals.

$$r_2 = A_{mc}\sqrt{\mathcal{E}_s}\sin(\phi - \hat{\phi}) + A_{ms}\sqrt{\mathcal{E}_s}\cos(\phi - \hat{\phi}) + n_c \sin\hat{\phi} - n_s \cos\hat{\phi}$$

We observe that the effect of an imperfect phase estimate is twofold. First, the desired signal components in r_1 and r_2 are reduced in amplitude by the factor $\cos(\phi - \hat{\phi})$. In turn, this reduces the SNR by the factor $\cos^2(\phi - \hat{\phi})$. Second, there is a leakage of the quadrature signal components into the desired signal. This signal leakage, which is scaled by $\sin(\phi - \hat{\phi})$, causes a significant performance degradation unless $\phi - \hat{\phi}$ is very small. This point serves to emphasize the importance of having an accurate carrier-phase estimate in order to demodulate the QAM signal.

The optimum detector computes the distance metrics

$$D(\mathbf{r}, \mathbf{s}_m) = |\mathbf{r} - \mathbf{s}_m|^2, \quad m = 1, 2, \ldots, M \qquad (9.3.10)$$

and selects the signal corresponding to the smallest value of $D(\mathbf{r}, \mathbf{s}_m)$. If a correlation metric is used in place of a distance metric, it is important to recognize that correlation metrics must employ bias correction because the QAM signals are not equal energy signals.

Symbol timing may be extracted from the received signal by using one of the methods for symbol synchronization described in Section 8.5. For QAM signals, the spectral line methods described in Section 8.5.1 have proved to be particularly suitable for timing recovery. Figure 9.24 illustrates a spectral line method which is based on filtering out a signal component at the frequency $\frac{1}{2T}$ and squaring the filter output to generate a sinusoidal signal at the desired symbol rate $\frac{1}{T}$. Because the demodulation of the QAM signal is accomplished, as described above, by multiplication of the input signal with the two quadrature-carrier signals $\psi_1(t)$ and $\psi_2(t)$, the in-phase and quadrature signal components at the outputs of the two correlators are used as the inputs to the two bandpass filters tuned to $1/2T$. The two filter outputs are squared (rectified), summed, and then filtered by a narrowband filter tuned to the clock frequency $1/T$. Thus, we generate a sinusoidal signal that is the appropriate clock signal for sampling the outputs of the correlators to recover the information.

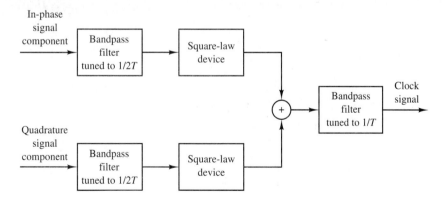

FIGURE 9.24.   Block diagram of timing recovery method for QAM.

Finally, we should consider the problem of linear channel distortion that may be encountered in a practical communication system employing QAM modulation. If the channel frequency response characteristic $C_{bp}(f)$ (or the equivalent baseband frequency response) is known, we may use the methodogy described in Section 9.1.3 to design optimum transmitting and receiving filters. On the other hand, if the channel frequency response is not known a priori, an adaptive equalizer may be used to compensate for the channel distortion. The equalizer may be implemented either as a bandpass equalizer or as a baseband equalizer.

9.3.2 Probability of Error for QAM in AWGN Channel

To determine the probability of error for QAM, we must specify the signal point constellation. We begin with QAM signal sets that have $M = 4$ points. Figure 9.25 illustrates two four-point signal sets. The first is a four-phase-modulated signal and the second is a QAM signal with two amplitude levels, labeled A_1 and A_2, and four phases. Because the probability of error is dominated by the minimum distance between pairs of signal points, let us impose the condition that $d_{\min} = 2A$ for both signal constellations and let us evaluate the average transmitter power, based on the premise that all signal points are equally probable. For the four-phase signal we have

$$P_{av} = \frac{1}{4}(4)A^2 = A^2 \qquad (9.3.11)$$

For the two-amplitude, four-phase QAM, we place the points on circles of radii A and $\sqrt{3}A$. Thus, $d_{\min} = 2A$, and

$$P_{av} = \frac{1}{4}\left[2\left(\frac{3}{2}\right)A^2 + 2\frac{A}{2}^2\right] = A^2 \qquad (9.3.12)$$

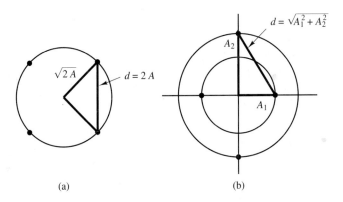

(a)   (b)

FIGURE 9.25.   Two 4-point signal constellations.

which is the same average power as the $M = 4$-phase signal constellation. Hence, for all practical purposes, the error rate performance of the two signal sets is the same. In other words, there is no advantage of the two-amplitude QAM signal set over $M = 4$ phase modulation.

Next, let us consider $M = 8$ QAM. In this case there are many possible signal constellations. We shall consider the four signal constellations shown in Figure 9.26, all of which consist of two amplitudes and have a minimum distance between signal points of $2A$. The coordinates (A_{mc}, A_{ms}) for each signal point, normalized by A, are given in the figure. Assuming that the signal points are equally

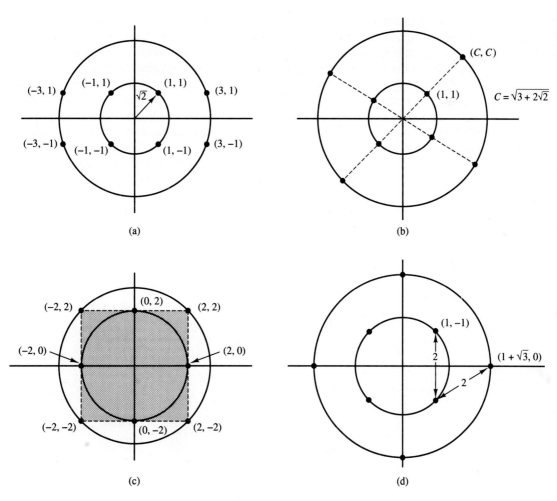

FIGURE 9.26.   Four 8-point QAM signal constellations.

probable, the average transmitted signal power is

$$P_{av} = \frac{1}{M} \sum_{m=1}^{M} \frac{1}{2} \left(A_{mc}^2 + A_{ms}^2 \right)$$

$$= \frac{A^2}{2M} \sum_{m=1}^{M} \left(a_{mc}^2 + a_{ms}^2 \right) \tag{9.3.13}$$

where (a_{mc}, a_{ms}) are the coordinates of the signal points, normalized by A.

The first two signal sets in Figure 9.26 contain signal points that fall on a rectangular grid and have $P_{av} = 3A^2$. The third signal set requires an average transmitted power $P_{av} = 3.41A^2$, and the fourth requires $P_{av} = 2.36A^2$. Therefore, the fourth signal set requires approximately 1 dB less power than the first two and 1.6 dB less power than the third to achieve the same probability of error. This signal constellation is known to be the best 8-point QAM constellation because it requires the least power for a given minimum distance between signal points.

For $M \geq 16$, there are many more possibilities for selecting the QAM signal points in the two-dimensional space. For example, we may choose a circular multiamplitude constellation for $M = 16$, as shown in Figure 9.27. In this case, the signal points at a given amplitude level are phase rotated by $\pi/4$ relative to the signal points at adjacent amplitude levels. This 16-QAM constellation is a generalization of the optimum 8-QAM constellation. However, the circular 16-QAM constellation is not the best 16-point QAM signal constellation for the AWGN channel.

Rectangular QAM signal constellations have the distinct advantage of being easily generated as two PAM signals impressed on phase-quadrature carriers. In addition, they are easily demodulated as previously described. Although they are

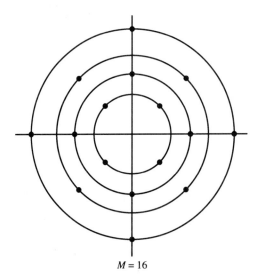

$M = 16$

FIGURE 9.27.   Circular 16-point QAM signal constellations.

not the best M-ary QAM signal constellations for $M \geq 16$, the average transmitted power required to achieve a given minimum distance is only slightly greater than the average power required for the best M-ary QAM signal constellation. For these reasons, rectangular M-ary QAM signals are most frequently used in practice.

For rectangular signal constellations in which $M = 2^k$ where k is even, the QAM signal constellation is equivalent to two PAM signals on quadrature carriers, each having $\sqrt{M} = 2^{k/2}$ signal points. Because the signals in the phase-quadrature components are perfectly separated by coherent detection when $\hat{\phi} = \phi$, the probability of error for QAM is easily determined from the probability of error for PAM. Specifically, the probability of a correct decision for the M-ary QAM system is

$$P_c = \left(1 - P_{\sqrt{M}}\right)^2 \tag{9.3.14}$$

where $P_{\sqrt{M}}$ is the probability of error of a $\sqrt{M}$-ary PAM with one-half the average power in each quadrature signal of the equivalent QAM system. By appropriately modifying the probability of error for M-ary PAM, we obtain

$$P_{\sqrt{M}} = 2\left(1 - \frac{1}{\sqrt{M}}\right)Q\left(\sqrt{\frac{3}{M-1}\frac{\mathcal{E}_{av}}{N_0}}\right) \tag{9.3.15}$$

where $\mathcal{E}_{av}/N_0$ is the average SNR per symbol. Therefore, the probability of a symbol error for the M-ary QAM is

$$P_M = 1 - \left(1 - P_{\sqrt{M}}\right)^2 \tag{9.3.16}$$

We note that this result is exact for $M = 2^k$ when k is even. On the other hand, when k is odd there is no equivalent $\sqrt{M}$-ary PAM system. This is no problem, however, because it is rather easy to determine the error rate for a rectangular signal set. If we employ the optimum detector that bases its decisions on the optimum distance metrics given by (9.3.10), it is relatively straightforward to show that the symbol error probability is tightly upper-bounded as

$$P_M \leq 1 - \left[1 - 2Q\left(\sqrt{\frac{3\mathcal{E}_{av}}{(M-1)N_0}}\right)\right]^2$$

$$\leq 4Q\left(\sqrt{\frac{3k\mathcal{E}_{bav}}{(M-1)N_0}}\right) \tag{9.3.17}$$

for any $k \geq 1$, where $\mathcal{E}_{bav}/N_0$ is the average SNR per bit. The probability of a symbol error is plotted in Figure 9.28 as a function of the average SNR per bit.

It is interesting to compare the performance of QAM with that of phase modulation for any given signal size M, because both types of signals are two-dimensional. Recall that for M-ary phase modulation, the probability of a symbol

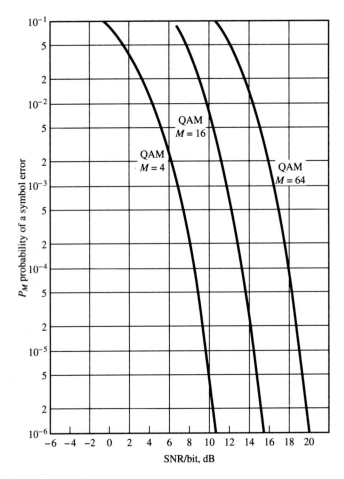

FIGURE 9.28.   Probability of a symbol error for QAM.

error is approximated as

$$P_M \approx 2Q \left(\sqrt{2\rho_s} \sin \frac{\pi}{M} \right) \tag{9.3.18}$$

where ρ_s is the SNR per symbol. For M-ary QAM, we may use the expression in (9.3.15). Because the error probability is dominated by the argument of the Q-function, we may simply compare the arguments of Q for the two signal formats. Thus, the ratio of these two arguments is

$$\mathcal{R}_M = \frac{3/(M-1)}{2 \sin^2 \pi / M} \tag{9.3.19}$$

For example, when $M = 4$, we have $\mathcal{R}_M = 1$. Hence, 4-PSK and 4-QAM yield comparable performance for the same SNR per symbol. On the other hand, when

TABLE 9-1 SNR ADVANTAGE OF M-ARY QAM OVER M-ARY PSK.

M	$10\log_{10}\mathcal{R}_M$
8	1.65
16	4.20
32	7.02
64	9.95

$M > 4$ we find that $\mathcal{R}_M > 1$, so that M-ary QAM yields better performance than M-ary PSK. Table 9.1 illustrates the SNR advantage of QAM over PSK for several values of M. For example, we observe that 32-QAM has a 7 dB SNR advantage over 32-PSK.

Finally, we note that QAM, PAM, and PSK are linear modulation methods that allow us to increase the bit-rate-to-bandwidth ratio (R_b/W) by simply increasing the number of signal points $M = 2^k$ in the constellation, where k is the number of information bits conveyed by each transmitted symbol. The rate of increase of R_b/W is logarithmic in M (linear in k). We have already observed that the cost of doubling the number of phases (increasing the number of bits per symbol by one bit) in PSK approaches 6 dB (a factor of 4) in additional transmitted power for large M. A similar comparison for QAM (same as PAM) indicates that the increase in transmitted power is approximately 3 dB per additional bit per symbol. Table 9.2 gives the factor $10\log 2(M-1)/3$, which represents the increase in average power required to maintain a given level of performance for QAM as the number of signal points in the rectangular constellation increases. Thus, we observe that QAM (and PAM) is preferable to PSK for large signal constellation sizes.

TABLE 9-2 QAM SIGNAL CONSTELLATIONS.

No. of signal points M	Increase in average power (dB) relative to $M = 2$
4	3
8	6.7
16	10.0
32	13.2
64	16.2
128	19.2

9.3.3 Carrier-Phase Estimation

As we have indicated in Section 9.3.1, the demodulation of a QAM signal requires a carrier that is phase-locked to the phase of the received carrier signal. Carrier-phase estimation for QAM can be accomplished in a number of different ways, depending on the signal point constellation and the phase relationships of the various signal points.

For example, let us consider the 8-point QAM signal constellations shown in Figure 9.26(c) and (d). The signal points in these constellations have one of two possible amplitude values and eight possible phases. The phases are spaced 45° apart. This phase symmetry allows us to use a PLL driven by the output of an 8th power-law device that generates a carrier component at $8f_c$, where f_c is the carrier frequency. Thus, the method illustrated in Figure 9.16, and described in Section 9.2.3, may be used in general for any QAM signal constellation that contains signal points with phases that are multiples of some phase angle θ, where $L\theta = 360^o$ for some integer L.

Another method for extracting a carrier-phase estimate $\hat{\phi}$ from the received M-ary QAM signal is the DFPLL previously described in Section 9.2.3. The basic idea in the DFPLL is to estimate the phase of the QAM signal in each signal interval and remove the phase modulation from the carrier. The DFPLL may be used with any QAM signal, irrespective of the phase relationships among the signal points.

To be specific, let us express the received QAM signal in the form

$$r(t) = A_m g_T(t) \cos(2\pi f_c t + \theta_n + \phi) + n(t), \qquad (9.3.20)$$

where θ_n is the phase of the signal point and ϕ is the carrier phase. This signal is demodulated by crosscorrelating $r(t)$ with $\psi_1(t)$ and $\psi_2(t)$, which are given by (9.3.8). The sampled values at the output of the correlators are

$$r_1 = A_m \sqrt{\mathcal{E}_s} \cos(\theta_n + \phi - \hat{\phi}) + n_c \cos(\theta_n + \phi - \hat{\phi}) - n_s \sin(\theta_n + \phi - \hat{\phi})$$

$$r_2 = A_m \sqrt{\mathcal{E}_s} \sin(\theta_n + \phi - \hat{\phi}) + n_c \sin(\theta_n + \phi - \hat{\phi}) - n_s \cos(\theta_n + \phi - \hat{\phi}) \qquad (9.3.21)$$

Now suppose that the detector, based on r_1 and r_2, has made the correct decision on the transmitted signal point. Then we multiply r_1 by $-\sin\theta_n$ and r_2 by $\cos\theta_n$. Thus, we obtain

$$-r_1 \sin\theta_n = -A_m \sqrt{\mathcal{E}_s} \cos(\theta_n + \phi - \hat{\phi}) \sin\theta_n + \text{noise component}$$

$$= A_m \sqrt{\mathcal{E}_s} [-\sin\theta_n \cos\theta_n \cos(\phi - \hat{\phi}) + \sin^2\theta_n \sin(\phi - \hat{\phi})]$$

$$+ \text{noise component}$$

$$r_2 \cos\theta_n = A_m \sqrt{\mathcal{E}_s} \sin(\theta_n + \phi - \hat{\phi}) \cos\theta_n + \text{noise component}$$

$$= A_m \sqrt{\mathcal{E}_s} [\sin\theta_n \cos\theta_n \cos(\phi - \hat{\phi}) + \cos^2\theta_n \sin(\phi - \hat{\phi})]$$

$$+ \text{noise component} \qquad (9.3.22)$$

By adding these two terms we obtain an error signal $e(t)$, given as

$$e(t) = r_2 \cos \theta_n - r_1 \sin \theta_n$$

$$= A_m \sqrt{\mathcal{E}_s} \sin(\phi - \hat{\phi}) + \text{noise components} \qquad (9.3.23)$$

This error signal is now passed to the loop filter that drives the VCO. Thus, only the phase of the QAM signal is used in obtaining an estimate of the carrier phase. Consequently, the general block diagram for the DFPLL, given in Figure 9.17, also applies to carrier-phase estimation for an M-ary QAM signal.

As in the case of digitally phase-modulated signals, the method described above for carrier-phase recovery results in phase ambiguities. This problem is solved generally by differential encoding of the data sequence at the input to the modulator.

9.4 CARRIER-FREQUENCY MODULATION

Above, we described methods for transmitting digital information by modulating either the amplitude of the carrier, or the phase of the carrier, or the combined amplitude and phase. Digital information can also be transmitted by modulating the frequency of the carrier.

As we will observe from our treatment below, digital transmission by frequency modulation is a nonlinear modulation method that is appropriate for channels that lack the phase stability that is necessary to perform carrier-phase estimation. In contrast, the linear modulation methods that we have introduced, namely PAM, PM and QAM, require the estimation of the carrier phase to perform phase-coherent detection.

9.4.1 Frequency-Shift Keying

The simplest form of frequency modulation is binary frequency-shift keying (FSK). In binary FSK we employ two different frequencies, say f_1 and $f_2 = f_1 + \Delta f$, to transmit a binary information sequence. The choice of frequency separation $\Delta f = f_2 - f_1$ is considered below. Thus the two signal waveforms may be expressed as

$$u_1(t) = \sqrt{\frac{2\mathcal{E}_b}{T_b}} \cos 2\pi f_1 t, \quad 0 \leq t \leq T_b$$

$$u_2(t) = \sqrt{\frac{2\mathcal{E}_b}{T_b}} \cos 2\pi f_2 t, \quad 0 \leq t \leq T_b \qquad (9.4.1)$$

where $\mathcal{E}_b$ is the signal energy per bit and T_b is the duration of the bit interval.

More generally, M-ary FSK may be used to transmit a block of $k = \log_2 M$ bits per signal waveform. In this case, the M signal waveforms may be expressed as

$$u_m(t) = \sqrt{\frac{2\mathcal{E}_s}{T}} \cos(2\pi f_c t + 2\pi m \Delta f t), \qquad m = 0, 1, \ldots, M-1 \qquad 0 \le t \le T$$

$$(9.4.2)$$

where $\mathcal{E}_s = k\mathcal{E}_b$ is the energy per symbol, $T = kT_b$ is the symbol interval, and Δf is the frequency separation between successive frequencies, i.e., $\Delta f = f_m - f_{m-1}$ for all $m = 1, 2, \ldots, M-1$, where $f_m = f_c + m\Delta f$.

Note that the M FSK waveforms have equal energy $\mathcal{E}_s$. The frequency separation Δf determines the degree to which we can discriminate among the M possible transmitted signals. As a measure of the similarity (or dissimilarity) between a pair of signal waveforms we use the correlation coefficient γ_{mn}

$$\gamma_{mn} = \frac{1}{\mathcal{E}_s} \int_0^T u_m(t) u_n(t) \, dt \qquad (9.4.3)$$

By substituting for $u_m(t)$ and $u_n(t)$ in (9.4.3), we obtain

$$\gamma_{mn} = \frac{1}{\mathcal{E}_s} \int_0^T \frac{2\mathcal{E}_s}{T} \cos(2\pi f_c t + 2\pi m \Delta f t) \cos(2\pi f_c t + 2\pi n \Delta f t) \, dt$$

$$= \frac{1}{T} \int_0^T \cos 2\pi (m-n) \Delta f t \, dt + \frac{1}{T} \int_0^T \cos \left[4\pi f_c t + 2\pi (m+n) \Delta f t \right] dt$$

$$= \frac{\sin 2\pi (m-n) \Delta f T}{2\pi (m-n) \Delta f T} \qquad (9.4.4)$$

where the second integral vanishes when $f_c \gg 1/T$. A plot of γ_{mn} as a function of the frequency separation Δf is given in Figure 9.29. We observe that the signal waveforms are orthogonal when Δf is a multiple of $1/2T$. Hence, the minimum frequency separation between successive frequencies for orthogonality is $1/2T$. We

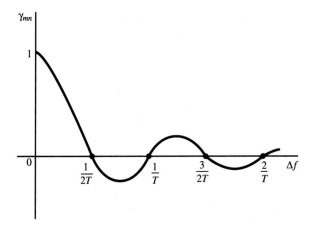

FIGURE 9.29. Crosscorrelation coefficient as a function of frequency separation for FSK signals.

also note that the minimum value of the correlation coefficient is $\gamma_{mn} = -0.217$, which occurs at the frequency separation $\Delta f = 0.715/T$.

M-ary orthogonal FSK waveforms have a geometric representation as M M-dimensional orthogonal vectors, given as

$$\mathbf{s}_1 = \left(\sqrt{\mathcal{E}_s}, 0, \dots, 0\right)$$

$$\mathbf{s}_2 = \left(0, \sqrt{\mathcal{E}_s}, 0, \dots, 0\right) \tag{9.4.5}$$

$$\vdots$$

$$\mathbf{s}_M = \left(0, 0, \dots, 0, \sqrt{\mathcal{E}_s}\right)$$

where the basis functions are $\psi_m(t) = \sqrt{2/T} \cos 2\pi (f_c + m\Delta f)t$. The distance between pairs of signal vectors is $d = \sqrt{2\mathcal{E}_s}$ for all m, n, which is also the minimum distance among the M signals.

The demodulation and detection of the M-ary FSK signals is considered below.

9.4.2 Demodulation and Detection of FSK Signals

Let us assume that the FSK signals are transmitted through an additive white Gaussian noise channel. Furthermore, we assume that each signal is delayed in the transmission through the channel. Consequently, the filtered received signal at the input to the demodulator may be expressed as

$$r(t) = \sqrt{\frac{2\mathcal{E}_s}{T}} \cos(2\pi f_c t + 2\pi m \Delta f t + \phi_m) + n(t) \tag{9.4.6}$$

where ϕ_m denotes the phase shift of the mth signal (due to the transmission delay) and $n(t)$ represents the additive bandpass noise, which may be expressed as

$$n(t) = n_c(t) \cos 2\pi f_c t - n_s(t) \sin 2\pi f_c t \tag{9.4.7}$$

The demodulation and detection of the M FSK signals may be accomplished by one of two methods. One approach is to estimate the M carrier-phase shifts $\{\phi_m\}$ and perform *phase-coherent demodulation and detection*. As an alternative method, the carrier phases may be ignored in the demodulation and detection.

In phase-coherent demodulation, the received signal $r(t)$ is correlated with each of the M possible received signals $\cos(2\pi f_c t + 2\pi m \Delta f t + \hat{\phi}_m)$, $m = 0, 1, \dots, M - 1$, where $\{\hat{\phi}_m\}$ are the carrier-phase estimates. A block diagram illustrating this type of demodulation is shown in Figure 9.30. It is interesting to note that when $\hat{\phi}_m \neq \phi_m$, $m = 0, 1, \dots, M - 1$ (imperfect phase estimates), the frequency separation required for signal orthogonality at the demodulator is $\Delta f = 1/T$ (see Problem 9.26), which is twice the minimum separation for orthogonality when $\phi_m = \hat{\phi}_m$.

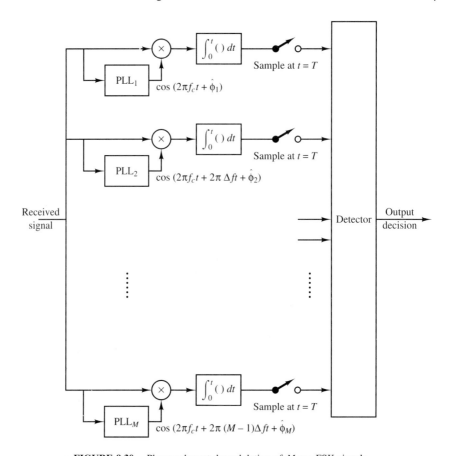

FIGURE 9.30. Phase-coherent demodulation of M-ary FSK signals.

The requirement for estimating M carrier phases makes coherent demodulation of FSK signals extremely complex and impractical, especially when the number of signals is large. Therefore, we shall not consider coherent detection of FSK signals.

Instead we consider a method for demodulation and detection that does not require knowledge of the carrier phases. The demodulation may be accomplished as shown in Figure 9.31. In this case there are two correlators per signal waveform, or a total of $2M$ correlators, in general. The received signal is correlated with the basis functions (quadrature carriers) $\sqrt{\frac{2}{T}}\cos(2\pi f_c t + 2\pi m\Delta f t)$ and $\sqrt{\frac{2}{T}}\sin(2\pi f_c t + 2\pi m\Delta f t)$ for $m = 0, 1, \ldots, M-1$. The $2M$ outputs of the correlators are sampled at the end of the signal interval and the $2M$ samples are passed to the detector. Thus, if the mth signal is transmitted, the $2M$ samples at the detector may be expressed as

$$r_{kc} = \sqrt{\mathcal{E}_s}\left[\frac{\sin 2\pi(k-m)\Delta f T}{2\pi(k-m)\Delta f T}\cos\phi_m - \frac{\cos 2\pi(k-m)\Delta f T - 1}{2\pi(k-m)\Delta f T}\sin\phi_m\right] + n_{kc}$$

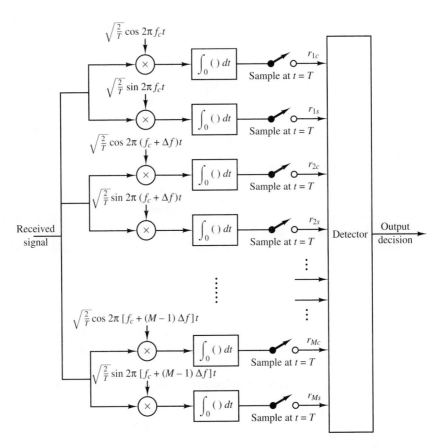

FIGURE 9.31.   Demodulation of M-ary FSK signals for noncoherent detection.

$$r_{ks} = \sqrt{\mathcal{E}_s} \left[\frac{\cos 2\pi (k-m)\Delta f T - 1}{2\pi (k-m)\Delta f T} \cos \phi_m + \frac{\sin 2\pi (k-m)\Delta f T}{2\pi (k-m)\Delta f T} \sin \phi_m \right] + n_{ks}$$

$$(9.4.8)$$

where n_{kc} and n_{ks} denote the Gaussian noise components in the sampled outputs.

We observe that when $k = m$, the sampled values to the detector are

$$r_{mc} = \sqrt{\mathcal{E}_s} \cos \phi_m + n_{mc}$$
$$r_{ms} = \sqrt{\mathcal{E}_s} \sin \phi_m + n_{ms} \qquad (9.4.9)$$

Furthermore, we observe that when $k \neq m$, the signal components in the samples r_{kc} and r_{ks} will vanish, independent of the values of the phase shift ϕ_k, provided that the frequency separation between successive frequencies is $\Delta f = 1/T$. In such a case, the other $2(M-1)$ correlator outputs consist of noise only, i.e.,

$$r_{kc} = n_{kc}, \qquad r_{ks} = n_{ks}, \qquad k \neq m \qquad (9.4.10)$$

In the following development we assume that $\Delta f = 1/T$ so that the signals are orthogonal.

It is shown easily (see Problem 9.27) that the $2M$ noise samples $\{n_{kc}\}$ and $\{n_{ks}\}$ are zero-mean, mutually uncorrelated Gaussian random variables with equal variance $\sigma^2 = N_0/2$. Consequently, the joint p.d.f. for r_{mc} and r_{ms} conditioned on ϕ_m is

$$f_{\mathbf{r}_m}(r_{mc}, r_{ms}|\phi_m) = \frac{1}{2\pi\sigma^2} e^{-[(r_{mc}-\sqrt{\mathcal{E}_s}\cos\phi_m)^2+(r_{ms}-\sqrt{\mathcal{E}_s}\sin\phi_m)^2]/2\sigma^2} \tag{9.4.11}$$

and for $m \neq k$, we have

$$f_{\mathbf{r}_k}(r_{kc}, r_{ks}) = \frac{1}{2\pi\sigma^2} e^{-(r_{kc}^2+r_{ks}^2)/2\sigma^2} \tag{9.4.12}$$

Given the $2M$ observed random variables $\{r_{kc}, r_{ks}, k = 1, 2, \dots, M\}$, the optimum detector selects the signal that corresponds to the maximum of the posterior probabilities, i.e.,

$$P[\mathbf{s}_m \text{ was transmitted}|\mathbf{r}] \equiv P(\mathbf{s}_m|\mathbf{r}), \quad m = 1, 2, \dots, M \tag{9.4.13}$$

where r is the $2M$ dimensional vector with elements $\{r_{kc}, r_{ks}, k = 1, 2, \dots, M\}$. Let us derive the form for the optimum noncoherent detector for the case of binary FSK. The generalization to M-ary FSK is straightforward.

Optimum detector for binary FSK. In binary orthogonal FSK, the two posterior probabilities are

$$P(\mathbf{s}_1|\mathbf{r}) = \frac{f_{\mathbf{r}}(\mathbf{r}|\mathbf{s}_1)P(\mathbf{s}_1)}{f_{\mathbf{r}}(\mathbf{r})}$$

$$P(\mathbf{s}_2|\mathbf{r}) = \frac{f_{\mathbf{r}}(\mathbf{r}|\mathbf{s}_2)P(\mathbf{s}_2)}{f_{\mathbf{r}}(\mathbf{r})} \tag{9.4.14}$$

and, hence, the optimum detection rule may be expressed as

$$P(\mathbf{s}_1|\mathbf{r}) \underset{s_2}{\overset{s_1}{\gtrless}} P(\mathbf{s}_2|\mathbf{r}) \tag{9.4.15}$$

or, equivalently,

$$\frac{f_{\mathbf{r}}(\mathbf{r}|\mathbf{s}_1)P(\mathbf{s}_1)}{f_{\mathbf{r}}(\mathbf{r})} \underset{s_2}{\overset{s_1}{\gtrless}} \frac{f_{\mathbf{r}}(\mathbf{r}|\mathbf{s}_2)P(\mathbf{s}_2)}{f_{\mathbf{r}}(\mathbf{r})} \tag{9.4.16}$$

where $\mathbf{r}$ is the four-dimensional vector $\mathbf{r} = (r_{1c}, r_{1s}, r_{2c}, r_{2s})$. The relation in (9.4.16) simplifies to the detection rule

$$\frac{f_{\mathbf{r}}(\mathbf{r}|\mathbf{s}_1)}{f_{\mathbf{r}}(\mathbf{r}|\mathbf{s}_2)} \underset{s_2}{\overset{s_1}{\gtrless}} \frac{P(\mathbf{s}_2)}{P(\mathbf{s}_1)} \tag{9.4.17}$$

The ratio of p.d.f.'s in the left-hand side of (9.4.17) is the likelihood ratio, which we denote as

$$\Lambda(\mathbf{r}) = \frac{f_{\mathbf{r}}(\mathbf{r}|\mathbf{s}_1)}{f_{\mathbf{r}}(\mathbf{r}|\mathbf{s}_2)} \qquad (9.4.18)$$

The right-hand side of (9.4.17) is the ratio of the two prior probabilities, which takes the value of unity when the two signals are equally probable.

The p.d.f.'s $f_{\mathbf{r}}(\mathbf{r}|\mathbf{s}_1)$ and $f_{\mathbf{r}}(\mathbf{r}|\mathbf{s}_2)$ in the likelihood ratio may be expressed as

$$f_{\mathbf{r}}(\mathbf{r}|\mathbf{s}_1) = f_{\mathbf{r}_2}(r_{2c}, r_{2s}) \int_0^{2\pi} f_{\mathbf{r}_1}(r_{1c}, r_{1s}|\phi_1) f_{\phi_1}(\phi_1) \, d\phi_1$$

$$f_{\mathbf{r}}(\mathbf{r}|\mathbf{s}_2) = f_{\mathbf{r}_1}(r_{1c}, r_{1s}) \int_0^{2\pi} f_{\mathbf{r}_2}(r_{2c}, r_{2s}|\phi_2) f_{\phi_2}(\phi_2) \, d\phi_2 \qquad (9.4.19)$$

where $f_{\mathbf{r}_m}(r_{mc}, r_{ms}|\phi_m)$ and $f_{\mathbf{r}_k}(r_{kc}, r_{ks})$, $m \neq k$, are given by (9.4.11) and (9.4.12), respectively. Thus, the carrier phases ϕ_1 and ϕ_2 are eliminated by simply averaging $f_{\mathbf{r}_m}(r_{mc}, r_{ms}|\phi_m)$.

The uniform p.d.f. for ϕ_m represents the most ignorance regarding the phases of the carriers. This is called the *least favorable p.d.f.* for ϕ_m. With $f_{\phi_m}(\phi_m) = 1/2\pi, 0 \le \phi_m < 2\pi$, substituted into the integrals given in (9.4.19), we obtain

$$\frac{1}{2\pi} \int_0^{2\pi} f_{\mathbf{r}_m}(r_{mc}, r_{ms}|\phi_m) \, d\phi_m$$

$$= \frac{1}{2\pi\sigma^2} e^{-(r_{mc}^2 + r_{ms}^2 + \mathcal{E}_s)/2\sigma^2} \frac{1}{2\pi} \int_0^{2\pi} e^{\sqrt{\mathcal{E}_s}(r_{mc}\cos\phi_m + r_{ms}\sin\phi_m)/\sigma^2} \, d\phi_m \quad (9.4.20)$$

But

$$\frac{1}{2\pi} \int_0^{2\pi} e^{\sqrt{\mathcal{E}_s}(r_{mc}\cos\phi_m + r_{ms}\sin\phi_m)/\sigma^2} \, d\phi_m = I_0\left(\frac{\sqrt{\mathcal{E}_s(r_{mc}^2 + r_{ms}^2)}}{\sigma^2} \right) \qquad (9.4.21)$$

where $I_0(x)$ is the modified Bessel function of order zero. This function is a monotonically increasing function of its argument as illustrated in Figure 9.32. $I_0(x)$ has the power series expansion

$$I_0(x) = \sum_{k=0}^{\infty} \frac{x^{2k}}{2^{2k}(k!)^2} \qquad (9.4.22)$$

From (9.4.19)–(9.4.21), we obtain the likelihood ratio $\Lambda(\mathbf{r})$ in the form

$$\Lambda(\mathbf{r}) = \frac{I_0\left(\sqrt{\mathcal{E}_s(r_{1c}^2 + r_{1s}^2)}/\sigma^2 \right)}{I_0\left(\sqrt{\mathcal{E}_s(r_{2c}^2 + r_{2s}^2)}/\sigma^2 \right)} \underset{s_2}{\overset{s_1}{\gtrless}} \frac{P(\mathbf{s}_2)}{P(\mathbf{s}_1)} \qquad (9.4.23)$$

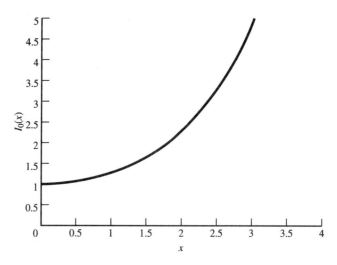

FIGURE 9.32. Graph of $I_0(x)$.

Thus, the optimum detector computes the two envelopes $r_1 = \sqrt{r_{1c}^2 + r_{1s}^2}$ and $r_2 = \sqrt{r_{2c}^2 + r_{2s}^2}$ and the corresponding values of the Bessel function $I_0\left(\sqrt{\mathcal{E}_s r_1^2}/\sigma^2\right)$ and $I_0\left(\sqrt{\mathcal{E}_s r_2^2}/\sigma^2\right)$ to form the likelihood ratio. We observe that this computation requires knowledge of the noise variance σ^2 and the signal energy $\mathcal{E}_s$. The likelihood ratio is then compared with the threshold $P(\mathbf{s}_2)/P(\mathbf{s}_1)$ to determine which signal was transmitted.

A significant simplification in the implementation of the optimum detector occurs when the two signals are equally probable. In such a case the threshold becomes unity, and, due to the monotonicity of the Bessel function, the optimum detector rule simplifies to

$$\sqrt{r_{1c}^2 + r_{1s}^2} \underset{s_2}{\overset{s_1}{\gtrless}} \sqrt{r_{2c}^2 + r_{2s}^2} \tag{9.4.24}$$

Thus, the optimum detector bases its decision on the two envelopes $r_1 = \sqrt{r_{1c}^2 + r_{1s}^2}$ and $r_2 = \sqrt{r_{2c}^2 + r_{2s}^2}$, and hence it is called an *envelope detector.*

We observe that the computation of the envelopes of the received signal samples at the output of the demodulator renders the carrier signal phases $\{\phi_m\}$ irrelevant in the decision as to which signal was transmitted. Equivalently, the decision may be based on the computation of the squared envelopes r_1^2 and r_2^2, in which case the detector is called a *square-law detector.* Figure 9.33 shows the block diagram of the demodulator and the square-law detector.

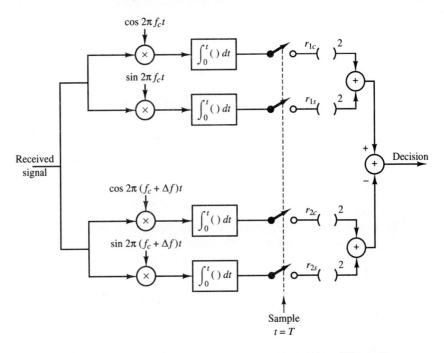

FIGURE 9.33. Demodulation and square-law detection of binary FSK signals.

The generalization of the optimum demodulator and detector to M-ary orthogonal FSK signals is straightforward. As illustrated in Figure 9.31, the output of the optimum demodulator at the sampling instant consists of the 2M vector components $(r_{1c}, r_{1s}, r_{2c}, r_{2s}, \ldots, r_{Mc}, r_{Ms})$. Then the optimum noncoherent detector computes the M envelopes as

$$r_m = \sqrt{r_{mc}^2 + r_{ms}^2} \quad , \quad m = 1, 2, \ldots, M \tag{9.4.25}$$

Thus, the unknown carrier phases of the received signals are rendered irrelevant to the decision as to which signal was transmitted. When all the M signals are equally likely to be transmitted, the optimum detector selects the signal corresponding to the largest envelope (or squared envelope). In the case of nonequally probable transmitted signals, the optimum detector must compute the M posterior probabilities in (9.4.13) and then select the signal corresponding to the largest posterior probability.

9.4.3 Probability of Error for Noncoherent Detection of FSK

Let us consider M-ary orthogonal FSK signals that are detected noncoherently. Let us also assume that the M signals are equally probable a priori and that $s_1(t)$ was transmitted in the interval $0 \leq t \leq T$.

The M decision metrics at the detector are the M envelopes

$$r_m = \sqrt{r_{mc}^2 + r_{ms}^2} \quad m = 1, 2, \ldots, M \tag{9.4.26}$$

where

$$
\begin{aligned}
r_{1c} &= \sqrt{\mathcal{E}_s} \cos \phi_1 + n_{1c} \\
r_{1s} &= \sqrt{\mathcal{E}_s} \sin \phi_1 + n_{1s}
\end{aligned}
\tag{9.4.27}
$$

and

$$
\begin{aligned}
r_{mc} &= n_{mc}, \quad m = 2, 3, \ldots, M \\
r_{ms} &= n_{ms}, \quad m = 2, 3, \ldots, M
\end{aligned}
\tag{9.4.28}
$$

The additive noise components $\{n_{mc}\}$ and $\{n_{ms}\}$ are mutually statistically independent zero-mean Gaussian variables with equal variance $\sigma^2 = N_0/2$. Thus the p.d.f.'s of the random variables at the input to the detector are

$$f_{\mathbf{r}_1}(r_{1c}, r_{1s}) = \frac{1}{2\pi\sigma^2} e^{-(r_{1c}^2 + r_{1s}^2 + \mathcal{E}_s)/2\sigma^2} I_0 \left(\frac{\sqrt{\mathcal{E}_s(r_{1c}^2 + r_{1s}^2)}}{\sigma^2} \right) \tag{9.4.29}$$

$$f_{\mathbf{r}_m}(r_{mc}, r_{ms}) = \frac{1}{2\pi\sigma^2} e^{-(r_{mc}^2 + r_{ms}^2)/2\sigma^2}, \quad m = 2, 3, \ldots, M \tag{9.4.30}$$

Let us make a change in variables in the joint p.d.f.'s given by (9.4.29) and (9.4.30). We define the normalized variables

$$R_m = \frac{\sqrt{r_{mc}^2 + r_{ms}^2}}{\sigma}$$

$$\Theta_m = \tan^{-1} \frac{r_{ms}}{r_{mc}} \tag{9.4.31}$$

Clearly, $r_{mc} = \sigma R_m \cos \Theta_m$ and $r_{ms} = \sigma R_m \sin \Theta_m$. The Jacobian of this transformation is

$$|\mathbf{J}| = \begin{vmatrix} \sigma \cos \Theta_m & \sigma \sin \Theta_m \\ -\sigma R_m \sin \Theta_m & \sigma R_m \cos \Theta_m \end{vmatrix} = \sigma^2 R_m \tag{9.4.32}$$

Consequently,

$$f_{R_1 \Theta_1}(R_1, \Theta_1) = \frac{R_1}{2\pi} e^{-(R_1^2 + 2\mathcal{E}_s/N_0)/2} I_0 \left(\sqrt{\frac{2\mathcal{E}_s}{N_0}} R_1 \right) \tag{9.4.33}$$

$$f_{R_m \Theta_m}(R_m, \Theta_m) = \frac{R_m}{2\pi} e^{-R_m^2/2}, \quad m = 2, 3, \ldots, M \tag{9.4.34}$$

Finally, by averaging $f_{R_m \Theta_m}(R_m, \Theta_m)$ over Θ_m, the factor of 2π is (see Problem 3.31) eliminated from (9.4.33) and (9.4.34). Thus, we find that R_1 has a Rice probability distribution and R_m, $m = 2, 3, \ldots, M$ are each Rayleigh distributed.

The probability of a correct decision is simply the probability that $R_1 > R_2$ and $R_1 > R_3$, ... and $R_1 > R_m$. Hence,

$$P_c = P(R_2 < R_1, R_3 < R_1, \ldots, R_M < R_1)$$

$$= \int_0^\infty P(R_2 < R_1, R_3 < R_1, \ldots, R_M < R_1 | R_1 = x) f_{R_1}(x) dx \qquad (9.4.35)$$

Because the random variables R_m, $m = 2, 3, \ldots, M$ are statistically i.i.d., the joint probability in (9.4.35) conditioned on R_1 factors into a product of $M - 1$ identical terms. Thus,

$$P_c = \int_0^\infty [P(R_2 < R_1 | R_1 = x)]^{M-1} f_{R_1}(x) \, dx \qquad (9.4.36)$$

where

$$P(R_2 < R_1 | R_1 = x) = \int_0^x f_{R_2}(r_2) \, dr_2$$

$$= 1 - e^{-x^2/2} \qquad (9.4.37)$$

The $(M - 1)$st power of (9.4.37) may be expressed as

$$\left[1 - e^{-x^2/2}\right]^{M-1} = \sum_{n=0}^{M-1} (-1)^n \binom{M-1}{n} e^{-nx^2/2} \qquad (9.4.38)$$

Substitution of this result in (9.4.36) and integration over x yields the probability of a correct decision as

$$P_c = \sum_{n=0}^{M-1} (-1)^n \binom{M-1}{n} \frac{1}{n+1} e^{-n\rho_s/(n+1)} \qquad (9.4.39)$$

where $\rho_s = \mathcal{E}_s / N_0$ is the SNR per symbol. Then, the probability of a symbol error, which is $P_M = 1 - P_c$, becomes

$$P_M = \sum_{n=1}^{M-1} (-1)^{n+1} \binom{M-1}{n} \frac{1}{n+1} e^{-nk\rho_b/(n+1)} \qquad (9.4.40)$$

where $\rho_b = \mathcal{E}_b / N_0$ is the SNR per bit.

For binary FSK ($M = 2$), (9.4.40) reduces to the simple form

$$P_2 = \frac{1}{2} e^{-\rho_b/2} \qquad (9.4.41)$$

We observe that the performance of noncoherent FSK is 3 dB worse than binary DPSK.

For $M > 2$, we may compute the probability of a bit error by making use of the relationship

$$P_b = \frac{2^{k-1}}{2^k - 1} P_M \qquad (9.4.42)$$

which was established in Section 7.3.2. Figure 9.34 shows the bit-error probability as function of the SNR per bit ρ_b for $M = 2, 4, 8, 16, 32$. Just as in the case of coherent detection of M-ary orthogonal signals (see Section 7.3.2), we observe that for any given bit-error probability, the SNR per bit decreases as M increases. It will be shown in Chapter 10 that, in the limit as $M \to \infty$ (or $k = \log_2 M \to \infty$), the probability of a bit-error P_b can be made arbitrarily small provided that the SNR per bit is greater than the Shannon limit of -1.6 dB. The cost for increasing M

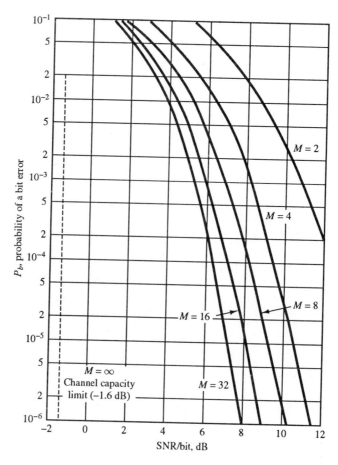

FIGURE 9.34. Probability of a bit error for noncoherent detection of orthogonal FSK signals.

is the bandwidth required to transmit the signals. Because the frequency separation between adjacent frequencies is $\Delta f = 1/T$ for signal orthogonality, the bandwidth required for the M signals is $W = M \Delta f = M/T$. Also, the bit rate is $R_b = k/T$ where $k = \log_2 M$. Therefore, the bit-rate-to-bandwidth ratio is

$$\frac{R_b}{W} = \frac{\log_2 M}{M} \qquad (9.4.43)$$

Thus, we observe that $R_b/W \to 0$ as $M \to \infty$.

9.4.4 Continuous-Phase FSK

Ordinary FSK signals as described above may be generated by having $M = 2^k$ separate oscillators tuned to the desired frequencies $f_c + m\Delta f \equiv f_m$, and selecting one of the M frequencies according to the particular k-bit symbol that is to be transmitted in a signal interval. However, such abrupt switching from one oscillator output to another in successive signaling intervals results in relatively large spectral sidelobes outside of the main spectral band of the signal, which decay slowly with frequency separation. Consequently, this method is wasteful of bandwidth.

To avoid the use of signals having large spectral sidelobes, we may use the information-bearing signal to phase modulate a single carrier whose phase is changed in a continuous manner. The resulting frequency-modulated signal is phase continuous and, hence, it is called *continuous-phase* FSK (CPFSK).

In order to represent a CPFSK signal, we begin with a PAM signal

$$v(t) = \sum_n a_n g_T(t - nT) \qquad (9.4.44)$$

where the amplitudes are obtained by mapping k-bit blocks of binary digits from the information sequence into the amplitude levels $\pm 1, \pm 3, \ldots, \pm(M-1)$, and $g_T(t)$ is a rectangular pulse of amplitude $1/2T$ and duration T. The signal $v(t)$ is used to frequency-modulate the carrier. Consequently, the frequency-modulated carrier is

$$u(t) = \sqrt{\frac{2\mathcal{E}_s}{T}} \cos\left[2\pi f_c t + 4\pi T f_d \int_{-\infty}^{t} v(\tau) d\tau + \phi_0\right] \qquad (9.4.45)$$

where f_d is the *peak frequency deviation* and ϕ_0 is an arbitrary initial phase of the carrier. Note that the instantaneous frequency of the carrier is $f_c + 2T f_d v(t)$.

We observe that, although $v(t)$ contains discontinuities, the integral of $v(t)$ is continuous. We may denote the phase of the carrier as

$$\theta(t; \mathbf{a}) = 4\pi T f_d \int_{-\infty}^{t} v(\tau) d\tau \qquad (9.4.46)$$

where $\mathbf{a}$ denotes the sequence of signal amplitudes. Because $\theta(t; \mathbf{a})$ is a continuous function of t, we have a continuous-phase signal.

The phase of the carrier in the interval $nT \leq t \leq (n+1)T$ is determined by integrating (9.4.46). Thus,

$$\theta(t; \mathbf{a}) = 2\pi f_d T \sum_{k=-\infty}^{n-1} a_k + 2\pi(t - nT) f_d a_n$$

$$= \theta_n + 2\pi h a_n q(t - nT) \tag{9.4.47}$$

where $h, \theta_n,$ and $q(t)$ are defined as

$$h = 2 f_d T \tag{9.4.48}$$

$$\theta_n = \pi h \sum_{k=-\infty}^{n-1} a_k \tag{9.4.49}$$

$$q(t) = \begin{cases} 0, & t < 0 \\ t/2T, & 0 \leq t \leq T \\ 1/2, & t > T \end{cases} \tag{9.4.50}$$

The parameter h is called the *modulation index*. We observe that θ_n represents the phase accumulation (memory) from all symbols up to time $(n-1)T$. The signal $q(t)$ is simply the integral of the rectangular pulse, as illustrated in Figure 9.35.

It is instructive to sketch the set of all phase trajectories $\theta(t; \mathbf{a})$ generated by all possible values of the information sequence $\{a_n\}$. For example, with binary symbols, $a_n = \pm 1$, the set of phase trajectories beginning at time $t = 0$ are shown in Figure 9.36. For comparison, the phase trajectories for quaternary CPFSK ($a_n = \pm 1, \pm 3$) are illustrated in Figure 9.37. These phase diagrams are called *phase trees*. We observe that the phase trees are piecewise linear as a consequence of the fact that the pulse $g_T(t)$ is rectangular. Smoother phase trajectories and phase trees may be obtained by using pulses that do not contain discontinuities.

The phase trees shown in these figures grow with time. However, the phase of the carrier is unique only in the range $\theta = 0$ to $\theta = 2\pi$ or, equivalently, from $\theta = -\pi$ to $\theta = \pi$. When the phase trajectories are plotted modulo 2π, say in the range $(-\pi, \pi)$, the phase tree collapses into a structure called a *phase trellis*. One way to view the phase trellis as a function of time is to plot the two quadrature

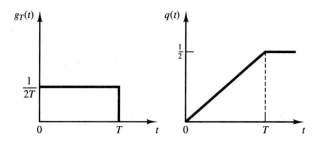

FIGURE 9.35. The signal pulse $g^T(t)$ and its integral $q(t)$.

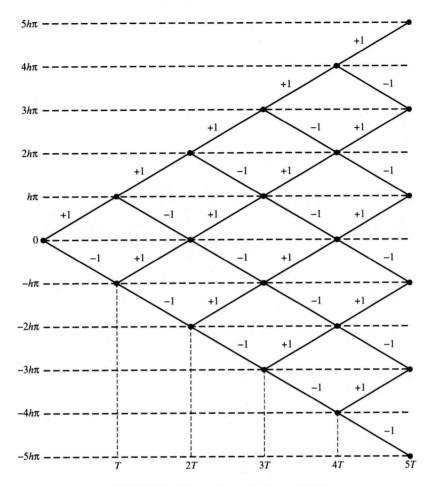

FIGURE 9.36. Phase trajectory for binary CPFSK.

components $x_c(t; \mathbf{a}) = \cos\theta(t; \mathbf{a})$ and $x_s(t; \mathbf{a}) = \sin\theta(t; \mathbf{a})$ as (x, y) coordinates and to let time vary in a third dimension. Because $x_c^2 + x_s^2 = 1$ at any time instant, the three-dimensional plot generated by $x_c(t; \mathbf{a})$ and $x_s(t; \mathbf{a})$ appears as a trajectory on the surface of a cylinder of unit radius.

Simpler representations for the phase trajectories can be obtained by displaying only the terminal values of the signal phase at the time instants $t = nT$. In this case, we restrict the modulation index h to be rational. In particular, let us assume that $h = m/p$, where m and p are relatively prime integers. Then, at the time instants $t = nT$, the terminal phase states for even m are

$$\Theta_s = \left\{ 0, \quad \frac{\pi m}{p}, \quad \frac{2\pi m}{p}, \dots, \frac{(p-1)\pi m}{p} \right\} \tag{9.4.51}$$

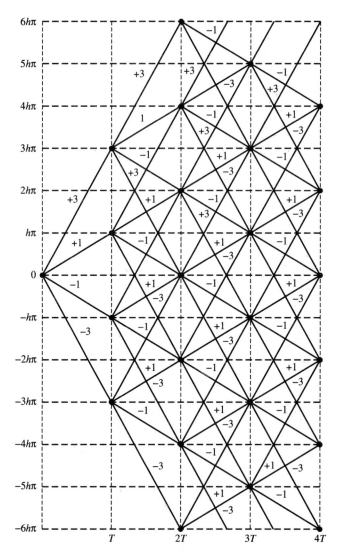

FIGURE 9.37. Phase trajectory for quaternary CPFSK.

and for odd m are

$$\Theta_s = \left\{ 0, \quad \frac{\pi m}{p}, \quad \frac{2\pi m}{p}, \ldots, \frac{(2p-1)\pi m}{p} \right\} \tag{9.4.52}$$

Hence, there are p terminal phase states when m is even and $2p$ terminal phase states when m is odd. For example, binary CPFSK with $h = 1/2$ has four terminal phase states. The *state trellis* for this signal is illustrated in Figure 9.38. We emphasize that the phase transitions from one state to another are not true phase

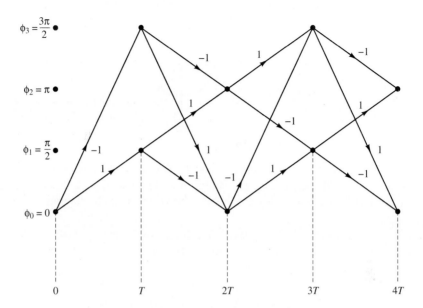

$\phi_3 = \dfrac{3\pi}{2}$

$\phi_2 = \pi$

$\phi_1 = \dfrac{\pi}{2}$

$\phi_0 = 0$

$0 \qquad T \qquad 2T \qquad 3T \qquad 4T$

FIGURE 9.38.   State trellis for binary CPFSK with $h = 1/2$.

trajectories. They represent phase transitions to the terminal states at the time instants $t = nT$.

An alternative representation to the state trellis is the *state diagram,* which also illustrates the state transitions at the time instants $t = nT$. This is an even more compact representation of the CPFSK signal. Only the possible terminal phase states and their transitions are displayed in the state diagram. Time does not appear explicitly as a variable. For example, the state diagram for the CPFSK signal with $h = 1/2$ is shown in Figure 9.39.

We should emphasize that a CPFSK signal cannot be represented by discrete points in signal space as in the case of PAM, PSK, and QAM, because the phase of the carrier is time-variant. Instead, the constant amplitude CPFSK signal may be represented in two-dimensional space by a circle, where points on the circle represent the combined amplitude and phase trajectory of the carrier as a function of time. For example, Figure 9.40 illustrates the signal space diagrams for binary

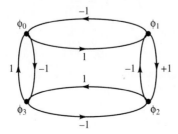

FIGURE 9.39.   Stae diagram for binary CPFSK with $h = 1/2$.

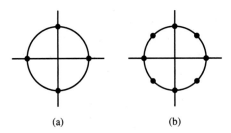

(a)   (b)

FIGURE 9.40. Signal-space diagrams for binary FSK with (a) $h = 1/2$ and (b) $h = 1/4$.

CPFSK with $h = 1/2$ and $h = 1/4$. The four dots at $\theta = 0, \pi/2, \pi, 3\pi/2$ and $\theta = 0, \pm\pi/4, \pm\pi/2, \pm3\pi/M, \pi$ for $h = 1/2$ and $h = 1/4$, respectively, represent the terminal phase states previously shown in the state diagram.

Minimum-shift keying. Minimum-shift keying (MSK) is a special form of binary CPFSK in which the modulation index $h = 1/2$. Thus, the phase of the carrier for the MSK signal is

$$\theta(t; \mathbf{a}) = \frac{\pi}{2} \sum_{k=-\infty}^{n-1} a_k + \pi a_n q(t - nT_b)$$

$$= \theta_n + \frac{\pi}{2} \left(\frac{t - nT_b}{T_b} \right) a_n, \quad nT_b \leq t \leq (n+1)T_b \qquad (9.4.53)$$

which follows from (9.4.47). The corresponding carrier-modulated signal is

$$u(t) = \sqrt{\frac{2\mathcal{E}_b}{T_b}} \cos[2\pi f_c t + \theta_n + \pi(t - nT_b)a_n/2T_b]$$

$$= \sqrt{\frac{2\mathcal{E}_b}{T_b}} \cos\left[2\pi(f_c + \frac{1}{4T_b} a_n)t - \frac{n\pi}{2} a_n + \theta_n \right] \qquad (9.4.54)$$

The expression in (9.4.54) indicates that the MSK (binary CPFSK) signal is basically a sinusoid consisting of one of two possible frequencies in the interval $nT_b \leq t \leq (n+1)T_b$, namely,

$$f_1 = f_c - \frac{1}{4T_b}$$

$$f_2 = f_c + \frac{1}{4T_b} \qquad (9.4.55)$$

Hence, the two sinusoidal signals may be expressed as

$$u_i(t) = \sqrt{\frac{2\mathcal{E}_b}{T_b}} \cos\left[2\pi f_i t + \theta_n + \frac{n\pi}{2}(-1)^{i-1} \right], \quad i = 1, 2 \qquad (9.4.56)$$

The frequency separation is $\Delta f = f_2 - f_1 = 1/2T_b$. Recall that this is the minimum frequency separation for orthogonality of the two sinusoids provided the signals are detected coherently. This explains why binary CPFSK with $h = 1/2$ is called *minimum-shift keying*. Note that the phase of the carrier in the nth signaling interval is the phase state of the signal that results in phase continuity between adjacent intervals.

It is interesting to demonstrate that MSK is also a form of four-phase PSK. To prove this point, let us begin with a four-phase PSK signal, which has the form

$$u(t) = \sqrt{\frac{2\mathcal{E}_b}{T_b}} \left\{ \left[\sum_{n=-\infty}^{\infty} a_{2n} g_T(t - 2nT_b) \right] \cos 2\pi f_c t \right.$$

$$\left. + \left[\sum_{n=-\infty}^{\infty} a_{2n+1} g_T(t - 2nT_b - T_b) \right] \sin 2\pi f_c t \right\} \qquad (9.4.57)$$

where $g_T(t)$ is a sinusoidal pulse defined as

$$g_T(t) = \begin{cases} \sin \frac{\pi t}{2T_b}, & 0 \le t \le 2T_b \\ 0, & \text{otherwise} \end{cases} \qquad (9.4.58)$$

and illustrated in Figure 9.41. First, we observe that the four-phase PSK signal consists of two quadrature carriers, $\cos 2\pi f_c t$ and $\sin 2\pi f_c t$, which are amplitude modulated at a rate of one bit per $2T_b$ interval. The even-numbered information bits $\{a_{2n}\}$ are transmitted by modulating the cosine carrier, while the odd-numbered information bits $\{a_{2n+1}\}$ are transmitted by amplitude modulating the sine carrier. Note that the modulation of the two quadrature carriers is staggered in time by T_b and that the transmission rate for each carrier is $1/2T_b$. This type of four-phase modulation is called *offset quadrature PSK* (OQPSK) or *staggered quadrature PSK* (SQPSK).

Figure 9.42 illustrates the SQPSK signal in terms of the two staggered quadrature-modulated binary PSK signals. The corresponding sum of the two quadrature signals is a constant-amplitude, continuous-phase FSK signal, as shown in Figure 9.42.

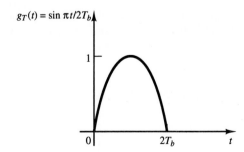

FIGURE 9.41.   Sinusoidal pulse shape.

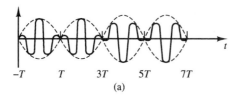

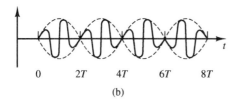
(a)

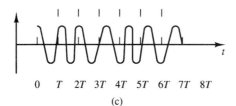
(b)

(c)

FIGURE 9.42. Representation of MSK signal as a form of two staggered binary PSK signals, each with a sinusoidal envelope. (a) In-phase signal component, (b) quadrature signal component, and (c) MSK signal $(a + b)$.

It is also interesting to compare the waveforms for MSK with the waveforms for SQPSK in which the pulse $g_T(t)$ is rectangular for $0 \le t \le 2T_b$, and with conventional QPSK in which the baseband pulse is rectangular in the interval $0 \le t \le 2T_b$. We emphasize that all three of these modulation methods result in identical data rates. The MSK signal is phase continuous. The SQPSK signal with a rectangular baseband pulse is basically two binary PSK signals for which the phase transitions are staggered in time by T_b seconds. Consequently, this signal contains phase jumps of $\pm 90^o$ that may occur as often as every T_b seconds. On the other hand, in conventional QPSK with constant envelope, one or both of the information symbols may cause phase transitions as often as every $2T_b$ seconds. These phase jumps may be $\pm 180^o$ or $\pm 90^o$. An illustration of these three types of four-phase PSK signals is shown in Figure 9.43.

Below we describe the spectral characteristics of CPFSK signals and compare them with other QPSK signals. The demodulation and detection of CPFSK is described in Section 9.5.1, which deals with the demodulation of the more general class of continuous-phase-modulated signals.

9.4.5 Spectral Characteristics of CPFSK Signals

In this section we consider the spectral characteristics of CPFSK signals and present some results on their power-spectral density. The derivation of these results may be

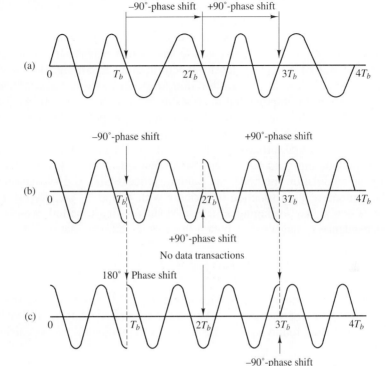

FIGURE 9.43. Signal waveforms for (a) MSK, (b) offset QPSK (rectangular pulse), and (c) conventional QPSK (rectanglar pulse). (From Gronemeyer and McBride, 1976; © 1976 IEEE).

found in more advanced digital communication textbooks, e.g., Anderson et al. (1986) and Proakis (1989).

A CPSFK signal may be expressed in the general form

$$u(t; \mathbf{a}) = \sqrt{\frac{2\mathcal{E}_s}{T}} \, \cos\left[2\pi f_c t + \theta(t; \mathbf{a})\right]$$

$$= \text{Re}\left\{\sqrt{\frac{2\mathcal{E}_s}{T}} \, e^{j\left[2\pi f_c t + \theta(t;\mathbf{a})\right]}\right\} \qquad (9.4.59)$$

where the carrier phase $\theta(t; \mathbf{a})$ is given by (9.4.47). Because the complex exponential $e^{j2\pi f_c t}$ serves as a frequency translation of the complex-valued baseband signal

$$s(t) = \sqrt{\frac{2\mathcal{E}_s}{T}} \, e^{j\theta(t;\mathbf{a})} \qquad (9.4.60)$$

it is sufficient to focus our attention on the spectral characteristics of the information-bearing signal $s(t)$.

Unlike the computation of the power-spectral density for linear modulation methods such as PAM, PSK, and QAM, which was relatively straightforward, the computation of the power density spectrum of a CPFSK signal is much more tedious and involved. The difficulties are due to the memory in the continuous-phase FSK signal and to the exponential relationship between $s(t)$ and $\theta(t; \mathbf{a})$. Nevertheless, the general procedure used in the derivation for the power density spectrum of PAM, given in Section 8.1.2, may be followed.

The general procedure begins with the computation of the autocorrelation function of the random process $S(t)$, denoted as $R_S(t + \tau, t)$. As in the case of linear modulation methods, the random process $S(t)$ is cyclostationary and, hence, the autocorrelation function is periodic with period T, i.e., $R_S(t + T + \tau, t + T) = R_S(t + \tau, t)$. By averaging $R_S(t + \tau, t)$ over a single period, we eliminate the time dependence t, and we thus obtain the average autocorrelation function

$$\bar{R}_S(\tau) = \frac{1}{T} \int_0^T R_S(t + \tau, t) \, dt \tag{9.4.61}$$

Finally, we compute the Fourier transform of $\bar{R}_S(\tau)$ to obtain the power-spectral density $S_S(f)$ of the signal. The power-spectral density of the real-valued signal $u(t; \mathbf{a})$ is then found by translating $S_S(f)$ in frequency by the carrier f_c. Thus, we obtain

$$S_U(f) = \frac{1}{4} \left[S_S(f - f_c) + S_S(f + f_c) \right] \tag{9.4.62}$$

In the case of the CPFSK signal, this procedure yields the following expression for the power-spectral density:

$$S_S(f) = T \left[\frac{1}{M} \sum_{n=1}^{M} A_n^2(f) + \frac{2}{M^2} \sum_{n=1}^{M} \sum_{m=1}^{M} B_{nm}(f) A_n(f) A_m(f) \right] \tag{9.4.63}$$

where

$$A_n(f) = \frac{\sin \pi \left[fT - (2n - 1 - M)h/2 \right]}{\pi \left[fT - (2n - 1 - M)h/2 \right]} \tag{9.4.64}$$

$$= \operatorname{sinc}\left(fT - \frac{h(2n - 1 - M)}{2} \right)$$

$$B_{nm}(f) = \frac{\cos\left(2\pi fT - \alpha_{nm} \right) - \beta \cos \alpha_{nm}}{1 + \beta^2 - 2\beta \cos 2\pi fT} \tag{9.4.65}$$

$$\alpha_{nm} = \pi h(m + n - 1 - M) \tag{9.4.66}$$

$$\beta = \frac{\sin M\pi h}{M \sin \pi h} \tag{9.4.67}$$

The power-spectral density of CPFSK for $M = 2$ is plotted in Figure 9.44 as a function of the normalized frequency fT, with the modulation index $h = 2f_dT$ as a parameter. Note that only one-half of the spectrum is shown in these graphs, because the spectrum is symmetric in frequency. The origin $fT = 0$ corresponds to the carrier frequency f_c in the spectrum of the real-valued signal.

These graphs show that the spectrum of the CPFSK signal is relatively smooth and well-confined for $h < 1$. As h approaches unity, the spectra become very peaked, and, for $h = 1$ where $|\beta| = 1$, we find that impulses occur at M frequencies. When $h > 1$, the spectrum becomes much broader. In communication systems that employ CPFSK, the modulation index is selected to conserve bandwidth, so that $h < 1$.

The special case of binary CPFSK with $h = 1/2$ (or $f_d = 1/4T_s$) and $\beta = 0$ corresponds to MSK. In this case, the power-spectral density obtained from (9.4.63)–(9.4.67) is

$$S_S(f) = \frac{32\mathcal{E}_s}{\pi^2} \left[\frac{\cos 2\pi f T}{1 - 16 f^2 T^2} \right]^2 \tag{9.4.68}$$

In contrast, the power-density spectrum of SQPSK with a rectangular pulse $g_T(t)$ of duration $2T_b$ is

$$S_S(f) = 4\mathcal{E}_s \left(\frac{\sin 2\pi f T_b}{2\pi f T_b} \right)^2 \tag{9.4.69}$$

The power-density spectra in (9.4.68) and (9.4.69) are illustrated in Figure 9.45. Note that the main lobe of MSK is 50% wider than that of SQPSK. However, the sidelobes of MSK fall off considerably faster. As a consequence, MSK is significantly more bandwidth efficient than SQPSK.

9.5 CONTINUOUS-PHASE CARRIER MODULATION

The digital PAM, PSK, and QAM carrier modulation methods described in Sections 9.1, 9.2, and 9.3 are basically memoryless modulation methods. In other words, the transmitted carrier phase in digital-phase modulation, the transmitted amplitude in PAM, and the transmitted (combined) amplitude and phase in QAM are statistically independent over successive signaling intervals, unless the channel introduces memory (intersymbol interference).

In contrast, CPFSK is a modulation method with memory. The memory results from the phase continuity of the transmitted carrier phase from one signal interval to the next. When the phase is expressed in the form of (9.4.47), CPFSK becomes a special case of a general class of continuous-phase modulated (CPM) signals in which the carrier phase is

$$\theta(t; \mathbf{a}) = 2\pi \sum_{k=-\infty}^{n} a_k h_k q(t - kT), \quad nT \leq t \leq (n+1)T \tag{9.5.1}$$

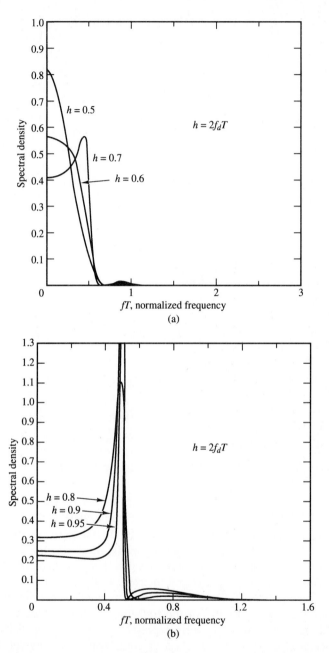

FIGURE 9.44. Power-density spectrum of binary CPFSK. (From *Digital Communications,* Sec. Ed., by J.G. Proakis; © 1989 by the McGraw-Hill Book Co. Reprinted with permission of the publisher.)

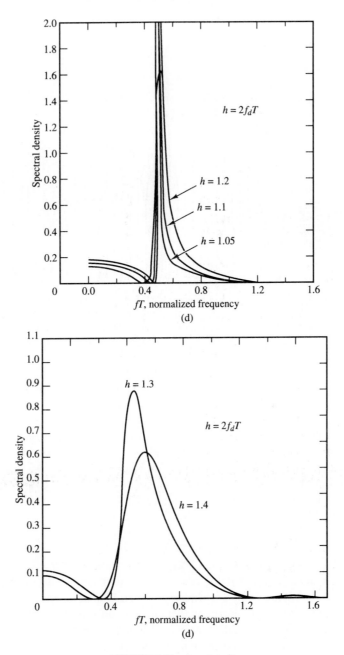

FIGURE 9.44.   (*continued*)

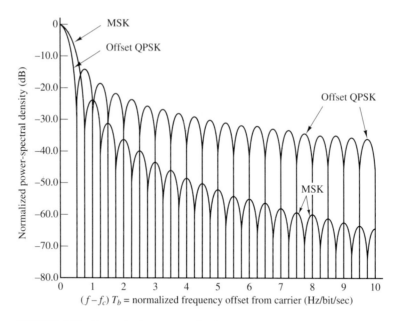

FIGURE 9.45. Power-density spectra of MSK and QPSK. (From Gronemeyer and McBride, 1976; © 1976 IEEE).

where $\{a_k\}$ is the sequence of M-ary information symbols with possible values $\pm1, \pm3, \ldots, \pm(M-1)$, $\{h_k\}$ is a sequence of modulation indices, and $q(t)$ is some arbitrary normalized waveform. Recall that for CPFSK, $q(t) = t/2T$ for $0 \le t \le T$, $q(t) = 0$ for $t < 0$, and $q(t) = 1/2$ for $t > T$.

When $h_k = h$ for all k, the modulation index is fixed for all symbols. When the modulation index varies from one symbol to another, the CPM signal is called *multi-h*. In such a case, the $\{h_k\}$ are usually selected to vary in a cyclic pattern through the set of indices.

The waveform $q(t)$ is the integral of a pulse $g_T(t)$ of arbitrary shape, i.e.,

$$q(t) = \int_0^t g_T(\tau)\,d\tau \qquad (9.5.2)$$

If $g_T(t) = 0$ for $t > T$, the CPM signal is called a *full-response CPM signal.* If the signal pulse $g_T(t) \ne 0$ for $t > T$, the modulated signal is called *partial-response CPM.* In Figure 9.46 we illustrate several pulse shapes for $g_T(t)$ and the corresponding $q(t)$. It is apparent that there is an infinite number of CPM signals that can be obtained by selecting different pulse shapes for $g_T(t)$ and by varying the modulation index h and the number of symbols M.

In general, CPM signals cannot be represented by discrete points in signal space as in the case of PAM, PSK, and QAM, because the phase of the carrier is time-variant. As we observed in the case of CPFSK, we may employ phase trajectories (or phase trees) to illustrate the signal phase as a function of time.

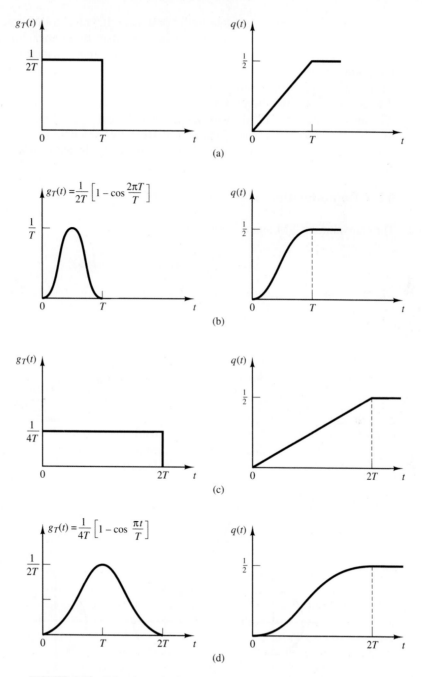

FIGURE 9.46.   Pulse shapes for full-response (a,b) and partial response (c,d) CPM.

Alternatively, we may use a state trellis or a state diagram to illustrate the terminal phase states and the phase transitions from one state to another. Finally, as in the case of CPFSK, we may represent a CPM signal in signal space by a circle, where the points on the circle correspond to the combined amplitude and phase of the carrier as a function of time. The terminal phase states are usually identified as discrete points on the circle. For example, Figure 9.47 illustrates the signal space diagrams for binary CPFSK with $h = 1/4, 1/3$, and $2/3$. Note that the length of the phase trajectory (lengths of the arc) between two terminal phase states increases with an increase in h. An increase in h also results in an increase of the signal bandwidth, as previously shown in Section 9.4.5.

9.5.1 Demodulation and Detection of CPM Signals

The transmitted CPM signal may be expressed as

$$u(t) = \sqrt{\frac{2\mathcal{E}_s}{T}} \cos[2\pi f_c t + \theta(t; \mathbf{a})] \tag{9.5.3}$$

where $\theta(t; \mathbf{a})$ is the carrier phase. The filtered received signal for an additive Gaussian noise channel is

$$r(t) = u(t) + n(t) \tag{9.5.4}$$

where

$$n(t) = n_c(t) \cos 2\pi f_c t - n_s(t) \sin 2\pi f_c t \tag{9.5.5}$$

The optimum receiver for this signal consists of a correlator followed by a maximum-likelihood sequence detector that searches the paths through the state trellis for the minimum Euclidean distance path. The Viterbi algorithm is an efficient method for performing this search. Let us establish the general state-trellis structure for CPM and then describe the metric computations.

The carrier phase for a CPM signal having a fixed modulation index h may be expressed as

$$\theta(t; \mathbf{a}) = 2\pi h \sum_{h=-\infty}^{n} a_k q(t - kT), \quad nT \leq t \leq (n+1)T$$

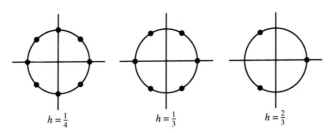

$$h = \tfrac{1}{4} \qquad\qquad h = \tfrac{1}{3} \qquad\qquad h = \tfrac{2}{3}$$

FIGURE 9.47. Signal-space diagram for CPFSK.

$$= \pi h \sum_{k=-\infty}^{n-L} a_k + 2\pi h \sum_{k=n-L+1}^{n} a_k q(t - kT)$$

$$= \theta_n + \phi(t; \mathbf{a}), \quad nT \le t \le (n+1)T \tag{9.5.6}$$

where, by definition,

$$\theta_n = \pi h \sum_{k=-\infty}^{n-L} a_k \tag{9.5.7}$$

$$\phi(t; \mathbf{a}) = 2\pi h \sum_{k=n-L+1}^{n} a_k q(t - kT) \tag{9.5.8}$$

and $q(t) = 0$ for $t < 0$, $q(t) = 1/2$ for $t \ge LT$, and

$$q(t) = \int_0^t g_T(\tau) d\tau \tag{9.5.9}$$

where L is a positive integer. The signal pulse $g_T(t) = 0$ for $t < 0$ and $t \ge LT$. For $L = 1$ we have a full-response CPM signal, and for $L > 1$ we have a partial-response CPM signal. Note that θ_n represents the phase of the carrier at $t = nT$ and $\phi(t; \mathbf{a})$ represents the additional phase accumulation in the interval $nT \le t \le (n+1)T$.

When h is rational, i.e., $h = m/p$ where m and p are relatively prime integers, the phases of the CPM signal at $t = nT$ may be represented by a trellis. The number of terminal phase states for $L = 1$ is p for even m and $2p$ for odd m, as previously indicated by (9.4.51) and (9.4.52). However, for $L > 1$ we have additional phase states due to the partial-response characteristic of the signal pulse $g_T(t)$. These additional terminal phase states can be determined by expressing the phase $\phi(t; \mathbf{a})$ in (9.5.8) as

$$\phi(t; \mathbf{a}) = 2\pi h \sum_{k=n-L+1}^{n-1} a_k q(t - kT) + 2\pi h a_n q(t - nT) \tag{9.5.10}$$

The first term on the right-hand side of (9.5.10) is a function of the information symbols $(a_{n-1}, a_{n-2}, \ldots, a_{n-L+1})$, which is called the *correlative state vector,* and represents the phase contribution of signal pulses that have not reached their final value. The second term on the right-hand side of (9.5.10) represents the phase contribution due to the most recent symbol a_n. Therefore, the state of the CPM signal at $t = nT$ may be expressed as the combined phase state θ_n and the correlative state vector $(a_{n-1}, a_{n-2}, \ldots, a_{n-L+1})$, which we denote as

$$S_n = \{\theta_n, a_{n-1}, \ldots, a_{n-L+1}\} \tag{9.5.11}$$

for the partial-response signal of length LT. Thus, the number of states is

$$N_s = \begin{cases} pM^{L-1} & m \text{ even} \\ 2pM^{L-1} & m \text{ odd} \end{cases} \tag{9.5.12}$$

when $h = m/p$.

The state transition from S_n to S_{n+1} can be determined by the effect of the new symbol a_n in the time interval $nT \leq t \leq (n+1)T$. In general, we have

$$S_{n+1} = \{\theta_{n+1}, a_n, a_{n-1}, \ldots, a_{n-L+2}\} \qquad (9.5.13)$$

where

$$\theta_{n+1} = \theta_n + \pi h a_{n-L+1} \qquad (9.5.14)$$

Example 9.5.1

Determine the terminal states of a binary signal having a modulation index $h = 3/4$ and a partial-response pulse of $L = 2$. Also, sketch the state trellis.

Solution   Because $p = 4$, m is odd ($m = 3$) and $M = 2$, we have $N_s = 16$ phase states. The $2p$ phase states corresponding to θ_n are

$$\Theta_s = \left\{ 0, \pm\frac{\pi}{4}, \pm\frac{\pi}{2}, \pm\frac{3\pi}{4}, \pi \right\}$$

For each of these phase states, there are two states that result from the memory due to the partial-response signal. Hence, the 16 states $S_n = (\theta_n, a_{n-1})$ are

$$(0, +1), \quad (0, -1), \quad (\pi/4, 1), \quad (\pi/4, -1), \quad (-\pi/4, 1), \quad (-\pi/4, -1),$$

$$(\pi/2, 1), \quad (\pi/2, -1), \quad (-\pi/2, 1), \quad (-\pi/2, -1), \quad (3\pi/4, 1), \quad (3\pi/4, -1),$$

$$(-3\pi/4, 1), \quad (-3\pi/4, -1), \quad (\pi, 1), \quad (\pi, -1)$$

Suppose the system is in phase state $\theta_n = 3\pi/4$ and $a_{n-1} = -1$. Then $S_n = (3\pi/4, -1)$ and

$$\theta_{n+1} = \theta_n + \pi h a_{n-1}$$

$$= 3\pi/4 + (3\pi/4)(-1) = 0$$

Hence, $S_{n+1} = (\theta_{n+1}, a_n) = (0, a_n)$. If $a_n = 1$, then $S_{n+1} = (0, 1)$. If $a_n = -1$, then $S_{n+1} = (0, -1)$. The state trellis is illustrated in Figure 9.48. A path through the state trellis corresponding to the data sequence $(1, -1, -1, -1, 1, 1)$ is illustrated in Figure 9.49.

Let us now focus on the demodulation and detection of CPM signals. Because all signal waveforms have the same energy, we may base decisions at the detector on correlation metrics. The crosscorrelation of the received signal $r(t)$ with each of the possible transmitted signals yields the metrics

$$C_n(\mathbf{a}, \mathbf{r}) = \int_{-\infty}^{(n+1)T} r(t) \cos[2\pi f_c t + \theta(t; \mathbf{a})]dt$$

$$= C_{n-1}(\mathbf{a}, \mathbf{r}) + \int_{nT}^{(n+1)T} r(t) \cos[2\pi f_c t + \theta_n + \phi(t; \mathbf{a})]dt \qquad (9.5.15)$$

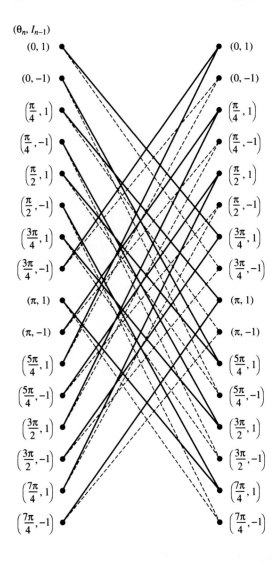

(θ_n, I_{n-1})

FIGURE 9.48. State trellis for partial-response ($L = 2$) CPM with $h = 3/4$.

In the maximum-likelihood sequence detector, the term $C_{n-1}(\mathbf{a}, \mathbf{r})$ represents the metrics of the surviving sequences up to time nT. The term

$$C_n(\theta_n; \mathbf{a}, \mathbf{r}) = \int_{nT}^{(n+1)T} r(t) \cos[2\pi f_c t + \theta_n + \phi(t; \mathbf{a})] dt \qquad (9.5.16)$$

represents the additional increments to the correlation metrics contributed by the received signal in the time interval $nT \leq t \leq (n + 1)T$. We observe that there are M^L possible sequences $\mathbf{a} = (a_n, a_{n-1}, \ldots, a_{n-L+1})$ of symbols and either p or $2p$ possible phase states $\{\theta_n\}$. Therefore, there are either pM^L or $2pM^L$ different values of the correlation metrics $C_n(\theta_n; \mathbf{a}, \mathbf{r})$ computed in each signal interval,

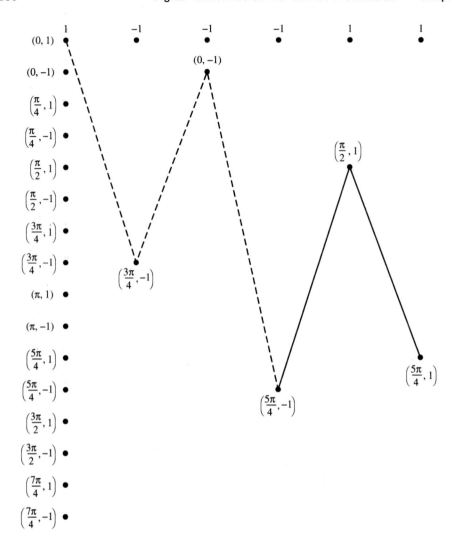

FIGURE 9.49. A signal path through the trellis.

and each value is used to increment the metrics corresponding to the surviving sequences from the previous signal interval. A general block diagram that illustrates the computations of $C_n(\theta_n; \mathbf{a}, \mathbf{r})$ for the Viterbi detector is shown in Figure 9.50.

In the maximum-likelihood sequence detector, the number of surviving sequences at each state of the Viterbi search process is either pM^{L-1} or $2pM^{L-1}$. For each surviving sequence, we have M new correlation increments $C_n(\theta_n; \mathbf{a}, \mathbf{r})$ that are added to the existing metrics to yield either pM^L or $2pM^L$ sequences with corresponding metrics. However, this number is then reduced back to either pM^{L-1}

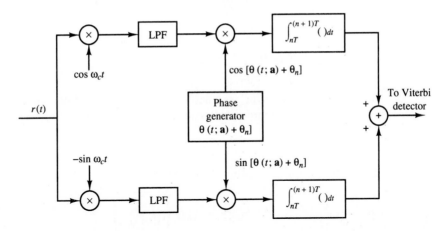

FIGURE 9.50.   Computation of metric increments $C_n(\theta_n; \mathbf{a}, \mathbf{r})$.

or $2pM^{L-1}$ survivors with corresponding metrics by selecting the most probable sequence of the M sequences merging at each node of the trellis and discarding the other $M - 1$ sequences.

9.5.2 Performance of CPM in an AWGN Channel

In determining the probability of error for the Viterbi detector of CPM signals, we must determine the minimum Euclidean distance of paths through the trellis that separate at a node and remerge at a later time at the same node. The distance between two paths through the trellis depends on the phase difference between the paths, as we now demonstrate.

Suppose that we have two signals $u_i(t)$ and $u_j(t)$ corresponding to two phase trajectories $\theta(t; \mathbf{a}_i)$ and $\theta(t; \mathbf{a}_j)$. The sequences $\mathbf{a}_i$ and $\mathbf{a}_j$ must be different in their first symbol. Then, the Euclidean squared distance between the two signals over an interval of length NT, where $1/T$ is the symbol rate, is defined as

$$d_{ij}^2 = \int_0^{NT} [u_i(t) - u_j(t)]^2 dt$$

$$= \int_0^{NT} u_i^2(t)dt + \int_0^{NT} u_j^2(t)dt - 2\int_0^{NT} u_i(t)u_j(t)dt$$

$$= 2N\mathcal{E}_s - 2\left(\frac{2\mathcal{E}_s}{T}\right)\int_0^{NT} \cos[2\pi f_c t + \theta(t; \mathbf{a}_i)]\cos[2\pi f_c t + \theta(t, \mathbf{a}_j)]dt$$

$$= \frac{2\mathcal{E}_s}{T}\int_0^{NT} \{1 - \cos[\theta(t; \mathbf{a}_i) - \theta(t; \mathbf{a}_j)]\}dt \qquad (9.5.17)$$

Because the symbol energy is related to the energy per bit by the expression $\mathcal{E}_s = \mathcal{E}_b \log_2 M$, the squared Euclidean distance d_{ij}^2 may also be expressed as

$$d_{ij}^2 = \frac{2\mathcal{E}_b \log_2 M}{T} \int_0^{NT} \{1 - \cos[\theta(t; \mathbf{a}_i) - \theta(t; \mathbf{a}_j)]\} dt \qquad (9.5.18)$$

The probability of error for the Viterbi detector is dominated by the term corresponding to the minimum Euclidean distance, which is defined as

$$d_{\min}^2 = \lim_{N \to \infty} \min_{i,j} \left[d_{ij}^2 \right] \qquad (9.5.19)$$

Then, the symbol-error probability is approximated as

$$P_M \simeq N_{d \min} Q \left(\sqrt{\frac{d_{\min}^2}{2N_0}} \right) \qquad (9.5.20)$$

where $N_{d \min}$ is the number of minimum distance paths. We note that for conventional binary PSK with no memory, $N = 1$ and $d_{\min}^2 = d_{12}^2 = 4\mathcal{E}_b$. Hence, (9.5.20) agrees with our previous result.

In general, $d_{\min}^2$ is a function of the modulation index h, the number of symbols M, and the pulse shape $g_T(t)$. The choice of the modulation index h and the pulse shape have a significant impact in determining the bandwidth occupied by the transmitted signal. The class of raised cosine pulses of the form

$$g_T(t) = \begin{cases} \dfrac{1}{2LT} \left(1 - \cos \dfrac{2\pi t}{LT} \right), & 0 \leq t \leq LT \\ 0, & \text{otherwise} \end{cases} \qquad (9.5.21)$$

are especially suitable for use in the design of bandwidth efficient CPM signals.

The value of $d_{\min}^2$ has been evaluated for a variety of CPM signals, including full response and partial response, by Aulin and Sundberg (1981, 1984). For example, Figure 9.51 illustrates the value of $d_{\min}^2$ normalized by $4\mathcal{E}_b$ as a function of the time-bandwidth product $2WT_b$, where W is the 99% in-band power bandwidth. Because MSK results in a $d_{\min}^2 = 4\mathcal{E}_b$, the MSK signal serves as a point of reference (0 dB) in this graph. The notation LRC denotes that the signal pulse $g_T(t)$ is a raised cosine pulse with $L = 3, 4$, or 6, as defined by (9.5.21). Along any curve in Figure 9.51, the bandwidth W increases as the modulation index h increases, and the SNR gain increases as h increases. Furthermore, the bandwidth efficiency increases with an increase in the symbol size M for fixed LRC pulse shape. We observe from this figure that there are several decibels to be gained by using partial-response signals and higher signal alphabets. The major price paid for this performance gain is the added exponentially increasing complexity in the implementation of the Viterbi detector for searching for the most probable path through the trellis.

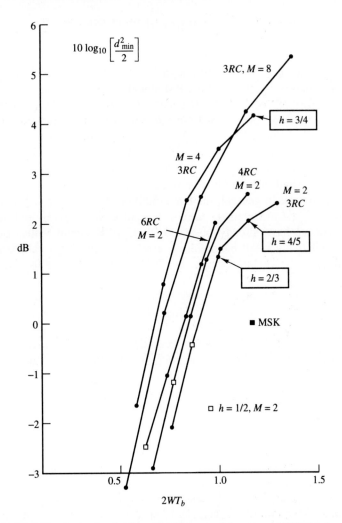

FIGURE 9.51. Power bandwidth trade-off for partial-response CPM signals with raised cosine pulses; W is the 99% in-band power bandwidth. (From Sundberg, 1986; © 1986 IEEE).

The performance results in Figure 9.51 illustrate that 3–4-dB gain relative to MSK can be easily obtained with relatively no increase in bandwidth by use of raised-cosine partial-response CPM and $M = 4$. Although these results are for raised cosine signal pulses, similar gains can be achieved with other partial-response signal shapes. We emphasize that this gain in SNR is achieved by introducing memory into the signal modulation and exploiting this memory in the detection of the signal.

9.5.3 Spectral Characteristics of CPM Signals

In Section 9.4.5 we described the spectral characteristics of CPFSK signals, which are a special class of CPM signals. The general procedure for determining the power-density spectrum of a CPM signal is basically the same as the procedure that was described for CPFSK. Because CPM is more general, however, the derivation for the average autocorrelation function and the power-spectral density is more complicated and will not be given here. The interested reader may refer to more advanced books, e.g., Anderson et al. (1989) and Proakis (1989), for a detailed derivation of the power-spectral density of a general CPM signal.

We shall describe briefly some general trends in the spectral characteristics of CPM signals. In general, the bandwidth occupancy of a CPM signal depends on the choice of the modulation index h, the pulse shape $g_T(t)$, and the number of signals M. As we have already observed for CPFSK, small values of h result in signals with relatively small bandwidth occupancy, while large values of h result in signals with large bandwidth occupancy. This is also the case for the more general CPM signals.

The use of smooth pulses such as raised cosine pulses of the form given by (9.5.21) where $L = 1$ for full response and $L > 1$ for partial response, result in smaller bandwidth occupancy and, hence, in greater bandwidth efficiency than the use of rectangular pulses. For example, Figure 9.52 illustrates the power-spectral density for binary CPM with different partial-response raised cosine (LRC) pulses

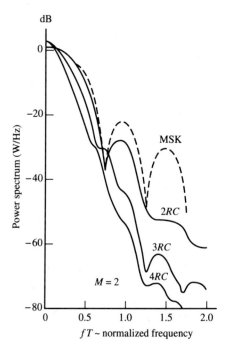

FIGURE 9.52. Power-spectral density for binary CPM with $h = 1/2$ and different pulse shapes (From Aulin et al., 1981; © 1981 IEEE).

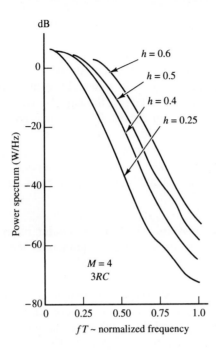

FIGURE 9.53. Power-spectral density for $M = 4$ CPM for $3RC$ different modulation indices (From Aulin et al., 1981; © 1981 IEEE).

and $h = 1/2$. For comparison, the spectrum of binary CPFSK with $h = 1/2$ (MSK) is also shown. We note that as L increases, the pulse $g_T(t)$ becomes smoother and the corresponding spectral occupancy of the signal is reduced.

The effect of varying the modulation index in a CPM signal is illustrated in Figure 9.53 for the case of $M = 4$ and a raised cosine pulse of the form given (9.5.21) with $L = 3$. Note that these spectral characteristics are similar to the ones illustrated previously in Section 9.4.5 for CPFSK, except that these power spectra for CPM are narrower due to the smoother raised cosine pulse shape.

9.6 DIGITAL TRANSMISSION ON FADING MULTIPATH CHANNELS

In this and the previous two chapters we have described digital modulation and demodulation methods for transmission of information over two types of channels, namely, an additive Gaussian noise channel and a linear filter channel. These channel models are appropriate for a large variety of physical channels. However, these two channel models are inadequate in characterizing signal transmission over radio channels whose transmission characteristics change with time. In such cases, we must develop more general mathematical models that characterize the time-varying behavior of the channel. We cite the following examples of communication channels that require a different type of channel model.

Signal Transmission via Ionospheric Propagation in the HF Band. We recall from our discussion in Chapter 1 that sky-wave propagation, as illustrated in Figure 1.6, results from transmitted signals (in the HF frequency band) being bent or refracted by the ionosphere, which consists of several layers of charged particles ranging in altitude from 30 to 250 miles above the surface of the earth. As a consequence of these ionospheric layers, the signal arrives at the receiver via different propagation paths at different delays. These signal components are called *multipath components*. The signal multipath components generally have different carrier-phase offsets and, hence, they may add destructively at times, resulting in a phenomenon called *signal fading*. Hence, signal fading is a result of multipath signal propagation. To characterize such channel behavior, we must adopt a time-varying impulse response model.

Mobile Cellular Transmission. In mobile cellular radio transmission between a base station and a telephone-equipped automobile, the signal transmitted from the base station to the automobile is usually reflected from surrounding buildings, hills, and other obstructions. As a consequence, we observe multiple propagation paths arriving at the receiver at different delays. Hence, the received signal has characteristics similar to those for ionospheric propagation. The same is true of transmission from the automobile to the base station. Moreover, the speed that the automobile is traveling results in frequency offsets, called *Doppler shifts,* of the various frequency components (see Problem 9.35) of the signal.

Line-of-sight Microwave Radio Transmission. In line-of-sight (LOS) radio transmission of signals, the transmitting and receiving antennas generally are mounted on high towers, to avoid obstructions such as buildings and hills in the path of signal propagation. However, when there are tall obstructions or hilly terrain in the path of propagation, it is likely that signals will be reflected from the ground to the receiving antenna as illustrated in Figure 9.54. This is especially a problem under severe weather conditions. In this case there is a received signal component that arrives via the direct path and an ensemble of secondary paths that are reflected from the ground terrain. The latter arrive at the receiver with various delays and constitute multipath propagation. Relatively narrow-beamwidth antennas are employed in microwave LOS transmission to reduce the occurrence of secondary

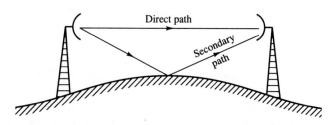

FIGURE 9.54. Illustration of multipath propagation in LOS microwave transmission.

reflections. Nevertheless, some secondary signal reflections are frequently observed in practice.

Airplane-to-airplane Radio Communications.   In radio communications between two aircrafts, it is possible for secondary signal components to be received from ground reflections, as illustrated in Figure 9.55. This is especially the case when omnidirectional antennas are employed in the communication system. The ensemble of ground-reflected signal components generally arrive at the receiver with different delays and different attenuations. In addition, the motions of the aircraft result in Doppler frequency offsets in the various signal components. In many respects, this situation is similar to that in mobile cellular communications.

The four channels briefly described above may be generally characterized as linear systems with time-varying impulse responses. Because it is generally difficult, if not impossible, to characterize the microscopic effects of signal transmission on channels as the ones described above in a deterministic fashion, it is logical to adopt a statistical characterization. Such an approach is described below.

9.6.1 Channel Model for Time-Variant Multipath Channels

As we have observed, there are basically two distinct characteristics of the types of radio channels described above. One characteristic is that the transmitted signal arrives at the receiver via multiple propagation paths, each of which has an associated time delay. For example, if we transmit an extremely short pulse, ideally an impulse, the channel response due to multiple scatterers (such as ionized particles

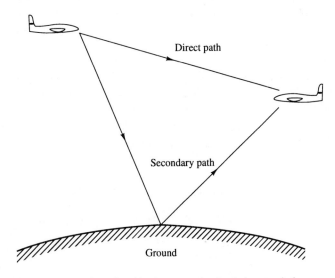

FIGURE 9.55.   Illustration of multipath propagation in airplane-to-airplane communications.

Transmitted signal   *Received signal*

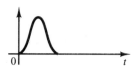

$t = t_0$

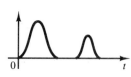

$t = t_1$

$t = t_2$

FIGURE 9.56. Illustration of time-variant channel response characteristics.

in the ionosphere) might appear as shown in Figure 9.56. Because the received signal is spread in time due to the multiple scatterers at different delays, we say that the channel is *time dispersive.*

A second characteristic of the types of radio channels described above is concerned with the time variations in the structure of the medium. As a result of such time variations, the response of the channel to any signal transmitted through it will change with time. Hence, if we repeat the short pulse transmission experiment over and over, we would observe changes in the received signal, which are due to physical changes in the medium. Such changes include variations in the relative delays of signals from the multiple scatterers. Hence, the received signal might appear as illustrated in Figure 9.56. In other words, the impulse response of the channel is varying with time. In general, the time variations in the received signal appear to be unpredictable to the user of the channel. This leads us to characterize the time-variant multipath channel statistically.

To obtain a statistical description of the channel, let us consider the transmission of an unmodulated carrier

$$c(t) = A \cos 2\pi f_c t \tag{9.6.1}$$

The received signal in the absence of noise may be expressed as

$$x(t) = A \sum_n \alpha_n(t) \cos[2\pi f_c(t - \tau_n(t))]$$

$$= A \operatorname{Re}\left[\sum_n \alpha_n(t) e^{-j2\pi f_c \tau_n(t)} e^{j2\pi f_c t}\right] \tag{9.6.2}$$

where $\alpha_n(t)$ is the time-variant attenuation factor associated with the nth propagation path and $\tau_n(t)$ is the corresponding propagation delay. The complex-valued signal,

$$z(t) = \sum_n \alpha_n(t) e^{-j2\pi f_c \tau_n(t)}$$

$$= \sum_n \alpha_n(t) e^{-j\phi_n(t)} \qquad (9.6.3)$$

represents the response of the channel to the complex exponential $\exp(j2\pi f_c t)$. We note that, although the input to the channel is a monochromatic signal, i.e., a signal at a single frequency, the output of the channel consists of a signal that contains many different frequency components. These new components are generated as a result of the time variations in the channel response. The r.m.s. (root-mean-square) spectral width of $z(t)$ is called the *Doppler frequency spread* of the channel and is denoted as B_d. This quantity is a measure of how rapidly the signal $z(t)$ is changing with time. If $z(t)$ changes slowly, the Doppler frequency spread is relatively small, while if $z(t)$ changes rapidly, the Doppler frequency spread is large.

We may view the received complex-valued signal $z(t)$ in (9.6.3) as the sum of a number of vectors (phasors), each of which has a time-variant amplitude $\alpha_n(t)$ and phase $\phi_n(t)$. In general, it takes large dynamic changes in the physical medium to cause a large change in $\{\alpha_n(t)\}$. On the other hand, the phases $\{\phi_n(t)\}$ will change by 2π radians whenever $\{\tau_n(t)\}$ change by $1/f_c$. But $1/f_c$ is a small number and, hence, the phases $\{\phi_n(t)\}$ change by 2π or more radians with relatively small changes of the medium characteristics. We also expect the delays $\{\tau_n(t)\}$ associated with the different signal paths to change at different rates and in an unpredictable (random) manner. This implies that the complex-valued signal $z(t)$ in (9.6.3) can be modeled as a random process. When there are a large number of signal propagation paths, the central limit theorem can be applied. Thus, $z(t)$ can be modeled as a complex-valued Gaussian random process.

The multipath propagation model for the channel, embodied in the received signal $x(t)$ or, equivalently, $z(t)$ given by (9.6.3), results in signal fading. The fading phenomenon is primarily a result of the time-variant phase factors $\{\phi_n(t)\}$. At times the complex-valued vectors in $z(t)$ add destructively to reduce the power level of the received signal. At other times the vectors in $z(t)$ add constructively and, thus, produce a large signal value. The amplitude variations in the received signal due to the time-variant multipath propagation in the channel are usually called *signal fading*.

A general model for a time-variant multipath channel is illustrated in Figure 9.57. The channel model consists of a tapped delay line with uniformly spaced taps. The tap spacing between adjacent taps is $1/W$, where W is the bandwidth of the signal transmitted through the channel. The tap coefficients, denoted as $\{c_n(t) \equiv \alpha_n(t) e^{j\phi_n(t)}\}$ are usually modeled as complex-valued, Gaussian random processes that are mutually uncorrelated. The length of the delay line corresponds to the amount of time dispersion in the multipath channel, which is usually called

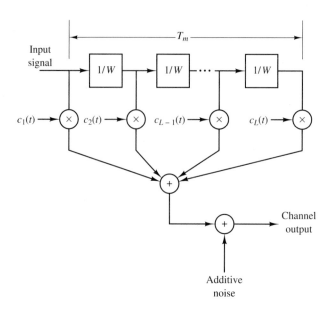

FIGURE 9.57. Model for time-variant multipath channel.

the *multipath spread*. We denote the multipath spread as $T_m = L/W$, where L represents the maximum number of possible multipath signal components.

Example 9.6.1

Determine an appropriate channel model for two-path ionospheric propagation where the relative time delay between the two received signal paths is 1 msec and the transmitted signal bandwidth W is 10 kHz.

Solution A 10-kHz signal can provide a time resolution of $1/W = 0.1$ msec. Because the relative time delay between the two received signal paths is 1 msec, the tapped delay line model consists of 10 taps, with only the first tap and the last tap having nonzero, time-varying coefficients, denoted as $c_1(t)$ and $c_2(t)$, as shown in Figure 9.58. Because $c_1(t)$ and $c_2(t)$ represent the signal response of a large number of ionized particles from two different regions of the ionosphere, we may characterize $c_1(t)$ and $c_2(t)$ as complex-valued, uncorrelated Gaussian random processes. The speed of variation of the tap coefficients determine the value of the Doppler spread for each path.

Example 9.6.2

Determine an appropriate channel model for an airplane-to-airplane communication link in which there is a direct signal propagation path and a secondary propagation resulting from signal scattering due to the surrounding ground terrain. The secondary path has a propagation delay of $\tau_0 = 10\mu\text{sec}$ relative to the propagation delay of the direct path. The signal bandwidth is $W = 100$ kHz.

Solution A 100 kHz signal provides a time resolution of $1/W = 1\mu\text{sec}$. Consequently, the secondary signal path is resolvable, because its relative time delay is 10 μsec.

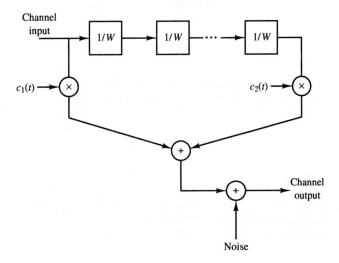

FIGURE 9.58. Channel model for two-path channel in Example 9.6.1.

In this case, a channel model that has two taps with tap coefficients $c_1 = \alpha_1 e^{j\phi_1}$ and $c_2(t) = \alpha_2(t)e^{j\phi_2(t)}$, as shown in Figure 9.59, is appropriate. Notice that the direct path is modeled as having a fixed (time-invariant) attenuation α_1 and phase shift ϕ_1. On the other hand, the secondary path that results from ground terrain is modeled as having a time-varying attenuation and phase shift, because the ground terrain is changing with time due to the motion of the airplanes. In this case, it is appropriate to model $c_2(t)$ as a complex-valued, Gaussian random process. The speed of variation of the tap coefficient $c_2(t)$ determines the value of the Doppler frequency spread for this path. Note that in both the direct and the reflected signal paths there will be Doppler frequency shifts resulting from motion of the aircraft.

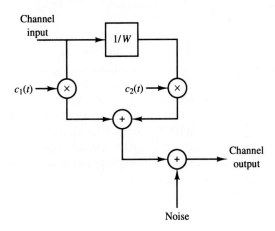

FIGURE 9.59. Channel model for two-path channel in Example 9.6.2.

Example 9.6.3

Determine the appropriate channel model for the airplane-to-airplane communication link described in Example 9.6.2, but now assume that the transmitted signal bandwidth is 10 kHz.

Solution   A 10 kHz signal provides a time resolution of 100 μsec. Because the relative delay between the two signal components is 10 μsec, the two signal paths are not resolvable in time. Consequently the channel appears as a single path channel with tap coefficient $c(t)$ that includes the direct path and the secondary path. Because the secondary path results from signal scattering from a large number of scatterers, we may characterize $c(t)$ as a complex-valued, Gaussian random process with a mean value of $\alpha_1 e^{j\phi_1}$, which is due to the direct path.

In the channel models described in the above examples, the tap coefficients in the tapped delay line model were characterized as complex-valued Gaussian random processes. We may express each of the tap coefficients as

$$c(t) = c_r(t) + jc_i(t) \qquad (9.6.4)$$

where $c_r(t)$ and $c_i(t)$ represent real-valued Gaussian random processes. We assume that $c_r(t)$ and $c_i(t)$ are stationary and statistically independent. This assumption generally holds for the tapped delay model of physical channels.

We may also express $c(t)$ in the form as

$$c(t) = v(t)e^{j\phi(t)} \qquad (9.6.5)$$

where

$$v(t) = \sqrt{c_r^2(t) + c_i^2(t)}$$

$$\phi(t) = \tan^{-1}\frac{c_i(t)}{c_r(t)} \qquad (9.6.6)$$

In this representation, if $c_r(t)$ and $c_i(t)$ are Gaussian with zero-mean values, the amplitude $v(t)$ is characterized statistically by the Rayleigh probability distribution and $\phi(t)$ is uniformly distributed over the interval $[0, 2\pi)$. As a consequence, the channel is called a *Rayleigh fading channel*. On the other hand, if $c_r(t)$ and $c_i(t)$ are Gaussian with nonzero mean as in the airplane-to-airplane communication link described in the examples above, the amplitude $v(t)$ is characterized statistically by the Rice probability distribution and the phase $\phi(t)$ is also nonzero mean. In this case the channel is called a *Ricean fading channel* (see Problem 3.31).

Besides the multipath (time) spread T_m and the Doppler (frequency) spread B_d, there are two other parameters that are useful in characterizing fading, multipath channels. One parameter is the reciprocal of the Doppler spread. This quantity is a measure of the time interval over which the channel characteristics will change very little. We call this parameter the *coherence time* of the channel and define it as

$$T_{ct} = \frac{1}{2B_d} \qquad (9.6.7)$$

The other useful parameter is the reciprocal of the multipath spread, which has units of frequency. This quantity is a measure of the bandwidth over which the channel characteristics, the magnitude $v(t)$, and the phase $\phi(t)$ are highly correlated. In other words, all frequency components of a signal within this bandwidth will fade simultaneously. We call this parameter the *coherence bandwidth* of the channel and define it as

$$B_{cb} = \frac{1}{2T_m} \qquad (9.6.8)$$

Example 9.6.4

Consider the communication link described in Example 9.6.2. Determine the coherence bandwidth of this channel.

Solution The multipath spread for this channel is $T_m = 10$ μsec. Hence, the coherence bandwidth of the channel is $1/2T_m$ 50 kHz. Within this bandwidth, all frequency components of a signal will be affected similarly by the channel. For example, all frequency components of a signal that fall within the coherence bandwidth will fade simultaneously.

Examples 9.6.1, 9.6.3, and 9.6.4 provide some indication on the role that the signal bandwidth plays in relation to the characteristics of the channel model. For example, if the transmitted signal has a bandwidth W that is larger than the coherence bandwidth B_{cb} of the channel ($W > B_{cb}$), as is the case in Examples 9.6.1 and 9.6.2, the multipath components are resolvable, and as a consequence frequency components of the transmitted signal that are separated by more that B_{cb} are affected (attenuated and phase shifted) differently by the channel. In such a case, we say that the channel is *frequency selective*. On the other hand, if the signal bandwidth W is smaller that $B_{cb}(W < B_{cb})$, as is the case in Example 9.6.3, all the frequency components in the signal are affected (attenuated and phase shifted) similarly by the channel at any instant in time. In such a case, we say that the channel is *frequency nonselective*.

Below we describe how these channel parameters are used in the design of modulation signals for transmitting digital information over the channel. Table 9.3 lists the values of these channel parameters for several multipath channels.

9.6.2 Signal Design for Fading Multipath Channels

There are different approaches that one can take in the design of signals for transmitting digital information through a fading multipath channel. Although, in principle, any of the carrier modulation methods (PAM, QAM, PM, FSK, and CPM) are applicable, signal fading generally causes large fluctuations in the received signal amplitude, and consequently it is extremely difficult to discriminate among multiple amplitude levels in the received signal. For this reason, PAM and

TABLE 9-3   MULTIPATH SPREAD AND DOPPLER SPREAD FOR SEVERAL TIME-VARIANT MULTIPATH CHANNELS.

Channel	Multipath spread T_m (sec)	Doppler spread B_d (Hz)
Shortwave ionospheric propagation (HF)	10^{-3}–10^{-2}	10^{-1}–1
Ionospheric forward scatter (VHF)	10^{-4}	10
Tropospheric scatter (SHF)	10^{-6}	10
Moon at maximum libration	10^{-2}	10

QAM are generally avoided for digital communications through fading channels. In our discussion, we shall focus our attention on PSK and FSK.

As indicated in the previous section, the available channel bandwidth plays an important role in the design of the modulation. If the bandwidth that is allocated to the user exceeds the coherence bandwidth B_{cb} of the channel, we have basically two options. One option is to transmit the information on a single sinusoidal carrier, using a signal that occupies the entire available channel bandwidth. In this case, the multipath signal components are resolvable to within a time resolution of $1/W$. Hence, signal propagation paths that differ in time-of-arrival at the receiver by a time interval greater than or equal to $1/W$ will be distinguishable as separate propagating signal components. The number of distinct signal components depends on the channel characteristics. For example, in the airplane-to-airplane communication link, we would expect to see the direct path and the ground-reflected path. In radio signal transmission via ionospheric propagation in the HF-frequency band, the number of received signal components depends on the number of ionospheric reflecting layers and the possibility of paths arriving by multiple skips.

We observe that when $W > B_{cb}$, the received signal may be corrupted by intersymbol interference unless the symbol rate is selected to be significantly smaller than the multipath spread T_m. We recall that the symbol rate $1/T$ satisfies the condition

$$\frac{1}{T} \le 2W \tag{9.6.9}$$

For example, if we select the symbol rate $1/T = W$, then because $T_m > 1/W$ it follows that $T_m > T$. In this case, the receiver encounters ISI and must employ some type of equalization, as previously described in Section 8.4. On the other hand, if the symbol rate is selected to be very small, i.e., $T \gg T_m$, than the ISI becomes insignificant and no equalization is necessary. Such signal design is usually accomplished by the use of spread spectrum signals, as described in Chapter 11. Further discussion on this topic is deferred to Section 11.2.2.

Example 9.6.5

Suppose that an HF channel with a nominal bandwidth allocation of 3200 Hz is to be used for transmitting digital information at a rate of either (1) 4800 bits/sec or (2) 20 bits/sec. The channel multipath spread is $T_m = 5$ msec. Specify a modulation method for achieving the desired data rates and indicate whether or not an equalizer is necessary at the receiver for the intersymbol interference.

Solution   For the 4800 bits/sec system, we select a symbol rate of 2400 symbol/sec and four-phase modulation (either coherent PSK or DPSK). A symbol rate of 2400 can be accommodated in a channel bandwidth of 3200 Hz, because there is an excess bandwidth of 800 Hz. A transmitting filter with a raised cosine spectral characteristic may be used to obtain the desired signal shaping. The symbol rate $1/T = 2400$ means that the received signal will be corrupted by ISI ($T_m \gg T$), which will require the use of an equalizer to reduce its effect on performance.

For the low rate of 20 bits/sec we may again employ four-phase modulation (PSK or DPSK). Thus, the symbol rate is 10 symbol/sec and hence $T = 100$ msec. Because $T \gg T_m$, the effect of the multipath spread is negligible and no equalization is necessary at the receiver. To use the entire channel bandwidth, we may employ a phase-modulated spread spectrum signal as described in Chapter 11.

The second option in the design of modulation when the available channel bandwidth exceeds the coherence bandwidth ($W > B_{cb}$) is to employ multicarrier transmission. In this method, the available channel bandwidth is subdivided into N subchannels such that the bandwidth of each subchannel is much less than B_{cb}. The primary objective in this subdivision of the channel bandwidth is to transmit information in each subchannel at a rate of $1/T_{sc} \ll B_{cb}$, such that ISI is negligible. This is clearly the case when $T_{sc} \gg T_m$. With N parallel subchannels, the overall symbol rate is N/T_{sc}.

Example 9.6.6

Consider an HF channel that has a nominal bandwidth of 3200 Hz and a multipath spread of $T_m = 1$ msec. Design a multiple-carrier modulator that achieves a data rate of 4800 bits/sec.

Solution   We may select the number N of subcarriers to be as large as we like to achieve the desired condition, $T_{sc} \gg T_m$. However, the complexity of the demodulator increases (linearly) with N, and the demodulation delay for delivering the information to the user also increases (linearly) with N. Therefore, it is desirable to keep N as small as possible. Suppose we select N such that $T_{sc} = 100$ msec. Then each subchannel may be as narrow[†] as $W_{ac} \approx \frac{1}{T_{sc}} = 10$ Hz. Note that $W_{sc} \ll B_{cb} = 500$ Hz as desired. If we employ four-phase (PSK or DPSK) modulation in each subchannel, we achieve a bit rate of 20 bits/sec subchannel. With $N = 240$ subchannels, we achieve the desired data rate of 4800 bits/sec.

[†]In practice, it will be necessary to have some excess bandwidth in each subchannel. The bandwidth may be in the range of 25% to 50%.

Of the two options described, the multiple subcarrier system requires no equalization and may be preferable from the viewpoint of reducing receiver complexity, especially when the channel multipath spread is large, necessitating the use of a long equalizer at the receiver. On the other hand, if the multipath spread is relatively small and the ISI is confined to a few symbols (two or three, for example), the complexity of the equalizer is also relatively small. In such a case, the multicarrier design method may not be as attractive.

In the above discussion we assumed that the available channel bandwidth exceeds the coherence bandwidth B_{cb} of the channel. On the other hand, if the channel bandwidth is much smaller than B_{cb}, there is no point in designing a multicarrier system. A single carrier system can be designed using the entire bandwidth, with a symbol rate of $W \leq 1/T \leq 2W$. In this case $T \gg T_m$, so that the effect of ISI on the performance of the system is negligible.

Below we evaluate the performance of the demodulator and detector for PSK, DPSK, and FSK modulations in a Rayleigh fading channel under the condition that ISI is negligible.

9.6.3 Performance of Binary Modulation in Rayleigh Fading Channels

In this section we determine the probability of error at the receiver of a binary digital communication system that transmits information through a Rayleigh fading channel. The signal bandwidth W is assumed to be much smaller than the coherence bandwidth B_{cb} of the channel, as in Example 9.6.3. Because the multipath components are not resolvable, the channel is frequency nonselective and, hence, the channel impulse response is represented as

$$h(\tau; t) = \alpha(t)\delta(\tau - \tau_0(t)) \tag{9.6.10}$$

where $\alpha(t)$ has a Rayleigh distribution at any instant in time.

We assume that the time variations of $\alpha(t)$ and $\tau_0(t)$ are very slow compared to the symbol interval, so that within the time interval $0 \leq t \leq T$, the channel impulse response is constant, i.e.,

$$h(\tau; t) \equiv h(\tau) = \alpha\delta(\tau - \tau_0) \tag{9.6.11}$$

where the amplitude α is Rayleigh distributed, i.e.,

$$p(\alpha) = \begin{cases} \dfrac{\alpha}{\sigma_\alpha^2} e^{-\alpha^2/2\sigma_\alpha^2}, & \alpha > 0 \\ 0, & \text{otherwise} \end{cases} \tag{9.6.12}$$

and $\sigma_\alpha^2 = E(\alpha^2)/2$.

Now suppose that binary antipodal signals, e.g., binary PSK, are used to transmit the information through the channel. Hence, the two possible signals are

$$u_m(t) = \sqrt{\frac{2E_b}{T}} \cos(2\pi f_c t + m\pi) + n(t), \quad m = 0, 1 \tag{9.6.13}$$

The received signal in the interval $0 \le t \le T$ is

$$r(t) = \alpha \sqrt{\frac{2E_b}{T}} \cos(2\pi f_c t + m\pi + \phi) + n(t) \tag{9.6.14}$$

where ϕ is the carrier-phase offset. Let us assume that ϕ is known to the demodulator, which crosscorrelates $r(t)$ with

$$\psi(t) = \sqrt{\frac{2}{T}} \cos(2\pi f_c t + \phi), \quad 0 \le t \le T \tag{9.6.15}$$

Hence, the input to the detector at the sampling instant is

$$r = \alpha \sqrt{E_b} \cos m\pi + n, \quad m = 0, 1 \tag{9.6.16}$$

For a fixed value of α, the probability of error is the familiar form,

$$P_2(\alpha) = Q\left(\sqrt{\frac{2\alpha^2 E_b}{N_0}}\right) \tag{9.6.17}$$

We view $P_2(\alpha)$ as a conditional probability of error for a given value of the channel attenuation α. To determine the probability of error averaged over all possible values of α, we compute the integral

$$P_2 = \int_0^\infty P_2(\alpha) p(\alpha) d\alpha \tag{9.6.18}$$

where $p(\alpha)$ is the Rayleigh p.d.f. given by (9.6.12). This integral has the simple closed-form expression

$$P_2 = \frac{1}{2}\left[1 - \sqrt{\frac{\bar{\rho}_b}{1 + \bar{\rho}_b}}\right] \tag{9.6.19}$$

where, by definition,

$$\bar{\rho}_b = \frac{E_b}{N_0} E(\alpha^2) \tag{9.6.20}$$

Hence, $\bar{\rho}_b$ is the average received SNR/bit and $E[\alpha^2] = 2\sigma_\alpha^2$.

If the binary signals are orthogonal, as in orthogonal FSK, where the two possible transmitted signals are given as

$$u_m(t) = \sqrt{\frac{2E_b}{N_0}} \cos\left[2\pi\left(f_c + \frac{m}{2T}\right)t\right], \quad m = 0, 1, \tag{9.6.21}$$

the received signal is

$$r(t) = \alpha \sqrt{\frac{2\mathcal{E}_b}{T}} \cos \left[2\pi \left(f_c + \frac{m}{2T} \right) t + \phi \right] + n(t) \qquad (9.6.22)$$

In this case, the received signal is crosscorrelated with the two signals

$$\psi_1(t) = \sqrt{\frac{2}{T}} \cos(2\pi f_c t + \phi)$$

$$\psi_2(t) = \sqrt{\frac{2}{T}} \cos \left[2\pi \left(f_c + \frac{1}{2T} \right) t + \phi \right] \qquad (9.6.23)$$

If $m = 0$, for example, the two correlator outputs are

$$r_1 = \alpha \sqrt{\mathcal{E}_b} + n_1$$

$$r_2 = n_2 \qquad (9.6.24)$$

where n_1 and n_2 are the additive noise components at the outputs of the two correlators. Hence the probability of error is simply the probability that $r_2 > r_1$. Because the signals are orthogonal, the probability of error for a fixed value of α has the familiar form

$$P_2(\alpha) = Q \left(\sqrt{\frac{\alpha^2 \mathcal{E}_b}{N_0}} \right) \qquad (9.6.25)$$

As in the case of antipodal signals, the average probability of error over all values of α is determined by evaluating the integral in (9.6.18). Thus, we obtain

$$P_2 = \frac{1}{2} \left[1 - \sqrt{\frac{\bar{\rho}_b}{2 + \bar{\rho}_b}} \right] \qquad (9.6.26)$$

where $\bar{\rho}_b$ is the average SNR/bit defined by (9.6.20).

Figure 9.60 illustrates the average probability of error for binary antipodal and orthogonal signals. The striking aspects of these graphs is the slow decay of the probability of error as a function of SNR. In fact, for large values of $\bar{\rho}_b$, the probability of error for binary signals is

$$P_2 \approx \frac{1}{4\bar{\rho}_b}, \quad \text{antipodal signals}$$

$$P_2 \approx \frac{1}{2\bar{\rho}_b}, \quad \text{orthogonal signals} \qquad (9.6.27)$$

Hence, the probability of error in both cases decreases only inversely as the SNR. This is in contrast to the exponential decrease in the case of the AWGN channel. We also note that the difference in SNR between antipodal signals (binary PSK) and orthogonal signals (binary FSK) is 3 dB.

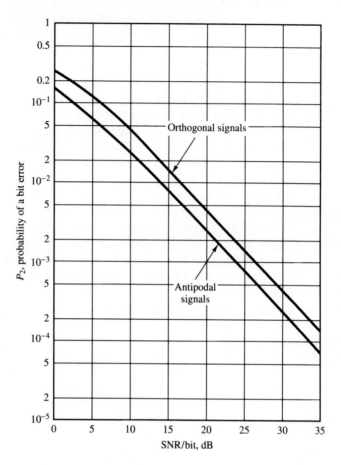

FIGURE 9.60. Performance of binary signaling on a Rayleigh fading channel.

Two other types of signal demodulation are DPSK and noncoherent FSK. For completeness, we state that the average probability of error for these signals (see Problem 9.38) is

$$P_2 = \frac{1}{2(1 + \bar{\rho}_b)}, \quad \text{DPSK} \tag{9.6.28}$$

$$P_2 = \frac{1}{2 + \bar{\rho}_b}, \quad \text{noncoherent FSK} \tag{9.6.29}$$

Performance improvement through signal diversity.   The basic problem in digital communication through a fading channel is that a large number of errors occur when the channel attenuation is large, i.e., when the channel is in a deep fade. If we can supply to the receiver two or more replicas of the same information signal transmitted through independently fading channels, the probability that all the signal components will fade simultaneously is reduced considerably. If p is the probability that any one signal will fade below some critical value, than p^D is the probability that all D independently fading replicas of the same signal will fade below the critical value. There are several ways to provide the receiver with D independently fading replicas of the same information-bearing signal.

One method for achieving D independently fading versions of the same information-bearing signal is to transmit the same information on D carrier frequencies, where the separation between successive carriers equals or exceeds the coherence bandwidth B_{cb} of the channel. This method is called *frequency diversity*.

A second method for achieving D independently fading versions of the same information-bearing signal is to transmit the same information in D different time slots, where the time separation between successive time slots equals or exceeds the coherence time T_{ct} of the channel. This method is called *time diversity*.

Another commonly used method for achieving diversity is via use of multiple receiving antennas, but only one transmitting antenna. The receiving antennas must be spaced sufficiently far apart so that the multipath components in the signal have significantly different propagation paths, as illustrated in Figure 9.61. Usually, a separation of 10 or more wavelengths is required between two receiving antennas in order to obtain signals that fade independently.

There are other diversity transmission and reception techniques that have been tried in practice, such as angle-of-arrival diversity and polarization diversity. However, these methods have not been as widely used as the methods described above.

Given that the information is transmitted to the receiver via D independently fading channels, there are several ways that the receiver may extract the transmitted information from the received signal. The simplest method is for the receiver to monitor the received power level in the D received signals and to select for demodulation and detection the strongest signals. In general, this approach results in frequent switching from one signal to another. A slight modification that leads to a simpler implementation is to use a signal for demodulation and detection as long as the received power level in that signal is above a preset threshold. When the signal falls below the threshold, a switch is made to the channel that has the

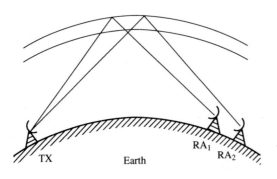

FIGURE 9.61. Illustration of diversity reception using two receiving antennas.

largest received power level. This method of signal combining and detection is called *selection combining.*

There are several more complex methods for combining the independently fading received signals. One that is appropriate for coherent demodulation and detection requires that the receiver estimate and correct for the different phase offsets on each of the D received signals after demodulation. Then, the phase-corrected signals at the outputs of the D demodulators are summed and fed to the detector. This type of signal combining is called *equal-gain combining.* If, in addition, the received signal power level is estimated for each of the D received signals and the phase-corrected demodulator outputs are weighted in direct proportion of the received signal strength (square-root of power level) and then fed to the detector, the combiner is called a *maximal ratio combiner.* On the other hand, if orthogonal signals are used for transmitting the information through D independently fading channels, the receiver may employ noncoherent demodulation. In such a case, the outputs from the D demodulators may be squared, summed, and then fed to detector. This combiner is called a *square-law combiner.*

All these types of combining methods lead to performance characteristics that result in a probability of error, which behaves as $K_D/\bar{\rho}_b^D$ where K_D is a constant that depends on D, and $\bar{\rho}_b$ is the average SNR. Thus, we achieve an exponential decrease in the error probability. Without providing a detailed derivation, we simply state that for antipodal signals with maximal ratio combining, the probability of error has the general form,

$$P_2 \approx \frac{K_D}{(4\bar{\rho}_b)^D}, \qquad \bar{\rho}_b \gg 1 \tag{9.6.30}$$

where K_D is defined as

$$K_D = \frac{(2D-1)!}{D!(D-1)!} \tag{9.6.31}$$

For binary orthogonal signals with square-law combining, the probability of error has the asymptotic form,

$$P_2 \approx \frac{K_D}{\bar{\rho}_b^D}, \qquad \bar{\rho}_b \gg 1 \tag{9.6.32}$$

Finally, for binary DPSK with equal gain combining, the probability of error has the asymptotic form,

$$P_2 \approx \frac{K_D}{(2\bar{\rho})^D}, \qquad \bar{\rho}_b \gg 1 \tag{9.6.33}$$

These error probabilities are plotted in Figure 9.62 for $D = 1, 2, 4$. It is apparent that a large reduction in SNR/bit is achieved in having $D = 2$ (dual diversity) compared to no diversity. A further reduction in SNR is achieved by increasing the order of diversity to $D = 4$, although the additional gain from $D = 2$ to $D = 4$ is smaller than going from $D = 1$ to $D = 2$. Beyond $D = 4$, the additional

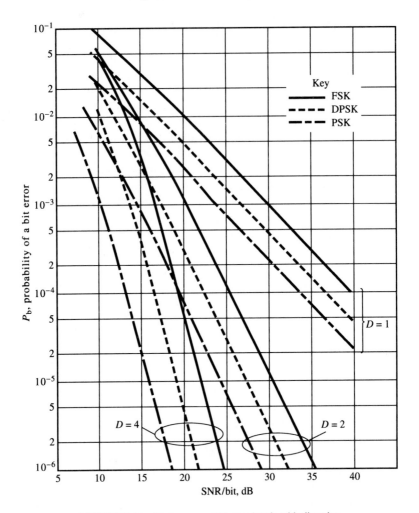

FIGURE 9.62. Performace of binary signals with diversity.

reduction in SNR is not as significant. Hence, one must question the performance advantages in increasing the diversity beyond $D = 4$, when the cost of the increase is to use additional channel bandwidth.

In conclusion, we have demonstrated that efficient use of transmitter power in a Rayleigh fading channel can be achieved by using some form of diversity to provide the receiver with several independently fading signals all carrying the same information. The types of diversity that we described (time, frequency, etc.) constitute a form of channel coding usually called *repetition coding*. In the following chapter, we describe channel codes that are considerably more bandwidth efficient than repetition coding and are generally more attractive for digital communication through Rayleigh fading channels.

9.7 FURTHER READING

In this chapter we have focused on the transmission of digital information via carrier modulation. We have described conventional carrier modulation methods that include pulse-amplitude modulation, phase modulation, including phase-shift keying and differential phase-shift keying, quadrature-amplitude modulation, frequency modulation, and continuous-phase modulation. The demodulation and detection of these carrier-modulated signals were described for an AWGN channel and their performance characteristics were evaluated.

More extensive treatments of carrier-modulation techniques may be found in more advanced textbooks on digital communications, such as the books by Proakis (1989), Biglieri et al. (1987), and Blahut (1990), and in the technical journals, especially the *IEEE Transactions on Communications.*

CPFSK and CPM are modulation techniques that have been treated extensively in the technical journals and more recently in textbooks. A thorough treatment of CPM can be found in the book by Anderson et al. (1986). The journal papers by Aulin and Sundberg (1981, 1982, 1984) and by Aulin et al. (1981) provide detailed analysis of the performance characteristics of CPM. The tutorial paper by Sundberg (1986) gives a very readable overview of CPM, its demodulation and its performance characteristics. This paper also contains a comprehensive list of references.

Symbol synchronization for carrier-modulated signals is a topic that has been treated thoroughly and analyzed in many journal articles. Of particular importance are the journal papers by Lyon (1975a,b) that treat timing recovery for QAM signals and the paper by Mueller and Muller (1976) that describes symbol timing methods based on digital processing of signal samples.

Finally, our treatment of the topic of digital communication through fading multipath channels was indeed very brief. More extensive and advanced treatments are found in books by Schwartz et al. (1966) and Proakis (1989). The pioneering work on the characterization of fading multipath channels and on signal and receiver design for reliable digital communications over such channels was done by Price (1954, 1956). This early work was followed by additional significant contributions from Price and Green (1958, 1960), Kailath (1960, 1961), and Green (1962). Diversity transmission and diversity combining techniques under a variety of channel conditions have been treated in the papers by Pierce (1958), Brennan (1959), Turin (1961, 1962), Pierce and Stein (1960), Barrow (1963), Bello and Nelin (1962, 1963), Price (1962), and Lindsey (1964).

Our treatment of digital communications over fading multipath channels was necessarily restricted to the Rayleigh fading channel model. For the most part, this is due to the wide acceptance of this model for describing the fading effects on many radio channels and to its mathematical tractability. Although other statistical models, such as the Ricean fading model, may be more appropriate for characterizing fading on some real channels, such as microwave LOS, the general approach for providing reliable communications by means of signal diversity is still applicable.

PROBLEMS

9.1 A single sideband PAM signal may be represented as

$$u_m(t) = A_m \left[g_T(t) \cos 2\pi f_c t - \widehat{g}_T(t) \sin 2\pi f_c t \right]$$

where $\widehat{g}_T(t)$ is the Hilbert transform of $g_T(t)$ and A_m is the amplitude level that conveys the information. Demonstrate mathematically that a Costas loop (see Figure 6.10) can be used to demodulate the SSB PAM signal.

9.2 Two quadrature carriers $\cos 2\pi f_c t$ and $\sin 2\pi f_c t$ are used to transmit digital information through an AWGN channel at two different data rates, 10 kbits/sec and 100 kbits/sec. Determine the relative amplitudes of the signals for the two carriers so that the $\mathcal{E}_b/N_b$ for the two channels is identical.

9.3 An ideal voice-band telephone line channel has a bandpass frequency response characteristic spanning the frequency range 600–3000 Hz.

1. Design an $M = 4$ PSK (quadrature PSK or QPSK) system for transmitting data at a rate of 2400 bits/sec and a carrier frequency $f_c = 1800$. For spectral shaping, use a raised-cosine frequency-response characteristic. Sketch a block diagram of the system and describe their functional operation.

2. Repeat part 1 if the bit rate $R = 4800$ bits per second.

9.4 Suppose that binary PSK is used for transmitting information over an AWGN with a power-spectral density of $N_0/2 = 10^{-10}$ W/Hz. The transmitted signal energy is $\mathcal{E}_b = A^2 T/2$, where T is the bit interval and A is the signal amplitude. Determine the signal amplitude required to achieve an error probability of 10^{-6} if the data rate is (a) 10 kbits/sec, (b) 100 kbits/sec, (c) 1 Mbits/sec.

9.5 Consider the signal

$$u(t) = \begin{cases} \frac{A}{T} t \cos 2\pi f_c t, & 0 \le t \le T \\ 0, & \text{otherwise} \end{cases}$$

1. Determine the impulse response of the matched filter for the signal.

2. Determine the output of the matched filter at $t = T$.

3. Suppose the signal $u(t)$ is passed through a correlator that correlates the input $u(t)$ with $u(t)$. Determine the value of the correlator output at $t = T$. Compare your result with that in part 2.

9.6 A carrier component is transmitted on the quadrature carrier in a communication system that transmits information via binary PSK. Hence, the received signal has the form

$$r(t) = \pm \sqrt{2P_s} \cos \left(2\pi f_c + \phi \right) + \sqrt{2P_c} \sin \left(2\pi f_c + \phi \right) + n(t)$$

where ϕ is the carrier phase and $n(t)$ is AWGN. The unmodulated carrier component is used as a pilot signal at the receiver to estimate the carrier phase.

1. Sketch a block diagram of the receiver, including the carrier-phase estimator.

2. Illustrate mathematically the operations involved in the estimation of the carrier-phase ϕ.

3. Express the probability of error for the detection of the binary PSK signal as a function of the total transmitted power $P_T = P_s + P_c$. What is the loss in performance due to the allocation of a portion of the transmitted power to the pilot signal? Evaluate the loss for $P_c/P_T = 0.1$.

9.7 Consider a four-phase PSK signal that is represented by the equivalent lowpass signal

$$v(t) = \sum_n a_n g(t - nT)$$

where a_n takes on one of the four possible values $\frac{\pm 1 \pm j}{\sqrt{2}}$ with equal probability. The sequence of information symbols $\{a_n\}$ is statistically independent.

1. Determine and sketch the power-spectral density of $v(t)$ when

$$g(t) = \begin{cases} A, & 0 \le t \le T \\ 0, & \text{otherwise} \end{cases}$$

2. Repeat part 1 when

$$g(t) = \begin{cases} A \sin \dfrac{\pi t}{2}, & 0 \le t \le T \\ 0, & \text{otherwise} \end{cases}$$

3. Compare the spectra obtained in parts 1 and 2 in terms of the 3-dB bandwidth and the bandwidth to the first spectral null.

9.8 In the demodulation of a binary PSK signal received in white Gaussian noise, a PLL is used to estimate the carrier-phase ϕ.

1. Determine the effect of a phase error $\phi - \widehat{\phi}$ on the probability of error.

2. What is the loss in SNR if the phase error $\phi - \widehat{\phi} = 45°$?

9.9 The probability of error for binary PSK demodulation and detection when there is a carrier phase error ϕ_e is

$$P_2(\phi_e) = Q\left(\sqrt{\frac{2\mathcal{E}_b}{N_0}}\cos^2\phi_e\right)$$

Suppose that the phase error is modeled as a zero-mean, Gaussian random variable with variance $\sigma_\phi^2 \ll \pi$. Determine the expression for the average probability of error (in integral form).

9.10 Suppose that the loop filter (see 6.2.4) for a PLL has the transfer function

$$G(s) = \frac{1}{s + \sqrt{2}}$$

1. Determine the closed-loop transfer function $H(s)$ and indicate if the loop is stable.
2. Determine the damping factor and the natural frequency of the loop.

9.11 Consider the PLL for estimating the carrier phase of a signal in which the loop filter is specified as

$$G(s) = \frac{K}{1 + \tau_1 s}$$

1. Determine the closed-loop transfer function $H(s)$ and its gain at $f = 0$.
2. For what range of value of τ_1 and K is the loop stable?

9.12 The loop filter $G(s)$ in a PLL is implemented by the circuit shown in Figure P-9.12. Determine the system function $G(s)$ and express the time constants τ_1 and τ_2 (see 6.2.4) in terms of the circuit parameters.

9.13 The loop filter $G(s)$ in a PLL is implemented with the active filter shown in Figure P-9.13. Determine the system function $G(s)$ and express the time constants τ_1 and τ_2 (see 6.2.4) in terms of the circuit parameters.

9.14 Consider the four-phase and eight-phase signal constellations shown in Figure P-9.14. Determine the radii r_1 and r_2 of the circles such that the

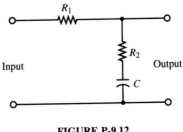

FIGURE P-9.12

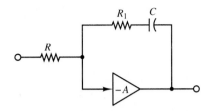

FIGURE P-9.13

distance between two adjacent points in the two constellations is d. From this result determine the additional transmitted energy required in the 8-PSK signal to achieve the same error probability as the four-phase signal at high SNR, where the probability of error is determined by errors in selecting adjacent points.

9.15 A voice-band telephone channel passes the frequencies in the band from 300 to 3300 Hz. It is desired to design a modem that transmits at a symbol rate of 2400 symbols/sec, with the objective of achieving 9600 bits/sec. Select an appropriate QAM signal constellation, carrier frequency, and the roll-off factor of a pulse with a raised cosine spectrum that utilizes the entire frequency band. Sketch the spectrum of the transmitted signal pulse and indicate the important frequencies.

9.16 Consider the two 8-point QAM signal constellation shown in Figure P-9.16. The minimum distance between adjacent points is $2A$. Determine the average transmitted power for each constellation assuming that the signal points are equally probable. Which constellation is more power efficient?

9.17 The 16-QAM signal constellation shown in Figure P-9.17 is an international standard for telephone line modems (called V.29). Determine the optimum decision boundaries for the detector, assuming that the SNR is sufficiently high so that errors only occur between adjacent points.

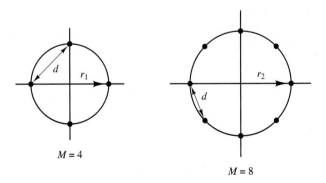

$M = 4$

$M = 8$

FIGURE P-9.14

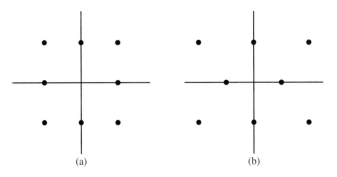

(a)   (b)

FIGURE P-9.16

9.18 Specify a Gray code for the 16-QAM V.29 signal constellation shown in Problem 9.17.

9.19 A 4 kHz bandpass channel is to be used for transmission of data at a rate of 9600 bits/sec. If $N_0/2 = 10^{-10}$ W/Hz is the spectral density of the additive, zero-mean Gaussian noise in the channel, design a QAM modulation and determine the average power that achieves a bit error probability of 10^{-6}. Use a signal pulse with a raised-cosine spectrum having a roll-off factor of at least 50%.

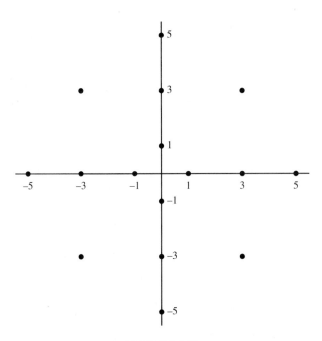

FIGURE P-9.17

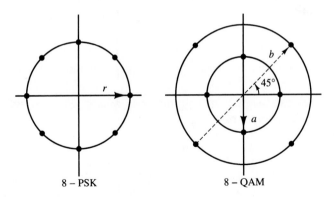

8 – PSK   8 – QAM

FIGURE P-9.20

9.20 Consider the octal signal point constellations in Figure P-9.20.

1. The nearest neighbor signal points in the 8-QAM signal constellation are separated in distance by A units. Determine the radii a and b of the inner and outer circles.
2. The adjacent signal points in the 8-PSK are separated by a distance of A units. Determine the radius r of the circle.
3. Determine the average transmitter powers for the two signal constellations and compare the two powers. What is the relative power advantage of one constellation over the other? (Assume that all signal points are equally probable.)

9.21 Consider a digital communication system that transmits information via QAM over a voice-band telephone channel at a rate 2400 symbols/sec. The additive noise is assumed to be white and Gaussian.

1. Determine the $\mathcal{E}_b/N_0$ required to achieve an error probability of 10^{-5} at 4800 bits/sec.
2. Repeat part 1 for a rate of 9600 bits/sec.
3. Repeat part 1 for a rate of 19,200 bits/sec.
4. What conclusions do you reach from these results.

9.22 Consider the 8-point QAM signal constellation shown in Figure P-9.20.

1. Is it possible to assign three data bits to each point of the signal constellation such that nearest (adjacent) points differ in only one bit position? Explain.
2. Determine the symbol rate if the desired bit rate is 90 Mbits/sec.

3. Compare the SNR required for the 8-point QAM modulation with that of an 8-point PSK modulation having the same error probability.

4. Which signal constellation, 8-point QAM or 8-point PSK, is more immune to phase errors? Explain the reason for your answer.

9.23 The two signal waveforms for binary FSK signal transmission with discontinuous phase are

$$s_0(t) = \sqrt{\frac{2E_b}{T_b}} \cos\left[2\pi\left(f - \frac{\Delta f}{2}\right)t + \theta_0\right], \quad 0 \le t \le T$$

$$s_1(t) = \sqrt{\frac{2E_b}{T_b}} \cos\left[2\pi\left(f + \frac{\Delta f}{2}\right)t + \theta_1\right], \quad 0 \le t \le T$$

where $\Delta f = 1/T \ll f_c$, and θ_0 and θ_1 are uniformly distributed random variables on the interval $(0, 2\pi)$. The signals $s_0(t)$ and $s_1(t)$ are equally probable.

1. Determine the power-spectral density of the FSK signal.
2. Show that the power-spectral density decays as $1/f^2$ for $f \gg f_c$.

9.24 In Section 9.4.1 it was shown that the minimum frequency separation for orthogonality of binary FSK signals with coherent detection is $\Delta f = \frac{1}{2T}$. However, a lower error probability is possible with coherent detection of FSK if Δf is increased beyond $\frac{1}{2T}$. Show that the minimum value of Δf is $\frac{0.715}{T}$ and determine the probability of error for this choice of Δf.

9.25 The lowpass equivalent signal waveforms for three signal sets are shown in Figure P-9.25. Each set may be used to transmit one of four equally probable messages over an additive white Gaussian noise channel with noise power-spectral density $\frac{N_0}{2}$.

1. Classify the signal waveforms in set I, set II, and set III. In other words, state the category or class to which each signal set belongs.
2. What is the average transmitted energy for each signal set?
3. For signal set I, specify the average probability of error if the signals are detected coherently.
4. For signal set II, give a union bound on the probability of a symbol error if the detection is performed (a) coherently and (b) noncoherently.
5. Is it possible to use noncoherent detection on signal set III? Explain.
6. Which signal set or signal sets would you select if you wished to achieve a bit rate to bandwidth ($\frac{R}{W}$) ratio of at least 2. Explain your answer.

9.26 Consider the phase-coherent demodulator for M-ary FSK signals as shown in Figure 9.30.

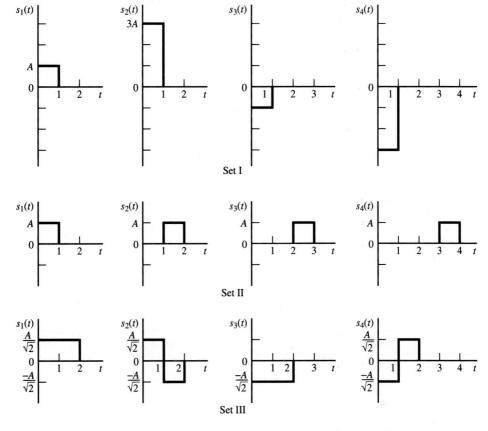

FIGURE P-9.25

1. Assume that the signal

$$u_0(t) = \sqrt{\frac{2\mathcal{E}_s}{T}} \cos 2\pi f_c t, \qquad 0 \leq t \leq T$$

was transmitted and determine the output of the $M - 1$ correlators at $t = T$, corresponding to the signals $u_m(t)$, $m = 1, 2, \ldots, M - 1$, given by (9.4.2) when $\hat{\phi}_m \neq \phi_m$.

2. Show that the minimum frequency separation required for the signal orthogonality at the demodulator when $\hat{\phi}_m \neq \phi_m$ is $\Delta f = \frac{1}{T}$.

9.27 In the demodulation and noncoherent detection of M-ary FSK signals, as illustrated in Figure 9.31, show that the $2M$ noise samples given in (9.4.9) and (9.4.10) are zero-mean, mutually independent Gaussian random variables with equal variance $\sigma^2 = \frac{N_0}{2}$.

9.28 In on–off keying of a carrier-modulated signal, the two possible signals are

$$s_0(t) = 0, \qquad\qquad\qquad 0 \le t \le T_b$$
$$s_1(t) = \sqrt{\tfrac{2E_b}{T_b}} \cos 2\pi f_c t, \qquad 0 \le t \le T_b$$

The corresponding received signals are

$$r(t) = n(t), \qquad\qquad\qquad 0 \le t \le T_b$$
$$r(t) = \sqrt{\tfrac{2E_b}{T_b}} \cos\left(2\pi f_c t + \phi\right) + n(t), \qquad 0 \le t \le T_b$$

where ϕ is the carrier phase and $n(t)$ is AWGN.

1. Sketch a block diagram of the receiver (demodulator and detector) that employs noncoherent (envelope) detection.
2. Determine the p.d.f.'s for the two possible decision variables at the detector corresponding to the two possible received signals.
3. Derive the probability of error for the detector.

9.29 Determine the bit rate that can be transmitted through a 4 kHz voice-band telephone (bandpass) channel if the following modulation methods are used: (1) binary PSK, (2) four-phase PSK, (3) 8-point QAM, (4) binary orthogonal FSK, with noncoherent detection, (5) orthogonal four-FSK with noncoherent detection, and (6) orthogonal 8-FSK with noncoherent detection. For parts 1–3 assume that the transmitter pulse shape has a raised-cosine spectrum with a 50% roll-off.

9.30 Digital information is to be transmitted by carrier modulation through an additive Gaussian noise channel with a bandwidth of 100 kHz and $N_0 = 10^{-10}$ W/Hz. Determine the maximum rate that can be transmitted through the channel for four-phase PSK, binary FSK, and four-frequency orthogonal FSK, which is detected noncoherently.

9.31 In a MSK signal, the initial state for the phase is either 0 or π radians. Determine the terminal phase state for the following four input pairs of input data: (a) 00, (b) 01, (c) 10, (d) 11.

9.32 A continuous-phase FSK signal with $h = 1/2$ is represented as

$$s(t) = \pm\sqrt{\tfrac{2E_b}{T_b}} \cos\left(\tfrac{\pi t}{2T_b}\right)\cos 2\pi f_c t \pm \sqrt{\tfrac{2E_b}{T_b}} \sin\left(\tfrac{\pi t}{2T_b}\right)\sin 2\pi f_c t, \qquad 0 \le t \le 2T_b$$

where the $\pm$ signs depend on the information bits transmitted.

1. Show that this signal has a constant amplitude.
2. Sketch a block diagram of the modulator for synthesizing the signal.
3. Sketch a block diagram of the demodulator and detector for recovering the information.

9.33 Sketch the phase tree, the state trellis, and the state diagram for partial-response CPM with $h = \frac{1}{2}$ and

$$
u(t) = \begin{cases} \dfrac{1}{4T}, & 0 \le t \le 2T \\[2mm] 0, & \text{otherwise} \end{cases}
$$

9.34 Determine the number of terminal phase states in the state trellis diagram for (1) a full response binary CPFSK with either $h = \frac{2}{3}$ or $\frac{3}{4}$ and (2) a partial-response $L = 3$ binary CPFSK with either $h = \frac{2}{3}$ or $\frac{3}{4}$.

9.35 In the transmission and reception of signals to and from moving vehicles, the transmitted signal frequency is shifted in direct proportion to the speed of the vehicle. The so-called *Doppler frequency shift* imparted to a signal that is received in a vehicle traveling at a velocity v relative to a (fixed) transmitter is given by the formula

$$
f_D = \pm \frac{v}{\lambda}
$$

where λ is the wavelength, and the sign depends on the direction (moving toward or moving away) that the vehicle is traveling relative to the transmitter. Suppose that a vehicle is traveling at a speed of 100 km/hr relative to a base station in a mobile cellular communication system. The signal is a narrowband signal transmitted at a carrier frequency of 1 GHz.

1. Determine the Doppler frequency shift.
2. What should be the bandwidth of a Doppler frequency tracking loop if the loop is designed to track Doppler frequency shifts for vehicles traveling at speeds up to 100 km/hr?
3. Suppose the transmitted signal bandwidth is 2 MHz centered at 1 GHz. Determine the Doppler frequency spread between the upper and lower frequencies in the signal.

9.36 Suppose the binary antipodal signals $\pm s(t)$ are transmitted over a fading channel and the received signal is

$$
r(t) = \pm a s(t) + n(t), \quad 0 \le t \le T
$$

where $n(t)$ is zero-mean white Gaussian noise with autocorrelation function $\frac{N_0}{2} \delta(\tau)$. The energy in the transmitted signal is $\mathcal{E} = \int_0^T |s(t)|^2 \, dt$. The channel gain a is specified by the p.d.f. $p(a) = 0.1\delta(a) + 0.9\delta(a - 2)$

1. Determine the average probability of error P_e for the demodulator which employs a filter matched to $s(t)$.
2. What value does P_e approach as $\frac{\mathcal{E}}{N_0}$ approaches infinity?

3. Suppose the same signal is transmitted over two statistically independent fading channels with gains a_1 and a_2 where

$$p(a_k) = 0.1\delta(a_k) + 0.9\delta(a_k - 2), \qquad k = 1, 2$$

The noises on the two channels are statistically independent and identically distributed. The demodulator employs a matched filter for each channel and simply adds the two filter outputs to form the decision variable. Determine the average P_e.

4. For the case in part 3 what value does P_e approach as $\frac{\mathcal{E}}{N_0}$ approaches infinity?

9.37 A multipath fading channel has a multipath spread of $T_m = 1$ sec and a Doppler spread $B_d = 0.01$ Hz. The total channel bandwidth at bandpass available for signal transmission is $W = 5$ Hz. To reduce the effect of ISI the signal designer selects a pulse duration of $T = 10$ sec.

1. Determine the coherence bandwidth and the coherence time.
2. Is the channel frequency selective? Explain.
3. Is the channel fading slowly or rapidly? Explain.
4. Suppose that the channel is used to transmit binary data via (antipodal) coherently detected PSK in a frequency diversity mode. Explain how you would use the available channel bandwidth to obtain frequency diversity and determine how much diversity is available.
5. For the case in part 4, what is the approximate SNR required per diversity to achieve an error probability of 10^{-6}?

9.38 The probability of error for binary DPSK and binary FSK with noncoherent detection in an AWGN channel is

$$P_2 = \frac{1}{2} e^{-c\rho_b}$$

where $c = 1$ for DPSK and $c = \frac{1}{2}$ for FSK, and $\rho_b = \frac{\alpha^2 \mathcal{E}_b}{N_0}$, where α is the attenuation factor. By averaging P_2 over the Rayleigh distributed variable α, as indicated by (9.6.18), verify the expression for the probability of error for DPSK and FSK in a Rayleigh fading channel.

9.39 A communication system employs dual antenna diversity and binary orthogonal FSK modulation. The received signals at the two antennas is

$$r_1(t) = \alpha_1 s(t) + n_1(t)$$
$$r_2(t) = \alpha_2 s(t) + n_2(t)$$

where α_1 and α_2 are statistically i.i.d. Raleigh random variables and $n_1(t)$ and $n_2(t)$ are statistically independent, zero-mean white Gaussian random processes with power-spectral density $N_0/2$ W/Hz. The two signals are demodulated, squared and then combined (summed) prior to detection.

1. Sketch the functional block diagram of the entire receiver including the demodulator, the combiner and the detector.
2. Plot the probability of error for the detector and compare the result with the case of no diversity.

9.40 A binary communication system transmits the same information on two diversity channels. The two received signals are

$$r_1 = \pm\sqrt{\mathcal{E}_b} + n_1$$
$$r_2 = \pm\sqrt{\mathcal{E}_b} + n_2$$

where $E(n_1) = E(n_2) = 0$, $E\left(n_1^2\right) = \sigma_1^2$ and $E\left(n_2^2\right) = \sigma_2^2$, and n_1 and n_2 are uncorrelated Gaussian variables. The detector bases its decision on the linear combination of r_1 and r_2, i.e.,

$$r = r_1 + kr_2$$

1. Determine the value of k that minimizes the probability of error.
2. Plot the probability of error for $\sigma_1^2 = 1$, $\sigma_2^2 = 3$, and either $k = 1$ or k is the optimum value found in part 1. Compare the results.

10

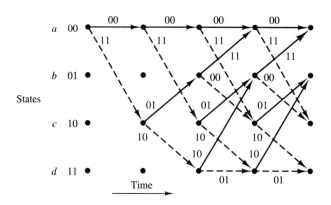

Channel Capacity and Coding

The purpose of any communication system is to transmit information from an *information source* to a *destination* via a *communication channel.* A communications engineer usually has very little control over these three components. His role is to design transmitters and receivers that send the source output over the channel to the destination with high fidelity (low distortion, that can be error probability or any other distortion measure, as discussed in Chapter 4). The mathematical model for information sources together with a quantitative measure of information were presented in Chapter 4. In this chapter we study the other important component of a communication system, i.e., the communication channel. We also introduce the concept of coding for protection of messages against channel errors.

10.1 MODELING OF COMMUNICATION CHANNELS

As defined in Chapter 1, a communication channel is any medium over which information can be transmitted or in which information can be stored. Coaxial cables, ionospheric propagation, free space, fiber optic cables, and magnetic and optical disks are examples of communication channels. What is common among all these is that they accept signals at their inputs and deliver signals at their outputs at a later time (storage case) or at another location (transmission case). Therefore, each communication channel is characterized by a relation between its input and output. In this sense a communication channel is a system as defined in Chapter 2.

There are many factors that cause the output of a communication channel to be different from its input. Among these factors are attenuation, nonlinearities, bandwidth limitations, multipath propagation, and noise. All these factors contribute to a usually complex input–output relation in a communication channel. Due to the presence of fading and noise, the input–output relation in a communication channel is generally a stochastic relation.

Channels encountered in practice are generally waveform channels that accept (continuous-time) waveforms as their inputs and produce waveforms as their outputs. Because the bandwidth of any practical channel is limited, by using the sampling theorem a waveform channel becomes equivalent to a discrete-time channel. In a discrete-time channel both input and output are discrete-time signals.

In a discrete-time channel, if the values that the input and output variables can take are finite, or countably infinite, the channel is called a *discrete channel.* An example of a discrete-channel is a binary-input binary-output channel. In general, a discrete channel is defined by X, the input alphabet, $\mathcal{Y}$, the output alphabet, and $p(\mathbf{y}|\mathbf{x})$ the conditional p.m.f. of the output sequence given the input sequence. A schematic representation of a discrete channel is given in Figure 10.1. In general, the output y_i depends not only on the input at the same time x_i but also on the previous inputs (channels with ISI, see Chapter 8), or even previous and future inputs (in storage channels). Therefore, a channel can have *memory.* However, if a discrete channel does not have memory, it is called a *discrete-memoryless channel,* and for such a channel, for any $\mathbf{y} \in \mathcal{Y}^n$ and $\mathbf{x} \in X^n$, we have

$$p(\mathbf{y}|\mathbf{x}) = \prod_{i=1}^{n} p(y_i|x_i) \tag{10.1.1}$$

All channel models that we will discuss here are memoryless.

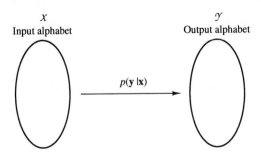

X
Input alphabet

$\mathcal{Y}$
Output alphabet

$p(\mathbf{y}|\mathbf{x})$

FIGURE 10.1. A discrete channel.

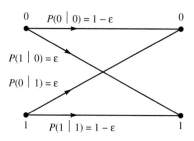

FIGURE 10.2. The binary-symmetric channel.

A special case of a discrete-memoryless channel is the *binary-symmetric channel* (BSC) shown in Figure 10.2. In a binary-symmetric channel, $\epsilon = p(0|1) = p(1|0)$ is called the *crossover probability*.

Example 10.1.1

Let us assume that we are dealing with an additive white Gaussian noise channel with binary antipodal signaling. We have already seen in Chapter 7 that in such a channel the error probability of a one being detected as zero or a zero being detected as one are given by

$$\epsilon = p(1|0) = p(0|1) = Q\left(\sqrt{\frac{2\mathcal{E}_b}{N_0}}\right) \tag{10.1.2}$$

where N_0 is the noise power-spectral density and $\mathcal{E}_b$ denotes the energy content of each of the antipodal signals representing 0 and 1. This discrete channel is an example of a binary-symmetric channel.

The most important continuous alphabet channel is the discrete-time, additive white Gaussian noise channel with an input power constraint. In this channel both X and $\mathcal{Y}$ are the set of real numbers, and the input–output relation is given by

$$Y = X + Z \tag{10.1.3}$$

where Z denotes the channel noise, which is assumed to be Gaussian, with mean equal to 0 and variance equal to P_N. It is further assumed that inputs to this channel satisfy some power constraint. For example, for large n, input blocks of length n satisfy

$$\frac{1}{n}\sum_{i=1}^{n} x_i^2 \leq P \tag{10.1.4}$$

where P is some fixed power constraint. This channel model is shown in Figure 10.3.

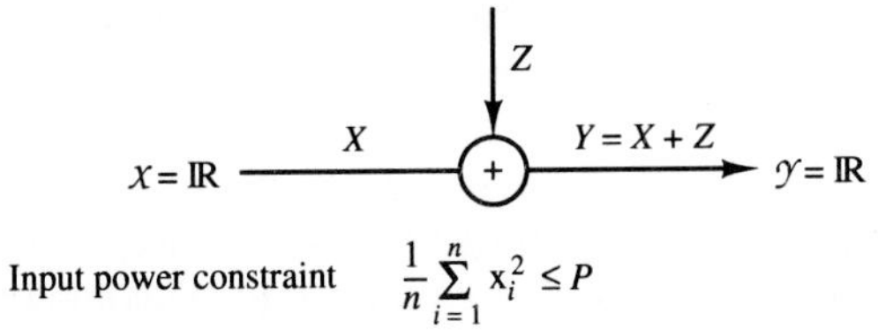

$$Z$$

$$X = \mathbb{R} \xrightarrow{\quad X \quad} \boxed{+} \xrightarrow{\quad Y = X + Z \quad} \mathcal{Y} = \mathbb{R}$$

Input power constraint $\dfrac{1}{n} \displaystyle\sum_{i=1}^{n} x_i^2 \leq P$

FIGURE 10.3.   Additive white Gaussian noise channel with power constraint.

10.2 CHANNEL CAPACITY

We have already seen in Chapter 4 that $H(X)$ defines a fundamental limit on the rate at which a discrete source can be encoded without errors in its reconstruction, and $R(D)$ gives a fundamental rate for reproduction of the source output with distortion less than or equal to D. A similar "fundamental limit" exists also for information transmission over communication channels.

Of course, the main objective when transmitting information over any communication channel is *reliability,* which is measured by the probability of correct reception at the receiver. Due to the presence of noise, at first glance it seems that error probability is always bounded away from zero by some positive number. However, a fundamental result of information theory is that reliable transmission (that is, transmission with error probability less any given value) is possible even over noisy channels as long as the transmission rate is less than some number called the *channel capacity.* This remarkable result, first shown by Shannon (1948), is known as the *noisy channel coding theorem.* What the noisy channel coding theorem says is that *the basic limitation that noise causes in a communication channel is not on the reliability of communication, but on the speed of communication.*

Figure 10.4 shows a discrete memoryless channel with four inputs and outputs. If the receiver receives a, it does not know whether an a or a d was transmitted; if it receives a b, it does not know whether an a or b was transmitted, etc.; therefore, there always exists a possibility of error. But if the transmitter and the receiver agree that the transmitter only uses letters a and c, then there exists no ambiguity.

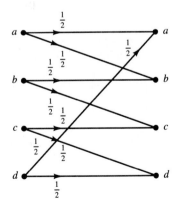

FIGURE 10.4.   An example of a discrete channel.

In this case, if the receiver receives a or b, it knows that an a was transmitted; and if it receives a c or d, it knows that a c was transmitted. This means that the two symbols a and c can be transmitted over this channel with no error, i.e., we are able to avoid errors by using only a subset of the possible inputs to the channel. Admittedly, using a smaller subset of possible inputs reduces the number of possible inputs, but this is the price that must be paid for reliable communication. This is the essence of the noisy channel coding theorem, i.e., using only those inputs whose corresponding possible outputs are disjoint. The chosen inputs should be in some sense "far apart" such that their "images" under the channel operation are nonoverlapping (or have negligible overlaps).

Looking at the binary-symmetric channel and trying to apply the same approach as above, we observe that there is no way that one can have nonoverlapping outputs. In fact, this is the case with most channels. To use the results of the above argument for the binary-symmetric channel, one has to apply it not to the channel itself but to the *extension channel*. The nth extension of a channel with input and output alphabets X and Y and conditional probabilities $p(y|x)$ is a channel with input and output alphabets X^n and Y^n and conditional probability $p(\mathbf{y}|\mathbf{x}) = \prod_{i=1}^{n} p(y_i|x_i)$. The nth extension of a binary-symmetric channel takes binary blocks of length n as its input and its output. This channel is shown in Figure 10.5. By the law of large numbers as discussed in Chapter 3, for n large enough, if a binary sequence of length n is transmitted over the channel, the output will disagree with the input with high probability at $n\epsilon$ positions. The number of possible sequences that disagree with a sequence of length n at $n\epsilon$ positions is given by

$$\binom{n}{n\epsilon}$$

Using Stirling's approximation $n! \approx n^n e^{-n} \sqrt{2\pi n}$, we obtain

$$\binom{n}{n\epsilon} \approx 2^{n H_b(\epsilon)} \tag{10.2.1}$$

$X^n = \{0, 1\}^n$ $Y^n = \{0, 1\}^n$

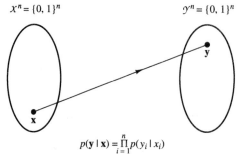

$$p(\mathbf{y}|\mathbf{x}) = \prod_{i=1}^{n} p(y_i|x_i)$$

$\mathbf{x}$: A binary sequence of length n.
$\mathbf{y}$: A binary sequence of length n.

FIGURE 10.5. nth extension of a binary-symmetric channel.

where $H_b(\epsilon)$ is the binary entropy function as defined in Chapter 4. This means that for any input block there exist roughly $2^{nH_b(\epsilon)}$ highly probable corresponding outputs. On the other hand, the total number of highly probable output sequences is roughly $2^{nH(Y)}$. The maximum number of input sequences that produce almost nonoverlapping output sequences, therefore, is at most equal to

$$M = \frac{2^{nH(Y)}}{2^{nH_b(\epsilon)}} = 2^{n(H(Y)-H_b(\epsilon))} \tag{10.2.2}$$

and the transmission rate per channel use is

$$R = \frac{\log M}{n} = H(Y) - H_b(\epsilon) \tag{10.2.3}$$

Figure 10.6 gives a schematic representation of this case.

In the relation $R = H(Y) - H_b(\epsilon)$ depends on the channel and we can not control it. However, the probability distribution of the random variable Y depends both on the input distribution $p(x)$ and the channel properties characterized by ϵ. To maximize the transmission rate over the channel, one has to choose $p(x)$ that maximizes $H(Y)$. If X is chosen to be a uniformly distributed random variable, so will be Y, and $H(Y)$ will be maximized. The maximum value of $H(Y)$ is 1, and we therefore obtain

$$R = 1 - H_b(\epsilon) \tag{10.2.4}$$

It can be proved that the above rate is the maximum rate at which reliable transmission over the BSC is possible. By *reliable transmission* we mean that the error probability can be made to tend to zero as the block length n tends to infinity. A plot of the channel capacity in this case is given in Figure 10.7. It is interesting to note that the cases $\epsilon = 0$ and $\epsilon = 1$ both result in $C = 1$. This means that a channel that always flips the input is as good as the channel that transmits the input with no errors. The worst case, of course, happens when the channel flips the input with $\epsilon = .5$.

The maximum rate at which one can communicate over a discrete-memoryless channel and still make the error probability approach zero as the code block length increases is called the *channel capacity* and is denoted by C. The noisy channel

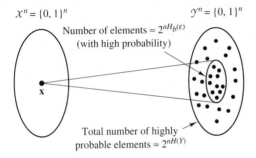

$X^n = \{0, 1\}^n$ $\mathcal{Y}^n = \{0, 1\}^n$

Number of elements $\approx 2^{nH_b(\epsilon)}$ (with high probability)

x

Total number of highly probable elements $\approx 2^{nH(Y)}$

FIGURE 10.6.   Schematic representation of a BSC.

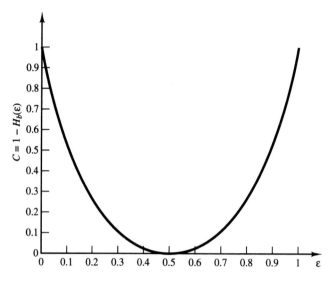

FIGURE 10.7.   The capacity of a BSC.

coding theorem stated below gives the capacity of a general discrete-memoryless channel.

Theorem 10.2.1 [Noisy Channel Coding Theorem].   The capacity of a discrete-memoryless channel is given by

$$C = \max_{p(x)} I(X; Y) \qquad (10.2.5)$$

where $I(X; Y)$ is the mutual information between the channel input X and the output Y, previously defined in Chapter 4. If the transmission rate R is less than C, then for any $\epsilon > 0$ there exists a code with block length n large enough whose error probability is less than ϵ. If $R > C$, the error probability of any code with any block length is bounded away from zero.

This theorem is one of the fundamental results in information theory and gives a fundamental limit on the possibility of reliable communication over a noisy channel. According to this theorem, regardless of all other properties, any communication channel is characterized by a number called capacity that determines how much information can be transmitted over it. Therefore, to compare two channels from an information transmission point of view, it is enough to compare their capacities.

Example 10.2.1

Find the capacity of the channel shown in Figure 10.8.

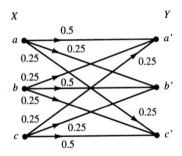

FIGURE 10.8. The DMC of Example 10.2.1.

Solution   We have to find the input distribution that maximizes $I(X;Y)$. We have

$$I(X;Y) = H(Y) - H(Y|X)$$

But

$$H(Y|X) = p(X = a)H(Y|X = a) + p(X = b)H(Y|X = b)$$
$$+ p(X = c)H(Y|X = c)$$

From the channel input–output relation it is seen that for all the three cases $X = a$, $X = b$, and $X = c$, Y is a ternary random variable with probabilities .25, .25, and .5. Therefore,

$$H(Y|X = a) = H(Y|X = b) = H(Y|X = c) = 1.5$$

Then

$$H(Y|X) = 1.5$$

and

$$I(X;Y) = H(Y) - 1.5$$

To maximize $I(X;Y)$, it remains to maximize $H(Y)$, which is maximized when Y is an equiprobable random variable. But it is not clear if there exists an input distribution that results in a uniform distribution on the output. However, in this special case a uniform input distribution results in a uniform output distribution and for this distribution,

$$H(Y) = \log 3 = 1.585$$

This means that the capacity of this channel is given by

$$C = 1.585 - 1.5 = .085 \text{ bits/transmission}$$

10.2.1 Gaussian Channel Capacity

A discrete-time Gaussian channel with input power constraint is characterized by the input–output relation

$$Y = X + Z \tag{10.2.6}$$

where Z is a zero-mean Gaussian random variable with variance P_N, and for n large enough, an input power constraint of the form

$$\frac{1}{n}\sum_{i=1}^{n} x_i^2 \leq P \tag{10.2.7}$$

applies to any input sequence of length n. If we look at blocks of length n of the input, the output, and the noise, we have

$$\mathbf{y} = \mathbf{x} + \mathbf{z} \tag{10.2.8}$$

If n is large, by the law of large numbers, we have

$$\frac{1}{n}\sum_{i=1}^{n} z_i^2 = \frac{1}{n}\sum_{i=1}^{n} |y_i - x_i|^2 \leq P_N \tag{10.2.9}$$

or

$$|\mathbf{y} - \mathbf{x}|^2 \leq nP_N \tag{10.2.10}$$

This means that with probability approaching one (as n increases), $\mathbf{y}$ will be located in an n-dimensional sphere (hypersphere) of radius $\sqrt{nP_N}$ and centered at $\mathbf{x}$. On the other hand, due to the power constraint of P on the input and the independence of the input and noise, the output power is the sum of the input power and the noise power, i.e.,

$$\frac{1}{n}\sum_{i=1}^{n} y_i^2 \leq P + P_N \tag{10.2.11}$$

or

$$|\mathbf{y}|^2 \leq n(P + P_N) \tag{10.2.12}$$

This implies that the output sequences (again, asymptotically and with high probability) will be inside an n-dimensional hypersphere of radius $\sqrt{n(P + P_N)}$ and centered at the origin. Figure 10.9 shows the sequences in the output space. The question now is: How many $\mathbf{x}$ sequences can we transmit over this channel such that the hyperspheres corresponding to these sequences do not overlap in the output space? Obviously if this condition is satisfied, then the input sequences can be decoded reliably. An equivalent question is: How many hyperspheres of radius $\sqrt{nP_N}$ can we pack in a hypersphere of radius $\sqrt{n(P_N + P)}$? The answer is roughly the ratio of the volumes of two hyperspheres. If we denote the volume of an n-dimensional hypersphere by

$$V_n = K_n R^n \tag{10.2.13}$$

where R denotes the radius and K_n is independent of R, we see that the number of messages that can be reliably transmitted over this channel is equal to

$$M = \frac{K_n(n(P_N + P))^{\frac{n}{2}}}{K_n(nP_N)^{\frac{n}{2}}}$$

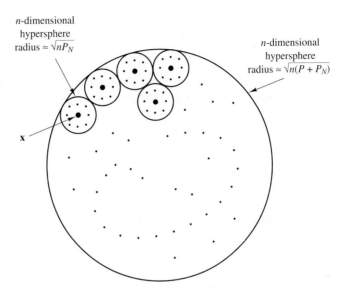

n-dimensional
hypersphere
radius $\approx \sqrt{nP_N}$

n-dimensional
hypersphere
radius $\approx \sqrt{n(P + P_N)}$

FIGURE 10.9.   The output sequences of a Gaussian channel with power constraint.

$$= \left(\frac{P_N + P}{P_N} \right)^{\frac{n}{2}}$$

$$= \left(1 + \frac{P}{P_N} \right)^{\frac{n}{2}} \tag{10.2.14}$$

Therefore, the capacity of a discrete-time additive white Gaussian noise channel with input power constraint P is given by

$$C = \frac{1}{n} \log M$$

$$= \frac{1}{n} \cdot \frac{n}{2} \log \left(1 + \frac{P}{P_N} \right)$$

$$= \frac{1}{2} \log \left(1 + \frac{P}{P_N} \right) \tag{10.2.15}$$

When dealing with a continuous-time, bandlimited, additive white Gaussian noise channel with noise power-spectral density $\frac{N_0}{2}$, input power constraint P, and bandwidth W, one can sample at the Nyquist rate and obtain a discrete-time channel. The power per sample will be P and the noise power per sample will be

$$P_N = \int_{-W}^{+W} \frac{N_0}{2} \, df = W N_0$$

Substituting these results in 10.2.15, we obtain

$$C = \frac{1}{2} \log(1 + \frac{P}{N_0 W}) \text{ bits/transmission} \qquad (10.2.16)$$

If we multiply this result by the number of transmissions per second, which is $2W$, we obtain the channel capacity in bits/sec.

$$C = W \log(1 + \frac{P}{N_0 W}) \text{ bits/sec} \qquad (10.2.17)$$

This is the celebrated Shannon's formula for the capacity of an additive white Gaussian noise channel.

Example 10.2.2

Find the capacity of a telephone channel with bandwidth $W = 3000$ Hz and SNR of 30 dB.

Solution   The SNR of 30 dB is equivalent to 1000. Using Shannon's relation we have

$$C = 3000 \log(1 + 1000) \approx 30,000 \text{ bits/sec}$$

10.3 BOUNDS ON COMMUNICATION

From the previous section, the capacity of an additive white Gaussian noise channel is given by

$$C = W \log(1 + \frac{P}{N_0 W})$$

From this result, the basic factors that determine the channel capacity are the channel bandwidth W, the noise power spectrum N_0, and the signal power P. There exists a tradeoff between P and W in the sense that one can compensate for the other.

Increasing the input signal power obviously increases the channel capacity, because when one has more power to spend, one can choose a larger number of input levels that are far apart, and therefore more information bits per transmission are possible. However, the increase in capacity as a function of power is logarithmic and slow. This is because if one is transmitting with a certain number of input levels that are Δ apart to allow a certain level of immunity against noise and wants to increase the number of input levels, one has to introduce new levels with amplitudes higher than the existing levels, and this requires a lot of power. This fact notwithstanding, the capacity of the channel can be increased to any value by increasing the input power.

The effect of the channel bandwidth, however, is quite different. Increasing W has two contrasting effects. On one hand, on a higher bandwidth channel one can transmit more samples per second and therefore increase the transmission rate. On the other hand, a higher bandwidth means higher input noise to the receiver and this degrades its performance. This is seen from the two W's that appear in the relation that describes the channel capacity. To see the effect of increasing the bandwidth, we let the bandwidth W tend to infinity and using L'Hospital's rule, we obtain

$$\lim_{W \to \infty} C = \frac{P}{N_0} \log e = 1.44 \frac{P}{N_0} \tag{10.3.1}$$

This means that, contrary to the power case, by increasing the bandwidth alone one can not increase the capacity to any desired value. Figure 10.10 shows C plotted versus W.

In any practical communication system we must have $R < C$. If an AWGN channel is employed, we have

$$R < W \log(1 + \frac{P}{N_0 W}) \tag{10.3.2}$$

By dividing both sides by W and defining $r = \frac{R}{W}$, the spectral bit rate, we obtain

$$r < \log(1 + \frac{P}{N_0 W}) \tag{10.3.3}$$

If $\mathcal{E}_b$ is the energy per bit, then $\mathcal{E}_b = \frac{P}{R}$. By substituting in the previous relation, we obtain

$$r < \log(1 + r \frac{\mathcal{E}_b}{N_0}) \tag{10.3.4}$$

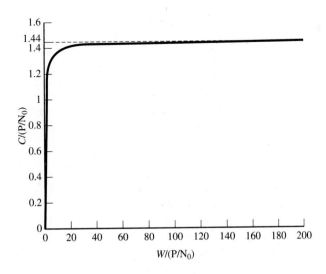

FIGURE 10.10.   Plot of channel capacity versus bandwidth.

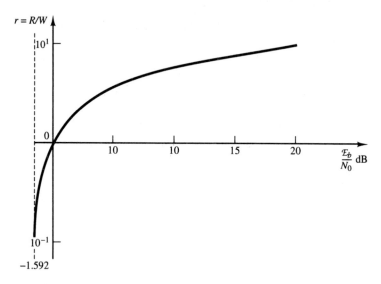

FIGURE 10.11.   Spectral bit rate versus SNR/bit in an optimal system.

This relation is plotted in Figure 10.11. The curve defined by

$$r = \log(1 + r\frac{\mathcal{E}_b}{N_0})$$ (10.3.5)

divides the plane into two regions. In one region (below the curve) reliable communication is possible, and in the other region (above the curve) reliable communication is not possible. The performance of any communication system can be denoted by a point in this plane and the closer the point is to this curve, the closer is the performance of the system to an optimal system. From this curve it is seen that

$$\frac{\mathcal{E}_b}{N_0} = \ln 2 = 0.693 \sim -1.592 \text{ dB}$$ (10.3.6)

is an absolute minimum for reliable communication. In other words, for reliable communication, we must have

$$\frac{\mathcal{E}_b}{N_0} > 0.693$$ (10.3.7)

In Figure 10.11, when $r \ll 1$, we are dealing with a case where bandwidth is large and the main concern is limitation on power. This case is usually referred to as the *power limited case*. Signaling schemes with high dimensionality, such as orthogonal, biorthogonal, and simplex, are frequently used in these cases. The case where $r \gg 1$ happens when the bandwidth of the channel is small, and therefore is referred to as the *bandwidth limited case*. Low-dimensional signaling schemes with crowded constellations are implemented in these cases.

In Chapter 4 we introduced the fundamental limits that exist in coding of information sources. These fundamental limits are expressed in terms of the entropy of the source, the rate-distortion function associated with the source, and the corresponding distortion measure. Entropy gives a lower bound on the rate of the codes that are capable of reproducing the source with no error, and the rate-distortion function gives a lower bound on the rate of the codes capable of reproducing the source with distortion D. If we want to transmit a source U reliably via a channel with capacity C, we require that

$$H(U) < C \tag{10.3.8}$$

If transmission with a maximum distortion equal to D is desired, then the condition is

$$R(D) < C \tag{10.3.9}$$

These two relations define fundamental limits on the transmission of information.

Example 10.3.1

A zero-mean Gaussian source with power-spectral density

$$S_X(f) = \Pi\left(\frac{f}{20,000}\right)$$

is to be transmitted via a telephone channel described in Example 10.2.2. What is the minimum distortion achievable if mean-squared distortion is used?

Solution   The bandwidth of the source is 10,000 Hz, and it therefore can be sampled at a rate of 20,000 samples/sec. The power of the source is

$$P_X = \int_{-\infty}^{+\infty} S_X(f) df = 20000$$

The variance of each sample is given by

$$\sigma^2 = P_X = 20,000$$

and the rate-distortion function for $D < 20,000$ is given by

$$R(D) = \frac{1}{2} \log \frac{\sigma^2}{D} = \frac{1}{2} \log \frac{20,000}{D} \quad \text{bits/sample}$$

which is equivalent to

$$R(D) = 10,000 \log \frac{20,000}{D} \quad \text{bits/sec}$$

Because the capacity of the channel in question, derived in Example 10.2.2, is 30,000 bits/sec, we can derive the least possible distortion by solving

$$30,000 = 10,000 \log \frac{20,000}{D}$$

for D, which results in $D = 2500$.

In general, it can be shown that the rate-distortion function for a waveform Gaussian source with power-spectral density

$$S_X(f) = \begin{cases} A, & |f| < W_s \\ 0, & \text{otherwise} \end{cases} \tag{10.3.10}$$

and with distortion measure

$$d(u(t), v(t)) = \lim_{T \to \infty} \frac{1}{T} \int_{-T/2}^{T/2} \left[u(t) - v(t) \right]^2 dt \tag{10.3.11}$$

is given by

$$R(D) = \begin{cases} W_s \log(\frac{2AW_s}{D}), & D < 2AW_s \\ 0, & D \geq 2AW_s \end{cases} \tag{10.3.12}$$

If this source is to be transmitted via an additive white Gaussian noise channel with bandwidth W_c, power P, and noise-spectral density N_0, the minimum achievable distortion is obtained by solving the equation

$$W_c \log(1 + \frac{P}{N_0 W_c}) = W_s \log(\frac{2AW_s}{D}) \tag{10.3.13}$$

From this equation we obtain

$$D = 2AW_s(1 + \frac{P}{N_0 W_c})^{-\frac{W_c}{W_s}} \tag{10.3.14}$$

It is seen that in an optimal system, distortion decreases exponentially as $\frac{W_c}{W_s}$ increases. The ratio $\frac{W_c}{W_s}$ is called the *bandwidth expansion factor*. We can also find the SQNR as

$$\text{SQNR} = \frac{2AW_s}{D} = (1 + \frac{P}{N_0 W_c})^{\frac{W_c}{W_s}} \tag{10.3.15}$$

The important conclusion is that in an optimal system, the final signal-to-noise (distortion) ratio increases exponentially with the bandwidth expansion factor. This is a criterion by which we can compare the performance of various waveform transmission systems.

10.3.1 Transmission of Analog Sources by PCM

We have seen in Chapter 4 that PCM is one of the most commonly used schemes for transmission of analog data. Here we find the performance characteristics of a PCM system when transmitting data over a noisy channel and compare the result with that of an optimal system.

We have already seen in Section 4.6 that in a PCM system the quantization noise (distortion) is given by

$$E[\tilde{X}^2] = \frac{x_{\max}^2}{3 \times 4^\nu}$$

where $x_{\max}$ denotes the maximum input amplitude and ν is the number of bits per sample. If the source bandwidth is W_s, then the sampling frequency in an optimal system is $f_s = 2W_s$, and the bit rate is $R = 2\nu W_s$. The minimum channel bandwidth that can accommodate this bit rate (Nyquist criterion, see Chapter 8) is

$$W_c = \frac{R}{2} = \nu W_s \tag{10.3.16}$$

Therefore, the quantization noise can be expressed as

$$E[\tilde{X}^2] = \frac{x_{\max}^2}{3 \times 4^{\frac{W_c}{W_s}}} \tag{10.3.17}$$

Now, if the output of a PCM system is transmitted over a noisy channel with error probability P_b, some bits will be received in error and another type of distortion, due to transmission, will also be introduced. This distortion will be independent from the quantization distortion; and because the quantization error is zero mean, the total distortion will be the sum of these distortions (see Section 3.3.2). To simplify the calculation of the transmission distortion, we assume that P_b is small enough such that in a block of length ν, either no error, or one error can occur. This one error can occur at any of the ν locations and, therefore, its contribution to the total distortion varies accordingly. If an error occurs in the least significant bit, it results in a change of Δ in the output level; if it happens in the next bit, it results in 2Δ change in the output; and if it occurs at the most significant bit, it causes a level change equal to $2^{\nu-1}\Delta$ at the output. Assuming that these are the only possible errors one obtains the following expression for the transmission distortion

$$E[D_T^2] = \sum_{i=1}^{\nu} P_b (2^{i-1}\Delta)^2 = P_b \Delta^2 \frac{4^\nu - 1}{3} \tag{10.3.18}$$

and the total distortion will be

$$D_{\text{total}} = \frac{x_{\max}^2}{3 \times 4^{\frac{W_c}{W_s}}} + P_b \Delta^2 \frac{4^{\frac{W_c}{W_s}} - 1}{3} \tag{10.3.19}$$

We note that

$$\Delta = \frac{2x_{\max}}{2^\nu} = \frac{x_{\max}}{2^{\nu-1}} \tag{10.3.20}$$

Substituting for Δ, and assuming $4^\nu \gg 1$, the expression for D_{total} simplifies to

$$D_{\text{total}} = \frac{x_{\max}^2}{3} \left(4^{-\frac{W_c}{W_s}} + 4P_b \right) \tag{10.3.21}$$

The signal-to-noise (distortion) ratio at the receiving end is, therefore,

$$\text{SNR} = \frac{3\overline{X^2}}{x_{max}^2 (4^{-\frac{W_c}{W_s}} + 4P_b)} \tag{10.3.22}$$

The value of P_b, of course, depends on the modulation scheme employed to transmit the outputs of the PCM system. If binary antipodal signaling is employed, then

$$P_b = Q\left(\sqrt{\frac{2\mathcal{E}_b}{N_0}}\right) \tag{10.3.23}$$

and if binary orthogonal signaling with coherent detection is used, then

$$P_b = Q\left(\sqrt{\frac{\mathcal{E}_b}{N_0}}\right) \tag{10.3.24}$$

From the expression for the SNR, it is seen that for small P_b, the SNR grows almost exponentially with $\frac{W_c}{W_s}$, the bandwidth expansion factor, and in this sense, a PCM system uses the available bandwidth efficiently. Recall from Section 6.3 that another frequently used system that trades bandwidth for noise immunity is an FM system. However, the SNR of an FM system is a quadratic function of the bandwidth expansion factor (see 6.3.26) and therefore an FM system is not as bandwidth efficient as a PCM system.

10.4 CODING FOR RELIABLE COMMUNICATION

In Chapters 7 and 9 it was shown that both in baseband and carrier modulation schemes the error probability is a function of the distance between the points in the signal constellation. In fact, for binary equiprobable signals the error probability can be expressed as (see 7.3.6)

$$p_e = Q\left(\frac{d_{12}}{\sqrt{2N_0}}\right) \tag{10.4.1}$$

where d_{12} is the Euclidean distance between $s_1(t)$ and $s_2(t)$, given by

$$d_{12}^2 = \int_{-\infty}^{+\infty} [s_1(t) - s_2(t)]^2 \, dt \tag{10.4.2}$$

On the other hand, we have seen that in the binary case, with coherent demodulation, antipodal signaling performs best and has a Euclidean distance $d_{12} = 2\sqrt{\mathcal{E}_b}$. Therefore,

$$p_e = Q\left(\sqrt{\frac{2\mathcal{E}_b}{N_0}}\right) \tag{10.4.3}$$

From the above it is seen that to decrease the error probability, one has to increase the signal energy. Increasing the signal energy can be done in two ways, either by increasing the transmitter power or by increasing the transmission duration. Increasing the transmitter power is not always feasible because each transmitter has a limitation on its average power. Increasing the transmission duration, in turn, decreases the transmission rate, and therefore it seems that the only way to make the error probability vanish is to let transmission rate vanish. In fact, this was the communication engineers' viewpoint in the pre-Shannon era.

We have seen also in Chapter 7 that by employing orthogonal signals one can achieve reliable transmission, at nonzero rate, as long as $\mathcal{E}_b/N_0 > 2\ln 2$. Here we show that $\mathcal{E}_b/N_0 > \ln 2$ is sufficient to achieve reliable transmission. Note that this is the same condition that guarantees reliable transmission over an additive white Gaussian noise channel when the bandwidth goes to infinity (see Figure 10.11 and Equation 10.3.6).

10.4.1 A Tight Bound on Error Probability of Orthogonal Signals

To find a tighter bound on the error probability of orthogonal signals compared to the one obtained in Chapter 7, we use another bounding technique. The problem with the union bound employed in Chapter 7 is that it is not tight for small SNR's. To compensate for this, we use two bounds, one for large SNR's, which is essentially the union bound, and another bound for small SNR's.

In Chapter 7 it was shown that the message-error probability for orthogonal signals is given by

$$P_M = \frac{1}{\sqrt{2\pi}} \int_{-\infty}^{+\infty} \left[1 - \left(1 - Q(u)\right)^{M-1}\right] \exp\left[-\frac{(u - \sqrt{2\mathcal{E}_s/N_0})^2}{2}\right] du \quad (10.4.4)$$

Defining

$$a = \sqrt{\frac{2\mathcal{E}_s}{N_0}} \quad (10.4.5)$$

and

$$f(u) = \frac{1}{\sqrt{2\pi}} e^{-\frac{u^2}{2}} \quad (10.4.6)$$

we have

$$P_M = \int_{-\infty}^{+\infty} \left[1 - \left(1 - Q(u)\right)^{M-1}\right] f(u - a) du \quad (10.4.7)$$

Using the inequality $nx + (1 - x)^n \geq 1$, which holds for all $0 \leq x \leq 1$ and $n \geq 1$,[†] we obtain

$$(1 - Q(u))^{M-1} \geq 1 - (M - 1)Q(u) \qquad (10.4.8)$$

and, therefore,

$$1 - (1 - Q(u))^{M-1} \leq (M - 1)Q(u) < M Q(u) < M e^{-u^2/2} \qquad (10.4.9)$$

where in the last step we have used the well-known bound on $Q(x)$ introduced in Section 3.1. This bound is basically the union bound, which we will use for large SNR's (large u). For small SNR's, we use the obvious bound

$$1 - (1 - Q(u))^{M-1} \leq 1 \qquad (10.4.10)$$

Substituting these bounds into (10.4.7) we obtain

$$P_M < \int_{-\infty}^{y_0} f(u - a)\, du + M \int_{y_0}^{\infty} e^{-u^2/2} f(u - a)\, du = P_1 + M P_2 \qquad (10.4.11)$$

where y_0 can be any arbitrary positive number. To obtain the tightest bound, we differentiate the right-hand side of (10.4.11) with respect to y_0 and set the derivative equal to zero. Thus we obtain

$$f(y_0 - a) - M e^{-y_0^2/2} f(y_0 - a) = 0 \qquad (10.4.12)$$

or

$$e^{y_0^2/2} = M \qquad (10.4.13)$$

from which we have

$$y_0 = \sqrt{2 \ln M} = \sqrt{2 \ln 2 \log_2 M} = \sqrt{2k \ln 2} \qquad (10.4.14)$$

The next step is to find bounds on P_1 and P_2, and therefore on P_M. For P_1 we have

$$P_1 = \int_{-\infty}^{y_0} f(u - a)\, du = \int_{-\infty}^{y_0-a} f(x)\, dx = \int_{a-y_0}^{\infty} f(x)\, dx = Q(a - y_0) \qquad (10.4.15)$$

By using bounds on Q, we obtain

$$P_1 \leq \frac{1}{2} e^{-(a-y_0)^2/2} < e^{-(a-y_0)^2/2} \quad \text{for } y_0 \leq a \qquad (10.4.16)$$

To bound P_2 we have

$$P_2 = \int_{y_0}^{\infty} e^{-u^2/2} f(u - a)\, du$$

[†]To see this, define $g(x) = nx + (1 - x)^n$, then $g(0) = 1$ and $g'(x) = n - n(1 - x)^{n-1}$. But for $0 \leq x \leq 1$ we have $0 \leq 1 - x \leq 1$, and therefore $(1 - x)^{n-1} \leq 1$, for all $n \geq 1$. This means that for $0 \leq x \leq 1$, $g'(x) \geq 0$ and therefore $g(x)$ is an increasing function in this interval. This in turn means that $g(x) \leq 1$ for $0 \leq x \leq 1$ and $n \geq 1$.

$$= \int_{y_0}^{\infty} e^{-\frac{u^2}{2}} \times \frac{1}{\sqrt{2\pi}} e^{-(u-a)^2/2} \, du$$

$$= e^{-a^2/4} \int_{y_0}^{\infty} \frac{1}{\sqrt{2\pi}} e^{-(u-a/2)^2} \, du$$

$$= e^{-a^2/4} \frac{1}{\sqrt{2}} \int_{\sqrt{2}(y_0-a/2)}^{\infty} \frac{1}{\sqrt{2\pi}} e^{-x^2/2} \, dx$$

$$= \frac{1}{\sqrt{2}} e^{-a^2/4} Q\left[\sqrt{2}\left(y_0 - \frac{a}{2}\right)\right] \tag{10.4.17}$$

Using the fact that for all x, $Q(x) \leq 1$, and that for $x \geq 0$, we have $Q(x) \leq \frac{1}{2} e^{-x^2/2}$, we obtain

$$P_2 \leq \begin{cases} e^{-a^2/4} e^{-(y_0-a/2)^2} & \text{for } y_0 \geq \frac{a}{2} \\ e^{-a^2/4} & \text{for } y_0 \leq \frac{a}{2} \end{cases} \tag{10.4.18}$$

Substituting the derived bounds on P_1 and P_2 in (10.4.11) and using $M = e^{y_0^2/2}$, we obtain

$$P_M < \begin{cases} e^{-(a-y_0)^2/2} + e^{y_0^2/2} e^{-a^2/4} & \text{for } 0 \leq y_0 \leq \frac{a}{2} \\ e^{-(a-y_0)^2/2} + e^{y_0^2/2} e^{-a^2/4-(y_0-a/2)^2} & \text{for } \frac{a}{2} \leq y_0 \leq a \end{cases} \tag{10.4.19}$$

For $0 \leq y_0 \leq \frac{a}{2}$, we have

$$e^{-(a-y_0)^2/2} + e^{y_0^2/2} e^{-a^2/4} = e^{y_0^2/2-a^2/4}\left[1 + e^{-(y_0-a/2)^2}\right] \leq 2e^{y_0^2/2-a^2/4} \tag{10.4.20}$$

and for $\frac{a}{2} \leq y_0 \leq a$, the two exponents in (10.4.19) are the same. Using these, the bound simplifies to

$$P_M < \begin{cases} 2e^{y_0^2/2-a^2/4} & \text{for } 0 \leq y_0 \leq \frac{a}{2} \\ 2e^{-(a-y_0)^2/2} & \text{for } \frac{a}{2} \leq y_0 \leq a \end{cases} \tag{10.4.21}$$

Now, using the relations

$$y_0 = \sqrt{2k \ln 2} \tag{10.4.22}$$

and

$$a = \sqrt{\frac{2\mathcal{E}_s}{N_0}} = \sqrt{\frac{2k\mathcal{E}_b}{N_0}} \tag{10.4.23}$$

we obtain the final result

$$
P_M < \begin{cases} 2e^{-\frac{k}{2}\left(\mathcal{E}_b/N_0 - 2\ln 2\right)} & \text{for } 0 \le \ln M \le \mathcal{E}_s/4N_0 \\ 2e^{-k\left(\sqrt{\mathcal{E}_b/N_0} - \sqrt{\ln 2}\right)^2} & \text{for } \frac{\mathcal{E}_s}{4N_0} \le \ln M \le \frac{\mathcal{E}_s}{N_0} \end{cases} \tag{10.4.24}
$$

The first bound coincides with the union bound derived in Chapter 7. The second bound is better than the union bound for large values of M (smaller $\mathcal{E}_b/N_0$). This bound shows that the condition for reliable communications is $\mathcal{E}_b/N_0 > \ln 2$ rather than $\frac{\mathcal{E}_b}{N_0} > 2\ln 2$ derived in Chapter 7. We can also express these bounds in terms of the transmission rate and the channel capacity. By using the relations

$$
k = RT \tag{10.4.25}
$$

$$
\mathcal{E}_b = \frac{P}{R} \tag{10.4.26}
$$

$$
R = \frac{1}{T} \log M \tag{10.4.27}
$$

$$
C_\infty = \frac{P}{N_0} \ln 2 \tag{10.4.28}
$$

we obtain

$$
P_M < \begin{cases} 2 \times 2^{-T\left(\frac{1}{2} C_\infty - R\right)} & \text{for } 0 \le R \le \frac{1}{4} C_\infty \\ 2 \times 2^{-T\left(\sqrt{C_\infty} - \sqrt{R}\right)^2} & \text{for } \frac{1}{4} C_\infty \le R \le C_\infty \end{cases} \tag{10.4.29}
$$

or, equivalently,

$$
P_M < 2 \times 2^{-T E^*(R)} \tag{10.4.30}
$$

where

$$
E^*(R) = \begin{cases} \frac{1}{2} C_\infty - R & \text{for } 0 \le R \le \frac{1}{4} C_\infty \\ \left(\sqrt{C_\infty} - \sqrt{R}\right)^2 & \text{for } \frac{1}{4} C_\infty \le R \le C_\infty \end{cases} \tag{10.4.31}
$$

The function $E^*(R)$ is called the *channel reliability function* and is plotted in Figure 10.12.

10.4.2 The Promise of Coding

From the above discussion it is seen that orthogonal signals are capable of achieving the capacity at nonvanishing rates. However, the price paid for this is quite high. To achieve this performance we have to let $k \to \infty$, or $T \to \infty$. On the other hand,

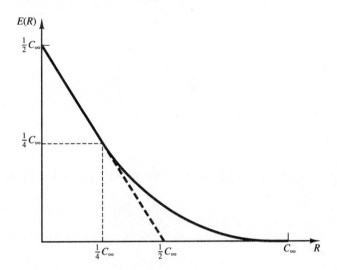

FIGURE 10.12.   The reliability function for AWGN channel with infinite bandwidth.

the number of orthogonal signals is given by $M = 2^k = 2^{RT}$. If one wants, for instance, to employ PPM signaling, this means that the time duration T has to be divided into 2^{RT} slots each of duration $T/2^{RT}$. This in turn means that the width of the pulses approaches zero exponentially, and therefore the bandwidth required to transmit them increases exponentially as well.[†] Now the question is: does there exist a way to transmit messages reliably over a noisy channel at a positive rate without an exponential increase in the bandwidth? The answer is positive and the tool to achieve this goal is coding. The following example clarifies the point.

Example 10.4.1

In a digital communication system, the transmitter power is P and the rate of the source is R. The system employs a $M = 4$ PSK signaling where a pair of information bits are mapped into any of the four signals shown in the constellation depicted in Figure 10.13. It is readily seen that $\mathcal{E}_b = \frac{P}{R}$, and the minimum Euclidean distance between any two signals is given by

$$d_{min}^2 = 4\mathcal{E}_b = 4\frac{P}{R} \qquad (10.4.32)$$

Now let us assume that instead of transmitting an $M = 4$ PSK signal (which is two dimensional), *three* orthonormal signals are employed to transmit the same *two* bits. For example, we can assume that the orthonormal signals are given by $\psi(t)$, $\psi(t-T)$,

---

[†]This exponential increase in bandwidth requirement is a characteristic of all orthogonal signaling schemes and not a peculiar property of the PPM signaling employed here. It can be shown that the number of orthogonal signals that are "almost" bandlimited to a bandwidth of W and "almost" time-limited to a duration T is $2WT$.

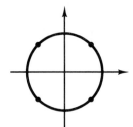

FIGURE 10.13. Signal constellation for a 4 PSK scheme.

and $\psi(t - 2T)$ where $\psi(t)$ is equal to zero outside the interval $[0, T]$ and

$$\int_0^T \psi^2(t) \, dt = 1 \qquad (10.4.33)$$

Let us assume that instead of the 4-PSK signals, we transmit the following four signals

$$s_1(t) = \sqrt{\mathcal{E}}\left(+\psi(t) + \psi(t - T) + \psi(t - 2T)\right) \qquad (10.4.34)$$

$$s_2(t) = \sqrt{\mathcal{E}}(+\psi(t) - \psi(t - T) - \psi(t - 2T)) \qquad (10.4.35)$$

$$s_3(t) = \sqrt{\mathcal{E}}(-\psi(t) - \psi(t - T) + \psi(t - 2T)) \qquad (10.4.36)$$

$$s_4(t) = \sqrt{\mathcal{E}}(-\psi(t) + \psi(t - T) - \psi(t - 2T)) \qquad (10.4.37)$$

or equivalently, in vector notation

$$\mathbf{s}_1 = \sqrt{\mathcal{E}}(+1, +1, +1) \qquad (10.4.38)$$

$$\mathbf{s}_2 = \sqrt{\mathcal{E}}(+1, -1, -1) \qquad (10.4.39)$$

$$\mathbf{s}_3 = \sqrt{\mathcal{E}}(-1, -1, +1) \qquad (10.4.40)$$

$$\mathbf{s}_4 = \sqrt{\mathcal{E}}(-1, +1, -1) \qquad (10.4.41)$$

The corresponding constellation is shown in the three-dimensional space in Figure 10.14. It is seen that with this choice of code words, each code word differs from any other code word at two components. Therefore, the Euclidean distance

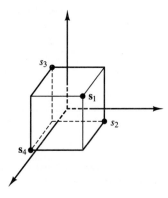

FIGURE 10.14. Code words on the vertices of a cube.

between any two signals is given by

$$d_{ij}^2 = |\mathbf{s}_i - \mathbf{s}_j|^2 = 8\mathcal{E} \quad \text{for } i \neq j \tag{10.4.42}$$

To find $\mathcal{E}$ in terms of P, we note that in a time interval of T, two bits are transmitted. Therefore,

$$2\mathcal{E}_b = 3\mathcal{E} \tag{10.4.43}$$

and

$$\mathcal{E} = \frac{2}{3}\mathcal{E}_b = \frac{2}{3}\frac{P}{R} \tag{10.4.44}$$

Thus,

$$d_{ij}^2 = \frac{16}{3}\frac{P}{R} \quad \text{for } i \neq j \tag{10.4.45}$$

Comparing this with the minimum distance in the 4-PSK signal, we observe that the minimum-distance squared has increased by a factor of

$$\frac{d_{ij}^2}{d_{4\text{PSK}}^2} = \frac{\frac{16}{3}\frac{P}{R}}{4\frac{P}{R}} = \frac{4}{3} \tag{10.4.46}$$

Because the error probability is a decreasing function of the minimum Euclidean distance, we have reduced the error probability by employing this new scheme. In fact, it can be said that the resulting reduction in error probability is equivalent to the reduction in error probability due to an increase in power by a factor of $\frac{4}{3}$. This, in turn is equivalent to a 1.25-dB power gain. This power gain, of course, has not been obtained for free. It is seen that with this coding scheme in a time duration of $\frac{2}{R}$, which is the time duration to transmit two bits, we have to transmit three signals. Therefore, the width of these signals is reduced by a factor of $\frac{2}{3}$, and the bandwidth required to transmit them is increased by a factor of $\frac{3}{2}$. Another problem with this second scheme is that it obviously is more elaborate and requires a more complex decoding scheme.

The above exercise basically describes what a coding scheme does. Coding results in a lower error probability (which is equivalent to a higher effective SNR) at the price of increasing the bandwidth and the complexity of the system. It should be mentioned here that, although employing coding increases the bandwidth,[†] this increase is not exponential, as was the case with orthogonal signaling.

In a general signaling scheme with coded waveforms, sequences of length $k = RT$ of the source output are mapped into sequences of length n of the form

$$\mathbf{s}_i = \sqrt{\mathcal{E}}\underbrace{(\pm 1, \pm 1, \ldots, \pm 1)}_{n} \tag{10.4.47}$$

[†]There exist coding-modulation schemes that increase the Euclidean distance between code words but do not increase the bandwidth. These schemes will be treated later in this chapter.

These points are located on the vertices of a hypercube of edge length $2\sqrt{E}$. The ratio

$$R_c = \frac{k}{n} \tag{10.4.48}$$

is defined to be the *code rate*. There exist a total of 2^n vertices of an n-dimensional hypercube of which we have to choose $M = 2^k$ as codewords. Obviously, one has to select these 2^k vertices in such a way to be as far apart from each other as possible. This makes the Euclidean distance between them large and, thus, reduces the error probability.

Let us assume that we have chosen 2^k vertices of the hypercube as the code words and each code word differs from another code word in at least $d_{\min}^{H}$ components. This parameter is called the *minimum Hamming distance of the code* and will be defined more precisely in Section 10.5. The relation between Euclidean distance and Hamming distance is very simple. If the sequences $\mathbf{s}_i$ and $\mathbf{s}_j$ differ in d_{ij}^{H} locations, then their Euclidean distance d_{ij}^{E} is related to d_{ij}^{H} by

$$\left(d_{ij}^{E}\right)^2 = \sum_{\substack{1 \le l \le n \\ l:s_i^l \ne s_j^l}} \left(\pm 2\sqrt{E}\right)^2 = 4 d_{ij}^{H} E \tag{10.4.49}$$

This means that the minimum Euclidean distance can be expressed in terms of the minimum Hamming distance as

$$(d_{\min}^{E})^2 = 4 d_{\min}^{H} E \tag{10.4.50}$$

Now if we assume that $\mathbf{s}_i$ is transmitted and use the union bound (see Chapter 7), we have

$$p_{ei} \le M Q \left(\sqrt{\frac{4 d_{\min}^{H} E}{2 N_0}} \right)$$

$$\le \frac{M}{2} e^{-d_{\min}^{H} E / N_0} \tag{10.4.51}$$

where in the last step we have used the bound on the Q function introduced in Section 3.1. Now, noting that the energy content of each codeword is nE and this has to be equal to PT, we have

$$E = \frac{PT}{n} = \frac{RT}{n} E_b = \frac{k}{n} E_b = R_c E_b \tag{10.4.52}$$

where we have used the relation $E_b = \frac{P}{R}$. Hence,

$$p_e \le \frac{M}{2} e^{-d_{\min}^{H} R_c E_b / N_0} \tag{10.4.53}$$

(the index i has been deleted because the bound is independent of i). If no coding were employed, that is, if we used all the vertices of a k-dimensional hypercube rather than 2^k vertices of an n-dimensional hypercube, we would have the following union bound on the error probability

$$p_e \leq M Q \left(\sqrt{\frac{2\mathcal{E}_b}{N_0}} \right)$$

$$\leq \frac{M}{2} e^{-\mathcal{E}_b/N_0} \qquad (10.4.54)$$

Comparing the two bounds one concludes that coding has resulted in a power gain equivalent to

$$G_{\text{coding}} = d_{\min}^{\text{H}} R_c \qquad (10.4.55)$$

which is called the *coding gain*. As seen here, the coding gain is a function of two main code parameters, the minimum Hamming distance and the code rate. Note that, in general, $R_c < 1$ and $d_{\min}^{\text{H}} \geq 1$, and therefore the coding gain can be greater or less than 1.[†] It turns out that there exist many codes that can provide good coding gains. The relation defining the coding gain once again emphasizes that for a given n and k, the best code is the code that can provide the highest minimum Hamming distance.

To study the bandwidth requirements of coding, we observe that, when no coding is used, the width of the pulses employed to transmit one bit is given by

$$T_b = \frac{1}{R} \qquad (10.4.56)$$

After using coding, in the same time duration that k pulses were transmitted, we must now transmit n pulses, which means that the duration of each pulse is reduced by a factor of $\frac{k}{n} = R_c$. Therefore, the *bandwidth expansion ratio* is given by

$$B = \frac{W_{\text{coding}}}{W_{\text{no coding}}} = \frac{1}{R_c} = \frac{n}{k} \qquad (10.4.57)$$

Thus, the bandwidth has increased linearly. It can be proved that in an AWGN channel, there exists a sequence of codes with parameters (n_i, k_i) with fixed rate ($\frac{k_i}{n_i} = R_c$ independent of i), satisfying

$$R_c = \frac{k}{n} < \frac{1}{2} \log \left(1 + \frac{P}{N_0 W} \right) \qquad (10.4.58)$$

where $\frac{1}{2} \log \left(1 + \frac{P}{N_0 W} \right)$ is the capacity of the channel in bits per transmission,[‡] for which the error probability goes to zero as n_i becomes larger and larger. Obviously

---

[†] Although in a very bad code design we can even have $d_{\min}^{\text{H}} = 0$, we will ignore such cases.

[‡] Recall that the capacity of this channel in bits per second is $W \log \left(1 + \frac{P}{N_0 W} \right)$.

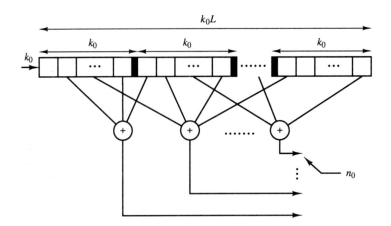

FIGURE 10.15. A convolutional encoder.

for such a scheme the bandwidth expands by a modest factor and does not, as in orthogonal signaling, grow exponentially.

In this chapter we study two major types of codes, *block codes* and *convolutional codes*. Block codes are codes that have already been described. In a block code the information sequence is divided into blocks of length k, and each block is mapped into channel inputs of length n. This mapping is independent from the previous blocks, i.e., there exists no memory from one block to another block. In convolutional codes there exists a shift register of length $k_0 L$ as shown in Figure 10.15. The information bits enter the shift register k_0 bits at a time and then n_0 bits, which are linear combinations of various shift register bits, are transmitted over the channel. These n_0 bits depend not only on the recent k_0 bits that just entered the shift register but also on the $(L-1)k_0$ previous contents of the shift register that constitute its *state*. The quantity

$$m = Lk_0 \qquad (10.4.59)$$

is defined as the *constraint length* of the convolutional code and the number of states of the convolutional code is equal to 2^{m-k_0}. The rate of a convolutional code is defined as

$$R_c = \frac{k_0}{n_0} \qquad (10.4.60)$$

The main difference between block codes and convolutional codes is the existence of memory in convolutional codes.

10.5 LINEAR BLOCK CODES

A (n, k) block code[†] is completely defined by $M = 2^k$ binary sequences of length n called *code words*. A code C consists of M codewords c_i for $1 \le i \le 2^k$.

$$C = \{c_1, c_2, \ldots, c_M\}$$

where each c_i is a sequence of length n with components equal to 0 or 1.

Definition 10.5.1. A block code is *linear* if any linear combination of two code words is also a code word. In the binary case this requires that if c_i and c_j are code words then $c_i \oplus c_j$ is also a code word, where $\oplus$ denotes componentwise modulo-2 addition.

With this definition it is readily seen that a linear block code is a k-dimensional subspace of an n-dimensional space. It is also obvious that the all zero sequence $\mathbf{0}$ is a code word of any linear block code because it can be written as $c_i \oplus c_i$ for any code word c_i. Note that, according to Definition 10.5.1, linearity of a code only depends on the code words and not on the way that the information sequences (messages) are mapped to the code words. However, it is natural to assume that if the information sequence x_1 (of length k) is mapped into the code word c_1 (of length n) and the information sequence x_2 is mapped into c_2, then $x_1 \oplus x_2$ is mapped into $c_1 \oplus c_2$. From now on we will assume that the linear codes that we study possess this special property.

Example 10.5.1

A $(5, 2)$ code is defined by

$$C = \{00000, 10100, 01111, 11011\}$$

It is very easy to verify that this code is linear. If the mapping between the information sequences and code words is given by

$$00 \rightarrow 00000$$
$$01 \rightarrow 10100$$
$$10 \rightarrow 01111$$
$$11 \rightarrow 11011$$

the special property mentioned above is satisfied as well. If the mapping is given by

$$00 \rightarrow 10100$$
$$01 \rightarrow 01111$$
$$10 \rightarrow 00000$$
$$11 \rightarrow 11011$$

[†]From now on we will deal with binary codes unless otherwise specified.

the special property is not satisfied. However, in both cases the code is linear.

Now we define some of the basic parameters that characterize a code.

Definition 10.5.2.   The *Hamming distance* between two code words $\mathbf{c}_i$ and $\mathbf{c}_j$ is the number of components at which the two code words differ and is denoted by $d(\mathbf{c}_i, \mathbf{c}_j)$.[†]

Definition 10.5.3.   The *Hamming weight,* or simply the *weight* of a code word $\mathbf{c}_i$ is the number of nonzero components of the code word and is denoted by $w(\mathbf{c}_i)$.

Definition 10.5.4.   The minimum distance of a code is the minimum Hamming distance between any two different code words, i.e.,

$$d_{\min} = \min_{\substack{i \neq j \\ \mathbf{c}_i, \mathbf{c}_j}} d(\mathbf{c}_i, \mathbf{c}_j) \tag{10.5.1}$$

Definition 10.5.5.   The minimum weight of a code is the minimum of the weights of the code words except the all-zero code word.

$$w_{\min} = \min_{\mathbf{c}_i \neq 0} w(\mathbf{c}_i) \tag{10.5.2}$$

Theorem 10.5.1.   In any linear code, $d_{\min} = w_{\min}$.

Proof.   If $\mathbf{c}$ is a code word, then $w(\mathbf{c}) = d(\mathbf{c}, \mathbf{0})$. Also, if $\mathbf{c}_i$ and $\mathbf{c}_j$ are code words, so is $\mathbf{c} = \mathbf{c}_i \oplus \mathbf{c}_j$ and, moreover, $d(\mathbf{c}_i, \mathbf{c}_j) = w(\mathbf{c})$. This implies that in a linear code corresponding to any weight of a code word there exists a Hamming distance between two code words, and corresponding to any Hamming distance there exists a weight of a code word. In particular, it shows that $d_{\min} = w_{\min}$.

Generator and parity check matrices.   In a linear block (n, k) code let the code words corresponding to the information sequences $\mathbf{e}_1 = (1000\ldots0)$, $\mathbf{e}_2 = (0100\ldots0)$, $\mathbf{e}_3 = (0010\ldots0)$, $\ldots$, $\mathbf{e}_k = (0000\ldots1)$ be denoted by $\mathbf{g}_1, \mathbf{g}_2, \mathbf{g}_3, \ldots, \mathbf{g}_k$, respectively, where each of the $\mathbf{g}_i$ sequences is binary sequence of length n. Now, any information sequence $\mathbf{x} = (x_1, x_2, x_3, \ldots, x_k)$ can be written as

$$\mathbf{x} = \sum_{i=1}^{k} x_i \mathbf{e}_i \tag{10.5.3}$$

and therefore the corresponding code word will be

$$\mathbf{c} = \sum_{i=1}^{k} x_i \mathbf{g}_i \tag{10.5.4}$$

[†]From now on Hamming distance is denoted by d and Euclidean distance is denoted by d^{E}.

If we define the *generator matrix* for this code as

$$
\mathbf{G} = \begin{bmatrix} \mathbf{g}_1 \\ \mathbf{g}_2 \\ \vdots \\ \mathbf{g}_k \end{bmatrix} = \begin{bmatrix} g_{11} & g_{12} & \cdots & g_{1n} \\ g_{21} & g_{22} & \cdots & g_{2n} \\ \vdots & \vdots & \ddots & \vdots \\ g_{k1} & g_{k2} & \cdots & g_{kn} \end{bmatrix}
\tag{10.5.5}
$$

then we can write

$$
\mathbf{c} = \mathbf{xG}
\tag{10.5.6}
$$

This shows that any linear combination of the rows of the generator matrix is a code word. The generator matrix for any linear block code is a $k \times n$ matrix of rank k (because, by definition, the dimension of the subspace is k). The generator matrix of a code completely describes the code. When the generator matrix is given, the structure of an encoder is quite simple.

Example 10.5.2

Find the generator matrix for the first code given in Example 10.5.1.

Solution   We have to find the code words corresponding to information sequences (01) and (10). These are (10100) and (01111), respectively. Therefore,

$$
\mathbf{G} = \begin{bmatrix} 10100 \\ 01111 \end{bmatrix}
\tag{10.5.7}
$$

It is seen that, for the information sequence (x_1, x_2), the code word is given by

$$
(c_1, c_2, c_3, c_4, c_5) = (x_1, x_2)\mathbf{G}
\tag{10.5.8}
$$

or

$$
c_1 = x_1
$$

$$
c_2 = x_2
$$

$$
c_3 = x_1 \oplus x_2
$$

$$
c_4 = x_2
$$

$$
c_5 = x_2
$$

The above code has the property that the code word corresponding to each information sequence starts with a replica of the information sequence itself followed by some extra bits. Such a code is called a *systematic code,* and the extra bits following the information sequence in the code word are called the *parity check bits.* A necessary and sufficient condition for a code to be systematic is that the generator matrix be in the form

$$
\mathbf{G} = \begin{bmatrix} \mathbf{I}_k & | & \mathbf{P} \end{bmatrix}
\tag{10.5.9}
$$

where $\mathbf{I}_k$ denotes a $k \times k$ identity matrix and $\mathbf{P}$ is a $k \times (n-k)$ binary matrix. In a systematic code we have

$$c_i = \begin{cases} x_i & 1 \leq i \leq k \\ \displaystyle\sum_{j=1}^{k} p_{ji} x_j & k+1 \leq i \leq n \end{cases} \tag{10.5.10}$$

where all summations are modulo-2.

By definition, a linear block code C is a k-dimensional linear subspace of the n-dimensional space. From linear algebra we know that if we take all sequences of length n that are orthogonal to all vectors of this k-dimensional linear subspace, the result will be an $(n-k)$-dimensional linear subspace called the *orthogonal complement* of the k-dimensional subspace. This $(n-k)$-dimensional subspace naturally defines a $(n, n-k)$ linear code, which is known as the *dual* of the original (n, k) code C. The dual code is denoted by $C^\top$. Obviously the code words of the original code C and the dual code $C^\top$ are orthogonal to each other. In particular, if we denote the generator matrix of the dual code by $\mathbf{H}$, which is an $(n-k) \times n$ matrix, then any code word of the original code is orthogonal to all rows of $\mathbf{H}$, i.e.,

$$\mathbf{cH}^t = \mathbf{0} \quad \text{for all } \mathbf{c} \in C \tag{10.5.11}$$

The matrix $\mathbf{H}$, which is the generator matrix of the dual code $C^\top$, is called the *parity check matrix* of the original code C. Because all rows of the generator matrix are code words, we conclude that

$$\mathbf{GH}^t = \mathbf{0} \tag{10.5.12}$$

In the special case of a systematic code, where

$$\mathbf{G} = \begin{bmatrix} \mathbf{I}_k & | & \mathbf{P} \end{bmatrix} \tag{10.5.13}$$

the parity check matrix has the following form

$$\mathbf{H} = \begin{bmatrix} -\mathbf{P}^t & | & \mathbf{I}_{n-k} \end{bmatrix} \tag{10.5.14}$$

Example 10.5.3

Find the parity check matrix for the code given in Example 10.5.1.

Solution Here

$$\mathbf{G} = \begin{bmatrix} 10100 \\ 01111 \end{bmatrix}$$

$$\mathbf{I} = \begin{bmatrix} 10 \\ 01 \end{bmatrix}$$

$$\mathbf{P} = \begin{bmatrix} 100 \\ 111 \end{bmatrix}$$

Noting that in the binary case $-\mathbf{P}^t = \mathbf{P}^t$, we conclude that

$$\mathbf{P}^t = \begin{bmatrix} 11 \\ 01 \\ 01 \end{bmatrix}$$

and therefore

$$\mathbf{H} = \left[\begin{array}{c|c} \begin{matrix} 11 \\ 01 \\ 01 \end{matrix} & \begin{matrix} 100 \\ 010 \\ 001 \end{matrix} \end{array} \right]$$

Hamming codes. Hamming codes are a class of linear block codes with $n = 2^m - 1$, $k = 2^m - m - 1$ and $d_{min} = 3$, for some integer $m \geq 2$. As we will see later, with this minimum distance, these codes are capable of providing error-correction capabilities for single errors. The parity check matrix for these codes has a very simple structure. It consists of all binary sequences of length m except the all-zero sequence. The rate of these codes is given by

$$R_c = \frac{2^m - m - 1}{2^m - 1} \tag{10.5.15}$$

which is close to 1 for large values of m. Therefore, Hamming codes are high rate codes with relatively small minimum distance ($d_{min} = 3$). We will see later that the minimum distance of a code is closely related to its error-correcting capabilities. Therefore, Hamming codes have limited error-correcting capability.

Example 10.5.4

Find the parity check matrix and the generator matrix of a $(7, 4)$ Hamming code in the systematic form.

Solution   In this case $m = 3$, and therefore $\mathbf{H}$ consists of all binary sequences of length 3 except the all-zero sequence. We generate the parity check matrix in the systematic form as

$$\mathbf{H} = \left[\begin{array}{cccc|ccc} 1 & 0 & 1 & 1 & 1 & 0 & 0 \\ 1 & 1 & 0 & 1 & 0 & 1 & 0 \\ 0 & 1 & 1 & 1 & 0 & 0 & 1 \end{array} \right]$$

and the generator matrix is obtained to be

$$\mathbf{G} = \left[\begin{array}{cccc|ccc} 1 & 0 & 0 & 0 & 1 & 1 & 0 \\ 0 & 1 & 0 & 0 & 0 & 1 & 1 \\ 0 & 0 & 1 & 0 & 1 & 0 & 1 \\ 0 & 0 & 0 & 1 & 1 & 1 & 1 \end{array} \right]$$

10.5.1 Decoding and Performance of Linear Block Codes

The main purpose of using coding in communication systems is to increase the Euclidean distance between the transmitted signals and, hence, to reduce the error probability at a given transmitted power. This was shown by an example in the previous section. Referring to Figure 10.14, we see that this goal is achieved by choosing the code words to be as far apart on the vertices of the cube as possible. This means that a good measure for comparing the performance of two codes is the Hamming distance between code words. Keeping track of all distances between any two code words is difficult and, in some cases, impossible. Therefore, the comparison between various codes is usually done based on the minimum distance of the code, which for linear codes is equal to the minimum weight. From this it follows that for a given n and k, a code with a larger $d_{\min}$ (or $w_{\min}$) usually performs better compared with a code with a smaller minimum distance.

Soft decision decoding.   In Chapters 7 and 8 we have seen that the optimum signal detection scheme on an additive white Gaussian noise channel is detection based on minimizing the Euclidean distance between the received signal and the transmitted signal. This means that after receiving the output of the channel and passing it through the matched filters, we choose one of the message signals that is closest to the received signal in the Euclidean distance sense. In using coded waveforms, the situation is the same. Assuming that we are employing binary PSK for transmission of the coded message, a code word $\mathbf{c}_i = (c_{i1}, c_{i2}, \ldots, c_{in})$ is mapped into the sequence $s_i(t) = \sum_{k=1}^{n} \psi_{ik}(t - (k-1)T)$, where

$$\psi_{ik}(t) = \begin{cases} \psi(t) & \text{If } c_{ik} = 1 \\ -\psi(t) & \text{If } c_{ik} = 0 \end{cases} \tag{10.5.16}$$

and $\psi(t)$ is a signal of duration T and energy $\mathcal{E}$, which is equal to zero outside the interval $[0, T]$. Now the Euclidean distance between two arbitrary signal waveforms is

$$\left(d_{ij}^{\text{E}}\right)^2 = \sum_{\substack{1 \le k \le n \\ k: c_{ik} \ne c_{jk}}} \left(\pm 2\sqrt{\mathcal{E}}\right)^2 = 4 d_{ij}^{\text{H}} \mathcal{E} \tag{10.5.17}$$

This gives a simple relation between the Euclidean and the Hamming distance when a binary PSK signaling scheme (or any antipodal signaling scheme) is employed. For orthogonal signalling, where $\psi_1(t)$ and $\psi_2(t)$ are orthogonal, the equivalent relation is

$$\left(d_{ij}^{\text{E}}\right)^2 = \sum_{\substack{1 \le k \le n \\ k: c_{ik} \ne c_{jk}}} \int_0^T \left[\psi_1(t) - \psi_2(t)\right]^2 dt = 2 d_{ij}^{\text{H}} \mathcal{E} \tag{10.5.18}$$

Now, using the general relation

$$p_e = Q\left(\frac{d^{\text{E}}}{\sqrt{2N_0}}\right) \tag{10.5.19}$$

we obtain

$$
p(j \text{ received}|i \text{ sent}) = \begin{cases} Q\left(\sqrt{\dfrac{d_{ij}\mathcal{E}}{N_0}}\right) & \text{for orthogonal signaling} \\[3em] Q\left(\sqrt{\dfrac{2d_{ij}\mathcal{E}}{N_0}}\right) & \text{for antipodal signaling} \end{cases} \tag{10.5.20}
$$

Because $d_{ij} \geq d_{\min}$ and because $Q(x)$ is a decreasing function of x, we conclude that

$$
p(j \text{ received}|i \text{ sent}) \leq \begin{cases} Q\left(\sqrt{\dfrac{d_{\min}\mathcal{E}}{N_0}}\right) & \text{for orthogonal signaling} \\[3em] Q\left(\sqrt{\dfrac{2d_{\min}\mathcal{E}}{N_0}}\right) & \text{for antipodal signaling} \end{cases} \tag{10.5.21}
$$

Now, using the union bound (see Chapter 7), we obtain

$$
p(\text{error}|i \text{ sent}) \leq \begin{cases} (M-1)Q\left(\sqrt{\dfrac{d_{\min}\mathcal{E}}{N_0}}\right) & \text{for orthogonal signaling} \\[3em] (M-1)Q\left(\sqrt{\dfrac{2d_{\min}\mathcal{E}}{N_0}}\right) & \text{for antipodal signaling} \end{cases}
$$

$$\tag{10.5.22}$$

and assuming equiprobable messages, we finally conclude that

$$
p_e \leq \begin{cases} (M-1)Q\left(\sqrt{\dfrac{d_{\min}\mathcal{E}}{N_0}}\right) & \text{for orthogonal signaling} \\[3em] (M-1)Q\left(\sqrt{\dfrac{2d_{\min}\mathcal{E}}{N_0}}\right) & \text{for antipodal signaling} \end{cases} \tag{10.5.23}
$$

These are bounds on the error probability of a coded communication system when optimal demodulation is employed. By optimal demodulation we mean passing the received signal $r(t)$ through a bank of matched filters to obtain the received vector $\mathbf{r}$ and then finding the closest point in the constellation to $\mathbf{r}$ in the Euclidean distance sense. This type of decoding that involves finding the minimum Euclidean distance is called *soft-decision decoding* and requires real number computations.

Example 10.5.5

Compare the performance of an uncoded data transmission system with the performance of a coded system using the $(7, 4)$ Hamming code given in Example 10.5.4 when applied to the transmission of a binary source with rate $R = 10^4$ bits/sec. The channel is assumed to be an additive white Gaussian noise channel, the received power is 1 μW and the noise power-spectral density is $\frac{N_0}{2} = 10^{-11}$. The modulation scheme is binary PSK.

Solution

1. If no coding is employed, we have

$$p_b = Q\left(\sqrt{\frac{2\mathcal{E}_b}{N_0}}\right) = Q\left(\sqrt{\frac{2P}{RN_0}}\right) \tag{10.5.24}$$

But $\frac{2P}{RN_0} = \frac{10^{-6}}{10^4 \times 10^{-11}} = 10$ and, therefore,

$$p_b = Q(\sqrt{10}) = Q(3.16) \approx 7.86 \times 10^{-4} \tag{10.5.25}$$

The error probability for four bits will be

$$p_{\text{error in 4 bits}} = 1 - (1 - p_b)^4 \approx 3.1 \times 10^{-3} \tag{10.5.26}$$

2. If coding is employed, we have $d_{\min} = 3$ and

$$\frac{\mathcal{E}}{N_0} = R_c \frac{\mathcal{E}_b}{N_0} = R_c \frac{P}{RN_0} = \frac{4}{7} \times 5 = \frac{20}{7}$$

Therefore, the *message*, error probability is given by

$$p_e \leq 15Q\left(\sqrt{\frac{2d_{\min}\mathcal{E}}{N_0}}\right) = 15Q\left(\sqrt{3\frac{40}{7}}\right) = 15Q(4.14) \approx 2.6 \times 10^{-4}$$

It is seen that using this simple code decreases the error probability by a factor of 12. Of course the price that has been paid is an increase in the bandwidth required for transmission of the messages. This bandwidth expansion ratio is given by

$$\frac{W_{\text{coded}}}{W_{\text{uncoded}}} = \frac{1}{R_c} = \frac{7}{4} = 1.75$$

Hard-decision decoding.   A simpler and more frequently used decoding scheme is to make hard binary decisions on the components of the received vector **r**, and then to find the code word that is closest to it in the Hamming distance sense. The following example clarifies the distinction between soft and hard decisions.

Example 10.5.6

A $(3,1)$ code consists of the two code words 000 and 111. The code words are transmitted using binary PSK modulation with $\mathcal{E} = 1$. The received vector (the sampled outputs of the matched filters) is $\mathbf{r} = (.5, .5, -3)$. If soft decision is employed, we have to compare the Euclidean distance between $\mathbf{r}$ and the two constellation points $(1, 1, 1)$ and $(-1, -1, -1)$ and choose the smaller one. We have $(d^{\mathrm{E}}(\mathbf{r}, (1, 1, 1)))^2 = .5^2 + .5^2 + 4^2 = 16.6$ and $(d^{\mathrm{E}}(\mathbf{r}, (-1, -1, -1)))^2 = 1.5^2 + 1.5^2 + (-2)^2 = 8.5$ and therefore a soft-decision decoder would decode $\mathbf{r}$ as $(-1, -1, -1)$ or equivalently $(0, 0, 0)$. However, if hard-decision decoding is employed, $\mathbf{r}$ is first componentwise detected as $+1$ or 0. This requires a comparison of the components of $\mathbf{r}$ with the zero threshold. The resulting vector $\mathbf{y}$ is therefore $\mathbf{y} = (1, 1, 0)$. Now, we have to compare $\mathbf{y}$ with the $(1, 1, 1)$ and $(0, 0, 0)$ and find the closer one in the Hamming distance sense. The result is of course $(1, 1, 1)$. As seen in this example the results of soft-decision decoding and hard decision decoding can be quite different. Of course, soft-decision decoding is the optimal detection method and achieves a lower probability of error.

There are three basic steps involved in hard-decision decoding. First, we perform demodulation by passing the received $r(t)$ through the matched filters and sampling the output to obtain the $\mathbf{r}$ vector. Second, we compare the components of $\mathbf{r}$ with the threshold and quantize each component to one of the two levels to obtain the $\mathbf{y}$ vector. Finally, we perform decoding by finding the code word that is closest to $\mathbf{y}$ in the Hamming distance sense. In this section we present a systematic approach to perform hard-decision decoding.

First we will define the notion of a *standard array*. Let the code words of the code in question be denoted by $\mathbf{c}_1, \mathbf{c}_2, \ldots, \mathbf{c}_M$, where each of the code words is of length n, $M = 2^k$ and let $\mathbf{c}_1$ denote the all-zero code word. A standard array is a $2^{n-k} \times 2^k$ array whose elements are binary sequences of length n and is generated by writing all the code words in a row starting with the all-zero code word. This constitutes the first row of the standard array. To write the second row, we look among all the binary sequences of length n that are not in the first row of the array. Choose one of these code words that has the minimum weight and call it $\mathbf{e}_1$. Write it under[†] $\mathbf{c}_1$ and write $\mathbf{e}_1 \oplus \mathbf{c}_i$ under $\mathbf{c}_i$ for $2 \le i \le M$. The third row of the array is completed in a similar way. From the binary n-tuples that have not been used in the first two rows, we choose the one with minimum weight and call it $\mathbf{e}_2$. Then, the elements of the third row become $\mathbf{c}_i \oplus \mathbf{e}_2$. This process is continued until no binary n-tuples remain to start a new row. Figure 10.16 shows the standard array generated

$$
\begin{array}{cccc}
\mathbf{c}_1 & \mathbf{c}_2 & \mathbf{c}_3 & \mathbf{c}_M \\
\mathbf{e}_1 & \mathbf{e}_1 \oplus \mathbf{c}_2 & \mathbf{e}_1 \oplus \mathbf{c}_3 & \mathbf{e}_1 \oplus \mathbf{c}_M \\
\mathbf{e}_2 & \mathbf{e}_2 \oplus \mathbf{c}_2 & \mathbf{e}_2 \oplus \mathbf{c}_3 & \mathbf{e}_2 \oplus \mathbf{c}_M \\
\vdots & \vdots & \vdots & \vdots \\
\mathbf{e}_2{}^{(n-k)}{-}1 & \mathbf{e}_2{}^{(n-k)}{-}1 \oplus \mathbf{c}_2 & \mathbf{e}_2{}^{(n-k)}{-}1 \oplus \mathbf{c}_3 & \mathbf{e}_2{}^{(n-k)}{-}1 \oplus \mathbf{c}_M
\end{array}
$$

FIGURE 10.16.   The standard array.

[†]Note that $\mathbf{c}_1 \oplus \mathbf{e}_1 = \mathbf{e}_1$, because $\mathbf{c}_1 = (0, 0, \ldots, 0)$.

as explained above. Each row of the standard array is called a *coset,* and the first element of each coset ($\mathbf{e}_l$ in general) is called the *coset leader.*

The standard array has the following important properties.

Theorem 10.5.2.   All elements of the standard array are different.

Proof.   Let us assume two elements of the standard array are equal. This can happen in two ways.

1. The two equal elements belong to the same coset. In this case we have $\mathbf{e}_l \oplus \mathbf{c}_i = \mathbf{e}_l \oplus \mathbf{c}_j$, from which we conclude $\mathbf{c}_i = \mathbf{c}_j$ which is impossible.
2. The two equal elements belong to two different cosets. Here we have $\mathbf{e}_l \oplus \mathbf{c}_i = \mathbf{e}_k \oplus \mathbf{c}_j$ for $l \neq k$, which means $\mathbf{e}_l = \mathbf{e}_k \oplus (\mathbf{c}_i \oplus \mathbf{c}_j)$. By linearity of the code, $\mathbf{c}_i \oplus \mathbf{c}_j$ is also a code word; let us call it $\mathbf{c}_m$. This means that $\mathbf{e}_l = \mathbf{e}_k \oplus \mathbf{c}_m$ and therefore $\mathbf{e}_l$ and $\mathbf{e}_k$ belong to the same coset, which is impossible because by assumption, $k \neq l$.

From this theorem we conclude that the standard array contains exactly 2^{n-k} rows.

Theorem 10.5.3.   If $\mathbf{y}_1$ and $\mathbf{y}_2$ are elements of the same coset, we have $\mathbf{y}_1 \mathbf{H}^t = \mathbf{y}_2 \mathbf{H}^t$.

Proof.   It is enough to note that because $\mathbf{y}_1$ and $\mathbf{y}_2$ are in the same coset, $\mathbf{y}_1 = \mathbf{e}_l \oplus \mathbf{c}_i$ and $\mathbf{y}_2 = \mathbf{e}_l \oplus \mathbf{c}_j$, therefore,

$$\mathbf{y}_1 \mathbf{H}^t = (\mathbf{e}_l \oplus \mathbf{c}_i)\,\mathbf{H}^t = \mathbf{e}_l \mathbf{H}^t + \mathbf{0} = \left(\mathbf{e}_l \oplus \mathbf{c}_j\right)\mathbf{H}^t = \mathbf{y}_2 \mathbf{H}^t$$

From this theorem we conclude that each coset of the standard array can be uniquely identified by the product $\mathbf{e}_l \mathbf{H}^t$. In general, for any binary sequence $\mathbf{y}$ of length n, we define the *syndrome* $\mathbf{s}$ as

$$\mathbf{s} = \mathbf{y}\mathbf{H}^t \tag{10.5.27}$$

If $\mathbf{y} = \mathbf{e}_l \oplus \mathbf{c}_i$, i.e., $\mathbf{y}$ belongs to the $(l+1)^{\text{st}}$ coset, then obviously $\mathbf{s} = \mathbf{e}_l \mathbf{H}^t$. The syndrome is a binary sequence of length $n - k$ and corresponding to each coset there exists a unique syndrome. Obviously, the syndrome corresponding to the first coset, which consists of the code words, is $\mathbf{s} = \mathbf{0}$.

Example 10.5.7

Find the standard array for the $(5, 2)$ code with code words 00000, 01011, 10101, 11110. Also find the syndromes corresponding to each coset.

Solution   The generator polynomial of the code is

$$G = \begin{bmatrix} 1 & 0 & 1 & 0 & 1 \\ 0 & 1 & 0 & 1 & 1 \end{bmatrix}$$

and the parity check matrix corresponding to **G** is

$$\mathbf{H} = \begin{bmatrix} 1 & 0 & 1 & 0 & 0 \\ 0 & 1 & 0 & 1 & 0 \\ 1 & 1 & 0 & 0 & 1 \end{bmatrix}$$

Using the construction explained above, we obtain the standard array

00000	01011	10101	11110	syndrome = 000
10000	11011	00101	01110	syndrome = 101
01000	00011	11101	10110	syndrome = 011
00100	01111	10001	11010	syndrome = 100
00010	01001	10111	11100	syndrome = 010
00001	01010	10100	11111	syndrome = 001
11000	10011	01101	00110	syndrome = 110
10010	11001	00111	01100	syndrome = 111

Assuming that the received vector **r** has been componentwise compared with a threshold and the resulting binary vector is **y**, we have to find the code word that is at minimum Hamming distance from **y**. First, we find in which coset **y** is located. To do this, we find the syndrome of **y** by calculating $\mathbf{s} = \mathbf{yH}^t$. After finding **s**, we refer to the standard array and find the coset corresponding to **s**. Let us assume that the coset leader corresponding to this coset is $\mathbf{e}_l$. Because **y** belongs to this coset, it is of the form $\mathbf{e}_l \oplus \mathbf{c}_i$ for some i. The Hamming distance of **y** from any code word $\mathbf{c}_j$ is therefore

$$d(\mathbf{y}, \mathbf{c}_j) = w(\mathbf{y} \oplus \mathbf{c}_j) = w(\mathbf{e}_l \oplus \mathbf{c}_i \oplus \mathbf{c}_j) \tag{10.5.28}$$

Because the code is linear, $\mathbf{c}_i \oplus \mathbf{c}_j = \mathbf{c}_k$ for some $1 \le k \le M$. This means that

$$d(\mathbf{y}, \mathbf{c}_j) = w(\mathbf{c}_k \oplus \mathbf{e}_l) \tag{10.5.29}$$

but $\mathbf{c}_k \oplus \mathbf{e}_l$ belongs to the same coset that **y** belongs to. Therefore, to minimize $d(\mathbf{y}, \mathbf{c}_j)$, we have to find the minimum weight element in the coset to which **y** belongs. By construction of the standard array, this element is the coset leader, i.e., we choose $\mathbf{c}_k = \mathbf{0}$ and therefore $\mathbf{c}_j = \mathbf{c}_i$. This means that **y** is decoded into $\mathbf{c}_i$ by finding

$$\mathbf{c}_i = \mathbf{y} \oplus \mathbf{e}_l \tag{10.5.30}$$

Therefore, the procedure for hard-decision decoding can be summarized as follows.

1. Find **r**, the vector representation of the received signal.
2. Compare each component of **r** to the optimal threshold and make a binary decision on it to obtain the binary vector **y**.
3. Find $\mathbf{s} = \mathbf{yH}^t$ the syndrome of **y**.
4. Find the coset corresponding to **s** by using the standard array.
5. Find the coset leader **e** and decode **y** as $\mathbf{c} = \mathbf{y} \oplus \mathbf{e}$.

Because in this decoding scheme the difference between the vector $\mathbf{y}$ and the decoded vector $\mathbf{c}$ is $\mathbf{e}$, the binary n-tuple $\mathbf{e}$ is frequently referred to as the *error pattern*. This means that *the coset leaders constitute the set of all correctable error patterns*.

To obtain error bounds in hard-decision decoding, we note that, since a decision is made on each individual bit, the error probability for each bit for antipodal signaling is

$$p_b = Q\left(\sqrt{\frac{2\mathcal{E}_b}{N_0}}\right) \tag{10.5.31}$$

and for orthogonal signaling is

$$p_b = Q\left(\sqrt{\frac{\mathcal{E}_b}{N_0}}\right) \tag{10.5.32}$$

The channel between the input code word $\mathbf{c}$ and the output of the hard limiter $\mathbf{y}$ is a binary-input binary-output channel that can be modeled by a binary-symmetric channel with crossover probability p_b. Because the code is linear, the distance between any two code words $\mathbf{c}_i$ and $\mathbf{c}_j$ is equal to the distance between the all zero code word $\mathbf{0}$ and the code word $\mathbf{c}_i \oplus \mathbf{c}_j = \mathbf{c}_k$. Without loss of generality we can assume that the all-zero code word is transmitted. If $\mathbf{0}$ is transmitted, then the error probability, by the union bound, can not exceed $(M-1)$ times the probability of decoding the code word that is closest to $\mathbf{0}$ in the Hamming distance sense. For this code word $\mathbf{c}$, which is at distance $d_{\min}$ from $\mathbf{0}$, we have

$$p(\mathbf{c}|\mathbf{0} \text{ sent}) = \begin{cases} \displaystyle\sum_{i=(d_{\min}+1)/2}^{d_{\min}} \binom{d_{\min}}{i} p_b^i (1-p_b)^{d_{\min}-i} & d_{\min} \text{ odd} \\[3mm] \displaystyle\sum_{i=(d_{\min}+2)/2}^{d_{\min}} \binom{d_{\min}}{i} p_b^i (1-p_b)^{d_{\min}-i} + \\[3mm] \qquad + \dfrac{1}{2}\binom{d_{\min}}{d_{\min}/2} p_b^{\frac{d_{\min}}{2}} (1-p_b)^{\frac{d_{\min}}{2}} & d_{\min} \text{ even} \end{cases}$$

or, in general,

$$p(\mathbf{c}|\mathbf{0} \text{ sent}) \leq \sum_{i=[d_{\min}/2]+1}^{d_{\min}} \binom{d_{\min}}{i} p_b^i (1-p_b)^{d_{\min}-i} \tag{10.5.33}$$

Therefore,

$$p_e \leq (M-1) \sum_{i=[d_{\min}/2]+1}^{d_{\min}} \binom{d_{\min}}{i} p_b^i (1-p_b)^{d_{\min}-i} \tag{10.5.34}$$

This gives an upper bound on the error probability of a linear block code using hard decision decoding. As seen here, both in soft-decision and hard-decision decoding, $d_{\min}$ plays a major role in bounding the error probability. This means that, at for a given (n, k), it is desirable to have codes with large $d_{\min}$.

It can be shown that the difference between the performance of soft- and hard-decision decoding is roughly 2 dB for an additive white Gaussian noise channel. That is, the error probability of a soft-decision decoding scheme is comparable to the error probability of a hard-decision scheme whose power is 2-dB higher than the soft-decision scheme. It can also be shown, that if instead of quantization of each component of $\mathbf{r}$ to two levels, an eight-level quantizer (three bits per component) is employed, the performance difference with soft decision (infinite precision) reduces to 0.1 dB. This multilevel quantization scheme, which is a compromise between soft (infinite precision) and hard decision, is also referred to as soft decision in the literature.

Example 10.5.8

If hard-decision decoding is employed in Example 10.5.5, how will the results change?

Solution   Here $p_b = Q\left(\sqrt{\frac{40}{7}}\right) = Q(2.39) = 0.0084$ and $d_{\min} = 3$. Therefore,

$$p_e \leq \binom{7}{2} p_b^2 (1 - p_b)^2$$

$$+ \binom{7}{3} p_b^3 (1 - p_b)^3 + \cdots + p_b^7$$

$$\approx 21 p_b^2 \approx 1.5 \times 10^{-3}$$

In this case coding has decreased the error probability by a factor of 2, compared to 12 in the soft-decision case.

Error detection versus error correction.   Let C be a linear block code with minimum distance $d_{\min}$. Then, if $\mathbf{c}$ is transmitted and hard-decision decoding is employed, any code word will be decoded correctly if the received $\mathbf{y}$ is closer to $\mathbf{c}$ than any other code word. This situation is shown in Figure 10.17. As shown, around each code word there is a "Hamming sphere" of radius e_c, where e_c denotes the number of correctable errors. As long as these spheres are disjoint, the code is capable of correcting e_c errors. A little thinking shows that the condition for nonoverlapping spheres is

$$\begin{cases} 2e_c + 1 \leq d_{\min} & \text{for } d_{\min} \text{ odd} \\ 2e_c + 2 \leq d_{\min} & \text{for } d_{\min} \text{ even} \end{cases}$$

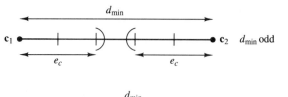

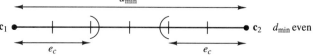

FIGURE 10.17. Relation between e_c and d_{min}.

or

$$e_c \leq \begin{cases} \dfrac{d_{min} - 1}{2} & \text{for } d_{min} \text{ odd} \\[2ex] \dfrac{d_{min} - 2}{2} & \text{for } d_{min} \text{ even} \end{cases} \qquad (10.5.35)$$

which can be summarized as

$$e_c \leq \left[\frac{d_{min} - 1}{2} \right] \qquad (10.5.36)$$

In some cases we are interested in decoding procedures that can *detect* errors rather than correct them. For example, in a communication system where a feedback link is available from the receiver to the transmitter, it might be desirable to detect if an error has occurred and, if so, to ask the transmitter via the feedback channel to retransmit the message. If we denote the error detection capability of a code by e_d, then obviously $e_d \leq d_{min} - 1$, because if $d_{min} - 1$ or fewer errors occur, the transmitted code word will be converted to a non-code word sequence and therefore an error is detected. If both error correction and error detection are desirable, then there is naturally a trade-off between these. Figure 10.18 demonstrates this. From this picture we see that

$$e_c + e_d \leq d_{min} - 1 \qquad (10.5.37)$$

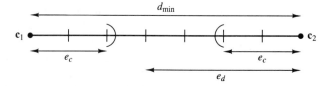

FIGURE 10.18. Relation between e_c, e_d and d_{min}.

with the extra condition $e_c \le e_d$.

10.5.2 Burst-Error-Correcting-Codes

Most of the linear block codes are designed for correcting *random errors,* i.e., errors that occur independently from the location of other channel errors. Certain channel models, including the additive white Gaussian noise channel, can be well modeled as channels with random errors. In some other physical channels, however, the assumption of independently generated errors is not a valid assumption. One such example is a fading channel, as discussed in Section 9.6. In such a channel, if the channel is in deep fade, a large number of errors occur in sequence, i.e., the errors have a *bursty nature.* Obviously in this channel the probability of error at a certain location or time depends on whether or not its adjacent bits are received correctly. Another example of a channel with bursty errors is a compact disc (see Sections 7.1.4 and 10.9). Any physical damage to a compact disc, such as a scratch, damages a sequence of bits and therefore the errors tend to occur in *bursts.* Of course, any random error correcting code can be used to correct bursts of errors as long as the number of errors is less than half of the minimum distance of the code. But the knowledge of the bursty nature of errors makes it possible to design more efficient coding schemes. Two particular codes that are designed to be used for burst error correction are Fire codes and Burton codes. The interested reader is referred to Lin and Costello (1983) for a discussion of these codes.

   An effective method for correction of error bursts is to *interleave* the coded data such that the location of errors looks random and is distributed over many code words rather than a few code words. In this way the number of errors that occur in each block are low and can be corrected by using a random-error-correcting code. At the receiver a *deinterleaver* is employed to undo the effect of the interleaver. A block diagram of a coding system employing interleaving/deinterleaving is shown in Figure 10.19.

   An *interleaver of depth m* reads m code words of length n each and arranges them in a block with m rows and n columns. Then this block is read by column and the output is sent to the digital modulator. At the receiver the output of the detector is supplied to the deinterleaver, which generates the same $m \times n$ block structure and then reads by row and sends the output to the channel decoder. This is shown in Figure 10.20.

   Let us assume that $m = 8$ and the code in use is a (15, 11) Hamming code capable of correcting one error per code word. Then the block generated by the interleaver is a 8×15 block containing 120 binary symbols. Obviously any burst of errors of length 8 or less will result in at most one error per code word and therefore can be corrected. If interleaving/deinterleaving was not employed an error burst of length 8 could possibly result in erroneous detection in two code words (up to 22 information bits).

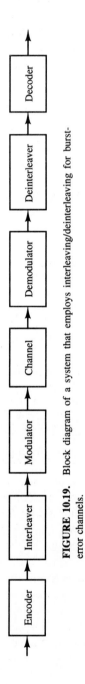

FIGURE 10.19. Block diagram of a system that employs interleaving/deinterleaving for burst-error channels.

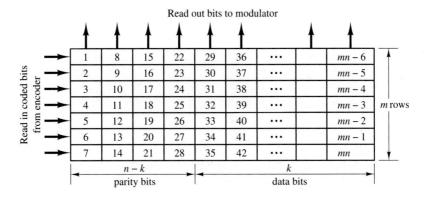

FIGURE 10.20.   A block interleaver for coded data.

10.6 CYCLIC CODES

Cyclic codes are a subset of linear block codes for which easily implementable encoders and decoders exist. In this section we will study the structure of cyclic codes.

Definition 10.6.1.   A *cyclic code* is a linear block code with the extra condition that if $\mathbf{c}$ is a code word, a cyclic shift[†] of it is also a code word.

Example 10.6.1

The code $\{000, 110, 101, 011\}$ is a cyclic code because it is easily verified to be linear, and a cyclic shift of any code word is also a code word. The code $\{000, 010, 101, 111\}$ is not cyclic because, although it is linear, a cyclic shift of 101 is not a code word.

10.6.1 The Structure of Cyclic Codes

To study the properties of cyclic codes it is easier to represent each code word as a polynomial called the *code word polynomial.* The code word polynomial corresponding to $\mathbf{c} = (c_1, c_2, \ldots, c_{n-1}, c_n)$ is simply defined to be

$$c(p) = \sum_{i=1}^{n} c_i p^{n-i} = c_1 p^{n-1} + c_2 p^{n-2} + \cdots + c_{n-1} p + c_n \qquad (10.6.1)$$

The code word polynomial of $\mathbf{c}^{(1)} = (c_2, c_3, \ldots, c_{n-2}, c_{n-1}, c_n, c_1)$, the cyclic shift of $\mathbf{c}$, is

$$c^{(1)}(p) = c_2 p^{n-1} + c_3 p^{n-2} + \cdots + c_{n-1} p^2 + c_n p + c_1 \qquad (10.6.2)$$

[†]A cyclic shift of the code word $\mathbf{c} = (c_1, c_2, \ldots, c_{n-1}, c_n)$ is defined to be $\mathbf{c}^{(1)} = (c_2, c_3 \ldots, c_{n-1}, c_n, c_1)$.

which can be written as

$$c^{(1)}(p) = pc(p) + c_1(p^n + 1) \tag{10.6.3}$$

Noting that in the binary field addition and subtraction are equivalent, this reduces to

$$pc(p) = c^{(1)}(p) + c_1(p^n + 1) \tag{10.6.4}$$

or

$$c^{(1)}(p) = pc(p) \quad (\text{mod } (p^n + 1)) \tag{10.6.5}$$

If we shift $c^{(1)}$ once more the result will also be a code word and its code word polynomial will be

$$c^{(2)}(p) = pc^{(1)}(p) \quad (\text{mod } (p^n + 1)) \tag{10.6.6}$$

$$= p^2 c(p) \quad (\text{mod } (p^n + 1)) \tag{10.6.7}$$

where

$$a(p) = b(p) \quad (\text{mod } d(p))$$

means that $a(p)$ and $b(p)$ have the same remainder when divided by $d(p)$. In general, for i shifts we have the code word polynomial

$$c^{(i)}(p) = p^i c(p) \quad (\text{mod } (p^n + 1)) \tag{10.6.8}$$

For $i = n$ we have

$$c^{(n)}(p) = p^n c(p) \quad (\text{mod } (p^n + 1)) \tag{10.6.9}$$

$$= (p^n + 1)c(p) + c(p) \quad (\text{mod } (p^n + 1)) \tag{10.6.10}$$

$$= c(p) \tag{10.6.11}$$

The above result is obvious because shifting any code word n times leaves it unchanged.

The following theorem, which we will state without proof,[†] is fundamental in further study of cyclic codes.

Theorem 10.6.1. In any (n, k) cyclic code all code word polynomials are multiples of a polynomial of degree $n - k$ of the form

$$g(p) = p^{n-k} + g_2 p^{n-k-1} + g_3 p^{n-k-2} + \cdots + g_{n-k} p + 1$$

called the *generator polynomial,* where $g(p)$ divides $p^n + 1$. Furthermore, for any information sequence $\mathbf{x} = (x_1, x_2, \ldots, x_{k-1}, x_k)$, we have the *information sequence polynomial* $X(p)$, defined by

$$X(p) = x_1 p^{k-1} + x_2 p^{k-2} + \cdots + x_{k-1} p + x_k$$

[†]See Lin and Costello (1983) for a proof.

and the code word polynomial corresponding to **x** is given by $c(p) = X(p)g(p)$.

The fact that any code word polynomial is the product of the generator polynomial and the information sequence polynomial implies that $\mathbf{c} = (c_1, c_2, \ldots, c_{n-1}, c_n)$ is the discrete convolution of the two sequences $\mathbf{x} = (x_1, x_2, \ldots, x_k)$ and $\mathbf{g} = (1, g_2, \ldots, g_{n-k}, 1)$. This fact is very important in designing cyclic encoders.

Example 10.6.2

To generate a $(7, 4)$ cyclic code we need a generator polynomial of degree $n - k = 3$. This polynomial has to divide $p^7 + 1$. Because

$$p^7 + 1 = (p + 1)(p^3 + p^2 + 1)(p^3 + p + 1)$$

the only third degree polynomials that divide $p^7 + 1$ are $p^3 + p^2 + 1$ and $p^3 + p + 1$. We can choose either of these polynomials to generate a cyclic code. If we choose

$$g(p) = p^3 + p^2 + 1$$

and multiply it by all $X(p)$ polynomials of the form

$$X(p) = x_1 p^3 + x_2 p^2 + x_3 p + x_4$$

where x_i is either one or zero, we generate 16 code word polynomials from which we can obtain the code words. The $X(p)$ polynomial is the *message polynomial* corresponding to the binary sequence to be encoded and in general,

$$c(p) = X(p)g(p)$$

The following table shows the input binary sequences and the corresponding code words:

Input	$X(p)$	$c(p) = X(p)g(p)$	Code word
0000	0	0	0000000
0001	1	$p^3 + p^2 + 1$	0001101
0010	p	$p^4 + p^3 + p$	0011010
0100	p^2	$p^5 + p^4 + p^2$	0110100
1000	p^3	$p^6 + p^5 + p^3$	1101000
0011	$p + 1$	$p^4 + p^2 + p + 1$	0010111
0110	$p^2 + p$	$p^5 + p^3 + p^2 + p$	0101110
1100	$p^3 + p^2$	$p^6 + p^4 + p^3 + p^2$	1011100
1001	$p^3 + 1$	$p^6 + p^5 + p^2 + 1$	1100101
0101	$p^2 + 1$	$p^5 + p^4 + p^3 + 1$	0111001
1010	$p^3 + p$	$p^6 + p^5 + p^4 + p$	1110010
0111	$p^2 + p + 1$	$p^5 + p + 1$	0100011
1110	$p^3 + p^2 + p$	$p^6 + p^2 + p$	1000110
1101	$p^3 + p^2 + 1$	$p^6 + p^4 + 1$	1010001
1011	$p^3 + p + 1$	$p^6 + p^5 + p^4 + p^3 + p^2 + p + 1$	1111111
1111	$p^3 + p^2 + p + 1$	$p^6 + p^3 + p + 1$	1001011

As seen from the above table, all cyclic shifts of a given code word are code words themselves.

The generator matrix. Because the generator matrix **G** for a linear block code is not unique, for any cyclic code with a given generator polynomial, there exist many generator matrices. Here we describe a method to obtain a generator matrix in the systematic form. A generator matrix is in the systematic form if it can be written in the form

$$\mathbf{G} = \begin{bmatrix} \mathbf{I}_k \mid \mathbf{P} \end{bmatrix} \qquad (10.6.12)$$

On the other hand, all rows of the generator matrix are code words. This means that the polynomials corresponding to different rows of the generator matrix are all multiples of the generator polynomial $g(x)$. On the other hand, the ith row of the matrix **G** is

$$\mathbf{g}_i = (0, 0, \ldots, 0, 1, 0, \ldots, 0, p_{i,1}, p_{i,2}, \ldots, p_{i,n-k}) \qquad 1 \le i \le k \qquad (10.6.13)$$

where the first k components of $\mathbf{g}_i$ are all zeros except the ith component, which is 1 and $p_{i,j}$ denotes the (i, j)th element of the matrix **P**. From above, the polynomial corresponding to $\mathbf{g}_i$ is

$$g_i(p) = p^{n-i} + p_{i,1} p^{n-k-1} + p_{i,2} p^{n-k-2} + \cdots + p_{i,n-k} \qquad (10.6.14)$$

Because $g_i(p)$ is the code word polynomial corresponding to the code word $\mathbf{g}_i$, it has to be a multiple of $g(p)$, the generator polynomial of the cyclic code under study. Therefore, we have

$$g_i(p) = p^{n-i} + p_{i,1} p^{n-k-1} + p_{i,2} p^{n-k-2} + \cdots + p_{i,n-k} = X(p)g(p) \qquad (10.6.15)$$

Now, since $g(p)$ is of degree $n - k$, we conclude that

$$p_{i,1} p^{n-k-1} + p_{i,2} p^{n-k-2} + \cdots + p_{i,n-k} = p^{n-i} \quad (\text{mod } g(p)) \qquad 1 \le i \le k \quad (10.6.16)$$

This relation shows that by having $g(p)$ we can find all $p_{i,j}$'s for all $1 \le i \le k$ and $1 \le j \le n - k$. Therefore, **P** can be obtained from $g(p)$ and, thus, we obtain **G**, the generator matrix.

Example 10.6.3

Find the generator matrix in the systematic form for the $(7, 4)$ cyclic code generated by $g(p) = p^3 + p^2 + 1$.

Solution We observe that

$$p^6 \bmod p^3 + p^2 + 1 = p^2 + p$$

$$p^5 \bmod p^3 + p^2 + 1 = p + 1$$

$$p^4 \bmod p^3 + p^2 + 1 = p^2 + p + 1$$

$$p^3 \bmod p^3 + p^2 + 1 = p^2 + 1$$

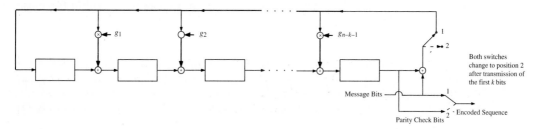

FIGURE 10.21.   Implementation of a cyclic encoder.

Now, having the $p_{i,j}$'s, we can find **G** as follows

$$\mathbf{G} = \begin{bmatrix} 1 & 0 & 0 & 0 & 1 & 1 & 0 \\ 0 & 1 & 0 & 0 & 0 & 1 & 1 \\ 0 & 0 & 1 & 0 & 1 & 1 & 1 \\ 0 & 0 & 0 & 1 & 1 & 0 & 1 \end{bmatrix}$$

Encoding of cyclic codes.   Compared to the general class of linear block codes, cyclic codes have more built-in structure and this extra structure makes the implementation of their encoders easier. The cyclic encoders can be implemented by shift registers. In this section we examine some basic structures for encoding cyclic codes. An easily implementable cyclic encoder is based on the observation that any code word polynomial can be obtained by multiplying the generator polynomial $g(p)$ by the input sequence polynomial $X(p)$, or equivalently by convolving sequences **x** and **p**. This can easily be done by a digital filter-like structure as shown in Figure 10.21.

Example 10.6.4

The $(7,4)$ cyclic code structure is shown in Figure 10.22.

BCH codes.   Bose, Chaudhuri, and Hocquenghem (BCH) codes are a subclass of cyclic codes that, in general, can be designed for correction of t errors. This

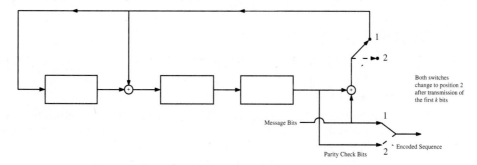

FIGURE 10.22.   Encoder structure for $(7,4)$ cyclic Hamming code.

versatility in design and the existence of an efficient decoding algorithm for these codes (the Berlekamp-Massey algorithm) makes these codes particularly attractive.

To be more specific for any m and t, there exists a BCH code with parameters

$$n = 2^m - 1 \tag{10.6.17}$$

$$n - k \le mt \tag{10.6.18}$$

$$d_{\min} = 2t + 1 \tag{10.6.19}$$

Because m and t are arbitrary, the designer of the communication system has a large number of selections in this family of codes. BCH codes are well tabulated. Table 10.1 shows a table of the coefficients of the generator polynomials for BCH codes of length $7 \le n \le 255$. The coefficients of $g(p)$ are given in octal form. Thus the coefficients of the generator polynomial of the $(15, 5)$ code are 2467, which in binary form is 10,100,110,111. Consequently, the generator polynomial for this code is $g(p) = p^{10} + p^8 + p^5 + p^4 + p^2 + p + 1$.

More information on the properties of BCH codes and their decoding algorithms can be found in Lin and Costello (1983) and Blahut (1983).

Reed-Solomon codes. Reed-Solomon codes are a subset of BCH codes and belong therefore to the family of cyclic codes. Reed-Solomon codes are nonbinary codes, i.e., in a code word $\mathbf{c} = (c_1, c_2, \ldots, c_N)$, the elements c_i, $1 \le i \le N$ are members of a q-ary alphabet. In most practical cases q is chosen to be a power of 2, say $q = 2^k$. In such a case k information bits are mapped into a single element from the q-ary alphabet and then, using an (N, K) Reed-Solomon code K, q-ary symbols are mapped into N q-ary symbols and transmitted over the channel. Figure 10.23 shows this procedure.

A Reed-Solomon code is a BCH code defined by the following parameters.

$$N = q - 1 = 2^k - 1 \tag{10.6.20}$$

$$K = 1, 2, 3, \ldots, N - 1 \tag{10.6.21}$$

$$D_{\min} = N - K + 1 \tag{10.6.22}$$

$$R_c = \frac{K}{N} \tag{10.6.23}$$

This code is capable of correcting up to

$$t = \left[\frac{D_{\min} - 1}{2} \right] \tag{10.6.24}$$

symbol errors.

Reed-Solomon codes have particularly good distance properties, are quite suitable for use in conjunction with q-ary modulation, and are useful in situations where errors tend to happen in "bursts" rather than randomly. This latter property is a consequence of the fact that bursts of errors cause only a few symbol errors in a Reed-Solomon code, which can be corrected easily.

TABLE 10-1   COEFFICIENTS OF THE GENERATOR POLYNOMIALS OF
BCH CODES

n	k	t	$g(p)$
7	4	1	13
15	11	1	23
	7	2	721
	5	3	2467
31	26	1	45
	21	2	3551
	16	3	107657
	11	5	5423325
	6	7	313365047
63	57	1	103
	51	2	12471
	45	3	1701317
	39	4	166623567
	36	5	1033500423
	30	6	157464165547
	24	7	17323260404441
	18	10	1363026512351725
	16	11	6331141367235453
	10	13	472622305527250155
	7	15	5231045543503271737
127	120	1	211
	113	2	41567
	106	3	11554743
	99	4	3447023271
	92	5	624730022327
	85	6	130704476322273
	78	7	26230002166130115
	71	9	6255010713253127753
	64	10	1206534025570773100045
	57	11	335265252505705053517721
	50	13	54446512523314012421501421
	43	14	17721772213651227521220574343
	36	15	31460746665220750447645757721735
	29	21	4031144613767060366753014117 6155
	22	23	123376070404722522435445626637647043
	15	27	22057042445604554770523013762217604353
	8	31	7047264052751030651476224271567733130217
255	247	1	435
	239	2	267543
	231	3	156720665
	223	4	75626641375
	215	5	23157564726421
	207	6	16176560567636227
	199	7	7633031270420722341
	191	8	2663470176115333714567

TABLE 10-1   (Continued)

n	k	t	$g(p)$
	187	9	52755313540001322236351
	179	10	226247107173404324163000455
	171	11	154162142123423560770616306037
	163	12	75004155100756025515747245514601
	155	13	3757513005407665015722506464677633
	147	14	1642130173537165525304165305441011711
	139	15	461401732060175561570722730247453567445
	131	18	2157133314715101512612502774421420241654471
	123	19	120614052242066003717210326516141226272506267
	115	21	6052666557210024726363640460027635255631347273
	107	22	2220577232206625631241730023534742017657475015444
	99	23	10656667253473174222741416201574332252411076432303431
	91	25	675026503032744417272363172473251107555076272072434456
	87	26	11013676341474323643523163430717204620672254527331172131
	79	27	6670003563765750002027034420736617462101532671176654134235
	71	29	2402471052064432151555417211233116320544425036255764322170603
	63	30	10754475055163544325315217357707003666111726455267613656702543301
	55	31	73154252035011001330152753060320543254143267550105570444260354 73617
	47	42	25335420170626465630330413774062331751233341454460450050660245 52543173
	45	43	15202056055234161131101346376423740156367002447076237303320215 025051541
	37	45	51363302550670074141774472454375304207357061743234323476443547 37403044003
	29	47	30257155366730714655270640123613771153422423242011741140602547 57410403565037
	21	55	12562152570603326560017731536076121032273414056530745425211531 21614466513473725
	13	59	46417320050525645444265737142500660043306774454765614031746772 1357026134460500547
	9	63	15726025217472463201031043255355134614162367212044074545112766 115547705561677516057

[†] From paper by Stenbit (1964). © 1964 IEEE; reprinted with permission.

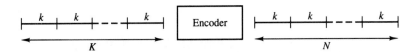

FIGURE 10.23.   A Reed-Solomon encoder.

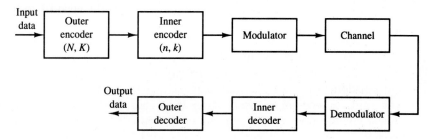

FIGURE 10.24.   Block diagram of a communication system with concatenated coding.

Reed-Solomon codes also can be concatenated with a binary code to provide higher levels of error protection. The binary code used in concatenation with the Reed-Solomon code could be either a block code or a convolutional code. The binary encoder and decoder are located right before the modulator and after the demodulator and are called the *inner* encoder–decoder pair. If the inner code is a (n, k) code, the combination of the inner encoder, digital modulator, waveform channel, digital demodulator, and the inner decoder can be considered as a channel whose input and output are binary blocks of length k, or equivalently, elements of a q-ary alphabet where $q = 2^k$. Now the Reed-Solomon code (usually called the *outer* code) can be used on this q-ary-input q-ary-output channel to provide further error protection. It can be seen easily that if the rates of the inner and the outer codes are r_c and R_c, respectively, the rate of the concatenated code will be

$$R = R_c r_c \qquad (10.6.25)$$

Also the minimum distance of the concatenated code is the product of the minimum distances of the inner and the outer codes. Figure 10.24 shows the block diagram of a digital communication system employing a concatenated coding scheme.

10.7   CONVOLUTIONAL CODES

Convolutional codes are different from block codes by the existence of memory in the encoding scheme. In block codes, each block of k input bits is mapped into a block of length n of output bits by a rule defined by the code (for example, by **G** or $g(p)$) and regardless of the previous inputs to the encoder. The rate of such a code is given by

$$R_c = \frac{k}{n} \qquad (10.7.1)$$

In convolutional codes each block of k bits is again mapped into a block of n bits to be transmitted over the channel, but these n bits are not only determined by the present k information bits but also by the previous information bits. This dependence on the previous information bits causes the encoder to be a finite state machine.

To be more specific, the block diagram of a convolutional encoder is given in Figure 10.25. The convolutional encoder consists of a shift register with kL stages where L is called the *constraint length* of the code. At each instant of time k information bits enter the shift register and the contents of the last k stages of the shift register are dropped. After the k bits have entered the shift register, n linear combinations of the contents of the shift register as shown in the figure are computed and used to generate the encoded waveform. From the above coding procedure it is obvious that the n encoder outputs not only depend on the most recent k bits that have entered the encoder but also on the $(L-1)k$ contents of the first $(L-1)k$ stages of the shift register before the k bits arrived. Therefore, the shift register is a finite state machine with $2^{(L-1)k}$ states. Because for each k input bits we have n output bits, the rate of this code is simply

$$R_c = \frac{k}{n} \tag{10.7.2}$$

Example 10.7.1

A convolutional encoder is shown in Figure 10.26. In this encoder $k = 1$, $n = 2$, and $L = 3$. Therefore, the rate of the code is $\frac{1}{2}$ and the number of states is $2^{(L-1)k} = 4$. One way to describe such a code (other than drawing the encoder) is to specify how the two output bits of the encoder depend on the contents of the shift register. This is usually done by specifying n vectors $\mathbf{g}_1, \mathbf{g}_2, \ldots, \mathbf{g}_n$, known as *generator sequences* of the convolutional code. The ith $1 \le i \le kL$ component of $\mathbf{g}_j$, $1 \le j \le n$, is one if the ith stage of the shift register is connected to the combiner corresponding to the jth bit in the output, and zero otherwise. In the above example, the generator sequences are

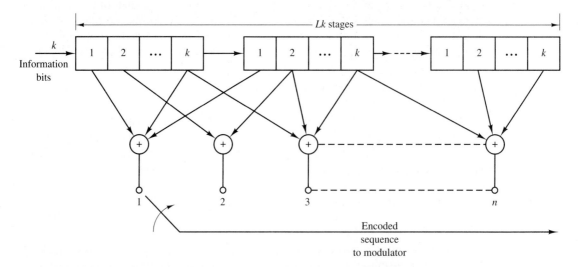

FIGURE 10.25.   The block diagram of a convolutional encoder.

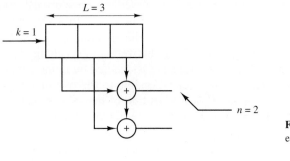

FIGURE 10.26.   A rate $\frac{1}{2}$ convolutional encoder.

given by

$$\mathbf{g}_1 = [1\ 0\ 1]$$

$$\mathbf{g}_2 = [1\ 1\ 1]$$

10.7.1 Basic Properties of Convolutional Codes

Because a convolutional encoder has finite memory, it can easily be represented by a *state-transition diagram.* In the state-transition diagram each state of the convolutional encoder is represented by a box and transitions between states are denoted by lines connecting these boxes. On each line both the input(s) causing that transition and the corresponding outputs are specified. The number of lines emerging from each state is therefore equal to the number of possible inputs to the encoder at that state, which is equal to 2^k. The number of lines merging at each state is equal to the number of states from which a transition is possible to this state. This is equal to the number of possible combinations of bits that leave the encoder as the k bits enter the encoder. This again is equal to 2^k. Figure 10.27 shows the state transition diagram for the convolutional code of Figure 10.26.

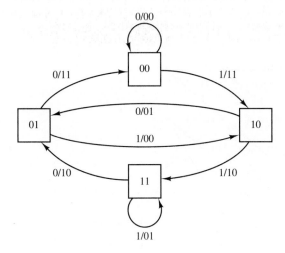

FIGURE 10.27.   State transition diagram for the encoder of Figure 10.26.

A second, and more popular method, to describe convolutional codes is to specify their *trellis diagram*. The trellis diagram is a way to show the transition between various states as the time evolves. The trellis diagram is obtained by specifying all states on a vertical axis and repeating this vertical axis along the time axis. Then each transition from a state to another state is denoted by a line connecting the two states on two adjacent vertical axes corresponding to the two time instances. In a sense, the trellis diagram is nothing but a repetition of the state-transition diagram along the time axes. As was the case with the state-transition diagram, we have 2^k branches of the trellis leaving each state and 2^k branches merging at each. In the case where $k = 1$, it is common to denote the branch corresponding to a 0 input to the encoder by a bold line and the branch corresponding to a 1 input to the encoder by a dashed line. Figure 10.28 shows the trellis diagram for the code described by the encoder of Figure 10.26.

Encoding. The encoding procedure in a convolutional code is very simple. We assume that the encoder, before the first information bit enters it, is loaded with zeros. The information bits enter the encoder k bits at a time and the corresponding n output bits are transmitted over the channel. This procedure is continued until the last group of k bits are loaded into the encoder and the corresponding n output bits are sent over the channel. We will assume, for simplicity, that after the last set of k bits another set of $k(L - 1)$ bits consisting of all zeros enter the encoder and the corresponding n outputs are transmitted over the channel. This makes the encoder ready to be used for the next transmission.

Example 10.7.2

In the convolutional code shown in Figure 10.26, what is the encoded sequence corresponding to the information sequence $\mathbf{x} = (1101011)$?

Solution   It is enough to note that the encoder is in state 0 before transmission, and after transmission of the last information bit two zero bits are transmitted. This means that the transmitted sequence is $\mathbf{x}_1 = (110101100)$. Using this transmission sequence we have the following code word $\mathbf{c} = (111010000100101011)$.

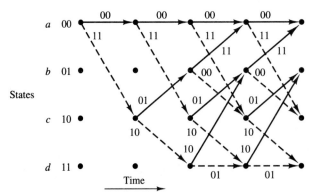

FIGURE 10.28.   Trellis diagram for the encoder of Figure 10.26.

The transfer function.   For every convolutional code the transfer function gives information about the various paths through the trellis that start from the all-zero state and return to this state for the first time. According to the coding convention described before, any code word of a convolutional encoder corresponds to a path through the trellis that starts from the all-zero state and returns to the all-zero state. As we will see later, the transfer function of a convolutional code plays a major role in bounding the error probability of the code. To obtain the transfer function of a convolutional code, we split the all-zero state into two states, one denoting the starting state and one denoting the first return to the all-zero state. All the other states are denoted as in-between states. Corresponding to each branch connecting two states a function of the form $D^\alpha N^\beta J$ is defined where α denotes the number of ones in the output bit sequence for that branch and β is the number of ones in the corresponding input sequence for that branch. The *transfer function* of the convolutional code is then the transfer function of the flow graph between the starting all-zero state and the final all-zero state and will be a function of the tree parameters D, N, J and denoted by $T(D, N, J)$. Each element of $T(D, N, J)$ corresponds to a path through the trellis, starting from the all-zero state and ending at the all-zero state. The exponent of J indicates the number of branches spanned by that path, the exponent of D shows the number of ones in the code word corresponding to that path (or, equivalently, the Hamming distance of the code word from the all-zero code word), and finally the exponent of N indicates the number of ones in the input information sequence. Because $T(D, N, J)$ indicates the properties of all paths through the trellis starting from the all-zero state and returning to it *for the first time,* then in deriving it, any self-loop at the all-zero state is ignored. To obtain the transfer function of the convolutional code, one can use all rules that can be used to obtain the transfer function of a flow graph.

Example 10.7.3

Find the transfer function of the convolutional code of Figure 10.26.

Solution   Figure 10.29 shows the diagram used to find the transfer function of this code. The code has a total of four states denoted by the contents of the first two stages of the shift register. We denote these states by the following letters

$$00 \rightarrow a$$

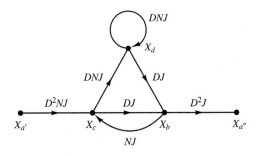

FIGURE 10.29. Flow graph for finding the transfer function.

$$01 \rightarrow b$$

$$10 \rightarrow c$$

$$11 \rightarrow d$$

As seen in the figure, state a is split into states a' and a'' denoting the starting and returning state. Using the flow graph relations we can write

$$X_c = X_{a'} D^2 N J + N J X_b$$

$$X_b = D J X_d + D J X_c$$

$$X_d = D N J X_c + D N J X_d$$

$$X_{a''} = D^2 J X_b$$

Eliminating X_b, X_c, and X_d results in

$$T(D, N, J) = \frac{X_{a''}}{X_{a'}} = \frac{D^5 N J^3}{1 - D N J - D N J^2} \tag{10.7.3}$$

Now, expanding $T(D, N, J)$ in a polynomial form we obtain

$$T(D, N, J) = D^5 N J^3 + D^6 N^2 J^4 + D^6 N^2 J^5 + D^7 N^3 J^5 + \cdots \tag{10.7.4}$$

This indicates that exactly one path exists through the trellis starting from the all-zero state and returning to the all-zero state for the first time, which spans three branches, corresponding to an input information sequence containing one 1 (and, therefore, two 0's), and the code word for this path has Hamming weight equal to 5. This path is indicated with bold lines in the Figure 10.30. This path is somewhat similar to the minimum-weight code word of block codes. In fact, this path corresponds to the code word that is at "minimum distance" from the all-zero code word. This minimum distance, which is equal to the minimum power of D in the expansion of $T(D, N, J)$ is called the *free distance of the code* and is denoted by d_{free}. The free distance of the above code is equal to 5. The general form of the transfer function is, therefore,

$$T(D, N, J) = \sum_{d=d_{\text{free}}}^{\infty} a_d D^d N^{f(d)} J^{g(d)} \tag{10.7.5}$$

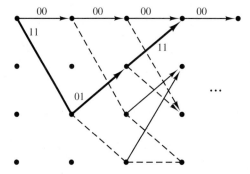

FIGURE 10.30. The path corresponding to $D^5 N J^3$ in the code represented in Figure 10.26.

A shorter form of transfer function, which only provides information about the weight of the code words, can be obtained from $T(D, N, J)$ by setting $N = J = 1$. This shorter form will be denoted by

$$T_1(D) = \sum_{d=d_{\text{free}}}^{\infty} a_d D^d \qquad (10.7.6)$$

and will later be used in deriving bounds on the error probabilities of the convolutional codes.

Example 10.7.4

For the code of Figure 10.26 we have

$$T_1(D) = \left. \frac{D^5 N J^3}{1 - DNJ - DNJ^2} \right|_{N=J=1}$$

$$= \frac{D^5}{1 - 2D}$$

$$= D^5 + 2D^6 + 4D^7 + \cdots$$

$$= \sum_{i=0}^{\infty} 2^i D^{5+i}$$

Catastrophic Codes.   A convolutional code maps a (usually long) sequence of input information bits into a code word to be transmitted over the channel. The purpose of coding is to provide higher levels of protection against channel noise. Obviously, a code that maps information sequences that are far apart into code words that are not far apart is not a good code because these two code words can be mistaken rather easily and the result would be a large number of bit errors in the information stream. A limiting case of this undesirable property happens when two information sequences that are different in infinitely many positions are mapped into code words that differ only in a finite number of positions. In such a case, since the code words differ in a finite number of bits, there always exists the probability that they will be erroneously decoded, and this in turn results in an infinite number of errors in detecting the input information sequence. Codes that exhibit this property are called *catastrophic codes* and should be avoided in practice.

As an example of a catastrophic code, let us consider the $(2, 1)$ code described by

$$\mathbf{g}_1 = [1 \ 1 \ 0]$$

$$\mathbf{g}_2 = [0 \ 1 \ 1]$$

The encoder and the state-transition diagram for this code are given in Figure 10.31. As seen in this diagram a self-loop exists in state "11" that corresponds to a "1" input to the encoder and the corresponding output consists of all-zeros. Therefore, if an input information stream consists of all ones, the corresponding

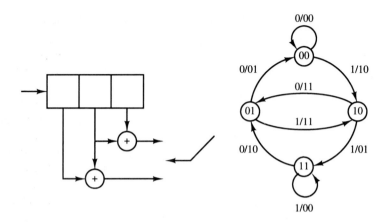

FIGURE 10.31. Encoder and the state-transition diagram for a catastrophic code.

output will be

$$\mathbf{c} = (10010000 \ldots 0001001)$$

If we compare this code word to the code word corresponding to the all-zero information sequence

$$\mathbf{c}_0 = (0000 \ldots 000)$$

we observe that, although the information sequences are different in a large number of positions, the corresponding output sequences are quite close (the Hamming distance being only 4) and, therefore, for a large number of transmissions, they can be mistaken very easily. The existence of such a self-loop corresponding to k inputs, which are not all zeros and for which the n output bits are all zeros, shows that a code is catastrophic and should therefore be avoided.

10.7.2 Optimum Decoding of Convolutional Codes—The Viterbi Algorithm

In our discussion of various decoding schemes for block codes, we saw that there exists the possibility of soft- and hard-decision decoding. In soft-decision decoding $\mathbf{r}$, the vector denoting the outputs of the matched filters, is compared to the various signal points in the constellation of the coded modulation system and the one closest to it in Euclidean distance is chosen. In hard-decision decoding $\mathbf{r}$ is first turned into a binary sequence $\mathbf{y}$ by making decisions on individual components of $\mathbf{r}$, and then the code word, which is closest to $\mathbf{y}$ in the Hamming distance sense, is chosen. It is seen that in both approaches a fundamental task is *to find a path through the trellis that is at minimum distance from a given sequence.* These fundamental problems arise in many areas of communications and other disciplines of electrical engineering. Particularly, the same problem is encountered in maximum-likelihood sequence estimation when transmitting over bandlimited channels with ISI (Chapter 8), demodulation of CPM schemes (Chapter 9), speech recognition, some pattern classification schemes, etc. All these problems are essentially the same and can be titled as *optimal trellis searching algorithms.* The well-known Viterbi algorithm, described previously, provides a satisfactory solution to all these problems.

In hard-decision decoding of convolutional codes, we want to choose a path through the trellis whose code word, denoted by **c**, is at minimum Hamming distance from the quantized received sequence **y**. In hard-decision decoding the channel is binary and memoryless (the fact that the channel is memoryless follows from the fact that the channel noise is assumed to be white). Because the desired path starts from the all-zero state and returns back to the all-zero state, we assume that this path spans a total of m branches, and because each branch corresponds to n bits of the encoder output, the total number of bits in **c** and **y** is mn. We denote the sequence of bits corresponding to the ith branch by $\mathbf{c}_i$ and $\mathbf{y}_i$, respectively, where $1 \leq i \leq m$ and each $\mathbf{c}_i$ and $\mathbf{y}_i$ is of length n. The Hamming distance between **c** and **y** is therefore

$$d(\mathbf{c}, \mathbf{y}) = \sum_{i=1}^{m} d(\mathbf{c}_i, \mathbf{y}_i) \qquad (10.7.7)$$

In soft-decision decoding we have a similar situation with three differences.

1. Instead of **y** we are dealing directly with the vector **r**, the vector output of the optimal (matched filter-type or correlators-type) digital demodulator.

2. Instead of the binary $(0, 1)$ sequence **c** we are dealing with the corresponding sequence **c**$'$ with (for antipodal signals)

$$c'_{ij} = \begin{cases} \sqrt{\mathcal{E}} & \text{if } c_{ij} = 1 \\ -\sqrt{\mathcal{E}} & \text{if } c_{ij} = 0 \end{cases} \quad \text{for } 1 \leq i \leq m \text{ and } 1 \leq j \leq n$$

3. Instead of Hamming distance we are using Euclidean distance. This is a consequence of the fact that the channel under study is an additive white Gaussian noise channel.

From the above we have

$$d_E^2(\mathbf{c}', \mathbf{r}) = \sum_{i=1}^{m} d_E^2(\mathbf{c}'_i, \mathbf{r}_i) \qquad (10.7.8)$$

From (10.7.7) and (10.7.8) it is seen that the generic form of the problem we have to solve is: Given a vector **a** to find a path through the trellis starting at the all-zero state and ending at the all-zero state so that some distance measure between

a and a sequence **b** corresponding to the desired path is minimized.[†] The important fact that makes this problem easy to solve is that the distance between **a** and **b** in both cases of interest can be written as the sum of distances corresponding to individual branches of the path. This is easily observed from (10.7.7) and (10.7.8).

Now let us assume that we are dealing with a convolutional code with $k = 1$. This means that there are only two branches entering each state in the trellis. If the optimal path at a certain point passes through state S, there are two paths that connect the previous states S_1 and S_2 to this state (see Figure 10.32). If we want to see which one of these two branches is a good candidate to minimize the overall distance, we have to add the overall (minimum) metrics at states S_1 and S_2 to the metrics of the branches connecting these two states to the state S. Then, obviously, the branch that has the minimum total metric accumulation up to state S is a candidate to be considered for the states after the state S. This branch is called a *survivor* at state S and the other branch is simply not a suitable candidate and is deleted. Now after the survivor at state S is determined, we also save the minimum metric up to this state and we can move to the next state. This procedure is continued until we reach the all-zero state at the end of the trellis. For cases where $k > 1$, the only difference is that at each stage we have to choose one survivor path from among 2^k branches leading to state S.

The above procedure can be summarized in the following algorithm known as *Viterbi algorithm.*

1. Parse the received sequence into m subsequences each of length n.
2. Draw a trellis of depth m for the code under study. For the last $L - 1$ stages of the trellis, draw only paths corresponding to the all-zero input sequences (this is done because we know that the input sequence has been padded with $k(L - 1)$ zeros).
3. Set $l = 1$ and set the metric of the initial all-zero state equal to zero.

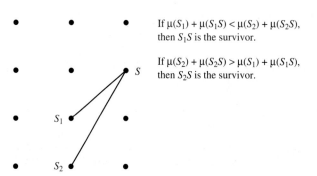

If $\mu(S_1) + \mu(S_1 S) < \mu(S_2) + \mu(S_2 S)$, then $S_1 S$ is the survivor.

If $\mu(S_2) + \mu(S_2 S) > \mu(S_1) + \mu(S_1 S)$, then $S_2 S$ is the survivor.

FIGURE 10.32.   The Viterbi algorithm.

[†]The problem can also be formulated as a maximization problem. For example, instead of minimizing the Euclidean distance, one could maximize the correlation.

4. Find the distance of the lth subsequence of the received sequence to all branches connecting lth stage states to the $(l+1)^{\text{st}}$ stage states of the trellis.

5. Add these distances to the metrics of the lth stage states to obtain the metric candidates for the $(l+1)^{\text{st}}$ stage states. For each state of the $(l+1)^{\text{st}}$ stage, there are 2^k metric candidates each corresponding to one branch ending at that state.

6. For each state at the $(l+1)^{\text{st}}$ stage, choose the minimum of the metric candidates and label the branch corresponding to this minimum value as the *survivor*, and assign the minimum of the metric candidates as the metrics of the $(l+1)^{\text{st}}$ stage states.

7. If $l = m$, go to the next step; otherwise increase l by 1 and go to step 4.

8. Starting with the all-zero state at the $(m+1)^{\text{st}}$ stage, go back through the trellis along the survivors to reach the initial all-zero state. This path is the optimal path and the input bit sequence corresponding to that is the maximum-likelihood decoded information sequence. To obtain the best guess about the input bit sequence, remove the last $k(L-1)$ zeros from this sequence.

As seen from the above algorithm, the decoding delay and the amount of memory required for decoding a long information sequence is unacceptable. The decoding can not be started until the whole sequence (which, in the case of convolutional codes, can be very long) is received, and the total surviving paths have to be stored. In practice, a suboptimal solution that does not cause these problems is desirable. One such approach, which is referred to as *path memory truncation*, is that the decoder at each stage only searches δ stages back in the trellis and not to the start of the trellis. With this approach at the $(\delta+1)$th stage, the decoder makes a decision on the input bits corresponding to the first stage of the trellis (the first k bits) and future received bits do not change this decision. This means that the decoding delay will be $k\delta$ bits and it is only required to keep the surviving paths corresponding to the last δ stages. Computer simulations have shown that, if $\delta \geq 5L$, the degradation in performance due to path memory truncation is negligible.

Example 10.7.5

Let us assume that in hard-decision decoding the quantized received sequence is

$$\mathbf{y} = (01101111010001)$$

The convolutional code is given in Figure 10.26. Find the maximum-likelihood information sequence and the number of errors.

Solution   The code is a $(2, 1)$ code with $L = 3$. The length of the received sequence $\mathbf{y}$ is 14. This means that $m = 7$ and we have to draw a trellis of depth 7. Also note that because the input information sequence is padded with $k(L-1) = 2$ zeros for the final two stages of the trellis, we will only draw the branches corresponding to all-zero inputs. This also means that the actual length of the input sequence is 5, which, after padding with two zeros, has increased to 7. The trellis diagram for this case is shown

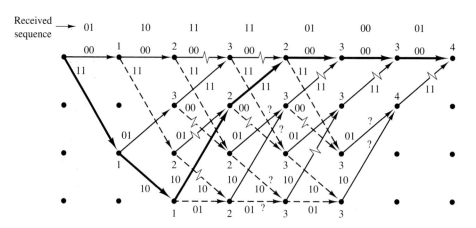

FIGURE 10.33.   The trellis diagram for Viterbi decoding of the sequence (01101111010001).

in Figure 10.33. The parsed received sequence **y** is also shown in this figure. Note that in drawing the trellis in the last two stages, we have only considered the zero inputs to the encoder (notice that in the final two stages, there exist no dashed lines corresponding to 1 inputs). Now the metric of the initial all-zero state is set to zero and the metrics of the next stage are computed. In this step there is only one branch entering each state, therefore there is no comparison, and the metrics (which are the Hamming distances between that part of the received sequence and the branches of the trellis) are added to the metric of the previous state. In the next stage there exists no comparison either. In the third stage, for the first time we have two branches entering each state. This means that a comparison has to be made here and survivors are to be chosen. From the two branches that enter each state, one that corresponds to the least total accumulated metric remains as a survivor and the other branches are deleted (marked with ∼ on the graph). If at any stage two paths result in the same metric, each one of them can be a survivor. Such cases have been marked by a "?" in the trellis diagram. The procedure is continued to the final all-zero state of the trellis and then starting from that state we move along the surviving paths to the initial all-zero state. This path, which is denoted by a heavy path through the trellis, is the optimal path. The input bit sequence corresponding to this path is 1100000, where the last two zeros are not information bits but were added to return the encoder to the all-zero state. Therefore, the information sequence is 11000. The corresponding code word for the selected path is 11101011000000, which is at Hamming distance 4 from the received sequence. All other paths through the trellis correspond to code words that are at greater Hamming distance from the received sequence.

For soft-decision decoding a similar procedure is followed with squared Euclidean distances substituted for Hamming distances.

10.7.3 Other Decoding Algorithms for Convolutional Codes

The Viterbi algorithm provides maximum-likelihood decoding for convolutional codes. However, as we have already seen, the complexity of the algorithm is proportional to the number of states in the trellis diagram. This means that the complexity of the algorithm increases exponentially with the constraint length of the convolutional codes. Therefore, the Viterbi algorithm can be applied only to codes with low-constraint lengths, usually below 10. For higher-constraint length codes other suboptimal decoding schemes have been proposed. These include the sequential decoding of Wozencraft (1957), the Fano algorithm (1963), the stack algorithm (Zigangirov, 1966, and Jelinek, 1969), the feedback-decoding algorithm (Heller, 1975), and majority logic decoding (Massey, 1963).

10.7.4 Bounds on Error Probability of Convolutional Codes

Finding bounds on the error performance of convolutional codes is different from the method used to find error bounds for block codes because here we are dealing with sequences of very large length and, because the free distance of these codes is usually small, some errors will eventually occur. The number of errors is a random variable that depends both on the channel characteristics (SNR in soft-decision decoding and crossover probability in hard-decision decoding) and the length of the input sequence. The longer the input sequence, the higher the probability of making errors. Therefore, it makes sense to normalize the number of bit errors to the length of the input sequence. A measure that is usually adopted for comparing the performance of convolutional codes is the expected number of bits received in error per input bit.

   To find a bound on the average number of bits in error for each input bit, we first derive a bound on the average number of bits in error for each input sequence of length k. To determine this, let us assume that the all-zero sequence is transmitted[†] and, up to stage l in the decoding, there has been no error. Now k information bits enter the encoder and result in moving to the next stage in the trellis. We are interested in finding a bound on the expected number of errors that can occur due to this input block of length k. Because we are assuming that up to stage l there has been no error, then up to this stage the all-zero path through the trellis has the minimum metric. Now, moving to the next stage (stage $(l + 1)$), it is possible that another path through the trellis will have a metric less than the all-zero path and therefore cause errors. If this happens, we must have a path through the trellis that merges with the all-zero path, for the first time, at the $(l + 1)^{st}$ stage and has a metric less than the all-zero path. Such an event is called the *first-error*

[†]Because of the linearity of convolutional codes we can, make this assumption without loss of generality.

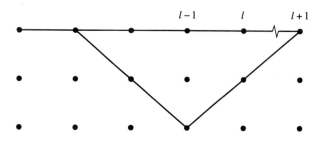

FIGURE 10.34. The path corresponding to the first-error event.

event and the corresponding probability is called *the first-error-event probability.* This situation is depicted in Figure 10.34.

Our first step would be bounding the first-error-event probability. Let $P_2(d)$ denote the probability that a path through the trellis, which is at Hamming distance d from the all-zero path, is the survivor at the $(l + 1)$st stage. Because d is larger than d_{free}, we can bound the first-error-event probability by

$$P_e \le \sum_{d=d_{\text{free}}}^{\infty} a_d P_2(d) \tag{10.7.9}$$

where on the right-hand side we have included all paths through the trellis that merge with the all-zero path at the $(l + 1)^{\text{st}}$ stage. The value of $P_2(d)$ depends on whether soft- or hard-decision decoding is employed.

For soft-decision decoding, if antipodal signaling (binary PSK) is used, we have

$$P_2(d) = Q\left(\frac{d^E}{\sqrt{2N_0}}\right)$$

$$= Q\left(\sqrt{\frac{2\mathcal{E}d}{N_0}}\right)$$

$$= Q\left(\sqrt{2R_c d \frac{\mathcal{E}_b}{N_0}}\right) \tag{10.7.10}$$

and therefore

$$P_e \le \sum_{d=d_{\text{free}}}^{\infty} a_d Q\left(\sqrt{2R_c d \frac{\mathcal{E}_b}{N_0}}\right) \tag{10.7.11}$$

Using the upper bound on the Q function, we have

$$Q\left(\sqrt{2R_c d \frac{\mathcal{E}_b}{N_0}}\right) \leq \frac{1}{2} e^{-R_c d \mathcal{E}_b/N_0} \qquad (10.7.12)$$

Now, noting that

$$e^{-R_c d \mathcal{E}_b/N_0} = D^d \Big|_{D=e^{-R_c \mathcal{E}_b/N_0}} \qquad (10.7.13)$$

we finally obtain

$$P_e \leq \frac{1}{2} \sum_{d=d_{\text{free}}}^{\infty} a_d D^d \Big|_{D=\exp(-R_c \mathcal{E}_b/N_0)} = \frac{1}{2} T_1(D) \Big|_{D=\exp(-R_c \mathcal{E}_b/N_0)} \qquad (10.7.14)$$

This is a bound on the first-error-event probability. To find a bound on the average number of bits in error for k input bits, $\bar{P}_b(k)$, we note that each path through the trellis causes a certain number of input bits to be decoded erroneously. For a general $D^d N^{f(d)} J^{g(d)}$ in the expansion of $T(D, N, J)$, there are a total of $f(d)$ nonzero input bits. This means that the average number of input bits in error can be obtained by multiplying the probability of choosing each path by the total number of input errors that would result if that path were chosen. Hence, the average number of bits in error, in the soft-decision case, can be bounded by

$$\bar{P}_b(k) \leq \sum_{d=d_{\text{free}}}^{\infty} f(d) P_2(d)$$

$$= \sum_{d=d_{\text{free}}}^{\infty} a_d f(d) Q\left(\sqrt{2R_c d \frac{\mathcal{E}_b}{N_0}}\right)$$

$$\leq \frac{1}{2} \sum_{d=d_{\text{free}}}^{\infty} a_d f(d) e^{-R_c d \mathcal{E}_b/N_0} \qquad (10.7.15)$$

If we define

$$T_2(D, N) = T(D, N, J)\big|_{J=1}$$

$$= \sum_{d=d_{\text{free}}}^{\infty} a_d D^d N^{f(d)} \qquad (10.7.16)$$

we have

$$\frac{\partial T_2(D, N)}{\partial N} = \sum_{d=d_{\text{free}}}^{\infty} a_d f(d) D^d \qquad (10.7.17)$$

Therefore, using (10.7.15) and (10.7.17), one obtains

$$\bar{P}_b(k) \le \frac{1}{2} \left. \frac{\partial T_2(D, N)}{\partial N} \right|_{N=1, D=\exp(-R_c \mathcal{E}_b / N_0)} \tag{10.7.18}$$

To obtain the average number of bits in error for each input bit, we have to divide this bound by k. Thus, the final result is

$$\bar{P}_b = \frac{1}{2k} \left. \frac{\partial T_2(D, N)}{\partial N} \right|_{N=1, D=\exp(-R_c \mathcal{E}_b / N_0)} \tag{10.7.19}$$

For hard-decision decoding, the basic procedure follows the above derivation. The only difference is the bound on $P_2(d)$. It can be shown that (see Problem 10.46) $P_2(d)$ can be bounded by

$$P_2(d) \le \left[4p(1 - p) \right]^{d/2} \tag{10.7.20}$$

Using this result, it is straightforward to show that in hard-decision decoding, the probability of error is upperbounded as

$$\bar{P}_b \le \frac{1}{k} \left. \frac{\partial T_2(D, N)}{\partial N} \right|_{N=1, D=\sqrt{4p(1-p)}} \tag{10.7.21}$$

A comparison of hard-decision decoding and soft-decision decoding for convolutional codes shows that here, as in the case for linear block codes, soft-decision decoding outperforms hard-decision decoding by a margin of roughly 2 dB in additive white Gaussian noise channels.

Convolutional codes with good distance properties.   From the analysis carried out above it is obvious that d_{free} plays a major role in the performance of convolutional codes. For a given n and k the free distance of a convolutional code depends on the constraint length of the code. Searching for convolutional codes with good distance properties has been extensively carried out in the literature. Tables 10.2 and 10.3 summarize the result of computer simulations carried out for rate $\frac{1}{2}$ and rate $\frac{1}{3}$ convolutional codes. In these tables, for each constraint length, the convolutional code that achieves the highest free distance is tabulated. For this code, the generators $\mathbf{g}_i$ are given in octal form. The resulting free distance of the code is also given in these tables.

10.8 CODING FOR BANDWIDTH CONSTRAINED CHANNELS

In the two major classes of codes studied so far, i.e., block and convolutional codes, an improvement in the performance of the communication system is achieved by expanding bandwidth. In both cases the Euclidean distance between the transmitted coded waveforms is increased by use of coding, but at the same time the bandwidth is increased by a factor of $\frac{n}{k} = \frac{1}{R_c}$. These type of codes have wide

TABLE 10-2   RATE $\frac{1}{2}$ MAXIMUM
FREE-DISTANCE CODES

Constraint length L	Generators in octal		d_{free}
3	5	7	5
4	15	17	6
5	23	35	7
6	53	75	8
7	133	171	10
8	247	371	10
9	561	753	12
10	1167	1545	12
11	2335	3661	14
12	4335	5723	15
13	10533	17661	16
14	21675	27123	16

Odenwalder (1970) and Larsen (1973).

applications in cases where there exists enough bandwidth and the communication system designer is not under tight bandwidth constraints. Examples of such cases are deep-space communication systems and storage on high-density media. However, in many practical applications we are dealing with communication channels with strict bandwidth constraints, and the bandwidth expansion due to coding may not be acceptable. For example, in transmission of digital data over telephone channels (modem design) we are dealing with a channel that has a restricted bandwidth,

TABLE 10-3   RATE $\frac{1}{3}$ MAXIMUM
FREE-DISTANCE CODES

Constraint length L	Generators in octal			d_{free}
3	5	7	7	8
4	13	15	17	10
5	25	33	37	12
6	47	53	75	13
7	133	145	175	15
8	225	331	367	16
9	557	663	711	18
10	1117	1365	1633	20
11	2353	2671	3175	22
12	4767	5723	6265	24
13	10533	10675	17661	24
14	21645	35661	37133	26

Odenwalder (1970) and Larsen (1973).

and the overhead due to coding imposes a major restriction on the transmission rate. In this section we will discuss an integral coding and modulation scheme called *trellis coded modulation* that is particularly useful for bandwidth constrained channels.

10.8.1 Combined Coding and Modulation

Use of block or convolutional codes introduces redundancy that in turn causes increased Euclidean distance between the coded waveforms. On the other hand, the dimensionality of the transmitted signal will increase from k dimensions per transmission to n dimensions per transmission if binary PSK modulation is employed. This increase in dimensionality results in an increase in bandwidth since bandwidth and dimensionality are proportional. If we want to reap the benefits of coding and at the same time not increase the bandwidth, we have to use a modulation scheme other than binary PSK, i.e., a scheme that is more bandwidth efficient. This means that we have to employ a multilevel/multiphase modulation scheme to reduce the bandwidth. Of course, using a multilevel/multiphase modulation scheme results in a more "crowded" constellation and, at a constant power level, decreases the minimum Euclidean distance within the constellation. This certainly has a negative effect on the error performance of the overall coding-modulation scheme. But, as we will see below, this reduction of the minimum Euclidean distance within the constellation can be well-compensated by the increase in the Hamming distance due to coding such that the overall performance shows considerable improvement.

As an example let us assume that in the coding stage we want to use a rate $\frac{2}{3}$ code. If the rate of the source is R bits/sec, the number of encoder output binary symbols/sec will be $\frac{3}{2}R$. If we want to use a constellation such that the bandwidth requirement is equal to the bandwidth requirement of the uncoded signal (no bandwidth expansion), we must assign m dimensions for each output binary symbol such that the resulting number of dimensions/sec is equal the number of dimensions/sec of the uncoded data which is R. Therefore, we must have

$$R = \frac{3}{2} Rm \tag{10.8.1}$$

and, hence,

$$m = \frac{2}{3} \text{ dimension/binary symbol} \tag{10.8.2}$$

This means that the constellation should be designed in such a way that we have two dimensions for every three binary symbols. But three binary symbols are equivalent to eight points in the constellation, and therefore the final conclusion is that we can achieve our goal with an eight-point constellation in the two-dimensional space. One such constellation is of course an 8-PSK modulation scheme. Therefore, the overall conclusion is that if we use a rate $\frac{2}{3}$ code in conjunction with an 8-PSK modulation scheme, there will be no bandwidth expansion.

Now let us examine how much coding gain we can obtain from such a scheme. Assuming that the available power is P, with no coding we have

$$\mathcal{E}_b = \frac{P}{R} \tag{10.8.3}$$

and therefore for the minimum Euclidean distance between two sequences we have

$$d^2 = \frac{4P}{R} \tag{10.8.4}$$

If two information bits are mapped into a point in an 8-PSK constellation, the energy of this point is

$$\mathcal{E}_s = \frac{2P}{R} \tag{10.8.5}$$

From this we can derive an expression for the minimum Euclidean distance within the constellation (see Figure 10.35) as

$$d_{\min}^2 = 4\frac{2P}{R}\sin^2\frac{\pi}{8} = 2(2-\sqrt{2})\frac{P}{R} \tag{10.8.6}$$

Obviously the minimum Euclidean distance has been decreased. To see this effect we derive the loss due to using this constellation.

$$\left(\frac{d_{\min}^2}{d^2}\right)^{-1} = \frac{2}{2-\sqrt{2}} = 2 + \sqrt{2} = 3.414 \sim 5.33 \text{ dB} \tag{10.8.7}$$

This loss has to be compensated by the code. Of course, the rate $\frac{2}{3}$ code employed here should not only compensate for this loss but should also provide additional gain to justify its use of the overall coding-modulation scheme. We can use any block or convolutional code that can provide the minimum distance required to achieve a certain overall coding gain. For example, if we need an overall coding gain of

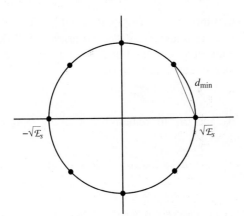

FIGURE 10.35. The 8-PSK constellation used for bandwidth efficient coding.

3 dB, the code must provide a coding gain of 8.33 dB to compensate for the 5.33-dB loss due to modulation and to provide an extra 3-dB coding gain. A code that can provide such a high coding gain is a very complex (long constraint length) code requiring a sophisticated encoder and decoder. However, by interpreting coding and modulation as a single entity as shown in the next section, we see that a comparable performance can be achieved using a much simpler coding scheme.

10.8.2 Trellis-Coded Modulation

Trellis-coded modulation, or TCM, is a simple method for designing coded modulation schemes that can achieve good overall performance. This coding-modulation scheme is based on the concept of *mapping by set partitioning* developed by Ungerboeck (1982). Mapping by set partitioning can be used in conjunction with both block and convolutional codes, but due to the existence of a simple optimal soft-decision decoding algorithm for convolutional codes (the Viterbi algorithm), it has been mostly used with convolutional codes. When used with convolutional codes, the resulting coded modulation scheme is known as *trellis coded modulation.*[†]

Set partitioning principles.   The key point in partitioning of a constellation is to find subsets of the constellation that are similar and the points inside each partition are maximally separated. Starting from the original constellation, we partition it into two subsets that are congruent and the points within each partition are separated maximally. Then apply the same principle to each partition and continue. The point at which the partitioning is stopped depends on the code that we are using and will be discussed shortly.

An example of set partitioning is shown in Figure 10.36. We start with an 8-PSK constellation with signal points on a circle of radius $\mathcal{E}_s$. The minimum distance within this constellation is

$$d_0 = \sqrt{(2 - \sqrt{2})\mathcal{E}_s} \qquad (10.8.8)$$

This constellation is partitioned into two partitions denoted by B_0 and B_1. Note that B_0 and B_1 are congruent. There are many ways that the original 8-PSK constellation can be partitioned into two congruent subsets, but B_0 and B_1 provide the maximum partition distance. This distance is easily seen to be

$$d_1 = \sqrt{2\mathcal{E}_s} \qquad (10.8.9)$$

Now we further partition B_0 and B_1 to obtain C_0, C_1, C_2, and C_3. The inner partition distance now has increased to

$$d_2 = 2\sqrt{\mathcal{E}_s} \qquad (10.8.10)$$

[†]To be more precise, the term *trellis coded modulation* is used when a general trellis code is employed.

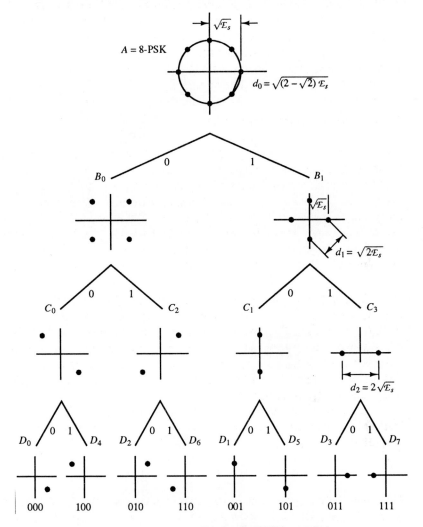

FIGURE 10.36.   Set partitioning for 8-PSK constellation.

We can still go one step further to obtain eight partitions, each containing a single point. The corresponding subsets are denoted by D_0 through D_7. Another example of set partitioning applied to a QAM constellation is given in Figure 10.37. The reader can verify that this partitioning follows the general rules for set partitioning as described above.

Coded modulation.   The block diagram of a coded modulation scheme is shown in Figure 10.38. A block of length k input bits is divided into two subblocks of lengths k_1 and k_2, respectively. The first k_1 bits are applied to a (n_1, k_1) binary

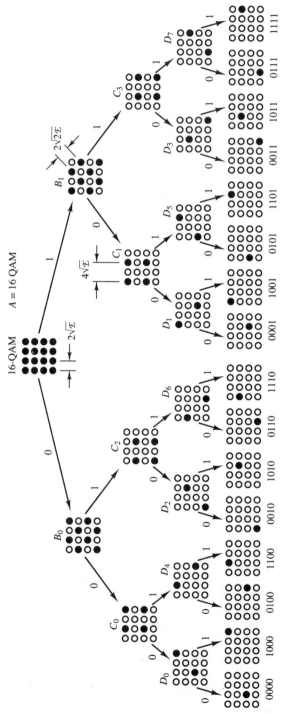

FIGURE 10.37. Set partitioning for 16-QAM constellation.

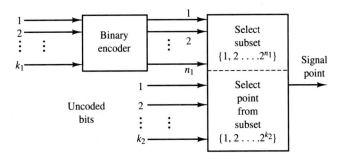

FIGURE 10.38.   The block diagram of a coded modulation system.

encoder. The output of the encoder consists of n_1 bits. These bits are used to choose one of 2^{n_1} partitions in the constellation. This means that the constellation has been partitioned into 2^{n_1} subsets. After the constellation is chosen, the remaining k_2 bits are used to choose one of the points in the chosen constellation. This means that there exist 2^{k_2} points in each partition. Therefore, the partitioning that is used contains 2^{n_1} subsets and each subset contains 2^{k_2} points. This gives us a rule for how large a constellation is required and how many steps in partitioning of this constellation must be taken.

Ungerboeck (1982) has shown that by choosing $n_1 = k_1 + 1$ and $k_2 = 1$ and using simple convolutional codes, we can design coded modulation schemes that achieve an overall coding gain between 3 and 6 dB. One such scheme is shown in Figure 10.39. In this coding scheme $k_1 = 1$, $n_1 = 2$, and $k_2 = 1$. The constellation contains $2^{n_1+k_2} = 8$ points, which are partitioned into $2^{n_1} = 4$ subsets each containing $2^{k_2} = 2$ points. The constellation chosen here is an 8-PSK constellation and it is partitioned as shown before (see Figure 10.36). The convolutional code employed here can be any rate $\frac{k_1}{n_1} = \frac{1}{2}$ code. The constraint length of this code is a design parameter and can be chosen to provide the desired coding gain. Higher constraint lengths, of course, provide higher coding gains at the price of increased encoder–decoder complexity. In this very simple example, the constraint length has been chosen to be equal to 3. The (one stage) trellis diagram of this code is also shown in Figure 10.39.

The trellis diagram shown in Figure 10.39 is the trellis diagram of an ordinary convolutional code. The main difference is that here we have two paths connecting two states. The reason for this is the existence of the extra $k_2 = 1$ bit, which chooses a point in each partition. In fact, the two parallel paths connecting two states correspond to a partition, and any single path corresponds to a point in the partition. One final question remains to be answered: What is the optimal mapping in the transition between the states of the convolutional code and the partitions? Extensive computer simulations as well as heuristic reasoning result in the following rules.

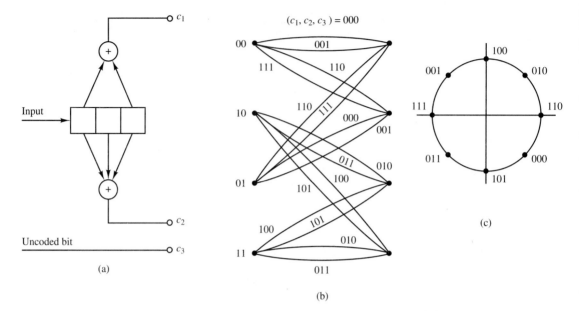

FIGURE 10.39. A simple TCM scheme: (a) encoder, (b) four-state trellis, and (c) mapping of coded bits (c_1, c_2, c_3) to signal points.

1. Parallel transitions (when they occur) correspond to signal points in a single partition at the next stage of partitioning. In the above example, $C_0 = \{D_0, D_4\}$, $C_2 = \{D_2, D_6\}$, $C_1 = \{D_1, D_5\}$ and $C_3 = \{D_3, D_7\}$ correspond to parallel transitions. These points are separated by the maximum Euclidean distance of $d_2 = 2\sqrt{\mathcal{E}_s}$.

2. The transitions originating from and merging into any state are assigned partitions in the next stage of partitioning that have a single parent partition in the preceding stage. In the above example, $B_0 = \{C_0, C_2\}$ and $B_1 = \{C_1, C_3\}$ are such partitions. The maximum distance in this case is $d_1 = \sqrt{2\mathcal{E}_s}$.

3. The signal points should occur with equal frequency.

To see how the trellis coded modulation scheme of Figure 10.39 performs, we have to find the minimum Euclidean distance between two paths originating from a node and merging into another node. This distance, known as the *free Euclidean distance* and denoted by D_{fed}, is an important characteristic of a trellis coded modulation scheme. One obvious candidate for D_{fed} is the Euclidean distance between two parallel transitions. The Euclidean distance between two parallel transitions is $d_2 = 2\sqrt{\mathcal{E}_s}$. Another candidate path is shown in Figure 10.40. However, the Euclidean distance between these two paths is $d^2 = d_0^2 + 2d_1^2 = 4.58\mathcal{E}_s$. Obviously, this is larger than the distance between two parallel transitions. It is easily verified that for this code the free Euclidean distance is $D_{\text{fed}} = d_2 = 2\sqrt{\mathcal{E}_s}$. To compare this

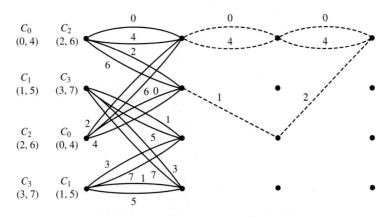

FIGURE 10.40. Two candidate minimum-distance paths.

result with an uncoded scheme, we note that in an uncoded scheme

$$d^2_{\text{uncoded}} = 4\mathcal{E}_b = 4\frac{P}{R}$$

(10.8.11)

and in the coded scheme

$$d^2_{\text{coded}} = 4\mathcal{E}_s = 8\mathcal{E}_b$$

(10.8.12)

Therefore, the coding gain is given by

$$G_{\text{coding}} = \frac{d^2_{\text{coded}}}{d^2_{\text{uncoded}}} = 2 \sim 3 \text{ dB}$$

(10.8.13)

Thus, this simple coding scheme is capable of achieving a 3-dB coding gain without increasing bandwidth. Of course, the price paid for this better performance is increased complexity in encoding and decoding.

Instead of a four-state trellis, a trellis with a higher number of states yields higher coding gains. Extensive computer simulations by Ungerboeck indicate that with 8, 16, 32, 64, 128, and 256 states, coding gains in the range of 3.6 dB to 5.75 dB can be achieved. The encoder for an eight-state trellis is shown in Figure 10.41.

Decoding of Trellis-Coded Modulation Codes.   The decoding of trellis-coded modulation is performed in two steps. Because each transition in the trellis corresponds to a partition of the signal set and each partition generally corresponds to a number of signal points, the first step is to find the most likely signal point in each partition. This is accomplished by finding the point in each partition that is closest in Euclidean distance to the received point. This first step in decoding of a trellis coded modulation scheme is called *subset decoding*. After this step, corresponding to each transition in the trellis there exists only one point (the most likely one) and only one Euclidean distance (the distance between the received

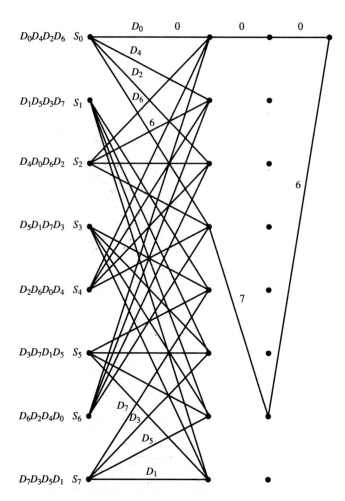

FIGURE 10.41. An eight-state trellis encoder for coded 8-PSK modulation.

point and this most likely point). The second step of the decoding procedure is to use this Euclidean distance to find a path through the trellis whose total Euclidean distance from the received sequence is minimum. This is done by applying the Viterbi algorithm.

Trellis-coded modulation is used widely in high-speed modems. Without coding, high-speed modems achieved data rates up to 9600 bits/sec with a $M = 16$ QAM signal constellation. The added coding gain provided by trellis coded modulation has made it possible to increase the speed of transmission by at least a factor of 2.

10.9 PRACTICAL APPLICATIONS OF CODING

In the previous sections we have seen that coding can be employed to improve the effective SNR and thus enhance the performance of the digital communication system. Block and convolutional codes and combinations of them in the form of concatenated codes as discussed earlier have been applied to communication system design where bandwidth is not a major concern, and thus some bandwidth expansion due to coding is allowed. On the other hand, in cases where bandwidth is a major concern, as in digital communication over telephone channels, coded modulation can be employed. By using coding, performance of practical digital communication systems has been improved by up to 9 dB, depending on the application and the type of the code employed. In this section we discuss applications of coding to deep-space communications, telephone-line modems, and compact disc players (CD's).

10.9.1 Coding for Deep-Space Communications

Deep-space communication channels are characterized by very low SNR's and practically no bandwidth limitations. The transmitter power is usually obtained from on-board solar cells and therefore is typically limited to 20–30 Watts. The physical dimensions of the transmitting antenna are also quite limited and therefore its gain is limited too. The enormous distance between the transmitter and the receiver and lack of repeaters results in a very low SNR at the receiver. The channel noise can be characterized by a white Gaussian random process. These channels are very well modeled as AWGN channels. Because bandwidth is not a major concern on these channels, both block and convolutional codes can be applied.

In 1975 the Viking orbiters and landers were launched to Mars. In this mission a (32,6) block code was employed that provided a coding gain a gain of approximately 4 dB with respect to an uncoded PSK system at an error rate of 10^{-6}. Later, in the Voyager space mission to the outer planets (Mars, Jupiter, and Saturn), convolutional codes with Viterbi decoding were employed. Two codes that were designed at the Jet Propulsion Laboratory (JPL) for that mission were a (2,1) convolutional code with a constraint length of $L = 7$ with

$$\mathbf{g}_1 = \begin{bmatrix} 1 & 1 & 0 & 1 & 1 & 0 & 1 \end{bmatrix}$$
$$\mathbf{g}_2 = \begin{bmatrix} 1 & 0 & 0 & 1 & 1 & 1 & 1 \end{bmatrix}$$

and a (3,1) convolutional code with $L = 7$ and

$$\mathbf{g}_1 = \begin{bmatrix} 1 & 1 & 0 & 1 & 1 & 0 & 1 \end{bmatrix}$$
$$\mathbf{g}_2 = \begin{bmatrix} 1 & 0 & 0 & 1 & 1 & 1 & 1 \end{bmatrix}$$
$$\mathbf{g}_3 = \begin{bmatrix} 1 & 0 & 1 & 0 & 1 & 1 & 1 \end{bmatrix}$$

The first code has a free distance of $d_{free} = 10$ and the second code has a free distance of $d_{free} = 15$. Both codes were decoded using the Viterbi algorithm and a soft-decoding scheme in which the output was quantized to $Q = 8$ levels. The first code provides a coding gain of 5.1 dB with respect to an uncoded PSK system operating at an error rate of 10^{-5}. The second code provides a gain of 5.7 dB. Both codes operate about 4.5 dB from the theoretical limit predicted by Shannon's formula.

In subsequent missions of the Voyager to Uranus in 1986, the (2,1) convolutional code with $L = 7$ was used as an inner code in a concatenated coding scheme where a (255,223) Reed-Solomon code served as the outer code. Viterbi decoding followed by a Reed-Solomon decoder at the earth terminal provided a total coding gain of 8 dB at an error rate of 10^{-6}. This system operated at a data rate of approximately 30 Kbits/sec.

Other decoding algorithms for convolutional codes also have been applied to certain deep space communication projects. For NASA's Pioneer 9 mission, a (2,1) convolutional code with a constraint length of $L = 21$ was designed with generator sequences (in octal representation)

$$\mathbf{g}_1 = \begin{bmatrix} 4 & 0 & 0 & 0 & 0 & 0 & 0 \end{bmatrix}$$
$$\mathbf{g}_2 = \begin{bmatrix} 7 & 1 & 5 & 4 & 7 & 3 & 7 \end{bmatrix}$$

which employed Fano's algorithm with a soft-decision decoding scheme and eight levels of output quantization. Another frequently used coding scheme, used in Pioneers 10, 11, and 12 and also in Helios A and B German solar orbiter missions, employed a (2,1) convolutional code with a constraint length of $L = 32$. The generator sequences for this code (in octal representation) are given below.

$$\mathbf{g}_1 = \begin{bmatrix} 7 & 3 & 3 & 5 & 3 & 3 & 6 & 7 & 6 & 7 & 2 \end{bmatrix}$$
$$\mathbf{g}_2 = \begin{bmatrix} 5 & 3 & 3 & 5 & 3 & 3 & 6 & 7 & 6 & 7 & 2 \end{bmatrix}$$

This code has a free distance of $d_{free} = 23$. For decoding, again, the Fano decoding algorithm with eight-level output quantization was employed. Majority logic decoding also has been used in a number of coding schemes designed for the INTELSAT communication satellites. As an example, a (8,7) code with $L = 48$ designed to operate at 64 kbits/sec on an INTELSAT satellite was capable of improving the error rate from 10^{-4} to 5×10^{-8}.

10.9.2 Coding for Telephone-Line Modems

Telephone-line channels are characterized by a limited bandwidth, typically 300–3300 Hz, and a rather high SNR, which is usually 28 dB or more. Therefore, in designing coding schemes for telephone-line channels, we are faced with a bandwidth limitation. This is in direct contrast to the deep-space communication channel, which is primarily power limited. This corresponds to the case of $r \gg 1$ in Figure 10.11. Because bandwidth is limited, we have to use low dimensional signaling schemes, and because power is rather abundant, we can employ multilevel modulation schemes. As we have already seen in Section 10.8, trellis coded-modulation is an appropriate scheme to be employed in such a case.

Historically, the first modems on telephone channels (prior to the 1960's) employed FSK with asynchronous detection and achieved bit rates in the range of 300–1200 bits/sec. Later, in the early 1960's, the first generation of synchronous modems employing 4-PSK modulation achieved bit rates of up to 2400 bits/sec. Advances in equalization techniques allowed for more sophisticated constellations, which resulted in higher bit rates. These included 8-PSK modems achieving a bit rate of 4800 bits/sec. and 16-point QAM modems that increased the bit rate to 9600 bits/sec. In the early 1980's, modems with a bit rate of 14400 bits/sec were introduced that employed a 64-point QAM signal constellation. All these improvements were results of advances in equalization and signal processing techniques and also in the characteristics of telephone lines.

The advent of trellis-coded modulation made it possible to design coded modulation systems that improved overall system performance without requiring excess bandwidth. Trellis-coded modulation schemes based on variations of the original Ungerboeck's codes and introduced by Wei (1984) were adopted as standard by the CCITT standard committees. These codes are based on linear or nonlinear convolutional codes to guarantee invariance to $180°$ or $90°$-phase rotations. This is crucial in applications where differential encoding is employed to avoid phase ambiguities when a PLL is employed for carrier-phase estimation at the receiver. These codes achieve a coding gain comparable to Ungerboeck's codes with the same number of states but at the same time provide the required phase invariance. In Figure 10.42 we have shown the combination of the differential encoder, the nonlinear convolutional encoder, and the signal mapping for the 8-state trellis-coded modulation system that is adopted in the CCITT V.32 standard.

10.9.3 Coding for Compact Discs

In Chapters 4 and 7 the problems of source coding and modulation codes for compact discs were addressed. In this section we consider error-correcting techniques that are employed for compact-disc digital audio recording.

The storage medium for digital audio recording on a compact disc is a plastic disc with a diameter of 120 mm, a thickness of 1.2 mm, and a track pitch of 1.6 μm. At playing time this disc is read by a laser beam at a velocity of 1.25 m/sec. Inside the spiral track on the disc are depressions called "pits" and flat areas between pits called "lands." The digital audio is stored by the length of these pits and lands. A 1 is represented by a transition from a pit to a land or vice versa, whereas a 0 corresponds to no transition (NRZI modulation). As we have seen previously in Chapter 7, constraints on the physical length of pits and lands make it necessary to employ runlength-limited (RLL) codes. The 8–14 modulation (EFM) code with $d = 2$ and $\kappa = 10$ is used in compact-disc recording.

The main source of errors in a compact disc are imperfections in the manufacturing of the disc, such as air bubbles in the plastic material or pit inaccuracies,

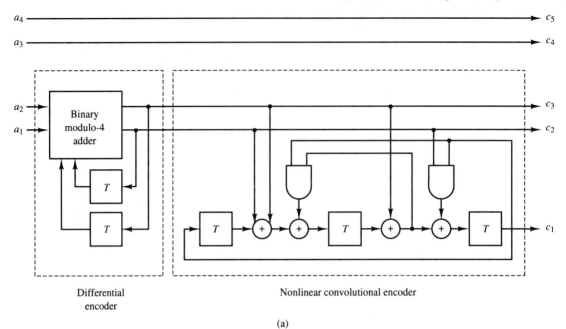

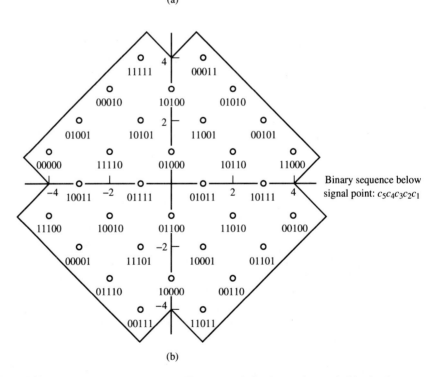

FIGURE 10.42. (a) Differential encoder, nonlinear convolutional encoder, and (b) signal constellation adopted in V.32 the standard.

and damages to the disc, such as fingerprints or scratches, dust, dirt, and surface abrasions. Because each pit is almost 0.5-μm wide and between 0.9- and 3.3-μm long, these sources of errors result in error bursts, affecting many adjacent information bits. This means that a good model for this storage channel is a channel with bursts of errors, and well-known techniques for correcting bursts of errors can be employed. As we have previously discussed, Reed-Solomon codes are particularly attractive for such applications.

Recall from Chapter 4 that the left and the right channels are sampled at a rate of 44.1 kHz and then each sample is quantized to 16 levels. Therefore, at each sampling instance there are 32 bits, or four 8-bit sequences to be encoded. Each 8-bit sequence is called a "symbol."

For error correction and detection two Reed-Solomon codes are employed,[†] as shown in Figure 10.43. The first code, denoted by C_1, is a (28,24) RS code, and the second code, C_2, is a (32,28) RS code. The alphabet on which these codes are defined consists of binary sequences of length 8, which coincide with our definition of a symbol.

The input sequence to the C_1 encoder consists of 24 symbols (usually known as a "frame"), which are encoded into 28 symbols. The 28 symbols at the output of the C_1 encoder are *interleaved* (see Section 10.5.2) to reduce the effect of error bursts and to spread them over a longer interval, making them look more "random." These are encoded by the C_2 encoder to 32 symbols. At the output of the C_2 encoder, the odd-numbered symbols of each frame are grouped with the even-numbered symbols of the next frame to form a new frame. At the output of the C_2 encoder corresponding to each set of six audio samples, we have 32 8-bit

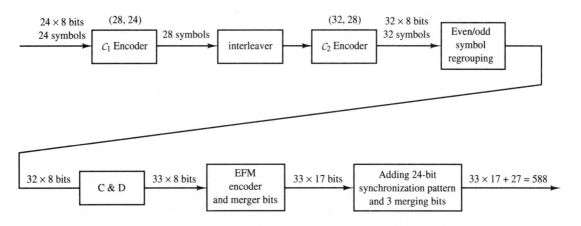

FIGURE 10.43.   Encoding for compact-disc recording.

[†]These are, in fact, *shortened* Reed-Solomon codes, obtained from Reed-Solomon codes by putting some information bits equal to zero, and therefore reducing both k and n by a constant while keeping the minimum distance intact.

symbols. One more 8-bit symbol is added that contains the control and display (C & D) information, bringing the total to 33 symbols/frame.

The output is then applied to the 8-14 (EFM) runlength-limited encoder, which maps each symbol into a binary sequence of length 14. We have seen in Chapter 7 that three more bits, called "merger" bits, are added for each symbol to make sure that the merger of code words satisfies the runlength constraint. These bring the length of each sequence to 17. Next, a frame is completed by adding a 24-bit synchronization pattern and 3 additional "merger" bits to guarantee the run-length constraint after a merger. This brings the total number of encoded bits/frame (six samples, or $6 \times 2 \times 16 = 192$ audio bits) to

$$33 \times 17 + 24 + 3 = 588$$

channel bits (see Figure 7.30). The number of channel bits/sec is given by

$$\frac{44100 \times 588}{6} = 4,321,800$$

For a compact disc that can store 67 minutes of music, this results in

$$67 \times 60 \times 4321800 = 17,373,636,000$$

channel bits. The number of information bits (audio bits) is given by

$$67 \times 60 \times 44100 \times 192 = 5,673,024,000$$

This means that an expansion of roughly three times has taken place. These extra bits are used to protect the digital audio information bits from errors (the RS codes) and also make sure that the runlength constraint is satisfied (the EFM code). The efficient use of the RLL code makes it possible to store more than 17 billion channel bits with less than 2 billion pits.

On the playback, first synchronization and merger bits are separated, and then the 32 symbols are de-interleaved. The result then enters the decoder for the code C_2. This code has a minimum distance of 5 and therefore is capable of correcting up to two errors. The decoder, however, is designed to correct only one error. Then, according to the relation (10.5.37),

$$e_c + e_d \leq d_{min} - 1$$

it can detect up to three errors with certainty and four or more errors with high probability. If a single error is encountered, it is corrected; if multiple errors are detected, then all 28 symbols are flagged as "unreliable." After de-interleaving, these symbols are passed to the decoder for the code C_1. Decoder C_1 tries single error, or two erasure corrections. If it fails, all output symbols are flagged. If three or more flags are at its input, it copies them to its output.

At the output of the second decoder, the symbol corresponding to "unreliable" positions are filled in by interpolation of the other positions. Using this rather complex encoding–decoding technique together with the signal processing methods, burst errors of up to 12,000 data bits, which correspond to a track length of 7.5 mm on the disk, can be concealed.

10.10 FURTHER READING

The noisy channel coding theorem, which plays a central role in information theory, and the concept of channel capacity were first proposed and proved by Shannon (1948). For detailed discussion of this theorem and its variations, the reader may refer to standard books on information theory such as Gallager (1968), Blahut (1987), and Cover and Thomas (1992).

Golay (1949), Hamming (1950), Hocquenghem (1959), Bose and Ray-Chaudhuri (1960a,b), and Reed and Solomon (1960) are landmark papers in the development of block codes. Concatenated codes were introduced and analyzed by Forney (1966). Convolutional codes were introduced by Elias (1955), and various methods for their decoding were developed by Wozencraft and Reiffen (1961), Fano (1963), Zigangirov (1966), Viterbi (1967), and Jelinek (1969). Trellis coded modulation was introduced by Ungerboeck (1982) and later developed by Forney (1988a,b). The interested reader is referred to books on coding theory by Berlekamp (1968), Peterson and Weldon (1972), MacWilliams and Sloane (1977), Lin and Costello (1983), and Blahut (1984).

PROBLEMS

10.1 Find the capacity of the channel shown in Figure P-10.1.

10.2 The channel shown in Figure P-10.2 is known as the *binary erasure channel*. Find the capacity of this channel and plot it as a function of ϵ.

10.3 Find the capacity of the cascade connection of n binary-symmetric channels with the same crossover probability ϵ. What is the capacity when the number of channels goes to infinity?

10.4 The matrix whose elements are the transition probabilities of a channel, i.e., $p(y_i|x_j)$'s, is called the channel probability transition matrix. A channel is called *symmetric* if all rows of the channel probability transition matrix are

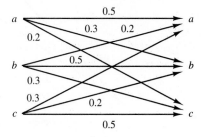

FIGURE P-10.1

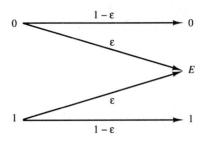

FIGURE P-10.2

permutations of each other, and all its columns are also permutations of each other. Show that in a symmetric channel the input probability distribution that achieves capacity is a uniform distribution. What is the capacity of this channel?

10.5 Channels 1, 2, and 3 are shown in Figure P-10.5.

    1. Find the capacity of channel 1. What input distribution achieves capacity?

    2. Find the capacity of channel 2. What input distribution achieves capacity?

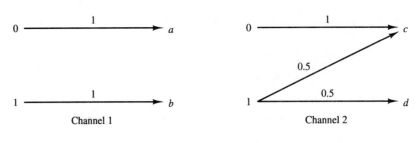

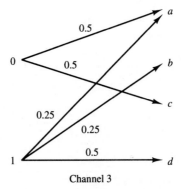

FIGURE P-10.5

3. Let C denote the capacity of the third channel and C_1 and C_2 represent the capacities of the first and second channel. Which of the following relations holds true and why?

   a. $C < \frac{1}{2}(C_1 + C_2)$.

   b. $C = \frac{1}{2}(C_1 + C_2)$.

   c. $C > \frac{1}{2}(C_1 + C_2)$.

10.6 Let C denote the capacity of a discrete memoryless channel with input alphabet $X = \{x_1, x_2, \ldots, x_N\}$ and output alphabet $\mathcal{Y} = \{y_1, y_2, \ldots, y_M\}$. Show that $C \leq \min\{\log M, \log N\}$.

10.7 The channel C is (known as the Z channel) shown in Figure P-10.7.

   1. Find the input probability distribution that achieves capacity.

   2. What is the input distribution and capacity for the special cases $\epsilon = 0$, $\epsilon = 1$, and $\epsilon = 0.5$?

   3. Show that if n such channels are cascaded, the resulting channel will be equivalent to a Z channel with $\epsilon_1 = \epsilon^n$.

   4. What is the capacity of the equivalent Z channel when $n \to \infty$.

10.8 Find the capacity of the channels A and B as shown in Figure P-10.8. What is the capacity of the cascade channel AB? (Hint: Look carefully at the channels, avoid lengthy math.)

10.9 Find the capacity of an additive white Gaussian noise channel with a bandwidth of 1 MHz, power of 10 W, and noise power-spectral density of $\frac{N_0}{2} = 10^{-9}$ W/Hz.

10.10 Channel C_1 is an additive white Gaussian noise channel with a bandwidth of W, transmitter power of P, and noise power-spectral density of $\frac{N_0}{2}$. Channel C_2 is an additive Gaussian noise channel with the same bandwidth and power as channel C_1 but with noise power-spectral density $S_n(f)$. It is further assumed that the total noise power for both channels is the same,

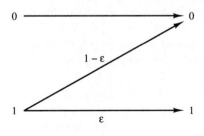

FIGURE P-10.7

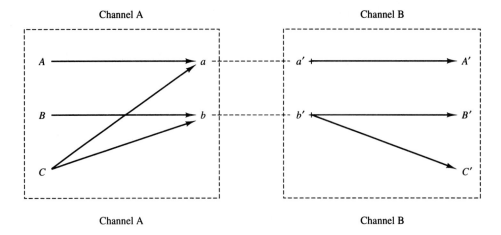

FIGURE P-10.8

that is

$$\int_{-W}^{W} S_n(f)\,df = \int_{-W}^{W} \frac{N_0}{2}\,df = N_0 W$$

Which channel do you think has a larger capacity? Give an intuitive reasoning.

10.11 A discrete time memoryless Gaussian source with mean 0 and variance σ^2 is to be transmitted over a binary-symmetric channel with crossover probability ϵ.

1. What is the minimum value of the distortion attainable at the destination (distortion is measured in mean-squared error)?
2. If the channel is a discrete time memoryless additive Gaussian noise channel with input power P and noise power σ_n^2, what is the minimum attainable distortion?
3. Now assume that the source has the same basic properties but is not memoryless. Do you expect that the distortion in transmission over the binary-symmetric channel to be decreased or increased? Why?

10.12 For the channel shown on page 813 find the channel capacity and the input distribution that achieves capacity.

10.13 Consider two discrete memoryless channels represented by $(X_1, p(y_1|x_1), \mathcal{Y}_1)$ and $(X_2, p(y_2|x_2), \mathcal{Y}_2)$ with corresponding (information) capacities C_1 and C_2. A new channel is defined by $(X_1 \times X_2, p(y_1|x_1)p(y_2|x_2), \mathcal{Y}_1 \times \mathcal{Y}_2)$. This channel is referred to as the "product channel" and models the case where $x_1 \in X_1$ and $x_2 \in X_2$ are simultaneously transmitted over the two channels

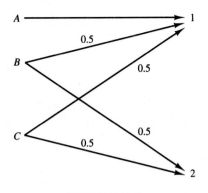

FIGURE P-10.12

with no interference. Prove that the capacity of this new channel is the sum of C_1 and C_2.

10.14 Let $(X_1, p(y_1|x_1), \mathcal{Y}_1)$ and $(X_2, p(y_2|x_2), \mathcal{Y}_2)$ represent two discrete memoryless communication channels with inputs X_i, outputs $\mathcal{Y}_i$, and conditional probabilities $p(y_i|x_i)$. Further assume that $X_1 \cap X_2 = \emptyset$ and $\mathcal{Y}_1 \cap \mathcal{Y}_2 = \emptyset$. We define the sum of these channels as a new channel with input alphabet $X_1 \cup X_2$, output alphabet $\mathcal{Y}_1 \cup \mathcal{Y}_2$, and conditional probability $p(y_i|x_i)$ where i denotes the index of X to which the input to the channel belongs. This models a communication situation where we have two channels in parallel, and at each transmission interval we can use one and only one of the channels, the input and output alphabets are disjoint, and therefore at the receiver there is no ambiguity which channel was being used.

1. Show that the capacity of the *sum* channel satisfies $2^C = 2^{C_1} + 2^{C_2}$ where C_1 and C_2 are the capacities of each channel.
2. Using the result of part 1 show that if $C_1 = C_2 = 0$, we still have $C = 1$; that is, using two channels with zero capacity, we are able to transmit one bit per transmission. How do you interpret this result?
3. Find the capacity of the channel shown in Figure P-10.14.

10.15 X is a binary memoryless source with $p(X = 0) = 0.3$. This source is transmitted over a binary-symmetric channel with crossover probability $\epsilon = 0.1$.

1. Assume that the source is directly connected to the channel, i.e., no coding is employed. What is the error probability at the destination?
2. If coding is allowed, what is the minimum possible error probability in reconstruction of the source.
3. For what values of ϵ is reliable transmission possible (with coding, of course)?

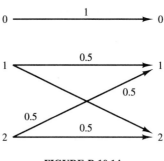

FIGURE P-10.14

10.16 Each sample of a Gaussian memoryless source has a variance equal to 4 and the source produces 8000 samples/sec. The source is to be transmitted via an additive white Gaussian noise channel with a bandwidth equal to 4000 Hz. It is desirable to have a distortion/sample not exceeding 1 at the destination (assume squared-error distortion).

1. What is the minimum required SNR of the channel?
2. If it is further assumed that, on the same channel, a binary PSK scheme is employed with hard-decision decoding, what will be the minimum required channel SNR?

(**Hint:**The SNR of the channel is defined by $\frac{P}{N_0 W}$.)

10.17 A certain source can be modeled as a stationary zero mean Gaussian process $X(t)$ with power-spectral density

$$S_x(f) = \begin{cases} 2, & |f| < 10 \\ 0, & \text{otherwise} \end{cases}$$

The distortion in reproducing $X(t)$ by $\hat{X}(t)$ is $D = E|X(t) - \hat{X}(t)|^2$. This source is to be transmitted over an additive Gaussian noise channel, in which the noise power-spectral density is given by

$$S_n(f) = \begin{cases} 1, & |f| < 4 \\ 0, & \text{otherwise} \end{cases}$$

1. What is $D_{\max}$ in the compression of the source?
2. Find the rate-distortion function for the source.
3. If we want to reproduce $X(t)$ with a distortion equal to 10, what transmission rate/sec is required?
4. Find the channel capacity-cost function, where cost is assumed to be the power. What is the required power such that the source can be transmitted via the channel with a distortion not exceeding 10?

10.18 It can be shown that the capacity of a discrete-time power constrained additive noise channel described by $Y = X + Z$, where X and Y are the input and the output and Z is the noise, satisfies the inequalities

$$\frac{1}{2} \log \left(1 + \frac{P}{N}\right) \leq C \leq \frac{1}{2} \log[2\pi e(P + N)] - h(Z)$$

where P is the input power constraint, N is the variance (power) of the noise process, and $h(\cdot)$ denotes the differential entropy as defined in Chapter 4. Using the result of Problem 4.36, plot the lower and upper bounds to the capacity for a channel with Laplacian noise (see Problem 4.36) as a function of the noise variance (noise power).

10.19 Plot the capacity of an AWGN channel that employs binary antipodal signaling, with optimal bit-by-bit detection at the receiver, as a function of $\mathcal{E}_b/N_0$. On the same axis plot the capacity of the same channel when binary orthogonal signaling is employed.

10.20 In Example 10.5.1 find the minimum distance of the code. Which code word(s) is(are) minimum weight?

10.21 In Example 10.5.3, verify that all code words of the original code satisfy

$$\mathbf{cH}^t = \mathbf{0}$$

10.22 By listing all code words of the $(7, 4)$ Hamming code, verify that its minimum distance is equal to 3.

10.23 Find the parity check matrix and the generator matrix of a $(15, 11)$ Hamming code in the systematic form.

10.24 Show that the minimum Hamming distance of a linear block code is equal to the minimum number of columns of its parity check matrix that are linearly dependent. From this conclude that the minimum Hamming distance of a Hamming code is always equal to 3.

10.25 A simple repetition code of blocklength n is a simple code consisting of only two code words one $\underbrace{(0, 0, \dots, 0)}_{n}$ and the other $\underbrace{(1, 1, \dots, 1)}_{n}$. Find the parity check matrix and the generator matrix of this code in the systematic form.

10.26 $\mathbf{G}$ is the generator matrix of a $(6, 3)$ linear code. This code is *extended* by adding an overall parity check bit to each code word so that the Hamming weight of each resulting code word is even.

$$\mathbf{G} = \begin{bmatrix} 1 & 0 & 0 & 1 & 1 & 0 \\ 0 & 1 & 0 & 1 & 0 & 1 \\ 0 & 0 & 1 & 0 & 1 & 1 \end{bmatrix}$$

    1. Find the parity check matrix of the extended code.

    2. What is the minimum distance of the extended code?

    3. Find the coding gain of the extended code?

10.27 Compare the block error probability of an uncoded system with a system that uses a $(15, 11)$ Hamming code. The transmission rate is $R = 10^4$ bits/sec and the channel is AWGN with a received power of 1 μW and noise power-spectral density of $\frac{N_0}{2}$. The modulation scheme is binary PSK and soft-decision decoding is employed. Answer the question when hard decision is employed.

10.28 Generate the standard array for a $(7,4)$ Hamming code and use it to decode the received sequence $(1, 1, 1, 0, 1, 0, 0)$.

10.29 For what values of k does a (n, k) cyclic code exist with $n = 6$? List all possible k's with corresponding generator polynomial(s).

10.30 Find a generator polynomial and the corresponding code words for a $(7, 3)$ cyclic code. What is the minimum distance of this code?

10.31 Design an encoder for a $(15, 11)$ cyclic code.

10.32 Using the generator polynomial $g(p) = 1 + p + p^4$ find the generator matrix and the parity check matrix (in systematic form) of a $(15,11)$ cyclic code.

10.33 Let $g(p) = p^8 + p^6 + p^4 + p^2 + 1$ denote a polynomial over the binary field.

    1. Find the lowest-rate cyclic code whose generator polynomial is $g(p)$. What is the rate of this code?

    2. Find the minimum distance of the code found in part 1.

    3. What is the coding gain for the code found in part 1?

10.34 The polynomial $g(p) = p + 1$ over the binary field is considered.

    1. Show that this polynomial can generate a cyclic code for any choice of n. Find the corresponding k.

    2. Find the systematic form of **G** and **H** for the code generated by $g(p)$.

    3. Can you say what type of code this generator polynomial generates?

10.35 Design a $(6, 2)$ cyclic code by choosing the shortest possible generator polynomial.

    1. Determine the generator matrix **G** (in the systematic form) for this code and find all possible code words.

    2. How many errors can be corrected by this code?

    3. If this code is used in conjunction with binary PSK over an AWGN channel with $P = 1$ W, $N_0 = 2 \times 10^{-6}$ W/Hz, and $W = 6 \times 10^4$ Hz and the information is transmitted at the maximum theoretically possible speed, while avoiding ISI, find the upper bound on the block error probability assuming the receiver employs a soft-decision scheme.

10.36 The Golay code is a cyclic $(23, 12)$ code with $d_{\min} = 7$. Show that if this code is used with interleaving/deinterleaving of depth 5, the resulting block of length 115 can correct single bursts of errors of length up to 15.

10.37 Let C_1 and C_2 denote two cyclic codes with the same block length n, with generator polynomials $g_1(p)$ and $g_2(p)$ and with minimum distances d_1 and d_2. Define $C_{\max} = C_1 \cup C_2$ and $C_{\min} = C_1 \cap C_2$.

1. Is $C_{\max}$ a cyclic code? Why? If yes, what are its generator polynomial and its minimum distance?

2. Is $C_{\min}$ a cyclic code? Why? If yes, find its generator polynomial. What can you say about its minimum distance?

10.38 In a coded communication system, M messages $1, 2, \ldots, M = 2^k$ are transmitted by M *baseband* signals $x_1(t), x_2(t), \ldots, x_M(t)$, each of duration nT. The general form of $x_i(t)$ is given by

$$x_i(t) = \sum_{j=0}^{n-1} \varphi_{ij}(t - jT)$$

where $\varphi_{ij}(t)$ can be either of the two signals $\phi_1(t)$ or $\phi_2(t)$, where $\phi_1(t) = \phi_2(t) \equiv 0$ for all $t \notin [0, T]$. We further assume that $\phi_1(t)$ and $\phi_2(t)$ have equal energy $\mathcal{E}$ and the channel is ideal (no attenuation) with additive white Gaussian noise of power-spectral density $\frac{N_0}{2}$. This means that the received signal is $r(t) = x(t) + n(t)$ where $x(t)$ is one of the $x_i(t)$'s and $n(t)$ represents the noise.

1. With $\phi_1(t) = -\phi_2(t)$, show that N, the dimensionality of the signal space, satisfies $N \leq n$.

2. Show that in general $N \leq 2n$.

3. With $M = 2$, show that for general $\phi_1(t)$ and $\phi_2(t)$

$$p(\text{error}|x_1(t) \text{ sent}) \leq \underset{R^N}{\int \cdots \int} \sqrt{p(\mathbf{r}|\mathbf{x}_1)p(\mathbf{r}|\mathbf{x}_2)} \, d\mathbf{r}$$

where $\mathbf{r}$, $\mathbf{x}_1$, and $\mathbf{x}_2$ are the vector representations of $r(t)$, $x_1(t)$, and $x_2(t)$ in the N-dimensional space.

4. Using the result of part 3, show that for general M,

$$p(\text{error}|x_m(t) \text{ sent}) \leq \sum_{\substack{1 \leq m' \leq M \\ m' \neq m}} \underset{R^N}{\int \cdots \int} \sqrt{p(\mathbf{r}|\mathbf{x}_m)p(\mathbf{r}|\mathbf{x}_{m'})} \, d\mathbf{r}$$

5. Show that

$$\int_{R^N} \cdots \int \sqrt{p(\mathbf{r}|\mathbf{x}_m) p(\mathbf{r}|\mathbf{x}_{m'})} \, d\mathbf{r} = \exp\left(-\frac{|\mathbf{x}_m - \mathbf{x}_{m'}|^2}{4N_0}\right)$$

and therefore

$$p(\text{error}|x_m(t) \text{ sent}) \leq \sum_{\substack{1 \leq m' \leq M \\ m' \neq m}} \exp\left(-\frac{|\mathbf{x}_m - \mathbf{x}_{m'}|^2}{4N_0}\right)$$

10.39 A convolutional code is described by

$$\mathbf{g}_1 = [1\ 0\ 0]$$
$$\mathbf{g}_2 = [1\ 0\ 1]$$
$$\mathbf{g}_3 = [1\ 1\ 1]$$

1. Draw the encoder corresponding to this code.
2. Draw the state-transition diagram for this code.
3. Draw the trellis diagram for this code.
4. Find the transfer function and the free distance of this code.
5. Verify whether this code is catastrophic or not.

10.40 The block diagram of a binary convolutional code is shown in Figure P-10.40.

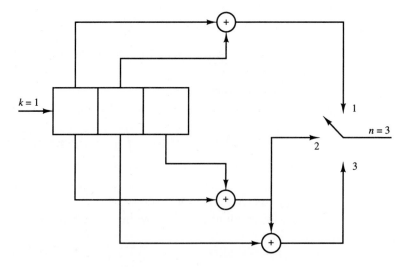

FIGURE P-10.40

1. Draw the state diagram for the code.

2. Find $T(D)$, the transfer function of the code.

3. What is d_{free}, the minimum free distance of the code?

4. Assume that a message has been encoded by this code and transmitted over a binary-symmetric channel with an error probability of $p = 10^{-5}$. If the received sequence is $\mathbf{r} = (110, 110, 110, 111, 010, 101, 101)$, using the Viterbi algorithm find the transmitted bit sequence.

5. Find an upper bound to bit error probability of the code when the above binary-symmetric channel is employed. Make any reasonable approximations.

10.41 The block diagram of a $(3, 1)$ convolutional code is shown in Figure P-10.41.

1. Draw the state diagram of the code.

2. Find the transfer function $T(D)$ of the code.

3. Find the minimum free distance (d_{free}) of the code and show the corresponding path (at distance d_{free} from the all-zero code word) on the trellis.

4. Assume that four information bits (x_1, x_2, x_3, x_4), followed by two zero bits, have been encoded and sent via a binary-symmetric channel with crossover probability equal to 0.1. The received sequence is $(111, 111, 111, 111, 111, 111)$. Use the Viterbi decoding algorithm to find the most likely data sequence.

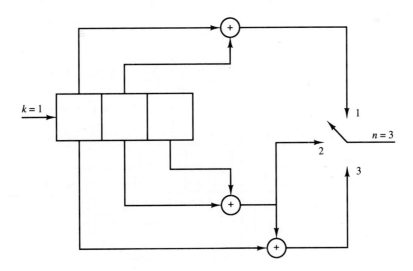

FIGURE P-10.41

10.42 The convolutional code of Problem 10.39 is used for transmission over a AWGN channel with hard-decision decoding. The output of the demodulator detector is (101001011110111). Using the Viterbi algorithm find the transmitted sequence.

10.43 Repeat Problem 10.39 for a code with

$$\mathbf{g}_1 = [1\ 1\ 0]$$

$$\mathbf{g}_2 = [1\ 0\ 1]$$

$$\mathbf{g}_3 = [1\ 1\ 1]$$

10.44 Show the paths corresponding to all code words of weight 6 in Example 10.7.3.

10.45 In the convolutional code generated by the encoder shown in Figure P-10.45:

1. Find the transfer function of the code in the form $T(N, D)$.
2. Find d_{free} of the code.
3. If the code is used on a channel using hard-decision Viterbi decoding, assuming the crossover probability of the channel is $p = 10^{-6}$, use the hard decision bound to find an upper bound on the average bit error probability of the code.

10.46 Let $\mathbf{x}_1$ and $\mathbf{x}_2$ be two code words of length n with distance d and assume that these two code words are transmitted via a binary-symmetric channel with crossover probability p. Let $P(d)$ denote the error probability in transmission of these two code words.

1. Show that

$$P(d) \leq \sum_{i=1}^{2^n} \sqrt{p(\mathbf{y}_i|\mathbf{x}_1)p(\mathbf{y}_i|\mathbf{x}_2)}$$

where the summation is over all binary sequences $\mathbf{y}_i$.

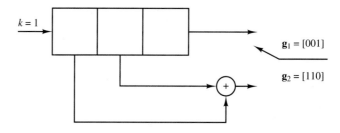

FIGURE P-10.45

2. From above conclude that

$$P(d) \le \left[4p(1-p)\right]^{\frac{d}{2}}$$

10.47 The complementary error function erfc(x) is defined by

$$\text{erfc}(x) = \frac{2}{\sqrt{\pi}} \int_x^\infty e^{-t^2}\, dt$$

1. Express $Q(x)$ in terms of erfc(x).
2. Using the inequality

$$\text{erfc}\left(\sqrt{x+y}\right) \le \text{erfc}\left(\sqrt{x}\right) e^{-y} \qquad x \ge 0,\; y \ge 0$$

prove the following bound on the average bit error probability of a convolutional code, assuming that soft-decision decoding is employed.

$$P_b \le \frac{1}{2k} \text{erfc}\left(\sqrt{d_{\text{free}} R_c \rho_b}\right) e^{d_{\text{free}} R_c \rho_b} \left. \frac{\partial T(D,N)}{\partial N} \right|_{N=1, D=e^{-R_c \rho_b}}$$

where ρ_b is the SNR per bit.

10.48 A trellis coded modulation system uses an 8-ary PAM signal set given by $\{\pm 1, \pm 3, \pm 5, \pm 7\}$ and the 4-state trellis encoder shown in Figure 10.39(a).

1. Using the set partitioning rules, partition the signal set into four subsets.
2. If the channel is additive white Gaussian noise and at the output of the matched filter the sequence $(-.2, 1.1, 6, 4, -3, -4.8, 3.3)$ is observed, what is the most likely transmitted sequence?

11

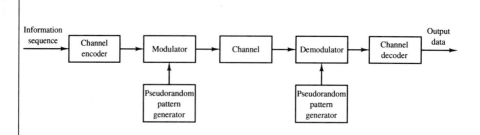

Spread-Spectrum
Communication Systems

In our treatment of signal design for digital communication over an AWGN channel, the major objective has been the efficient utilization of transmitter power and channel bandwidth. As we observed in the preceding chapter, channel coding allows us to reduce the transmitter power by increasing the transmitted signal bandwidth through code redundancy and, thus, to trade off transmitter power with channel bandwidth. This is the basic methodology for the design of digital communication systems for AWGN channels.

In practice, one often encounters other factors that influence the design of an efficient digital communication system. For example, in the design of a military communication system, the system designer is usually concerned with the actions that an adversary may take to disrupt normal communications. Intentional jamming of the transmitted signal is a common type of adversarial action. In such a case, the system designer must design a communication system that contains some built-in immunity to the presence of jamming signals. A similar problem arises in multiple-access communications when two or more transmitters use the same common channel to transmit information. In such a case, the interference created by other users of the channel is unintentional. Nevertheless, the system designer must take into account the existence of such interference in the design of a reliable digital communication system.

Even in these more complex design problems cited above, the basic system design parameters are transmitter power and channel bandwidth. To overcome the problems of intentional or unintentional interference, we may further increase the bandwidth of the transmitted signal, as described below, so that the bandwidth expansion factor $B_e = W/R$ is much greater than unity. This is one characteristic of a *spread-spectrum signal*. A second important characteristic is that the information signal at the modulator is spread in bandwidth by means of a code that is independent of the information sequence. This code has the property of being *pseudorandom*, i.e., it appears to be random to receivers other than the intended receiver that uses the knowledge of the code to demodulate the signal. It is this second characteristic property that distinguishes a spread spectrum communication system from the conventional communication system that expands the transmitted signal bandwidth by means of channel code redundancy. However, we should emphasize that channel coding is an important element in the design of an efficient spread spectrum communication system.

Spread-spectrum signals for digital communications were originally developed and used for military communications either (1) to provide resistance to jamming (antijam protection), or (2) to hide the signal by transmitting it at low power and, thus, making it difficult for an unintended listener to detect its presence in noise (low probability of intercept). However, spread-spectrum signals may also be used to provide reliable communications in a variety of civilian applications, including mobile vehicular communications and interoffice communications.

In this chapter we present the basic characteristics of spread-spectrum signals and assess their performance in terms of probability of error. We concentrate our discussion on two methods for spreading the signal bandwidth, namely, by direct sequence modulation and by frequency hopping. Both methods require the use of pseudorandom code sequences whose generation is also described. Several applications of spread-spectrum signals are presented.

11.1 MODEL OF A SPREAD-SPECTRUM DIGITAL COMMUNICATION SYSTEMS

The basic elements of a spread spectrum digital communication system are illustrated in Figure 11.1. We observe that the channel encoder and decoder and the modulator and demodulator are the basic elements of a conventional digital communication system. In addition to these elements, a spread-spectrum system employs two identical pseudorandom sequence generators, one which interfaces with the modulator at the transmitting end and the second which interfaces with the demodulator at the receiving end. These two generators produce a pseudorandom or pseudonoise (PN) binary-valued sequence that is used to spread the transmitted

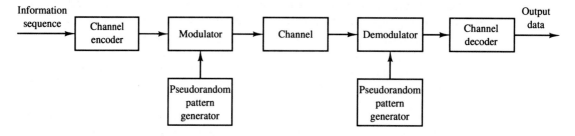

FIGURE 11.1.   Model of spread-spectrum digital communications system.

signal in frequency at the modulator and to despread the received signal at the demodulator.

Time synchronization of the PN sequence generated at the receiver with the PN sequence contained in the received signal is required to properly despread the received spread-spectrum signal. In a practical system, synchronization is established prior to the transmission of information by transmitting a fixed PN bit pattern that is designed so that the receiver will detect it with high probability in the presence of interference. After time synchronization of the PN sequence generators is established, the transmission of information commences. In the data mode, the communication system usually tracks the timing of the incoming received signal and keeps the PN sequence generator in synchronism. The synchronization of the PN sequence generators is treated in Section 11.5.

Interference is introduced in the transmission of the spread-spectrum signal through the channel. The characteristics of the interference depend to a large extent on its origin. The interference may be generally categorized as being either broadband or narrowband (partial band) relative to the bandwidth of the information-bearing signal, and either continuous in time or pulsed (discontinuous) in time. For example, a jamming signal may consist of a high-power sinusoid in the bandwidth occupied by the information-bearing signal; such a jamming signal is narrowband. As a second example, the interference generated by other users in a multiple-access channel depends on the type of spread-spectrum signals that are employed by the various users to transmit their information. If all users employ broadband signals, the interference may be characterized as an equivalent broadband noise. If the users employ frequency hopping to generate spread-spectrum signals, the interference from other users may be characterized as narrowband. We shall consider these types of interference and some others in the following sections.

Our discussion will focus on the performance of spread-spectrum signals for digital communication in the presence of narrowband and broadband interference. Two types of digital modulation are considered, namely, PSK and FSK. PSK modulation is appropriate for applications where phase coherence between the transmitted signal and the received signal can be maintained over a time interval that spans several symbol (or bit) intervals. On the other hand, FSK modulation is appropriate in applications where phase coherence of the carrier cannot be maintained due to time variations in the transmission characteristics of the communications channel. For example, this may be the case in a communications link between two high-speed aircraft or between a high-speed aircraft and a ground-based terminal.

The PN sequence generated at the modulator is used in conjunction with the PSK modulation to shift the phase of the PSK signal pseudorandomly, as described below, at a rate that is an integer multiple of the bit rate. The resulting modulated signal is called a *direct-sequence* (DS) *spread-spectrum signal*. When used in conjunction with binary or M-ary ($M > 2$) FSK, the PN sequence is used to select the frequency of the transmitted signal pseudorandomly. The resulting signal is called a *frequency-hopped* (FH) *spread-spectrum signal*. Although a number of other types of spread-spectrum signals will be described briefly, our treatment will emphasize DS and FH spread-spectrum communication systems.

11.2 DIRECT-SEQUENCE SPREAD SPECTRUM SYSTEMS

Let us consider the transmission of a binary information sequence by means of binary PSK. The information rate is R bits per second and the bit interval is $T_b = 1/R$ seconds. The available channel bandwidth is B_c Hz, where $B_c \gg R$. At the modulator the bandwidth of the information signal is expanded to $W = B_c$ Hz by shifting the phase of the carrier pseudorandomly at a rate of W times per second according to the pattern of the PN generator. The basic method for accomplishing the spreading is shown in Figure 11.2.

The information-bearing baseband signal is denoted as $v(t)$ and is expressed as

$$v(t) = \sum_{n=-\infty}^{\infty} a_n g_T(t - nT_b) \tag{11.2.1}$$

where $\{a_n = \pm 1, -\infty < n < \infty\}$ and $g_T(t)$ is a rectangular pulse of duration T_b. This signal is multiplied by the signal from the PN sequence generator, which may be expressed as

$$c(t) = \sum_{n=-\infty}^{\infty} c_n p(t - nT_c) \tag{11.2.2}$$

where $\{c_n\}$ represents the binary PN code sequence of ± 1's and $p(t)$ is a rectangular pulse of duration T_c, as illustrated in Figure 11.2. This multiplication operation serves to spread the bandwidth of the information-bearing signal (whose bandwidth is R_b Hz, approximately) into the wider bandwidth occupied by PN generator signal $c(t)$ (whose bandwidth is $1/T_c$, approximately). The spectrum spreading is illustrated in Figure 11.3, which shows, in simple terms, using rectangular spectra, the convolution of the two spectra, the narrow spectrum corresponding to the information-bearing signal and the wide spectrum corresponding to the signal from the PN generator.

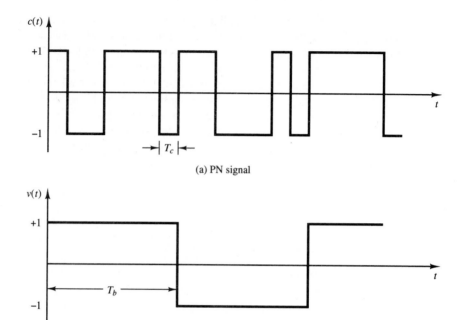

(a) PN signal

(b) Data signal

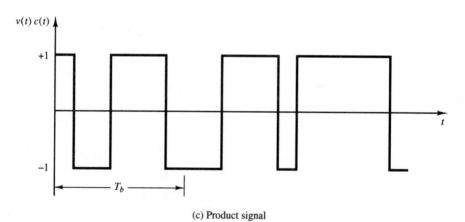

(c) Product signal

FIGURE 11.2. Generation of a DS spread-spectrum signal.

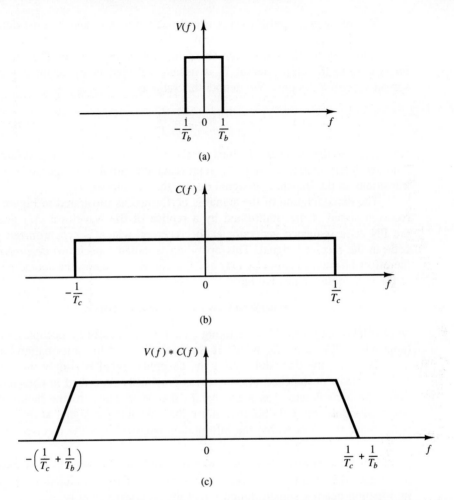

FIGURE 11.3. Convolution of spectra of the (a) data signal with the (b) PN code signal.

The product signal $v(t)c(t)$, also illustrated in Figure 11.2, is used to amplitude modulate the carrier $A_c \cos 2\pi f_c t$ and, thus, to generate the DSB-SC signal

$$u(t) = A_c v(t) c(t) \cos 2\pi f_c t \qquad (11.2.3)$$

Since $v(t)c(t) = \pm 1$ for any t, it follows that the carrier-modulated transmitted signal may also be expressed as

$$u(t) = A_c \cos[2\pi f_c t + \theta(t)] \qquad (11.2.4)$$

where $\theta(t) = 0$ when $v(t)c(T) = 1$ and $\theta(t) = \pi$ when $v(t)c(t) = -1$. Therefore, the transmitted signal is a binary PSK signal.

The rectangular pulse $p(t)$ is usually called a *chip*, and its time duration T_c is called the *chip interval*. The reciprocal $1/T_c$ is called the *chip rate* and corresponds (approximately) to the bandwidth W of the transmitted signal. The ratio of the bit interval T_b to the chip interval T_c is usually selected to be an integer in practical spread-spectrum systems. We denote this ratio as

$$L_c = \frac{T_b}{T_c} \tag{11.2.5}$$

Hence, L_c is the number of chips of the PN code sequence per information bit. Another interpretation is that L_c represents the number of possible $180°$ phase transitions in the transmitted signal during the bit interval T_b.

The demodulation of the signal is performed as illustrated in Figure 11.4. The received signal is first multiplied by a replica of the waveform $c(t)$ generated by the PN code sequence generator at the receiver, which is synchronized to the PN code in the received signal. This operation is called (spectrum) *despreading*, since the effect of multiplication by $c(t)$ at the receiver is to undo the spreading operation at the transmitter. Thus, we have

$$A_c v(t) c^2(t) \cos 2\pi f_c t = A_c v(t) \cos 2\pi f_c t \tag{11.2.6}$$

since $c^2(t) = 1$ for all t. The resulting signal $A_c v(t) \cos 2\pi f_c t$ occupies a bandwidth (approximately) of R Hz, which is the bandwidth of the information-bearing signal. Therefore, the demodulator for the despread signal is simply the conventional crosscorrelator or matched filter that was previously described in Chapters 7 and 9. Since the demodulator has a bandwidth that is identical to the bandwidth of the despread signal, the only additive noise that corrupts the signal at the demodulator is the noise that falls within the information bandwidth of the received signal.

Effect of despreading on a narrowband interference. It is interesting to investigate the effect of an interfering signal on the demodulation of the desired information-bearing signal. Suppose that the received signal is

$$r(t) = A_c v(t) c(t) \cos 2\pi f_c t + i(t) \tag{11.2.7}$$

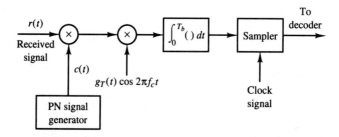

FIGURE 11.4. Demodulation of DS spread-spectrum signal.

where $i(t)$ denotes the interference. The despreading operation at the receiver yields

$$r(t)c(t) = A_c v(t) \cos 2\pi f_c t + i(t)c(t) \tag{11.2.8}$$

The effect of multiplying the interference $i(t)$ with $c(t)$ is to spread the bandwidth of $i(t)$ to W Hz.

As an example, let us consider a sinusoidal interfering signal of the form

$$i(t) = A_J \cos 2\pi f_J t, \tag{11.2.9}$$

where f_J is a frequency within the bandwidth of the transmitted signal. Its multiplication with $c(t)$ results in a wideband interference with power-spectral density $J_0 = P_J / W$, where $P_J = A_J^2 / 2$ is the average power of the interference. Since the desired signal is demodulated by a matched filter (or correlator) that has a bandwidth R, the total power in the interference at the output of the demodulator is

$$J_0 R = P_J R / W = \frac{P_J}{W/R} = \frac{P_J}{T_b / T_c} = \frac{P_J}{L_c} \tag{11.2.10}$$

Therefore, the power in the interfering signal is reduced by an amount equal to the bandwidth expansion factor W/R. The factor $W/R = T_b/T_c = L_c$ is called the *processing gain* of the spread-spectrum system. The reduction in interference power is the basic reason for using spread-spectrum signals to transmit digital information over channels with interference.

In summary, the PN code sequence is used at the transmitter to spread the information-bearing signal into a wide bandwidth for transmission over the channel. By multiplying the received signal with a synchronized replica of the PN code signal, the desired signal is despread back to a narrow bandwidth while any interference signals are spread over a wide bandwidth. The net effect is a reduction in the interference power by the factor W/R, which is the processing gain of the spread-spectrum system.

The PN code sequence $\{c_n\}$ is assumed to be known only to the intended receiver. Any other receiver that does not have knowledge of the PN code sequence cannot demodulate the signal. Consequently, the use of a PN code sequence provides a degree of privacy (or security) that is not possible to achieve with conventional modulation. The primary cost for this security and performance gain against interference is an increase in channel bandwidth utilization and in the complexity of the communication system.

11.2.1 Probability of Error

To derive the probability of error for a DS spread-spectrum system, we assume that the information is transmitted via binary PSK. Within the bit interval $0 \le t \le T_b$, the transmitted signal is

$$s(t) = a_0 g_T(t)c(t) \cos 2\pi f_c t, \qquad 0 \le t \le T_b \tag{11.2.11}$$

where $a_o = \pm 1$ is the information symbol, the pulse $g_T(t)$ is defined as

$$
g_T(t) = \begin{cases} \sqrt{\dfrac{2\mathcal{E}_b}{T_b}}, & 0 \le t \le T_b \\ 0, & \text{otherwise} \end{cases} \tag{11.2.12}
$$

and $c(t)$ is the output of the PN code generator that, over a bit interval, is expressed as

$$
c(t) = \sum_{n=0}^{L_c-1} c_n p(t - nT_c) \tag{11.2.13}
$$

where L_c is the number of chips per bit, T_c is the chip interval, and $\{c_n\}$ denotes the PN code sequence. The code chip sequence $\{c_n\}$ is uncorrelated (white), i.e.,

$$
E[c_n c_m] = E[c_n]E[c_m] \quad \text{for } n \ne m \tag{11.2.14}
$$

and each chip is $+1$ or -1 with equal probability. These conditions imply that $E[c_n] = 0$ and $E[c_n^2] = 1$.

The received signal is assumed to be corrupted by an additive interfering signal $i(t)$. Hence,

$$
r(t) = a_o g_T(t - t_d)c(t - t_d)\cos(2\pi f_c t + \phi) + i(t) \tag{11.2.15}
$$

where t_d represents the propagation delay through the channel and ϕ represents the carrier phase shift. Since the received signal $r(t)$ is the output of an ideal bandpass filter in the front end of the receiver, the interference $i(t)$ is also a bandpass signal and may be represented as

$$
i(t) = i_c(t)\cos 2\pi f_c t - i_s(t)\sin 2\pi f_c t \tag{11.2.16}
$$

where $i_c(t)$ and $i_s(t)$ are the two quadrature components.

Assuming that the receiver is perfectly synchronized to the received signal, we may set $t_d = 0$ for convenience. In addition, the carrier phase is assumed to be perfectly estimated by a PLL. Then, the signal $r(t)$ is demodulated by first despreading through multiplication by $c(t)$ and then crosscorrelation with $g_T(t)\cos(2\pi f_c t + \phi)$, as shown in Figure 11.5. At the sampling instant $t = T_b$, the output of the correlator is

$$
y(T_b) = \mathcal{E}_b + y_i(T_b) \tag{11.2.17}
$$

where $y_i(T_b)$ represents the interference component, which has the form

$$
y_i(T_b) = \int_0^{T_b} c(t)i(t)g_T(t)\cos(2\pi f_c t + \phi)\,dt
$$

$$
= \sum_{n=0}^{L_c-1} c_n \int_0^{T_b} p(t - nT_c)i(t)g_T(t)\cos(2\pi f_c t + \phi)\,dt
$$

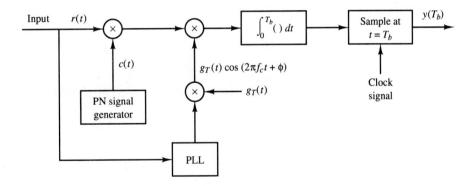

FIGURE 11.5. DS spread-spectrum signal demodulator.

$$= \sqrt{\frac{2\mathcal{E}_b}{T_b}} \sum_{n=0}^{L_c-1} c_n v_n \qquad (11.2.18)$$

where, by definition,

$$v_n = \int_{nT_c}^{(n+1)T_c} i(t)\cos(2\pi f_c t + \phi)dt \qquad (11.2.19)$$

The probability of error depends on the statistical characteristics of the interference component. Clearly, its mean value is

$$E[y_i(T_b)] = 0 \qquad (11.2.20)$$

Its variance is

$$E[y_i^2(T_b)] = \frac{2\mathcal{E}_b}{T_b} \sum_{n=0}^{L_c-1} \sum_{m=0}^{L_c-1} E[c_n c_m] E[v_n v_m]$$

But $E[c_n c_m] = \delta_{mn}$. Therefore,

$$E[y_i^2(T_b)] = \frac{2\mathcal{E}_b}{T_b} \sum_{n=0}^{L_c-1} E(v_n^2)$$

$$= \frac{2\mathcal{E}_b}{T_b} L_c E(v^2) \qquad (11.2.21)$$

where $v = v_n$, as given by (11.2.19). To determine the variance of v, we must postulate the form of the interference.

First, let us assume that the interference is sinusoidal. Specifically, we assume that the interference is at the carrier frequency and has the form

$$i(t) = \sqrt{2P_J} \cos(2\pi f_c t + \Theta_J) \qquad (11.2.22)$$

where P_J is the average power and Θ_J is the phase of the interference, which we assume to be random and uniformly distributed over the interval $(0, 2\pi)$. If we substitute for $i(t)$ in (11.2.19), we obtain

$$v_n = \int_{nT_c}^{(n+1)T_c} \sqrt{2P_J} \cos(2\pi f_c t + \Theta_J) \cos(2\pi f_c t + \phi)dt$$

$$= \frac{1}{2}\sqrt{2P_J} \int_{nT_c}^{(n+1)T_c} \cos(\Theta_J - \phi)dt = \frac{T_c}{2}\sqrt{2P_J}\cos(\Theta_J - \phi) \quad (11.2.23)$$

Since Θ_J is a random variable, v_n is also random. Its mean value is zero, i.e.,

$$E[v_n] = \frac{T_c}{2}\sqrt{2P_J} \int_0^{2\pi} \frac{1}{2\pi} \cos(\Theta_j - \phi)d\Theta_J = 0 \quad (11.2.24)$$

Its mean square value is

$$E(v_n^2) = \frac{T_c^2 P_J}{2} \frac{1}{2\pi} \int_0^{2\pi} \cos^2(\Theta_j - \phi)d\Theta_J$$

$$= \frac{T_c^2 P_J}{4} \quad (11.2.25)$$

We may now substitute for $E(v^2)$ into (11.2.21). Thus, we obtain

$$E\left[y_i^2(T_b)\right] = \frac{\mathcal{E}_b P_J T_c}{2} \quad (11.2.26)$$

The ratio of $\{E[y(T_b)]\}^2$ to $E\left[y_i^2(T_b)\right]$ is the SNR at the detector. In this case we have

$$(\text{SNR})_D = \frac{\mathcal{E}_b^2}{\mathcal{E}_b P_J T_c/2} = \frac{2\mathcal{E}_b}{P_J T_c} \quad (11.2.27)$$

To see the effect of the spread-spectrum signal, we express the transmitted energy $\mathcal{E}_b$ as

$$\mathcal{E}_b = P_S T_b \quad (11.2.28)$$

where P_S is the average signal power. Then, if we substitute for $\mathcal{E}_b$ in (11.2.27), we obtain

$$(\text{SNR})_D = \frac{2P_S T_b}{P_J T_c} = \frac{2P_S}{P_J/L_c} \quad (11.2.29)$$

where $L_c = T_b/T_c$ is the processing gain. Therefore, the spread-spectrum signal has reduced the power of the interference by the factor L_c.

Another interpretation of the effect of the spread-spectrum signal on the sinusoidal interference is obtained if we express $P_J T_c$ in (11.2.29) as follows. Since $T_c \simeq 1/W$, we have

$$P_J T_c = P_J/W \equiv J_0 \quad (11.2.30)$$

where J_0 is the power-spectral density of an equivalent interference in a bandwidth W. Therefore, in effect, the spread-spectrum signal has spread the sinusoidal interference over the wide bandwidth W, creating an equivalent spectrally flat noise with power-spectral density J_0. Hence,

$$(\text{SNR})_D = \frac{2\mathcal{E}_b}{J_0} \tag{11.2.31}$$

Example 11.2.1

The SNR required at the detector to achieve reliable communication in a DS spread-spectrum communication system is 13 dB. If the jammer-to-signal power at the receiver is 20 dB, determine the processing gain required to achieve reliable communication.

Solution We are given $(P_J/P_S)_{\text{dB}} = 20$ dB or, equivalently, $P_J/P_S = 100$. We are also given $(\text{SNR})_D = 13$ dB, or equivalently, $(\text{SNR})_D = 20$. The relation in (10.2.29) may be used to solve for L_c. Thus,

$$L_c = \frac{1}{2}\left(\frac{P_J}{P_S}\right)(\text{SNR})_D = 1000$$

Therefore, the processing gain required is 1000 or, equivalently, 30 dB.

As a second case, let us consider the effect of an interference $i(t)$ that is a zero-mean broadband random process with a constant power-spectral density over the bandwidth W of the spread-spectrum signal, as illustrated in Figure 11.6. Note that the total interference power is

$$P_J = \int_{-\infty}^{\infty} S_{ii}(f)df = W J_0 \tag{11.2.32}$$

The variance of the interference component at the input to the detector is given by (11.2.21). To evaluate the moment $E(v^2)$, we substitute the bandpass representation of $i(t)$, given by (11.2.16), into (11.2.19). By neglecting the double frequency terms (terms involving $\cos 4\pi f_c t$) and making use of the statistical properties of the quadrature components, namely $E[i_c(t)] = E(i_s(t)] = 0$ and

$$R_{i_c}(\tau) = E[i_c(t)i_c(t+\tau)] = E[i_s(t)i_s(t+\tau)]$$

$$= J_0 \frac{\sin \pi W \tau}{\pi \tau} \tag{11.2.33}$$

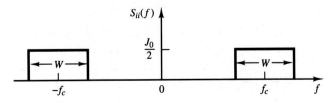

FIGURE 11.6. Power-spectral density of broadband interference.

we obtain the result

$$E(v^2) = \frac{1}{4} \int_0^{T_c} \int_0^{T_c} R_{i_c}(t_1 - t_2) dt_1 dt_2 \qquad (11.2.34)$$

This is an integral of the autocorrelation function over the square defined by the region $0 \le t_1 \le T_c$ and $0 \le t_2 \le T_c$, as shown in Figure 11.7. If we let $\tau = t_1 - t_2$, $E(v^2)$ can be reduced to the single integral (see Figure 11.7),

$$
\begin{aligned}
E(v^2) &= \frac{1}{4} \int_{-T_c}^{T_c} (T_c - |\tau|) R_{i_c}(\tau)\, dt \\
&= \frac{J_0 T_c}{4} \int_{-T_c}^{T_c} \left(1 - \frac{|\tau|}{T_c}\right) \frac{\sin \pi W \tau}{\pi \tau}\, d\tau \\
&= \frac{P_J T_c}{4W} \int_{-T_c}^{T_c} \left(1 - \frac{|\tau|}{T_c}\right) \frac{\sin \pi W \tau}{\pi \tau}\, d\tau \qquad (11.2.35)
\end{aligned}
$$

Since $T_c \approx \alpha/W$, where $\alpha \ge 1$, the above integral may be expressed as

$$I(\alpha) = 2 \int_0^{\alpha} (1 - \frac{x}{\alpha}) \frac{\sin \pi x}{x}\, dx \qquad (11.2.36)$$

and it can be numerically evaluated for any value of α. Figure 11.8 illustrates $I(\alpha)$. Note that $I(\alpha) \le 1$ for any value of α and that $I(\alpha) \to 1$ as $\alpha \to \infty$.

By combining the results in (11.2.21), (11.2.35), and (11.2.36), we conclude that the variance of the interference component $y_i(T_b)$ for the broadband interference

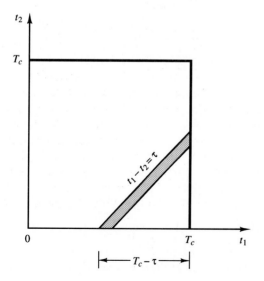

FIGURE 11.7. Region of integration of the autocorrelation function $R_{cc}(t_1, t_2)$.

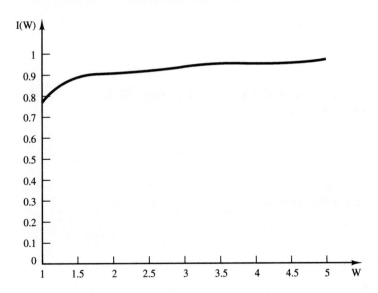

FIGURE 11.8.   Plot of the function $I(\alpha)$ given in (11.2.36).

is

$$E\left[y_i^2(T_b)\right] = \frac{\mathcal{E}_b P_J T_c}{2} I(\alpha) \qquad (11.2.37)$$

Therefore, the SNR at the detector is

$$(SNR)_D = \frac{2\mathcal{E}_b}{J_0 I(\alpha)} \qquad (11.2.38)$$

If we compare (11.2.31) with (11.2.38), we observe that the SNR for the case of the broadband interference is larger due to the factor $I(\alpha)$. Hence, the sinusoidal interference results in a somewhat larger degradation in the performance of the DS spread-spectrum system compared to that of a broadband interference.

The probability of error for a DS spread-spectrum system with binary PSK modulation is easily obtained from the SNR at the detector, if we make an assumption on the probability distribution of the sample $y_i(T_b)$. From (11.2.18) we note that $y_i(T_b)$ consists of a sum L_c of uncorrelated random variables $\{c_n v_n, \quad 0 \le n \le L_c - 1\}$, all of which are identically distributed. Since the processing gain L_c is usually large in any practical system, we may use the Central Limit Theorem to justify a Gaussian probability distribution for $y_i(T)$. Under this assumption, the probability of error for the sinusoidal interference is

$$P_2 = Q\left(\sqrt{\frac{2\mathcal{E}_b}{J_0}}\right) \qquad (11.2.39)$$

where J_0 is the power-spectral density of an equivalent broadband interference. A similar expression holds for the case of a broadband jamming signal, where the SNR at the detector is increased by the factor $1/I(\alpha)$.

The jamming margin. When the interference signal is a jamming signal, we may express $\frac{\mathcal{E}_b}{J_0}$ as

$$\frac{\mathcal{E}_b}{J_0} = \frac{P_S T_b}{P_J/W} = \frac{P_S/R}{P_J/W} = \frac{W/R}{P_J/P_S} \qquad (11.2.40)$$

Now, suppose we specify a required $\mathcal{E}_b/J_0$ to achieve a desired performance. Then, using a logarithmic scale, we may express (11.2.40) as

$$10\log\frac{P_J}{P_S} = 10\log\frac{W}{R} - 10\log\left(\frac{\mathcal{E}_b}{J_0}\right)$$

$$\left(\frac{P_J}{P_S}\right)_{dB} = \left(\frac{W}{R}\right)_{dB} - \left(\frac{\mathcal{E}_b}{J_0}\right)_{dB} \qquad (11.2.41)$$

The ratio $(P_J/P_S)_{dB}$ is called the *jamming margin*. This is the relative power advantage that a jammer may have without disrupting the communication system.

Example 11.2.2

Suppose we require an $(\mathcal{E}_b/J_0)_{dB} = 10$ dB to achieve reliable communication. What is the processing gain that is necessary to provide a jamming margin of 20 dB?

Solution   Clearly, if $W/R = 1000$, then $(W/R)_{dB} = 30$ dB and the jamming margin is $(P_J/P_S)_{dB} = 20$ dB. This means that the average jamming power at the receiver may be 100 times the power P_S of the desired signal and we can still maintain reliable communication.

Performance of coded spread-spectrum signals. As shown in Chapter 10, when the transmitted information is coded by a binary linear (block or convolutional) code, the SNR at the output of a soft-decision decoder is increased by the coding gain, defined as

$$\text{coding gain} = R_c d_{min}^H \qquad (11.2.42)$$

where R_c is the code rate and d_{min}^H is the minimum Hamming distance of the code. Therefore, the effect of coding is to increase the jamming margin by the coding gain. Thus, (10.2.41) may be modified as

$$\left(\frac{P_J}{P_S}\right)_{dB} = \left(\frac{W}{R}\right)_{dB} + (CG)_{dB} - \left(\frac{\mathcal{E}_b}{J_0}\right)_{dB} \qquad (11.2.43)$$

11.2.2 Some Applications of DS Spread-Spectrum Signals

In this subsection we briefly describe the use of DS spread-spectrum signals in four applications. First, we consider an application in which the signal is transmitted at very low power, so that a listener trying to detect the presence of the signal would encounter great difficulty. A second application is multiple-access radio communications. A third application involves the use of a DS spread-spectrum signal to resolve the multipath in a time-dispersive radio channel. Finally, we consider the use of DS spread-spectrum to overcome jamming.

Low-detectability signal transmission.   In this application the information-bearing signal is transmitted at a very low power level relative to the background channel noise and thermal noise that is generated in the front end of a receiver. If the DS spread-spectrum signal occupies a bandwidth W and the power-spectral density of the additive noise is N_0 W/Hz, the average noise power in the bandwidth W is $P_N = W N_0$.

The average received signal power at the intended receiver is P_R. If we wish to hide the presence of the signal from receivers that are in the vicinity of the intended receiver, the signal is transmitted at a power level such that $P_R/P_N \ll 1$. The intended receiver can recover the weak information-bearing signal from the background noise with the aid of the processing gain and the coding gain. However, any other receiver that has no knowledge of the PN code sequence is unable to take advantage of the processing gain and the coding gain. Consequently, the presence of the information-bearing signal is difficult to detect. We say that the transmitted signal has a *low probability of being intercepted* (LPI), and it is called an *LPI signal.*

The probability of error given in Section 11.2.1 applies as well to the demodulation and decoding of LPI signals at the intended receiver.

Example 11.2.3

A DS spread-spectrum signal is designed so that the power ratio P_R/P_N at the intended receiver is 10^{-2}. (a) If the desired $\mathcal{E}_b/N_0 = 10$ for acceptable performance, determine the minimum value of the processing gain. (b) Suppose that the DS spread-spectrum signal is transmitted via radio to a receiver at a distance of 2000 km. The transmitter antenna has a gain of 20 dB, while the receiver antenna is omnidirectional. The carrier frequency is 3 MHz, the available channel bandwidth $W = 10^5$ Hz, and the receiver has a noise temperature of 300° K. Determine the required transmitter power and the bit rate of the DS spread-spectrum system.

Solution   (a) We may write $\mathcal{E}_b/N_0$ as

$$\frac{\mathcal{E}_b}{N_0} = \frac{P_R T_b}{N_0} = \frac{P_R L_c T_c}{N_0} = \left(\frac{P_R}{W N_0}\right) L_c = \left(\frac{P_R}{P_N}\right) L_c$$

Since $\mathcal{E}_b/N_0 = 10$ and $P_R/P_N = 10^{-2}$, it follows that the necessary processing gain is $L_c = 1000$.

(b) The expression for the received signal power is

$$P_{R\mathrm{dB}} = P_{T\mathrm{dB}} - L_{s\mathrm{dB}} + G_{T\mathrm{dB}}$$

where $L_{s\mathrm{dB}}$ is the free-space path loss and $G_{T\mathrm{dB}}$ is the antenna gain. The path loss is

$$L_{s\mathrm{dB}} = 20 \log \left(\frac{4\pi d}{\lambda} \right)$$

where the wavelength $\lambda = 100$ m. Hence,

$$L_{s\mathrm{dB}} = 20 \log(8\pi \times 10^4) = 108 \mathrm{dB}$$

Therefore,

$$P_{T\mathrm{dB}} = P_{R\mathrm{dB}} + 108 - 20$$

$$= P_{R\mathrm{dB}} + 88$$

The received power level can be obtained from the condition $P_R/P_N = 10^{-2}$. First of all, $P_N = W N_0$, where $N_0 = kT = 4. \times 10^{-21}$ W/Hz and $W = 10^5$ Hz. Hence,

$$P_N = 4.1 \times 10^{-16} \text{ W.}$$

and

$$P_R = 4.1 \times 10^{-18} \text{ W.}$$

or, equivalently, $P_{R\mathrm{dB}} = -174$ dBw. Therefore,

$$P_{T\mathrm{dB}} = P_{R\mathrm{dB}} + 88 = -86 \mathrm{dBw}$$

or, equivalently, $P_T = 2.5 \times 10^{-9}$ W. The bit rate is $R = W/L_c = 10^5/10^3 = 100$ bits/sec.

Code division multiple access. The enhancement in performance obtained from a DS spread-spectrum signal through the processing gain, and the coding gain can be used to enable many DS spread-spectrum signals to occupy the same channel bandwidth provided that each signal has its own pseudorandom (signature) sequence. Thus, it is possible to have several users transmit messages simultaneously over the same channel bandwidth. This type of digital communication in which each transmitter/receiver user pair has its own distinct signature code for transmitting over a common channel bandwidth is called *code division multiple access* (CDMA).

In the demodulation of each DS spread-spectrum signal, the signals from the other simultaneous users of the channel appear as additive interference. The level of interference varies as a function of the number of users of the channel at any given time. A major advantage of CDMA is that a large number of users can be accommodated if each user transmits messages for a short period of time. In such a multiple access system, it is relatively easy either to add new users or to decrease the number of users without reconfiguring the system.

We determine below the number of simultaneous signals that can be accommodated in a CDMA system. For simplicity, we assume that all signals have identical average powers. In many practical systems the received signal power level from each user is monitored at a central station and power control is exercised over all simultaneous users by use of a control channel that instructs the users on whether to increase or decrease their power level. With such power control, if there are N_u simultaneous users, the desired signal-to-noise interference power ratio at a given receiver is

$$\frac{P_S}{P_N} = \frac{P_S}{(N_u - 1)P_S} = \frac{1}{N_u - 1} \qquad (11.2.44)$$

From this relation we can determine the number of users that can be accommodated simultaneously. The following example illustrates the computation.

Example 11.2.4

Suppose that the desired level of performance for a user in a CDMA system is an error probability of 10^{-6}, which is achieved when $\mathcal{E}_b/J_0 = 20$ (13 dB). Determine the maximum number of simultaneous users that can be accommodated in a CDMA system if the bandwidth-to-bit-rate ratio is 1000 and the coding gain is $R_c d_{min}^H = 4$ (6 dB).

Solution   From the basic relationship given in (11.2.43), we have

$$\frac{\mathcal{E}_b}{J_0} = \frac{W/R}{N_u - 1} R_c d_{min}^H = 20$$

If we solve for N_u, we obtain

$$N_u = \frac{W/R}{20} R_c d_{min}^H + 1$$

For $W/R = 1000$ and $R_c d_{min}^H = 4$, we obtain the result that $N_u = 201$.

In determining the maximum number of simultaneous users of the channel, we implicitly assumed that the pseudorandom code sequences used by the various users are uncorrelated and that the interference from other users adds on a power basis only. However, orthogonality of the pseudorandom sequences among the N_u users generally is difficult to achieve, especially if N_u is large. In fact, the design of a large set of pseudorandom sequences with good correlation properties is an important problem that has received considerable attention in the technical literature. We shall briefly treat this problem in Section 11.2.3.

CDMA is a viable method for providing digital cellular telephone service to mobile users. As described in Chapter 4, linear predictive coding of speech signals may be used to compress the bit rate of digitized speech to the range of 4800–7200 bits/sec. With channel coding, the digitized speech signal can be transmitted by means of four-phase PSK in a channel bandwidth of 10 KHz or less. The signal from each user may then be spread via DS to a wider bandwidth for transmission over the multiple access channel. For example, if we spread the signal over a bandwidth of $W = 1$ MHz, we have a processing gain of $\frac{W}{R} \approx 100$. This would allow a relatively large number of users to access the wideband channel simultaneously.

Communication over channels with multipath. In Section 9.6 we described the characteristics of fading multipath channels and the design of signals for effective communication through such channels. Examples of fading multipath channels include ionospheric propagation in the HF frequency band (3–30 MHz), where the ionospheric layers serve as signal reflectors, and in mobile radio communication systems, where the multipath propagation is due to reflection from buildings, trees, and other obstacles located between the transmitter and the receiver.

Our discussion on signal design in Section 9.6 focused primarily on frequency nonselective channels, where the signal bandwidth W is smaller than the coherence bandwidth B_{cb} of the channel. On the other hand, if $W > B_{cb}$, we considered two approaches to signal design. One approach is to subdivide the available bandwidth W into N subchannels such that the bandwidth per channel $W/N < B_{cb}$. In this way, each subchannel is frequency nonselective and the signals in each subchannel satisfy the condition that the symbol interval $T \gg T_m$, where T_m is the multipath spread of the channel. Thus, ISI is avoided. A second approach is to design the signal to utilize the entire signal bandwidth W and transmit it on a single carrier. In this case, the channel is frequency selective and the multipath components with differential delays of $1/W$ or greater become resolvable.

DS spread-spectrum is a particularly effective way to generate a wideband signal for resolving multipath signal components. By separating the multipath components, we may also reduce the effects of fading. For example, in LOS communication systems, where there is a direct path and a secondary propagation path resulting from signals reflecting from buildings and surrounding terrain, the demodulator at the receiver may synchronize to the direct signal component and ignore the existence of the multipath component. In such a case, the multipath component becomes a form of ISI on the demodulation of subsequent transmitted signals.

ISI can be avoided if we are willing to reduce the symbol rate $1/T$ such that $T \gg T_m$. In this case, we employ a DS spread-spectrum signal with bandwidth W to resolve the multipath. Thus, the channel is frequency selective and the appropriate channel model is the tapped-delay-line model with time-varying coefficients as shown in Figure 9.58 The optimum demodulator for this channel is a filter matched to the tapped-delay channel model. That is, the matched filter is a tapped-delay-line of length T_m, as shown in Figure 11.9, that employs a set of correlators, one for each tap, to form the crosscorrelation of the received signal with each of the possible resolvable multipath signal components. For coherent detection, each of the correlator outputs are phase-corrected by the carrier-phase shift for each signal component, possibly weighted by the corresponding signal strength of each signal component (for maximal ratio combining) and then all phase-corrected and weighted correlator outputs are summed and fed to the detector. Alternatively, for noncoherent detection of orthogonal signals, each correlator output may be squared and the sum of squares of all tap correlators are fed to the detector. Thus, for both coherent and noncoherent detection, all the energy in the resolvable signal components is effectively extracted from the received signal by the matched filter. The matched filter demodulator is called a RAKE correlator because of the resemblance of the

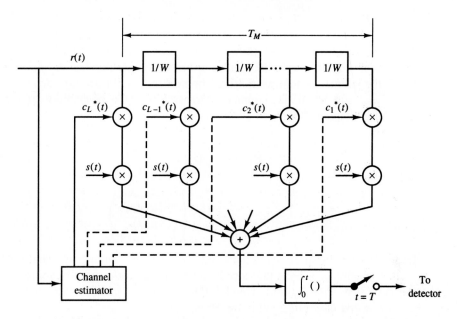

FIGURE 11.9.   RAKE correlation-type demodulator for coherent detection.

tapped-delay-line matched filter to an ordinary garden rake. That is, the RAKE matched filter/correlator resembles a garden rake in the way it collects the signal energy from all the resolvable multipath signal components. Thus, if there are D multipath components, the combined output consists of D signal elements. Since the multipath components that are separated in propagation delay by $\frac{1}{W}$ from one another exhibit independent fading characteristics, the RAKE demodulator/detector achieves the performance of a system with D-order diversity. Thus, the effects of fading that may occur within each signal path are significantly reduced. For more details on the characteristics and performance of the RAKE correlator, the interested reader is referred to Proakis (1989) and to the original works of Price (1954, 1956) and Price and Green (1958).

Resistance to intentional jamming.   In Section 11.2.1 we evaluated the effectiveness of a DS spread-spectrum system in the presence of a narrowband (sinusoidal) and a broadband interference. We have observed that the processing gain and the coding gain provide a means for overcoming the detrimental effects of these types of interference. Below we describe another jamming threat that has a dramatic effect on the performance of a DS spread-spectrum system.

Let us consider a jamming signal that consists of pulses of spectrally flat Gaussian noise that covers the entire signal bandwidth. This type of jamming is usually called *pulsed interference* or *partial-time jamming*. Suppose that the jammer is average-power limited with an average power P_J in the signal bandwidth W.

Hence, $J_0 = P_J/W$. Instead of transmitting the jamming signal continuously, the jammer transmits pulses at a power level P_J/α for α percent of the time. Thus, the probability that the jammer is transmitting at a given instant is α. For simplicity, we assume that an interference pulse spans an integer number of bits (or symbols). When the jammer is not transmitting, the transmitted bits are assumed to be received error free, and when the jammer is transmitting, the probability of error for an uncoded DS spread-spectrum system is

$$P(\alpha) = \frac{\alpha}{2} Q \left(\sqrt{\frac{2\mathcal{E}_b}{J_0/\alpha}} \right)$$

$$= \frac{\alpha}{2} Q \left(\sqrt{\frac{2\alpha W/R}{P_J/P_S}} \right) \qquad (11.2.45)$$

where W/R is the processing gain and P_J/P_S is the jammer-to-signal power ratio.

The jammer selects the duty cycle α to maximize the probability of error for the communication system. Upon differentiating (11.2.45) with respect to α, we find the worst-case pulse jamming occurs when

$$\alpha^\star = \begin{cases} \dfrac{0.71}{\mathcal{E}_b/J_0}, & \frac{\mathcal{E}_b}{J_0} \geq 0.71 \\[4mm] 1, & \frac{\mathcal{E}_b}{J_0} < 0.71 \end{cases} \qquad (11.2.46)$$

and the corresponding probability of error is

$$P_2 = \begin{cases} \dfrac{0.082}{\mathcal{E}_b/J_0} = \dfrac{0.082 P_J/P_S}{W/R}, & \frac{\mathcal{E}_b}{J_0} \geq 0.71 \\[4mm] Q \left(\sqrt{\dfrac{2\mathcal{E}_b}{J_0}} \right) = Q \left(\sqrt{\dfrac{2W/R}{P_J/P_S}} \right), & \frac{\mathcal{E}_b}{J_0} < 0.71 \end{cases} \qquad (11.2.47)$$

The error rate performance given by (11.2.45) for $\alpha = 1.0, 0.1$ and 0.01 along with the worst-case performance based on $\alpha^\star$ is plotted in Figure 11.10. When we compare the error rate for continuous wideband Gaussian noise jamming ($\alpha = 1$) with worst-case pulse jamming, we find a large difference in performance, e.g., approximately 40 dB at an error rate of 10^{-6}. This is indeed a large penalty.

We should point out that practical consideration may prohibit the jammer from achieving small values of α at a high peak power. Nevertheless, the error probability given by (11.2.47) serves as an upper bound on the performance of uncoded binary PSK with worst-case pulse jamming. Clearly, the performance of the DS spread-spectrum system in the presence of pulse jamming is poor.

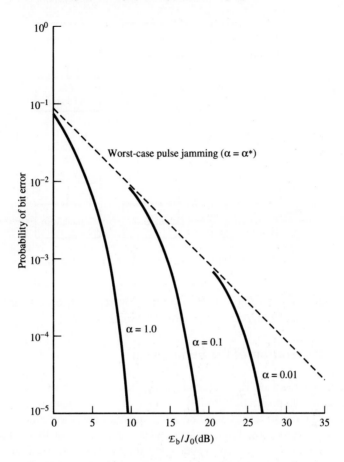

FIGURE 11.10.   Performance of DS binary PSK with pulse jamming.

If we simply add coding to the DS spread-spectrum system, the performance in SNR is improved by an amount equal to the coding gain, which in most cases is limited to less than 10 dB. The reason that the addition of coding does not improve the performance significantly is that the jamming signal pulse duration (duty cycle) may be selected to affect many consecutive coded bits. Consequently, the code word error probability is high due to the burst characteristics of the jammer.

To improve the performance of the coded DS spread-spectrum system, we should interleave the coded bits prior to transmission over the channel. The effect of interleaving is to make the coded bits that are hit by the jammer statistically independent. Figure 11.11 illustrates a block diagram of a DS spread-spectrum system that employs coding and interleaving. By selecting a sufficiently long interleaver so that the burst characteristics of the jammer are eliminated, the penalty in

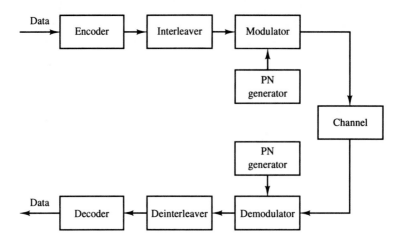

FIGURE 11.11. Block diagram of AJ communication system.

performance due to jamming is significantly reduced, e.g., to the range of 3–5 dB for conventional binary block or convolutional codes.

11.2.3 Generation of PN Sequences

A pseudorandom or PN sequence is a code sequence of 1's and 0's whose autocorrelation has properties similar to those of white noise. In this section, we briefly describe the construction of some PN sequences and their autocorrelation and cross-correlation properties. For a comprehensive treatment of this subject, the interested reader may refer to Golomb (1967) and the paper by Sarwate and Pursley (1980).

By far, the most widely known binary PN code sequences are the maximum-length shift-register sequences. A maximum-length shift register sequence, or m-sequence for short, has a length $L = 2^m - 1$ bits and is generated by an m-stage shift register with linear feedback as illustrated in Figure 11.12. The sequence is periodic with period L. Each period has a sequence of 2^{m-1} ones and $2^{m-1} - 1$

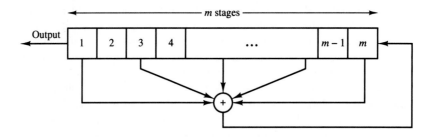

FIGURE 11.12. General m-stage shift register with linear feedback.

TABLE 11-1   SHIFT-REGISTER CONNECTIONS FOR GENERATING ML SEQUENCES.

m	Stages connected to modulo-2 adder	m	Stages connected to modulo-2 adder	m	Stages connected to modulo-2 adder
2	1, 2	13	1, 10, 11, 13	24	1, 18, 23, 24
3	1, 3	14	1, 5, 9, 14	25	1, 23
4	1, 4	15	1, 15	26	1, 21, 25, 26
5	1, 4	16	1, 5, 14, 16	27	1, 23, 26, 27
6	1, 6	17	1, 15	28	1, 26
7	1, 7	18	1, 12	29	1, 28
8	1, 5, 6, 7	19	1, 15, 18, 19	30	1, 8, 29, 30
9	1, 6	20	1, 18	31	1, 29
10	1, 8	21	1, 20	32	1, 11, 31, 32
11	1, 10	22	1, 22	33	1, 21
12	1, 7, 9, 12	23	1, 19	34	1, 8, 33, 34

[†] Forney [1970].

zeros. Table 11.1 lists shift register connections for generating maximum-length sequences.

In DS spread-spectrum applications, the binary sequence with elements $\{0, 1\}$ is mapped into a corresponding binary sequence with elements $\{-1, 1\}$. We shall call the equivalent sequence $\{c_n\}$ with elements $\{-1, 1\}$ a *bipolar sequence*.

An important characteristic of a periodic PN sequence is its autocorrelation function which is usually defined in terms of the bipolar sequences $\{c_n\}$ as

$$R_c(m) = \sum_{n=1}^{L} c_n c_{n+m}, \qquad 0 \le m \le L - 1 \qquad (11.2.48)$$

where L is the period of the sequence. Since the sequence $\{c_n\}$ is periodic with period L, the autocorrelation sequence $\{R_c(m)\}$ is also periodic with period L.

Ideally, a PN sequence should have an autocorrelation function that has correlation properties similar to white noise. That is, the ideal autocorrelation sequence for $\{c_n\}$ is $R_c(0) = L$ and $R_c(m) = 0$ for $1 \le m \le L-1$. In the case of m-sequences, the autocorrelation sequence is

$$R_c(m) = \begin{cases} L, & m = 0 \\ -1, & 1 \le m \le L - 1 \end{cases} \qquad (11.2.49)$$

For long m-sequences, the size of the off-peak values of $R_c(m)$ relative to the peak value $R_c(0)$, i.e., the ratio $R_c(m)/R_c(0) = -1/L$ is small and, from a practical viewpoint, inconsequential. Therefore, m-sequences are very close to ideal PN sequences when viewed in terms of their autocorrelation function.

In antijamming applications of DS spread-spectrum signals, the period of the PN sequence must be large to prevent the jammer from learning the feedback connections of the PN code sequence from observation of the received signal.

However, this requirement is impractical in most cases because, due to the linearity property of the PN code, the jammer can determine the feedback connections by observing only $2m$ chips from the PN sequence. To reduce this vulnerability to a jammer, the output sequences from several stages of the shift register or the outputs from several distinct m-sequences are combined in a nonlinear way to produce a nonlinear sequence that is considerably more difficult for the jammer to learn. Further reduction in vulnerability is achieved by frequently changing the feedback connections and/or the number of stages in the shift register according to some prearranged plan formulated between the transmitter and the intended receiver.

In some applications the cross correlation properties of PN sequences are as important as the autocorrelation properties. For example, in CDMA each user is assigned a particular PN sequence. Ideally, the PN sequences among users should be mutually uncorrelated so that the level of interference experienced by one user from transmissions of other users adds on a power basis. However, the PN sequences used in practice by different users exhibit some correlation.

To be specific, let us consider the class of m-sequences. It is known that the periodic crosscorrelation function between a pair of m-sequences of the same period can have relatively large peaks. Table 11.2 lists the peak magnitude R_{max} for the periodic crosscorrelation between pairs of m-sequences for $3 \le m \le 12$. Also listed in Table 11.2 is the number of m-sequences of length $L = 2^m - 1$ for $3 \le m \le 12$. We observe that the number of m-sequences of length L increases rapidly with m. We also observe that, for most sequences, the peak magnitude R_{max} of the cross-correlation function is a large percentage of the peak value of the autocorrelation function. Consequently, m-sequences are not suitable for CDMA communication systems. Although it is possible to select a small subset of m-sequences that have relatively smaller crosscorrelation peak values than R_{max}, the number of sequences in the set is usually too small for CDMA applications.

TABLE 11-2   PEAK CROSSCORRELATIONS OF M SEQUENCES AND GOLD SEQUENCES

		m sequences			Gold sequences	
m	$L = 2^{m-1}$	Number	Peak cross-correlation R_{max}	$R_{max}/R(0)$	R_{max}	$R_{min}/R(0)$
3	7	2	5	0.71	5	0.71
4	15	2	9	0.60	9	0.60
5	31	6	11	0.35	9	0.29
6	63	6	23	0.36	17	0.27
7	127	18	41	0.32	17	0.13
8	255	16	95	0.37	33	0.13
9	511	48	113	0.22	33	0.06
10	1023	60	383	0.37	65	0.06
11	2047	176	287	0.14	65	0.03
12	4095	144	1407	0.34	129	0.03

Methods for generating PN sequences with better periodic crosscorrelation properties than m-sequences have been developed by Gold (1967, 1968) and by Kasami (1966). Gold sequences are constructed by taking a pair of specially selected m-sequences, called *preferred m-sequences,* and forming the modulo-2 sum of the two sequences, for each of L cyclically shifted versions of one sequence relative to the other sequence. Thus, L Gold sequences are generated as illustrated in Figure 11.13. For m odd, the maximum value of the crosscorrelation function between any pair of Gold sequences is $R_{\max} = \sqrt{2L}$. For m even, $R_{\max} = \sqrt{L}$.

Kasami (1966) described a method for constructing PN sequences by decimating an m-sequence. In Kasami's method of construction, every $2^{m/2} + 1$ bit of an m-sequence is selected. This method of construction yields a smaller set of PN sequences compared with Gold sequences, but their maximum crosscorrelation value is $R_{\max} = \sqrt{L}$.

It is interesting to compare the peak value of the crosscorrelation function for Gold sequences and for Kasami sequences with a known lower bound for the maximum crosscorrelation between any pair of binary sequences of length L. Given a set of N sequences of period L, a lower bound on their maximum crosscorrelation is

$$R_{\max} \geq L\sqrt{\frac{N-1}{NL-1}} \qquad (11.2.50)$$

which, for large values of L and N, is well approximated as $R_{\max} \geq \sqrt{L}$. Hence, we observe that Kasami sequences satisfy the lower bound and, hence, they are optimal. On the other hand, Gold sequences with m odd have a $R_{\max} = \sqrt{2L}$. Hence, they are slightly suboptimal.

Besides the well-known Gold sequences and Kasami sequences, there are other binary sequences that are appropriate for CDMA applications. The interested

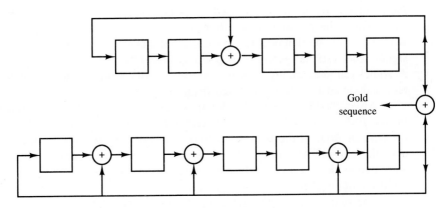

FIGURE 11.13.   Generation of Gold sequences of length 31.

reader is referred to the papers by Scholtz (1979), Olsen (1977), and Sarwate and Pursley (1980).

Finally, we should point out that although we discussed the periodic cross-correlation function between pairs of periodic sequences, many practical CDMA systems use an information bit duration that encompasses only a fraction of a periodic sequence. In such a case it is the partial period crosscorrelation between two sequences that is important. The partial period crosscorrelation properties of periodic PN sequences has been widely investigated and discussed in the technical literature.

11.3 FREQUENCY-HOPPED SPREAD SPECTRUM

In frequency-hopped (FH) spread-spectrum the available channel bandwidth W is subdivided into a large number of nonoverlapping frequency slots. In any signaling interval the transmitted signal occupies one or more of the available frequency slots. The selection of the frequency slot(s) in each signal interval is made pseudorandomly according to the output from a PN generator.

A block diagram of the transmitter and receiver for a FH spread-spectrum system is shown in Figure 11.14. The modulation is either binary or M-ary FSK (MFSK). For example, if binary FSK is employed, the modulator selects one of two frequencies, say f_0 or f_1, corresponding to the transmission of a 0 for a 1. The resulting binary FSK signal is translated in frequency by an amount that is determined by the output sequence from a PN generator, which is used to select a frequency f_c that is synthesized by the frequency synthesizer. This frequency is mixed with the output of the FSK modulator and the resultant frequency-translated signal is transmitted over the channel. For example, by taking m bits from the PN generator, we may specify $2^m - 1$ possible carrier frequencies. Figure 11.15 illustrates a FH signal pattern.

At the receiver, there is an identical PN sequence generator, synchronized with the received signal, which is used to control the output of the frequency synthesizer. Thus, the pseudorandom frequency translation introduced at the transmitter is removed at the demodulator by mixing the synthesizer output with the received signal. The resultant signal is then demodulated by means of an FSK demodulator. A signal for maintaining synchronism of the PN sequence generator with the FH received signal is usually extracted from the received signal.

Although binary PSK modulation generally yields better performance than binary FSK, it is difficult to maintain phase coherence in the synthesis of the frequencies used in the hopping pattern and, also, in the propagation of the signal over the channel as the signal is hopped from one frequency to the other over a wide bandwidth. Consequently, FSK modulation with noncoherent demodulation is usually employed in FH spread-spectrum systems.

The frequency-hopping rate, denoted as R_h, may be selected to be either equal to the symbol rate, or lower than the symbol rate, or higher than the symbol rate.

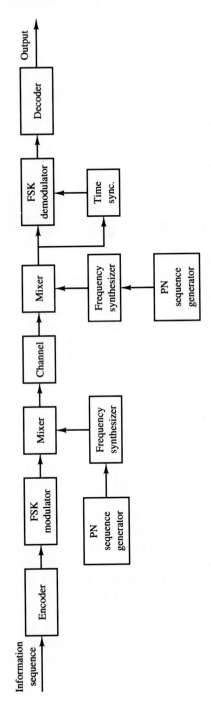

FIGURE 11.14. Block diagram of a FH spread-spectrum system.

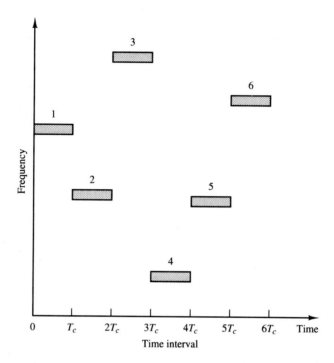

FIGURE 11.15.   An example of a FH pattern.

If R_h is equal to or lower than the symbol rate, the FH system is called a *slow-hopping* system. If R_h is higher than the symbol rate, i.e., there are multiple hops per symbol, the FH system is called a *fast-hopping* system.

Fast frequency hopping is usually employed in antijam (AJ) applications when it is necessary to prevent a type of jammer, called a *follower jammer,* from having sufficient time to intercept the frequency and retransmit it along with adjacent frequencies so as to create interfering signal components. However, there is a penalty incurred in subdividing an information symbol into several FH elements, because the energy from these separate elements is combined noncoherently.

FH spread-spectrum signals are used primarily in digital communication systems that require AJ protection and in CDMA where many users share a common bandwidth. In most cases, a FH signal is preferred over a DS spread-spectrum signal because of the stringent synchronization requirements inherent in DS spread-spectrum signals. Specifically, in a DS system, timing and synchronization must be established to within a fraction of a chip interval $T_c = 1/W$. On the other hand, in an FH system, the chip interval T_c is the time spent in transmitting a signal in a particular frequency slot of bandwidth $B \ll W$. But this interval is approximately $1/B$, which is much larger than $1/W$. Hence, the timing requirements in an FH system are not as stringent as in a DS system.

Below we shall evaluate the performance of FH spread-spectrum systems under the condition that the system is either slow-hopping or fast-hopping.

11.3.1 Slow Frequency-Hopping Systems

Let us consider a slow frequency hopping system in which the hop rate $R_h = 1$ hop/bit. We assume that the interference on the channel is broadband and is characterized as AWGN with power-spectral density J_0. Under these conditions, the probability of error for the detection of noncoherently demodulated binary FSK is

$$P_2 = \frac{1}{2} e^{-\rho_b/2} \tag{11.3.1}$$

where $\rho_b = \mathcal{E}_b/J_0$ is the SNR/bit.

As in the case of a DS spread-spectrum system, we observe that $\mathcal{E}_b$, the energy/bit, can be expressed as $\mathcal{E}_b = P_S T_b = P_S/R$, where P_S is the average transmitted power and R is the bit rate. Similarly, $J_0 = P_J/W$, where P_J is the average power of the broadband interference and W is the available channel bandwidth. Therefore, the SNR ρ_b can be expressed as

$$\rho_b = \frac{\mathcal{E}_b}{J_0} = \frac{W/R}{P_J/P_S} \tag{11.3.2}$$

where W/R is the processing gain and P_J/P_S is the jamming margin for the FH spread-spectrum signal.

Slow FH spread-spectrum systems are particularly vulnerable to partial-band interference that may result from either intentional jamming or in FH CDMA systems. To be specific, suppose that the partial band interference is modeled as a zero-mean Gaussian random process with a flat power-spectral density over a fraction of the total bandwidth W and zero in the remainder of the frequency band. In the region or regions where the power-spectral density is nonzero, its value is $S_J(f) = J_0/\alpha$, where $0 < \alpha \le 1$. In other words, the interference average power P_J is assumed to be constant.

Suppose that the partial-band interference comes from a jammer who selects α to optimize the effect on the communications system. In an uncoded slow-hopping system with binary FSK modulation and noncoherent detection, the transmitted frequencies are selected with uniform probability in the frequency band W. Consequently, the received signal will be jammed with probability α and it will not be jammed with probability $1 - \alpha$. When it is jammed, the probability of error is $1/2 \exp(-\alpha\rho_b/2)$ and when it is not jammed, the detection of the signal is assumed to be error free. Therefore, the average probability of error is

$$P_2(\alpha) = \frac{\alpha}{2} e^{-\alpha\rho_b/2}$$

$$= \frac{\alpha}{2} \exp\left(-\frac{\alpha W/R}{2P_J/P_S}\right) \tag{11.3.3}$$

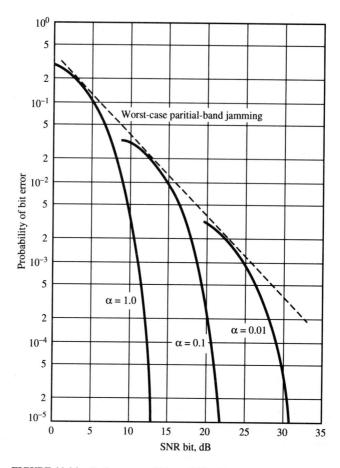

FIGURE 11.16. Performance of binary FSK with partial-band interference.

Figure 11.16 illustrates the error rate as a function of ρ_b for several values of α. The jammer is assumed to optimize its strategy by selecting α to maximize the probability of error. By differentiating $P_2(\alpha)$, and solving for the value of α that maximizes $P_2(\alpha)$, we find that the jammers best choice of α is

$$\alpha^\star = \begin{cases} 2/\rho_b, & \rho_b \geq 2 \\ 1, & \rho_b < 2 \end{cases} \tag{11.3.4}$$

The corresponding error probability for the worst-case partial-band jammer is

$$P_2 = \begin{cases} e^{-1}/\rho_b, & \rho_b \geq 2 \\ \dfrac{1}{2} e^{-\rho_b/2}, & \rho_b < 2 \end{cases} \tag{11.3.5}$$

which is also shown in Figure 11.16. Whereas the error probability decreases exponentially for full-band jamming as given by (11.3.3), the error probability for worst-case partial-band jamming decreases only inversely with $\mathcal{E}_b/J_0$. This result is similar to the error probability for DS spread-spectrum signals in the presence of pulse jamming. It is also similar to the error probability for binary FSK in a Rayleigh fading channel.

In our discussion of signal design for efficient and reliable communication over a fading channel in Section 9.6, we found that diversity, which can be obtained by simple repetition of the transmitted information bit on different frequencies (or by means of block or convolutional coding), provides a significant improvement in performance relative to uncoded signal transmission. It should not be surprising that the same type of signal coding is also effective on partial-band interference channels. In fact, it has been shown by Viterbi and Jacobs (1975) that, by optimizing the code design for the partial-band jamming scenario, the communication system can achieve an average bit error probability of

$$P_2 = e^{-\rho_b/4} \tag{11.3.6}$$

Therefore, the probability of error achieved with the optimum code design decreases exponentially with an increase in SNR and is within 3 dB of the performance obtained in an AWGN channel. Thus, the penalty due to partial-band jamming is reduced significantly.

11.3.2 Fast Frequency-Hopping Systems

In fast FH systems, the frequency-hop rate R_h is some multiple of the symbol rate. Basically, each (M-ary) symbol interval is subdivided into N subintervals, which are called *chips* and one of M frequencies is transmitted in each subinterval. The hop rate R_h is selected sufficiently high so that a potential jammer does not have sufficient time to detect the presence of the transmitted frequency and to synthesize a jamming signal that occupies the same bandwidth.

To recover the information at the receiver, the received signal is first dehopped by mixing it with the hopped carrier frequency. This operation removes the hopping pattern and brings the received signal in all subintervals (chips) to a common frequency band that encompasses the M possible transmitted frequencies. The signal in each subinterval is then passed through the M matched filters (or correlators) tuned to the M possible transmitted frequencies, which are sampled at the end of each subinterval and passed to the detector. The detection of the FSK signals is noncoherent. Hence, decisions are based on the magnitude of the matched filter (or correlator) outputs.

Since each symbol is transmitted over N chips, the decoding may be performed either on the basis of hard decisions or soft decisions. The following example illustrates the decoding based on hard decisions.

Example 11.3.1

Suppose that binary FSK is used to transmit binary symbols, and each symbol is transmitted over N frequency hops, where N is odd. Determine the probability of error for an AWGN channel if hard-decision decoding is used.

Solution   The probability of error for noncoherent detection of binary FSK for each hop is

$$p = \frac{1}{2} e^{-\rho_b/2N} \tag{11.3.7}$$

where

$$\frac{\rho_b}{N} = \frac{\mathcal{E}_b/N}{N_0} \tag{11.3.8}$$

is the SNR/chip and $\mathcal{E}_b$ is the total bit energy. The decoder decides in favor of the transmitted frequency that is larger in at least $(N+1)/2$ chips. Thus, the decision is made on the basis of a majority vote given the decisions on the N chips. Consequently, the probability of a bit error is

$$P_2 = \sum_{m=(N+1)/2}^{N} \binom{N}{m} p^m (1-p)^{N-m} \tag{11.3.9}$$

where p is given by (11.3.7). We should note that the error probability P_2 for hard-decision decoding of the N chips will be higher than the error probability for a single hop/bit FSK system, which is given by (11.3.1), when the SNR/bit ρ_b is the same in the two systems (see Problem 11.10). However the opposite is expected to be true in the presence of a jammer.

The alternative to hard-decision decoding is soft-decision decoding in which the magnitudes (or magnitudes squared) of the corresponding matched filter outputs are summed over the N chips and a single decision is made based on the frequency giving the largest output. For example, if binary orthogonal FSK is used to transmit the information, the two soft-decision metrics for the N chips based on square-law combining are

$$DM_1 = \sum_{k=1}^{N} \left| \frac{\mathcal{E}_b}{N} + \nu_{1k} \right|^2$$

$$DM_2 = \sum_{k=1}^{N} |\nu_{2k}|^2 \tag{11.3.10}$$

where $\{\nu_{1k}\}$ and $\{\nu_{2k}\}$ are the noise components from the two matched filters for the N chips. Since frequency f_1 is assumed to have been transmitted, a decision error occurs when $DM_2 > DM_1$. The probability of this event error for additive Gaussian noise may be obtained in closed form, although its derivation is cumbersome. The final result is

$$P_2 = \frac{1}{2^{2N-1}} e^{-\rho_b/2} \sum_{i=0}^{N-1} K_i \left(\frac{\rho_b}{2} \right)^i \tag{11.3.11}$$

where the set $\{K_i\}$ are constants that may be expressed as

$$K_i = \frac{1}{i!} \sum_{r=0}^{N-1-i} \binom{2N-1}{r} \qquad (11.3.12)$$

The error probability for soft-decision decoding given by (11.3.11) is lower than that for hard-decision decoding given by (11.3.9) for the same $\mathcal{E}_b/N_0$. The difference in performance is the loss in hard-decision decoding. However, (11.3.11) is higher than the error probability for single-hop FSK, which is given by (11.3.1) for the AWGN channel. The difference in performance between (11.3.1) and (11.3.11) for the same SNR is due to the noncoherent (square-law) combining at the decoder. This loss is usually called the *noncoherent combining loss* of the system.

If soft-decision decoding is used in the presence of partial-band jamming or interference, it is important to scale (or normalize) the matched filter outputs in each hop, so that a strong jammer or interference that falls within the transmitted signal band in any hop does not dominate the output of the combiner. A good strategy in such a case is to normalize or clip the matched filter outputs from each hop if their values exceed some threshold that is set near (slightly above) the mean of the signal-plus-noise power level. Alternatively, we may monitor the noise power level and scale the matched filter outputs for each hop by the reciprocal of the noise power level. Thus, the noise power levels from the matched filter outputs are normalized. Therefore, with proper scaling, a fast FH spread-spectrum system will not be as vulnerable to partial-band jamming or interference because the transmitted information per bit is distributed (or spread) over N frequency hops.

11.3.3 Applications of FH Spread Spectrum

FH spread-spectrum is a viable alternative to DS spread-spectrum for protection against jamming and for CDMA. In CDMA systems based on frequency hopping, each transmitter/receiver pair is assigned its own pseudorandom FH pattern. Aside from this distinguishing feature, the transmitters and receivers of all users may be identical, i.e., they have identical encoders, decoders, modulators and demodulators.

CDMA systems based on FH spread-spectrum signals are particularly attractive for mobile (land, air, sea) users because timing (synchronization) requirements are not as stringent as in a DS spread-spectrum system. In addition, frequency synthesis techniques and associated hardware have been developed that make it possible to frequency-hop over bandwidths that are significantly larger, by one or more orders of magnitude, than those currently possible with DS spread-spectrum signals. Consequently, larger processing gains are possible by FH, which more than offset the loss in performance inherent in noncoherent detection of the FSK-type signals.

FH is also effective against jamming signals. As we have described in the previous section, a FH M-ary ($M \geq 2$) FSK system that employs coding, or simply repeats the information symbol on multiple hops (repetition coding), is very effective against a partial-band jammer. As a consequence, the jammer's threat is reduced to that of an equivalent broadband noise jammer whose transmitter power is spread across the channel bandwidth W.

11.4 OTHER TYPES OF SPREAD-SPECTRUM SIGNALS

DS and FH are the most common forms of spread-spectrum signals used in practice. However, other methods may be used to pseudorandomly spread the bandwidth of the information signal.

One method that is analogous to FH is time hopping (TH). In TH, a time interval, which is selected to be much larger than the reciprocal of the information rate, is subdivided into a large number of time slots. The coded information symbols are transmitted in a pseudorandomly selected time slot as a block of one or more code words. PSK, QAM, or PAM are suitable modulation methods for transmitting the information.

A block diagram of a transmitter and a receiver for a TH spread-spectrum system is shown in Figure 11.17. Due to the burst characteristics of the transmitted signal, buffer storage must be provided at the transmitter in a TH system, as shown in Figure 11.17. A buffer may also be used at the receiver to provide a uniform data stream to the user.

Example 11.4.1

Suppose that we wish to transmit the output of an information source whose bit rate 10,000 bits/sec. The information is encoded by a rate $R_c = 1/2$ code. Determine the channel bandwidth required to transmit the coded information using a TH spread-

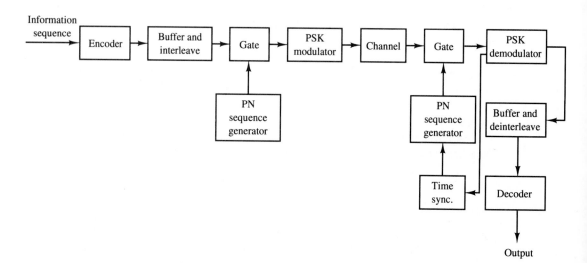

FIGURE 11.17.   Block diagram of a TH spread-spectrum system.

spectrum system with a processing gain of 1000. Assume that the data is transmitted by four-phase PSK, i.e., QPSK.

Solution   To achieve a processing gain of 1000, we may assume that we have a time interval T, which is divided into 1000 time slots of width T/1000 each. The coded bit rate 20,000 bits/sec. With the use of QPSK, the corresponding symbol rate is 10,000 symbols/sec. Consequently, in a time interval of T/1000, we must transmit 10,000 T symbols. With QPSK, the channel bandwidth required is

$$W = \frac{10,000T}{T/1000}$$

$$= 10\text{MHz}$$

Other types of spread-spectrum signals can be obtained by combining DS, FH, and TH. For example, we may have a hybrid DS/FH, which means that a DS spread-spectrum signal is generated and then frequency hopped. Thus, the signal transmitted on a single hop consists of a DS spread-spectrum signal that is demodulated coherently. Another possible hybrid spread-spectrum signal is DS/TH. Such a spread-spectrum signal may not be practical because of an increase in system complexity at the receiver and more stringent timing requirements.

11.5 SYNCHRONIZATION OF SPREAD-SPECTRUM SYSTEMS

Time synchronization of the receiver to the received spread-spectrum signal may be separated into two distinct phases. There is an initial acquisition phase, during which time the receiver establishes time synchronization by detecting the presence of a special initial acquisition sequence. The initial acquisition phase is followed by the transmission of data, during which period the receiver must track the signal timing.

Acquisition phase.   In a DS spread-spectrum system, the PN code sequence must be synchronized in time to within a small fraction of the chip interval $T_c = 1/W$. The problem of initial synchronization may be viewed as one in which we attempt to synchronize the receiver clock to the transmitter clock. Usually, extremely accurate and stable time clocks are used in spread-spectrum systems to reduce the time uncertainty between the receiver clock and the transmitter clock. Nevertheless, there is always an initial timing uncertainty that is due to propagation delay in the transmission of the signal through the channel. This is especially a problem when communication is taking place between two mobile users. In any case, the usual procedure for establishing initial synchronization is for the transmitter to send a known pseudorandom sequence to the receiver. The receiver is continuously in a search mode looking for this sequence to establish initial synchronization.

Suppose that the initial timing uncertainty is T_u seconds and the chip duration is T_c. Since initial synchronization takes place in the presence of additive noise and, perhaps other interference, it is necessary to dwell for $T_d = NT_c$ seconds to test synchronism at each time instant, where N is some positive integer. If we search over the time uncertainty interval in (coarse) time steps of $T_c/2$, then the time required to establish initial synchronization is

$$T_{\text{initsync}} = \frac{T_u}{T_c/2} T_d = 2NT_u$$

Clearly, the synchronization sequence transmitted to the receiver must be at least as long as $2NT_c$ seconds for the receiver to have sufficient time to perform the necessary search in a serial fashion.

In principle, matched filtering or crosscorrelation are optimum methods for establishing initial synchronization in the presence of additive Gaussian noise. A filter matched to the known data waveform generated from the known pseudorandom sequence continuously compares its output with a predetermined threshold. When the threshold is exceeded, initial synchronization is established and the demodulator enters the "data receive" mode.

Alternatively, we may implement a *sliding correlator* as shown in Figure 11.18. The correlator cycles through the time uncertainty, usually in discrete time intervals of $T_c/2$ seconds or less. The crosscorrelation is performed over the time interval NT_c, where N is the number of chips in the synchronization sequence, and the correlator output is compared with a threshold to determine if the known signal sequence is present. If the threshold is not exceeded, the known reference sequence is advanced by $T_c/2$ seconds and the correlation process is repeated. These operations are performed until a signal is detected or until the search has been performed over the time uncertainty interval T_u. In the case of the latter outcome, the search process is repeated.

A similar procedure may be used for FH signals. In this case, the problem is to synchronize the PN code sequence generated at the receiver that controls the hopped frequency pattern. To accomplish this initial synchronization, a known FH signal is transmitted to the receiver. The initial acquisition system at the receiver looks for this known FH signal pattern. For example, a bank of matched filters tuned to the transmitted frequencies in the known pattern may be employed. Their outputs

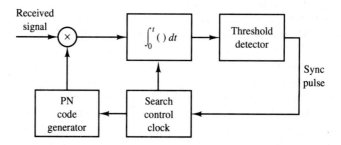

FIGURE 11.18.   A sliding correlator for DS signal acquisition.

must be properly delayed, envelope or square-law detected, weighted, if necessary, and added to produce the signal output that is compared with a threshold. A signal present (signal acquisition) is declared when the threshold is exceeded. The search process is usually performed continuously in time until a threshold is exceeded. A block diagram illustrating this signal acquisition scheme is given in Figure 11.19. As an alternative, a single matched-filter and envelope detector may be used preceded by a FH pattern generator and followed by a threshold detector. This configuration, which is shown in Figure 11.20, is based on a serial search and is akin to the sliding correlator for DS spread-spectrum signals.

The sliding correlator for DS signals and its counterpart shown in Figure 11.20 for FH signals basically perform a serial search that is generally time consuming. As an alternative, one may employ some degree of parallelism by having two or more such correlators operating in parallel and searching over nonoverlapping time slots. In such a case, the search time is reduced at the expense of a more complex and costly implementation. Figure 11.19 depicts such a parallel implementation for FH signals.

During the search mode there may be false alarms that occur occasionally due to additive noise and other interference. To handle the occasional false alarms, it is

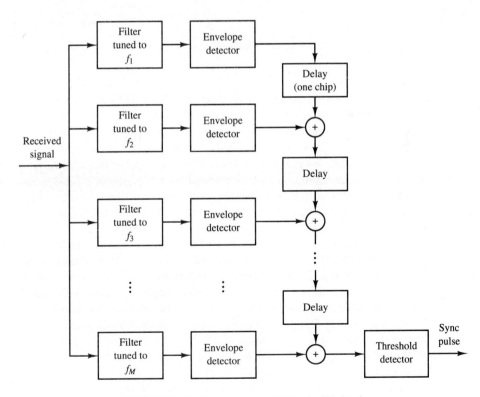

FIGURE 11.19.   System for acquisition of a FH signal.

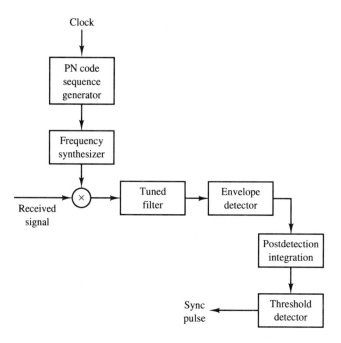

FIGURE 11.20.   Alternative system for acquisition of a FH signal.

necessary to have an additional method or circuit that checks to confirm that the received signal at the output of the correlator remains above the threshold. With such a detection strategy, a large noise pulse that causes the matched filter output to exceed the threshold will have only a transient effect on synchronization, since the matched filter output will fall below the threshold once the large noise pulse passes through the filter. On the other hand, when a signal is present the correlator or matched filter output will remain above the threshold for the duration of the transmitted signal. Thus, if confirmation fails, the search for signal synchronization is resumed.

In the above discussion we considered only time uncertainty in establishing initial synchronization. However, another aspect of initial synchronization is frequency uncertainty. If the transmitter and/or the receiver are mobile, the relative velocity between them results in a Doppler frequency shift in the received signal relative to the transmitted signal. Since the receiver does not know the relative velocity, a priori, the Doppler frequency shift is unknown and must be determined by means of a frequency search method. Such a search is usually accomplished in parallel over a suitably quantized frequency uncertainty interval and serially over the time uncertainty interval. A block diagram of this scheme for DS spread-spectrum signals is shown in Figure 11.21. Appropriate Doppler frequency search methods can also be devised for FH signals.

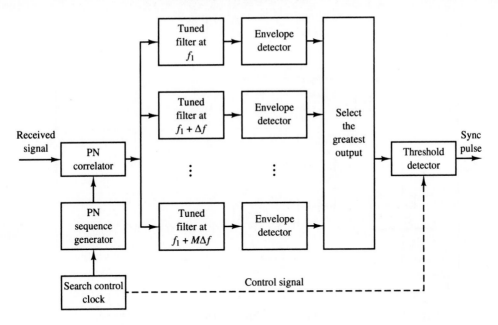

FIGURE 11.21. Initial search for the Doppler frequency offset in a DS system.

Tracking.   Once the signal is acquired, the initial synchronization process is stopped and fine synchronization and tracking begins. The tracking maintains the PN code generator at the receiver in synchronism with the received signal. Tracking includes fine chip synchronization.

For a DS spread-spectrum signal, tracking is usually performed by means of a tracking loop, called a *delay-locked loop* (DLL), as shown in Figure 11.22. In this tracking loop, the received signal is applied to two multipliers, where it is multiplied by two outputs from the local PN code generator, which are delayed relative to each other by an amount of $2\delta \leq T_c$. Thus, the product signals are the crosscorrelations between the received signal and the PN sequence at the two values of delay. These products are bandpass filtered, envelope (or square-law) detected and then subtracted. This difference signal is applied to the loop filter that drives the clock VCO. The VCO output serves as the clock for the PN code signal generator.

If the synchronism is not exact, the filtered output from one correlator will exceed the other and the VCO will be appropriately advanced or delayed. At the equilibrium point, the two filtered correlator outputs will be equally displaced from the peak value, and the PN code generator output will be exactly synchronized to the received signal, which is fed to the demodulator. We observe that this implementation of the DLL for tracking the DS signal is equivalent to the early–late gate bit tracking synchronizer previously described in Section 8.5.

An alternative method for time tracking a DS signal is to use a tau-dither loop (TDL), which is illustrated by the block diagram in Figure 11.23. The TDL employs

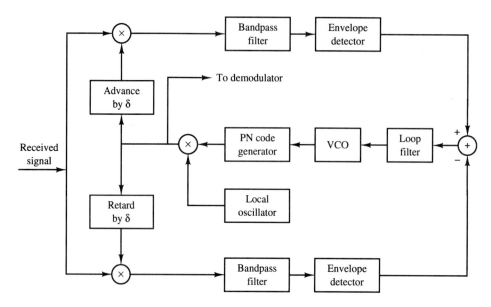

FIGURE 11.22.   DLL for PN code tracking.

only a single "arm" instead of the two "arms" shown in Figure 11.22. By providing a suitable gating waveform it is possible to make this single "arm" implementation appear to be equivalent to the two "arm" realization. In this case, the crosscorrelator output is regularly sampled at two values of delay, by stepping the code clock forward and backward in time by an amount δ. The envelope of the crosscorrelation that is sampled at $\pm\delta$ has an amplitude modulation whose phase relative to the tau-dither modulator determines the sign of the tracking error.

One advantage of the TDL is the less costly implementation resulting from elimination of one of the two arms that are employed in the conventional DLL. A

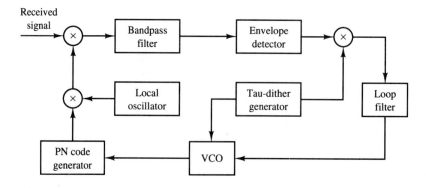

FIGURE 11.23.   Tau-dither loop.

second and less apparent advantage is that the TDL does not suffer from performance degradation that is inherent in the DLL when the amplitude gain in the two
arms is not properly balanced.

Both the DLL and the TDL generate an error signal by sampling the signal
correlation function at $\pm\delta$ off the peak, as shown in Figure 11.24(a). This generates
an error signal as shown in Figure 11.24(b). The analysis of the performance of
the DLL is similar to that for the PLL, previously described in Section 6.2. If not
for the envelope detectors in the two arms of the DLL, the loop would resemble
a Costas loop. In general, the variance of the time estimation error in the DLL
is inversely proportional to the loop SNR, which depends on the input SNR to
the loop and on the loop bandwidth. Its performance is somewhat degraded as in
the squaring PLL by the nonlinearities inherent in the envelope detectors, but this
degradation is relatively small.

A tracking method for FH spread-spectrum signals is illustrated in Figure 11.25. This method is based on the premise that, although initial acquisition

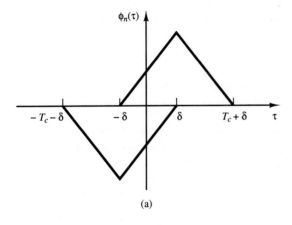

(a)

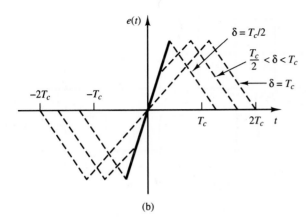

(b)

FIGURE 11.24. Autocorrelation
function (a) and tracking error signal (b)
for DLL.

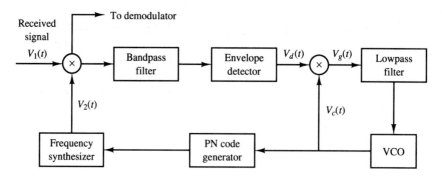

FIGURE 11.25. Tracking method for FH signals [from paper by Pickholtz et al. (1982). © IEEE. Reprinted with permission.]

has been achieved, there is a small timing error between the received signal and the received clock. The bandpass filter is tuned to a single intermediate frequency and its bandwidth is of the order of $1/T_c$, where T_c is the chip interval. Its output is envelope detected and then multiplied by the clock signal to produce a three-level signal, as shown in Figure 11.26, which drives the loop filter. Note that when the chip transitions from the locally generated sinusoidal waveform do not occur at the

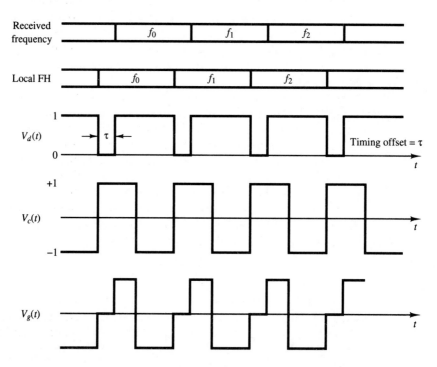

FIGURE 11.26. Waveforms for the tracking method of FH signals shown in Figure 11.25.

same time as the transitions in the incoming received signal, the output of the loop filter will be either negative or positive, depending on whether the VCO is lagging or advanced relative to the timing of the input signal. This error signal from the loop filter will provide the control signal for adjusting the VCO timing signal so as to drive the frequency synthesized FH signal to proper synchronism with the received signal.

11.6 FURTHER READING

In this chapter we have given a brief, introductory treatment of spread-spectrum modulation and demodulation and evaluated the performance of this type of signalling in the presence of several different interference conditions. Historically, the primary application of spread-spectrum modulation has been in the design of secure digital communication systems for military use. However, in the last decade we have seen a trend toward commercial use of spread-spectrum modulation, especially its use in mobile cellular communications, in multiple access communications via satellites, and in interoffice radio communications.

A historical account on the development of spread-spectrum communications covering the period 1920–1960 is given in a paper by Scholtz (1982). Tutorial treatments of spread-spectrum modulation that deal with basic concepts are found in papers by Scholtz (1977) and Pickholtz et al. (1982). These papers also contain a large number of references to the previous work. Two tutorial papers by Viterbi (1979, 1985) contain a basic analysis of the performance characteristics of DS and FH spread-spectrum signals.

Comprehensive treatments of various aspects concerning the analysis and design of spread-spectrum signals and systems, including synchronization techniques, are found in books by Simon et al (1985), Ziemer and Peterson (1985), and Holmes (1982). Also, special issues of the *IEEE Transactions on Communications* (August 1977, May 1982) are devoted to spread-spectrum communications. These special issues contain a collection of papers devoted to a variety of topics, including multiple-access techniques, synchronization techniques, and performance analysis with various types of channel interference. A number of important papers that have been published in IEEE journals have been reprinted in book form by the IEEE press (see Dixon, ed. [1977] and Cook et al. [1983]).

PROBLEMS

11.1 A rate 1/2 convolutional code with $d_{\text{free}} = 10$ is used to encode a data sequence occurring at a rate of 1000 bits/sec. The modulation is binary PSK. The DS spread-spectrum sequence has a chip rate of 10 MHz.

1. Determine the coding gain.
2. Determine the processing gain.
3. Determine the jamming margin assuming an $\mathcal{E}_b/J_0 = 10$.

11.2 Demonstrate that a DS spread-spectrum signal without coding provides no improvement in performance against additive white Gaussian noise.

11.3 A total of 30 equal-power users are to share a common communication channel by CDMA. Each user transmits information at a rate of 10 kbit/sec via DS spread-spectrum and binary PSK. Determine the minimum chip rate to obtain a bit error probability of 10^{-5}. Additive noise at the receiver may be ignored in this computation.

11.4 A CDMA system is designed based on DS spread-spectrum with a processing gain of 1000 and binary PSK modulation. Determine the number of users if each user has equal power and the desired level of performance is an error probability of 10^{-6}. Repeat the computation if the processing gain is changed to 500.

11.5 A DS spread-spectrum system transmits at a rate of 1000 bit/sec in the presence of a tone jammer. The jammer power is 20 dB greater than the desired signal and the required $\mathcal{E}_b/J_0$ to achieve satisfactory performance is 10 dB.

   1. Determine the spreading bandwidth required to meet the specifications.
   2. If the jammer is a pulse jammer, determine the pulse duty cycle that results in worst-case jamming and the corresponding probability of error.

11.6 A DS spread-spectrum system is used to resolve the multipath signal component in a two-path radio signal propagation scenario. If the path length of the secondary path is 300 m longer than that of the direct path, determine the minimum chip rate necessary to resolve the multipath component.

11.7 A CDMA system consists of 15 equal-power users that transmit information at a rate of 10,000 bit/sec, each using a DS spread-spectrum signal operating at a chip rate of 1 MHz. The modulation is binary PSK.

   1. Determine the $\mathcal{E}_b/J_0$, where J_0 is the spectral density of the combined interference.
   2. What is the processing gain?
   3. How much should the processing gain be increased to allow for doubling the number of users without affecting the output SNR?

11.8 A FH binary orthogonal FSK system employs a $m = 15$ stage linear feedback shift register that generates a ML sequence. Each state of the shift register selects one of N nonoverlapping frequency bands in the hopping pattern. The bit rate is 100 bit/sec and the hop rate is once per bit. The demodulator employs noncoherent detection.

   1. Determine the hopping bandwidth for this channel.

2. What is the processing gain?

3. What is the probability of error in the presence of AWGN?

11.9 Consider the FH binary orthogonal FSK system described in Problem 11.8. Suppose that the hop rate is increased to two hops per bit. The receiver uses square-law combining to combine the signal over the two hops.

1. Determine the hopping bandwidth for the channel.

2. What is the processing gain?

3. What is the error probability in the presence of AWGN?

11.10 In a fast FH spread-spectrum system, the information is transmitted via FSK, with noncoherent detection. Suppose there are $N = 3$ hops per bit, with hard-decision decoding of the signal in each hop.

1. Determine the probability of error for this system in an AWGN channel with power-spectral density $\frac{N_0}{2}$ and an SNR = 13 dB (total SNR over the three hops).

2. Compare the result in part 1 with the error probability of a FH spread-spectrum system that hops once per bit.

11.11 A slow FH binary FSK system with noncoherent detection operates at an $\mathcal{E}_b/J_0 = 10$, with a hopping bandwidth of 2 GHz, and a bit rate of 10 kbit/sec.

1. What is the processing gain for the system?

2. If the jammer operates as a partial-band jammer, what is the bandwidth occupancy for worst-case jamming?

3. What is the probability of error for the worst-case partial-band jammer?

11.12 A DS binary PSK spread-spectrum has a processing gain of 500. What is the jamming margin against a continuous tone jammer if the desired error probability is 10^{-5}?

11.13 Repeat Problem 11.12 if the jammer is a pulsed-noise jammer with a duty cycle of 1%.

11.14 Consider the DS spread-spectrum signal

$$c(t) = \sum_{n=-\infty}^{\infty} c_n p(t - nT_c)$$

where c_n is a periodic m-sequence with a period $L = 127$ and $p(t)$ is a rectangular pulse of duration $T_c = 1$ μsec. Determine the power-spectral density of the signal $c(t)$.

11.15 Suppose that $\{c_{1i}\}$ and $\{c_{2i}\}$ are two binary (0, 1) periodic sequences with periods L_1 and L_2, respectively. Determine the period of the sequence obtained by forming the modulo-2 sum of $\{c_{1i}\}$ and $\{c_{2i}\}$.

11.16 An $m = 10$ ML shift register is used to generate the pseudorandom sequence in a DS spread-spectrum system. The chip duration is $T_c = 1$ μsec, and the bit duration is $T_b = LT_c$, where L is the length (period) of the m-sequence.

1. Determine the processing gain of the system in decibels.
2. Determine the jamming margin if the required $E_b/J_0 = 10$ and the jammer is a tone jammer with an average power J_{av}.

A

The Probability of Error for Multichannel Reception of Binary Signals

In multichannel communication systems that employ binary signals for transmitting information over an AWGN channel, the decision metric can be expressed as a special case of the general quadratic form

$$Q = \sum_{k=1}^{N} \left(A\,|X_k|^2 + B\,|Y_k|^2 + C X_k Y_k^* + C^* X_k^* Y_k \right) \qquad (A.1)$$

in complex-valued Gaussian random variables. A, B, and C are constants satisfying the condition $|C|^2 - AB > 0$; X_k and Y_k are a pair of (possibly) correlated, complex-valued Gaussian random variables of the form $X_k = X_{kr} + jX_{ki}$ and $Y_k = Y_{kr} + jY_{ki}$; and N represents the number of independent channels over which the signal is received.

For example, in binary DPSK detection, which is described in Section 9.2.5, the decision metric is given as

$$Re\left(r_k r_k^*\right) = \frac{1}{2}\left(r_k r_{k-1}^* + r_k^* r_{k-1}\right) \qquad (A.2)$$

where r_k is the output of the demodulation filter. We note that (A.2) is a special case of (A.1) which is obtained from (A.1) by setting $N = 1, A = B = 0, X_k = r_k, Y_k = r_{k-1},$ and $C = 1/2$.

As another example, we have the case of fast frequency hopped, binary FSK, in which case the output of the two square-law combiners is given by (11.3.10), where N is the number of frequency-hopped chips per symbol. The probability of error, is simply the probability that $DM_1 - DM_2 < 0$, where

$$DM_1 - DM_2 = \sum_{k=1}^{N} \left[\left| v_{1k} + \frac{E_b}{N_0} \right|^2 - |v_{2k}|^2 \right] \tag{A.3}$$

where v_{1k} and v_{2k} are complex-valued Gaussian random variables that represent the noises of the outputs of the correlators. We again note that (A.3) is a special case of the general quadratic form in (A.1), where $A = --B = 1, C = 0, X_k = \left(v_{1k} + \frac{E_b}{N_0} \right)$ and $Y_k = v_{2k}$.

The probability of error is just the probability that $Q < 0$, i.e.,

$$P_2 = P(Q < 0) = \int_{-\infty}^{0} f_Q(q) \, dq \tag{A.4}$$

where $f_Q(q)$ is the p.d.f. of Q.

The derivation for the error probability P_2 may be found in the technical literature and in the book by Proakis (1989). This derivation is lengthy and is not repeated here. The expression for the probability of error resulting from this derivation is

$$P_2 = Q(a, b) - I_0(ab) \exp\left(-\frac{a^2 + b^2}{2} \right)$$

$$+ \frac{I_0(ab) \exp\left[-(a^2 + b^2)/2 \right]}{(1 + v_2/v_1)^{2N-1}} \sum_{k=0}^{N-1} \binom{2N-1}{k} \left(\frac{v_2}{v_1} \right)^k$$

$$+ \frac{\exp\left[-(a^2 + b^2)/2 \right]}{(1 + v_2/v_1)^{2N-1}} \sum_{n=1}^{N-1} I_n(ab)$$

$$\times \left\{ \sum_{k=0}^{N-1-n} \binom{2N-1}{k} \left[\left(\frac{b}{a} \right)^n \left(\frac{v_2}{v_1} \right)^k - \left(\frac{a}{b} \right)^n \left(\frac{v_2}{v_1} \right)^{2N-1-k} \right] \right\} \quad N > 1$$

$$P_2 = Q(a, b) - \frac{v_2/v_1}{1 + v_2/v_1} I_0(ab) \exp\left(-\frac{a^2 + b^2}{2} \right) \quad N = 1 \tag{A.5}$$

where the parameters a, b, v_1, v_2 are defined as

$$a = \left[\frac{2v_1^2 v_2 (\alpha_1 v_2 - \alpha_2)}{(v_1 + v_2)^2} \right]^{1/2}$$

$$b = \left[\frac{2v_1 v_2^2 (\alpha_1 v_1 + \alpha_2)}{(v_1 + v_2)^2} \right]^{1/2}$$

$$v_1 = \sqrt{w^2 + \frac{1}{\left(\mu_{xx}\mu_{yy} - |\mu_{xy}|^2 \right) \left(|C|^2 - AB \right)} - w}$$

$$v_2 = \sqrt{w^2 + \frac{1}{\left(\mu_{xx}\mu_{yy} - |\mu_{xy}|^2 \right) \left(|C|^2 - AB \right)} + w}$$

$$w = \frac{A\mu_{xx} + B\mu_{yy} + C\mu_{xy}^* + C^*\mu_{xy}}{2 \left(\mu_{xx}\mu_{yy} - |\mu_{xy}|^2 \right) \left(|C|^2 - AB \right)}$$

$$\alpha_{1k} = \left(|C|^2 - AB \right) \left(|\overline{X}_k|^2 \mu_{yy} + |\overline{Y}_k|^2 \mu_{xx} - \overline{X}_k^* \overline{Y}_k \mu_{xy} - \overline{X}_k \overline{Y}_k^* \mu_{xy}^* \right)$$

$$\alpha_{2k} = A|\overline{X}_k|^2 + B|\overline{Y}_k|^2 + C\overline{X}_k^* \overline{Y}_k + C^*\overline{X}_k \overline{Y}_k^*$$

$$\overline{X}_k = E(X_k); \overline{Y}_k = E(Y_k)$$

$$\mu_{xx} = E\left[\left(X_k - \overline{X}_k \right)^2 \right], \mu_{yy} = E\left[\left(Y_k - \overline{Y}_k \right)^2 \right]$$

$$\mu_{xy} = E\left[\left(X_k - \overline{X}_k \right) \left(Y_k - \overline{Y}_k \right) \right] \tag{A.6}$$

The function $Q(a, b)$ is defined as

$$Q(a, b) = \int_b^\infty x e^{-\frac{x^2 + a^2}{2}} I_0(ax) \, dx \tag{A.7}$$

We note that in the special case where $a = 0$,

$$Q(0, b) = \frac{1}{2} e^{-b^2/2} \tag{A.8}$$

Finally, $I_n(x)$ is the modified Bessel function of order n.

R

References

ADLER, R.L., COPPERSMITH, D., AND HASSNER, M. (1983), "Algorithms for Sliding Block Codes," *IEEE Trans. Inform. Theory*, vol. IT-29, pp. 5–22. January.

ANDERSON, J. B., AULIN, T., AND SUNDBERG, C. W. (1986), *Digital Phase Modulation*, Plenum, New York.

AULIN, T., RYDBECK, N., AND SUNDBERG, C. W. (1981), "Continuous Phase Modulation—Part II: Partial Response Signalling," *IEEE Trans. Commun.*, vol. COM-29, pp. 210–225, March.

AULIN, T., AND SUNDBERG, C. W., (1981), "Continuous Phase Modulation—Part I: Full Response Signalling," *IEEE Trans. Commun.*, vol. COM-29, pp. 196–209, March.

AULIN, T., AND SUNDBERG, C. W., (1982a), "On the Minimum Euclidean Distance for a Class of Signal Space Codes," *IEEE Trans. Inform. Theory*, vol. IT-28, pp. 43–55, January.

AULIN, T., AND SUNDBERG, C. W., (1982b), "Minimum Euclidean Distance and Power Spectrum for a Class of Smoothed Phase Modulation Codes with Constant Envelope," *IEEE Trans. Commun.*, vol. COM-30, pp. 1721–1729, July.

AULIN, T., AND SUNDBERG, C. W., (1984), "CPM—An Efficient Constant Amplitude Modulation Scheme," *Int. J. Satellite Commun.*, vol. 2, pp. 161–186.

BARROW, B. (1963), "Diversity Combining of Fading Signals with Unequal Mean Strengths," *IEEE Trans. Commun. Sys.*, vol. CS-11, pp. 73–78, March.

BELLO, P. A., AND NELIN, B. D. (1962a), "Predetection Diversity Combining with Selectively Fading Channels," *IRE Trans. Commun. Sys.*, vol. CS-10, pp. 32–44, March.

BELLO, P. A., AND NELIN, B. D. (1962b), "The Influence of Fading Spectrum on the Binary Error Probabilities of Incoherent and Differentially Coherent Matched Filter Receivers," *IRE Trans. Commun. Sys.*, vol. CS-11, pp. 170–186, June.

BELLO, P. A., AND NELIN, B. D. (1963), "The Effect of Frequency Selective Fading on the Binary Error Probabilities of Incoherent and Differentially Coherent Matched Filter Receivers," *IEEE Trans. Commun. Sys.*, vol. CS-11, pp. 170–186, June.

BENEDETTO, S., BIGLIERI, E., AND CASTELLANI, V. (1987), *Digital Transmission Theory*, Prentice-Hall, Englewood Cliffs, N.J.

BERGER, T, AND TUFTS, D. W. (1967), "Optimum Pulse Amplitude Modulation, Part I: Transmitter-Receiver Design and Bounds from Information Theory," *IEEE Trans. Inform. Theory*, vol. IT-13, pp. 196–208.

BERGER, T. (1971), *Rate Distortion Theory: A Mathematical Basis for Data Compression*, Prentice-Hall, Englewood Cliffs, N.J.

BERLEKAMP, E. R. (1968), *Algebraic Coding Theory*, McGraw-Hill, New York.

BLAHUT, R. E. (1983), *Theory and Practice of Error Control Codes*, Addison-Wesley, Reading, Mass.

BLAHUT, R. E. (1987), *Principles and Practice of Information Theory*, Addison-Wesley, Reading, Mass.

BLAHUT, R. E. (1990), *Digital Transmission of Information*, Addison-Wesley, Reading, Mass.

BOSE, R. C., AND RAY-CHAUDHURI, D. K. (1960a), "On a Class of Error Correcting Binary Group Codes," *Inform. Control*, vol. 3, pp. 68–79, March.

BOSE, R. C., AND RAY-CHAUDHURI, D. K. (1960b), "Further Results in Error Correcting Binary Group Codes," *Inform. Control*, vol. 3, pp. 279–290, September.

BRACEWELL, R. (1965), *The Fourier Transform and Its Applications*, 2nd Ed., McGraw-Hill, New York.

BRENNAN, D. G. (1959), "Linear Diversity Combining Techniques," *Proc. IRE*, vol. 47, pp. 1075–1102, June.

CARLSON, A. B. (1986), *Communication Systems*, 3rd Ed., McGraw-Hill, New York.

CLARKE, K. K., AND HESS, D. T. (1971), *Communication Circuits: Analysis and Design*, Addison-Wesley, Reading, Mass.

COOK, C. E., ELLERSICK, F. W., MILSTEIN, L. B., AND SCHILLING, D. L. (1983), *Spread Spectrum Communications*, IEEE Press, New York.

COUCH, L. W. II (1993), *Digital and Analog Communication Systems*, 4th Ed., Macmillan, New York.

COVER, T. M., AND THOMAS, J. A. (1992), *Elements of Information Theory*, Wiley-Interscience, New York.

DAVENPORT, W. B. JR., AND ROOT, W. L. (1958), *Random Signals and Noise*, McGraw-Hill, New York.

DELLER, J. P., PROAKIS, J. G., AND HANSEN, H. L. (1993), *Discrete-Time Processing of Speech Signals*, Macmillan, New York.

DIXON, R. C. (1976), *Spread Spectrum Techniques*, IEEE Press, New York.

ELIAS, P. (1955), "Coding for Noisy Channels," *IRE Conv. Rec.*, vol. 3, pt. 4, pp. 37–46.

FORNEY, G. D. JR. (1966), *Concatenated Codes*, MIT Press, Cambridge, Mass.

FORNEY, G. D. JR. (1972), "Maximum-Likelihood Sequence Estimation of Digital Sequences in the Presence of Intersymbol Interference," *IEEE Trans. Inform. Theory*, vol. 18, pp. 363–378, May.

FORNEY, G. D. JR. (1988a), "Coset Codes I: Introduction and Geometrical Classification," *IEEE Trans. Inform. Theory*, vol. IT-34, pp. 671–680, September.

FORNEY, G. D. JR. (1988b), "Coset Codes II: Binary Lattices and Related Codes," *IEEE Trans. Inform. Theory*, vol. IT-34, pp. 671–680, September.

FRANASZEK, P. A. (1968), "Sequence-State Coding for Digital Transmission," *Bell Sys. Tech. J.*, vol. 27, p. 143.

FRANASZEK, P. A. (1969), "On Synchronous Variable Length Coding for Discrete Noiseless Channels," *Information and Control*, vol. 15, pp. 155–164.

FRANASZEK, P. A. (1970), "Sequence-State Methods for Run-Length-Limited Coding," *IBM J. Res. Dev.*, pp. 376–383, July.

FRANKS, L. E. (1969), *Signal Theory*, Prentice-Hall, Englewood Cliffs, N.J.

FRANKS, L. E. (1980), "Carrier and Bit Synchronization in Data Communication—A Tutorial Review," *IEEE Trans. Commun.*, vol. COM-28, pp. 1107–1121, August.

FREIMAN, C. E., AND WYNER, A. D. (1964), "Optimum Block Codes for Noiseless Input Restricted Channels," *Inform. Control*, vol. 7, pp. 398–415.

GABOR, A. (1967), "Adaptive Coding for Self Clocking Recording," *IEEE Trans. on Electronic Computers*, vol. EC-16, p. 866.

GALLAGER, R. G. (1968), *Information Theory and Reliable Communication*, Wiley, New York.

GARDNER, F. M. (1979), *Phaselock Techniques*, Wiley, New York.

GERSHO, A., AND GRAY, R. M (1992), *Vector Quantization and Signal Compression*, Kluwer, Boston.

GERST, I., AND DIAMOND, J., (1961), "The Elimination of Intersymbol Interference by Input Pulse Shaping," *Proc. IRE*, vol. 53, July.

GIBSON, J. D. (1993), *Principles of Digital and Analog Communications*, 2nd Ed., Macmillan, New York.

GOLAY, M. J. E. (1949), "Notes on Digital Coding," *Proc. IRE*, vol. 37, p. 657, June.

GOLD, R. (1967), "Maximal Recursive Sequences with 3-Valued Recursive Cross Correlation Functions", *IEEE Trans. Inform. Theory*, vol. IT-14, pp. 154–156, January.

GOLOMB, S. W., (1967), *Shift Register Sequences*, Holden-Day, San Francisco, Calif.

GRAY, R. M., AND DAVISSON L. D. (1986), *Random Processes: A Mathematical Approach for Engineers*, Prentice-Hall, Englewood Cliffs, N.J.

GREEN, P. E. JR., (1962), "Radar Astronomy Measurement Techniques," MIT Lincoln Laboratory, Lexington, Mass., Tech. Report No. 282, December.

GUPTA, S. C. (1975), "Phase-Locked Loops," *Proc. IEEE*, vol. 63, pp. 291–306, February.

HAMMING, R. W. (1950), "Error Detecting and Error Correcting Codes," *Bell Sys. Tech. Journal*, vol. 29, pp. 147–160, April.

HARTLEY, R. V. (1928), "Transmission of Information," *Bell Sys. Tech. Journal*, vol. 7, p. 535.

HEEMSKERK, J. P. J., AND SCHOUHAMER IMMINK, K. A. (1982), "Compact Disc: system aspects and modulation," *Philips Tech. Rev.*, vol. 40, pp. 157–164.

HELLER, J. A. (1975), "Feedback Decoding of Convolutional Codes." in *Advances in Communication Systems*, vol. 4, A. J. Viterbi (Ed.), Academic, New York.

HELSTROM, C. W. (1991), *Probability and Stochastic Processes for Engineers*, Macmillan, New York.

HOCQUENGHEM, A. (1959), "Codes Correcteurs d'Erreurs," *Chiffers*, vol. 2, pp. 147–156.

HOLMES, J. K. (1982), *Coherent Spread Spectrum Systems*, Wiley-Interscience, New York.

HUFFMAN, D. A. (1952), "A Method for the Construction of Minimum Redundancy Codes," *Proc. IRE*, vol. 40, pp. 1098–1101, September.

JACOBY, G. V. (1977), "A New Look-Ahead Code for Increased Data Density," *IEEE Transactions on Magnetics*, vol. MAG-13, pp. 1202–1204.

JAYANT, N. S., AND NOLL. P. (1984), *Digital Coding of Waveforms*, Prentice-Hall, Englewood Cliffs, N.J.

JELINEK, F. (1969), "Fast Sequential Decoding Algorithm Using a Stack," *IBM J. Res. Dev.*, vol. 13, pp. 675–685, November.

KAILATH, T. (1960), "Correlation Detection of Signals Perturbed by a Random Channel," *IRE Trans. Inform. Theory*, vol. IT-6, pp. 361–366, June.

KAILATH, T. (1961), "Channel Characterization: Time-Variant Dispersive Channels," in *Lectures on Communication Theory*, Chapter 6, E. Baghdadi (Ed.), McGraw-Hill, New York.

KARABED, R., AND SIEGEL, P. H. (1991), "Matched-Spectral Null Codes for Partial-Response Channels,"*IEEE TRans. Inform. Theory*, vol. IT-37, pp. 818-855, May.

KASAMI, T. (1966), "Weight Distribution Formula for Some Class of Cyclic Codes," Coordinated Science Laboratory, University of Illinois, Urbana, Ill., Tech. Report No. R-285, April.

KOTELNIKOV, V. A. (1947), "The Theory of Optimum Noise Immunity," Ph.D. Dissertation, Molotov Energy Institute, Moscow. Translated by R. A. Silverman, McGraw Hill, New York, 1959.

KRETZMER, E. R. (1966), "Generalization of a Technique for Binary Data Communication," *IEEE Trans. Commun. Tech.*, vol. COM-14, pp. 67–68, February.

LLOYD, S. P. (1957), "Least Square Quantization in PCM," Reprinted in *IEEE Trans. Inform. Theory*, vol. IT-28, pp. 129–137, March 1982.

LENDER, A. (1963), "The Duobinary Technique for High Speed Data Transmission," *AIEE Trans. Commun. Electronics*, vol. 82, pp. 214–218.

LEON-GARCIA, A. (1989), *Probability and Random Processes for Electrical Engineering*, Addison-Wesley, Reading Mass.

LIN, S., AND COSTELLO, D. J. JR. (1983), *Error Control Coding: Fundamentals and Applications*, Prentice-Hall, Englewood Cliffs, N.J.

LINDE, J., BUZO, A., AND GRAY, R. M. (1980), "An Algorithm for Vector Quantizer Design," *IEEE Trans. Commun.*, vol. COM-28, pp. 84–95, January.

LINDSEY, W. C. (1964), "Error Probabilities for Ricean Fading Multichannel Reception of Binary and *N*-ary Signals," *IEEE Trans. Inform. Theory*, vol. IT-10, pp. 339–350, October.

LINDSEY, W. C. (1972), *Synchronization Systems in Communications*, Prentice-Hall, Englewood Cliffs, N.J.

LINDSEY, W. C., AND CHIE, C. M. (1981), "A Survey of Digital Phase-Locked Loops," *Proc. IEEE*, vol. 69, pp. 410–432.

LINDSEY, W. C., AND SIMON, M. K. (1973), *Telecommunication Systems Engineering*, Prentice-Hall, Englewood Cliffs, N.J.

LUCKY, R. W. (1965), "Automatic Equalization for Digital Communication," *Bell Sys. Tech. Journal*, vol. 45, pp. 255–286, April.

LUCKY, R. W. (1966), "Techniques for Adaptive Equalization for Digital Communication," *Bell Sys. Tech. Journal*, vol. 45, pp. 255–286.

LUCKY, R.W., SALZ, J., AND WELDON, E. J. JR. (1968), *Principles of Data Communication*, McGraw-Hill, New York.

LYON, D L. (1975a), "Timing Recovery in Synchronous Equalized Data Communication," *IEEE Trans. on Communication*, vol. COM-23, pp. 269–274, February.

LYON, D. L. (1975b), "Envelope-Derived Timing Recovery in QAM and SQAM Systems," *IEEE Trans. on Communications*, vol. COM-23, pp. 1327–1331, November.

MAC WILLIAMS, F. J., AND SLOANE, J. J. (1977), *The Theory of Error Correcting Codes*, North Holland, New York.

MARKEL, J.D, AND GRAY, A. H. JR. (1976), *Linear Prediction of Speech*, Springer-Verlag, New York.

MASSEY, J. L. (1963), *Threshold Decoding*, MIT Press, Cambridge, Mass.

MAX, J. (1960), "Quantizing for Minimum Distortion," *IRE Trans. Inform. Theory*, vol. IT-6, pp. 7–12, March.

MCMAHON, M. A. (1984), *The Making of a Profession—A Century of Electrical Engineering in America*, IEEE Press.

MEYR, H., AND ASCHEID, G. (1990), *Synchronization in Digital Communication*, Wiley-Interscience, New York.

MILLMAN, S. ED. (1984), *A History of Engineering and Science in the Bell System—Communications Sciences (1925-1980)*, AT&T Bell Laboratories.

MONZIGO, R. A., AND MILLER, T. W. (1980), "*Introduction to Adaptive Arrays*," Wiley, New York.

MUELLER, K. H., AND MULLER, M. S. (1976), "Timing Recovery in Digital Synchronous Data Receivers," *IEEE Trans. on Communications*, vol. COM-24, pp. 516–531, May.

NORTH, D. O. (1943), "An Analysis of the Factors Which Determine Signal/Noise Discrimination in Pulse-Carrier Systems," RCA Tech. Report No. 6, PTR-6C.

NYQUIST, (1924), "Certain Factors Affecting Telegraph Speed," *Bell System Tech. Journal*, vol. 3, p.324.

NYQUIST, H. (1928), "Certain Topics in Telegraph Transmission Theory," *AIEE Trans.*, vol. 47, pp. 617–644.

OLSEN, J. D. (1979), "Nonlinear Binary Sequences with Asymptotically Optimum Periodic Cross Correlation," Ph.D. Dissertation, University of Southern California, Los Angeles.

OMURA, J. (1971), "Optimal Receiver Design for Convolutional Codes and Channels with Memory via Control Theoretical Concepts," *Inform. Sci.*, vol. 3, pp. 243–266.

OPPENHEIM, A. V., WILLSKY, A. S., AND YOUNG, I. T. (1983), *Signals and Systems*, Prentice-Hall, Englewood Cliffs, N.J.

PAPOULIS, A. (1962), *The Fourier Integral and Its Applications*, McGraw-Hill, New York.

PAPOULIS, A. (1991), *Probability, Random Variables, and Stochastic Processes*, 3rd Ed., McGraw-Hill, New York.

PETERSON, W. W., AND WELDON, E. J. JR. (1972), *Error Correcting Codes*, 2nd Ed., MIT Press, Cambridge, Mass.

PICKHOLTZ, R. L., SCHILLING, D. L., AND MILSTEIN, L. B. (1982), "Theory of Spread Spectrum Communications—A Tutorial," *IEEE Trans. Commun.*, vol. COM-30, pp. 855–884, May.

PIERCE, J. N. (1958), "Theoretical Diversity Improvement in Frequency Shift Keying," *Proc. IRE*, vol. 46, pp. 903–910, May.

PIERCE, J. N., AND STEIN, S. (1960), "Multiple Diversity with Non-Independent Fading." *Proc. IRE*, vol. 48, pp. 89–104, January.

PRICE, R. (1954), "The Detection of Signals Perturbed by Scatter and Noise", *IRE Trans. Inform. Theory*, vol. PGIT-4, pp. 163–170, September.

PRICE, R. (1956), "Optimum Detection of Random Signals in Noise with Application to Scatter Multipath Communication," *IRE Trans. Inform. Theory*, vol. IT-2, pp. 125–135, December.

PRICE, R. (1962a), "Error Probabilities for Adaptive Multichannel Reception of Binary Signals," MIT Lincoln Laboratory, Lexington, Mass., Tech. Report No. 258, July.

PRICE, R. (1962b), "Error Probabilities for Adaptive Multichannel Reception of Binary Signals," *IRE Trans. Inform. Theory*, vol. IT-8, pp. 308–316, September.

PRICE, R., AND GREEN, P. E. JR., (1958), "A Communication Technique for Multipath Channels", *Proc. IRE*, vol. 46, pp. 555–570, March.

PROAKIS, JOHN G. (1989), *Digital Communications*, 2nd Ed., McGarw Hill, New York.

RABINER, L. R., AND SCHAFER, R. W. (1978), *Digital Processing of Speech Signals*, Prentice-Hall, Englewood Cliffs, N.J.

REED, I. S., AND SOLOMON, G. (1960), "Polynomial Codes over Certain Finite Fields," *SIAM J.* vol. 8, pp. 300–304, June.

RYDER, J. D., AND FINK, D. G. (1984), *Engineers and Electronics*, IEEE Press.

SAKRISON, D. J. (1968), *Communication Theory: Transmission of Waveforms and Digital Information*, New York, Wiley.

SARWATE, D. V., AND PURSLEY, M. B. (1980), "Crosscorrelation Properties of Pseudorandom and Related Sequences," *Proc. IEEE*, vol. 68, pp. 2399–2419, September.

SCHOLTZ, R. A. (1977), "The Spread Spectrum Concept," *IEEE Trans. Commun.*, vol. COM-25, pp. 748–755, August.

SCHOLTZ, R. A. (1979), "Optimal CDMA Codes," *1979 National Telecommunications Conf. Record*, Washington, D.C., pp. 54.2.1–54.2.4, November.

SCHOLTZ, R. A. (1982), "The Origins of Spread Spectrum," *IEEE Trans. Commun.*, vol. COM-30, pp. 822–854, May.

SCHOUHAMER IMMINK, K. A. (1990), "Run length-limited sequences", *Proc. IEEE*, vol. 78, pp. 1745–1759, November.

SCHOUHAMER IMMINK, K. A. (1991), *Coding Techniques for Digital Recorders*, Prentice-Hall, Englewood-Cliffs, N.J.

SCHWARTZ, M., BENNETT, W. R., AND STEIN, S. (1966), *Communication Systems and Techniques*, McGraw-Hill, New York.

SHANMUGAM, K. S. (1979), *Digital and Analog Communication Systems*, Wiley, New York.

SHANNON, C. E. (1948a), "A Mathematical Theory of Communication," *Bell Sys. Tech. Journal*, vol. 27, pp. 379–423, July.

SHANNON, C. E. (1948b), "A Mathematical Theory of Communication," *Bell Sys. Tech. Journal*, vol. 27, pp. 623–656, October.

SIMON, M. K., OMURA, J. K., SCHOLTZ, R. A., AND LEVITT, B. K. (1985), *Spread Spectrum Communications vol. I, II, III*, Computer Science Press, Rockville, Md.

SMITH, J. W. (1965), "The Joint Optimization of Transmitted Signal and Receiving Filter for Data Transmission Systems," *Bell Sys. Tech. Journal*, vol. 44, pp. 1921–1942, December.

STIFFLER, J. J. (1971), *Theory of Synchronous Communications*, Prentice-Hall, Englewood Cliffs, N.J.

STREMLER, F. G. (1990), *Introduction to Communication Systems*, 3rd Ed., Addison-Wesley, Reading, Mass.

SUNDBERG, C. W. (1986), "Continuous Phase Modulation," *IEEE Commun. Magazine*, vol. 24, pp. 25–38, April.

TANG, D. L., AND BAHL, L. R. (1970), "Block Codes for a Class of Constrained Noiseless Channels," *Information and Control*, vol. 17, pp. 436–461.

TAUB, H., AND SCHILLING, D. L. (1986), *Principles of Communication Systems*, 2nd Ed., McGraw-Hill, New York.

TUFTS, D. W. (1965), "Nyquist's Problem—The Joint Optimization of Transmitter and Receiver in Pulse Amplitude Modulation," *Proc. IEEE*, vol. 53, pp. 248–259, March.

TURIN, G. L. (1961), "On Optimal Diversity Reception," *IRE Trans. Inform. Theory*, vol. IT-7, pp. 154–166, March.

TURIN, G. L. (1962), "On Optimal Diversity Reception II," *IRE Trans. Commun. Sys.*, vol. CS-12, pp. 22–31, March.

UNGERBOECK, G. (1974), "Adaptive Maximum Likelihood Receiver for Carrier Modulated Data Transmission Systems," *IEEE Trans.. Commun.*, vol. COM-22, pp. 624–636, May.

UNGERBOECK, G. (1982), "Channel Coding with Multilevel/Phase Signals," *IEEE Trans. Inform. Theory*, vol. IT-28, pp. 55–67, January.

VITERBI, A. J. (1966), *Principles of Coherent Communication*, McGraw-Hill, New York.

VITERBI, A. J. (1967), "Error Bounds for Convolutional Codes and an Asymptotically Optimum Decoding Algorithm," *IEEE Trans. Inform. Theory*, vol. IT-13, pp. 260-269, April.

VITERBI, A. J. (1979), "Spread Spectrum Communication—Myths and Realities," *IEEE Commun. Magazine*, vol. 17, pp. 11–18, May.

VITERBI, A. J. (1985), "When not to Spread Spectrum—A Sequel," *IEEE Commun. Magazine*, vol. 23, pp. 12–17, April.

VITERBI, A. J., AND JACOBS, I. M., (1975), "Advances in Coding and Modulation for Non-coherent Channels Affected by Fading, Partial Band, and Multiple-Access Interference," in *Advances in Communication Systems*, vol. 4, A. J. Viterbi, (Ed.), Academic, New York.

WEI, L. F. (1984), "Rotationally Invariant Convolutional Channel Coding with Expanded Signal Space. Part I: 180°, Part II: Nonlinear Codes," *IEEE J. Selected Areas Commun.*, vol. SAC-2, pp. 659–687. September.

WIDROW, B. (1966), "Adaptive Filters, I: Fundamentals," Tech. Report No. 6764-6, Stanford Electronic Laboratories, Stanford University, Stanford, Calif., December.

WIENER, N. (1949), *The Extrapolation, Interpolation, and Smoothing of Stationary Time-Series with Engineering Applications*, Wiley, New York. (The original work appeared in an MIT Radiation Laboratory Report, 1942).

WONG, E., AND HAJEK, B. (1985), *Stochastic Processes in Engineering Systems*, Springer-Verlag, New York.

WOZENCRAFT, J. M. (1957), "Sequential Decoding for Reliable Communication," *IRE Natl. Conv. Rec.*, vol. 5, pt. 2, pp. 11–25.

WOZENCRAFT, J. M., AND JACOBS, I. M. (1965), *Principles of Communication Engineering*, Wiley, New York.

WOZENCRAFT, J. M., AND REIFFEN, B. (1961), *Sequential Decoding*, MIT Press, Cambridge, Mass.

ZIEMER, R. E., AND PETERSON, R. L. (1985), *Digital Communications and Spread Spectrum Systems*, Macmillan, New York.

ZIEMER, R. E., AND TRANTER, W. H. (1990), *Principles of Communications Systems, Modulation, and Noise*, Houghton Mifflin, Boston.

ZIGANGIROV, K. S. (1966), "Some Sequential Decoding Procedures," *Problemy Peredachi Informatsii*, vol. 2, pp. 13–25.

ZIV, J., AND LEMPEL, A. (1978), "Compression of Individual Sequences via Variable Rate Coding," *IEEE Trans. Inform. Theory*, vol. IT-24, pp. 530–536.

Index

A posteriori probabilities, 481
A priori probabilities, 481
AMI code, 517
Aliasing, 108
  error, 108
Amplitude modulation, 298–328
  amplitude shift keying (ASK),
    621
  broadcasting, 351–353
  conventional AM, 306–310
  DSB-SC, 298–305
  PAM, 618–628
  SSB-AM, 310–317
  VSB-AM, 317–319
Amplitude-shift keying, 621
Analog repeaters, 428–430
Analog signal, 8
Analog source, 8
Analysis-synthesis coding, 246,
  270–273
Analytic signal, 116

Angle modulation, 328–351
  by a Gaussian process, 343
  indirect generation, 345
  narrowband, 332
  noise in, 404–417
  by periodic signals, 338
  by sinusoidal signal, 334
  time-domain representation,
    329
Antenna beamwidth, 510
Antenna gain, 509
Anti-causal signals, 33
Anti-imaging filter, 279
Antipodal signals, 488
Applications of coding,
  803–809
  compact discs, 805–808
  deep space communications,
    803–804
  telephone-line modems,
    804–805

Autocorrelation function, 87,
  166
  at output of LTI systems, 102,
    177
  statistical average, 166
  time-average, 101

Balanced discriminator, 347
Balanced modulator, 322
Bandlimited channels, 531
Bandlimited processes, 193–196
Bandpass channel, 437, 617
Bandpass processes, 196–204
  envelope of, 199
  in-phase component, 197
  phase of, 199
  quadrature component, 197
Bandpass signals, 112–125
  envelope of, 119
  in-phase component, 117
  lowpass representation of, 117

Bandpass signals (*cont.*)
  phase of, 119
  quadrature component, 117
Bandpass system, 114, 120–122
Bandpass to lowpass
    transformation, 118
  table of, 121
Bandwidth expansion factor,
    740, 751
Baseband channel, 437
Baseband signal, 297
Basis, 53
Bessel functions of the first
    kind, 334
  table of, 336
Binary symmetric channel, 728
  discrete-memoryless channel,
    727
Binary symmetric source
    (BSS), 223
Biorthogonal signals, 442
  error probability, 500–501
  geometrical representation,
    450
Bipolar sequences, 845
Block codes, 753–776
  for burst error correction,
    767–768
  cyclic codes, 769–777
  decoding of, 757–763
  dual code, 756
  error detection capability,
    765–767
  error correction capability,
    765–767
  generator matrix, 755
  Hamming distance, 754
  Hamming weight, 754
  linearity, 753
  parity check matrix, 755
  systematic, 755
Block length, 443
Boltzmann's constant, 191
Bounds on error probability
  for orthogonal signals,
    743–746
  union bound, 499

Carrier demodulation, 7
Carrier phase estimation,
    394–403
  with costas loop, 402–403
  with PLL, 394–402

  with square-law device, 401
Carrier modulation
  Analog, 298
    amplitude, 6, 298–328
    angle, 328–351
    binary, 10
    conventional AM, 306
    double-sideband suppressed
      carrier (DSB-SC), 298
    frequency, 329–351
    implementation of, 320
    phase, 330
    Single-sideband AM
      (SSB-AM), 310
    Vestigial-sideband AM
      (VSB-AM), 317
  Digital, 617–695
    amplitude, 618–628
    comparison of, 504–506
    continuous-phase, 681–695
    frequency, 659–681
    phase, 628–646
    quadrature AM, 646–659
Carrier phase recovery
  for PAM signals, 621–623
  for PSK signals, 640–642
  for QAM signals, 658–659
Carson's rule, 340
Catastrophic codes, 783–784
Cauchy-Schwartz inequality,
    49, 130
Causal
  signals, 32
  systems, 45
Center frequency, 114
Central limit theorem, 159
Channel capacity, 729–736
  Gaussian channel, 733–736
Channel coding, 9, 742–752
  applications of, 803–809
  block codes, 752, 753–776
  code rate, 750
  coded modulation, 792–802
  coding gain, 751
  convolutional codes, 752,
    777–792
  trellis coded modulation,
    796–802
Channel distortion
  amplitude distortion, 571
  delay distortion, 571
  envelope distortion, 571

  in PAM signal transmission,
    626–628
  phase distortion, 571
  in QAM signal transmission,
    652
Channel equalization
  adaptive, 579, 585
  decision directed mode, 587
  decision-feedback, 587
  fractionally spaced, 582
  linear, 579–589
  ML sequence estimation,
    578–579
  preset, 579
  symbol spaced, 581
  training mode, 587
  zero-forcing, 580
Channels, 7
  acoustic, 20
  AWGN, 15
  bandlimited, 531
  binary symmetric (BSC), 728
  discrete-memoryless (DMC),
    727
  electromagnetic, 15
  fading, 695
  fiber-optic, 14
  linear-filter, 22
  linear time-variant filter, 22
  mathematical modeling of, 21
  storage, 20
  underwater acoustic, 20
Characteristic function, 154
Chebychev inequality, 207
Chip, 828
  chip interval, 828
  chip rate, 828
Code division multiple access
    (CDMA), 838
Code rate, 9, 750
Codes (see block codes;
    convolutional codes;
    modulation codes)
Coding gain, 751
Coherence bandwidth, 703
Coherence time, 702
Communication system, 5
Compact disc, 277–281
  error correcting codes for,
    805–808
  modulation codes for,
    466–467

Companding, 251–263
  A-law, 262
  μ-law, 261, 294–295
  optimal, 262–264
Comparison of analog
    modulations, 421–422
  bandwidth efficiency, 421
  ease of implementation, 422
  power efficiency, 422
Complete basis, 53
Concatenated codes, 777
Conditional entropy, 225
Conditional p.d.f., 155
Conditional probability, 145
Continuous-phase FSK
    (CPFSK), 671–681
  minimum-shift keying (MSK),
    676–677
  modulation index, 672
  peak frequency deviation, 671
  spectral characteristics,
    678–681
Continuous-phase modulation
    (CPM), 681–695
  correlative state vector, 687
  demodulation, 686–691
  detection, 691
  error probability, 691–693
  full response, 684
  multi-h, 684
  partial response, 684
  power spectrum, 694–695
Controlled ISI, 555
  data detection for, 558
  maximum likelihood sequence
    detection, 561
  symbol-by-symbol detection,
    558
Conventional AM, 306–310
  demodulation, 309–310
  modulation efficiency,
    308–309
  modulation by deterministic
    signals, 306
  modulation by random
    signals, 307
  modulation index, 306
  q power, 308
  SNR in, 390
Convolution integral, 46
Convolution theorem, 82–83
Convolutional codes, 777–792
  code rate, 777

constraint length, 778
  decoding of, 784–786, 789
  distance properties, 792–793
  error probability, 789–792
  free distance, 782
  properties of, 779–784
  state transition diagram, 779
  tables of free distances, 793
  transfer function of, 781
  trellis diagram, 780
  Viterbi algorithm, 786
Correlation coefficient, 155
Correlation metrics, 482
Correlation-type demodulator,
    470
Coset, 762
  coset leader, 763
Covariance, 155
Crosscorrelation coefficient, 451
Crosscorrelation function, 175
  at output of LTI systems,
    185–186
Cumulative distribution function
    (c.d.f), 146
  joint c.d.f., 154
Cyclic codes, 769–777
  BCH codes, 773–774
  code word polynomial of, 769
  encoder for, 773
  generator matrix, 772
  generator polynomial, 770
  Reed-Solomon codes, 774
  table of generator
    polynomials, 775–776
Cyclostationary processes, 168,
    538
  power spectral density of, 182

Data compression, 9, 229
Data translation codes, 453
Decision feedback equalizer
    (DFE), 592–595
  feedback filter, 592
  feedforward filter, 592
Decoding of convolutional
    codes, 784–786, 789
Decoding of linear block codes,
    758–766
Deinterleaver, 767
Delay-locked loop, 861
Delay modulation, 464
Delta modulation (ΔM), 267
  adaptive, 269

granular noise, 268
  slope-overload distortion, 269
Demodulation gain, 434
Demodulation
  Analog
    of conventional AM,
      309–310
    of DSB-SC AM, 303
    of FM, 347–351
    implementation of, 320
    of PM, 347–351
    of SSB-AM, 315, 325
    of VSB-AM, 325
  Digital
    of CPM, 686–691
    of FSK signals, 661–664
    of PAM signals, 621,
      623–624
    in presence of channel
      distortion, 626–628
    of PSK signals, 633–636
    of QAM signals, 649–651
Demodulator
  coherent, 387
  correlation-type, 470
  matched-filter type, 474
  synchronous, 387
Despreading, 828
Detection
  of CPM, 690–691
  of DPSK, 644
  of FSK, 664–667
  of PAM, 623–624
  of PSK, 633–636
  of QAM, 649–651
Detector
  maximum likelihood, 481
  minimum distance, 482
  optimum, 480–488
  Viterbi algorithm, 486
Deviation constant, 330
Differential PCM, 264–267
Differential PSK (DPSK),
    644–646
  demodulator, 643–644
  error probability, 645–646
  modulator, 643
Differential encoding, 463
Differential entropy, 239
Differentially encoded PSK,
    643
Differentiator, 43

Digital audio
 recording, 276–281
 in telephone transmission,
  274–276
Digital interpolator, 279
Direct sequence spread
  spectrum systems,
  825–844
 for anti-jamming, 841–842
 applications of, 837–844
 CDMA, 838–839
 chip rate, 828
 coding gain, 836
 error probability, 829–836
 jamming margin, 836
 LPI, 837
 for multipath channels, 840
 processing gain, 829
 synchronization of, 857
Dirichlet conditions
 for Fourier series, 55
 for Fourier transform, 72
Discrete memoryless source
  (DMS), 223
Discrete source, 8
Discrete spectrum, 56
Distance metrics, 482
Distortion measure, 241
Doppler frequency spread, 699
Doppler shift, 696
Double-sideband
  suppressed-carrier AM,
  298–305
 demodulation, 303, 324
 modulation by deterministic
  signal, 298
 modulation by random signal,
  300
 power spectral density of, 302
 SNR, 388
Dual code, 756

Early-late gate synchronizer,
  601
Effective area of antenna, 509
Effective bandwidth, 329, 335,
  337–338
Effective isotropic radiated
  power, 509
Effective noise temperature, 425
Eigenvectors, 50
Electromagnetic channels, 15

Energy and power relations,
  105
Energy of signals, 35, 97–101
Energy spectral density, 99
Energy-type signals, 98–101
Ensemble average, 170
Entropy rate, 227
Entropy, 224
 conditional, 225–226
 differential, 239
 joint, 225–226
 rate, 227
Envelope delay, 141
Envelope detector, 310, 324,
  666
Equal-gain combiner, 711
Equalizers (see channel
  equalization)
Equivalent discrete-time
  channel filter, 578
Ergodic processes, 170–172
 Gaussian, 189–190
Error event, 502
Error probability, 481–506
 for binary modulation,
  488–491
 for biorthogonal signals,
  500–501
 for CPM, 691–693
 for DPSK, 645–646
 for FSK, 667–670
 for M-ary modulation,
  492–502
 for M-ary orthogonal signals,
  495–498
 for M-PAM, 494–495
 for ML detection, 502–504
 for PAM, 565
 for partial response signals,
  566–570
 for PSK signals, 636–640
 for QAM signals, 652–657
 Shannon's limit, 500
 for simplex signals, 502
 union bound on, 499
Euclidean distance metric, 485
Excess bandwidth, 553
Expectation
 of a random process, 164
 of a random variable, 153
Exponential signal, 36

Fading multipath
 channel model, 697–703
 channels, 695–706
 diversity, 709–712
 error probability for Rayleigh
  fading, 703–709
 examples of, 696–697
 signal design for, 703–706
Fiber optic channels, 14
Filtered process, 194
Filters
 bandpass, 96
 bandwidth of, 96
 highpass, 96
 ideal, 96
 lowpass, 96
First-error-event probability,
  790
Follower jammer, 850
Fourier series, 47, 55–63
 of even harmonic signals,
  63–65
 of even signals, 63
 of odd harmonic signals,
  63–64
 of odd signals, 63
 properties of, 55–63
 of real signals, 60–67
Fourier transform, 70–94
 basic properties, 78–96
 of even signals, 76–78
 from Fourier series, 70–72
 of impulse, 74–75
 of odd signal, 76–78
 of periodic signals, 91–94
 properties of, 78–91
  autocorrelation, 87
  convolution, 82–83
  differentiation, 87
  differentiation in frequency,
   88
  duality, 79
  integration, 89
  linearity, 78
  modulation, 83
  moments, 90
  Parseval's relation, 85
  scaling, 82
  shift in time, 80
 of real signals, 76
 of rectangular pulse, 74
 of signum, 75–76
 table of, 92

Free distance, 782
Free Euclidean distance, 800
Frequency diversity, 710
Frequency-division multiplexing
    (FDM), 326–329
Frequency hopped spread
    spectrum, 848–856
  applications of, 855–856
  error probability, 852–855
  fast-hopping, 850, 853–855
  noncoherent combining loss
    of, 855
  slow-hopping, 851–853
  system model, 849
Frequency modulation (FM)
  Analog, 329–351
    bandwidth of, 337–340
    broadcast, 354–356
    comparison with AM,
      421–422
    de-emphasis, 418
    demodulation, 347–351
    deviation constant, 330
    effect of noise, 404–414
    by Gaussian processes, 340
    implementation of, 343–351
    narrowband, 332–333
    by periodic signals, 338
    pre-emphasis, 418
    by sinusoidal, 334
    spectral characteristics,
      333–343
    stereo, 355–356
    threshold, 392, 414–415
    threshold SNR
  Digital, 659–681
    demodulation, 661–663
    detection, 664–667
    error probability, 667–671
    FSK, 659–661
    modulator, 659–660
Frequency response, 66
Frequency-selective channel,
    703
Frequency-shift keying (FSK),
    659–681
  CPFSK, 671–681
Fundamental frequency, 56

Gaussian processes, 188–190
Gaussian random variables,
    149–151
  joint p.d.f., 148

Generalized functions, 39
Generator polynomial, 770
Geometric representation of
    signals, 448–452
  biorthogonal signals, 450
  coded signals, 452
  orthogonal signals, 450
  PAM, 448–449
  PPM, 449–450
  simplex, 451
Gold sequences, 847
Gram-Schmidt
    orthogonalization, 50–53,
    445
Granular noise, 268
Gray encoding, 631
Group delay, 141
Guard band, 109

Hadamard matrix, 525
Hamming codes, 757
Hamming distortion, 241
Hard decision decoding,
    760–766
Hermitian (Hermitic) symmetry,
    34
Hilbert transform, 116
  properties of, 123–125
Horn antenna, 510
Huffman coding, 230–235
Hybrid spread-spectrum signals,
    857

Illumination efficiency factor,
    510
Image frequency, 352
Impulse response, 45
Impulse signal, 39–42
  generalized derivatives of, 41
  important properties of, 41
  as a limit, 39–40
  sifting property of, 39
Impulse train, 60
  Fourier series of, 60
  Fourier transform of, 81
In-phase signal, 117
Indirect method
  for generating FM, 345
  for generating PM, 345
Information sequence
    polynomial, 770
Information sources, 222
Inner product, 49

Innovation process, 270
Instantaneous decoding, 231
Interleaver, 767
Intersymbol interference (ISI),
    537
Inverse Fourier transform, 73
Inverse channel filter, 580

Jacobian, 156

Kasami sequences, 847

LMS algorithm, 587
LPI, 837
LTI systems, 65–68
  additivity, 128
  frequency characteristics of,
    66–68
  frequency-domain analysis, 94
  homogeneity, 128
  response to periodic signals,
    65–68
  time-domain analysis, 45–47
Least favorable p.d.f., 665
Lempel-Ziv coding, 235–237
Levinson-Durbin algorithm, 272
Line coders, 453
Linear equalizer, 579–589
Linear estimation, 216–218
Linear predictive coding (LPC),
    270
Linear space, 53
Link budget, 509–512
Lower sideband, 299

MMSE equalizer, 484–585
Magnetic storage systems,
    454–456
  normalized density, 455
Mapping by set partitioning,
    796
  principles of, 796, 800
Markov inequality, 207
Matched filter, 474
  frequency domain
    interpretation, 477–478
  output SNR, 478
  properties of, 475–477
Maximal ratio combiner, 711
Maximum-likelihood criterion,
    481
Mean
  of a random process, 164

Mean (*cont.*)
  of a random variable, 153
Method of steepest descent, 585
Miller code, 461
Minimum Hamming distance,
  750
Minimum-shift keying (MSK),
  676–678
Mixer, 323
Mobile radio, 367–369
  base station, 368
  MTSO, 368
Modified Bessel function, 211
Modulated signals, 114
Modulation (see carrier
  modulation)
Modulation codes, 453
  for the compact disc, 466–467
  runlength-limited codes,
    456–466
Modulation efficiency, 308
Modulation index
  for AM, 306
  for CPFSK, 672
  for CPM, 684
  for FM, 331–332
  for PM, 331–332
Modulation signals
  biorthogonal, 442
  geometric representation, 444
  orthogonal, 442
  simplex, 442
Modulator, 129
Monochromatic signals, 115
Morse code, 2
Multichannel transmission,
  869–871
  error probability, 870–871
Multipath, 696
Multipath spread, 700
Multiple random processes, 174
  jointly stationary, 175
  uncorrelated
Multiple random variables, 154
  functions of, 156
Mutual information, 238

NRZ signal, 454
NRZI signal, 454
Narrowband process, 197
Narrowband signal, 112–125
Noise equivalent bandwidth,
  192

Noise figure, 426
Noise
  effect on analog systems,
    386–394
  effect on angle modulation,
    404–421
  effect on baseband systems,
    386
  effect on conventional AM,
    389–390
  effect on DSB-SC AM, 386
  effect on SSB-SC AM,
    388–389
Noncausal signal, 32
Noncoherent combining loss,
  855
Norm, 49
Nyquist criterion for zero ISI,
  550
Nyquist sampling rate, 109

Odd harmonic signals, 63
Offset quadrature PSK
  (OQPSK), 677
On-off keying (OOK), 439
On-off signals, 519
Optimum linear filter, 12
Optimum receiver, 467–488
Optimum receiver filter,
  572–575
Optimum transmitter filter,
  572–575
Orthogonal signals, 442
Orthogonality condition, 218

**PN sequences (see
  pseudorandom sequences)**
Parabolic antenna, 510
Parseval's relation
  for Fourier series, 68–70
  for Fourier transform, 85
Partial response signalling,
  555–558
  duobinary signal pulse, 556
  error probability, 566–570
  modified duobinary signal
    pulse, 556
  ML sequence detection,
    569–570
  symbol-by-symbol detection,
    566–568
Partial-time hopping, 856
Partial-time jamming, 841

Path memory truncation, 787
Peak frequency deviation, 671
Period, 31
Phase delay, 141
Phase trellis, 672
Phase-locked loop (PLL), 305,
  394–402
  decision-feedback PLL,
    641–642
  effect of additive noise, 398
  linear model, 396, 399–400
  loop damping factor, 397
  model of, 396
  natural frequency, 397
Phase modulation (PM)
  Analog, 330–351
    deviation constant, 330
    implementation of, 343–351
    spectral characteristics of,
      333
  Digital, 628–646
    carrier phase recovery,
      460–462
    demodulator, 633–635
    detector, 634
    error probability, 636–639
    modulator, 633–635
    PSK, 630
Phase-shift keying (PSK), 630
Phasor, 114
Planck's constant, 191
Posterior probabilities, 481
Power of signals, 35, 97
Power relation, 105
Power spectral density,
  101–104, 179–185
  of CPFSK signals, 678–684
  of CPM signals, 694–695
  of cyclostationary processes,
    182
  of delay modulation, 545
  of the output of LTI systems,
    102
  of NRZ signal, 545
  of PAM signals, 537–541, 619
  of periodic signals, 103
  of PSK signals, 632–633
  of QAM signals, 649
  of random processes, 179–185
  of signals with memory,
    542–547
  of stationary processes, 182
  of sum of processes, 187

Power spectrum (see power
    spectral density)
Power-law modulator, 320
Power-type signals, 101–104
Practical sampling, 110–112
    switched sampling, 110–112
    zero-order hold sampling, 112
Pre-envelope, 116
Precoding, 558
Preferred *m*-sequences, 847
Prefix condition, 231
Preset equalizers, 579
Probability density function
    (p.d.f.), 149
Probability measure, 144–145
Probability space, 144
Processing gain, 829
Propagation
    fading, 16
    line-of-sight (LOS), 18
    multipath, 16
Pseudorandom code, 823
Pseudorandom sequences,
    844–848
    correlation properties,
        845–847
    generation of, 844–848
    Gold sequences, 847
    Kasami sequences, 847
Pulse amplitude modulation
    (PAM), 438, 618–628
    carrier-phase recovery, 621
    with channel distortion,
        626–628
    demodulation, 623–624
    detection, 624
    error probability, 624–626
    modulator, 618–619
Pulse code modulation (PCM),
    259
    nonuniform, 260–261
    in presence of channel noise,
        740–742
    uniform, 257
Pulse modulation, 438–444
Pulse position modulation
    (PPM), 439
Pulsed interference, 841
    error probability for, 842–843
Pythagorean relation, 49

Q-function
    approximation of, 209

bounds on, 150–151
definition, 149
table, 152
Quadrature amplitude
    modulation (QAM)
    comparison with PSK, 657
    demodulator, 649–650
    detector, 651
    error probability, 652–656
    modulator, 647–648
    phase estimation, 658
    synchronization, 651
Quadrature carrier multiplexing,
    328
Quadrature component, 117
Quadrature filter, 117
Quadrature forms in Gaussian
    variables, 869
Quadrature signal, 115
Quantization, 247–258
    noise, 250
    nonuniform, 253–256
    optimality conditions, 254
    scalar, 246–256
    tables for Gaussian sources,
        252, 255
    uniform, 250–253
    vector, 256–258

RAKE correlator, 840–841
Radio broadcast
    AM, 351–353
    FM, 354–356
    FM Stereo, 355–356
    TV, 356–366
Raised cosine spectrum, 553
Random errors, 767
Random processes, 160–203
    analytic description, 162
    complete statistical
        description, 163
    continuous-time, 160
    discrete-time, 161
    in frequency domain, 178–188
    passage through LTI systems,
        185
    RMS bandwidth of, 214
    statistical averages, 164–169
    statistical description, 161
Random variables, 146
    Bernoulli, 148
    Binomial, 148
    continuous, 147

discrete, 147
    functions of, 151–162
    Gaussian, 149
    Rayleigh, 158
    Rice, 211
    sum of, 159
    uniform, 149
Random
    event, 144
    experiment, 144
    process, 160
    variable, 144, 146
Rate-distortion function, 242
    for binary symmetric source,
        244
    for Gaussian source, 245
    Shannon's lower bound, 289
Rayleigh fading channel, 702
Rayleigh p.d.f., 158
Rayleigh's theorem, 86
Reactance tube, 344
Rectangular pulse, 37
Regenerative repeaters, 506–508
Repeaters
    analog, 428–430
    regenerative, 506–508
Repetition coding, 712
Rician distribution, 211
Rician fading channel, 702
Ring modulator, 323
Rolloff factor, 553
Runlength-limited codes (RLL),
    456–466
    capacity of, 459
    efficiency, 459
    Miller code, 461
    state-dependent, 461–462
    state-independent, 460
    state transition matrix, 457
    trellis representation, 461
Running digital sum (RDS),
    467, 516

Sample function, 160
Sample space, 144
Sample waveform, 27
Sampling, 104–114, 193–196
    ideal, 11, 104–110
    Nyquist rate, 109
    practical, 110–112
    of random processes, 193–196
    switched, 110
    zero-order hold, 112

Self synchronizing codes, 231
Self-information, 224
Sigma-delta modulator
    $(\Sigma\text{-}\Delta M)$, 279
Signal constellation
    for biorthogonal, 451
    for orthogonal, 450
    for PAM, 449
    for PSK, 631
    for QAM, 647–654
    for simplex, 452
Signal design
    for AWGN channel, 438–452
    for bandlimited channel,
        547–558
    for zero ISI, 548–554
Signal energy, 34–36
Signal fading, 16
Signal multiplexing, 329
    FDM, 326–327
    quadrature carrier, 328
    TDM, 275
Signal power, 34–36
Signal space, 48
Signal transformation
    bandpass to lowpass, 118
    table of, 121
Signal-to-noise ratio (SNR),
        490
    in baseband, 386
    in conventional AM, 390
    in DSB-AM, 388
    in FM, 412
    in PM, 412
    in SSB-AM, 389
Signal-to-quantization noise
        ratio (SQNR), 250
Signals, 26
    anti-causal, 33
    causal, 32
    classification of, 27–36
    complex, 29
    continuous-time, 27
    deterministic, 31
    discrete-time, 27
    distorted, 26
    energy of, 34–36
    energy-type, 35
    even, 33
    information-bearing, 26
    magnitude of, 29
    modulus of, 29
    noncausal, 32

nonperiodic, 31
odd, 31
orthogonal expansion of,
    53–45
periodic, 31
power of, 34–36
power-type, 35
random, 31
real, 29
received, 26
sinusoidal, 28, 36
stochastic, 31
Signum signal, 38
Simplex signals, 442
Sinc signal, 37
Single-sideband AM
    (SSB-AM), 310–317
    demodulation, 315, 325
    modulation by deterministic
        signal, 311
    modulation by random signal,
        313
    Weaver's modulator, 347
Sinusoidal signal, 36
Sliding correlator, 858
Slope-overload distortion, 268
Soft-decision decoding,
    758–760
Source
    analog, 8
    discrete, 8
    encoding, 9, 228–230
Source coding
    algorithms, 230–237
    theorem, 228–230
Spread spectrum signals, 823
Spread spectrum systems,
    823–863
    CDMA, 838–839
    channels with multipath, 840
    direct sequence, 825–848
    frequency-hopped, 825,
        848–856
    LPI, 837
    model of, 823–825
    other types, 856–857
Square-law combiner, 711
Square-law detector, 666
Squared error distortion, 241,
    250
Staggered quadrature PSK
    (SQPSK), 677
Standard array, 761

Standard deviation, 153
State diagram, 675
State transition diagram, 779
State trellis, 674
Stationary processes, 167
    cyclostationary, 168
    power of, 172–174
    strictly, 167
    wide-sense, 168
Stationary state probabilities,
    464–465
Step-size parameter, 580
Stochastic gradient algorithm,
    587
Stochastic processes (see
    random processes)
Sufficient statistics, 472
Superheterodyne AM radio
    receiver, 4, 351–353
Suppressed-carrier AM, 300
Surviving sequence, 787
Switched sampling, 110
Switching modulator, 321
Symbol synchronization,
    595–605
    via early-late gate, 600–601
    via ML method, 603–605
    via minimum MSE, 603
    via spectral line method,
        597–599
Synchronization
    of PAM signals, 624
    of QAM signals, 651
    of spread-spectrum systems,
        857–865
    acquisition phase, 857–860
    for DS, 857–858
    for FH, 858–859
    tracking, 861–865
Syndrome, 762
Systems, 42
    causal, 45
    classification of, 42
    continuous-time, 43
    discrete-time, 43
    linear, 43
    noncausal, 45
    nonlinear, 43
    time-invariant, 44
    time-varying, 44

Tau-dither loop (TDL), 861

Telegraphy, 2
  wireless, 4
Television, 356
  black-and-white, 357–361
  chrominance signal, 363
  color, 361–366
  color burst, 366
  luminance signal, 363
Thermal noise, 7, 13, 191, 423
Time diversity, 710
Time-division multiplexing
    (TDM), 275
Timing phase, 596
Timing recovery, 596
Toeplitz matrix, 272
Total probability theorem, 146
Transition probability matrix,
    465
Transmission losses, 422–431
  free space path loss, 428
Transmitter, 6
Trellis, 462, 672, 780
  for convolutional codes, 780

  for CPFSK, 672
  for NRZI signals, 486
Trellis coded modulation
    (TCM), 792–808
  decoding, 801–802
  encoding, 799
Triangle inequality, 49
Triangular signal, 37
Typical sequences, 228–229

Uniquely decodable codes, 231
Unit-step signal, 36
Universal source coding, 236
Upper sideband, 299

Varactor diode, 343
Variance, 153
Vector quantization, 256–258
Vector space, 48, 53
Vestigial-sideband AM,
    317–319
  demodulation, 325
  modulation, 317

Video signal, 357
Viterbi algorithm, 486, 579,
    690, 786
Voltage-controlled oscillator
    (VCO), 343

Waveform coding, 246, 259
Weak law of large numbers
    (WLLN), 159
Weaver's SSB modulator, 374
White processes, 190
Wiener filter, 12
Wiener-Khinchin theorem, 180
  for cyclostationary processes,
    182
  for stationary processes, 182
Wireline channels, 14

Yule-Walker equation, 266,
    272

Zero-forcing equalizer, 580
Zero-order hold sampling, 112